A TEXTBOOK OF COMPLEX ANALYSIS

Prasun Kumar Nayak
Mijanur Rahaman Seikh

Universities Press

TEXTBOOK OF COMPLEX ANALYSIS

UNIVERSITIES PRESS (INDIA) PRIVATE LIMITED

Registered Office
3-6-747/1/A & 3-6-754/1, Himayatnagar, Hyderabad 500 029, Telangana, India
info@universitiespress.com; www.universitiespress.com

Distributed by
Orient Blackswan Private Limited

Registered Office
3-6-752 Himayatnagar, Hyderabad 500 029, Telangana, India

Other Offices
Bengaluru / Bhopal / Chennai / Guwahati / Hyderabad / Jaipur / Kolkata / Lucknow / Mumbai / New Delhi / Noida / Patna / Vijayawada

First published 2018

ISBN 978 93 86235 65 7

501471

Typeset in Times Roman 10/12 *by*
TEXTSOFT Solutions Pvt. Ltd., Hyderabad 500 033

Printed at
Yash Printographics, Noida

Published by
Universities Press (India) Private Limited
3-6-747/1/A & 3-6-754/1, Himayatnagar, Hyderabad 500 029, Telangana, India

Contents

Preface

We have written this book primarily to meet the requirements of undergraduate (Hons) and postgraduate courses in mathematics, physics, statistics and engineering offered by universities in India. It follows the latest UGC syllabus under the choice-based credit system (CBCS) and covers all the prescribed topics in detail. We have included numerous worked-out examples, supplementary examples and extensive review exercises, including multiple choice questions (MCQs) in all the chapters for enabling the readers a thorough review the concepts learnt. The MCQs, in particular, would be of interest to those preparing for competitive examinations such as the NET, SET and GATE. Researchers, too, would find the organisation of the content useful as preparatory material.

This book comprises ten chapters. In Chapter 1, we begin the discussions with how the real number system is inadequate for many purposes, such as solving all polynomial equations, and therefore the need for complex numbers. We have introduced stereographic projections and defined the Riemannian sphere to ensure the understanding that in the complex plane there is only one point at infinity. We also discuss the algebraic and analytic properties of complex numbers followed by some topological notions of sets in the complex plane.

In Chapter 2 we have explained the sequence and series representation of complex numbers in detail. The complex power series and the behaviour of power series on the circle of convergence are also included.

In Chapter 3 we have covered the concepts of limit and continuity for functions of complex variables. The notion of differentiability for complex variable functions, concepts of analytic functions, and the ideas of branch cut and branch point for multi-valued functions are explored.

In Chapter 4 we introduce some of the elementary transcendental functions such as exponential, logarithmic, trigonometric, and hyperbolic functions of the complex variable. Further, the branches of logarithmic functions are explained with the help of multi-valued functions.

In Chapter 5 we have covered the theory of complex integration. Few important theorems such as the Cauchy–Goursat theorem, Cauchy integral formula, derivative of analytic functions in terms of contour integrals, Morera's theorem, Liouville's theorem, maximum modulus principle and their applications are also included. Further, the series expansions of an analytic complex function are discussed.

In chapter 6 we have described the more global aspects of analytic functions by considering a geometrical view point. Specifically, the theory of linear fractional transformation and the concept of conformal mapping are incorporated. The notion of Schawarz–Cristoffel transformations, widely used in fluid dynamics, is explored.

In Chapter 7 we present the notion of evaluating residue at an isolated singular point and establish the Cauchy's residue theorem using Laurent series. The Argument Principle is described and used to establish the Rouché's theorem, local mapping theorem, inverse function theorem and open mapping theorem. In the second part of the chapter, we show how certain types of real definite integrals that have a finite number of isolated singularities inside a contour can be evaluated rather easily. In the last part, we show how the sum of infinite series can be identified as the sum of residues of some meromorphic functions.

In Chapter 8, some relevant theorems such as the Ascoli–Arzelà theorem, Montel's theorem, and Riemann mapping theorem are established. These play an important role in advanced complex analysis.

In Chapter 9 we cover entire functions, their growth properties and meromorphic functions. Infinite products and their convergence are also discussed, as also are the properties of the zeros of the entire function.

In Chapter 10 we introduce the ideas of direct and indirect analytic continuation of an analytic function. The concepts of analytic continuation by means of power series, complete analytic function, natural boundary, and analytic continuation along a curve are explained with the help of illustrative examples. Analytic continuation of multi-valued function and Monodromy theorem are also investigated.

In preparing this book, we have taken great care to ensure clarity of all concepts using logical arguments, without, however, sacrificing any rigour. We thank the publisher, Universities Press, for their cooperation in bringing out the book in the present form.

Prasun Kumar Nayak and Mijanur Rahaman Seikh
prasunkrnayak@gmail.com; mrseikh@ymail.com

1 Complex Numbers

The complex number system is a natural extension of the real number system. The terminology 'complex number' was introduced by German mathematician C. F. Gauss in 1831. Later, in 1837, Irish mathematician William Rowan Hamilton contributed greatly to the development of the arithmetic theory of complex numbers. In this chapter, we will first learn about complex numbers and then develop the basic algebraic and geometric properties of the complex number system $\mathbb{C}$. Three representations of complex numbers, namely, the Cartesian representation, the polar representation, and the spherical representation have been discussed.

Any collection of points in the complex plane is called a *two-dimensional point set,* and each point is called a *member* or an *element* of the set.

We begin with a quick review of the algebraic and analytic properties of complex numbers, followed by some topological notions of sets in the complex plane.

1.1 Complex Numbers

There is no real number x which satisfies the polynomial equation $x^2 + 1 = 0$. To permit solutions of this and similar kinds of equations, the set of complex numbers was introduced.

Definition 1.1.1. A *complex number* is an expression of the form $z = a + ib$, where $a, b \in \mathbb{R}$ and i (iota) is an *imaginary unit* with the property $i^2 = -1$.

The components a and b are called respectively the *real* and *imaginary* parts of z, and we write

$$a = \text{Re } z \text{ and } b = \text{Im } z. \tag{1.1}$$

Let $\mathbb{C}$ denotes the set of all complex numbers, then $\mathbb{C} = \{a + ib : a, b \in \mathbb{R}\}$ which forms the complex plane.

The system of *complex numbers* can be obtained by appending to the system of real numbers consisting of ordered pair (a, b) a special element $i = (0, 1)$ which has the property that $i^2 = -1$. Euler called this number i as the 'imaginary number'.

Two complex numbers $z_1 = (a_1, b_1)$ and $z_2 = (a_2, b_2)$ are said to be *equal* , denoted by $z_1 = z_2$ if and only if they have the same real parts and the same imaginary parts, i.e.,

$$(a_1, b_1) = (a_2, b_2) \quad \text{if and only if} \quad a_1 = a_2 \text{ and } b_1 = b_2. \tag{1.2}$$

The symbol z, which can stand for any of a set of complex numbers, is called a *complex variable.* A complex number $z = (a, b)$ is said to be *non-zero* if at least one of the real numbers a and b is not equal to 0.

A complex number with zero real part ($a = 0$) is said to be *purely imaginary.* The real numbers are precisely those complex numbers with zero imaginary parts, i.e., $b = 0$.

1.1.1 Fundamental Operations

While performing operations on complex numbers, we follow the same rules as in algebra and replace i^2 by -1 whenever it occurs. We define the binary operations in $\mathbb{C}$ as follows: Let $z_1 = a_1 + ib_1 = (a_1, b_1)$ and $z_2 = a_2 + ib_2 = (a_2, b_2)$ be in $\mathbb{C}$, then

$$z_1 + z_2 = (a_1 + a_2) + i(b_1 + b_2) = (a_1 + a_2, b_1 + b_2) \tag{1.3}$$

$$z_1 - z_2 = (a_1 - a_2) + i(b_1 - b_2) = (a_1 - a_2, b_1 - b_2) \tag{1.4}$$

$$z_1 \cdot z_2 = (a_1a_2 - b_1b_2) + i(a_1b_2 + a_2b_1) = (a_1a_2 - b_1b_2, a_1b_2 + a_2b_1), \tag{1.5}$$

$$\text{and if } z_2 \neq 0, \text{ then } \frac{z_1}{z_2} = \frac{a_1a_2 + b_1b_2}{a_2^2 + b_2^2} + i\frac{b_1a_2 - a_1b_2}{a_2^2 + b_2^2} \tag{1.6}$$

where a_1, b_1, a_2, b_2 are all real numbers.

RESULT **1.** It is incorrect to write $\sqrt{-2} \cdot \sqrt{-3} = \sqrt{(-2) \cdot (-3)} = \sqrt{6}$. Correctly, we have $\sqrt{-2} \cdot \sqrt{-3} = i^2\sqrt{2 \cdot 3} = -\sqrt{6}$.

Theorem 1.1. *Let* $\mathcal{D} = \{a + ib : a, b \in \mathbb{R}\}$. *With the operations of addition and multiplication, as defined in equations* (1.3) *and* (1.5), $\mathcal{D}$ *is a field.*

Proof: Let $z_1 = a_1 + ib_1 = (a_1, b_1)$, $z_2 = a_2 + ib_2 = (a_2, b_2)$ and $z_3 = a_3 + ib_3 = (a_3, b_3)$ are three elements of $\mathcal{D}$, where $a_1, b_1, a_2, b_2, a_3, b_3 \in \mathbb{R}$.
(i) Closure: Since $a_1, b_1, a_2, b_2 \in \mathbb{R}$, so $a_1 + a_2, b_1 + b_2 \in \mathbb{R}$, consequently, $z_1 + z_2 \in \mathcal{D}$, for every $z_1, z_2 \in \mathcal{D}$.
(ii) Associative: Now,

$$\begin{aligned}(z_1 + z_2) + z_3 &= ([a_1 + a_2] + a_3, [b_1 + b_2] + b_3) \quad \text{by Eq.(1.3)}\\ &= (a_1 + [a_2 + a_3], b_1 + [b_2 + b_3]) \quad \text{by an associative law in } \mathbb{R}\\ &= z_1 + (z_2 + z_3) \quad \text{by Eq.(1.3)},\end{aligned}$$

for every $z_1, z_2, z_3 \in \mathcal{D}$.

(iii) Commutative: Here, $\forall\, z_1, z_2 \in \mathcal{D}$.

$$\begin{aligned}z_1 + z_2 &= (a_1, b_1) + (a_2, b_2) = (a_1 + a_2, b_1 + b_2)\\ &= (a_2 + a_1, b_2 + b_1)z_2 + z_1, \quad \text{as } a_k, b_k \in \mathbb{R}.\end{aligned}$$

(iv) Identity: Since $0 \in \mathbb{R}$, so $\underline{0} = 0 + i0 \in \mathcal{D}$ and

$$\underline{0} + z_1 = z_1 = z_1 + \underline{0}; \text{ for all } z_1 \in \mathcal{D}.$$

Thereforc $\underline{0}$ is the additive identity in $\mathcal{D}$.

(v) Inverse: We know, if $a_1, b_1 \in \mathbb{R}$; $-a_1, -b_1 \in \mathbb{R}$ and so $-z_1 = (-a_1, -b_1) \in \mathcal{D}$. Now

$$z_1 + (-a_1, -b_1) = \underline{0} = (-a_1, -b_1) + z_1.$$

Thus $-z_1$ is the additive inverse of z_1 in $\mathcal{D}$.

Hence $(\mathcal{D}, +)$ is an abelian group.

(vi) Closure: Since $a_1, b_1, a_2, b_2 \in \mathbb{R}$, so $a_1a_2 - b_1b_2, a_1b_2 + a_2b_1 \in \mathbb{R}$. Consequently using Eq.(1.5) $z_1 \cdot z_2 \in \mathcal{D}$, for all $z_1, z_2 \in \mathcal{D}$.

(vii) Associative: Using Eq.(1.5) we get

$$\begin{aligned}(z_1 \cdot z_2) \cdot z_3 &= [(a_1, b_1) \cdot (a_2, b_2)] \cdot (a_3, b_3)\\ &= (a_1a_2 - b_1b_2, a_1b_2 + a_2b_1) \cdot (a_3, b_3)\\ &= ((a_1a_2 - b_1b_2)a_3 - (a_1b_2 + a_2b_1)b_3, (a_1a_2 - b_1b_2)b_3 + (a_1b_2 + a_2b_1)a_3)\end{aligned}$$

(viii) Commutative: Using Eq.(1.5) and properties of $\mathbb{R}$, we get

$$\begin{aligned}z_1 \cdot z_2 &= (a_1a_2 - b_1b_2) + i(a_1b_2 + a_2b_1) = (a_1a_2 - b_1b_2, a_1b_2 + a_2b_1)\\ &= (a_2a_1 - b_2b_1, a_2b_1 + a_1b_2);\ a_k, b_k \in \mathbb{R}\\ &= z_2 \cdot z_1 \text{ for all } z_1, z_2 \in \mathcal{D}.\end{aligned}$$

(ix) Identity: Since $1, 0 \in \mathbb{R}$ so $z_0 = 1 + i0 = (1, 0) \in \mathcal{D}$ and

$$z_1 \cdot z_0 = z_1 = (a_1, b_1) = z_0 \cdot z_1;\ \forall\, z_1 \in \mathcal{D}.$$

Therefore $z_0 \in \mathcal{D}$ is the multiplicative identity element.

(x) Inverse: Let $z_1 = a_1 + ib_1$ be a non-zero complex number belonging to $\mathcal{D}$. So, either $a_1 \neq 0$ or $b_1 \neq 0$. Hence $a_1^2 + b_1^2 > 0$. Now,

$$\frac{1}{z_1} = \frac{a_1 - ib_1}{(a_1 + ib_1)(a_1 - ib_1)} = \frac{a_1}{a_1^2 + b_1^2} + i\frac{-b_1}{a_1^2 + b_1^2}.$$

Since $a_1, a_2 \in \mathbb{R}$ so $\dfrac{a_1}{a_1^2 + b_1^2}, \dfrac{-b_1}{a_1^2 + b_1^2} \in \mathbb{R}$ and so $\frac{1}{z_1} \in \mathcal{D}$. Also

$$z_1 \cdot \frac{1}{z_1} = z_0 = \frac{1}{z_1} \cdot z_1\ \forall\, z_1 \in \mathcal{D}.$$

Hence $\frac{1}{z_1} \in \mathcal{D}$ is the reciprocal or inverse of the complex number $z_1 = (a_1, b_1)$; where $a_1, b_1 \in \mathbb{R}$, not both zero.

Hence $(\mathcal{D}, \cdot)$ is an abelian group.

(xi) Distributive: We observe that

$$\begin{aligned}z_1 \cdot (z_2 + z_3) &= (a_1, b_1)(a_2 + a_3, b_2 + b_3); \qquad \text{by Eq.(1.3)}\\ &= (a_1[a_2 + a_3] - b_1[b_2 + b_3], a_1[b_2 + b_3] + [a_2 + a_3]b_1)\\ &= (a_1a_2 - b_1b_2, a_1b_2 + a_2b_1) + (a_1a_3 - b_1b_3, a_1b_3 + a_3b_1)\\ &= z_1 \cdot z_2 + z_1 \cdot z_3;\ \forall\, z_1, z_2, z_3 \in \mathcal{D}.\end{aligned}$$

Since the associative, commutative and distributive laws for addition and multiplication hold in $\mathcal{D}$, so $(\mathcal{D}, +, \cdot)$ is a field, in which the complex numbers $(0,0)$ and $(1,0)$ are identities for addition and multiplication respectively. □

RESULT **2.** We now observed that by Eq. (1.3) and (1.5), the following equations hold

$$(a_1, 0) + (a_2, 0) = (a_1 + a_2, 0); \quad (a_1, 0) \cdot (a_2, 0) = (a_1 a_2, 0). \tag{1.7}$$

Equations (1.7) suggest that it is convenient and natural to identify each complex number of the form $(a, 0)$ with the real number a. Indeed, we make this identification and we write

$$(a, 0) = a \text{ for each real number } a.$$

If $f : \mathbb{R} \to \mathbb{C}$, the mapping $a \to (a, 0)$ then f is a field isomorphism of $\mathbb{R}$ into $\mathbb{C}$ and we consider $\mathbb{R}$ as a subset of $\mathbb{C}$. Thus we replace each pair $(a, 0)$ by the real number a.

RESULT **3.** Let $x \in \mathbb{R}$, then it can be written as $x = (x, 0)$. Using Eq. (1.5), we get

$$xz = (x, 0)(a, b) = (xa, xb). \tag{1.8}$$

Taking $x = 1$ in Eq. (1.8), it follows that $1z = z$. By Eq. (1.8),

$$y(0, 1) = (0, y) \text{ for each real number } y. \tag{1.9}$$

RESULT **4.** If we take $i = (0, 1)$, then using Eq. (1.5), we get

$$\begin{aligned} & ib = (0,1)(b,0) = (0,b); \; b \in \mathbb{R}. \\ \therefore \quad & a + ib = (a,0) + (0,b) = (a,b); \; \forall \, (a,b) \in \mathbb{R} \times \mathbb{R}. \end{aligned}$$

Henceforth, the complex number $a + ib$ can be represented in lieu of (a, b).

RESULT **5.** Note that $i = (0, 1)$ and so using Eq. (1.5), we get

$$\begin{aligned} & (0,1) \cdot (0,1) = (0 - 1, 0 + 0) = (-1, 0) = -1 \\ \text{or,} \quad & (0,1)^2 = -1, \; i.e., \; i^2 = -1. \end{aligned}$$

RESULT **6.** The most important difference between $\mathbb{R}$ and $\mathbb{C}$ is that $\mathbb{R}$ is an ordered field, whereas $\mathbb{C}$ is a field but there is no ordering relation in $\mathbb{C}$.

If possible, let $i > 0$ or $i < 0$, then

$$i > 0 \Rightarrow -1 > 0; \; i < 0 \Rightarrow 0 < -i \Rightarrow 0 < -1.$$

But $-1 > 0$ is impossible. Hence there is no order structure in the complex number system so that we cannot compare two complex numbers.

Example 1.1.1. *If z and w are complex numbers, then there exists a unique p in $\mathbb{C}$ such that $p + w = z$; namely $p = z + (-w)$.*

Solution. First, if $p = z + (-w)$, then

$$p + w = [z + (-w)] + w = z; \text{ by associative law.}$$

Conversely, if $p + w = z$, then

$$p = (p + w) + (-w) = z + (-w).$$

Because of this theorem, it is natural to use $z-w$ to denote $z+(-w)$. Thus if $z=(x,y)$ and $w=(u,v)$ then

$$z-w=z+(-w)=(x-u,y-v).$$

1.1.2 Conjugation

Let $z=a+ib=(a,b);\ a,b\in\mathbb{R}$ be a complex number in $\mathbb{C}$. The *complex conjugate* of z, denoted by $\bar{z}$, is defined by

$$\bar{z}=a-ib=(a,-b) \tag{1.10}$$

and it is obtained by a reflection across the real axis in the plane. A complex number z is real if and only if $z=\bar{z}$, and it is purely imaginary if and only if $z=-\bar{z}$.

The following are the important properties of conjugation.

(i)	$\bar{\bar{z}}=z$	(ii)	$z+\bar{z}=2\,\mathrm{Re}\,z=2a; a=\frac{1}{2}(z+\bar{z})$
(iii)	$z-\bar{z}=2\,\mathrm{Im}\,z=2bi; b=\frac{1}{2i}(z-\bar{z})$	(iv)	$\overline{z_1+z_2}=\bar{z_1}+\bar{z_2}$
(v)	$\overline{z_1\cdot z_2}=\bar{z_1}\cdot\bar{z_2}$	(vi)	$\overline{(z_1/z_2)}=\bar{z_1}/\bar{z_2}$ with $z_2\neq(0,0)$

The mapping $f:\mathbb{C}\to\mathbb{C}$ defined by $f(z)=\bar{z}$ is called the *conjugation map*. It preserves both addition as well as multiplication operations of complex numbers. The conjunction map f is clearly one-to-one and onto.

Example 1.1.2. *The non-real roots of a polynomial equation with real coefficients occur is conjugate pairs.*

Solution. Let α be a real root of the polynomial equation $f(z)=a_0z^n+a_1z^{n-1}+\cdots+a_{n-1}z+a_n=0$, where $a_0,a_1,\cdots,a_{n-1},a_n\in\mathbb{R}$ with $a_0\neq0$. A number is called an *algebraic number* if it is a solution of the polynomial equation $f(x)=0$. Therefore $f(\alpha)=0$ and hence

$$\begin{aligned}
& a_0\alpha^n+a_1\alpha^{n-1}+\cdots+a_{n-1}\alpha+a_n=0\\
\text{or, }\ & \overline{a_0\alpha^n+a_1\alpha^{n-1}+\cdots+a_{n-1}\alpha+a_n}=\bar{0}\\
\text{or, }\ & \bar{a}_0\bar{\alpha}^n+\bar{a}_1\bar{\alpha}^{n-1}+\cdots+\bar{a}_{n-1}\bar{\alpha}+\bar{a}_n=\bar{0}\\
\text{or, }\ & a_0\bar{\alpha}^n+a_1\bar{\alpha}^{n-1}+\cdots+a_{n-1}\bar{\alpha}+a_n=0 \text{ as } a_i\text{'s are real}\\
\text{or, }\ & a_0(\bar{\alpha})^n+a_1(\bar{\alpha})^{n-1}+\cdots+a_{n-1}(\bar{\alpha})+a_n=0\\
\text{or, }\ & f(\bar{\alpha})=0.
\end{aligned}$$

Therefore $\bar{\alpha}$ is also a root of the equation $f(z)=0$.

Example 1.1.3. *Prove that $\sqrt[3]{4}-2i$ is an algebraic number.*

Solution. Let $z=\sqrt[3]{4}-2i,\ i.e.,\ z+2i=\sqrt[3]{4}$. Cubing, we get

$$(z+2i)^3=4\Rightarrow z^3-12z-4=i(8-6z^2).$$

Further squaring we get

$$(z^3-12z-4)^2=-(8-6z^2)^2\Rightarrow z^6+12z^4-8z^3+48z^2+96z+80=0,$$

a polynomial equation with integer coefficients having $\sqrt[3]{4}-2i$ as a root. Hence $\sqrt[3]{4}-2i$ is an algebraic number.

1.1.3 Modulus

The Euclidean scalar product in the real vector space $\mathbb{C} = \mathbb{R}^2$ with respect to the basis $\{(1,0) \equiv 1, (0,1) \equiv i\}$ is

$$\begin{aligned} \langle z_1, z_2 \rangle &= \text{Re}\,(z_1 \cdot \bar{z}_2) = \text{Re}\,\{(a_1 + ib_1) \cdot (a_2 - ib_2)\} \\ &= \text{Re}\,\{(a_1a_2 + b_1b_2) + i(a_2b_1 - a_1b_2)\} = a_1a_2 + b_1b_2. \end{aligned}$$

Observe that

$$\begin{aligned} \langle z_1, z_2 \rangle &= \text{Re}\,\{(a_1a_2 + b_1b_2) - i(a_2b_1 - a_1b_2)\} \\ &= \text{Re}\,\{(a_2 + ib_2) \cdot (a_1 - ib_1)\} = \text{Re}\,(z_2 \cdot \bar{z}_1) = \langle z_2, z_1 \rangle \\ \therefore \langle z_1, z_1 \rangle &= a_1^2 + b_1^2 = \begin{vmatrix} a_1 & -b_1 \\ b_1 & a_1 \end{vmatrix} \geq 0. \end{aligned}$$

Hence, the Euclidean scalar product of each complex number with itself is a non-negative real number, the positive square root of which is called its *modulus.*

Let $z = a + ib = (a, b);\ a, b \in \mathbb{R}$ be a complex number in $\mathbb{C}$. The *modulus* or *absolute value* of z, denoted by $|z|$, is defined by

$$|z| = |a + ib| = \sqrt{a^2 + b^2} = +\sqrt{\langle z, z \rangle}. \tag{1.11}$$

Geometrically, $|z|$ represents the distance (metric) between $z = (a, b)$ and the origin $O(0, 0)$. The following are the important properties of modulus:

(i) $|(a, 0)| = |a| = \sqrt{a^2}$

(ii) $|a| \leq |z|$ and $|b| \leq |z|$, equality holds if and only if $b = 0$ and $a = 0$ respectively.

(iii) $|z| \geq 0$ and $|z| = 0$ iff $z = 0$ and $|z| = |\bar{z}|$

(iv) $z \cdot \bar{z} = |z|^2 = \bar{z} \cdot z$

(v) $|z_1 \cdot z_2| = |z_1| \cdot |z_2|$ and $|\frac{z_1}{z_2}| = \frac{|z_1|}{|z_2|}$; provided $z_2 \neq 0$.

(vi) $\frac{1}{z} = \frac{1}{|z|^2}\,\bar{z} = \frac{1}{|\bar{z}|^2}\,\bar{z}$; provided $z \neq 0$.

(vii) $|z_1 \pm z_2|^2 = |z_1|^2 + |z_2|^2 \pm 2\,\text{Re}\,(z_1\bar{z}_2)$

(viii) $|\bar{z}| = |(a, -b)| = |(a, b)| = |z| = |-z|$

(ix) $|z_1 + z_2| \leq |z_1| + |z_2|;\ |z_1 + z_2| \geq |z_1| - |z_2|$

(x) $||z_1| - |z_2|| \leq |z_1 - z_2|;\ |z_1 - z_2| \geq |z_1| - |z_2|$

(xi) If $z_1 - \bar{z}_2$ and z_1z_2 are both real, then either z_1 and z_2 are both real or one is the conjugate of the other.

(xii) $\overline{z_1 \cdot z_2} = \bar{z}_1 \cdot \bar{z}_2$.

Example 1.1.4. *Find the condition under which the equation $az + b\bar{z} + c = 0$ in one complex unknown has exactly one solution and compute that solution.*

Solution. Taking conjugate of the given equation, we get

$$\overline{az + b\bar{z} + c} = \bar{0} \Rightarrow \bar{a}\bar{z} + \bar{b}z + \bar{c} = 0.$$

Eliminating $\bar{z}$ between these equations we get

$$(a\bar{a} - b\bar{b})z = b\bar{c} - \bar{a}c \Rightarrow (|a|^2 - |b|^2)z = b\bar{c} - \bar{a}c.$$

Hence, if $|a| \neq |b|$ the given equation has unique solution and the solution is given by

$$z = \frac{1}{|a|^2 - |b|^2}(b\bar{c} - \bar{a}c).$$

Example 1.1.5. *If z_1 and z_2 be two complex numbers, prove that* $\left|\frac{z_1 - z_2}{1 - \bar{z}_1 z_2}\right| = 1$ *if either* $|z_1| = 1$ *or* $|z_2| = 1$. *What happens if* $|z_1| = 1$ *and* $|z_2| = 1$?

Solution. Suppose $|z_1| = 1$. Hence $|\bar{z}_1| = 1$ and $z_1\bar{z}_1 = |z_1|^2 = 1$. Now

$$\left|\frac{z_1 - z_2}{1 - \bar{z}_1 z_2}\right| = \left|\frac{z_1 - z_2}{z_1\bar{z}_1 - \bar{z}_1 z_2}\right| = \left|\frac{z_1 - z_2}{\bar{z}_1(z_1 - z_2)}\right| = \frac{1}{|z_1|} = 1.$$

similarly if $|z_2| = 1$, i.e., $|\bar{z}_2| = 1$, we have the required result. If $|z_1| = 1$ and $|z_2| = 1$, then the result is true provided $1 - \bar{z}_1 z_2 \neq 0$,

$$\text{i.e., if } z_1 - z_1\bar{z}_1 z_2 \neq 0, \text{ i.e., if } z_1 \neq |z_1|^2 z_2, \text{ i.e., if } z_1 \neq z_2.$$

1.2 Point Representation of Complex Numbers

We can represent any complex number $z = a + ib$ by the point (a, b) in the xy-plane (Fig. 1.1) whose rectangular Cartesian coordinate are (a, b) relative to the given set of axes.

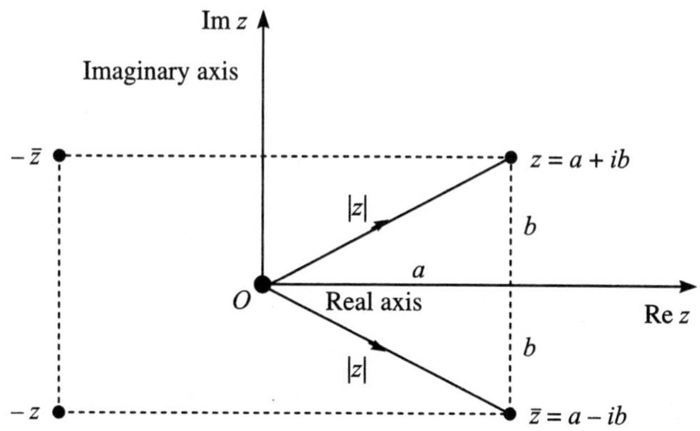

Figure 1.1 *Representation of complex numbers z, $\bar{z}$, $-z$, $-\bar{z}$ in the Argand plane*

The complex numbers belonging to the set $\mathbb{R} \times \{0\} \subset \mathbb{C}$ are plotted along the x-axis, called the *real axis*, which we denote by Re z. The complex numbers belonging to the set $\{0\} \times \mathbb{R} \subset \mathbb{C}$ are plotted along the y-axis, called the *imaginary axis*, which we denote by Im z. The origin represents $z = 0$.

From the geometrical viewpoint, the map $\phi : z \to \bar{z}$ can be thought of as the 'reflection with respect to the real axis'.

$|z|$ represents the distance between the complex number z and the origin.

The plane whose points represent complex numbers is known as *complex plane* or *Argand plane* or the *Gaussian plane.*

The coordinates$(z, \bar{z})$ which locate a point are called *complex conjugate coordinates* or briefly *conjugate coordinates* of the point.

RESULT **7.** $z_1 - z_2$ is constructed as the sum of vectors z_1 and $-z_2$. Let $z_1 = a_1 + ib_1$ and $z_2 = a_2 + ib_2$, then the distance (metric) between two points z_1 and z_2 in $\mathbb{C}$ is given by

$$|z_1 - z_2| = \sqrt{(a_1 - a_2)^2 + (b_1 - b_2)^2}$$

confirms the fact that the distance $|z_1 - z_2|$ is the length of the line segment with end points z_2 and z_1. This fact is useful in describing certain curves in the plane. For example:

(i) The equation $|z - 1 + 3i| = 2$ represents a circle whose centre is $z_0 = 1 - 3i$ and radius $\rho = 2$ (Fig. 1.2a).

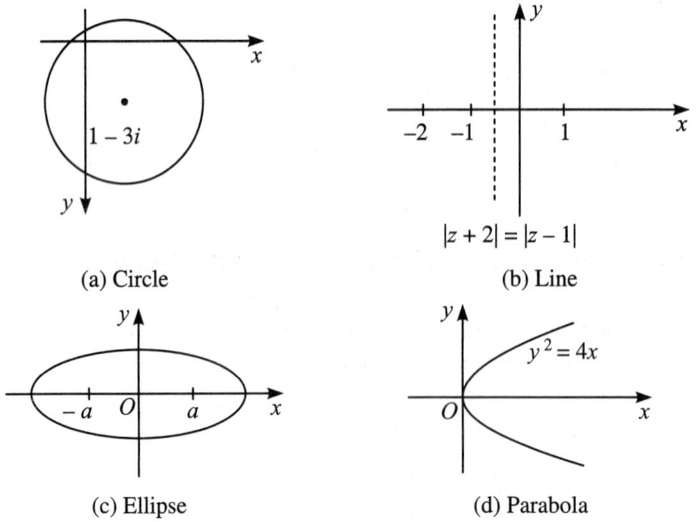

Figure 1.2 *Geometrical representations*

(ii) Consider the equation $|z + 2| = |z - 1|$. A point z satisfies that relation if and only if it is equidistant from the points -2 and 1. Hence it represents the perpendicular bisectors of the line segment joining -2 and 1; i.e., the line $x = -1/2$ (Fig. 1.2b).

(iii) The set $S = \{z : |z - a| + |z + a| = 2c\}$ represents the equation of an ellipse with centre at origin, foci at $\pm a$ and semi-major and semi-minor axes equal to c and $\sqrt{c^2 - a^2}$ (Fig. 1.2c).

(iv) Consider the set $S = \{z \in \mathbb{C} : |z - 1| = \operatorname{Re} z + 1\}$. It represents the curve $y^2 = 4x$, which is the equation of the parabola depicted in the Fig. 1.2d.

Example 1.2.1. *If z_1, z_2, z_3 are the vertices of an isosceles triangle, right angled at the vertex z_2, prove that $z_1^2 + 2z_2^2 + z_3^2 = 2z_2(z_1 + z_3)$.*

Solution. Let the points A, B and C in ΔABC represent the complex numbers z_1, z_2 and z_3 respectively, then $\angle ABC = 90° = \pi/2$. Therefore,

$$\arg\left(\frac{z_2 - z_1}{z_2 - z_3}\right) = \frac{\pi}{2} \Rightarrow \frac{z_2 - z_1}{z_2 - z_3} \text{ is an imaginary number}$$

so that its real part is zero. As Re $z = (z + \bar{z})/2$, so

$$\frac{1}{2}\left(\frac{z_2 - z_1}{z_2 - z_3} + \frac{\bar{z}_2 - \bar{z}_1}{\bar{z}_2 - \bar{z}_3}\right) = 0 \Rightarrow \frac{z_2 - z_1}{z_2 - z_3} = -\frac{\bar{z}_2 - \bar{z}_1}{\bar{z}_2 - \bar{z}_3}.$$

Now, as $\triangle ABC$ is isosceles, so $AB = CB$ and so

$$\begin{aligned} & |z_2 - z_1| = |z_2 - z_3| \Rightarrow |z_2 - z_1|^2 = |z_2 - z_3|^2 \\ \Rightarrow \quad & (z_2 - z_1)(\bar{z}_2 - \bar{z}_1) = (z_2 - z_3)(\bar{z}_2 - \bar{z}_3) \\ \Rightarrow \quad & \frac{z_2 - z_1}{z_2 - z_3} = \frac{\bar{z}_2 - \bar{z}_3}{\bar{z}_2 - \bar{z}_1}. \end{aligned}$$

Multiplying the results, we get

$$\left(\frac{z_2 - z_1}{z_2 - z_3}\right)^2 = -1 \Rightarrow z_1^2 + 2z_2^2 + z_3^2 = 2z_2(z_1 + z_3).$$

Lines and Half Planes in $\mathbb{C}$

Let L denote a straight line in the complex plane $\mathbb{C}$ passing through a point z0 in C, whose slope is the argument of a non-zero z1 in C (called direction vector). Then,

$$L = \{z = z_0 + tz_1; t \in \mathbb{R}, -\infty < t < \infty\}.$$

As $z_1 \neq 0$, z lies on the straight line L, iff

$$\frac{z - z_0}{z_1} \in \mathbb{R} \Rightarrow \text{Im}\left(\frac{z - z_0}{z_1}\right) = 0 \Rightarrow t = \frac{z - z_0}{z_1}.$$

Thus the equation of the straight line L can be written as

$$L = \left\{z \in \mathbb{C} : \text{Im}\left(\frac{z - z_0}{z_1}\right) = 0\right\}. \tag{1.12}$$

Therefore, $z \notin L$ iff $\text{Im}\left(\frac{z-z_0}{z_1}\right) \neq 0$. Hence

$$\mathcal{H}_{z_0} = \left\{z \in \mathbb{C} : \text{Im}\left(\frac{z - z_0}{z_1}\right) > 0\right\}; \mathcal{K}_{z_0} = \left\{z \in \mathbb{C} : \text{Im}\left(\frac{z - z_0}{z_1}\right) < 0\right\}$$

are the two half planes determined by the straight line L. The half plane $\mathcal{H}_{z_0}$ lies to the left of the line L if we are 'walking along L in the direction of z_1'.

Example 1.2.2. *Find the equation of a straight line joining two points z_1 and z_2 in the Argand plane.*

Solution. Take any point z on the straight line joining the points z_1, z_2 (Fig 1.3).

Figure 1.3 *Equation of a straight line*

In this figure, let z lies between A and B. Note that $\arg(\frac{z-z_1}{z-z_2})$ represents the angle between the lines joining z_1 to z and z_2 to z. Hence, if z, z_1, z_2 are collinear, then

$$\arg\left(\frac{z - z_1}{z - z_2}\right) = 0 \text{ or } \pi.$$

In any case, $(z - z_1)/(z - z_2)$ is purely real, i.e.,

$$\operatorname{Im}\left(\frac{z - z_1}{z - z_2}\right) = 0 \Rightarrow \frac{1}{2i}\left[\frac{z - z_1}{z - z_2} - \overline{\left(\frac{z - z_1}{z - z_2}\right)}\right] = 0$$

$$\Rightarrow \quad \frac{z - z_1}{z - z_2} = \overline{\left(\frac{z - z_1}{z - z_2}\right)} = \frac{\overline{z - z_1}}{\overline{z - z_2}} = \frac{\bar{z} - \bar{z}_1}{\bar{z} - \bar{z}_2}$$

$$\Rightarrow \quad (z - z_1)(\bar{z} - \bar{z}_2) = (z - z_2)(\bar{z} - \bar{z}_1)$$

$$\Rightarrow \quad z\bar{z} - z\bar{z}_2 - z_1\bar{z} + z_1\bar{z}_2 = z\bar{z} - z\bar{z}_1 - z_2\bar{z} + z_2\bar{z}_1$$

$$\Rightarrow \quad (\bar{z}_1 - \bar{z}_2)z - (z_1 - z_2)\bar{z} + (z_1\bar{z}_2 - z_2\bar{z}) = 0.$$

Multiplying by i, we get

$$(\bar{z}_1 - \bar{z}_2)zi - (z_1 - z_2)\bar{z}\, i + (z_1\bar{z}_2 - z_2\bar{z})i = 0.$$

Since $\bar{z}_1\, z_2$ is conjugate to $z_1\, \bar{z}_2$, the number $z_1\bar{z}_2 - z_2\bar{z}$ is purely imaginary and hence $i(z_1\bar{z}_2 - z_2\bar{z})$ is real, say $i(z_1\bar{z}_2 - z_2\bar{z}) = b$, where $b \in \mathbb{R}$. Denoting $i(z_1 - z_2)$ by a, we get $(\bar{z}_1 - \bar{z}_2)i = \bar{a}$. Thus the equation reduces to

$$\bar{a}\, z + a\, \bar{z} + b = 0, \text{ where } a \neq 0, b \in \mathbb{R},$$

which is the required general equation of a straight line.

Definition 1.2.1. [Reflection point]: Two points $P(z_1)$ and $Q(z_2)$ are called the *reflection points* for a given straight line l (Fig. 1.4) if the line l is the perpendicular bisector of the line segment PQ.

Theorem 1.2. *Two points z_1 and z_2 are the reflection points for the straight line $l : \bar{a}z + a\bar{z} + b = 0$ if and only if $\bar{a}z_1 + a\bar{z}_2 + b = 0$.*

Proof [Necessary condition]: Let z_1 and z_2 be reflection points for the straight line l (Fig. 1.4) such that for any point z on the line l, we have $|z - z_1| = |z - z_2|$. Therefore,

$$\begin{aligned} & |z - z_1|^2 = |z - z_2|^2 \\ \Rightarrow \quad & (z - z_1)\overline{(z - z_1)} = (z - z_2)\overline{(z - z_2)} \\ \Rightarrow \quad & (z - z_1)(\bar{z} - \bar{z}_1) = (z - z_2)(\bar{z} - \bar{z}_2) \\ \Rightarrow \quad & (\bar{z}_2 - \bar{z}_1)z + (z_2 - z_1)\bar{z} + z_1\bar{z}_1 - z_2\bar{z}_2 = 0. \end{aligned}$$

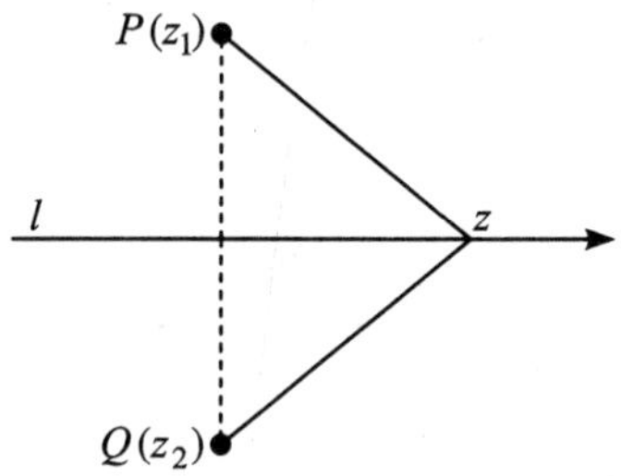

Figure 1.4 *Reflection points for a straight line*

Since the equation is true for any point z on the given line l, it may be regarded as the equation of the line l. Thus, comparing, we get

$$\frac{\bar{a}}{\bar{z}_2 - \bar{z}_1} = \frac{a}{z_2 - z_1} = \frac{b}{z_1\bar{z}_1 - z_2\bar{z}_2} = k \text{ (say)}.$$

$$\Rightarrow \quad a = k(z_2 - z_1);\ \bar{a} = k(\bar{z}_2 - \bar{z}_1) \text{ and } b = k(z_1\bar{z}_1 - z_2\bar{z}_2).$$

$$\begin{aligned} \therefore \quad \bar{a}z_1 + a\bar{z}_2 + b &= k\Big[z_1(\bar{z}_2 - \bar{z}_1) + \bar{z}_2(z_2 - z_1) + (z_1\bar{z}_1 - z_2\bar{z}_2)\Big] \\ &= k\Big[z_1\bar{z}_2 - z_1\bar{z}_1 + z_2\bar{z}_2 - z_1\bar{z}_2 + z_1\bar{z}_1 - z_2\bar{z}_2\Big] = 0. \end{aligned}$$

Sufficient condition: Conversely, suppose $\bar{a}z_1 + a\bar{z}_2 + b = 0$ is satisfied. Subtracting from l we get

$$\begin{aligned} &\bar{a}(z - z_1) + a(\bar{z} - \bar{z}_2) = 0 \\ \Rightarrow\ &\bar{a}(z - z_1) = -a(\bar{z} - \bar{z}_2) \\ \Rightarrow\ &|\bar{a}||z - z_1| = |a||\bar{z} - \bar{z}_2| \\ \Rightarrow\ &|z - z_1| = |\bar{z} - \bar{z}_2| = |\overline{z - z_2}| = |z - z_2| \end{aligned}$$

for any point z on the given line. Therefore, z_1 and z_2 are reflection points for the line l. □

Example 1.2.3. *Find the equation of the circle whose center is z_0 and radius is R.*

Solution. Let z be a complex number lying on the circle (Fig. 1.5); obviously the line joining z and z_0 is the radius R of the circle, i.e., $|z - z_0| = R$. Its parametric representation is given by

$$z = z_0 + R\,e^{i\theta}, \text{ where } 0 \le \theta \le 2\pi.$$

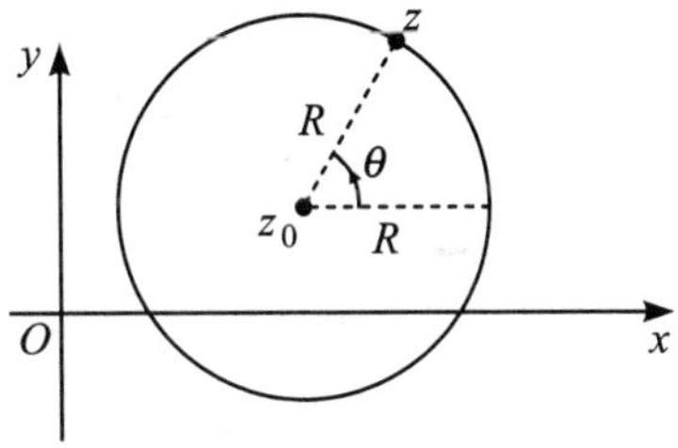

Figure 1.5 *Equation of a circle in terms of complex numbers*

Now, $|z - z_0| = R$ can be written as

$$\begin{aligned} &|z - z_0|^2 = R^2 \Rightarrow (z - z_0)(\overline{z - z_0}) = R^2 \\ \Rightarrow\ &(z - z_0)(\bar{z} - \bar{z}_0) = R^2; \text{ as } R \in \mathbb{R} \\ \Rightarrow\ &z\bar{z} - \bar{z}_0 z - z_0\bar{z} + (z_0\bar{z}_0 - R^2) = 0, \end{aligned}$$

which is the required equation of the circle. The general equation of a circle is of the form

$$z\bar{z} - \bar{z}_0 z - z_0\bar{z} + \beta = 0; \quad \beta = |z_0|^2 - R^2 \in \mathbb{R}$$
$$\Big|z - z_0\Big|^2 = z_0\bar{z}_0 - \beta = |z_0|^2 - \beta$$

provided $z_0\bar{z}_0 - \beta > 0$, with center $z = z_0$ and radius $\sqrt{z_0\bar{z}_0 - \beta}$.

Definition 1.2.2. [Inverse points with respect to the circle]: Two points $P(z_1)$ and $Q(z_2)$ are said to be *inverse points* with respect to a circle (Fig. 1.6) with center $O(z_0)$ and radius R if Q lies on the ray OP and

$$OP \cdot OQ = |z_1 - z_0| \cdot |z_2 - z_0| = R^2. \tag{1.13}$$

Example 1.2.4. *If z_1, z_2 are inverse points with respect to the circle $\Gamma : z\bar{z} + z_0\bar{z} + \bar{z}_0 z + \beta = 0$, then show that $z_1\bar{z}_2 + z_0\bar{z}_2 + \bar{z}_0 z_1 + \beta = 0$.*

Solution. The equation of the circle Γ can be written in the form

$$|z + z_0|^2 = z_0\bar{z}_0 - \beta = R^2$$

with centre at $O(-z_0)$ and radius $R = \sqrt{z_0\bar{z}_0 - \beta}$ (Fig. 1.6). Let P represents z_1 and Q represents z_2. Suppose P and Q are inverse points with respect to the circle Γ, then O, P, Q are collinear, so by definition

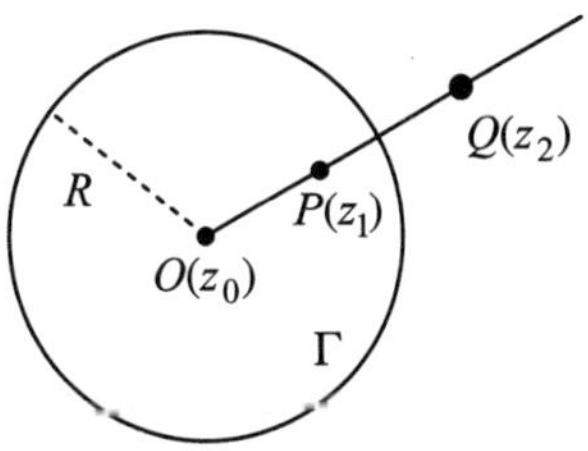

Figure 1.6 *Inverse points with respect to a circle*

$$\arg(z_1 + z_0) = \arg(z_2 + z_0) \text{ and } |z_1 + z_0||z_2 + z_0| = z_0\bar{z}_0 - \beta.$$

Therefore,

$$\begin{aligned} \arg(z_1 + z_0)(\overline{z_2 + z_0}) &= \arg(z_1 + z_0) + \arg(\overline{z_2 + z_0}) \\ &= \arg(z_1 + z_0) - \arg(z_2 + z_0) = 0, \end{aligned}$$

giving $(z_1 + z_0)(\overline{z_2 + z_0}) = (z_1 + z_0)(\bar{z}_2 + \bar{z}_0)$ to be a positive real number.

$$
\begin{aligned}
\therefore \quad & |(z_1 + z_0)(\bar{z}_2 + \bar{z}_0)| = (z_1 + \alpha)(\bar{z}_2 + \bar{z}_0) \\
\text{or,} \quad & (z_1 + z_0)(\bar{z}_2 + \bar{z}_0) = |z_1 + z_0||z_2 + z_0| \\
\text{or,} \quad & z_1 z_2 + z_0 \bar{z}_2 + \bar{z}_0 z_1 + z_0 \bar{z}_0 = z_0 \bar{z}_0 - \beta \\
\text{or,} \quad & z_1 \bar{z}_2 + z_0 \bar{z}_2 + \bar{z}_0 z_1 + \beta = 0.
\end{aligned}
$$

The converse can be similarly proved.

(i) If z_2 is the inverse point of z_1 with respect to the circle $|z - z_0| = R$, then z_0, z_1, z_2 lie on the same straight line. Hence,

$$z_1 = z_0 + R_1 e^{i\theta} \text{ and } z_2 = z_0 + R_2 e^{i\theta}.$$

Also, as z_1 and z_2 are inverse points, so $R_1 R_2 = R^2$ and $(z_1 - z_0)(\bar{z}_2 - \bar{z}_0) = R^2$.

(ii) For any point z_1, its inverse with respect to the unit circle $|z| = 1$ is $1/\bar{z}_1$.

(iii) Two points z_1 and z_2 are inverse points or the reflection points for the joint family of straight line and circle $\alpha z\bar{z} + \bar{\beta} z + \beta \bar{z} + \gamma = 0$ if and only if $\alpha z_1 \bar{z}_2 + \bar{\beta} z_1 + \beta \bar{z}_2 + \gamma = 0$.

Two Families of Circles

The equation $|\frac{z-z_1}{z-z_2}| = \lambda$, where λ is positive constant, represents a family of circles such that z_1 and z_2 are inverse points for each member of the family. Now

$$
\begin{aligned}
& \left|\frac{z - z_1}{z - z_2}\right| = \lambda \Rightarrow \left|\frac{z - z_1}{z - z_2}\right|^2 = \lambda^2 \\
\Rightarrow \quad & \left(\frac{z - z_1}{z - z_2}\right)\overline{\left(\frac{z - z_1}{z - z_2}\right)} = \lambda^2 \Rightarrow \left(\frac{z - z_1}{z - z_2}\right)\left(\frac{\bar{z} - \bar{z}_1}{\bar{z} - \bar{z}_2}\right) = \lambda^2 \\
\Rightarrow \quad & (1 - \lambda^2)\bar{z}z + (\bar{z}_2\lambda^2 - \bar{z}_1)z + (z_2\lambda^2 - z_1)\bar{z} + (z_1\bar{z}_1 - \lambda^2 z_2\bar{z}_2) = 0, \qquad (1.14)
\end{aligned}
$$

which represents the equation of a circle for every positive value of λ. Equation (1.14) can be written as

$$
\begin{aligned}
& |z|^2 - \frac{\bar{z}_1 - \bar{z}_2\lambda^2}{1 - \lambda^2} z - \frac{z_1 - z_2\lambda^2}{1 - \lambda^2} \bar{z} + \frac{|z_1|^2 - \lambda^2|z_2|^2}{1 - \lambda^2} = 0 \\
\text{or,} \quad & \left(z - \frac{z_1 - z_2\lambda^2}{1 - \lambda^2}\right)^2 = \frac{|z_1 - z_2\lambda^2|^2}{(1 - \lambda^2)^2} - \frac{|z_1|^2 - \lambda^2|z_2|^2}{(1 - \lambda^2)^2} \\
\text{or,} \quad & \left(z - \frac{z_1 - z_2\lambda^2}{1 - \lambda^2}\right)^2 = \frac{\lambda^2}{(1 - \lambda^2)^2}\left[|z_1|^2 + |z_2|^2 - \bar{z}_1 z_2 - \bar{z}_2 z_1\right] \\
\text{or,} \quad & \left(z - \frac{z_1 - z_2\lambda^2}{1 - \lambda^2}\right)^2 = \frac{\lambda^2}{(1 - \lambda^2)^2} |z_1 - z_2|^2,
\end{aligned}
$$

which shows that the circle (1.14) has its centre at $z_0 = (z_1 - z_2\lambda^2)/(1-\lambda^2)$ and its radius $R = \lambda(|z_1 - z_2|)/(1-\lambda^2)$. Now,

$$z_1 - z_0 = \frac{\lambda^2(z_2 - z_1)}{1-\lambda^2}, \bar{z}_2 - \bar{z}_0 = \frac{\bar{z}_2 - \bar{z}_1}{1-\lambda^2}$$

$$\Rightarrow \quad (z_1 - z_0)(\bar{z}_2 - \bar{z}_0) = \frac{\lambda^2}{(1-\lambda^2)^2}|z_1 - z_2|^2 = R^2,$$

which is the definition of the inverse points. Thus, the condition for z_1 and z_2 to be inverse points is satisfied for the circle (1.14) for all values of λ. In particular, the centre of every circle of the family must be collinear with z_1 and z_2.

(i) If $\lambda = 1$, the equation $|\frac{z-z_1}{z-z_2}| = \lambda = 1$ reduces to $|z - z_1| = |z - z_2|$, which means that z is equidistant from z_1 and z_2. The right bisector of the line segment joining z_1 and z_2 is also a member of the given family. The points z_1 and z_2 are inverse points for this line also.

(ii) The equation $\arg(\frac{z-z_1}{z-z_2}) = \mu$(constant) represents a family of circles and its every member passes through the points z_1 and z_2 (Fig. 1.7). The locus of the point z such that $\arg(\frac{z-z_1}{z-z_2}) = \mu$, where μ represents an arc of a circle through the points z_1 and z_2. The remaining part of the circle whose part is this arc is given by $\arg(\frac{z-z_1}{z-z_2}) = \mu + \pi$.

The line that joins the points z_1 and z_2 is itself a member of the given family, arising as it does for $\mu = 0$ and π.

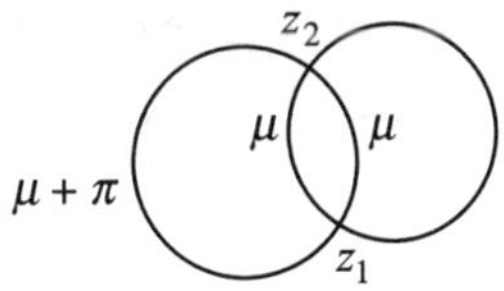

Figure 1.7 *Two families of circles number*

Example 1.2.5. *Find the family of circles which are orthogonal to the circles* $|z| = 1$ *and* $|z-1| = 4$.

Solution. The two circles cut orthogonally if the sum of the squares of their radii is equal to the square of the distance between their centres. Let the circle $|z - z_0| = R$, where $z_0 = a + ib$, be orthogonal to the circles $|z| = 1$ and $|z-1| = 4$. Then,

$$R^2 + 1 = |z_0 - 0|^2 = |z_0|^2 = z_0\bar{z}_0$$

$$\text{and} \quad R^2 + 16 = |z_0 - 1|^2 = (z_0 - 1)(\bar{z}_0 - 1) = z_0\bar{z}_0 - z_0 - \bar{z}_0 + 1$$

$$\Rightarrow z_0\bar{z}_0 - R^2 = 1 \text{ and } z_0\bar{z}_0 - R^2 = z_0 + \bar{z}_0 + 15.$$

From these two relations, we get

$$1 - (z_0 + \bar{z}_0 + 15) = 0 \Rightarrow z_0 + \bar{z}_0 = -14 \Rightarrow 2a = -14 \Rightarrow a = -7$$

$$\therefore \quad z_0 = -7 + ib \text{ and } R^2 = |z_0|^2 - 1 = (-7)^2 + b^2 - 1 = 48 + b^2.$$

Thus the required family of of circles is $|z + 7 - ib| = \sqrt{48 + b^2}; b \in \mathbb{R}$.

1.2.1 Polar Form of Complex Number

Consider a complex number $z = a + ib \in \mathbb{C}/\{0\}$. Let $(r, \theta) : x = r\cos\theta, y = r\sin\theta$ be polar coordinates of the point (a, b), where the polar axis is the non-negative real axis (Fig. 1.8). Then,

$$z = (a, b) = (r\cos\theta, r\sin\theta) = r(\cos\theta, \sin\theta) \tag{1.15}$$

is called the *polar form of the complex number z*. For convenience, if $e^{i\theta} = \cos\theta + i\sin\theta$, then the polar form of z can be written as

$$z = r\, e^{i\theta}, \quad \text{where } r = |z| = \sqrt{a^2 + b^2}. \tag{1.16}$$

r is called the *absolute value* or the *magnitude of the complex number* and θ, the angle measured in radians that z makes with x-axis is called the *amplitude* or *argument* of z and is denoted by arg z or amp z.

Now $r = |z|$ is uniquely determined by z, but θ is not unique. If $\theta = \arg z$, then $\theta + 2k\pi$, where k is any integer, is also a value of $\arg z$. However, we have the following results:

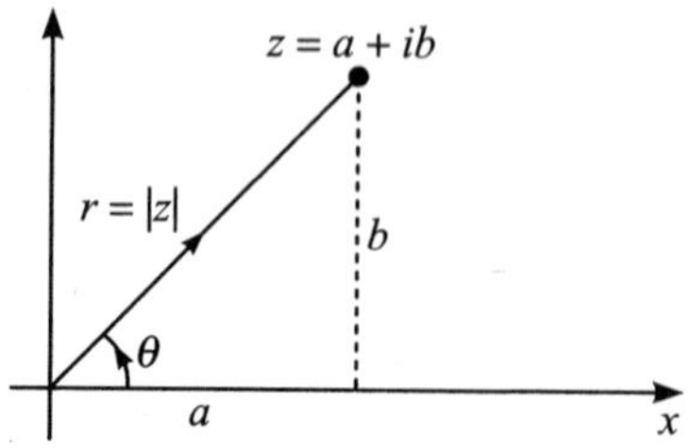

Figure 1.8 *Polar form of a complex number*

(i) If $z = |z|e^{i\theta} = |z|e^{i\phi}$, then $\theta - \phi = 2k\pi$ for some integer k.

(ii) If $z = re^{i\theta} \neq 0$, then there is a unique argument of z, called the *principal argument* of z and denoted by $\arg z$, satisfying the condition

$$-\pi < \arg z \leq \pi; \quad \text{i.e., } \arg z \in (-\pi, \pi].$$

This unique value of θ or $\arg z$ is called the *principle value* of θ and is denoted by Arg z. Thus,

$$\arg z = \{ \text{Arg } z + 2\pi k : k \in \mathbb{Z}\}; z \neq 0.$$

(iii) If $z = re^{i\theta}$, then $\bar{z} = re^{-i\theta}$. For

$$\bar{z} = \overline{r(\cos\theta, \sin\theta)} = r(\cos\theta, -\sin\theta) = r(\cos[-\theta], \sin[-\theta]) = re^{-i\theta}.$$

Therefore, arg $\bar{z} = -\theta = -\arg z$.

If $z = a + ib$, then

(i) $\arg z = \tan^{-1}\frac{b}{a}$ if $a > 0$;

(ii) $\arg z = \tan^{-1}\frac{b}{a} + \pi$ if $a < 0, b \geq 0$; and $\arg z = \tan^{-1}\frac{b}{a} - \pi$ if $a < 0, b < 0$;

(iii) $\arg z = \frac{\pi}{2}$ if $a = 0, b > 0$; and $\arg z = -\frac{\pi}{2}$ if $a = 0, b < 0$.

Example 1.2.6. *If z_1 and z_2 are any two non-zero complex numbers, then*
(a) $\arg z_1 z_2 = \arg z_1 + \arg z_2$; *(b)* $\arg(z_1/z_2) = \arg z_1 - \arg z_2$.

Solution. Let $z_1 = r_1 e^{i\theta_1} = r_1(\cos\theta_1, \sin\theta_1)$ and $z_2 = r_2 e^{i\theta_2} = r_2(\cos\theta_2, \sin\theta_2)$, where $r_1 = |z_1|$, $r_2 = |z_2|$ and $\theta_1, \theta_2 \in (-\pi, \pi]$.

(a) By Eq. (1.5), we have

$$\begin{aligned} z_1 z_2 &= (r_1\cos\theta_1, r_1\sin\theta_1)\cdot(r_2\cos\theta_2, r_2\sin\theta_2) \\ &= (r_1 r_2[\cos\theta_1\cos\theta_2 - \sin\theta_1\sin\theta_2], r_1 r_2[\cos\theta_1\sin\theta_2 + \sin\theta_1\cos\theta_2]) \\ &= r_1 r_2(\cos[\theta_1+\theta_2], \sin[\theta_1+\theta_2]) = r_1 r_2 e^{i(\theta_1+\theta_2)}. \end{aligned} \quad (1.17)$$

Therefore, $\arg z_1 \cdot z_2 = \theta_1 + \theta_2 = \arg z_1 + \arg z_2$. Since $z_1 \cdot z_2 = z_2 \cdot z_1$, so $r_1 r_2 = r_2 r_1$ and $\theta_1 + \theta_2 = \theta_2 + \theta_1$. From the definition of a quotient, we obtain the usual rule for multiplying fractions:

$$\frac{z_1}{z_2}\frac{p}{q} = \frac{z_1 p}{z_2 q}; \text{ if } z_2 q \neq 0.$$

(b) By Eq.(1.6), the quotient $\frac{z_1}{z_2}$ is given by

$$\frac{z_1}{z_2} = \frac{r_1 e^{i\theta_1}}{r_2 e^{i\theta_2}} = \frac{r_1}{r_2} e^{\theta_1 - \theta_2}, \; z_2 \neq 0. \tag{1.18}$$

Thus, if $z_1 = r_1 e^{i\theta_1}$ and $z_2 = r_2 e^{i\theta_2} \neq 0$, then there is a unique $q \in \mathcal{D}$ such that $z_2 q = z_1$. From this definition of $\frac{z_1}{z_2}$, we observe that the modulus of the quotient of two complex numbers is equal to the quotient of their noduli, and

$$\arg\left(\frac{z_1}{z_2}\right) = \theta_1 - \theta_2 = \arg z_1 - \arg z_2.$$

Now, if $z_2 \neq 0$, then

$$\frac{z_1}{z_2} = \frac{z_1\,\bar{z}_2}{z_2\,\bar{z}_2} = \frac{1}{|z_2|^2} z_1 \bar{z}_2. \tag{1.19}$$

This equation gives the extremely useful rule for writing the quotient $\frac{z_1}{z_2}$ in the form of $A + iB$.

Example 1.2.7. *If* $|\frac{z_1 - z_2}{1 - \bar{z}_1 z_2}| < 1$, *then* $|z_1| < 1$ *and* $|z_2| < 1$.

Solution. Let $z_1 = r_1 e^{i\theta_1}$ and $z_2 = r_2 e^{i\theta_2}$, where $|z_1| = r_1$ and $|z_2| = r_2$. Now, consider

$$\begin{aligned}
& z_1 - z_2 = r_1 e^{i\theta_1} - r_2 e^{i\theta_2} = (r_1 \cos\theta_1 - r_2 \cos\theta_2) + i(r_1 \sin\theta_1 - r_2 \sin\theta_2) \\
\Rightarrow \quad & |z_1 - z_2|^2 = (r_1 \cos\theta_1 - r_2 \cos\theta_2)^2 + (r_1 \sin\theta_1 - r_2 \sin\theta_2)^2 \\
& \qquad = r_1^2 + r_2^2 - 2r_1 r_2 \cos(\theta_1 - \theta_2),
\end{aligned}$$

and

$$\begin{aligned}
& 1 - \bar{z}_1 z_2 = 1 - r_2 r_1 e^{-(\theta_1 - \theta_2)} = 1 - r_1 r_2 \cos(\theta_1 - \theta_2) + i r_1 r_2 \sin(\theta_1 - \theta_2) \\
\Rightarrow \quad & |1 - \bar{z}_1 z_2|^2 = \{1 - r_1 r_2 \cos(\theta_1 - \theta_2)\}^2 + \{r_1 r_2 \sin(\theta_1 - \theta_2)\}^2 \\
& \qquad = 1 + r_1^2 r_2^2 - 2r_1 r_2 \cos(\theta_1 - \theta_2).
\end{aligned}$$

Also let $|\frac{z_1 - z_2}{1 - \bar{z}_1 z_2}| < 1$, i.e., $|z_1 - z_2|^2 < |1 - \bar{z}_1 z_2|^2$, then

$$\begin{aligned}
& r_1^2 + r_2^2 - 2r_1 r_2 \cos(\theta_1 - \theta_2) < 1 + r_1^2 r_2^2 - 2r_1 r_2 \cos(\theta_1 - \theta_2) \\
\Rightarrow \quad & r_1^2 + r_2^2 - 1 - r_1^2 r_2^2 < 0 \\
\Rightarrow \quad & (r_1^2 - 1)(1 - r_2^2) < 0.
\end{aligned}$$

This is possible only if $r_1 < 1$ and $r_2 < 1$, i.e., $|z_1| < 1$ and $|z_2| < 1$.

Example 1.2.8. *Sketch the set of points in the complex plane satisfying the given inequality:* $\pi/6 < \arg(z - 2 - 3i) \leq \pi$.

Solution. The set of all points given by the inequality $\pi/6 < \arg(z - 2 - 3i) \leq \pi$ is geometrically represented in Fig. 1.9.

RESULT **8.** Suppose that z_1 and z_2 are non-zero complex numbers and $\operatorname{Re}(z_1\,\bar{z}_2) = 0$. Then

$$\begin{aligned}
\operatorname{Re}(z_1\,\bar{z}_2) = 0 \quad & \Leftrightarrow \quad z_1 \bar{z}_2 + \bar{z}_1 z_2 = 0 \Leftrightarrow \frac{z_1}{\bar{z}_1} = -\frac{z_2}{\bar{z}_2}, \text{ i.e., } \frac{z_1^2}{|z_1|^2} = -\frac{z_2^2}{|z_2|^2} \\
& \Leftrightarrow \quad 2\arg z_1 = \pi + 2\arg z_2 + 2k\pi; \;\; k \in \mathbb{Z} \\
& \Leftrightarrow \quad \arg z_1 = \arg z_2 + (2k+1)\pi/2.
\end{aligned}$$

This means that the vectors z_1 and z_2 are perpendicular if and only if $\text{Re}(z_1\, \bar{z}_2) = 0$. This can be written as $z_1 = itz_2$ for some $t \in \mathbb{R}$.

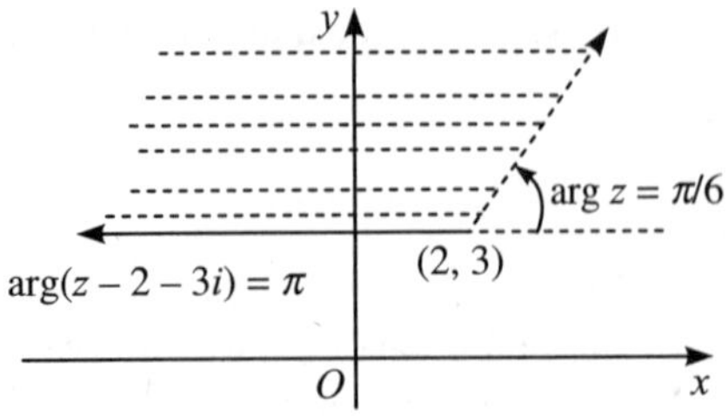

Figure 1.9 $\pi/6 < \arg(z - 2 - 3i) \le \pi$

RESULT **9.** Suppose that z_1 and z_2 are non-zero complex numbers and $\text{Im}(z_1\, \bar{z}_2) = 0$. Then

$$\begin{aligned}\text{Im}(z_1\, \bar{z}_2) = 0 \quad &\Leftrightarrow \quad z_1\bar{z}_2 - \bar{z}_1 z_2 = 0 \Leftrightarrow 2\,\text{Im}\,(z_1\, \bar{z}_2) = 0 \\ &\Leftrightarrow \quad \text{Im}\,(z_1/z_2) = 0;\ \text{i.e., } z_1/z_2 \text{ is real} \\ &\Leftrightarrow \quad \arg z_1 = \arg z_2 + k\pi, \quad k \in \mathbb{Z}.\end{aligned}$$

This means that the vectors z_1 is parallel to the vector z_2 if and only if $\text{Im}\,(z_1\, \bar{z}_2) = 0$. This can be written as $z_1 = tz_2$ for some $t \in \mathbb{R}$.

1.2.2 n^{th} Root of Complex Numbers

For $z \in \mathbb{C}$ and an integer n, we define the nth power of z as follows:

$$\begin{aligned} &z^0 = 1, \quad z^{n+1} = z^n \cdot z \text{ for } n \ge 0 \\ &z^{-n} = (z^{-1})^n \ \text{ if } z \ne 0 \text{ and } n > 0. \end{aligned} \tag{1.20}$$

When m, n are integers,

$$z^m \cdot z^n = z^{m+n} \text{ and } (z_1 \cdot z_2)^n = z_1^n \cdot z_2^n.$$

Theorem 1.3. [DeMoivre's theorem]: *If n is a positive integer,*

$$(\cos\theta + i\sin\theta)^n = \cos n\theta + i \sin n\theta.$$

Proof: Here we use the principle of mathematical induction. The result is true for $n = 1$. For $n = 2$,

$$\begin{aligned}(\cos\theta + i\sin\theta)^2 &= \cos^2\theta - \sin^2\theta + 2i\sin\theta\cos\theta \\ &= \cos 2\theta + i\sin 2\theta.\end{aligned}$$

Thus the result is true for $n = 2$. Let the result be true for $n = k$. Now,

$$\begin{aligned}(\cos\theta + i\sin\theta)^{k+1} &= (\cos\theta + i\sin\theta)^k \cdot (\cos\theta + i\sin\theta) \\ &= (\cos k\theta + i\sin k\theta) \cdot (\cos\theta + i\sin\theta) \\ &= \cos(k+1)\theta + i\sin(k+1)\theta.\end{aligned}$$

Hence the result is true for $n = k + 1$ if it true for $n = k$. But it is true for $n = 1, 2$ and so must be true for all positive integers. □

RESULT **10.** If $z \ne 0$ and if n is a positive integer, then there are exactly n distinct complex numbers $z_0, z_1, \cdots, z_{n-1}$ called the nth root of z, such that $z_k^n = z$ for each $k = 0, 1, \cdots, n-1$. Furthermore, the roots are given by the formula $z_k = re^{i\theta_k}$, where

$$r = |z|^{1/n} \text{ and } \theta_k - \frac{\theta}{n} + \frac{2\pi k}{n};\ k = 0, 1, \cdots, n-1.$$

From this we have the following results:

(i) When $z = 1$, we have $r = 1$ and $\theta = 0$. Thus the nth roots of unity are given by

$$\cos\frac{2k\pi}{n} + i\sin\frac{2k\pi}{n};\ k = 0, 1, \cdots, n-1.$$

Let $\omega = \cos\frac{2\pi}{n} + i\sin\frac{2\pi}{n} = e^{2\pi i/n}$, then the nth roots of unity are given by $1, \omega, \omega^2, \cdots, \omega^{n-1}$. The *primitive* n^{th} *root of unity* is a complex number ω such that $1, \omega, \omega^2, \cdots, \omega^{n-1}$ are distinct nth roots of unity.

(ii) Since $1, \omega, \omega^2, \cdots, \omega^{n-1}$, are the roots of the equation $z^n = 1$ we have $\omega^n = 1$ and $1 + \omega + \omega^2 + \cdots + \omega^{n-1} = 0$.

(iii) If z is any non-zero complex number and z_0 is any one of the nth root of z, then the set of all n^{th} roots of z are given by $z_0, z_0\omega, z_0\omega^2, \cdots, z_0\omega^{n-1}$.

1.2.3 Square Roots

Let the complex number $x + iy$ be the square root of a given complex number $a + ib$. Then

$$(x + iy)^2 = a + ib \Rightarrow (x^2 - y^2) + i2xy = a + ib.$$

Equating real and imaginary parts, we get $x^2 - y^2 = a, 2xy = b$. Now,

$$(x^2 + y^2)^2 = (x^2 - y^2)^2 + 4x^2y^2 = a^2 + b^2 \Rightarrow x^2 + y^2 = \sqrt{a^2 + b^2}.$$

Solving, we get,

$$x^2 = \frac{1}{2}(a + \sqrt{a^2 + b^2});\quad y^2 = \frac{1}{2}(-a + \sqrt{a^2 + b^2})$$

$$\Rightarrow\quad x = \pm\frac{1}{\sqrt{2}}(a + \sqrt{a^2 + b^2})^{1/2};\quad y = \pm\frac{1}{\sqrt{2}}(-a + \sqrt{a^2 + b^2})^{1/2}.$$

From $2xy = b$ we notice that if $b > 0$, x and y must be of the same sign and if $b < 0$, x and y must be of opposite signs. Thus, by choosing appropriate signs for x and y we get two square roots of $a + ib$. If $b = 0$, then z is real and we get two square roots as $\pm\sqrt{a}$ provided $a > 0$, and $\pm i\sqrt{a}$ provided $a < 0$.

Example 1.2.9. *Find the cube roots of unity and show that they form an equilateral triangle.*

Solution. Let ω be a cube root of unity. Then we have

$$\omega = (1)^{1/3} = (\cos 0 + i\sin 0)^{1/3} = (\cos 2k\pi)^{1/3} = \cos\frac{2k\pi}{3};\quad k = 0, 1, 2.$$

The three values of ω are given by

$$k = 0;\qquad \omega_1 = \cos 0 = 1,$$

$$k = 1;\qquad \omega_2 = \cos\frac{2\pi}{3} = \cos 120^\circ + i\sin 120^\circ = -\frac{1}{2} + i\frac{\sqrt{3}}{2},$$

$$k = 2;\qquad \omega_2 = \cos\frac{4\pi}{3} = \cos 240^\circ + i\sin 240^\circ = -\frac{1}{2} - i\frac{\sqrt{3}}{2}.$$

These three points are represented by the points P, Q, R on the Argand diagram such that $OP = OQ = OR$ and $\angle POQ = 120^\circ$ and $\angle POR = 240^\circ$ (Fig. 1.10). Therefore, these points lie on a circle with centre O and unit radius such that $\angle POQ = \angle QOR = \angle ROP = 120^\circ$, i.e., $PQ = QR = RP$. Hence, $\triangle PQR$ is an equilateral triangle.

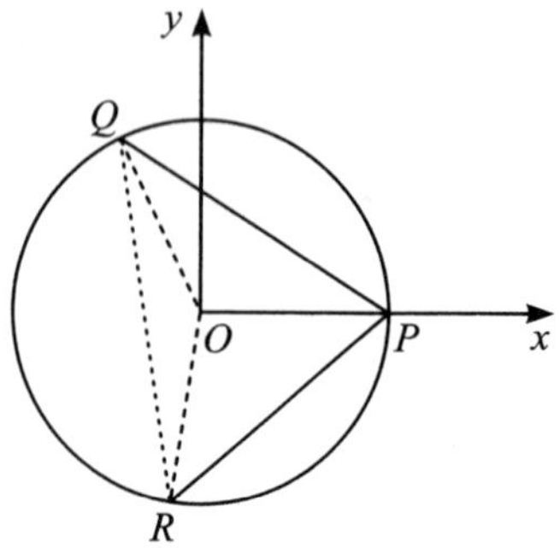

Figure 1.10 *Cube roots of unity*

1.2.4 Inequalities

From $|z| = \sqrt{[\text{Re}(z)]^2 + [\text{Im}(z)]^2}$, we have

$$|z| \geq |\text{Re}(z)| \geq \text{Re}(z) \text{ and } |z| \geq |\text{Im}(z)| \geq \text{Im}(z). \tag{1.21}$$

Theorem 1.4. [Triangle inequalities]: *If z, w and p are complex numbers, then*
(a) $|z + w| \leq |z| + |w|$; (b) $|z - w| \leq |z - p| + |p - w|$; (c) $|z - w| \geq ||z| - |w||$.

Proof: (a) If z and w are complex numbers, then $\overline{z\bar{w}} = \bar{z}w$. Now

$$\begin{aligned}
|z+w|^2 &= (z+w)(\overline{z+w}) = (z+w)(\bar{z}+\bar{w}) \\
&= z \cdot \bar{z} + (z \cdot \bar{w} + \bar{z} \cdot w) + w \cdot \bar{w} \\
&= |z|^2 + 2\text{Re}(z \cdot \bar{w}) + |w|^2 \\
&\leq |z|^2 + 2|z \cdot \bar{w}| + |w|^2; \text{ as } |\bar{w}| = |w| \\
&\leq |z|^2 + 2|z| \cdot |w| + |w|^2 = (|z| + |w|)^2.
\end{aligned}$$

Therefore, $|z + w| \leq |z| + |w|$.
(b) Let $Z = z - p$ and $W = p - w$, then using (a) we get

$$|z - w| \leq |z - p| + |p - w|.$$

(c) Now

$$\begin{aligned}
|z| &= |(z-w) + w| \leq |z-w| + |w|, \text{ by part } (a). \\
&\Rightarrow |z - w| \geq |z| - |w|.
\end{aligned} \tag{1.22}$$

Also,

$$\begin{aligned}
|w| = |-w| &= |(z-w) + (-z)| \leq |z-w| + |z| \\
&\Rightarrow |z - w| \geq |w| - |z|.
\end{aligned} \tag{1.23}$$

Using inequalities (1.22) and (1.23), we get the relation (c). □

RESULT **11.** For two complex numbers z_1, z_2, using the inequality $|z_1 + z_2| \leq |z_1| + |z_2|$, we get

$$|2z| = |z + a + (z - a)| \leq |z + a| + |z - a| = 2c$$

i. e., $\max_{z \in S} |z| = c.$

Here the maximum is attained at $z = ce^{-i\theta}$, where $\theta = \arg(c/a)$.

Example 1.2.10. *If* $|z - (4 - 3i)| = 2$, *find the greatest and least value of* $|z|$.

Solution. Using the inequality $|z_1 + z_2| \leq |z_1| + |z_2|$, we get

$$\begin{aligned} |2z| &= |z + z_0 + (z - z_0)| \leq |z + z_0| + |z - z_0|; \quad z_0 = 4 - 3i \\ &\leq 2 + |z| + |4 - 3i| = 2 + |z| + 5 \\ &\Rightarrow \max |z| = 7. \end{aligned}$$

Thus the greatest value of $|z|$ is 7. Using relation (1.23), we get

$$|z| \geq |z_0| - |z - z_0| = 5 - 2 = 3.$$

1.3 Stereographic Projection

Complex numbers can be represented by points on a sphere. For this it is sufficient to establish a one-one correspondence between the points on a sphere and the points on the plane. This correspondence is established through the projection of a sphere on a plane (Argand plane) at the origin. This projection is called the *stereographic projection.*

The complex plane $\mathbb{C}$, together with this point at ∞ is called the *extended complex plane*, denoted by set-theoretically $\mathbb{C}_\infty$, i.e., $\mathbb{C}_\infty = \mathbb{C} \cup \{\infty\}$.

(a) **Stereographic projection where the complex plane passes through the equator:** Let us consider the three dimensional Euclidean space $\mathbb{R}^3$ and identify $z = x + iy \in \mathbb{C}$ with the point P on the equilateral plane whose coordinatesare $(x, y, 0)$ with respect to the system Ox_1, Ox_2, Ox_3 of rectangular axes (Fig. 1.11) where we presume that the axis Ox_1 coincides with the real axis, Ox_2 coincides with the imaginary axis of the complex plane.

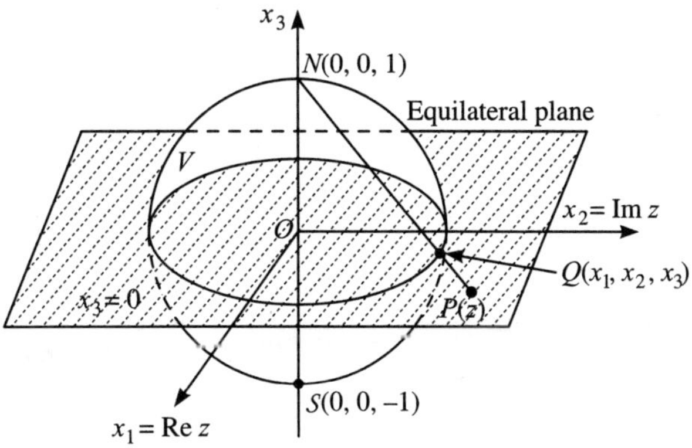

Figure 1.11 *Stereographic projection (complex plane passes through the equator)*

Let the equation of the sphere V of unit radius and centre at $(0,0,0)$ be

$$V : \{(x_1, x_2, x_3) \in \mathbb{R}^3 : x_1^2 + x_2^2 + x_3^2 = 1\}. \tag{1.24}$$

We identify $\mathbb{C}$ with $\{(x_1, x_2, 0) : x_1, x_2 \in \mathbb{R}\}$ so that $\mathbb{C}$ cuts V along the equator.

Let $N = (0,0,1)$, i.e., N is the north pole on V. Let (x_1, x_2, x_3) be the coordinatesof the point Q on V other than N and $(x, y, 0)$, the coordinatesof P, where the straight line NQ meets the coordinate plane $x_3 = 0$.

The point $N(0,0,1)$ $Q(x_1, x_2, x_3)$ and $P(x, y, 0)$ are lying on a straight line. So,

$$\frac{x_1 - 0}{x - 0} = \frac{x_2 - 0}{y - 0} = \frac{x_3 - 1}{0 - 1} = r \text{ (say)}$$

$$\Rightarrow \quad x_1 = rx, \quad x_2 = ry, \quad x_3 = 1 - r$$

$$\Rightarrow \quad x = \frac{x_1}{r} = \frac{x_1}{1 - x_3}, \quad y = \frac{x_2}{r} = \frac{x_2}{1 - x_3} \tag{1.25}$$

This geometric construction gives a bijective correspondence between points on the punctured sphere $V - \{N\}$ and the complex plane; it is described analytically by the formulas (1.25). Clearly, as Q approaches N, we identify N and the point ∞ in $\mathbb{C}_\infty$. Thus $\mathbb{C}_\infty$ is represented as the sphere V.

Using Eq. (1.24), we get

$$r^2x^2 + r^2y^2 + (1 - r)^2 = 1 \Rightarrow r = \frac{2}{x^2 + y^2 + 1} = \frac{2}{|z|^2 + 1}.$$

Hence the point $Q(x_1, x_2, x_3)$ on the sphere V corresponds to the points on the complex plane $x_3 = 0$ is given by

$$\left.\begin{aligned} x_1 = rx &= \frac{2x}{|z|^2 + 1} = \frac{z + \bar{z}}{|z|^2 + 1} \\ x_2 = ry &= \frac{2y}{|z|^2 + 1} = \frac{z - \bar{z}}{i(|z|^2 + 1)} \\ x_3 = 1 - r &= 1 - \frac{2}{|z|^2 + 1} = \frac{|z|^2 - 1}{|z|^2 + 1} \end{aligned}\right\}. \tag{1.26}$$

The point Q, where line NP cuts the sphere V is called the *stereographic image of* P and conversely, the point P is known as the *stereographic projection* of Q. The sphere V is sometimes called the *Riemann sphere*.

Analytically, one may sum up the results of stereographic projection as follows:

$$\left.\begin{aligned} z \in \mathbb{C} \leftrightarrow \left(\frac{z + \bar{z}}{|z|^2 + 1}, \frac{z - \bar{z}}{i(|z|^2 + 1)}, \frac{|z|^2 - 1}{|z|^2 + 1}\right) \in V - \{N\} \\ z = \infty \leftrightarrow (0,0,1) \equiv N. \end{aligned}\right\} \tag{1.27}$$

Continuing with this geographical interpretation of the Riemann sphere, we note that under streographic projection points on the circle $|z| = 1$ (in the z-plane) remain fixed, forming the equator. Points outside the unit circle $(|z| > 1)$ project to points in the northern hemisphere, while those inside the unit circle $(|z| < 1)$ project to points in the southern hemisphere. In particular, the origin of the z-plane projects to the south pole of the Riemann sphere. Since this construction takes the unbounded set $\mathbb{C}$ into the compact set V by adding one point, the Riemann sphere is sometimes called the *one-point compactification of* $\mathbb{C}$.

(b) **Stereographic projection where the complex plane is tangential to the Riemann sphere at the south pole:** Let us consider an unit sphere V with centre at the origin (Fig. 1.12). The equation of the sphere V of unit radius and centre at $(0,0,0)$ is Eq. (1.24), and the plane of projection is $x_3 = -1$. Let $Q(x_1, x_2, x_3)$ be the coordinates of the point Q on the unit sphere V other than N and $(x, y, 0)$ the coordinates of P where the straight line NQ meets the coordinate plane $x_3 = 0$. By the similar argument, we have

$$x_1 = \frac{2(z+\bar{z})}{|z|^2+4}, \quad x_2 = \frac{2(z-\bar{z})}{i(|z|^2+4)}, \quad x_3 = \frac{|z|^2-4}{|z|^2+4}. \tag{1.28}$$

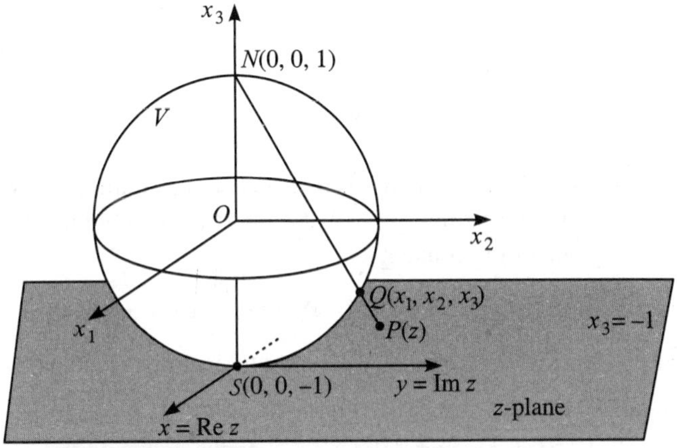

Figure 1.12 *Stereographic projection (complex plane tangential to the Riemann sphere at the south pole)*

RESULT **12.** Here we will show that the map $z \to (x_1, x_2, x_3)$ defined by Eq. (1.26) is injective. If possible, let

$$x_1' = \frac{w+\bar{w}}{|w|^2+1}, \quad x_2' = \frac{w-\bar{w}}{i(|w|^2+1)}, \quad x_3' = \frac{|w|^2-1}{|w|^2+1}.$$

If now, $(x_1, x_2, x_3) = (x_1', x_2', x_3')$, then $z = w$, since

$$x_3 = x_3' \Rightarrow |z| = |w|; \quad x_2 = x_2' \Rightarrow z - \bar{z} = w - \bar{w}; \quad x_1 = x_1' \Rightarrow z + \bar{z} = w + \bar{w}.$$

All these confirm injectivity. The same map is surjective if we have it restrict from $\mathbb{C}$ to V with its north pole N removed. Hence the stereographic map enables us to represent the points of the complex plane as points on an unit sphere V, known as *Riemann sphere*.

RESULT **13.** Any circle on $V - \{N\}$ is the intersection of some plane and the unit sphere V (with north pole removed). In rectangular Cartesian coordinate system, the equation of the circle is

$$\Gamma : x_1^2 + x_2^2 + x_3^2 = 1; \quad a_1x_1 + a_2x_2 + a_3x_3 = b, \tag{1.29}$$

where a_k's and b are all real. We know the stereographic map that builds up a one-to-one correspondence between $V - \{N\}$ and $\mathbb{C}$ is given by the equation (Eq. (1.26)). Using it, from equation (1.29) we get

$$\begin{aligned} & a_1(z+\bar{z}) - ia_2(z-\bar{z}) + a_3(|z|^2-1) - b(|z|^2+1) = 0 \\ \Rightarrow \quad & (a_3 - b)|z|^2 + (a_1 - ia_2)z + (a_1 + ia_2)\bar{z} - (a_3 + b) = 0. \end{aligned} \tag{1.30}$$

(i) If $a_3 = b$, Eq. (1.30) represents the equation of a straight line in the complex plane.

(ii) If $a_3 \neq b$ and $a_1^2 + a_2^2 + a_3^2 + b > 0$, Eq. (1.30) represents the equation of a straight line in the complex plane.

The the stereographic projection of any circle on $V - \{N\}$ is again a circle in the complex plane $\mathbb{C}$.

Example 1.3.1. *On the Riemann sphere, find the images of the points of the symmetrical points relative to: (i) the point $z = 0$ and (ii) the unit circle $|z| = 1$.*

Solution. The symmetrical pair of points relative to the origin is $(z, -z)$.

(i) If (x_1, x_2, x_3) is the image of z and (x_1', x_2', x_3') that of $-z$, then

$$x_1' = -x_1, \quad x_2' = -x_2, \quad x_3' = -x_3$$

and hence the stereographic images of z and $-z$ are directly opposite points on a circle lying paralle to the equator.

(ii) From the Apollonius circle it follows that if z is a given complex number, then its inverse point $1/\bar{z}$ is symmetrical relative to the unit circle $|z| = 1$. So the stereographic images are points mutually symmetrical about the plane of the equator.

Example 1.3.2. *Find the stereographic images of the points* $1, i, -1, \frac{1}{\sqrt{2}}(1 - i)$.

Solution. From Eq. (1.26), the stereographic map in the spherical representation

$$z \in \mathbb{C} \leftrightarrow \left(\frac{z + \bar{z}}{|z|^2 + 1}, \frac{z - \bar{z}}{i(|z|^2 + 1)}, \frac{z\bar{z} - 1}{|z|^2 + 1}\right).$$

Therefore, the images of the given points are as follows:

$$z = 1 \leftrightarrow \left(\frac{1 + 1}{1^2 + 1}, \frac{1 - 1}{i(1^2 + 1)}, \frac{1^2 - 1}{1^2 + 1}\right) = (1, 0, 0)$$

$$z = i \leftrightarrow \left(\frac{i + \bar{i}}{|i|^2 + 1}, \frac{i - \bar{i}}{i(|i|^2 + 1)}, \frac{|i|^2 - 1}{|i|^2 + 1}\right) = (0, 1, 0)$$

$$z = -1 \leftrightarrow \left(\frac{-1 - 1}{(-1)^2 + 1}, \frac{-1 + 1}{i(1^2 + 1)}, \frac{1^2 - 1}{1^2 + 1}\right) = (-1, 0, 0)$$

$$z = \frac{1}{\sqrt{2}}(1 - i) \leftrightarrow \left(\frac{\frac{1}{\sqrt{2}}(1 - i) + \frac{1}{\sqrt{2}}(1 + i)}{|\frac{1}{\sqrt{2}}(1 - i)|^2 + 1}, \frac{\frac{1}{\sqrt{2}}(1 - i) - \frac{1}{\sqrt{2}}(1 + i)}{|\frac{1}{\sqrt{2}}(1 - i)|^2 + 1}, \frac{|\frac{1}{\sqrt{2}}(1 - i)|^2 - 1}{|\frac{1}{\sqrt{2}}(1 - i)|^2 + 1}\right)$$

$$= \frac{1}{\sqrt{2}}(1, -1, 0).$$

Since, for all these points of the complex plane the modulus is unity, $x_3 = 0$ for each of the stereographic images—all lying on the equator of the Riemann sphere—the longitudes being respectively $0, \pi/2, \pi$ and $-\pi/4$.

1.4 Topological Aspects in $\mathbb{C}$

In the previous section, we have already discussed some of the algebraic, geometric and analytic properties of complex numbers $\mathbb{C}$ which, initially defined algebraically, can be identified with the points in the Euclidean plane $\mathbb{R}^2$. In this section we present some definitions and results concerning the topology of the complex plane, which we will use extensively throughout this book.

Definition 1.4.1. Let X be a non-empty set of elements $x, y, z, \cdots$. A mapping $d : X \times X \longrightarrow \mathbb{R}^+$ is said to be *metric* or *distance function* on X iff d satisfies the following axioms:

(i) **Non-negativity :** $d(x, y) \geq 0; \forall x, y \in X$.

(ii) **Identity :** $d(x, y) = 0$ iff $x = y; x, y \in X$

(iii) **Symmetry :** $d(x, y) = d(y, x); \forall x, y \in X$

(iv) **Triangle inequality :** $d(x, y) \leq d(x, z) + d(z, y); \forall x, y, z \in X$.

The set X together with the given metric d on X is called a *metric space* and is denoted by the ordered pair (X, d). The non-negative real number $d(x, y)$ is called the *distance between x and y*. The elements of the metric space are sometimes called *points*.

Definition 1.4.2. [Chordal distance]: Let z_1 and z_2 be two finite points in a complex plane $\mathbb{C}$, then the distance between the streographic projections is given by

$$d(z_1, z_2) = \frac{2|z_1 - z_2|}{\sqrt{1 + |z_1|^2}\, \sqrt{1 + |z_2|^2}}; z_1, z_2 \in \mathbb{C}. \tag{1.31}$$

This distance is called *spherical distance* or *choral distance* between z_1 and z_2.

Definition 1.4.3. [Usual metric]: The set $\mathbb{C}$ of complex numbers becomes a metric space with the metric $d : \mathbb{C} \times \mathbb{C} \to \mathbb{R}$ defined by

$$d(z_1, z_2) = |z_1 - z_2|;\ z_1, z_2 \in \mathbb{C}$$

having the above four properties. Thus $\mathbb{C}$ is a metric space with Euclidean metric or usual metric d, denoted by $(\mathbb{C}, d)$.

Therefore, we can talk about neighbourhood, interior point, open set, closed set, limit point, connected set in the complex plane on the basis of $(\mathbb{C}, d)$.

1.4.1 Diameter

Definition 1.4.4. [Diameter of a set]: Let (X, d) be a metric space and A be a non-empty subset of X. The *diameter* of A, denoted by diam(A) or $\delta(A)$, is defined by

$$\delta(A) = \sup\{d(x, y) : x, y \in A\}. \tag{1.32}$$

Definition 1.4.5. [Distances from points to sets]: Suppose (X, d) is a metric space, A a subset of X and $x \in X$. We define the distance from x to A to be

$$\text{dist}(x, A) = \inf\{d(x, a)a \in A\}. \tag{1.33}$$

It is, of course, dependent on the metric; if it is necessary to specify which metric is being used in order to avoid ambiguity, we may use a subscript, as in $\text{dist}_d(x, A)$.

Definition 1.4.6. [Distances between two sets]: The distance between two non-empty subsets A and B of a metric space (X, d), denoted by $\rho(A, B)$, is defined by

$$\rho(A, B) = \inf\{d(x, y); x \in A, y \in B\}. \tag{1.34}$$

Since $d(x, y) = d(y, x)$, we have $\rho(A, B) = \rho(B, A)$.

If A consists of a single point $\{a\}$, then

$$\rho(\{a\}, B) = \inf\{d(a, x); x \in B\}. \tag{1.35}$$

It is called the distance of a point $a \in A$ from the set B, and is denoted by $\rho(a, B)$. If $\rho(a, B) = 0$, then $a \in B$.

Example 1.4.1. *Let* $A = \{(x, y) \in \mathbb{R}^2 : x^2 + y^2 = 1\}$ *and* $B = \{(x, y) \in \mathbb{R}^2 : (x-1)^2 + y^2 = 1\}$. *Find the diameter of the set* $A \cup B$ *and* $A \cap B$ *with respect to the usual metric on* $\mathbb{R}^2$.

Solution. Here $A = \{(x, y) : x^2 + y^2 = 1\}$ and $B = \{(x, y) : (x-1)^2 + y^2 = 1\}$ as shown in Fig. 1.13. Thus diameter of $A \cup B$ is the distance between the points $(-1, 0)$ and $(2, 0)$.

$$\delta(A \cup B) = \sqrt{(-1-2)^2 + (0-0)^2} = \sqrt{3^2 + 0^2} = 3.$$

Also, the diameter of $A \cap B$ is the distance between the points $(\frac{1}{2}, \frac{\sqrt{3}}{2})$ and $(\frac{1}{2}, -\frac{\sqrt{3}}{2})$.

$$\delta(A \cap B) = \sqrt{\left(\frac{1}{2} - \frac{1}{2}\right)^2 + \left(\frac{\sqrt{3}}{2} + \frac{\sqrt{3}}{2}\right)^2} = \sqrt{0^2 + (\sqrt{3})^2} = \sqrt{3}.$$

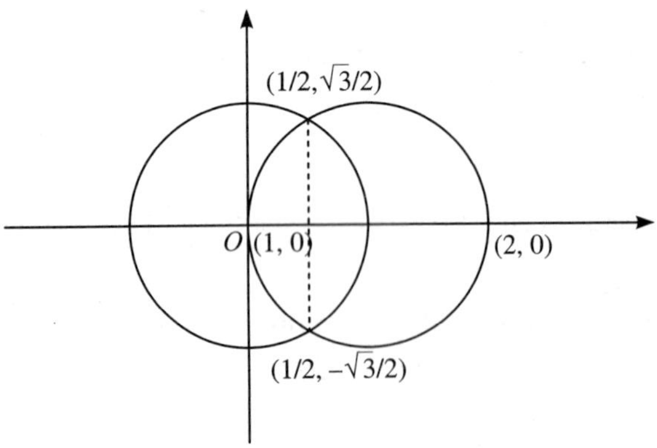

Figure 1.13 $A = \{z \in \mathbb{C} : |z| = 1\}$ *and* $B = \{z \in \mathbb{C} : |z-1| = 1\}$

1.4.2 Open and Closed Discs

The concept of a sphere or a ball will play a fundamental role in theory of complex metric spaces.

Definition 1.4.7. Let $(\mathbb{C}, d)$ be an usual metric space and $z_0 \in \mathbb{C}$. For any real number $\varepsilon > 0$, the non-empty set

$$\mathcal{S}(z_0, \varepsilon) = \{z \in \mathbb{C} : d(z, z_0) < \varepsilon\} \subseteq \mathbb{C} \tag{1.36}$$

is called an *open disc* with centre z_0 and radius ε and is denoted by $\mathcal{S}(z_0, \varepsilon)$ or $\mathcal{S}_\varepsilon(z_0)$. The *open ball* is defined by

$$B(z_0, \varepsilon) = B_\varepsilon(z_0) = \{z \in \mathbb{C} : d(z, z_0) = \varepsilon\}. \tag{1.37}$$

Definition 1.4.8. *A closed sphere* with centre with z_0 and radius $\varepsilon > 0$, denoted by $\mathcal{S}[z_0, \varepsilon]$ or $\mathcal{S}_\varepsilon[z_0]$, is the subset of $\mathbb{C}$ defined by

$$\mathcal{S}[z_0, \varepsilon] = \mathcal{S}_\varepsilon[z_0] = \{z \in \mathbb{C} : d(z, z_0) \le \varepsilon\}. \tag{1.38}$$

(i) The open and closed discs are always non-empty, since at least $z_0 \in \mathcal{S}(z_0, \varepsilon) \subseteq \mathcal{S}[z_0, \varepsilon]$.

(ii) Geometrically, $\mathcal{S}(z_0; \varepsilon)$ is just the disc centred at $z_0 \in \mathbb{C}$ consisting of all the points at a distance less than $\varepsilon > 0$ from z_0.

(iii) Evidently, $\mathcal{S}(z_0; \infty) = \mathbb{C}$ for any $z_0 \in \mathbb{C}$.

A few concrete examples are in order.

(i) In the usual metric space $\mathbb{C}$, let $z_0 \in \mathbb{C}$ and $\varepsilon > 0$. Then the open sphere centred at z_0 and radius ε is given by

$$\begin{aligned}\mathcal{S}(z_0, \varepsilon) &= \{z \in \mathbb{C} : d(z, z_0) = |z - z_0| < \varepsilon\} \\ &= \{(x, y) \in \mathbb{C} : (x - x_0)^2 + (y - y_0)^2 < \varepsilon^2\}.\end{aligned}$$

Thus every open spheres in the usual metric space $\mathbb{C}$ are circular discs. Also

$$\begin{aligned}\mathcal{S}[z_0, \varepsilon] &= \{z \in \mathbb{C} : d(z, z_0) = |z - z_0| \leq \varepsilon\} \\ &= \{(x, y) \in \mathbb{C} : (x - x_0)^2 + (y - y_0)^2 \leq \varepsilon^2\}.\end{aligned}$$

Thus the closed spheres on $\mathbb{C}$ are closed circular discs.

(ii) Let us consider any non-empty set $\mathbb{C}$ together with the discrete metric d on $\mathbb{C}$. Let $z_0 \in \mathbb{C}$ and take $0 < \varepsilon < 1$. Since d may assume only one of the two values, either 0 or 1, so

$$\begin{aligned}\mathcal{S}(z_0, \varepsilon) &= \{z \in \mathbb{C} : d(z, z_0) < \varepsilon\} \\ &= \{z \in \mathbb{C} : d(z, z_0) = 0\} \text{ as } 0 < \varepsilon < 1 \\ &= \{z_0\}.\end{aligned}$$

Thus for $0 < \varepsilon < 1$, as the only points in $\mathbb{C}$ whose distance from z_0 is less than 1 is z_0 itself. So $\mathcal{S}(z_0, \varepsilon) = \{z_0\}$. Using similar reasoning,

$$\mathcal{S}[z_0, r] = \{z \in \mathbb{C} : d(z, z_0) \leq \varepsilon\} = \{z_0\}.$$

If $\varepsilon = 1$, we get

$$\begin{aligned}\mathcal{S}(z_0, 1) &= \{z \in \mathbb{C} : d(z, z_0) < 1\} = \{z_0\} \\ \text{and } \mathcal{S}[z_0, 1] &= \{z \in \mathbb{C} : d(z, z_0) \leq 1\} \\ &= \{z \in \mathbb{C} : d(z, z_0) = 0 \text{ or } d(z, z_0) = 1\} = \mathbb{C}.\end{aligned}$$

If $\varepsilon > 1$, then, in a similar manner, $\mathcal{S}(z_0, \varepsilon) = \mathcal{S}[z_0, \varepsilon] = \mathbb{C}$. Therefore,

$$\mathcal{S}(z_0, \varepsilon) = \begin{cases} \{z_0\} & 0 < \varepsilon \leq 1 \\ \mathbb{C} & \varepsilon > 1 \end{cases} \quad \text{and } \mathcal{S}[z_0, \varepsilon] = \begin{cases} \{z_0\} & 0 < \varepsilon < 1 \\ \mathbb{C} & \varepsilon \geq 1. \end{cases}$$

Thus, the open spheres in a discrete metric space are either singletones, or coincides with the whole of $\mathbb{C}$.

Definition 1.4.9. [Open and closed annulus]: The set of points which satisfies the inequality $\varepsilon_1 < d(z, z_0) < \varepsilon_2$ defines an *open annulus*. If $\varepsilon_1 = 0$, then the set of points is called *punctured disc* of radius ε_2 around the point z_0.

The set of points which satisfies the inequality $\varepsilon_1 \leq d(z, z_0) \leq \varepsilon_2$ defines an *closed annulus*.

Example 1.4.2. *Describe the open spheres of radius* 1 *about* $(0,0)$ *for each of the of the following metrices in* $\mathbb{C}$*:*

(*i*) $d(z_1,z_2)=\sqrt{(x_1-x_2)^2+(y_1-y_2)^2}$

(*ii*) $d(z_1,z_2)=|x_1-x_2|+|y_1-y_2|$

(*iii*) $d(z_1,z_2)=\max\{|x_1-x_2|,|y_1-y_2|\}$

where $z_k=(x_k,y_k); k=1,2$ *are any two points in* $\mathbb{C}$.

Solution. (i) For the given metric, the open sphere with centre $z_0=(0,0)$ and radius 1 is given by

$$\begin{aligned}\mathcal{S}\{z_0,1\} &= \{z=(x,y)\in\mathbb{C}: d(z,z_0)<1\}\\ &= \{(x,y)\in\mathbb{R}^2:\sqrt{(x-0)^2+(y-0)^2}<1\}\\ &= \{(x,y)\in\mathbb{R}^2: x^2+y^2<1\}.\end{aligned}$$

The open unit sphere about $(0,0)$ is the interior of the circle $x^2+y^2=1$, as shown in the Fig. 1.14(a). Similarly, the closed sphere is given by

$$\mathcal{S}[(0,0),1]=\{(x,y)\in\mathbb{R}^2: x^2+y^2\leq 1\}.$$

Thus in the metric space $(\mathbb{R}^2,d)$, open and closed unit spheres are open and closed discs respectively, with centre $(0,0)$ and radius 1.

(ii) In this case, the open sphere with centre at $z_0=(0,0)$ and radius 1 is given by

$$\begin{aligned}\mathcal{S}\{z_0,1\} &= \{z\in\mathbb{C}: d(z,z_0)=|z-z_0|<1\}\\ &= \{(x,y)\in\mathbb{R}^2: |x|+|y|<1\}.\end{aligned}$$

Thus the open sphere in this case is the interior of the square bounded by the following four straight lines: $x+y=1$, $x-y=1$, $-x+y=1$, $-x-y=1$, as shown in the Fig. 1.14 (b). The open balls of this space are all squares with sides that are at 45° to the axes.

Similarly, the closed sphere is given by

$$\mathcal{S}[z_0,1]=\{(x,y)\in\mathbb{R}^2: |x|+|y|\leq 1\}.$$

Thus for the closed sphere $\mathcal{S}[z_0,1]$, we shall also consider all the points on the sides of this square which we do not consider in case of the open sphere $\mathcal{S}(z_0,1)$.

(iii) In this case, the open unit sphere with centre at $z_0=(0,0)$ is given by

$$\mathcal{S}\{z_0,1\}=\{z=(x,y)\in\mathbb{C}:\max\{|x|,|y|\}<1\}.$$

Thus the open sphere in this case is the interior of the square bounded by the following four straight lines $x=1$, $x=-1$, $y=1$, $y=-1$ as shown in the Fig. 1.14(c). Similarly, the closed sphere is given by

$$\mathcal{S}[z,1]=\{z=(x,y)\in\mathbb{R}^2:\max\{|x|,|y|\}\leq 1\}.$$

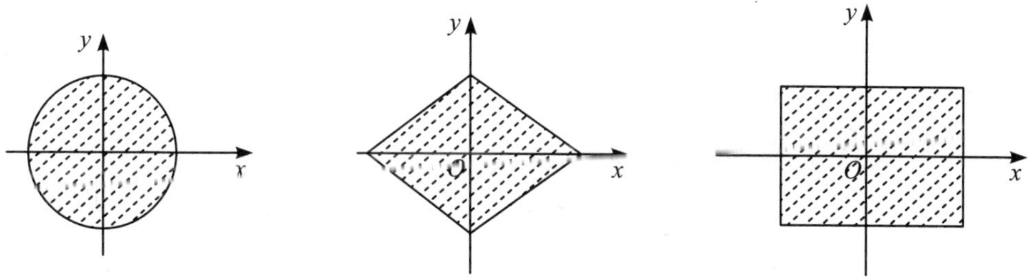

(a) Circle (b) Square (c) Square

Figure 1.14 $x^2 + y^2 = 1$

1.4.3 Neighbourhood of a Point

There are special types of sets that play a distinguished role in analysis: these are the open and closed sets. It is helpful to understand the notion of neighbourhood in complex metric space for this discussion.

Definition 1.4.10. Let $z_0 \in \mathbb{C}$ be any complex number in the Argand's plane and $\varepsilon > 0$ be a real number. Then the set of all points given by

$$\mathcal{N}(z_0, \varepsilon) = \{z \in \mathbb{C} : \mathcal{S}(z, z_0) < \varepsilon\}, \tag{1.39}$$

is called the *ε-neighbourhood* or simply *neighbourhood of* z_0 in the complex plane and is denoted by $\mathcal{N}_\varepsilon(z_0)$ or by $\mathcal{N}(z_0, \varepsilon)$.

Geometrically, $\mathcal{N}_\varepsilon(z_0)$ consists all points inside the circular disc (Fig. 1.15) with centre at z_0 and radius ε, not including the points of the boundary $\{z \in \mathbb{C} : |z - z_0| = \varepsilon\}$.

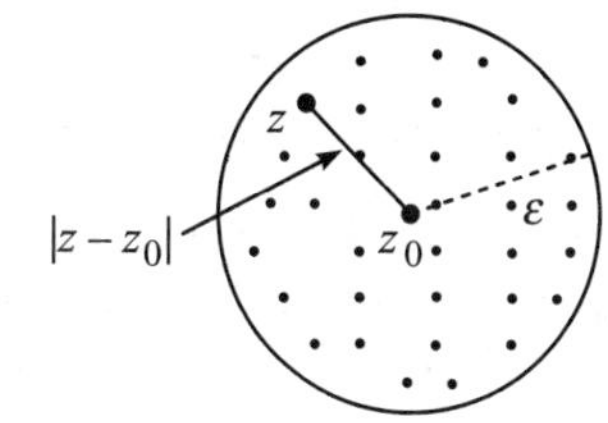

Figure 1.15 *Neighbourhood of a point*

The *deleted ε-neighbourhood of* z_0, denoted by $\mathcal{N}'_\varepsilon(z_0)$ or by $\mathcal{N}'(z_0, \varepsilon)$, is given by

$$\begin{aligned} \mathcal{N}'(z_0, \varepsilon) &= \mathcal{N}(z_0, \varepsilon) - \{z_0\} \\ &= \{z \in \mathbb{C} : 0 < d(z, z_0) < \varepsilon\}. \end{aligned} \tag{1.40}$$

The set of numbers z such that $|z| > k$, where k is any positive real number, is called a *neighbourhood of infinity*.

1.4.4 Interior, Exterior and Boundary Points

Definition 1.4.11. [Interior points]: Let $(\mathbb{C}, d)$ be a metric space and $\mathcal{D} \subseteq \mathbb{C}$ a subset of the Argand plane. A point $z_0 \in \mathcal{D}$ is said to be an *interior point of* $\mathcal{D}$ if $\exists$ an $\varepsilon > 0$ such that $\mathcal{S}(z_0, \varepsilon) \subseteq \mathcal{D}$.

The interior of $\mathcal{D}$ is the neighbourhood of each of its points (Fig. 1.16).

The set of all interior points of $\mathcal{D}$ is denoted by Int$\mathcal{D}$ or $\mathcal{D}^\circ$, i.e.,

$$\mathcal{D}^\circ = \{z \in \mathcal{D} : \mathcal{S}(z, \varepsilon) \subseteq \mathcal{D}, \text{ for some } \varepsilon > 0\}.$$

Definition 1.4.12. [Exterior points]: Let $(\mathbb{C}, d)$ be a metric space and $\mathcal{D} \subseteq \mathbb{C}$. A point $z_0 \in \mathbb{C}$ is called an *exterior point of* $\mathcal{D}$ if $\exists$ an $\varepsilon > 0$ such that

$$z_0 \in \mathcal{S}(z_0, \varepsilon) \subseteq \mathcal{D}^c, \text{ i.e., } \mathcal{S}(z_0, \varepsilon) \cap \mathcal{D} = \phi. \tag{1.41}$$

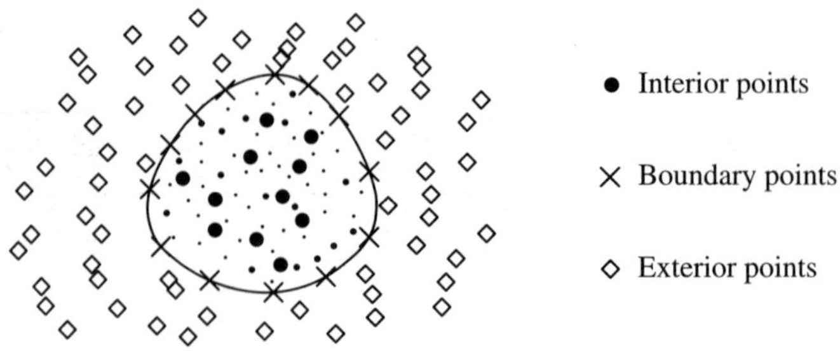

Figure 1.16 *Interior, exterior and boundary points*

The exterior of $\mathcal{D}$, denoted by Ext($\mathcal{D}$) or $\mathcal{D}^c$, is the collection of all exterior points of $\mathcal{D}$ and is defined by

$$\text{Ext}(\mathcal{D}) = (\mathcal{D}^c)^\circ = (\mathbb{C} - \mathcal{D})^\circ.$$

From the definition, it follows that:

(i) $z_0 \in \mathbb{C}$ is an exterior point of the set $\mathcal{D}$ in $(\mathbb{C}, d)$ if and only if we can find at least one sphere $\mathcal{S}(z_0, \varepsilon)$ with centre at z_0 which contains no point of the set $\mathcal{D}$.

(ii) Let $(\mathbb{C}, d)$ be a metric space, and $\mathcal{D}$ be a subset of $\mathbb{C}$. An interior point of $\mathbb{C} - \mathcal{D}$ is called an *exterior point* of $\mathcal{D}$.

(iii) Replacing $\mathcal{D}$ by $\mathcal{D}^c$, we get $\text{Ext}(\mathcal{D}^c) = \mathcal{D}^\circ$; as $(\mathcal{D}^c)^c = \mathcal{D}$.

(iv) From the definition of exterior of a set, it follows that Ext($\mathcal{D}$) is an open set and it is the largest open set contained in $\mathcal{D}^c$.

Definition 1.4.13. [Boundary points]: Let $\mathcal{D} \subseteq \mathbb{C}$ and let $z_0 \in \mathcal{D}$. Then z_0 is called a *boundary point* or *frontier point* of $\mathcal{D}$ if every neighbourhood of z_0 contains infinitely many points of $\mathcal{D}$ and infinitely many points of $\mathbb{C} - \mathcal{D}$. The boundary point is neither an interior point nor an exterior point of $\mathcal{D}$.

The set of all the boundary points of $\mathcal{D}$ is called the boundary of $\mathcal{D}$ and is denoted by $\partial\mathcal{D}$.

(i) Boundary point must belong to $\mathcal{D}$.

(ii) A boundary point of $\mathcal{D}$ is a frontier point of $\mathcal{D}$, but the converse is not necessarily true.

(iii) Obviously, $\text{Bd}(\mathcal{D}) \subseteq \text{Fr}(\mathcal{D})$.

(iv) From definition it follows that

$$\text{Fr}(\mathcal{D}) \cap \mathcal{D}^\circ = \phi, \quad \text{Fr}(\mathcal{D}) \cap \text{Ext}(\mathcal{D}) = \phi.$$

(v) Since $\text{Fr}(\mathcal{D}) = (\mathcal{D}^\circ \cup \text{Ext}(\mathcal{D}))^c$, it follows that

$$X = \mathcal{D}^\circ \cup \text{Ext}(\mathcal{D}) \cup \text{Fr}(\mathcal{D}).$$

(vi) The boundary of a subset of a metric space $(\mathbb{C}, d)$ is closed.

For example,

(i) the circle $|z| = 1$ is the boundary of the sets $|z| < 1$ and $|z| \leq 1$.

(ii) $z_0 = 2 + i$ is a boundary point but not in $S = \{z \in \mathbb{C} : |z - 2| < 1\}$.

1.4.5 Open Set

The notion of an open set is at the heart of metric space. It is basic in the study of continuous functions and all other notions like compactness, connectedness, bounded sets, totally bounded sets and so on in a metric space.

Definition 1.4.14. Let $(\mathbb{C}, d)$ be a metric space and $G \subseteq \mathbb{C}$. The subset G of the Argand plane is *open* if whenever $z_0 \in G$, $\exists$ a real number $\varepsilon > 0$ such that

$$S(z_0, \varepsilon) \subset G.$$

Thus the set G is called an open set if and only if every point of G is an interior point of G.

The open set does not contain any of its boundary points. In general, any set described by strict inequalities of continuous functions is open. For example,

(i) the open disc interior of a circle, i.e., $\{d(z, z_0) < \rho\}$.

(ii) the half-planes given by Re $z < a$, Im $z < b$ and Im $z > b$.

(iii) the sector $\{\theta_0 < \arg z < \theta_1\}$, the horizontal strip $\{-1 < \text{Im}\, z < 1\}$, punctured disc $\{0 < |z| < r\}$, the open annulus $\{r < |z| < s\}$.

(iv) In a discrete metric space $(\mathbb{C}, d)$, every subset of $G \subseteq \mathbb{C}$ is an open set. For any $z_0 \in G$, the open sphere $S(z_0, 1/2) = \{z_0\} \subseteq Y$. Therefore, G is a neighbourhood of z_0. Since z_0 is an arbitrary point in G, the assertion follows. A discrete metric space is richest in terms of open sets.

PROPOSITION **1.4.1.** Let $(\mathbb{C}, d)$ be a metric space. Then,

(i) The null set ϕ and the entire space $\mathbb{C}$ in a metric space $(\mathbb{C}, d)$ are open sets.

(ii) If $G_1, G_2, \cdots, G_n$ are open sets in $\mathbb{C}$, then so is $\bigcap_{k=1}^{n} G_k$.

(iii) If $\{G_k : k \in \bigwedge\}$ is an arbitrary collection of open sets in $\mathbb{C}$, $\bigwedge$ any indexing set, then $G = \bigcup\{G_k : k \in \bigwedge\}$ is also open.

1.4.6 Limit Point

Definition 1.4.15. Let $(\mathbb{C}, d)$ be a metric space and $\mathcal{D} \subseteq \mathbb{C}$. A point $z_0 \in \mathbb{C}$ is said to a *limit point* or *accumulation point* or *cluster point* or *condensation point* of $\mathcal{D}$, if

$$\mathcal{N}'(z_0, \varepsilon) \cap \mathcal{D} = \Big(\mathcal{S}(z_0, \varepsilon) - \{z_0\}\Big) \cap \mathcal{D} \neq \phi, \quad \forall \varepsilon > 0, \tag{1.42}$$

i.e., if each open sphere centred on z_0 contains at least one point of $\mathcal{D}$ other than z_0. Since ε can be any positive number, it follows that $\mathcal{D}$ must have infinitely many points.

Clearly, every interior point of a set is its limit point, but no exterior point of the set will be a limit point. The set of all limit points of $\mathcal{D}$ is called the *derived set* of $\mathcal{D}$ and is denoted by $\mathcal{D}'$. For example:

(i) Let $(\mathbb{C}, d)$ be any discrete metric space and $\mathcal{D}$ is any non-empty subset of $\mathbb{C}$. For all $z_0 \in \mathbb{C}$,

$$\mathcal{S}\Big(z_0, \frac{1}{2}\Big) \cap \Big[\mathcal{D} - \{z_0\}\Big] = \{z_0\} \cap \Big[\mathcal{D} - \{z_0\}\Big] = \phi.$$

i.e., every open sphere of radius 1 contains only the centre. This implies any point $z_0 \in \mathcal{D}$ is not a limit point of the set $\mathcal{D}$ in the metric space $(\mathbb{C}, d)$. Hence $\mathcal{D}' = \phi$.

(ii) In the usual metric space $(\mathbb{C}, d)$, if $\mathcal{D} = \{(-1 - \frac{1}{n}, 1 + \frac{1}{n}) : n \in \mathbb{N}\}$, then $\mathcal{D}' = \{(-1, 1)\}$.

(iii) Let $\mathcal{D} = \{z : |z| < 1\}$. Then each point on $|z| = 1$ is a limit point of $\mathcal{D}$ but does not belong to $\mathcal{D}$. Also, each $z \in \mathcal{D}$ being a limit point of $\mathcal{D}$ shows that a limit point may or may not belong to the set.

(iv) A finite set has no limit point.

Definition 1.4.16. [Closure of a set]: Let $(\mathbb{C}, d)$ be a metric space and $\mathcal{D} \subset \mathbb{C}$. The *closure of the set* $\mathcal{D}$, denoted by $\bar{\mathcal{D}}$ or cl$\mathcal{D}$, is defined as

$$\bar{\mathcal{D}} = \mathcal{D} \cup \mathcal{D}'.$$

From the definition it is clear that

$$z_0 \in \bar{\mathcal{D}} \Leftrightarrow \mathcal{S}(z_0, \varepsilon) \cap \mathcal{D} \neq \phi; \quad \forall \varepsilon > 0$$

$$\therefore \quad \bar{\mathcal{D}} = \{z \in \mathbb{C} : \mathcal{S}(z_0, \varepsilon) \cap \mathcal{D} \neq \phi; \quad \forall \varepsilon > 0\}.$$

Thus $z_0 \in \bar{\mathcal{D}}$ if and only if $\mathcal{S}(z_0, \varepsilon) \cap \mathcal{D} \neq \phi$ for every $\varepsilon > 0$.

1.4.7 Isolated Point

A point z_0 of a subset $\mathcal{D} \subseteq \mathbb{C}$ of a metric space $(\mathbb{C}, d)$ is necessarily at zero distance from $\mathcal{D}$. It may or may not, however, be at zero distance from the rest of $\mathcal{D}$; that is, from $\mathcal{D} - \{z_0\}$. Points that are not at zero distance from the rest of $\mathcal{D}$ are called *isolated points of S.*

Definition 1.4.17. Let $(\mathbb{C}, d)$ be a metric space and $\mathcal{D} \subseteq \mathbb{C}$. A point $z_0 \in \mathbb{C}$ is said to be an *isolated point* of $\mathbb{C}$ if there exists an open sphere centred on z_0 containing no point of $\mathcal{D}$ other than z_0 itself, i.e.,

$$\mathcal{S}(z_0, \varepsilon) \cap \mathcal{D} = \{z_0\}, \quad \text{for some } \varepsilon > 0. \tag{1.43}$$

The collection of all isolated points of $\mathcal{D}$ will be denoted by iso($\mathcal{D}$). If a point $z_0 \in \mathbb{C}$ is not a limit point of $\mathcal{D}$, then it is an isolated point. Hence every point of a metric space $\mathbb{C}$ is either a limit point or an isolated point of $\mathbb{C}$. For example,

(i) $1 + i$ is the isolated point of the set $\mathcal{D} = \{z \in \mathbb{C} : |z| = 1\} \cup \{1 + i\}$.

A point which is not the limit point of a set but belongs to the set is called an *isolated point.*

1.4.8 Closed Set

The dual of the concept of an open set is the concept of a closed set.

Definition 1.4.18. Let $(\mathbb{C}, d)$ be a metric space. A subset $\mathcal{D}(\subseteq \mathbb{C})$ is said to be *d-closed* or *closed*, if each limit point of $\mathcal{D}$ belongs to $\mathcal{D}$, i.e. $\mathcal{D}' \subset \mathcal{D}$.

A set $\mathcal{D}$ is closed if its complement $\mathcal{D}^c = \mathbb{C} - \mathcal{D}$ is open. For example,

(i) each closed sphere in $\mathbb{C}$ is a closed set

(ii) the set of points $\{z \in \mathbb{C} : |z| \leq 1\}$, $\{z \in \mathbb{C} : |z| \geq 1\}$, $\{z \in \mathbb{C} : |z| = 1\}$,

(iii) $\{z \in \mathbb{C} : \text{Re} \leq 0\}$, $\{z \in \mathbb{C} : \text{Im} = 1\}$, $\{z \in \mathbb{C} : |z - 2| \geq |z|\}$.

(iv) The closed disc $\{d(z, z_0) \leq \rho\}$ is not open set, as any open disc centred at a point on the boundary circle $\{|z - z_0| = \rho\}$ extends outside the closed disc. So the closed set contains all its boundary points.

(v) A closed set which cannot be split into two disjoint closed subsets is called a *continuum*.

A subset $\mathcal{D}$ of the complex plane $\mathbb{C}$ is closed if it contains the limit of every convergent sequence in $\mathcal{D}$. The closed disc $\{d(z, z_0) \leq \varepsilon\}$ is a closed set, as, if $|z_n - z_0| \leq \varepsilon$ and $z_n \to z$, then $|z - z_0| \leq \rho$. The closed upper half plane $H = \{z \in \mathbb{C} : \text{Re } z \geq 0\}$, is a closed set.

PROPOSITION **1.4.2.** Let $(\mathbb{C}, d)$ be a metric space. Then:

(i) Every finite set, empty set ϕ and whole space $\mathbb{C}$ are closed sets.

(ii) $\mathcal{D} \subset \mathbb{C}$ is closed if and only if $(\mathbb{C} - \mathcal{D})$ is open in $(\mathbb{C}, d)$.

(iii) A set $\mathcal{D}$ is closed if and only if it contains all its limit points.

(iv) The union of finite number of closed sets is a closed set.

(v) The intersection of a arbitrary collection of closed sets is a closed set.

(vi) The closure of every set is closed.

(vii) The set of all complex numbers is both open and closed.

Example 1.4.3. *State whether the given subset of $\mathcal{D}$ is open, closed and bounded:*

$$\begin{array}{rlrl} (i) & \{z : |z| \leq 5\} & (ii) & \{z : 0 < |z| < 5\} \\ (iii) & \{z : 1 < |z + i| \leq 2\} & (iv) & \{z : \text{Im}(z) = \text{Re}(z)\} \\ (v) & \{z : |\text{Re}(z)| > 4\} & (vi) & \{z : |z| \geq 4\} \\ (vii) & \{z : z = \frac{1}{n};\ n \in \mathbb{N}\} & (viii) & \mathbb{N} \\ (ix) & \{z : \text{Re}(z) \text{ is rational}\} & (x) & \text{Any non-empty finite set} \\ (xi) & \{z : \text{Re}(z)\text{Im}(z) = 0\} & (xii) & \{z : \text{Im}(z^2) < 0\} \\ (xiii) & \{z : \arg(z - 1 - i) = \frac{\pi}{3}\} & (xiv) & \{z;\ \frac{\pi}{6} < \arg z < \frac{\pi}{3}\} \end{array}$$

Solution. The solutions are given in the following tabular form:

	i	ii	iii	iv	v	vi	vii	viii	ix	x	xi	xii	xiii	xiv
Open	N	Y	N	N	Y	N	N	N	N	N	N	Y	N	N
Closed	Y	N	N	Y	N	Y	N	Y	N	Y	Y	N	Y	N
Bounded	Y	Y	Y	N	N	N	Y	N	N	Y	N	N	N	N

1.5 Compactness

In this section we shall study another important property called *compactness*. The concept of compactness is an extension of the benefits of finiteness to infinite sets. Most properties of compact sets are analogous to the properties of finite sets which are quite trivial.

Definition 1.5.1. Let (X, d) be a metric space and $\mathcal{D} \subset X$. Let $\mathcal{C} = \{G_\alpha : \alpha \in \Lambda\}$ be a family of open sets in (X, d) satisfying $\mathcal{D} \subset \bigcup\{G_\alpha : \alpha \in \Lambda\}$, called a *open cover* of $\mathcal{D}$. The subset $\mathcal{D}$ is called *compact* if there exists a finite family $\{G_{\alpha_1}, G_{\alpha_2}, \cdots, G_{\alpha_n}\}$ from $\mathcal{C}$ such that

$$\mathcal{D} \subset \bigcup_{k=1}^{n} \left\{G_{\alpha_k}\right\} = G_{\alpha_1} \cup G_{\alpha_2} \cup \cdots G_{\alpha_n}. \tag{1.44}$$

Thus a metric space (X, d) is said to be a *compact metric space* if and only if every open cover of X admits a finite subcover.

(i) The empty set ϕ and all finite sets are compact.

(ii) Let $\mathcal{D} = \{z \in \mathbb{C} : |z| < 1\}$. If $G_k = \{z \in \mathbb{C} : |z| < 1 - \frac{1}{k}\}$ for $k = 2, 3, \cdots$, then $\{G_2, G_3, \cdots\}$ is an open cover of $\mathcal{D}$ for which there is no finite subcover. Thus $\mathcal{D}$ is non compact.

Property 1.5.1. [Heine–Borel theorem]: A set $S \subseteq \mathbb{C}$ is compact if and only if it is closed and bounded in $\mathbb{C}$.

For example, unit circle in $\mathbb{C}$ is compact.

Property 1.5.2. [Cantor's theorem]: Let K_j's; $j \in \mathbb{N}$ be nested compact subset of a complex plane $\mathbb{C}$ with $K_1 \supset K_2 \supset K_3 \supset \cdots$. Then $\bigcap_{j=1}^{\infty} K_j$ is non-empty.

Proof: Suppose $\bigcap_{j=1}^{\infty} K_j$ is empty. Let us consider the collection

$$\mathcal{F} = \left\{K_j^c : j \in \mathbb{N}\right\}; \text{ and } \mathbb{C} = \bigcup_{j=1}^{\infty} K_j^c$$

In particular, $K_1 \subset \bigcup_{j=1}^{\infty} K_j^c$. Since K_1 is compact, there are numbers $j_1, j_2, \cdots, j_n$, where $j_1 < j_2 < \cdots < j_n$, such that

$$K_1 \subset \bigcup_{m=1}^{n} K_{j_m}^c \subset K_{j_n+1}^c \text{ as } K_{j_n+1} \subset K_{j_n} \subset \cdots \subset K_{j_1}$$

$$\Rightarrow \quad K_1 \bigcap K_{j_n+1} \neq \phi,$$

which is a contradiction. So $\bigcap_{j=1}^{\infty} K_j$ has to be non-empty. □

Property 1.5.3. A closed subset of a compact set is compact.

1.6 Connected Sets

Connectedness is an another important notion like compactness in the study of analysis in metric spaces.

Definition 1.6.1. [Separated sets]: Let (X, d) be a metric space. Two non-empty subsets A and B of X are said to be *d-separated* or simply *separated sets in* X if

$$A \cap \bar{B} = \phi \text{ and } \bar{A} \cap B = \phi, \tag{1.45}$$

where $\bar{B}, \bar{A}$ are closures of B and A respectively. Thus two sets $X, Y \subset \mathbb{C}$ are *separated* if there are disjoint open sets $A, B \subset \mathbb{C}$ so that $X \subseteq A$ and $Y \subseteq B$.

It follows from the above definition that separated sets are disjoint sets.

(i) Any two non-empty disjoint sets in a discrete metric space are separated sets.

(ii) In the complex plane $\mathbb{C}$, the sets $A = \{x + iy : x = 0 \text{ and } y \text{ is arbitrary }\}$, $B = \{x + iy : x \neq 0 \text{ and } y = 0\}$ are not separated sets.

Definition 1.6.2. [Disconnected sets]. Let (X, d) be a metric space. A subset A of X is said to be *d-disconnected* or simply *disconnected* if it is the union of two non-empty d-separated sets, i.e., if $\exists$ two sets C and D such that

$$A = C \cup D, \quad C \cap \bar{D} = \phi \text{ and } \bar{C} \cap D = \phi. \tag{1.46}$$

That is, the subsets must be non-empty, and together they must constitute the whole space; also, neither may contain a point of the closure of the other.

In particular, the space X is *disconnected* if there exists two non-empty sets A and B such that

$$X = A \cup B; \quad A \cap \bar{B} = \phi \text{ and } \bar{A} \cap B = \phi. \tag{1.47}$$

The pair (A, B) are said to form a *separation* of X. For example, let $X = A \cup B$, where

$$A = \{z \in \mathbb{C} : |z| \leq 1\}, B = \{z \in \mathbb{C} : |z - 3| < 1\}$$

with the metric induced from $\mathbb{C}$. Now, A is open, as, if $z_0 \in A$,

$$\mathcal{S}(z_0, 1) = \{z \in \mathbb{C} : |z - z_0| < 1\} \subset A$$

and it is closed because its complement $B = X - A$ is open. Thus the the set A is simultaneously open and closed. Similarly, B is both open and closed in X. Then X is disconnected.

Definition 1.6.3. [Connected sets]: A set $\mathcal{D} \subset \mathbb{C}$ is said to be *connected* if it is not contained in the union of two disjoint non-empty open subsets $\mathcal{D}_1$ and $\mathcal{D}_2$ of $\mathbb{C}$ which have a non-trivial intersection with $\mathcal{D}$.

Thus $\mathcal{D}$ is connected if

$$\mathcal{D} \subset \mathcal{D}_1 \cup \mathcal{D}_2; \mathcal{D}_1 \neq \phi, \mathcal{D}_2 \neq \phi \text{ and } \mathcal{D}_1 \cap \mathcal{D}_2 \neq \phi.$$

Equivalently, we say that $\mathcal{D} \subset \mathcal{D}_1 \cup \mathcal{D}_2$, where $\mathcal{D}_1$ and $\mathcal{D}_2$ are open, non-empty. Then, either

$$\mathcal{D}_1 \cap \mathcal{D} = \phi \text{ or } \mathcal{D}_2 \cap \mathcal{D} = \phi.$$

(i) A metric space $(\mathbb{C}, d)$ where $\mathbb{C}$ is a singleton is always connected.

(ii) $\mathbb{R}^n$ and $\mathbb{C}^n$ are connected metric spaces.

(iii) The empty set ϕ is taken to be connected by definition.

Definition 1.6.4. [Line segment]: Let z_1, z_2 be two points in $\mathbb{C}$. We call

$$L_{z_1,z_2} = \{z \in \mathbb{C} : z = (1 - t)z_1 + tz_2 : t \in [0, 1]\} \tag{1.48}$$

the line segment joining z_1 and z_2 (Fig. 1.17). The point z_1 is called the *initial point* of the line segment and z_2 is called the *end point* of the line segment. Every line segment is connected in $\mathbb{C}$.

Definition 1.6.5. [Polygon]: Let $\{z_k : k = 1, 2, \cdots, n+1\}$ be a finite set of complex numbers representing points in $\mathbb{C}$. For each $k = 1, 2, \cdots, n+1$, let $L_{z_k, z_{k+1}}$ denote the line segment joining z_k and z_{k+1} (Fig. 1.18). Then the successive line segments $L_{z_1,z_2}, L_{z_2,z_3}, \cdots L_{z_k,z_{k+1}}$ form a continuous chain known as a *polygonal path* joining z_1 and z_{n+1}.

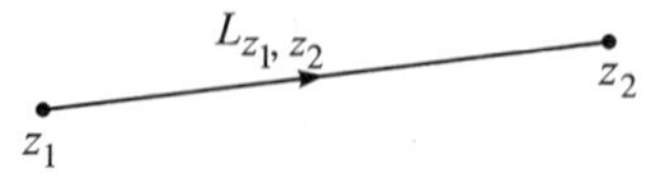

Figure 1.17 *Line segment*

The union $\bigcup_{k=1}^{n} L_{z_k,z_{k+1}}$ of the given finite number of line segments of the form $L_{z_k,z_{k+1}}; k = 1, 2, \cdots, n$ is called a *polygon*. Every polygon is connected.

A set P is said to be *polygonally connected* if any two points of P can be connected by a polygonal line contained in P. In other words, P is connected if and only if each pair of points z_1, z_2 of P can be connected by an arc lying in P.

For example, in $\mathbb{C}$, every open disc $\mathcal{S}(z_0, \varepsilon)$ is polygonally connected.

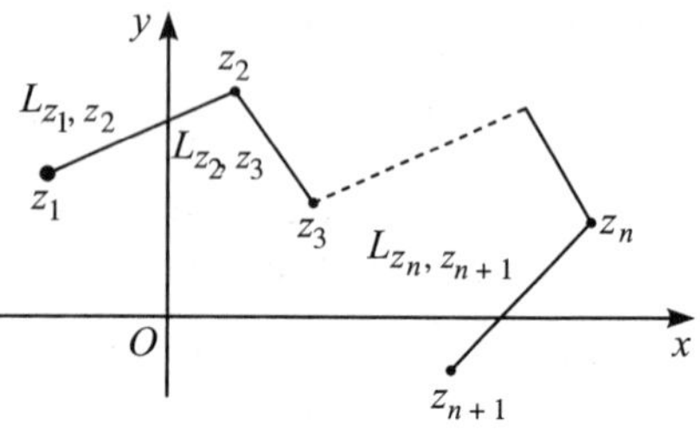

Figure 1.18 *Polygonal path*

RESULT **14.** A connected set can also be defined as an open set if any two points z_1 and z_2 of the set can be joined by a polygonal path, consisting of finite number of line segments joined end to end, which lies entirely in S. For example, the annulus $1 < |z| < 2$ is a open connected set.

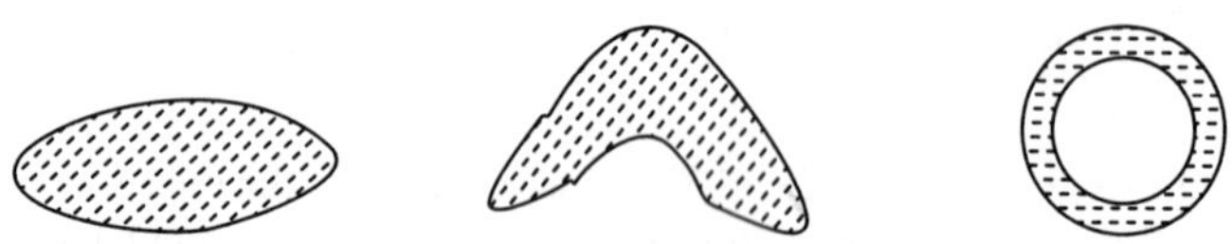

Figure 1.19 *Convex hull*

Definition 1.6.6. A set $G \subseteq \mathbb{C}$ is said to be *convex* if given any two points z_0 and z_1 in G, the line segment $L_{z_0 z_1}$ joining z_0 and z_1 lies entirely in G. An open and closed disc is convex, but a punctured disc is not convex.

Note: Clearly, every convex set is connected. Furthermore, it follows that the intersection of two or more convex sets is also convex.

Definition 1.6.7. The *convex hull* of a set $K \subseteq \mathbb{C}$ is a convex set $S \subseteq \mathbb{C}$ containing K such that no proper convex subset of S contains K.

Let $\mathcal{G}$ be a collection of all convex sets containing a given set K. Then $\bigcap_{S\in\mathcal{G}} S$ is a convex set containing K. This set $\bigcap_{S\in\mathcal{G}} S$ is the *convex hull of* K.

Property 1.6.1. (i) The convex hull of a finite number of points in $\mathbb{C}$ is a *polygon*.

(ii) Let $z_1, z_2, \cdots, z_n$ belong to a convex set K. Then $\exists$ scalars $c_1, c_2, \cdots, c_n \in \mathbb{R}$ such that

$$c_1 z_1 + c_2 z_2 + \cdots + c_n z_n \in K$$

whenever $0 \leq c_k \leq 1$ and $c_1 + c_2 + \cdots + c_n = 1$.

Theorem 1.5. [Gauss–Lucas theorem]: *Let $z_1, z_2, \cdots, z_n$ be zeros of a polynomial $p(z)$. Show that the zeros of the derivative of $p(z)$ lie within the convex hull of the points $z_1, z_2, \cdots, z_n$ in the complex plane $\mathbb{C}$.*

Proof: Let α be a zero of $p'(z)$. If α is one of the z_k's, then α lies in the convex hull of $z_1, z_2, \cdots, z_n$.
Suppose $p(\alpha) \neq 0$ and suppose $p(z) = k(z-z_1)(z-z_2)\cdots(z-z_n)$, where $k \in \mathbb{C}/\{0\}$. Therefore,

$$\frac{p'(z)}{p(z)} = \frac{1}{z-z_1} + \frac{1}{z-z_2} + \cdots + \frac{1}{z-z_n}$$

$$\Rightarrow \quad \frac{p'(\alpha)}{p(\alpha)} = \frac{1}{\alpha-z_1} + \frac{1}{\alpha-z_2} + \cdots + \frac{1}{\alpha-z_n} = 0$$

$$\Rightarrow \quad \frac{\overline{\alpha}-\overline{z}_1}{|\alpha-z_1|^2} + \frac{\overline{\alpha}-\overline{z}_2}{|\alpha-z_2|^2} + \cdots + \frac{\overline{\alpha}-\overline{z}_n}{|\alpha-z_n|^2} = 0$$

$$\Rightarrow \quad \bar{\alpha}\left[\frac{1}{|\alpha-z_1|^2} + \frac{1}{|\alpha-z_2|^2} + \cdots + \frac{1}{|\alpha-z_n|^2}\right] = \sum_{k=1}^{n} \frac{\bar{z}_k}{|\alpha-z_k|^2}$$

$$\Rightarrow \quad \bar{\alpha}\, A = \sum_{k=1}^{n}\left(\frac{A_k}{A}\right)\bar{z}_k; \text{ where } \sum A_k = A, 0 \leq \frac{A_k}{A} \leq 1 \text{ and } \sum \frac{A_k}{A} = 1$$

The right-hand side lies in the convex hull of $z_1, z_2, \cdots, z_n$ and hence by (ii) of Property 1.6.1, α lies in the convex hull of $z_1, z_2, \cdots, z_n$. □

Example 1.6.1. *Show that the zeros of* $p(z) = 1 + 2z + 3z^2 + \cdots + nz^{n-1}$ *lie inside the unit disc* $\overline{\mathcal{S}(0,1)}$.

Solution. Here $p(z) = (z + z^2 + \cdots + z^n)'$ and we use the Gauss–Lucas theorem 1.5. Now,

$$z + z^2 + \cdots + z^n = z\left(1 + z + \cdots + z^{n-1}\right) = z\left(\frac{z^n-1}{z-1}\right).$$

If $\zeta_k = e^{2\pi ik/n}; 1 \leq k \leq n-1$, then ζ_k is a root of the polynomial $z + z^2 + \cdots + z^n$. 0 is also a root of $z + z^2 + \cdots + z^n$. So the zeros of $p(z)$ lie in the convex hull of $\{\zeta_k : 1 \leq k \leq n-1\} \cup \{0\}$ as depicted in the Fig. 1.20(a). So the zeros of $p(z)$ lie on $\overline{\mathcal{S}(0,1)}$. In particular, when $n = 3$, then the convex hull, a triangular region as depicted in the Fig. 1.20b, lies on $\overline{\mathcal{S}(0,1)}$.

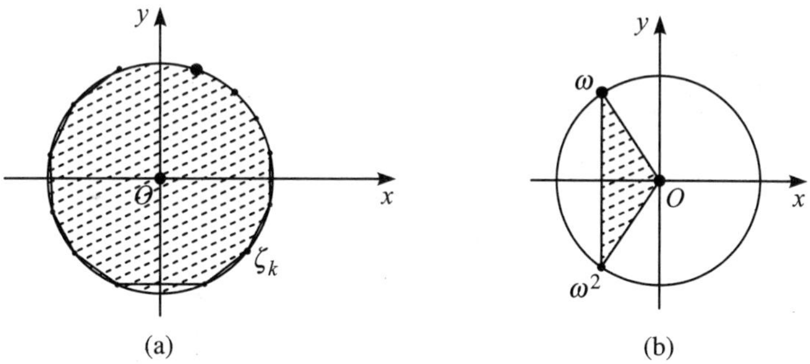

Figure 1.20 *Convex hull*

1.7 Domain and Region

Definition 1.7.1. A non-empty subset $\mathcal{D}$ of the complex plane $\mathbb{C}$ is a *domain* if $\mathcal{D}$ is open and if any two points of $\mathcal{D}$ can be connected by a broken line segment in $\mathcal{D}$. Clearly, all open discs are domains.

Open half planes, open discs, open sectors, open annuli, and open punctured discs, are all domains.

The closure of a domain, i.e., the union of domain and its boundary, is known as a *closed domain.*

An *exterior domain* is a domain $\mathcal{D}$ in the complex plane that includes all large z, i.e., $\mathcal{D}$ includes all z such that $\mathcal{S}^c = \{z \in \mathbb{C} : |z| \geq R\}$ for some R.

Definition 1.7.2. A domain together with some, none or all of its boundary points is known as a *region.* Thus a nonempty open connected subset of $\mathcal{D} \subset \mathbb{C}$ is called a region in $\mathbb{C}$. Every domain is a region, but the converse is not true.

Theorem 1.6. *Let G be a non-empty open subset of $\mathbb{C}$. Then G is a region if and only if any two points of G can be connected by a polygonal path.*

Proof: Suppose G is a region. Fix $a \in G$ (as G is non-empty) and let

$$G_1 = \{z \in G : \text{ there is a polygonal path from } a \text{ to } z \text{ in } G\}.$$

Let $G_2 = G/G_1$. Then both G_1 and G_2 are open sets.

Suppose $z \in G_1$. Then, since $G_1 \subset G$ and G is open, there is an $r > 0$ such that $\mathcal{S}(z;r) \subset G$. Since a and z are connected by a polygonal path, we can extend this path to any point $w \in \mathcal{S}(z;r)$ by joining an additional segment joining z and w. So, $w \in \mathcal{S}(z;r)$ is such that $w \in G_1$. Hence

$$\mathcal{S}(z;r) \subset G \Rightarrow G_1 \text{ is open.}$$

Likewise, G_2 is open. Thus $G_1 \cup G_2 = G$, G_1, G_2 is open and

$$G_1 \cap G_2 = \phi \Rightarrow \text{ either } G_1 = \phi \text{ or } G_2 = \phi.$$

We know G_1 is non-empty so $G_2 = \phi$, i.e., $G_1 = G$. Hence the theorem. □

For example,

(i) Let $\mathcal{D} = \{z : \operatorname{Re} z > 1\}$. Let $z = x + iy$. Then $\mathcal{D}$ is the set given by $x > 1$, which is the half plane as shown in the Fig. 1.21(a). Thus $\mathcal{D}$ which is a non-empty open connected subset of $\mathbb{C}$, is a region in $\mathbb{C}$.

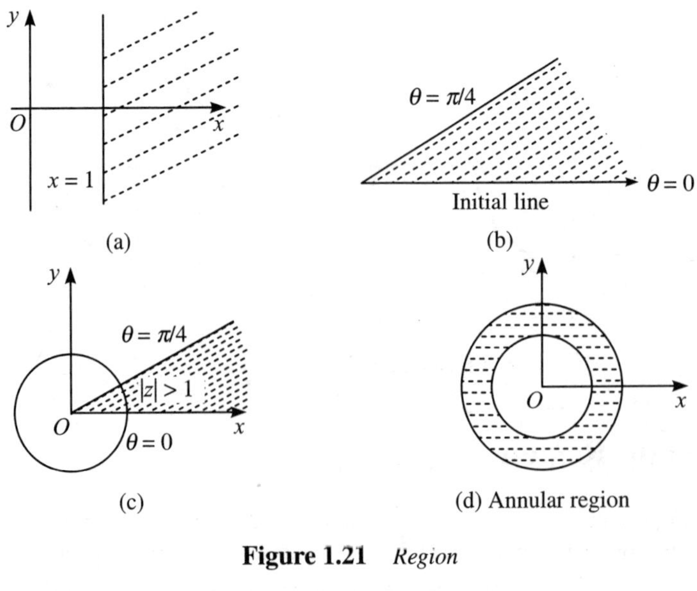

Figure 1.21 *Region*

(ii) Consider $\mathcal{D} = \{z : 0 < \arg z < \pi/4\}$ as shown in Fig. 1.21(b). Clearly, it is a region in $\mathbb{C}$.

(iii) Let $\mathcal{D} = \{z : 0 < \arg z < \pi/4 \text{ and } |z| > 1\}$ as shown in the Fig. 1.21(c). Clearly, $\mathcal{D}$ is an unbounded region in $\mathbb{C}$.

(iv) Let $\mathcal{D} = \{z : 1 < |z| < 2\}$ as shown in the Fig. 1.21(d). $\mathcal{D}$ is the region bounded by the circles $|z| = 1$ and $|z| = 2$. Such a region is called an annulus or annular region.

(v) Let $\mathcal{D} = \{z : |\text{Im } z| > 1\}$. Let $z = x + iy$. Then

$$\mathcal{D} = \{z : |y| > 1\} = \{z : y > 1\} \cup \{z : y < -1\},$$

which is the union of two half planes and is unbounded as shown in the Fig. 1.22 (a).

Obviously, if z_1 is any complex number with $\text{Im } z_1 > 1$ and z_2 is any complex number with $\text{Im } z_2 < -1$, then z_1 and z_2 cannot be joined by a polygon entirely lying in $\mathcal{D}$. Thus $\mathcal{D}$ is not connected and hence is not a region.

(vi) Let $\mathcal{D} = \{z : |z - 2 + i| \leq 1\}$, i.e., $\mathcal{D}$ is the set of complex numbers satisfying $|z - (2 - i)| \leq 1$. Clearly, $\mathcal{D}$ is a closed disc with centre $2 - i$ and 1 (Fig. 1.22b). Clearly, $\mathcal{D}$ is connected and bounded set. We observe that the points which lie on the circle $|z - (2 - i)| = 1$ are not interior points of $\mathcal{D}$. Hence $\mathcal{D}$ is not open and hence is not a region.

An example of an open set that is not a domain is the union of the open upper and lower half planes, $\mathcal{D} = \mathbb{C}/\mathbb{R}$. It is impossible to connect a point in the upper half-plane to a point in the lower half-plane by a broken line segment that does not cross the real line.

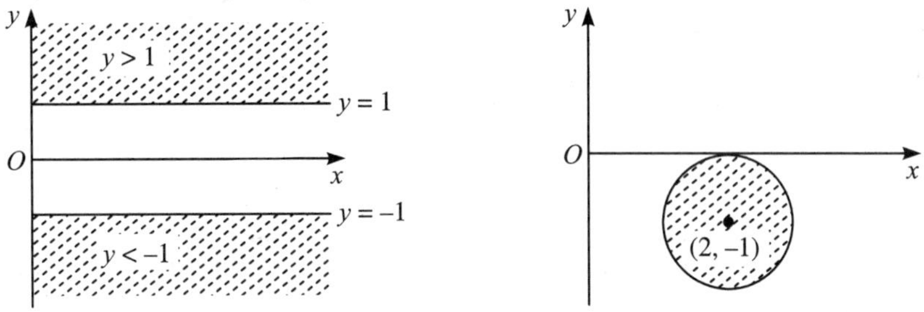

(a) (b) Disc

Figure 1.22

Exercises

Multiple-choice Questions

1. If z is a complex number such that $z^2 = (\bar{z})^2$, then
   (a) z is purely real (b) z is purely imaginary
   (c) either z is purely real or purely imaginary (d) none of these

2. The shaded region of Fig. 1.24 is given by
   (a) $|z + 2| \geq 6, 0 \leq \arg(z) \leq \pi/6$ (b) $|z + 2| \geq 6, 0 \leq \arg(z) \leq \pi/3$
   (c) $|z + 2| \geq 6, 0 \leq \arg(z) \leq \pi/2$ (d) none of these

3. How many elements does the set $\{z \in \mathbb{C} : z^{60} = -1 \text{ and } z^n \neq -1 \text{ for any } 0 < n < 60\}$ have?
   (a) 24 (b) 30 (c) 32 (d) 45

4. The area of the shaded region of Fig. 1.24 belongs to
(a) $\{z : |z+1| > 2, |\arg(z+1)| < \pi/4\}$ (b) $\{z : |z-1| > 2, |\arg(z-1)| < \pi/4\}$
(c) $\{z : |z+1| < 2, |\arg(z-1)| < \pi/2\}$ (d) $\{z : |z-1| < 2, |\arg(z-1)| < \pi/2\}$

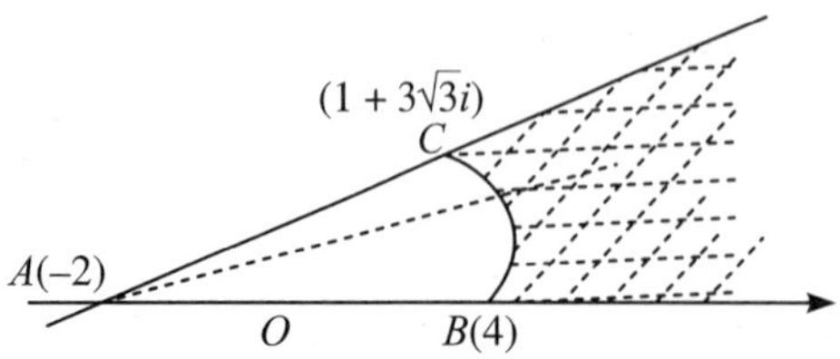

Figure 1.23

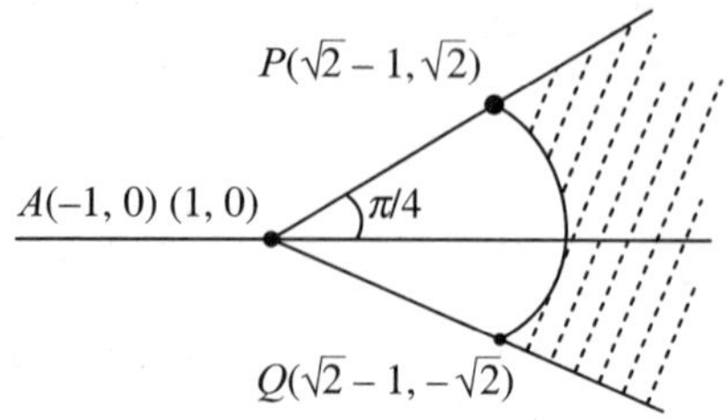

Figure 1.24

5. The inequality $|z-4| < |z-2|$ represents the region given by
(a) $\operatorname{Re} z > 0$ (b) $\operatorname{Re} z < 0$ (c) $\operatorname{Re} z > 3$ (d) none of these

6. The polar form of the complex number $-\sqrt{3}+i$ is:
(a) $2(\cos\frac{\pi}{6} + i\sin\frac{\pi}{6})$ (b) $2(\cos\frac{\pi}{3} + i\sin\frac{\pi}{3})$
(c) $2(\cos\frac{2\pi}{3} + i\sin\frac{2\pi}{3})$ (d) $2(\cos\frac{5\pi}{6} + i\sin\frac{5\pi}{6})$

7. If $z = re^{i\theta}$, then the value of $|e^{iz}|$ is equal to
(a) $e^{-r\cos\theta}$ (b) $e^{-r\cos\theta}$ (c) $e^{-r\sin\theta}$ (d) $e^{r\sin\theta}$

8. The conjugate (also called symmetric) point of $1+i$ with respect to the circle $\Gamma : |z-1| = 2$ is
(a) $1-i$ (b) $1+4i$ (c) $1+2i$ (d) $-1-i$
[**Hint:** All the other points lie outside Γ.]

9. If $x + \frac{1}{x} = 2\cos\theta$ and $x - \frac{1}{x} = 2i\sin\theta$, then $x^n + \frac{1}{x^n} =$
(a) $2^n\cos\theta$ (b) $2(\cos\theta)^n$ (c) $2\cos n\theta$ (d) $(2\cos\theta)^n$

10. Let $n \in \mathbb{N}$, then $e^{i\bar{z}} = \overline{e^{iz}}$
(a) for all z (b) only for $z = 2n\pi$ (c) for $z = n\pi$ (d) only for $z = (2n+1)\pi$

11. Primitive period of e^z is
(a) 2π (b) $2\pi i$ (c) $2n\pi$ (d) $2n\pi i$

12. If z_1, z_2 are two complex numbers such that $|z_1 + z_2| = |z_1| + |z_2|$, then it is necessary that
(a) $z_1 = z_2$, (b) $z_1 = \lambda z_2$, λ is some real number
(c) $z_2 = 0$ (d) $z_1 z_2 = 0$ or $z_1 = \lambda z_2$, λ is some real number

13. The locus of $z \in \mathbb{C}$ in the Argand plane satisfying the inequality $\log_{1/2}(\frac{|z-1|+4}{3|z-1|-2}) > 1$ (where $|z-1| \neq 2/3$) is
(a) a circle (b) interior of a circle (c) exterior of a circle (d) none of these

14. $\arg(z-i) = \pi/3$ represents
(a) a circle of radius $1/3$ and centre $(0,1)$ (b) a circle of radius $1/3$ and centre $(1,0)$
(c) a line with equation $y = \sqrt{3}x + 1$ (d) a point in the complex plane

15. If $\arg(z+3) = \pi/3$, then the least value of $|z|$ is
(a) 3 (b) $\sqrt{3}/2$ (c) $\sqrt{3}$ (d) $3\sqrt{3}/2$

16. The points which represents the root of the equation $(z+1)^n - z^n$ lie on
(a) a circle (b) a pair of straight lines (c) a straight line (d) a parabola

17. The principal value of $(1-i)^{1+i}$ is
    (a) $(1-i)e^{\pi/4}e^{i\log\sqrt{2}}$ (b) $(1+i)e^{\pi/4}e^{i\log\sqrt{2}}$
    (c) $(1-i)e^{-\pi/4}e^{i\log\sqrt{2}}$ (d) $(1+i)e^{-\pi/4}e^{i\log\sqrt{2}}$
18. n^{th} root of z: $\sqrt[n]{z}$ is equals to $(k=0,1,\cdots,n-1)$
    (a) $\cos\frac{\theta+\pi}{n}+i\sin\frac{\theta+\pi}{n}$ (b) $\sqrt[n]{r}(\cos\frac{\theta+2k\pi}{n}+i\sin\frac{\theta+2k\pi}{n})$
    (c) $\sqrt[n]{r}(\cos\frac{\theta+2k}{n}+i\sin\frac{\theta+2k}{n})$ (d) $\sqrt[n]{r}(\cos\frac{\theta+2k}{n\pi}+i\sin\frac{\theta+2k}{n\pi})$
19. The reciprocal of complex number $z=a+ib$ is
    (a) $\frac{1}{a}+i\frac{1}{b}$ (b) $\frac{1}{a^2+b^2}+i\frac{1}{a^2+b^2}$ (c) $\frac{a}{a^2+b^2}+i\frac{b}{a^2+b^2}$ (d) $\frac{a}{a^2+b^2}-i\frac{b}{a^2+b^2}$
20. What is the cardinality of the set $\{z\in\mathbb{C} : z^{98}=1 \text{ and } z^n\neq 1 \text{ for any } 0<n<98\}$?
    (a) 0 (b) 12 (c) 42 (d) 49
21. The real part of the principal value of 4^{4-i} is
    (a) $256\cos(\ln 4)$ (b) $164\cos(\ln 4)$ (c) $16\cos(\ln 4)$ (d) $4\cos(\ln 4)$
22. The point symmetric to the point $2+i$ with respect to the circle $|z-i|=3$ is
    (a) $9/2+i$ (b) $9/2-i$ (c) $2/9-i$ (d) none of these
23. The origin and the points representing the roots of the equation $z^2+pz+q=0$ form an equilateral triangle if
    (a) $p=3q^2$ (b) $p=3q$ (c) $p^2=q$ (d) $p^2=3q$
24. Given that $|z+\frac{1}{z}|=a$, the greatest possible value of $|z|$ is
    (a) $\frac{1}{2}(a+\sqrt{a^2+4})$ (b) $\frac{1}{2}(a+\sqrt{a+2})$ (c) a (d) $\sqrt{a^2+4}$
25. Let $S=S_1\cap S_2\cap S_3$, where $S_1=\{z\in\mathbb{C} : |z|<4\}$, $S_2=\{z\in\mathbb{C} : \text{Im}\,(\frac{z-1+\sqrt{3}i}{1-\sqrt{3}i})>0\}$ and $S_3=\{z\in\mathbb{C} : \text{Re}\,z>0\}$. Then $\min_{z\in S}|1-3i-z|=$
    (a) $(2-\sqrt{3})/2$ (b) $(2+\sqrt{3})/2$ (c) $(3-\sqrt{3})/2$ (d) $(3+\sqrt{3})/2$
26. For the problem 25, the area of S is
    (a) $10\pi/3$ (b) $20\pi/3$ (c) $16\pi/3$ (d) $32\pi/3$
27. Under the stereographic projection of points on the unit sphere $x^2+y^2+z^2=1$ in $\mathbb{R}^3$ to the extended plane $\mathbb{C}_\infty$, the point $z=x+iy$ corresponds to
    (a) $\left(\frac{x^2}{|z|^2+1},\frac{y^2}{|z|^2+1},\frac{|z|^2-1}{|z|^2+1}\right)$ (b) $\left(\frac{x}{|z|^2+1},\frac{y}{|z|^2+1},\frac{x+y+1}{|z|^2+1}\right)$
    (c) $\left(\frac{2x}{|z|^2+1},\frac{2y}{|z|^2+1},\frac{2x+2y+1}{|z|^2+1}\right)$ (d) $\left(\frac{2x}{|z|^2+1},\frac{2y}{|z|^2+1},\frac{|z|^2-1}{|z|^2+1}\right)$
28. The stereographic projection is
    (a) a homeomorphism of the complex plane into the sphere
    (b) a one-to-one continuous map of the complex plane into the sphere
    (c) a continuous map of the complex plane onto sphere which is not one-to-one
    (d) a one-to-one map of the extended complex plane onto sphere which is not continuous
29. By stereographic projection with the south pole at the origin $(0,0,0)$, the point $(1,0,0)$ goes to the complex number
    (a) $z=1$ (b) $z=1+i$ (c) $z=i$ (d) $z=-1$
30. The stereographic projection of i on the sphere touching the xy-plane and centre at $(0,0,1/2)$ is given by
    (a) $(1/2,1/2,0)$ (b) $(1/2,0,1/2)$ (c) $(0,1/2,1/2)$ (d) $(0,0,1)$

31. If the points on the complex plane are projected stereographically on the sphere $S : x^2 + y^2 + (z - 1/2)^2 = 1/4$ and $d(z, z')$ denotes the distance between the images of z and z' on S (the chordal distance), then $d(z, \infty)$, where $z = 1 + i$ is
    (a) $1/3$ (b) 1 (c) 0 (d) $1/\sqrt{3}$

32. If the point z on xy-plane has projection $(1, -1, 1)$ on the Riemannian sphere $x^2+y^2+(z-1/2)^2 = 1/4$, then the projection (stereographic) of iz is
    (a) $(1, 1, 1)$ (b) $(1, -1, 1)$ (c) $(-1, 1, -1)$ (d) $(-1, -1, -1)$

33. Let $\gamma : [0, 1] \to \mathbb{C}$ be defined by $\gamma(t) = (1 + i)(1 + t)^2$. Then the direction of the tangent at origin is
    (a) $\pi/3$ (b) $\pi/6$ (c) $\pi/4$ (d) cannot be determined

34. If $r = 2{+}4\cos\theta$ is a curve in the Argand plane, then the winding number of 1 and 3 are respectively
    (a) $1, 3$ (b) $2, 1$ (c) $1, 2$ (d) $2, 2$

35. A line in complex plane $\mathbb{C}$ is
    (a) $L = \left\{z = x + iy : \text{Im}\left(\frac{z - x}{y}\right) = 0\right\}$ (b) $L = \left\{z = x + iy : \text{Im}\left(\frac{z - x}{y}\right) > 0\right\}$
    (c) $L = \left\{z = x + iy : \text{Im}\left(\frac{z - x}{y}\right) < 0\right\}$ (d) $L = \left\{z = x + iy : \text{Re}\left(\frac{z - x}{y}\right) = 0\right\}$

36. The general equation of a line in the complex plane is (here $\alpha \neq 0$, α complex and n real):
    (a) $\alpha z + n = 0$ (b) $\alpha\bar{z} + n = 0$ (c) $\alpha z + \bar{\alpha}\bar{z} + n = 0$ (d) $z = n$

37. The half plane left to line L is
    (a) $H = \left\{z = x + iy : \text{Im}\left(\frac{z - x}{y}\right) = 0\right\}$ (b) $H = \left\{z = x + iy : \text{Im}\left(\frac{z - x}{y}\right) > 0\right\}$
    (c) $H = \left\{z = x + iy : \text{Im}\left(\frac{z - x}{y}\right) < 0\right\}$ (d) $H = \left\{z = x + iy : \text{Re}\left(\frac{z - x}{y}\right) = 0\right\}$

38. If P and Q are points on the unit circle corresponding to complex numbers $\frac{1}{2} + i\frac{\sqrt{3}}{2}$ and $-\frac{1}{2} + i\frac{\sqrt{3}}{2}$ respectively, what is the measure of $\angle POQ$?
    (a) 30° (b) 60° (c) 45° (d) 90°

39. A region of a plane is
    (a) an open connected subset (b) a disconnected subset
    (c) a connected subset (d) an open disconnected subset

Answers

1(c); 2(c); 3(c); 4(a); 5(c); 6(d); 7(b); 8(a); 9(c); 10(c); 11(b); 12(b); 13(c); 14(c); 15(d); 16(b); 17(a); 18(b); 19(c); 20(c); 21(a); 22(a); 23(d); 24(a); 25(c); 26(b); 27(d); 28(d); 29(a); 30(c); 31(d); 32(a); 33(c); 34(b); 35(a); 36(c); 37(c); 38(d); 39(a);

Review Questions [Short-answer Type]

1. Write the following numbers in the form $a + ib$.
   (a) $\cos(1 - 2i)$ (b) $e^{i(1+i)}$ (c) $e^{2+3i}\cos(i)$.

2. Graph the set of points $z \in \mathbb{C}$ which satisfy the equation
   (a) $|z - i| = 1$ (b) $z^2 + \bar{z}^2 = 2$.

3. If $z_1 = 1 - i$, $z_2 = -2 + 4i$, $z_3 = \sqrt{3} - 2i$, verify the following:

   (i) $\frac{1}{2}(\frac{z_3}{\bar{z}_3} + \frac{\bar{z}_3}{z_3}) = -\frac{1}{7}$ $\quad$ (ii) $|2z_2 - 3z_1|^2 = 170$

   (iii) $(z_3 - \bar{z}_3)^5 = 1024i$ $\quad$ $iv)$ $|z_1\bar{z}_2 + z_2\bar{z}_1| = 12$

   (v) $\text{Im}\{\frac{z_1 z_2}{z_3}\} = \frac{6\sqrt{3}+4}{7}$ $\quad$ (vi) $\text{Re}\{2z_1^3 + 3z_2^2 - 5z_3^2\} = 35$

   (vii) $|\frac{z_1+z_2+1}{z_1-z_2+i}| = \frac{3}{5}$ $\quad$ $(viii)$ $|z_1^2 + \bar{z}_2^2|^2 + |\bar{z}_3^2 - z_2^2| = 765 + 128\sqrt{3}$

4. What is the geometric interpretation of the multiplication of complex numbers?

5. Prove that the determinant of the real and imaginary parts of complex numbers z and w is $(\bar{z}w - z\bar{w})/2i$.

6. (a) Find i^{i^i}. Show that it does not coincide with $i^{i \cdot i} = i^{-1}$.

   (b) What is wrong in the following?

   $$1 = \sqrt{1} = \sqrt{(-1)(-1)} = \sqrt{-1}\,\sqrt{-1} = i \cdot i = -1.$$

   (c) Is it true that $i^i \cdot i^{-i} = i^0 = 1$? Give reasons.

7. Prove that for complex numbers a and b:

   (a) $|\frac{a-b}{1-\bar{a}b}| = 1$, with $|a| = 1$ or $|b| = 1$. $\quad$ (b) $|\frac{a-b}{1-\bar{a}b}| < 1$ if $|a| < 1$ or $|b| < 1$.

   (c) $|\frac{az+b}{\bar{a}+\bar{b}z}| = 1$, for $|z| = 1$. $\quad$ (d) $\frac{|z-a|}{1-\bar{a}z} = 1$ if $|z| = 1$ and $1 - \bar{a}z \neq 0$.

8. Find $\arg_I z$, if (a) $z = -i$ and $I = (-\pi, \pi]$, (b) $z = -i$ and $I = [0, 2\pi)$, (c) $z = 1$ and $I = [3\pi/2, 7\pi/2)$

9. Among the complex numbers z which satisfies $|z - 25i| \leq 15$, find the complex numbers z having (a) least positive argument, (b) maximum positive argument, (c) least modulus and (d) maximum modulus.

10. (a) Let $A = \{z : |z| \leq 2\}$ and $B = \{z : (1 - i)z + (1 + i)\bar{z} \geq 4\}$ be two sets. Find the area of the region $A \cap B$.

    (b) For all $x \in \mathbb{R}$, let the mapping $f(x) = 1/(x - i)$. If there exist $a, b, c, d \in \mathbb{R}$ for which $f(a), f(b), f(c)$ and $f(d)$ form a square on the complex plane, find the area of the square.

    (c) Let z_1 be a complex number with $|z_1| = 5$ and $\text{Im}\, z > 0$ and z_2 be a value of z_1 such that the distance between $(1 + 2i)z_1^3$ and z_1 is maximum. If $z_2^4 = c = id$, then find the value of $(c + d)$.

    (d) If z_1 and z_2 are two complex numbers such that $(z_1 - 2z_2)/(2 - z_1\bar{z}_2)$ is unimodular (whose modulus is one), while z_2 is not unimodular, find $|z_1|$.

    (e) Let $A = \{a \in \mathbb{R} : (1 + 2i)x^3 - 2(3 + i)x^2 + (5 - 4i)x + 2a^2 = 0\}$ have at leat one real root. Find the value of $\sum_{a \in A} a^2$.

11. Prove that

    (a) $\sqrt{3} + \sqrt{2}$ is an algebraic number. [**Hint:** Let $z = \sqrt{3} + \sqrt{2}$. Then $z^4 - 2z^2 + 1 = 8z^2 \Rightarrow z^4 - 10z^2 + 1 = 0$.]

    (b) $\sum_{p=1}^{32} (3p + 2)[\sum_{q=1}^{10} (\sin \frac{2q\pi}{11} - i \cos \frac{2q\pi}{11})]^p = 48(1 - i)$.

    (c) If $\alpha_k = \cos(k\pi/7) + i \sin(k\pi/7)$; $k \in \mathbb{Z}$, then

    $$\sum_{k=1}^{12} |\alpha_{k+1} - \alpha_k| \Big/ \sum_{k=1}^{3} |\alpha_{4k-1} - \alpha_{4k-2}| = 4.$$

(d) If $|z| =$, then $|\text{Im}\,(1-\bar{z}+z^2)| \leq 7$ and $|z^4-4z^2+3| \geq 3$.
(e) If $|z| = 3$, then $\frac{5}{13} \leq |\frac{2z-1}{4+z^2}| \leq \frac{7}{5}$.

12. Prove that every line or circle in $\mathbb{C}$ is the solution set of an equation of the form $a|z|^2 + \bar{z}_1 z + z_1\bar{z} + b = 0$, where a and b are real numbers and z_1 is a complex number. Conversely, show that every equation of this form has a line, circle, point, or empty set as its solution set.

13. Prove that
   (a) the equation $az + b\bar{z} + c = 0$ has exactly one solution if and only if $|a| \neq |b|$.
   (b) the set of points z such that $|z+ib| < |z+b|$ is the half plane $\{z = x+iy : x > y\}$.
   (c) the inequalities $\text{Re}\, z > 0$ and $|z-a| < |z+a|, a \in \mathbb{R}$ are equivalent.
   (d) the inequalities $|z| < 1$ and $\text{Re}\,\{(1+z)/(1-z)\} > 0$ are equivalent.
   (e) the inequality $|z_1 - z_2| < |\bar{z}_1 + z_2|$ holds provided either $\text{Re}\, z_1 > 0$ and $\text{Re}\, z_2 > 0$ or $\text{Re}\, z_1 < 0$ and $\text{Re}\, z_2 < 0$.
   (f) the minimum value of $|2z - 6 + 5i|$ is 5 if $|z - 3 - 2i| \leq 2$.

14. For any three distinct complex numbers z, z_1, z_2, the principal value of $\arg(\frac{z-z_1}{z-z_2})$ represents the angle between the line segment joining z and z_1 and the line segment joining z and z_2, taken in the appropriate sense.

15. If z is a variable point and $\text{Re}\,(\frac{z-4}{z-2i}) = 0$, prove that the locus of z is a circle.

16. Let R be a positive constant and let z_0 be a fixed complex number. Show that the equation of the circle with centre at z_0 and radius R may be written as

$$|z|^2 - 2\,\text{Re}\,(\bar{z}_0 z)| + |z_0|^2 = R^2.$$

17. Show that the inverse point of any point α with respect to the circle $|z| = 1$ is $1/|\alpha|$.

18. Find the inverse point of $-i$ with respect to the circle $2z\bar{z} + (i-1)z - (i+1)\bar{z} = 0$.

19. Identify and sketch the following sets in the complex plane:
   (i) $\Big\{z \in \mathbb{C} : |z-1+i| = 2\Big\}$ (ii) $\Big\{z \in \mathbb{C} : |z-1+i| \leq 2\Big\}$ (iii) $\Big\{z \in \mathbb{C} :\ \text{Im}\,(z^2) = 1\Big\}$
   (iv) $\Big\{z \in \mathbb{C} : |z-i|+|z+i| = 3\Big\}$ (v) $\Big\{z \in \mathbb{C} : |z| = |z+1|\Big\}$ (vi) $\Big\{z \in \mathbb{C} : |z-1| = 2|z+1|\Big\}$
   (vii) $\Big\{z \in \mathbb{C} :\ \text{Re}\,(z^2) = 1\Big\}$ (viii) $\Big\{z \in \mathbb{C} :\ \text{Re}\,(z+2-2i) = 3\Big\}$ (ix) $\Big\{z \in \mathbb{C} : |z-1| < |z|\Big\}$
   (x) $\Big\{z \in \mathbb{C} : |z-i|^2 + |z+i|^2 < 2\Big\}$

20. For a fixed complex number b, sketch the curve $\{e^{i\theta} + be^{-i\theta} : 0 \leq \theta \leq 2\pi\}$. Differentiate between the cases $|b| < 1$, $|b| = 1$ and $|b| > 1$.

21. Show that
   (a) $(\mathbb{C}, d)$ is a metric space, where $d(x+iy, a+ib) = |x-a| + |y-b|$.
   (b) $(\mathbb{C}, d)$ is a metric space, where $d(x+iy, a+ib) = \max\Big\{|x-a|, |y-b|\Big\}$.
   (c) the spherical distance $d(z_1, z_2)$ as in Eq.(1.31) is a metric.
   (d) if $z_2 = \infty$, then $d(z_1, \infty) = 2/\sqrt{1+|z_1|^2}$.

22. Show that $|z| \leq |\text{Re}\,(z)| + |\text{Im}\,(z)|$ and sketch the set of points for which equality holds.

23. Supply the geometric description of the following subsets of $\mathbb{C}$:

(i) $\{z : |z-1+i| = 1\}$
(ii) $\{z : -\frac{\pi}{6} < \arg(z - \frac{1}{k}) < \frac{13\pi}{6}; k = 1,2,3,4,5\}$
(iii) $\{z : |z+3i| > 1\}$
(iv) $\{z : |z-4i| \geq 1\}$
(v) $\{z : |z+i| \leq 3\}$
(vi) $\{z : |z+2| + |z-2| < 4\}$
(vii) $\{z : \operatorname{Re}(\bar{z} - i) = 2\}$
(viii) $S = \{z : |z-1| = 3|z+1|\}$
(ix) $\{z : 0 < \operatorname{Re}(iz) \leq 1\}$
(x) $S = \{z : \operatorname{Im}\left(\frac{z-z_1}{z-z_2}\right) = 0; z_1, z_2 \in \mathbb{C}\}$
(xi) $\{z : \operatorname{Re}(z^2) > 1\}$
(xii) $S = \{z : \operatorname{Re}\left(\frac{z-z_1}{z-z_2}\right) = 0; z_1, z_2 \in \mathbb{C}\}$
(xiii) $\{z : |z-1| = \operatorname{Re} z\}$
(xiv) $S = \{z : |z-1| > 2, 0 < \operatorname{Arg}(z-2) < \pi/4\}$
(xv) $\{z : |2\bar{z} + i| = 4\}$
(xvi) $\{z : -\frac{\pi}{4} < \arg(z-k) < \frac{3\pi}{2}; k = 1,2,3,4,5\}$

24. Sketch the sets defined by the following constraints and determine whether they are open, closed, or neither; bounded; connected; compact.

(i) $S = \{z : |z| > 2|z-1|\}$
(ii) $S = \{z : z = e^{2k\pi i/5}; k = 0,1,2,3,\cdots\}$
(iii) $S = \{z : \operatorname{Re} z < 1\}$
(iv) $S = \{z : |z+2| \geq |z|\}$
(v) $S = \{z : z^2 + \bar{z}^2 = 1\}$
(vi) $S = \{z : |z-1| + |z+1| = 2\}$
(vii) $S = \{z : |\operatorname{Im}(z)| < 1\}$
(viii) $S = \{z : 0 < |z-1| < 2\}$
(ix) $S = \{z : |z+3| < 2\}$
(x) $S = \{z : |z-1| + |z+1| < 3\}$
(xi) $S = \{z : |z| \geq \operatorname{Re}(z) + 1\}$

What are the boundaries of the sets in the previous exercise?

25. Which of the following subsets of $\mathbb{C}$ are open, closed, or neither open nor closed:

(i) $|z+i| + |z-i| \leq 4 \cap |\arg z| < \pi/4$
(ii) $\operatorname{Im} z > 1$
(iii) $0 \leq \arg z \leq \pi/4; z \neq 0$
(iv) $|z-4| \geq |z|$

26. Which of the following sets are bounded:

(i) $|z-4| \geq |z|$
(ii) $|z-2+i| \leq 1$
(iii) $\operatorname{Re}(z^2) > 0$
(iv) $\{z : |z-i| - |z+i| = 2\}$

27. Which of the following sets are connected?

(i) $\{z : |z| < 1\} \cup \{z : |z-1| < 1\}$
(ii) $\{z : -1 < |z| \leq 1\} \cup \{2\}$
(iii) $\{z : -\pi/4 < \arg z \leq \pi/4\}$
(iv) $\{z : |z^2 - 1| < 2\}$

28. Sketch the following sets and determine which of them are domains/regions:

(i) $|2z+3| > 4$
(ii) $\operatorname{Im} z = 1$
(iii) $\operatorname{Im} z > 1$
(iv) $|z-4| \geq |z|$
(v) $\{z : |z-2+i| \leq 1\}$
(vi) $S_1 = \{z : |\operatorname{Im}(z)| > 1\}$; $S_2 = \{z : |\operatorname{Re}(z)| > 1\}$
(vii) $\{z : 0 < \arg z < \frac{\pi}{4}\}$
(viii) $\{z : 0 < \arg z < \frac{\pi}{4} \text{ and } |z| > 1\}$
(ix) $\{z : 1 < |z| < 2\}$

29. Find the points on the Riemann sphere S corresponding to the complex numbers:
    $(i)\ -i$ $\quad (ii)\ 1+i$ $\quad (iii)\ 2-3i$ $\quad (iv)\ 1-i$
30. For each of the following points in $\mathbb{C}$, determine its stereographic projection on the Riemann sphere:
    (a) i; (b) $6-8i$; (c) $-\frac{3}{10}+\frac{2}{5}i$.
31. Show that the union of two regions with nonempty intersection is itself a region.

Answers

1. $\cos(1)\cosh(2)+i\sin(1)\sinh(2)$; $e^{-1}\cos(1)+ie^{-1}\sin(1)$; $\cosh(1)\cos(3)e^2+ie^2\cosh(1)\sin(3)$.
10. (a) $\pi-2$; (b) $1/2$; (c) 125; (d) $|z_1|=2$; (e) 18
18. $(2+i)/5$
25. (i) (iii) Neither open nor closed; (ii) Open; (iv) Closed
26. (i) (iii) Not bounded; (ii) (iv) Bounded
27. (i) (iii) Connected (ii) (iv) Not connected
29. (i) (0,-1,0); (ii) (2/3,2/3,1/2); (iii) (2/7,-3/7,6/7); (iv) (2/3,-2/3,1/2).

Review Questions [Long-answer Type]

1. Let $z_1, z_2 \in \mathbb{C}$. Prove that $|z_1+z_2|^2+|z_1-z_2|^2=2(|z_1|^2+|z_2|^2)$ and deduce that

$$|z_1+\sqrt{z_1^2-z_2^2}|+|z_1-\sqrt{z_1^2-z_2^2}|=|z_1+z_2|+|z_1-z_2|.$$

2. Let $z_1, z_2 \in \mathbb{C}$. Prove that the triangle inequality:

$$\Big||z_1|-|z_2|\Big| \le \Big|z_1+z_2\Big| \le |z_1|+|z_2|.$$

   Under what conditions does the equality hold? Generalise the inequality on the right to n complex numbers.

3. Show that the equation of a straight line in the Argand plane can be put in the form $\bar{b}z+b\bar{z}=c$, where b is a non-zero complex number and $c \in \mathbb{R}$.

4. Prove that the equation of the circle passing through the three points

   (a) z_1, z_2, z_3 is $\dfrac{(z-z_1)(z_3-z_2)}{(z-z_2)(z_3-z_1)}=\dfrac{(\bar{z}-\bar{z}_1)(\bar{z}_3-\bar{z}_2)}{(\bar{z}-\bar{z}_2)(\bar{z}_3-\bar{z}_1)}$.

   (b) $1, i, 1+i$ is $2z\bar{z}+(i-1)z-(i+1)\bar{z}=0$.

5. Find the centroid and circumcentre of the triangle whose vertices are given by complex numbers z_1, z_2, z_3.

6. The position vectors of A, B and C of triangle ABC are given by $z_1=1+2i$, $z_2=4-2i$, and $z_3=1-6i$ respectively. Prove that ABC is an isoscles triangle and find the lengths of the sides.

7. Prove that:

   (a) the reflection of z in the line $ax+by=c$; $a, b, c \in \mathbb{R}$ is given by $\frac{2ic+(b-ai)\bar{z}}{b+ai}$.

   (b) the equation $|\frac{z-z_1}{z-z_2}|=\lambda$, where λ is non-negative parameter, represents a family of circles such that z_1 and z_2 are inverse points for every member of the family.

   (c) the equation $\arg(\frac{z-z_1}{z-z_2})=\lambda$, where λ is non-negative real parameter, represents a family of circles, every member of which passes through z_1 and z_2.

(d) two points z_1 and z_2 are inverse points or the reflection points for the joint family of straight line and circle $\alpha z\bar{z} + \bar{\beta}z + \beta\bar{z} + \gamma = 0$ if and only if $\alpha z_1\bar{z}_2 + \bar{\beta}z_1 + \beta\bar{z}_2 + \gamma = 0$.

(e) $|\frac{z-1}{z+1}| =$ constant and $\arg(\frac{z-1}{z+1}) =$ constant are orthogonal circles.

8. (a) Let $P(z)$ be a point on the Argand plane and let the straight line NP intersects the unit sphere V at Q, where $N(0,0,1)$ is the north pole. Find the coordinatesof the point Q.

   (b) Let $P(z)$ be a point on the Argand plane $x_3 = 1/2$ and let the straight line NP intersects the sphere $V : x_1^2 + x_2^2 + (x_3 - \frac{1}{2})^2 = \frac{1}{4}$ at Q, where $N(0,0,1)$ is the north pole. Find the coordinatesof the point Q.

   (c) Suppose $(x, y, 0)$ is the spherical projection of (x_1, x_2, x_3). Show that the product of the distances from the north pole N to (x_1, x_2, x_3) and from N to $(x, y, 0)$ is 2. What is the situation when (x_1, x_2, x_3) lies on the equator of the sphere?

9. (a) Find the complex numbers represented by the points $(\frac{1}{\sqrt{2}}, \frac{1}{\sqrt{2}}, 0)$, $(1, 0, 0)$ and $(0, 1, 0)$.

   (b) For each of the following points in $\mathbb{C}$, give the corresponding point of V: $0, 1 + i, 3 + 2i$.

10. Describe the projections on the Riemann sphere of the following sets in the complex plane:
    (a) the right half plane $\{z \in \mathbb{C} : \text{Re } z > 0\}$; (b) the disc $\{z \in \mathbb{C} : |z| < \frac{1}{2}\}$
    (c) the annulus $\{z \in \mathbb{C} : 1 < |z| < 2\}$ (d) the set $\{z \in \mathbb{C} : |z| > 3\}$
    (e) the line $y = x$ (including the point ∞).

11. If a point Q on a sphere correspond to z under the streographic projection, show that the antipodal point $-Q$ on the sphere corresponds to $-\bar{z}^{-1}$.

12. Find the stereographic images of
    (a) a triangle with vertices z_1, z_2, z_3, (b) a pencil of parallel straight lines.
    (c) $\bar{z}$ and $-z$ if the equation of the Riemann sphere is $V : x_1^2 + x_2^2 + (x_3 - \frac{1}{2})^2 = \frac{1}{4}$.

13. Suppose that a cube has its vertices on the Riemann sphere and its edges parallel to the coordinate axes. Find the stereographic projections of the vertices.

14. Let S be an open set consisting of all points z such that $|z| < 1$ or $|z - 2| < 1$. Show that S is not connected.

15. Prove that if U is open and K is closed, then the set-theoretic difference U/K is open, while K/U is closed.

16. Show that if K is a compact subset of $\mathbb{C}$, then there is a point $z_0 \in K$ of minimum modulus, that is, a point $z_0 \in K$ such that

$$|z_0| \leq |z| \text{ for all } z \in K.$$

17. Prove that the union of a family of connected sets with a point in common is also a connected set.

18. Prove that if $\mathcal{S}[z_0; R]$ is a closed disc contained in an open set U, then there is an open disc $\mathcal{S}(z_0; r)$ such that $\mathcal{S}[z_0; R] \subset \mathcal{S}(z_0; r) \subset U$.

Answers

5. $(z_1 + z_2 + z_3)/3$, $\sum z_1\bar{z}_2(z_2 - z_3)/\sum(\bar{z}_2 z_3 - z_2\bar{z}_3)$.
6. $5, 5, 8$.
8. (b) $\alpha = \frac{2x}{4|z|^2+1}; \beta = \frac{2y}{4|z|^2+1}; \gamma = 1 - \frac{1}{4|z|^2+1}$.
9. (a) $(1 + i)/\sqrt{2}; 1; i$

2 Sequence and Series

In this chapter we will recall some definitions and results on sequences and series of complex numbers. Also, we will study about the sequence and series representation of analytic function in detail and prove the existence of such representation with the help of some theorems.

Next we will introduce the concept of power series and prove that there exists a circle of convergence associated with any power series. The power series converges within its circle of convergence, is analytic and its derivative can be obtained by termwise differentiation. We also define some standard elementary functions such as exponential function, trigonometric function, hyperbolic function and logarithmic function in terms of power series.

2.1 Complex Sequences

By a *sequence in* $\mathbb{C}$ we mean a function $z : \mathbb{N} \to \mathbb{C}$ such that $z(1) = z_1, z(2) = z_2, \ldots, z(n) = z_n, \ldots$ are all elements of $\mathbb{C}$. Each number in the sequence is called a *term* and z_n is called the nth *term*. The sequence $z_1, z_2 \ldots, z_n, \ldots$ is also designated briefly by $\{z_n\}$ or $\{z_1, z_2, z_3, \ldots\}$. For example, the set of complex numbers $1 + i, \frac{(1+i)^2}{2!}, \frac{(1+i)^3}{3!}, \cdots$ is an infinite sequence, the nth term given by $z_n = (1 + i)^n/n!$.

2.1.1 Convergent Sequences

A sequence of complex numbers $\{z_n\}$ in the Argand plane converges to a complex number $z_0 \in \mathbb{C}$ if corresponding to an $\varepsilon > 0$, there is an integer $M = M(\varepsilon) \in \mathbb{N}$ such that

$$d(z_n, z_0) = | z_n - z_0 | < \varepsilon \text{ whenever } n \geq M. \tag{2.1}$$

Formally, we write $z_n \to z_0$ as $n \to \infty$ or in brief, $\lim_{n\to\infty} z_n = z_0$. Geometrically, if every neighbourhood of z contains almost all terms of the sequence $\{z_n\}$, then the point z_0 is called a *limit* of the sequence.

If the limit of a sequence exists, the sequence is called *convergent*; otherwise it is called *divergent*. A sequence with limit zero is defined to be a *null sequence*.

We note that all the usual theorems about limits hold for complex numbers: limits of sums, limits of products, limits of quotients, limits of composite functions.

Definition 2.1.1. [Subsequence]: Let $z_n : \mathbb{N} \to \mathbb{C}$ be a sequence and $k : \mathbb{N} \to \mathbb{C}$ a monotone function such that $k(1) < k(2) < k(3) < \cdots$. Then the composite function $z \circ k$ is called a subsequence of z and is denoted by its elements $\{z_{k_n}\}_n$ in the complex plane. For instance, $\{z_{2k}\}, \{z_{2k+3}\}, \{z_{2^k}\}$, etc., are some subsequences of $\{z_n\}$.

A point z_0 is said to be a *limit point* of the sequence $\{z_n\}$ if there exists a subsequence that converges to z_0.

Example 2.1.1. *Find the limit points of the $\{z_n\}$, where $z_n = i^n$.*

Solution. Since $i = \cos(\pi/2) + i\sin(\pi/2) = e^{i\pi/2}$, so

$$z_n = i^n = e^{i\frac{n\pi}{2}}; \qquad n = 1, 2, \cdots.$$

Therefore, the different elements of the sequence are $z_1 = i, z_2 = i^2 = -1, z_3 = -i, z_4 = 1, z_5 = i, z_6 = -1, \ldots$. Hence $\{z_n\}$ has the following subsequences:

(a) $\{i, i, i, \ldots\}$ which converges to i (b) $\{-i, -i, -i, \ldots\}$ which converges to $-i$
(c) $\{1, 1, 1, \ldots\}$ which converges to 1 (d) $\{-1, -1, -1, \ldots\}$ which converges to -1

Therefore, the limit points of $\{z_n\}$ are i, $-i$, 1, -1.

Theorem 2.1. [Cauchy's general principle of convergence]: *A sequence $\{z_n\}$ converges iff for every $\varepsilon > 0$, there is a $p(\varepsilon) \in \mathbb{N}$ such that*

$$|z_n - z_{n+m}| < \varepsilon \text{ for } n \geq p(\varepsilon)$$

and for every positive integer $m > 1$.

Proof: The condition is necessary, for if $z_n \to z$ as $n \to \infty$, there exists an integer $p \in \mathbb{N}$ depending on ε only, such that

$$|z_n - z| < \frac{\varepsilon}{2} \text{ whenever } n \geq p.$$

Since m is any positive integer, we have

$$\begin{aligned} |z_{n+m} - z_n| &= |z_{n+m} - z + z - z_n| \\ &\leq |z_{n+m} - z| + |z - z_n| < \frac{\varepsilon}{2} + \frac{\varepsilon}{2} = \varepsilon \text{ whenever } n \geq p. \end{aligned}$$

The condition is also sufficient. For, if it is satisfied, the given sequence $\{z_n\}$ is bounded, since all save a finite number of its points lie in the neighbourhood of z_n of radius ε. Hence by the Bolzano–Weierstass theorem, the sequence possesses at least one limiting point.

Now, if the sequence $\{z_n\}$ is not convergent, it must possess at least two finite limiting points z' and z'', say. Accordingly, there exist integers q, r depending on ε, such that

$$|z' - z_{n+q}| < \frac{\varepsilon}{4}, |z'' - z_{n+r}| < \frac{\varepsilon}{4}.$$

With these values of p, q, r, we have

$$\begin{aligned} |z' - z''| &= |(z' - z_{n+q}) + (z_{n+q} - z_n) - (z_{n+r} - z_n) - (z'' - z_{n+r})| \\ &\leq |(z' - z_{n+q})| + |(z_{n+q} - z_n)| + |(z_{n+r} - z_n)| + |(z'' - z_{n+r})| \\ &< \frac{\varepsilon}{4} + \frac{\varepsilon}{4} + \frac{\varepsilon}{4} + \frac{\varepsilon}{4} = \varepsilon. \end{aligned}$$

Since $\varepsilon > 0$ is arbitrary, we conclude that $z' = z''$. Hence $\{z_n\}$ does not possess more than one limiting point; this completes the proof of the sufficiency of the condition. □

Example 2.1.2. *Show that* $\{z^n\}$ *is a null sequence for* $|z| < 1$.

Solution. Let $\varepsilon > 0$ be chosen arbitrary. We must find a value of p such that ,

$$|z^n| < |z|^n < \varepsilon \text{ for any } n > p$$

for all z with $|z| < 1$; this is certainly true when $z = 0$. So for $z \neq 0$ we simply need to find p such that for any $n > p$,

$$n \log|z| < \log\varepsilon \Rightarrow n > \frac{\log\varepsilon}{\log|z|}$$

as $\log|z|$ is negative for $|z| < 1$. This proves that if $|z| < 1$, then

$$|z|^n < \varepsilon \text{ for } n > p = \left[\frac{\log\varepsilon}{\log|z|}\right] + 1,$$

where $\left[u\right]$ is the integral part of $u = \log\varepsilon/\log|z|$. Thus $z^n \to 0$ as $n \to \infty$ if $|z| < 1$.

Example 2.1.3. *Investigate the convergence of* $\{z_n\}$, *where,* $z_n = i^n/n$.

Solution. The terms of the sequence are

$$i, \frac{i^2}{2}, \frac{i^3}{3}, \frac{i^4}{4}, \frac{i^5}{5}, \cdots \text{ or } i, \frac{-1}{2}, \frac{-i}{3}, \frac{1}{4}, \frac{i}{5}, \cdots.$$

On plotting the corresponding points in the z-plane, we suspect that the limit is zero. Let $\varepsilon > 0$ be chosen arbitrary. Now

$$d(z_n, 0) = \left|\frac{i^n}{n} - 0\right| = \frac{|i|^n}{n} = \frac{1}{n} < \varepsilon, \text{ when } n > \frac{1}{\varepsilon}$$

$$\therefore \quad d(z_n, 0) = \left|\frac{i^n}{n} - 0\right| < \varepsilon, \text{ when } n > M = [\frac{1}{\varepsilon}] + 1 \in \mathbb{N},$$

which shows that $\{z_n\}$ is a null sequence.

Theorem 2.2. *The limit of a convergent sequence* $\{z_n\}$ *is unique.*

Proof: If possible , let z_0 and z_0' be two limits of a convergent sequence $\{z_n\}$ such that $z_0 \neq z_0'$. Then for every $\varepsilon > 0$, $\exists\, M_1, M_2 \in \mathbb{N}$ such that

$$d(z_n, z_0) < \frac{\varepsilon}{2} \text{ for } n \geq M_1, \quad d(z_n, z_0') < \frac{\varepsilon}{2} \text{ for } n \geq M_2.$$

Let $M = \max\{M_1, M_2\} \in \mathbb{N}$. Now,

$$d(z_0, z_0') \;\leq\; d(z_0, z_n) + d(z_n, z_0') < \frac{\varepsilon}{2} + \frac{\varepsilon}{2} = \varepsilon \text{ for } n \geq M.$$

Therefore, $0 \leq d(z_0, z_0') < \varepsilon$ whenever $n > M$. Since ε is arbitrary, we have $z_0 = z_0'$, i.e., the limit of the convergent sequence is unique. □

Example 2.1.4. *Find the limit point(s) of* $\{z_n\}$, *where* $z_n = (-1)^n + \frac{n}{n+1}i;\ n = 1, 2, \ldots.$

Solution. The different elements of the sequence are

$$\begin{aligned} z_1 &= (-1)^1 + \frac{1}{1+1}i = -1 + \frac{1}{2}i \\ z_2 &= (-1)^2 + \frac{2}{2+1}i = 1 + \frac{2}{3}i \\ z_3 &= (-1)^3 + \frac{3}{3+1}i = -1 + \frac{3}{4}i \\ &\vdots \end{aligned}$$

From this we see that $\pm 1 + i$ are the limit points of $\{z_n\}$, so $\{z_n\}$ is not convergent.

Theorem 2.3. *Let $\{z_n\}$ be a sequence of complex numbers. If $z_n = a_n + ib_n$ for each $n \in \mathbb{N}$, then $\{z_n\}$ is convergent iff so are the sequences of real numbers $\{a_n\}$ and $\{b_n\}$.*

Proof: Let $\{z_n\}$ be a convergent sequence converging to $z_0 (= a + ib)$, say. Then corresponding to an $\varepsilon > 0$, $\exists\, M_0(\varepsilon) \in \mathbb{N}$ such that

$$\begin{aligned} & d(z_n, z_0) = | z_n - z_0 | < \varepsilon \quad \text{for all } n \geq M_0 \\ \Rightarrow \quad & | (a_n - a) + i(b_n - b) | < \varepsilon \quad \text{for all } n \geq M_0 \\ \Rightarrow \quad & | a_n - a | \leq | z_n - z_0 | < \varepsilon \text{ and } | b_n - b | \leq | z_n - z_0 | < \varepsilon \text{ for all } n \geq M_0. \end{aligned}$$

Therefore, $\lim_{n\to\infty} a_n = a$ and $\lim_{n\to\infty} b_n = b$. This completes proof of the necessary part.

Conversely, let $\lim_{n\to\infty} a_n = a$ and $\lim_{n\to\infty} b_n = b$. Then corresponding to $\varepsilon > 0$, there exists $M_1(\varepsilon)$, $M_2(\varepsilon) \in \mathbb{N}$ such that

$$|a_n - a| < \frac{\varepsilon}{2}; \forall n \geq M_1 \text{ and } |b_n - b| < \frac{\varepsilon}{2}; \forall n \geq M_2.$$

Let $M_0 = \max\{M_1, M_2\} \in \mathbb{N}$. Now both these is equalities hold simultaneously and

$$\begin{aligned} |z_n - z_0| &= |(a_n - a) + i(b_n - b)| \\ &\leq |a_n - a| + |b_n - b| < \frac{\varepsilon}{2} + \frac{\varepsilon}{2} = \varepsilon; \forall n \geq M_0. \end{aligned}$$

Consequently, $\lim_{n\to\infty} z_n = z_0$ i.e., $\{z_n\}$ converges to z_0. This proves the sufficiency of the condition. □

This theorem enables us to write

$$\lim_{n\to\infty} (a_n + ib_n) = \lim_{n\to\infty} a_n + i \lim_{n\to\infty} b_n.$$

Thus the study of convergence of $\{z_n\}$ is equivalent to the consideration of the real sequences $\{a_n\}$ and $\{b_n\}$.

Corollary 2.1.1. If $\lim_{n\to\infty} z_n = z_0$, then $\lim_{n\to\infty} \overline{z}_n = \overline{z}_0$ and $\lim_{n\to\infty} |z_n| = |z_0|$.

Proof: By the aforesaid Theorem 2.3,

$$\lim_{n\to\infty} z_n = z_0 \Leftrightarrow \lim_{n\to\infty} \operatorname{Re} z_n = \operatorname{Re} z_0 \text{ and } \lim_{n\to\infty} \operatorname{Im} z_n = \operatorname{Im} z_0.$$

$$\therefore \quad \lim_{n\to\infty} \operatorname{Re} z_n - i \lim_{n\to\infty} \operatorname{Im} z_n = \operatorname{Re} z_0 - \operatorname{Im} z_0 \Rightarrow \lim_{n\to\infty} \overline{z}_n = \overline{z}_0.$$

Using the property of modules, we have

$$\Big||z_n| - |z_0|\Big| \leq \Big|z_n - z_0\Big|$$

and with the fact that $\lim_{n\to\infty} z_n = z_0$, the second conclusion is immediate. □

Example 2.1.5. *Prove that the $\{z_n\}$ converges to i, where $z_n = \dfrac{1}{n^3} + i;\ n = 1, 2, \cdots$.*

Solution. Since $z_n = (1/n^3) + i$, so

$$\lim_{n\to\infty} z_n = \lim_{n\to\infty} \left(\frac{1}{n^3} + i\right) = \lim_{n\to\infty} \frac{1}{n^3} + i \lim_{n\to\infty} 1 = 0 + i \cdot 1 = i.$$

Theorem 2.1 can also be used to obtain the result. Let $\varepsilon > 0$ be chosen arbitrarily. Then

$$|z_n - i| = \frac{1}{n^3} < \varepsilon \text{ whenever } n \geq m = [\varepsilon^{-1/3}] + 1.$$

2.1.2 Bounded Sequence

Definition 2.1.2. A sequence $\{z_n\}$ of complex numbers is called *bounded* if there exists a positive number M such that

$$|z_n| < M, \qquad \forall\ n \in \mathbb{N}.$$

Now, like real analysis, every convergent sequence is bounded but not the converse.

2.1.3 Complex Cauchy's Sequence

Definition 2.1.3. A complex sequence $\{z_n\}$ is a *complex Cauchy's sequence* if for every preassigned $\varepsilon > 0$, $\exists\ M_0(\varepsilon) \in \mathbb{N}$ such that

$$d(z_n, z_m) = |z_n - z_m| < \varepsilon \text{ whenever } m, n > M_0$$

$$\text{or,} \quad d(z_{n+p}, z_n) = |z_{n+p} - z_n| < \varepsilon \text{ whenever } n > M_0 \text{ and } p \in \mathbb{N}. \tag{2.2}$$

Theorem 2.4. *A sequence $\{z_n\}$ is convergent, if and only if it is a Cauchy's sequence.*

***Proof*:** Let the sequence $\{z_n\}$ converge to z_0. Then, for $\varepsilon(> 0)$, $\exists\ n_0 \in \mathbb{N}$ such that

$$d(z_m, z_0) = |z_m - z_0| < \frac{\varepsilon}{2} \text{ whenever } n \geq n_0.$$

Hence, if $m \geq n_0, n \geq n_0$ then $|z_m - z_0| < \varepsilon/2$ and $|z_n - z_0| < \varepsilon/2$. Now,

$$\begin{aligned} |z_m - z_n| &= |z_m - z_0 + z_0 - z_n| \\ &\leq |z_m - z_0| + |z_0 - z_n| < \frac{\varepsilon}{2} + \frac{\varepsilon}{2} = \varepsilon, \text{ for } m, n \geq n_0. \end{aligned}$$

Therefore, $\{z_n\}$ is a Cauchy's sequence.

Conversely, let $\{z_n\}$ be a Cauchy's sequence. Then

$$|z_m - z_n| < \varepsilon \text{ for } m, n \geq n_0$$

$$\left|(a_m + ib_m) - (a_n + ib_n)\right| < \varepsilon \text{ for } m, n \geq n_0$$

$$\left|(a_m - a_n) + i(b_m - b_n)\right| < \varepsilon \text{ for } m, n \geq n_0.$$

Now, $\left|a_m - a_n\right| \leq \left|(a_m - a_n) + i(b_m - b_n)\right| < \varepsilon$ for $m, n \geq n_0$

$$\left|b_m - b_n\right| \leq \left|(a_m - a_n) + i(b_m - b_n)\right| < \varepsilon \text{ for } m, n \geq n_0.$$

Then, $\{a_n\}$and $\{b_n\}$ are Cauchy's sequence of real numbers, i.e., $\{a_n\}$ and $\{b_n\}$ are both convergent and converges to say a_0 and b_0 respectively. Then $\{z_n = a_n + ib_n\}$ converges to $a_0 + ib_0 = z_0$. Hence $\{z_n\}$ is convergent. □

2.2 Completeness

Definition 2.2.1. A metric space $(\mathbb{C}, d)$ is said to be *complete* if and only if every Cauchy sequence of elements of $\mathbb{C}$ converges to some element of $\mathbb{C}$ in the space.

(i) The definition of complete metric space at once suggests that if $d(z_n, z_m) \longrightarrow 0$ as $m, n \to \infty$, then $\exists$ a $z_0 \in \mathbb{C}$ such that $d(z_n, z_0) \longrightarrow 0$ as $n \to \infty$.

(ii) A metric space (X, d) is complete if and only if every Cauchy sequence in $\mathbb{C}$ has a convergent subsequence.

Example 2.2.1. *The set $\mathbb{C}$ with discrete metric forms a complete metric space.*

Solution. Let $(\mathbb{C}, d)$ be a discrete metric space and $\langle z_n \rangle$ be a Cauchy sequence in $(\mathbb{C}, d)$. Thus $d(z_m, z_n) = 0$ if $z_m = z_n$ and $d(z_n, z_m) = 1$ if $z_m \neq z_n$. Then for $\varepsilon = 1/2, \exists$ a natural number k such that

$$\begin{aligned} & d(z_m, z_n) < 1/2 \qquad \forall\, m, n \geq k \\ \Rightarrow\quad & d(z_m, z_n) = 0 \qquad \forall\, m, n \geq k, \qquad \text{since } d \text{ is discrete metric} \\ \Rightarrow\quad & z_k = z_{k+1} = z_{k+2} = \cdots = z* \qquad \text{(say).} \end{aligned}$$

Therefore, $d(z_n, z*) = 0\ \forall n \geq k$, i.e., $\langle z_n \rangle$ is of the form $\{z_1, z_2, \cdots, z_k, z*, z*, \cdots\}$. Hence $\langle z_n \rangle$ converges to $z*$, i.e., every Cauchy sequence is convergent in a discrete metric space. Thus the discrete metric space $(\mathbb{C}, d)$ is complete. From this we see that a sequence in a discrete metric space is Cauchy if and only if it is eventually constant.

Example 2.2.2. *Let $\mathbb{C}$ be the set of all convergent complex sequences with the function $d : \mathbb{C} \times \mathbb{C} \to \mathbb{R}$ is defined by*

$$d(y, z) = \sup_n \left\{|y_n - z_n|\right\}; \quad y = \{y_n\}, z = \{z_n\} \in \mathbb{C}.$$

Prove that $(\mathbb{C}, d)$ is a complete metric space.

Solution. We know $(\mathbb{C}, d)$ is a metric space. Let $\langle t_n \rangle$, where $t_n = \langle z_{n1}, z_{n2}, \cdots, z_{nk}, \cdots \rangle$ for all $n \in \mathbb{N}$, an element of $\mathbb{C}$, be a Cauchy sequence in $\mathbb{C}$. Let $\varepsilon > 0$ be a real number. Then $\exists$ a $p \in \mathbb{N}$ such that

$$\begin{aligned} & d(t_n, t_m) < \varepsilon \ \forall n, m \geq p \\ \Rightarrow \quad & \sup_{k \in \mathbb{N}} |z_{nk} - z_{mk}| < \varepsilon \ \forall n, m \geq p. \end{aligned} \tag{2.3}$$

Therefore, for each $k \in \mathbb{N}$, the sequence $\langle z_{nk} \rangle_{n=1}^{\infty}$ is a Cauchy sequence in $\mathbb{C}$, and using the completeness of $\mathbb{C}$, the sequence $\langle z_{nk} \rangle_{n=1}^{\infty}$ converges to a point $z_k \in \mathbb{C}$). Thus we get a sequence $t = \langle z_1, z_2, \cdots \rangle$. When we get $m \to \infty$ in Eq. (2.3), it follows that

$$|z_{nk} - z_k| < \varepsilon \text{ for all } n \geq p. \tag{2.4}$$

Now, by inequalities

$$\begin{aligned} |z_k| &\leq |z_k - z_{pk}| + |z_{pk}| \text{ for all } k \in \mathbb{N} \\ \Rightarrow \sup_{k \in \mathbb{N}} |z_k| &\leq \sup_{k \in \mathbb{N}} |z_k - z_{pk}| + \sup_{k \in \mathbb{N}} |z_{pk}| \\ &\leq \varepsilon + \sup_{k \in \mathbb{N}} \{|z_{pk}|\}. \end{aligned}$$

Since t_p is bounded, it follows that $t \in \mathbb{C}$. By taking supremum over k in Eq. (2.4), it can be seen that $\langle t_n \rangle$ converges to t, and $(\mathbb{C}, d)$ is complete.

2.3 Compact Sets

We shall now go through the basic results concerning compact sets.

Definition 2.3.1. Let $\mathcal{D} \subset \mathbb{C}$ be a set of complex numbers and $\{z_n\}$ be a complex sequence in $\mathcal{D}$. By a *point of accumulation* of $\{z_n\}$ we mean a complex number $z*$ such that, given $\varepsilon > 0$, there exist $n_0 \in \mathbb{N}$ such that

$$d(z_n, z*) = |z_n - z*| < \varepsilon \quad \text{for } n > n_0.$$

We may say that given an open set U containing $z*$, there exist infinitely many n such that $z_n \in U$.

We assume that the reader is acquainted with the *Bolzano–Weierstrass theorem* about sets of real numbers: *If S is an infinite bounded set of real numbers, then S has a point of accumulation.*

Definition 2.3.2. We define a set of complex numbers $\mathcal{D}$ to be *compact* if every sequence of elements of $\mathcal{D}$ has a point of accumulation in $\mathcal{D}$.

This property is equivalent to the following properties which can be taken as alternate definitions.

(i) Every infinite subset of $\mathcal{D}$ has a point of accumulation in $\mathcal{D}$.

(ii) Every sequence of elements of $\mathcal{D}$ has a convergent subsequence whose limit is in $\mathcal{D}$.

Theorem 2.5. *The set of complex numbers is compact if and only it is closed and bounded.*

Proof: Assume that $\mathcal{D}$ is compact. If $\mathcal{D}$ is not bounded, for each $n \in \mathbb{N}$, $\exists$ $z_n \subset \mathcal{D}$ such that

$$|z_n| > n.$$

Then, $\{z_n\}$ does not have a limit point. Indeed, if z^* is a limit point, pick $n_0 > 2|z^*|$ and note that $|z^*| > 0$. Then

$$|z_m - z^*| \geq |z_m| - |z^*| \geq m - |z^*| > |z^*|.$$

This contradicts the fact that for infinitely many m we must have z_m close to z^*. Hence $\mathcal{D}$ is bounded.

To show $\mathcal{D}$ is closed, let z^* be in its closure. Given n, there exists $z_n \in \mathcal{D}$ such that

$$\left|z_n - z^*\right| < 1/n.$$

The sequence $\{z_n\}$ converges to z^* and has a subsequence converging to a limit in $\mathcal{D}$ because $\mathcal{D}$ is assumed compact. This limit must be z^*, whence $z^* \in \mathcal{D}$ and $\mathcal{D}$ is closed.

Conversely, assume that $\mathcal{D}$ is closed and bounded, and let M be a bound. So $|z| \leq M$ for all $z \in \mathcal{D}$. If we write $z = x + iy$, then

$$|x| \leq M \text{ and } |y| \leq M.$$

Let $\{z_n\}$ be a sequence in $\mathcal{D}$, and let us write $z_n = x_n + iy_n$. There is a subsequence $\{z_{n_1}\}$ such that $\{x_{n_1}\}$ converges to $a \in \mathbb{R}$, and there is a subsequence $\{z_{n_2}\}$ such that $\{y_{n_2}\}$ converges to $b \in \mathbb{R}$. Then

$$\{z_{n_2} = x_{n_2} + iy_{n_2}\}$$

converges to $a + ib$, and $\mathcal{D}$ is compact. This proves the theorem. □

RESULT **15.** Let $\mathcal{D}$ be a compact set and let $\mathcal{D}_1 \supset \mathcal{D}_2 \supset \cdots$ be a sequence of non-empty closed subsets such that $\mathcal{D}_n \supset \mathcal{D}_{n+1}$. Let $z_n \in \mathcal{D}_n$. The sequence $\{z_n\}$ has a limit point, say z^* in $\mathcal{D}$. Then z^* is also a limit point for each subsequence $\{z_k\}$ with $k \geq n$, and hence it lies in the closure of $\mathcal{D}_n$ for each n. But $\mathcal{D}_n$ is assumed closed, and hence $z^* \in \mathcal{D}_n$ for all n. This shows that

$$\bigcap_{n=1}^{\infty} \mathcal{D}_n \neq \phi.$$

Theorem 2.6. *The continuous image of a complex compact set is compact.*

Proof: Let $\mathcal{D}$ be a compact set of complex numbers, and let f be a continuous function on $\mathcal{D}$. Let $\{w_n\}$ be a sequence in the image of f, so that

$$w_n = f(z_n) \quad \text{for} \quad z_n \in \mathcal{D}.$$

The sequence $\{z_n\}$ has a convergent subsequence $\{z_{n_k}\}$, with a limit $z^* \in \mathcal{D}$. Since f is continuous, we have

$$\lim_{k\to\infty} w_{n_k} = \lim_{k\to\infty} f(z_{n_k}) = f(z^*).$$

Hence the given sequence $\{w_n\}$ has a subsequence which converges in $f(\mathcal{D})$. This proves that the image $f(\mathcal{D})$ of f is compact. □

DEDUCTION **1**. Let $\mathcal{D}$ be a compact set of complex numbers, and let $f : \mathcal{D} \to \mathbb{R}$ be a continuous function. From the Theorem 2.6, we know that $f(\mathcal{D})$ is closed and bounded. Let b be its least upper bound. Then b is adherent to $f(\mathcal{D})$, whence b is in $f(\mathcal{D})$ because $f(\mathcal{D})$ is closed.

So there is some $z^* \in \mathcal{D}$ such that

$$f(z^*) = b \Rightarrow f(z) \leq b = f(z^*); \forall z \in \mathcal{D}.$$

Thus f has a maximum in $\mathcal{D}$.

Theorem 2.7. *Let $\mathcal{D}$ be a compact set, and let f be continuous function on $\mathcal{D}$. Then f is uniformly continuous.*

Proof: Here, we are to show that corresponding to an $\varepsilon > 0$, there exists $\delta > 0$ such that

$$z, w \in \mathcal{D} \text{ and } |z - w| < \delta \Rightarrow |f(z) - f(w)| < \varepsilon.$$

Suppose this assertion is not true. Then $\exists\, \varepsilon$, and for each n there exists a pair of elements $z_n, w_n \in \mathcal{D}$ such that

$$|z_n - w_n| < \frac{1}{n} \text{ but } |f(z_n) - f(w_n)| > \varepsilon.$$

There is an infinite subset $\mathbb{Z}_1$ of positive integers and some $z^* \in \mathcal{D}$ such that $z_n \to z^*$ for $n \to \infty$ and $n \in \mathbb{Z}_1$. There is an infinite subset $\mathbb{Z}_2$ of $\mathbb{Z}_1$ and some $z^{**} \in \mathcal{D}$ such that $w_n \to z^{**}$ for $n \to \infty$ and $n \in \mathbb{Z}_2$. Since

$$|z^* - z^{**}| \leq |z^* - z_n| + |z_n - w_n| + |w_n - z^{**}|.$$

taking the limit for $n \to \infty$ and $n \in \mathbb{Z}_2$ we obtain

$$|z^* - z^{**}| \Rightarrow z^* = z^{**}.$$

Hence $f(z^*) - f(z^{**}) = 0$. Furthermore,

$$|f(z_n) - f(w_n)| \leq |f(z^*) - f(z_n)| + |f(z^*) - f(z^{**})| + |f(w_n) - f(z^{**})|.$$

Again, taking the limit for $n \to \infty$ and $n \in \mathbb{Z}_2$, we conclude that

$$|f(z_n) - f(w_n)| \to 0.$$

This contradicts the assumption that $|f(z_n) - f(w_n)| > \varepsilon$ and proves the theorem. □

Theorem 2.8. *Let $\mathcal{D}$ be compact. Let $r > 0$ be a real number. There exists a finite number of open discs of radius r whose union contains $\mathcal{D}$.*

Proof: Suppose this is false. Let $z_1 \in \mathcal{D}$ and let $\mathcal{D}_1$ be the open disc of radius r centered at z_1. Then $\mathcal{D}_1$ does not contain $\mathcal{D}$, and there is some $z_2 \in \mathcal{D}$, $z_2 \neq z_1$. Proceeding inductively, suppose we have found open discs $\mathcal{D}_1, \cdots, \mathcal{D}_n$ of radius r centered at the points $z_1, \cdots, z_n$ respectively, such that z_{k+1} does not lie in $\mathcal{D}_1 \cup \cdots \cup \mathcal{D}_k$. We can then find z_{n+1} which does not lie in $\mathcal{D}_1 \cup \cdots \cup \mathcal{D}_n$, and we let D_{n+1} be the disc of radius r centered at z_{n+1}.

Let z^* be a limit point of $\{z_n\}$. By definition, there exist $m, k \in \mathbb{N}$ with $k > m$ such that

$$|z_k - z^*| < r/2 \quad \text{and} \quad |z_m - z^*| < r/2.$$

$$\therefore \quad |z_k - z_m| \leq |z_k - z^*| + |z_m - z^*| < r$$

and this contradicts the property of our sequence $\{z_n\}$ because z_k lies in the disc $\mathcal{D}_m$. This proves the theorem. □

2.4 Infinite Series

Definition 2.4.1. Let $\{z_n\}$ be a sequence of complex numbers. Then the formal expression involving the addition of an infinite number of complex number

$$z_1 + z_2 + \cdots + z_n + \cdots; \qquad z_n \in \mathbb{C} \tag{2.5}$$

is called an *infinite series of complex numbers* and is denoted by $\sum\limits_{n=1}^{\infty} z_n$, or simply, $\sum z_n$.

Form a new associated sequence $s_1, s_2, s_3, \ldots$ defined by

$$s_1 = z_1, s_2 = z_1 + z_2, s_3 = z_1 + z_2 + z_3, \ldots, s_n = z_1 + z_2 + z_3 + \cdots + z_n. \tag{2.6}$$

Then $\{s_n\}$ is called the *sequence of partial sums* of the series $\sum z_n$.

The series $\sum z_n$ is said to *converge* or *diverge* according as the sequence of partial sums $\{s_n\}$ converges or diverges. If $\lim\limits_{n\to\infty} s_n =$ finite, we say that $\sum u_n$ converges to the sum $s \in \mathbb{C}$ and we write

$$s = \sum z_n = z_1 + z_2 + z_3 + \cdots.$$

If $\lim\limits_{n\to\infty} s_n$ does not exist, then we say the series diverges.

Definition 2.4.2. The series $\sum z_n$ of complex numbers is said to *converge absolutely* or be *absolutely convergent* if the series $\sum_{n=1}^{\infty} |z_n|$ of real numbers converges.

Corollary 2.4.1. The absolute convergence of a series of complex numbers implies the convergence of the series.

The converse of this result is not always true. For example, consider the series $\sum z_n$, where $z_n = \frac{i^n}{\log n}$ is convergent but not absolutely convergent.

RESULT **16.** The harmonic series $\sum \frac{1}{n^p}$ converges if $p > 1$ and diverges if $p \leq 1$.

Theorem 2.9. *If* $z_n = a_n + ib_n$, $n = 1, 2, \ldots$, *then*

$$\sum_{n=1}^{\infty} z_n = \sum_{n=1}^{\infty} (a_n + ib_n) = a + ib \in \mathbb{C}$$

if and only if $\sum\limits_{n=1}^{\infty} a_n = a$ *and* $\sum\limits_{n=1}^{\infty} b_n = b$.

Proof: Let $\alpha_k = \sum\limits_{k=1}^{n} a_k$ and $\beta_k = \sum\limits_{k=1}^{n} b_k$ be the nth partial sum of the series $\sum a_n$ and $\sum b_n$ respectively.

Necessary part: Let $\lim\limits_{n\to\infty} \alpha_n = a$ and $\lim\limits_{n\to\infty} \beta_n = b$. Now

$$\begin{aligned} S_n &= n\text{th partial sum of } \sum z_n \\ &= \sum_{k=1}^{n} z_k = \sum_{k=1}^{\infty} (a_k + ib_k) = \sum_{k=1}^{n} a_k + i \sum_{k=1}^{n} b_k = \alpha_n + i\beta_n. \end{aligned}$$

Thus by using theorem 2.3, we get $\lim\limits_{n\to\infty} S_n = a + ib$.

Sufficient part: Suppose $\sum z_n = a + ib$. Then $\lim\limits_{n\to\infty} S_n = a + ib$, where S_n is the nth partial sum of the series $\sum z_n$. Thus, corresponding to an $\varepsilon > 0$, there is an integer $M (\in \mathbb{N})$ such that if $n > M$, then

$$\begin{aligned} \varepsilon > |S_n - (a + ib)| &= \sqrt{(\alpha_n - a)^2 + (\beta_n - b)^2} \\ &\geq \begin{cases} |\alpha_n - a| \\ |\beta_n - b| \end{cases}. \end{aligned}$$

Hence $\lim \alpha_n = a$ and $\lim \beta_n = b$. □

Corollary 2.4.2. If a series of complex numbers converges, the nth term converges to zero as $n \to \infty$.

Proof: Assuming that series (2.5) converges, we know from the theorem that if $z_n = a_n + ib_n$; $n = 1, 2, \ldots$ then each of the series $\sum a_n$ and $\sum b_n$ converges. We know, moreover, from calculus that the nth term of a convergent series of real numbers approaches zero as $n \to \infty$. Thus, by theorem 2.3,

$$\lim_{n\to\infty} z_n = \lim_{n\to\infty} a_n + i \lim_{n\to\infty} b_n = 0 + i \cdot 0 = 0.$$

Thus the proof of this corollary is complete. □

RESULT **17.** The terms of a convergent series are bounded, i.e. if the series (2.5) converges, there exists a positive constant M such that

$$|z_n| \leq M \;\; \forall n \in \mathbb{N}.$$

This follows from the Corollary 2.4.2.

Contrapositively, $\lim\limits_{n\to\infty} z_n \neq 0 \Rightarrow \sum z_n$ is divergent. For example, consider the series $\sum z_n$, where $z_n = (1 + in^2)/(1 + n^2)$. Then

$$\lim_{n\to\infty} z_n = \lim_{n\to\infty} \frac{1 + in^2}{1 + n^2} = \lim_{n\to\infty} \frac{i + (1/n^2)}{1 + (1/n^2)} = i \neq 0.$$

Therefore, $\sum z_n$ is divergent.

2.4.1 Test of Convergence

Theorem 2.10. [Comparison test]: *Let $\sum a_n$ and $\sum b_n$ be two series of non-zero complex numbers such that $\sum |b_n|$ is convergent, and if $|a_n| \leq |b_n|$ for $n \geq m (m \in \mathbb{N})$, then $\sum a_n$ is absolutely convergent.*

Example 2.4.1. *Prove that the series $\sum_{n=1}^{\infty} u_n(z)$ of complex variable, where $u_n(z) = \frac{z^n}{n(n+1)}$, converges absolutely for $|z| \leq 1$.*

Solution. If $|z| \leq 1$, then

$$|u_n(z)| = \left|\frac{z^n}{n(n+1)}\right| \leq \left|\frac{1}{n(n+1)}\right| \leq \frac{1}{n^2} = v_n \;\; \forall n \in \mathbb{N}.$$

We see that $\sum v_n$ is a hyper harmonic series with $p = 2 > 1$ and so convergent. Hence by comparison test $\sum |u_n|$ is convergent, i.e. $\sum u_n$ converges uniformly.

Theorem 2.11. [Ratio test]: *Let $\sum a_n$ be a series of non-zero complex numbers and let* $\lim \left|\frac{a_{n+1}}{\alpha_n}\right| = \rho$. *Then $\sum a_n$*

(i) converges absolutely if $\rho < 1$; (ii) diverges if $\rho > 1$; (iii) may converge or may diverge if $\rho = 1$.

Theorem 2.12. [Dedekind's test for non-absolute convergence]: *The series $\sum a_n b_n$ is convergent if*

(i) the complex series $\sum a_n$ has bounded partial sums,

(ii) the series $\sum (b_n - b_{n-1})$ is absolutely convergent for the complex sequence $\{b_n\}$, and

(iii) $b_n \to 0$ as $n \to \infty$.

Theorem 2.13. [Cauchy's general principle of convergence]: *Let $\sum_{n=1}^{\infty} z_n$ be an infinite series of complex number. The necessary and sufficient condition for the convergence of the series is that, corresponding to an $\varepsilon > 0$, $\exists\, m_0(\varepsilon) \in \mathbb{N}$ such that*

$$\sum z_n < \infty \Leftrightarrow |S_{n+p} - S_n| < \varepsilon; n \geq m_0 \text{ and } p \geq 1,$$

where S_n is the nth partial sum of the series $\sum z_n$.

DEDUCTION 2. Theorem 2.10 can easily be deduced from Cauchy's general principle of convergence.

Let $\{s_n\}$ and $\{t_n\}$ be the nth partial sum of the series $\sum a_n$ and $\sum |b_n|$ respectively. As $\sum |b_n|$ is convergent, its partial sum sequence $\{t_n\}$ satisfies Cauchy's criterion, i.e., corresponding to any pre-assigned $\varepsilon > 0$, $\exists\, n_0(\varepsilon) \in \mathbb{N}$ such that

$$|t_{n+p} - t_n| = \left|\sum_{k=1}^{p} b_{n+k}\right| < \varepsilon \text{ whenever } n \geq n_0 \text{ and } p \geq 1.$$

Using the property of modules,

$$\begin{aligned} |s_{n+p} - s_n| &= \left|\sum_{k=1}^{p} a_{n+k}\right| \leq \sum_{k=1}^{p} \left|a_{n+k}\right| \leq \left|\sum_{k=1}^{p} b_{n+k}\right| < \varepsilon \\ &< \varepsilon \text{ whenever } n \geq n_0 \text{ and } p \geq 1. \end{aligned}$$

This observation proves the theorem.

DEDUCTION 3. Corollary to Deduction 2 can be deduced from Cauchy's general principle of convergence.

Let $\{s_n\}$ and $\{t_n\}$ be the nth partial sum of the series $\sum z_n$ and $\sum |z_n|$ respectively. As $\sum |z_n|$ is convergent, its partial sum sequence $\{t_n\}$ satisfies Cauchy's criterion, i.e., corresponding to any pre-assigned $\varepsilon > 0$, $\exists\, n_0(\varepsilon) \in \mathbb{N}$ such that

$$|t_{n+p} - t_n| < \varepsilon \text{ whenever } n \geq n_0 \text{ and } p \geq 1.$$

Using the property of modules,

$$\begin{aligned} |s_{n+p} - s_n| &= \left|\sum_{k=1}^{p} z_{n+k}\right| \leq \sum_{k=1}^{p} \left|z_{n+k}\right| = |t_{n+p} - t_n| \\ &< \varepsilon \text{ whenever } n \geq n_0 \text{ and } p \geq 1. \end{aligned}$$

This ensures that $\{s_n\}$ is convergent and so also $\sum z_n$. Thus, if $\sum |z_n|$ is convergent, $\sum z_n$ converges.

DEDUCTION 4. Corollary to Deduction 3 can be deduced from Cauchy's general principle of convergence.

For $n \geq 2$, we write $z_n = s_n - s_{n-1}$, where $\{s_n\}$ is the nth partial sum sequence of the series $\sum z_n$. Let $\sum z_n$ be convergent. Then

$$\lim_{n\to\infty} s_n = \lim_{n\to\infty} s_{n-1} = l, \text{ say}$$

$$\Rightarrow \lim_{n\to\infty} z_n = \lim_{n\to\infty} s_n - \lim_{n\to\infty} s_{n-1} = l - l = 0.$$

Thus, if $\sum z_n$ converges, then $\lim\limits_{n\to\infty} z_n = 0$.

Example 2.4.2. *Show that the series $\sum z^n/n!$ is absolutely convergent.*

Solution. The given complex numbers is of the form $\sum z_n$, where

$$z_n = \frac{z^n}{n!}; \qquad z_{n+1} = \frac{z^{n+1}}{(n+1)!}$$

$$\Rightarrow \left|\frac{z_{n+1}}{z_n}\right| = \left|\frac{z^{n+1}}{(n+1)!} \times \frac{n!}{z^n}\right| = \frac{|z|}{n+1}$$

$$\Rightarrow \lim_{n\to\infty}\left|\frac{z_{n+1}}{z_n}\right| = \lim_{n\to\infty}\frac{|z|}{n+1} = 0\ \forall z \in \mathbb{C}.$$

This shows that $\sum z_n$ is convergent, no matter what $z \in \mathbb{C}$ is. The sum of the series is denoted by e^z.

Example 2.4.3. *Consider $\sum nz^n$ for testing its convergence or divergence.*

Solution. The given series of complex numbers is of the form $\sum z_n$, where

$$z_n = nz^n; \qquad z_{n+1} = (n+1)z^{n+1}$$

$$\Rightarrow \left|\frac{z_{n+1}}{z_n}\right| = \left(1+\frac{1}{n}\right)|z|$$

$$\Rightarrow \lim_{n\to\infty}\left|\frac{z_{n+1}}{z_n}\right| = |z|\lim_{n\to\infty}\left(1+\frac{1}{n}\right) = |z|.$$

Hence by ratio test $\sum z_n$ is convergent if $|z| < 1$ and is divergent if $|z| \geq 1$.

Theorem 2.14. [Heine's criterion]: *Let $f : \mathcal{D} \to \mathbb{C}$. Then f is continuous at $z_0 \in \mathcal{D}$ iff for any sequence $\{z_n\} \subset \mathcal{D}$ converging to z_0, the sequence $\{f(z_n)\}$ converges to $f(z_0)$.*

Proof *[Necessary part]*: Let f be continuous at $z_0 \in D$ and if there is a sequence $\{z_n\}$ in $\mathcal{D}$ such that $\lim_{n\to\infty} z_n = z_0$, then for any $\varepsilon > 0$, $\exists$ a $\delta > 0$ such that

$$|f(z) - f(z_0)| < \varepsilon \text{ whenever } 0 < |z - z_0| < \delta, z \in \mathcal{D}.$$

Further, as $\lim_{n\to\infty} z_n = z_0$, for $\delta > 0$, $\exists$ a positive integer $m \in \mathbb{N}$ such that $|z_n - z_0| < \delta$ for all $n \geq m$. Since $z_n \in D$,

$$|f(z) - f(z_0)| < \varepsilon\ \forall n \geq m \Rightarrow \lim_{n\to\infty} f(z_n) = f(z_0).$$

Sufficient part: Let $\lim_{n\to\infty} f(z_n) = f(z_0)$ for every sequence $\{z_n\}$ is $\mathcal{D}$ that converges to z_0. Suppose that f is not continuous at z_0. So $\exists$ some $\varepsilon > 0$ such that for each $\delta > 0$, $\exists\ z \in \mathcal{D}$ for which

$$0 < |z - z_0| < \delta \text{ but } |f(z) - f(z_0)| \geq \varepsilon.$$

Take $\delta = 1$. Then $\exists\ z_1 \in \mathcal{D}$ such that

$$0 < |z_1 - z_0| < 1 \text{ but } |f(z_1) - f(z_0)| \geq \varepsilon.$$

If we take $\delta = \frac{1}{2}$, $\exists\ z_2 \in \mathcal{D}$ such that

$$0 < |z_2 - z_0| < \frac{1}{2} \text{ but } |f(z_2) - f(z_0)| \geq \varepsilon.$$

If one continues with this process, one ultimately gets a sequence $\{z_n\}$ in $\mathcal{D}$ such that

$$0 < |z_n - z_0| < \frac{1}{n} \text{ but } |f(z_n) - f(z_0)| \geq \varepsilon \ \forall\, n.$$

Thus $\{z_n\}$ converges to z_0 but $\{f(z_n)\}$ does not converges to $f(z_0)$. This leads to a contradiction. Hence by method of contradiction we prove the proposition that f must be continuous at z_0. □

2.5 Complex Power Series

In this section the definition and basic properties of a power series will be given. The power series will then be used to give examples of analytic functions.

Definition 2.5.1. Let $\{a_n\}$ be a sequence of complex numbers independent of a complex variable $z \in \mathbb{C}$ and z_0 be a complex constant. An infinite series of the form

$$\sum_{n=0}^{\infty} a_n(z - z_0)^n = a_0 + a_1(z - z_0) + a_2(z - z_0)^2 + \cdots \infty \tag{2.7}$$

is called a *complex power series* in $z - z_0$ about z_0.

The sequence of nth *partial sums* of power series (2.7) is given by

$$S_n(z) = \sum_{k=0}^{n} a_k(z - z_0)^k = a_0 + a_1(z - z_0) + a_2(z - z_0)^2 + \cdots a_n(z - z_0)^n.$$

The convergence of the power series (2.7) depends on the sequence $\{a_n\}$ and the point z. So, it is sufficient to consider the following power series ($z_0 = 0$)

$$\sum_{n=0}^{\infty} a_n z^n = a_0 + a_1 z + a_2 z^2 + \cdots \infty. \tag{2.8}$$

One of the easiest examples of a power series is the infinite geometric power series $\sum z^n$.

Definition 2.5.2. The power series (2.8) is said to be *absolutely convergent* if the series $\sum_{n=0}^{\infty} |a_n|\ |z|^n$ is convergent. If the power series (2.8) is convergent but $\sum_{n=0}^{\infty} |a_n|\ |z|^n$ is not convergent, then (2.8) is said to be *conditionally convergent.*

Example 2.5.1. *Show that the series $\sum z_n$, where $z_n = \frac{(-1)^n}{n} + \frac{i}{n^2} \in \mathbb{C}$, is conditionally convergent but not absolutely.*

Solution. Let $z_n = a_n + ib_n \in \mathbb{C}$, where

$$a_n = \text{Re }(z_n) = \frac{(-1)^n}{n} \text{ and } b_n = \text{Im }(z_n) = \frac{1}{n^2}.$$

Now, $\sum a_n$ is an alternating series, and as in real numbers, by Leibniz's test, the series $\sum a_n$ is convergent, but $\sum |a_n| = \sum \frac{1}{n}$ is a hyperharmonic series with $p = 1$, and so divergent.

Therefore, $\sum a_n$ is conditionally convergent.

Now, $\sum b_n = \sum \frac{1}{n^2}$ is a hyperharmonic series with $p = 2 > 1$, and so convergent. Hence $\sum z_n$ is conditionally convergent.

Theorem 2.15. [Abel's theorem]: *If the power series $\sum\limits_{n=0}^{\infty} a_n z^n$ converges at a point $z = z_1 (\neq 0)$, then it converges absolutely for all z satisfying $|z| < |z_1|$ and if it diverges at $z = z_2$, then it diverges for all z satisfying $|z| > |z_2|$.*

Proof: It is given that $\sum\limits_{n=0}^{\infty} a_n z_1^n$ is convergent. So the terms of the series are bounded. Thus $\exists$ a real number $M (> 0)$ such that

$$|a_n z_1^n| < M \quad \text{for } n = 0, 1, 2, \ldots.$$

Let us choose $z_0 \in \mathbb{C}$ such that $|z_0| < |z_1|$, i.e., $\dfrac{z_0}{z_1} = r$ (say) < 1. For such z we have

$$\begin{aligned} |a_n z_0^n| &= \left|a_n z_1^n \left(\frac{z_0}{z_1}\right)^n\right| = \left|a_n z_1^n\right| \left|\left(\frac{z_0}{z_1}\right)\right|^n < Mr^n \ \forall n \in \mathbb{N} \\ \Rightarrow \sum_{n=0}^{\infty} |a_n z_0^n| &< M \sum_{n=0}^{\infty} r^n, \text{ where } 0 < r < 1. \end{aligned}$$

But $\sum_{n=0}^{\infty} r^n$ is an infinite geometric progression series of positive reals, which is convergent as $0 < r < 1$. By comparison test, the infinite series $\sum_{n=0}^{\infty} |a_n z_0^n|$ is convergent, i.e., $\sum_{n=0}^{\infty} a_n z_0^n$ is absolutely convergent. Since z_0 is arbitrary, we conclude that the power series $\sum_{n=0}^{\infty} a_n z^n$ is absolutely convergent (and hence convergent) for all z satisfying $|z| < |z_1|$ (Fig. 2.1).

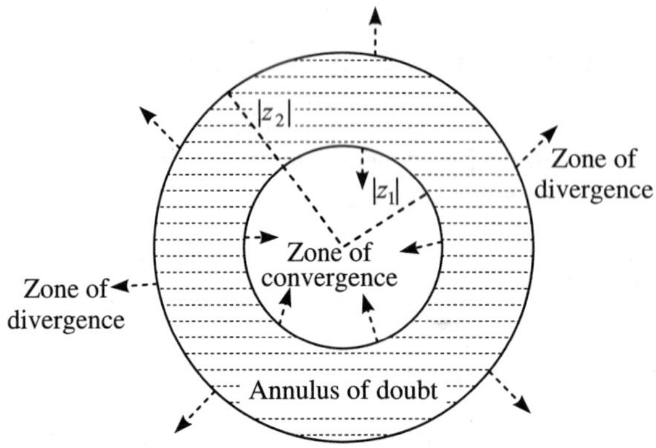

Figure 2.1 *Zones of convergence and divergence*

By hypothesis, if the given series $\sum_{n=0}^{\infty} a_n z^n$ diverges at $z = z_2$, then $\sum_{n=0}^{\infty} a_n z_2^n$ is divergent, and there is another value of z, say $z = z_0'$, with $|z_0'| > |z_2|$. If possible, let the series converge at $z = z_0'$. Then by the previous result, the series converges at z_0', where $|z_0'| < |z_2|$, which contradicts the hypothesis. Therefore, the series diverges for all z satisfying $|z_0'| > |z_2|$. □

RESULT **18.** From the above theorem we see that for the complex power series $\sum_{n=0}^{\infty} a_n z_2^n$, the annulus of doubt (Fig. 2.1) is given by

$$|z_1| < |z| < |z_2|.$$

Thus, irrespective of the fact that the power series $\sum_{n=0}^{\infty} a_n z^n$ converges or diverges, the annulus of doubt is slender in width.

RESULT **19.** Using the iterative procedure, the annulus of the doubt eventually shrinks to a limiting circle $|z| = R$ in the interior of which the power series $\sum_{n=0}^{\infty} a_n z^n$ converges. Thus we have the following theorem:

Existence theorem : If the power series $\sum_{n=0}^{\infty} a_n z^n$ converges at some point different from 0 and diverges at some other point, then $\exists$ a positive number R such that the series converges absolutely for all z satisfying $|z| < R$ and diverges for all z satisfying $|z| > R$.

2.5.1 Radius of Convergence

Definition 2.5.3. Let a complex power series be given by (2.7).

(i) If the power series (2.7) converges at some point other than z_0 and diverges at some point, then there is a real number $R \in [0, \infty]$ such that the series converges absolutely for $|z - z_0| \geq R$. This number R is defined as a *radius of convergence of the power series* and $\{z \in \mathbb{C} : |z - z_0| < R\}$ is defined as *interval of convergence.*

(ii) When the series converges for all z, the radius of convergence is defined as ∞ and the *interval of convergence* is $(-\infty, \infty)$.

(iii) When the complex power series does not converge for any z other than z_0, the radius of convergence is defined as zero and there is no interval of convergence.

When the series converges at any point at the interval of convergence, then the interval including the particular end point (or points) is known as *region of convergence.*

Theorem 2.16. [The Cauchy–Hadamard theorem]: *For a complex power series (2.7), define an extended real number $R \in [0, \infty]$ by*

$$\rho = \frac{1}{R} = \overline{\lim}(|a_n|)^{\frac{1}{n}} = \lim\sup(|a_n|)^{\frac{1}{n}}, \tag{2.9}$$

where ρ is any positive real number. Then

(i) *if $|z - z_0| < R$, the series converges absolutely for all $z \in \mathbb{C}$.*

(ii) *if $0 < \rho < R$, the series converges absolutely and uniformly on $\{z \in \mathbb{C} : |z - z_0| \leq \rho\}$.*

(iii) *if $0 < R < \infty$ and if $|z - z_0| > R$, the terms of the series become unbounded and so the series diverges.*

Moreover, the number R is the only member having the properties (i) and (iii).

Proof: We shall prove that for this choice of R, the assertions of the theorem are true.
(i) Let $|z - z_0| < R$, and $R \neq 0$. Choose a real number ρ_0 such that $|z - z_0| < \rho_0 < R$. By definition of upper limit, there is an integer $n_0 \in \mathbb{N}$ such that

$$|a_n|^{\frac{1}{n}} < \frac{1}{\rho_0}, \forall n \geq n_0, \quad \text{as } \frac{1}{\rho_0} > \frac{1}{R} = \overline{\lim}|a_n|^{\frac{1}{n}}$$

$$\Rightarrow \quad |a_n(z - z_0)^n|^{\frac{1}{n}} < \left(\frac{|z - z_0|}{\rho_0}\right)^n = r^n \text{ for all } n \geq n_0,$$

where $r = \frac{|z-z_0|}{\rho_0} < 1$. Now $\sum r^n$, being an infinite geometric series with $r < 1$, is convergent. Thus, by comparison test we get that the power series converges absolutely for each $|z - z_0| < R$. Let $R = 0$. Then

$$R = 0 \Rightarrow \overline{\lim}|a_n|^{\frac{1}{n}} = \infty.$$

So, $\{|a_n|^{\frac{1}{n}}\}$ is not bounded in the above. Therefore, for an infinite number of values of n,

$$|a_n|^{\frac{1}{n}} > \frac{2}{|z-z_0|} \Rightarrow |a_n(z-z_0)^n| > 2^n.$$

Hence the terms of the series are unbounded and the series is divergent.
(ii) Now, let $0 < \rho < R$. Choose $\rho_0 \in \mathbb{R}^+$ such that $\rho < \rho_0 < R$. Therefore,

$$\frac{1}{\rho_0} > \frac{1}{R} = \overline{\lim}|a_n|^{\frac{1}{n}} \Rightarrow \overline{\lim}|a_n|^{\frac{1}{n}} < \frac{1}{\rho_0}.$$

Also, $|z-z_0| \leq \rho$. By the definition of upper limit there exists a positive integer n_0 such that

$$|a_n|^{\frac{1}{n}} < \frac{1}{\rho_0} \quad \text{for all } n \geq n_0$$

$$\Rightarrow |a_n(z-z_0)^n|^{\frac{1}{n}} < \left(\frac{|z-z_0|}{\rho_0}\right)^n \leq \left(\frac{\rho}{\rho_0}\right)^n \quad \text{for all } n \geq n_0.$$

Since $|z-z_0| \leq \rho$, we have $\frac{\rho}{\rho_0} < 1$, so that $\sum_{n=0}^{\infty}\left(\frac{\rho}{\rho_0}\right)^n$ is a convergent series of constant terms. Hence by Weierstrass M-test, $\sum_{n=0}^{\infty} a_n(z-z_0)^n$ is uniformly and absolutely convergent on $\{z \in \mathbb{C} : |z-z_0| \leq \rho\}$.
(iii) Let $0 < R < \infty$. With the Cauchy's root test in mind, we observe that

$$\begin{aligned}
\overline{\lim}|a_n(z-z_0)^n|^{\frac{1}{n}} &= |z-z_0| \ \overline{\lim}|a_n|^{\frac{1}{n}} \\
&= |z-z_0| \times \rho = \begin{cases} 0 & \text{if } \rho = 0; \\ \infty & \text{if } \rho = \infty \text{ and } z \neq z_0. \end{cases} \\
&= \frac{1}{R}|z-z_0| \quad \text{if } 0 < \rho < \infty \\
&= \begin{cases} < 1 & \text{if } |z-z_0| < R; \\ > 1 & \text{if } |z-z_0| > R. \end{cases}
\end{aligned}$$

Thus the power series converges absolutely if $|z-z_0| < R$ and diverges if $|z-z_0| > R$.

Let $R = \infty$. We claim that the power series converges for all z. The series obviously converges when $z = 0$. Now,

$$R = \infty \Rightarrow \lim |a_n|^{\frac{1}{n}} = 0.$$

So, corresponding to an $\varepsilon > 0$, $\exists$ a $n_0 \in \mathbb{N}$ such that

$$|a_n|^{\frac{1}{n}} < \varepsilon \quad \text{for all } n \geq n_0.$$

In particular,

$$|a_n|^{\frac{1}{n}} < \frac{1}{2|z-z_0|} \quad \text{for all } n \geq n_0.$$

$$\Rightarrow \left|a_n(z-z_0)^n\right| < \frac{1}{2^n} \quad \text{for all } n \geq n_0.$$

Since $\sum_{n=0}^{\infty} \frac{1}{2^n}$ is a convergent series of positive terms, by comparison test $\sum_{n=0}^{n} a_n(z-z_0)^n$ is absolutely convergent for all z. □

Equation (2.9) is called Cauchy–Hadamard formula and it expresses the radius of convergence in terms of coefficients of power series.

Example 2.5.2. *Find the radii of convergence of the following power series: (i)* $\sum_{n=0}^{\infty}(3+4i)^n z^n$*; (ii)* $\sum_{n=0}^{\infty}\frac{z^n}{n^n}$.

Solution. Comparing the given series with (2.8), we get $a_n = (3+4i)^n$. Using the Cauchy's–Hadamard formula (2.10), if R is the radius convergence of the given power series, then it is given by

$$\frac{1}{R} = \overline{\lim}\sqrt[n]{|a_n|} = \overline{\lim}\sqrt[n]{|3+4i|^n} = 5.$$

Thus the radius of convergence is $1/5$.
(ii) Comparing the given series with (2.8), we get $a_n = 1/n^n$. Using the Cauchy's–Hadamard formula (2.10), if R be the radius convergence of the given power series, then it is given by

$$\frac{1}{R} = \overline{\lim}\sqrt[n]{|a_n|} = \overline{\lim}\sqrt[n]{n^n} = 0.$$

Thus the radius of convergence is ∞.

Example 2.5.3. *Let α, β be complex numbers with $|\alpha| < |\beta|$. Let $f(z) = \sum(3\alpha^n - 5\beta^n)z^n$. Determine the radius of convergence of $f(z)$.*

Solution. The given power series is of the form $\sum_{n=0}^{\infty} a_n z^n$, where $a_n = 3\alpha^n - 5\beta^n$. Thus

$$|a_n| = 5|\beta|^n\left|\frac{3}{5}\left(\frac{\alpha}{\beta}\right)^n - 1\right| \Rightarrow \lim_{n\to\infty}|a_n|^{\frac{1}{n}} = |\beta|;\ \text{as}\ \left|\frac{\alpha}{\beta}\right| < 1.$$

Thus the radius of convergence is $1/|\beta|$.

Theorem 2.17. [Ratio test]: *Let $\sum_{n=0}^{\infty} a_n(z-z_0)^n$ be a complex power series. If $\lim_{n\to\infty}\left|\frac{a_n}{a_{n+1}}\right|$ exists and is equal to R, then R is the radius of convergence of the complex power series.*

Proof: The given complex power series is $\sum_{n=0}^{\infty} a_n(z-z_0)^n$ of the form $\sum_{n=0}^{\infty} u_n$, where $u_n = a_n(z-z_0)^n$. Now

$$\begin{aligned}
\lim_{n\to\infty}\left|\frac{u_{n+1}}{u_n}\right| &= \lim_{n\to\infty}\left|\frac{a_{n+1}(z-z_0)^{n+1}}{a_n(z-z_0)^n}\right| \\
&= \lim_{n\to\infty}\left|\frac{a_{n+1}}{a_n}\right|.|z-z_0| = \frac{1}{R}|z-z_0| \\
&\lesseqgtr 1 \ \text{ according as } |z-z_0| \lesseqgtr R.
\end{aligned}$$

Hence the complex power series converges absolutely for all z satisfying $|z-z_0| < R$ and diverges for all z satisfying $|z-z_0| > R$. All these lead to the conclusion that

$$R = \lim_{n\to\infty}\left|\frac{a_n}{a_{n+1}}\right| \tag{2.10}$$

is the radius of convergence. The formula (2.10) express the radius of convergence in terms of the coefficients of the power series. □

Example 2.5.4. *Find the radius of convergence of the following complex power series:*

(i) $\sum_{n=0}^{\infty} \frac{z^n}{n!}$, (ii) $\sum_{n=0}^{\infty} \frac{n+1}{(n+2)(n+3)} z^n$, (iii) $\sum_{n=0}^{\infty} \frac{(n!)^2}{(2n)!} z^n$, (iv) $\sum_{n=0}^{\infty} \frac{z^n}{z^n+1}$

Solution. (i) Comparing the given series with (2.8), we get $a_n = z^n/n!$. Using formula (2.10), we get

$$R = \lim_{n\to\infty} \left|\frac{a_n}{a_{n+1}}\right| = \lim_{n\to\infty} \left|\frac{(n+1)!}{(n)!}\right| = \lim_{n\to\infty} (n+1) = \infty.$$

Hence radius of convergence is ∞.

(ii) Comparing the given series with (2.8), we get $a_n = (n+1)/(n+2)(n+3)$. Using formula (2.10), we get

$$R = \lim_{n\to\infty} \left|\frac{a_n}{a_{n+1}}\right| = \lim_{n\to\infty} \left|\frac{(n+1)(n+4)}{(n+2)^2}\right| = \lim_{n\to\infty} \frac{(1+\frac{1}{n})(1+\frac{1}{n})}{(1+\frac{2}{n})^2} = 1.$$

Hence radius of convergence of the power series is 1.

(iii) Comparing the given series with (2.8), we get $a_n = (n!)^2/2n$. Using formula (2.10), we get

$$\begin{aligned} R = \lim_{n\to\infty} \left|\frac{a_n}{a_{n+1}}\right| &= \lim_{n\to\infty} \frac{n!n!}{(2n)!} \times \frac{(2n+2)!}{(n+1)!(n+1)!} \\ &= \lim_{n\to\infty} \frac{(2n+2)(2n+1)}{(n+1)(n+1)} = \lim_{n\to\infty} \frac{(2+\frac{2}{n})(2+\frac{1}{n})}{(1+\frac{1}{n})(1+\frac{1}{n})} = 4. \end{aligned}$$

Hence radius of convergence of the power series is 4.

(iv) Comparing the given series with (2.8), we get $a_n = z^n/(z^n+1)$. Using formula (2.10), we get

$$R = \lim_{n\to\infty} \left|\frac{a_n}{a_{n+1}}\right| = \lim_{n\to\infty} \frac{2^{n+1}+1}{2^n+1} = \frac{2+0}{1+0} = 2.$$

Hence radius of convergence of the power series is 2.

Example 2.5.5. *Find the radius of convergence of the series*

$$\frac{1}{2} z + \frac{1 \cdot 3}{2 \cdot 5} z^2 + \frac{1 \cdot 3 \cdot 5}{2 \cdot 5 \cdot 8} z^3 + \cdots.$$

Solution. Comparing the given series with (2.8), we get

$$a_n = \frac{1 \cdot 3 \cdot 5 \cdots (2n-1)}{2 \cdot 5 \cdot 8 \cdots (3n-1)}; \quad a_{n+1} = \frac{1 \cdot 3 \cdot 5 \cdots (2n-1)(2n+1)}{2 \cdot 5 \cdot 8 \cdots (3n-1)(3n+2)}$$

$$\Rightarrow \quad \frac{1}{R} = \lim_{n\to\infty} \left|\frac{a_{n+1}}{a_n}\right| = \lim_{n\to\infty} \frac{2n+1}{3n+2} = \frac{2}{3} \lim_{n\to\infty} \frac{(1+\frac{1}{2n})(1+\frac{1}{n})}{1+\frac{2}{3n}} = \frac{2}{3}.$$

Hence radius of convergence of the power series is $3/2$.

Example 2.5.6. *Find the radius of convergence of the power series* $\sum_{n=0}^{\infty} \frac{n\sqrt{2}+i}{1+2in} z^n$.

Solution. Using formula (2.10), we get

$$\begin{aligned}
|a_n| &= \left|\frac{(n\sqrt{2}+i)(1-2in)}{1+4n^2}\right| = \left|\frac{(2+\sqrt{2})n + i(1-2\sqrt{2}n^2)}{1+4n^2}\right| \\
&= \frac{1}{4n^2+1}\sqrt{(2+\sqrt{2})^2n^2 + (1-2\sqrt{2}n^2)^2} \\
&= \frac{1}{4n^2+1}\sqrt{(6+4\sqrt{2})n^2 + 1 - 4\sqrt{2}n^2 + 8n^4} \\
&= \frac{1}{4n^2+1}\sqrt{8n^4+6n^2+1}.
\end{aligned}$$

Here the radius of convergence is given by

$$\begin{aligned}
R &= \lim_{n\to\infty}\left|\frac{a_n}{a_{n+1}}\right| = \lim_{n\to\infty}\frac{\left[1+4(n+1)^2\right]\sqrt{8n^4+6n^2+1}}{(1+4n^2)\sqrt{8(n+1)^4+6(n+1)^2+1}} \\
&= \lim_{n\to\infty}\frac{\left[\frac{1}{n^2}+4(1+\frac{1}{n})^2\right]\sqrt{8+\frac{6}{n^2}+\frac{1}{n^4}}}{(\frac{1}{n^2}+4)\sqrt{8(1+\frac{1}{n})^4+\frac{6}{n^2}(1+\frac{1}{n})^2+\frac{1}{n^4}}} \\
&= \frac{4\sqrt{8}}{4\sqrt{8+0}} = 1.
\end{aligned}$$

Thus the radius of convergence for the given power series is 1.

Example 2.5.7. *Find the domain of convergence of* $\sum_{n=1}^{\infty} u_n(z)$, *where*

$$u_n(z) = \frac{1\cdot 3\cdot 5\cdots(2n-1)}{n!}\left(\frac{1-z}{z}\right)^n.$$

Solution. Let $z = \xi^{-1}$. Then the given series is of the form $\sum\limits_{n=1}^{\infty} a_n(\xi-1)^n$, where

$$a_n = \frac{1\cdot 3\cdot 5\cdots(2n-1)}{n!};\quad a_{n+1} = \frac{1\cdot 3\cdot 5\cdots(2n-1)(2n+1)}{(n+1)!}$$

$$\Rightarrow\quad \frac{1}{R} = \lim_{n\to\infty}\left|\frac{a_{n+1}}{a_n}\right| = \lim_{n\to\infty}\frac{2n+1}{n+1} = \lim_{n\to\infty}\frac{2+\frac{1}{n}}{1+\frac{1}{n}} = 2.$$

Hence radius of convergence of the power series is $1/2$. Thus, the domain of convergence is given by

$$\begin{aligned}
& |\xi-1| < \frac{1}{2} \Rightarrow \left|\frac{1}{z}-1\right| < \frac{1}{2} \\
\Rightarrow\quad & |1-z|^2 < \frac{1}{4}|z|^2 \Rightarrow (1-z)(\overline{1-z}) < \frac{1}{4}z\,\bar{z} \\
\Rightarrow\quad & 3z\bar{z} - 4(z+\bar{z}) + 4 < 0 \\
\Rightarrow\quad & z\bar{z} - \frac{4}{3}(z+\bar{z}) + \frac{16}{9} < \frac{4}{9} \\
\Rightarrow\quad & \left(z-\frac{4}{3}\right)\left(\bar{z}-\frac{4}{3}\right) < \frac{4}{9} \Rightarrow \left|z-\frac{4}{3}\right| < \frac{2}{3}.
\end{aligned}$$

This shows that the series is convergent inside the circle of radius $2/3$ with centre at $4/3$.

Example 2.5.8. *Find the domain of convergence of the complex power series* $\sum\limits_{n=0}^{\infty} u_n(z)$, *where* $u_n(z) = \dfrac{n^2}{(1+i)^n}(z^2+1)^n$.

Solution. Let $z^2 = \xi$. Then the given series is of the form $\sum a_n(\xi+1)^n$, where

$$a_n = \frac{n^2}{(1+i)^n} \text{ and } a_{n+1} = \frac{(n+1)^2}{(1+i)^{n+1}}$$

$$\Rightarrow \left|\frac{a_n}{a_{n+1}}\right| = \frac{n^2}{(n+1)^2}|1+i| = \sqrt{2} \times \frac{n^2}{(n+1)^2}$$

$$\Rightarrow \lim_{n\to\infty}\left|\frac{a_n}{a_{n+1}}\right| = \sqrt{2} \times \lim_{n\to\infty}\frac{n^2}{(n+1)^2} = \sqrt{2}.$$

Thus the radius of the circle of convergence for this power series in ξ is $\sqrt{2}$ and the centre of the circle is -1, so that the domain of convergence is given by

$$|\xi+1| < \sqrt{2} \Rightarrow |z^2+1| < \sqrt{2}.$$

Example 2.5.9. *Find the region of convergence of the series* $\sum\limits_{n=0}^{\infty} u_n(z)$ *of complex variable, where* $u_n(z) = \dfrac{(z+2)^{n-1}}{4^n(n+1)^3}$.

Solution. The given series is of the form $\sum\limits_{n=0}^{\infty} a_n(z+2)^{n-1}$, where $a_n = 1/4^n(n+1)^3$. Excluding $z = -2$ for which the given series converges, we have

$$\lim_{n\to\infty}\left|\frac{u_{n+1}}{u_n}\right| = \lim_{n\to\infty}\left|\frac{z+2}{4}\;\frac{(n+1)^3}{(n+2)^3}\right| = \frac{1}{4}|z+2|.$$

Then the series converges absolutely for $\frac{1}{4}|z+2| < 1$, i.e., $|z+2| < 4$ as seen in the Fig (2.2).

The point $z = -2$ is included in $|z+2| < 4$. If $|z+2| = 4$, the ratio test fails. However it is seen that in this case

$$\left|u_n(z)\right| = \left|\frac{(z+2)^n}{4^n(n+2)^3}\right| = \frac{1}{4(n+1)^3} \le \frac{1}{n^3}$$

and since $\sum 1/n^3$ converges (hyperharmonic series with $p = 3 > 1$), the given series converges absolutely for $|z+2| \le 4$. Geometrically, this is the set of all points inside and on the circle of radius 4 with centre at $z = -2$, called the circle of convergence. The radius of convergence is 4.

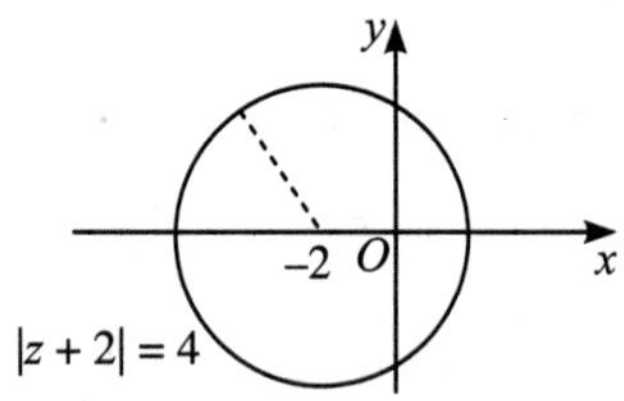

Figure 2.2 $|z+2| < 4$

Example 2.5.10. *Find the region of convergence of the series* $\sum\limits_{n=0}^{\infty} u_n(z)$ *of complex variable, where* $u_n(z) = (-1)^{n-1}\frac{1}{(2n+1)!}\, z^{2n-1}$.

Solution. The given series is of the form $\sum\limits_{n=0}^{\infty} a_n z^{2n-1}$, where $a_n = 1/(2n+1)!$. Excluding $z = 0$ for which the given series converges, we have

$$\lim_{n\to\infty} \left|\frac{u_{n+1}}{u_n}\right| = \lim_{n\to\infty} \left| -\frac{(2n-1)!}{(2n+1)!} z^2 \right| = \lim_{n\to\infty} \frac{|z|^2}{(2n+1)\cdot(2n)} = 0$$

for all finite z. Then the series converges absolutely for all z, and we say that the series converges for $|z| < \infty$. We can equivalently say that the circle of convergence is infinite, or that the radius of convergence is infinite.

2.5.2 Behaviour of Power Series on the Circle of Convergence

There is no information about the behaviour of the power series on the circle of convergence. We will now discuss the behaviour of some types of power series on the circle of convergence.

Property 2.5.1. There exist a power series which is always convergent on the circle of convergence.

Consider a complex power series $\sum\limits_{n=2}^{\infty} \frac{1}{n(\log n)^2}$. This series is of the form $\sum a_n z^n$, where

$$a_n = \frac{1}{n(\log n)^2};\ n \geq 2, a_{n+1} = \frac{1}{(n+1)[\log(n+1)]^2}$$

$$\Rightarrow \quad R = \lim_{n\to\infty} \frac{a_n}{a_{n+1}} = \lim_{n\to\infty} \frac{(n+1)[\log(n+1)]^2}{n(\log n)^2} = 1.$$

Therefore, the radius of convergence of the complex power series is 1 and the circle of convergence is $S(0,1) : |z| < 1$ (Fig. 2.3), having its centre at origin and radius of unity. Now, for any point z on the circle of convergence, where $|z| = 1$,

$$|a_n z^n| = \left|\frac{z^n}{n(\log n)^2}\right| = \frac{1}{n(\log n)^2} = v_n(\text{say});\ n \geq 2.$$

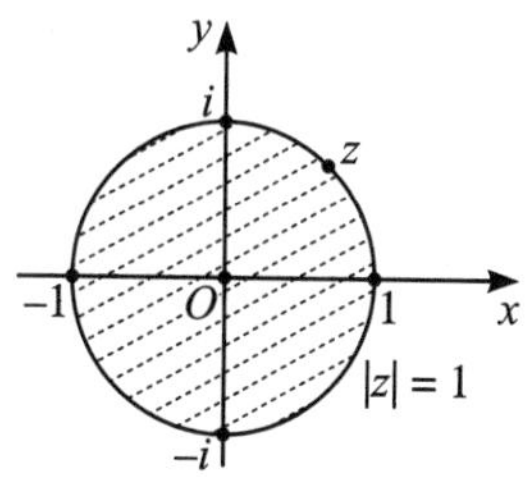

Figure 2.3 *Circle of convergence*

Now the series $\sum_{n=2}^{\infty} v_n$ is convergent by Cauchy condensation test of real numbers. Hence the series is always convergent for all points on the circle of convergence $S[0,1]$.

Property 2.5.2. It is not always true that a power series is always convergent at every point of the circle of convergence. It may also happen that it converges at all, barring a single point.

For this, we consider the complex power series $\sum\limits_{n=1}^{\infty} \frac{z^n}{n} = \sum b_n z^n$, where $b_n = \frac{1}{n}; n \geq 1$. If R is the radius of convergence then, it is given by

$$R = \lim_{n\to\infty} \left|\frac{b_n}{b_{n+1}}\right| = \lim_{n\to\infty} \left(\frac{n+1}{n}\right) = 1.$$

Hence, the radius of convergence is 1 and circle of convergence is $S(0,1) : |z| < 1$ (Fig. 2.3). At the point $z = 1$,

$$\sum b_n z^n = \sum_{n=1}^{\infty} \frac{z^n}{n} = \sum \frac{1}{n},$$

which is a hyperharmonic series with $p = 1$; so it is divergent. We now check what happens for $z \neq 1$ but $|z| = 1$. For this we take

$$a_n(z) = z^n \text{ and } b_n = \frac{1}{n}.$$

(i) Obviously the nth partial sum $S_n(z)$ of the complex power series $\sum a_n(z)$ is given by

$$\begin{aligned} |S_n(z)| &= \left|\sum_{k=1}^{n} a_k\right| = |1 + z + z^2 + \cdots + z^n| = \left|\frac{1 - z^n}{1 - z}\right|; z \neq 1 \\ &\leq \frac{1 + |z|^n}{|1 - z|} = \frac{1 + 1}{|1 - z|} = \frac{2}{|1 - z|}; z \neq 1 \end{aligned}$$

so that the sequence $\{S_n(z)\}$ of partial sums of $\sum a_n(z)$ is bounded.

(ii) Again, for the sequence $\{b_n\}$,

$$b_n - b_{n+1} = \frac{1}{n} - \frac{1}{n+1} = \frac{1}{n(n+1)} \leq \frac{1}{n^2}.$$

Now, $\sum \frac{1}{n^2}$ is a hyperharmonic series with $p = 2 > 1$; so it is convergent. Therefore, $\sum(b_n - b_{n+1})$ is absolutely convergent.

(iii) Also, $\lim_{n\to\infty} b_n = \lim_{n\to\infty} \frac{1}{n} = 0.$

Hence by Theorem 2.12, the complex power series $\sum a_n b_n = \sum_{n=1}^{\infty} \frac{z^n}{n}$ is convergent for every z other than 1 on the circle of convergence $S[0, 1]$.

Property 2.5.3. It is not always true that a power series is always convergent at every point of the circle of convergence; it may be so happen that it is not convergent at a single point.

For this we consider a power series $\sum\limits_{n=1}^{\infty} b_n z^n = \sum\limits_{n=1}^{\infty} \frac{(-1)^n}{n} z^n$. If R is the radius of convergence, then

$$\lim_{n\to\infty} \left|\frac{b_n}{b_{n+1}}\right| = \lim_{n\to\infty} \left| -\frac{n+1}{n}\right| = 1.$$

Hence, the radius of convergence is 1 and circle of convergence is $S(0, 1) : |z| < 1$ (Fig. 2.3). At the point $z = -1$,

$$\sum_{n=1}^{\infty} b_n\, z^n = \sum_{n=1}^{\infty} \frac{(-1)^n}{n} z^n = \sum_{n=1}^{\infty} \frac{(-1)^n}{n}(-1)^n = \sum_{n=1}^{\infty} \frac{1}{n},$$

which is a hyperharmonic series with $p = 1$, and hence divergent. We will now check behaviour of the series at all points on the circle of convergence $S(0, 1)$, except at $z = -1$. For this, we take

$$a_n = (-1)^n z^n; n \geq 1 \text{ and } b_n = \frac{1}{n}.$$

(i) The nth partial sum $S_n(z)$ of the complex power series $\sum a_n(z)$ is given by

$$|S_n(z)| \leq \frac{2}{|1 + z|}; \quad z \neq -1, |z| = 1$$

so that the sequence of the partial sums of the series $\sum a_n$ is bounded.

(ii) Again, for the sequence $\{b_n\}$,

$$b_n - b_{n+1} = \frac{1}{n} - \frac{1}{n+1} = \frac{1}{n(n+1)}$$

$$\Rightarrow \quad \lim_{n\to\infty} |b_n - b_{n+1}| = \lim_{n\to\infty} \frac{1}{n(n+1)} = 0,$$

so that $\sum |b_n - b_{n+1}|$ is convergent.

(iii) Also, $\lim\limits_{n\to\infty} b_n = \lim\limits_{n\to\infty} \frac{1}{n} = 0.$

Hence by Theorem 2.12, $\sum\limits_{n=1}^{\infty} a_n b_n = \sum \frac{(-1)^n}{n}$ is convergent for every z other than -1 on the circle of convergence $S[0,1]$.

Property 2.5.4. On the circle of convergence, a power series may diverge at more than one point.

For this we consider a complex power series $\sum\limits_{n=2}^{\infty} \frac{z^{4n}}{4n+1} = \sum\limits_{n=2}^{\infty} b_n z^{4n}$, where $b_n = 1/(4n+1)$. If R be the radius of convergence, it is given by

$$R = \lim_{n\to\infty} \left|\frac{b_n}{b_{n+1}}\right| = \lim_{n\to\infty} \frac{4n+5}{4n+1} = \lim_{n\to\infty} \left|\frac{4+(5/n)}{4+(1/n)}\right| = 1.$$

Hence the radius of convergence is 1 and circle of convergence is $S(0,1) : |z| < 1$ (Fig. 2.3). At the points $z = \pm 1, \pm i$

$$b_n z^{4n} = \frac{z^{4n}}{4n+1} = \frac{1}{4n+1} < \frac{1}{n},$$

where $\sum 1/(4n+1)$ is divergent on comparison with the divergent series $\sum 1/n$. So the given series is divergent at the points $z = \pm 1, \pm i$ on the circle of convergence (Fig. 2.3).

Therefore, we examine the behaviour of the given series for every value of z on the circle of convergence except at $z = \pm 1, \pm i$. For this, we take

$$a_n = z^{4n}; \quad b_n = \frac{1}{4n+1}, \qquad n(\geq 2) \in \mathbb{N}.$$

(i) Obviously, the nth partial sum $S_n(z)$ of the series $\sum a_n$ is given by

$$\begin{aligned} |S_n(z)| &= \left|\sum_{k=0}^{n-1} a_k\right| = |1 + z^4 + z^8 + \cdots + z^{4n}| \\ &= \left|\frac{1 - z^{4n+1}}{1 - z^4}\right| \leq \frac{1 + |z|^{4n+1}}{|1 - z^4|}; z \pm 1, \pm i \\ &\leq \frac{1+1}{|1 - z^4|} = \frac{2}{|1 - z^4|} \quad \text{on } |z| = 1 \end{aligned}$$

so that the sequence $\{S_n(z)\}$ of the partial sums is bounded.

(ii) Again, for the sequence $\{b_n\}$,

$$b_n - b_{n+1} = \frac{1}{4n+1} - \frac{1}{4n+5} = \frac{4}{(4n+1)(4n+5)} < \frac{1}{4n^2},$$

so that $\sum(b_n - b_{n+1})$ is absolutely convergent on comparison with the divergent hyperharmonic series $\sum 1/n^2$.

(iii) Also, $\lim\limits_{n\to\infty} b_n = \lim\limits_{n\to\infty} \frac{1}{4n+1} = 0.$

Hence by Theorem 2.12, the series $\sum a_n b_n$ converges for every z other than $\pm i, \pm 1$ on the circle of convergence (Fig. 2.3).

Theorem 2.18. [Uniform convergence of power series]: *If z_0 is any point inside the circle of convergence $|z| = R$ of the power series (2.8), then the power series converges uniformly in the closed disc $|z| \leq \rho$, where $\rho = |z_0|$.*

Proof: Consider the power series (2.8) with radius of convergence R given by Eq.(2.9) and circle of convergence $\mathcal{S}(0, R) : |z| < 1$ (Fig. 2.4).

Let $z_0 \in \mathcal{S}(0, R)$. Clearly, the series converges for the points inside the circle and these points are farther from the origin than z_0. Then by Abel's theorem 2.15, the series $\sum_{n=0}^{\infty} |a_n \, z_0^n|$ converges. For a positive integer $m(> n) \in \mathbb{N}$, the remainders $R_n(z)$ of the power series (2.8) and $T_n(z)$ of $\sum_{n=0}^{\infty} |a_n z_0^n|$ of real constants converges.

Let $\mathcal{R}_n(z)$ and T_n be respective nth partial sums of the series $\sum_{n=0}^{\infty} a_n z^n$ and $\sum_{n=0}^{\infty} |a_n z_0^n|$. Then they are respectively given by

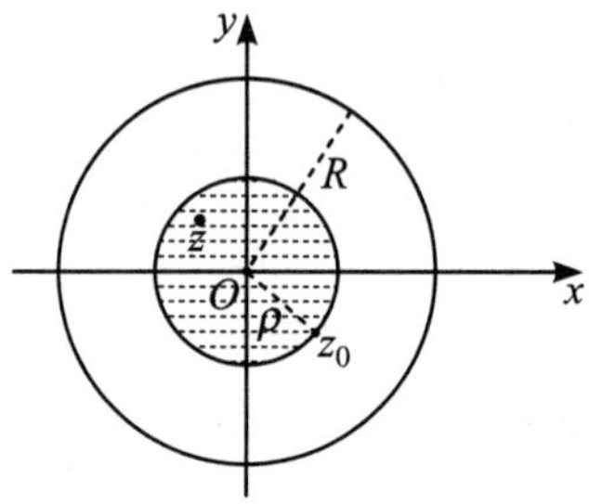

Figure 2.4 *Uniform convergence of power series*

$$\mathcal{R}_n(z) = \lim_{m\to\infty} \sum_{k=n}^{m} a_k z^k \text{ and } T_n = \lim_{m\to\infty} \sum_{k=n}^{m} |a_k z_0^k|.$$

Since $\sum\limits_{n=0}^{\infty} |a_n z_0^n|$ is convergent, so $T_n \to 0$ as $n \to \infty$. So, corresponding to an $\varepsilon > 0$, $\exists\, n_0 = n_0(\varepsilon) \in \mathbb{N}$ such that

$$|T_n - 0| = T_n < \varepsilon \text{ whenever } n \geq n_0.$$

We know that if $\lim\limits_{n\to\infty} z_n = z$, then $\lim\limits_{n\to\infty} |z_n| = |z|$. Therefore,

$$\begin{aligned}
|\mathcal{R}_n(z)| &= \lim_{m\to\infty} \left|\sum_{k=n}^{m} a_k z^k\right| \leq \lim_{m\to\infty} \sum_{k=n}^{m} \left|a_k z^k\right| \\
&\leq \lim_{m\to\infty} \sum_{k=n}^{m} |a_k|\,|z^k| \leq \lim_{m\to\infty} \sum_{k=n}^{m} |a_k|\,|z_0^k| \\
&\leq \lim_{m\to\infty} \sum_{k=n}^{m} \left|a_k \, z_0^k\right| = T_n \quad \text{for } |z| \leq z_0 \\
&< \varepsilon \text{ whenever } n \geq n_0,
\end{aligned}$$

which is true for all z in the closed disc $\mathcal{S}[0, \rho]$ and n_0 is independent of z. Thus the power series (2.8) converges uniformly in the closed disc $|z| \leq \rho$. □

Theorem 2.19. [Continuity of sum of power series]: *A complex power series (2.8) represents a continuous function $S(z)$ at each point inside the circle of convergence $|z| = R$.*

Proof: Let $S(z)$ be the sum of the complex power series (2.8) for each z in the circle of convergence $\mathcal{S}(0,\rho) : |z| < \rho$ of the series (Fig. 2.5). Also, let $S_n(z)$ denote the nth partial sum of the complex power series (2.8). Then $\lim\limits_{n\to\infty} S_n(z) = S(z)\ \forall z \in \mathcal{S}(0,\rho)$.

$$\left|S(z) - S_n(z)\right| < \frac{\varepsilon}{3} \text{ whenever } n > n_0(\varepsilon). \tag{2.11}$$

Also, let $z_1 \in \mathcal{S}(0,\rho)$ and let $S(z_1)$ be the sum of the power series (2.8) at $z = z_1$. Then

$$\left|S(z_1) - S_n(z_1)\right| < \frac{\varepsilon}{3} \text{ whenever } n > n_1(\varepsilon). \tag{2.12}$$

z_1 is inside the circle of convergence. Since the partial sum $S_n(z)$ is a polynomial, it is continuous at point z_1 for all values of $n \in \mathbb{N}$. So for small $\delta > 0$ and $n > n_2(\varepsilon) = n_0 + 1$,

$$\left|S_n(z) - S_n(z_1)\right| < \frac{\varepsilon}{3} \text{ for } |z - z_1| < \delta < \rho. \tag{2.13}$$

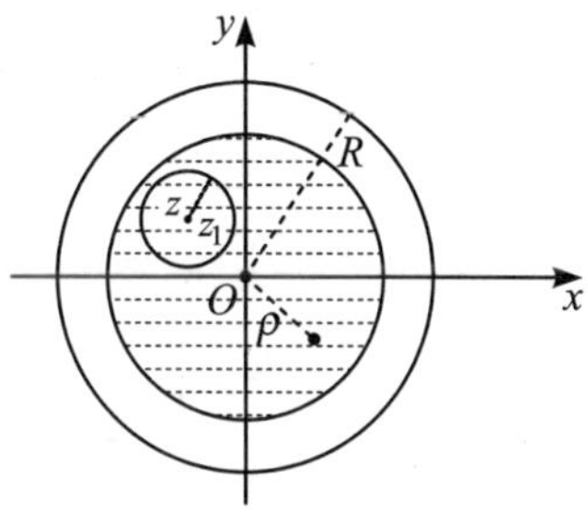

Figure 2.5 *Continuity of sum of power series*

Therefore, using Eqs. (2.11), (2.12) and (2.13) and for $n_2 = n_0 + 1$, we have

$$\begin{aligned}
|S(z) - S(z_1)| &= \left|S(z) - S_n(z) + S_n(z) - S_n(z_1) + S_n(z_1) - S(z_1)\right| \\
&\leq \left|S(z) - S_n(z)\right| + \left|S_n(z) - S_n(z_1)\right| + \left|S_n(z_1) - S(z_1)\right| \\
&< \frac{\varepsilon}{3} + \frac{\varepsilon}{3} + \frac{\varepsilon}{3} = \varepsilon \text{ whenever } |z - z_1| < \delta.
\end{aligned}$$

Thus the power series (2.8) represents a continuous function $S(z)$ at each point inside the circle of convergence $|z| = \rho$. □

Theorem 2.20. *The power series $\sum\limits_{n=0}^{\infty} a_n z^n$ and $\sum\limits_{n=1}^{\infty} n a_n z^{n-1}$, obtained by differentiating the first series term by term, have the same radius of convergence.*

Proof: Let R_1 and R_2 be the radii of convergence of the respective series $\sum\limits_{n=0}^{\infty} a_n z^n$ and $\sum\limits_{n=1}^{\infty} n a_n z^{n-1}$. Then they are given by the formula (2.9). First, we prove that $n^{1/n} \to 1$ as $n \to \infty$. To prove this, let $n^{1/n} = 1 + h_n; n \geq 2$ so that

$$\begin{aligned}
n &= \left(1 + h_n\right)^n = 1 + nh_n + \frac{n(n-1)}{2!}h_n^2 + \cdots + h_n^n > \frac{n(n-1)}{2!}h_n^2 \\
\Rightarrow\ & |h_n| < \sqrt{\frac{2}{n-1}} \to 0 \text{ as } n \to \infty.
\end{aligned}$$

It follows that $\lim\limits_{n\to\infty} n^{1/n} = 1$. Therefore,

$$\frac{1}{R_1} = \overline{\lim}|a_n|^{\frac{1}{n}} = \overline{\lim} n^{1/n}|a_n|^{\frac{1}{n}} = \frac{1}{R_2},$$

which shows that $R_1 = R_2$. □

Theorem 2.21. [Analyticity of power series]: *Every power series represents an analytic function at all points within the circle of convergence and its derivative can be obtained by termwise differentiation of the given power series.*

Proof: Let $f_1(z) = \sum_{n=0}^{\infty} a_n z^n$ be a given complex power series with the radius of convergence R_1 given by equation (2.9).

Consider another power series $f_2(z) = na_n z^{n-1}$ which is obtained by termwise differentiation of the given power series $f_1(z)$. Let R_2 denotes the radius of convergence of the power series $f_2(z)$. We claim that $R_1 = R_2$.

Let $S_n(z)$ denote the nth partial sum of the series of $f_1(z)$. Then,

$$f_1(z) = S_n(z) + \mathcal{R}_n(z),$$

where,
$$S_n(z) = a_0 + a_1 z + a_2 z^2 + \cdots + a_{n-1} z^{n-1}$$

$$\mathcal{R}_n(z) = a_n z^n + a_{n+1} z^{n+1} + \cdots \infty.$$

We claim that $f_1'(z) = f_2(z)$. Thus the nth partial sum of the derived series $f_2(z)$ is given by

$$S_n'(z) = a_1 + 2a_2 z + \cdots + (n-1)a_{n-1} z^{n-2}$$

$$\therefore \quad \lim_{n\to\infty} S_n'(z) = f_2(z) \text{ in } |z| < R_1.$$

Let $z_0(\neq 0)$ be any point within the circle of convergence, so that $|z_0| < R_1$. Then $\exists$ a $r \in \mathbb{R}^+$ such that $|z_0| < r < R_1$ (Fig. 2.6). Since $\lim_{n\to\infty} S_n'(z_0) = f_2(z_0)$, there exists $n_0 \in \mathbb{N}$ such that

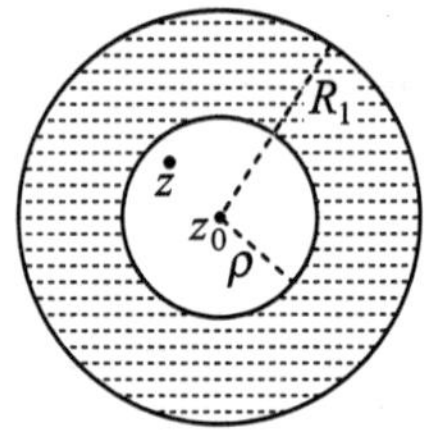

Figure 2.6 $|z_0| < r < R$

$$\left|S_n'(z_0) - f_2(z_0)\right| < \frac{\varepsilon}{3} \text{ for all } n \geq n_0.$$

From the definition of derivative $S_n'(z_0)$, corresponding to $\varepsilon > 0$, we can find $\delta > 0$ such that

$$\left|\frac{S_n(z) - S_n(z_0)}{z - z_0} - S_n'(z_0)\right| < \frac{\varepsilon}{3} \text{ whenever } 0 < |z - z_0| < \delta.$$

Also, for the remainder $\mathcal{R}_n(z)$, we have

$$\left|\frac{\mathcal{R}_n(z) - \mathcal{R}_n(z_0)}{z - z_0}\right| = \left|\sum_{k=n}^{\infty} \frac{a_k(z^k - z_0^k)}{z - z_0}\right| = \left|\sum_{k=n}^{\infty} a_k \Big[\sum_{m=1}^{k-1} z^{k-m-1} z_0^{m-1}\Big]\right|$$

$$\leq \sum_{k=n}^{\infty} |a_k| \Big[\sum_{m=1}^{k-1} |z|^{k-m-1} |z_0|^{m-1}\Big] \leq \sum_{k=n}^{\infty} |a_k|\, k\, r^{k-1} \quad (2.14)$$

The expression on the right side of equation (2.14) is the remainder after n terms of the convergent series $\sum_{n=1}^{\infty} na_n r^{n-1}$, and hence tends to $na_n r^{n-1} \to 0$ as $n \to \infty$. Hence we can find a positive integer $n_1 \in \mathbb{N}$ such that

$$\left|\frac{\mathcal{R}_n(z) - \mathcal{R}_n(z_0)}{z - z_0}\right| < \frac{\varepsilon}{3} \text{ for all } n \geq n_1.$$

Now chose a fixed a positive integer $n(\in \mathbb{N})$ such that $n > \max\{n_0, n_1\}$.

$$\left|\frac{f_1(z)-f_1(z_0)}{z-z_0} - f_2(z_0)\right| = \left|\frac{S_n(z)+\mathcal{R}_n(z)-S_n(z_0)-\mathcal{R}_n(z_0)}{z-z_0} - f_2(z_0)\right|$$

$$= \left|\frac{S_n(z)-S_n(z_0)}{z-z_0} - S_n'(z) + \frac{\mathcal{R}_n(z)-\mathcal{R}_n(z_0)}{z-z_0} + S_n'(z) - f_2(z_0)\right|$$

$$\leq \left|\frac{S_n(z)-S_n(z_0)}{z-z_0} - S_n'(z_0)\right| + \left|\frac{\mathcal{R}_n(z)-\mathcal{R}_n(z_0)}{z-z_0}\right| + \left|S_n'(z_0) - f_2(z_0)\right|. \tag{2.15}$$

Thus we get from the equation (2.15),

$$\left|\frac{f_1(z)-f_1(z_0)}{z-z_0} - f_2(z_0)\right| < \varepsilon \text{ whenever } 0 < |z-z_0| < \delta.$$

Hence $f_1'(z_0) = f_2(z_0)$. Since z_0 is an arbitrary point in $|z| < R_1$, we have $f_1'(z) = f_2(z)$ in $|z| < R_1$. Therefore the power series is an analytic function. □

DEDUCTION **5.**

(i) For each $k \geq 1$, the series $\sum\limits_{n=0}^{\infty} n(n-1)\cdots(n-k+1)a_k z^{n-k}$ has radius of convergence R.

(ii) Let the power series $\sum\limits_{n=0}^{\infty} a_n(z-z_0)^n$ converge in $\mathcal{S}(z_0, R) : |z-z_0| < R$ for some $R > 0$. Also, let

$$f(z) = \sum_{n=0}^{\infty} a_n(z-z_0)^n, \text{ for all } z \in \mathcal{S}(z_0, R).$$

Applying Theorem 2.21 repeatedly, first we note that $f^{(n)}(z)$ exists for every $n = 0, 1, 2, \cdots$ and

$$\begin{aligned} f^{(n)}(z) &= \frac{d^n}{dz^n}\Big[\sum_{k=0}^{n-1} a_k(z-z_0)^k + \sum_{k=n}^{\infty} a_k(z-z_0)^k\Big] \\ &= a_n\, n! + \sum_{k=n}^{\infty} k(k-1)\cdots(k-n+1)a_k(z-z_0)^{k-n}. \\ \therefore f^{(n)}(z_0) &= a_n\, n! \Rightarrow a_n = \frac{f^{(n)}(z_0)}{n!}; \forall n = 0, 1, 2, \cdots. \end{aligned}$$

(iii) Let $\sum\limits_{n=0}^{\infty} a_n(z-z_0)^n = \sum\limits_{n=0}^{\infty} b_n(z-z_0)^n \ \forall z \in \mathcal{S}(z_0, R)$. Let $f(z) = \sum\limits_{n=0}^{\infty} a_n(z-z_0)^n \ \forall z \in \mathcal{S}(z_0, R)$. Then, by the previous deduction (ii), we get

$$a_n = \frac{f^{(n)}(z_0)}{n!} = b_n; \forall n = 0, 1, 2, \cdots.$$

Conversely, if $a_n = b_n \ \forall n = 0, 1, 2, \cdots$, then, obviously, we get

$$\sum_{n=0}^{\infty} a_n(z-z_0)^n = \sum_{n=0}^{\infty} b_n(z-z_0)^n.$$

Corollary 2.5.1. A power series represents a function which is analytic in the interior of its circle of convergence.

Theorem 2.22. [Abel's limit theorem]: *Let* $\sum\limits_{n=0}^{\infty} a_n$ *converge. If* $f(z) = \sum\limits_{n=0}^{\infty} a_n z^n$, *then* $f(z) \to f(1)$ *as* $z \to 1$ *such that* $|1-z|/(1-|z|)$ *is bounded.*

Proof: First, we note that as $\sum\limits_{n=0}^{\infty} a_n$ converges to 1, the radius of convergence is at least 1 by Abel's theorem. If $R > 1$, then f is continuous at 1, and hence, the theorem follows. Therefore, assume that $R = 1$.

Case-I: Let $f(1) = \sum\limits_{n=0}^{\infty} a_n = 0$. Let α_n and $S_n(z)$ be respectively the nth partial sums of the series $\sum\limits_{n=0}^{\infty} a_n$ and $f(z)$. Then

$$\alpha_n = \sum_{k=0}^{n} a_k \text{ and } S_n(z) = \sum_{k=0}^{n} a_k z^k \ \ \forall z \in \mathcal{S}(0,1).$$

Since $|1-z|/(1-|z|)$ is bounded, there exists a $M \in \mathbb{R}^+$ such that

$$|1-z|/(1-|z|) \le M; \qquad M > 0.$$

Now, using the above notations

$$\begin{aligned}
S_n(z) &= \alpha_0 + \sum_{k=1}^{n}(\alpha_k - \alpha_{k-1})z^k = \alpha_0 + \sum_{k=1}^{n} \alpha_k z^k - \sum_{k=1}^{n} \alpha_{k-1} z^k \\
&= \alpha_0 + \sum_{k=1}^{n-1} \alpha_k z^k + a_n z^n - \alpha_0 z - \sum_{k=1}^{n-1} \alpha_k z^{k+1} \\
&= \alpha_0(1-z) + \sum_{k=1}^{n-1} \alpha_k \left[z^k - z^{k+1}\right] + \alpha_n z^n \\
&= \alpha_0(1-z) + \sum_{k=1}^{n-1} \alpha_k (1-z) z^k + \alpha_n z^n = (1-z)\sum_{k=0}^{n-1} \alpha_k z^k + \alpha_n z^n.
\end{aligned}$$

Allowing $n \to \infty$ on both sides, we get

$$f(z) = (1-z)\sum_{k=0}^{\infty} \alpha_k z^k + \lim_{n\to\infty} \alpha_n z^n = (1-z)\sum_{k=0}^{\infty} \alpha_k z^k$$

as $\alpha_n \to f(1) = 0$ as $n \to \infty$ and $|z| < 1$.

Again, using $\alpha_n \to 0$ as $n \to \infty$, given $\varepsilon > 0$, there exists $n_0 \in \mathbb{N}$ such that $|\alpha_n| < \varepsilon/2K$ $\forall n \ge n_0$. Choose $\delta > 0$ such that

$$0 < \delta < \frac{\varepsilon}{2\left(1 + \sum\limits_{k=1}^{n_0-1} |\alpha_n|\right)}.$$

If $0 < |z-1| < \delta$, and $|1-z|/(1-|z|) \leq M$, then

$$\begin{aligned}
|f(z)| &\leq \left|1-z\right|\Big[\sum_{k=0}^{n_0-1}\left|\alpha_k z^k\right| + \sum_{k=n_0}^{\infty}\left|\alpha_k z^k\right|\Big] \\
&< \left|1-z\right|\Big[\sum_{k=0}^{n_0-1}\left|\alpha_k\right| + \frac{\varepsilon}{2M}\sum_{k=n_0}^{\infty}\left|z^k\right|\Big] \\
&< \delta\sum_{k=0}^{n_0-1}\left|\alpha_k\right| + \frac{\varepsilon}{2M}|z^{n_0}|\left|1-z\right|\left|\sum_{k=0}^{\infty}\right|z^k\Big| \\
&< \frac{\varepsilon}{2} + \frac{\varepsilon}{2M}\frac{|1-z|}{1-|z|} \leq \frac{\varepsilon}{2} + \frac{\varepsilon}{2} = \varepsilon.
\end{aligned}$$

Hence, the theorem follows in this case.
Case-II: Let $f(1) \neq 0$. If

$$\phi(z) = f(z) - f(1) = a_0 - f(1) + \sum_{n=1}^{\infty} a_n z^n,$$

then $\phi(1) = 0$. Then, by Case-I, we obtain $\phi(z) \to \phi(1)$ as $z \to 1$ with $|1-z|/(1-|z|)$ bounded, and hence

$$f(z) - f(1) \to 0 \text{ as } z \to 1 \text{ with } |1-z|/(1-|z|) \text{bounded.}$$

Thus, the proof is complete. □

2.6 Sequence and Series of Functions

In this section we shall prove some results for complex sequences and series of functions.

Definition 2.6.1. (i) Let $z \in \mathbb{C}$. Then $f_1(z), f_2(z), \cdots, f_n(z), \cdots$ is called the *sequence of complex functions* defined in a domain $\mathcal{D} \subseteq \mathbb{C}$ denoted by $\langle f_n(z)\rangle$ and we write $f_n : \mathcal{D} \to \mathbb{C}$.

(ii) Suppose $\langle u_n(z)\rangle$ is a sequence of complex-valued functions defined in a domain $\mathcal{D} \subseteq \mathbb{C}$. Then $\sum\limits_{n=1}^{\infty} u_n(z)$ is called *series of functions.*

2.6.1 Pointwise Convergence

Definition 2.6.2. Suppose that for each $z \in \mathcal{D}$ the sequence $\langle f_n(z)\rangle$ is convergent. Define

$$\lim_{n\to\infty} f_n(z) = f(z); \text{ for any } z \in \mathcal{D}.$$

We say that $\langle f_n\rangle$ converges *pointwise* to f on $\mathcal{D}$ and f is called the *pointwise limit of* $\langle f_n\rangle$.

More explicitly, the sequence of functions $\langle f_n\rangle$ converges *pointwise to* f if for any fixed $z \in \mathcal{D}$ and any preassigned $\varepsilon > 0$, there exists $n_0(\varepsilon, z) \in \mathbb{N}$ such that

$$|f_n(z) - f(z)| < \varepsilon \text{ for all } n \geq n_0.$$

Domain $\mathcal{D}$ is called the *region of convergence of sequence of complex functions.*

Definition 2.6.3. Let $\langle u_n(z)\rangle$ be a sequence of complex functions defined in a domain $\mathcal{D} \subseteq \mathbb{C}$. The series of complex functions $\sum\limits_{n=1}^{\infty} u_n(z)$ is said to converge *pointwise to a function* $S(z)$, in $\mathcal{D}$, if

$$\lim_{n\to\infty} S_n(z) = \sum_{n=1}^{\infty} u_n(z) = S(z)$$

where, $S_n(z) = u_1(z) + u_2(z) + \cdots + u_n(z) =$ nth partial sum of the series. $S(z)$ is called the *sum of the series.*

When the sequence $\langle S_n(z)\rangle$ is divergent, the infinite series is also divergent. A necessary condition for the convergence of an infinite series of complex functions is that $\lim\limits_{n\to\infty} u_n(z) = 0$ for all $z \in \mathcal{D}$.

2.6.2 Uniform Convergence

Definition 2.6.4. The sequence of complex functions $\langle f_n(z)\rangle$, defined in a domain $\mathcal{D} \subseteq \mathbb{C}$, is said to *converge to $f(z)$ uniformly* on $\mathcal{D}$ if corresponding to an $\varepsilon > 0$ there exists a positive integer $n_0(\varepsilon) \in \mathbb{N}$ such that

$$\left|f_n(z) - f(z)\right| < \varepsilon \text{ for all } n \geq n_0 \text{ and } \forall z \in \mathcal{D}.$$

Definition 2.6.5. The series $\sum\limits_{n=1}^{\infty} u_n(z)$ is said to *converge uniformly to a function* $S(z)$ in $\mathcal{D}$ if the sequence of partial sums $\langle S_n(z)\rangle$ converges uniformly to $S(x)$ on $\mathcal{D}$.

For example, let $f_n(z) = z^n$, $f(0) = 0$. Then

(i) $f_n \to f$ pointwise on the open disc $S(0, 1)$

(ii) $f_n \to f$ uniformly on the closed disc $S[0, r]$, $0 < r < 1$, and

(iii) $f_n \not\to f$ uniformly on the closed disc $S[0, 1]$.

For the sequences and series of functions, we shall prove the following results.

Theorem 2.23. [Cauchy's criterion]:

(i) *The sequence $\langle f_n(z)\rangle$ converges uniformly on a domain $\mathcal{D}$ to $f(z)$ if and only if for any $\varepsilon > 0$ there exists an $n_0(\varepsilon) \in \mathbb{N}$ and any natural number m such that*

$$\left|f_{n+m}(z) - f_n(z)\right| < \varepsilon \ \forall z \in \mathcal{D} \text{ and } n \geq n_0(\varepsilon). \tag{2.16}$$

(ii) *The series $\sum\limits_{n=0}^{\infty} u(z)$ converges uniformly on a domain $\mathcal{D}$ to $S(z)$ if and only if for any $\varepsilon > 0$ there exists an $n_0(\varepsilon) \in \mathbb{N}$ and any natural number m such that*

$$\left|S_{n+m}(z) - S_n(z)\right| < \varepsilon \ \forall z \in \mathcal{D} \text{ and } n \geq n_0(\varepsilon) \tag{2.17}$$

holds, where $S_n(z)$ is the nth partial sum of the series.

Proof: The proof is simple and left as an exercise to the reader.

RESULT **20.** If a sequence/series of complex functions is uniformly convergent, then it is pointwise convergent, but not the converse. For example,

(i) Let $f_n(z) = 1/(1+nz)$, $f(z) = 0$. Then, for $|z| \geq 2$, we find

$$\left|\frac{1}{1+nz}\right| \leq \frac{1}{n|z|-1} \leq \frac{1}{2n-1} \leq \frac{1}{n},$$

i.e., the convergence is uniform. We also note that this sequence for $|z| \leq 1$ and Re $z \geq 0$ converges pointwise, but not uniformly. For this, clearly, $f(0) = 1$ and $f(z) = 0, z \neq 0$ is the pointwise limit of this sequence.

(ii) Let $u_n(z) = \left(1+\frac{1}{n}\right) z$, $z \in \mathbb{C}$. Obviously, if

$$\left|(1+\frac{1}{n})z - z\right| = \frac{1}{n}|z| < \varepsilon \text{ for } n \geq n_0,$$

it is necessary that $n_0 > \frac{1}{\varepsilon}|z|$. Such an n_0 exists for every z, but the requirement cannot be set simultaneously for every z. In this situation we say that the $\langle u_n \rangle$ converges pointwise but not uniformly.

Theorem 2.24. *If a series of complex functions $\sum\limits_{n=1}^{\infty} u_n(z)$ converges uniformly to $S(z)$ and if each $u_n(z)$ is continuous on $\mathcal{D}$, then S is continuous on $\mathcal{D}$.*

Proof: Let $S_n(z)$ be the nth partial sum of the series $\sum\limits_{n=1}^{\infty} u_n(z)$ of complex functions. Let $z_0 (\neq z) \in \mathcal{D}$. Since the series of functions converges uniformly to $S(z)$, so, corresponding to an $\varepsilon > 0$, $\exists\, M = M(\varepsilon) \in \mathbb{N}$ such that if $n \geq M$, then

$$\left|S_n(z) - S(z)\right| < \frac{\varepsilon}{3} \quad \text{for each } z \in \mathcal{D}.$$

In particular, at $z = z_0$ and $n = M$,

$$\left|S_M(z_0) - S(z_0)\right| < \frac{\varepsilon}{3}.$$

Now, $S_M(z)$, being the sum of a finite number of continuous functions on $\mathcal{D}$, is continuous on $\mathcal{D}$. Therefore, corresponding to $\varepsilon > 0$, $\exists\, \delta > 0$ such that

$$\left|S_M(z) - S(z_0)\right| < \frac{\varepsilon}{3}; \text{ for each } z \in \mathcal{N}(z_0,\delta) \bigcap \mathcal{D}.$$

Then for each $z \in \mathcal{N}(z_0,\delta) \bigcap \mathcal{D}$,

$$\begin{aligned} \left|S(z) - S(z_0)\right| &\leq \left|S(z) - S_M(z_0)\right| + \left|S_M(z) - S_M(z_0)\right| + \left|S_M(z_0) - S(z_0)\right| \\ &< \frac{\varepsilon}{3} + \frac{\varepsilon}{3} + \frac{\varepsilon}{3} = \varepsilon. \end{aligned}$$

Hence $S(z)$ is continuous at z_0. But z_0 is arbitrary. So we conclude that $S(z)$ is continuous in $\mathcal{D}$. Thus if a series of continuous functions is uniformly convergent in a region $\mathcal{D}$, then the sum function must be continuous in $\mathcal{D}$. If the sum function is not continuous, the series cannot be uniformly convergent.□

Corollary 2.6.1. If the sequence $\langle f_n(z)\rangle$ of complex functions converges uniformly on a domain $\mathcal{D} \subseteq \mathbb{C}$ to $f(z)$ and each $f_n(z)$ is continuous on $\mathcal{D}$, then $f(z)$ is continuous on $\mathcal{D}$.

Theorem 2.25. *Let Γ be a contour and let $\langle u_n(z)\rangle$ be a sequence of complex functions continuous on a domain $\mathcal{D} \subseteq \mathbb{C}$ containing the contour Γ. If the series of complex functions $\sum\limits_{n=0}^{\infty} u_n(z)$ converges uniformly to $S(z)$, then*

$$\lim_{n\to\infty} \int_\Gamma S_n(z)dz = \int_\Gamma \lim_{n\to\infty} S_n(z)dz = \int_\Gamma S(z)dz. \tag{2.18}$$

Proof: Let $S_n(z)$ be the nth partial sum of the series. Let $\varepsilon > 0$ and let $R_n(z) = S(z) - S_n(z)$ for each z on Γ and for each $n \in \mathbb{N}$. According to the definition of uniform convergence, $\exists\, M \in \mathbb{N}$ such that

$$|R_n(z)| = \Big|S(z) - S_n(z)\Big| < \frac{\varepsilon}{L} \text{ for each } z \text{ on } \Gamma$$

and for each $n \geq M$, where L is the length of Γ. Now $R_n(z)$ is a complex function continuous on a domain $\mathcal{D} \subseteq \mathbb{C}$ containing the contour Γ. Therefore,

$$\begin{aligned}
\Big|\int_\Gamma S(z)dz - \int_\Gamma S_n(z)dz\Big| &= \Big|\int\limits_\Gamma \Big[S(z)dz - S_n(z)\Big]dz\Big| \\
&\leq \int\limits_\Gamma \Big|S(z)dz - S_n(z)\Big|\,|dz| \\
&\leq \int_\Gamma \Big|R_n(z)\Big|\,|dz| \leq \frac{\varepsilon}{L}\cdot L = \varepsilon \text{ if } n \geq M.
\end{aligned}$$

By the definition of the sum of a series of complex number, relation (2.18) holds. □

Corollary 2.6.2. Let Γ be a contour and let $\langle f_n(z)\rangle$ be a sequence of complex functions continuous on a domain $\mathcal{D} \subseteq \mathbb{C}$ containing the contour Γ. If the sequence of functions converges uniformly to $f(z)$, then

$$\lim_{n\to\infty} \int_\Gamma f_n(z)dz = \int_\Gamma \lim_{n\to\infty} f_n(z)dz = \int_\Gamma f(z)dz. \tag{2.19}$$

Theorem 2.26. [Weierstrass M-test]: *Let $\langle u_n(z)\rangle$ be a sequence of complex functions in a domain $\mathcal{D} \subseteq \mathbb{C}$. Let $\sum\limits_{n=1}^{\infty} u_n(z)$ be a series of complex functions defined in $\mathcal{D}$ and*

(i) $|u_n(z)| \leq M_n$ *for each $n \in \mathbb{N}$ and $\forall z \in \mathcal{D}$,*

(ii) $\sum_{n=1}^{\infty} M_n$ *is a convergent series of positive real numbers,*

then the series converges uniformly and absolutely on $\mathcal{D}$.

Proof: Let $S_n(z)$ = the nth partial sum of the series $\sum_{n=0}^{\infty} u_n(z)$ of complex function. It is given that $|u_n(z)| \leq M_n\ \forall n \in \mathbb{N}$ and for each $z \in \mathcal{D}$. Thus by the comparison test in calculus, the series $\sum |u_n(z)|$ converges for each $z \in \mathcal{D}$, and so it has the limit $S(z)$, say, $z \in \mathcal{D}$.

Let $\varepsilon > 0$ be chosen arbitrarily. Since $\sum M_n$ is a convergent series of real numbers, $\exists\, n_0 \in \mathbb{N}$ such that

$$\sum_{k=n+1}^{\infty} M_k = \sum_{k=1}^{\infty} M_k - \sum_{k=1}^{n} M_k < \varepsilon \text{ wherever } n \geq n_0.$$

Thus for each $z \in \mathcal{D}$ and for each $n \geq n_0$,

$$\left|S(z) - S_n(z)\right| = \left|\sum_{k=n+1}^{\infty} u_k(z)\right| \leq \sum_{k=n+1}^{\infty} \left|u_k(z)\right| \leq \sum_{k=n+1}^{\infty} M_k < \varepsilon.$$

Therefore the infinite series converges uniformly to $S(z)$ on $\mathcal{D}$. □

Corollary 2.6.3. Suppose that $\langle f_n(z)\rangle$ converges to $f(z)$ pointwise on a domain $\mathcal{D}$. For each $n \geq 0$, let $\sup\{|f_n(z) - f(z)|; z \in \mathcal{D}\} = M_n$. Then $\langle f_n(z)\rangle$ converges to $f(z)$ uniformly on $\mathcal{D}$ if and only if $\lim\limits_{n\to\infty} M_n = 0$.

Example 2.6.1. *Discuss the convergence, uniformly convergence, and absolute convergence of the series of complex functions*

$$z(1-z) + z^2(1-z) + z^3(1-z) + \cdots$$

Solution. The given series of complex functions can be written as $\{u_n(z)\}$, where $u_n(z) = z^n(1-z)$. The nth partial sum of the given series is given by

$$\begin{aligned} S_n(z) &= u_1(z) + u_2(z) + \cdots + u_n(z) \\ &= z(1-z) + z^2(1-z) + z^3(1-z) + \cdots + z^n(1-z) \\ &= z - z^2 + z^2 - z^3 + \cdots + z^n - z^{n+1} = z - z^{n+1}. \end{aligned}$$

If $z = 0$, $S_n(0) = 0$ and $|S_n(z) - 0| \leq \varepsilon$ for all n. Now,

$$|S_n(z) - z| = |-z^{n+1}| = |z|^{n+1} < \varepsilon$$

$$\text{for } (n+1)\ln|z| < \ln\varepsilon, \text{ i.e., } n > \frac{\ln\varepsilon}{\ln|z|} - 1 \text{ if } z \neq 0.$$

Hence $\lim\limits_{n\to\infty} S_n(z) = z$, the required sum for all z such that $z < 1$. Thus if $|z| < 1$,

$$s(z) = \lim_{n\to\infty} S_n(z) = z.$$

For $z = 1$, $S_n(z) = S_1(1) = 0$ and $\lim\limits_{n\to\infty} S_n(1) = 0$. Hence $S(z)$ is discontinuous at $z = 1$. Also,

$$\max_{|z|\leq\frac{1}{2}} \left\{\frac{\ln\varepsilon}{\ln|z|} - 1\right\} = \left[\frac{\ln\varepsilon}{\ln|z|} - 1\right]_{|z|=\frac{1}{2}} = \frac{\ln\varepsilon}{\ln|z|} - 1.$$

Let us take $n_0 = \left[\frac{\ln\varepsilon}{\ln|z|} - 1\right] + 1 \in \mathbb{N}$. Then

$$|S_n(z) - z| < \varepsilon \text{ for all } n > n_0,$$

where $n_0 = n_0(\varepsilon)$ and not on the particular z in $|z| \leq \frac{1}{2}$. Thus the series converges uniformly to z for $|z| \leq \frac{1}{2}$. However, it is clear that we cannot extend the argument to $|z| \leq 1$ since this would require $n_0 = \left[\frac{\ln\varepsilon}{\ln|z|} - 1\right] + 1$, which is infinite, i.e., there is no finite value of n_0 which can be used in this case. Thus the series does not converge uniformly for $z \leq 1$.

For the absolute convergence of $\sum u_n(z)$, let $\sum t_n(z)$ be the nth partial sum of the series $\sum |u_n(z)|$. Then

$$\begin{aligned} t_n(z) &= |u_1(z)| + |u_2(z)| + \cdots + |u_n(z)| \\ &= |z(1-z)| + |z^2(1-z)| + \cdots + |z^n(1-z)| \\ &= |1-z|\Big[|z| + |z|^2 + \cdots + |z|^n\Big] \\ &= |1-z||z|\left[\frac{1-|z|^n}{1-|z|}\right]. \end{aligned}$$

If $|z| < 1$, $\lim_{n\to\infty} t_n(z)$ exists, so that the series converges absolutely. Thus if $|z| < 1$,

$$t(z) = \lim_{n\to\infty} t_n(z) = \frac{|1-z|}{1-|z|}|z|.$$

Example 2.6.2. *Test for uniform convergence of the series of complex variable in the indicated region:*

(i) $\sum_{n=1}^{\infty} u_n(z); u_n(z) = \frac{z^n}{n\sqrt{n+1}}, |z| \le 1$ *(ii)* $\sum_{n=1}^{\infty} u_n(z); u_n(z) = \frac{1}{n^2+z^2}, 1 < |z| < 2$

(iii) $\sum_{n=1}^{\infty} u_n(z); u_n(z) = \frac{\cos nz}{n^3}, |z| \le 1$

Solution. (i) For the given series of complex function, we have

$$|u_n(z)| \le \frac{|z|^n}{n\sqrt{n+1}} \le \frac{1}{n^{3/2}}, \quad \text{if } |z| \le 1.$$

Let $|M_n| = \frac{1}{n^{3/2}}$. We see that $\sum M_n$ converges (hyperharmonic series with $p = \frac{3}{2} > 1$). Hence by the Weierstrass M-test, the given series converges uniformly (and absolutely) for $|z| \le 1$.
(ii) The given series can be written as

$$\frac{1}{1^2+z^2} + \frac{1}{2^2+z^2} + \frac{1}{3^2+z^2} + \cdots$$

The first two terms can be omitted without affecting the uniform convergence of the series. For $n \ge 3$ and $1 < |z| < 2$, we have

$$\begin{aligned} & |n^2+z^2| \ge |n^2| - |z^2| \ge n^2 - 4 \ge \frac{1}{2}n^2 \\ \Rightarrow \quad & \frac{1}{|n^2+z^2|} \le \frac{2}{n^2} = M_n. \end{aligned}$$

Since $\sum M_n$ converges, it follows from Weierstrass M-test (hyperharmonic series with $p = 2 > 1$) that the given series converges uniformly for $1 < |z| < 2$.
(iii) Let $z = x + iy \in \mathbb{M}$. We have

$$\begin{aligned} \frac{\cos nz}{n^3} &= \frac{1}{2n^3}(e^{inz} + e^{-inz}) = \frac{1}{2n^3}(e^{inx-ny} + e^{-inx+ny}) \\ &= \frac{e^{-ny}}{2n^3}(\cos nx + i\sin nx) + \frac{e^{ny}}{2n^3}(\cos nx - i\sin nx) \\ &= p_n(x,y) + q_n(x,y). \end{aligned}$$

The series $p_n(x,y)$ and $q_n(x,y)$ cannot converge for $y > 0$ and $y < 0$ respectively (since in these cases the nth term does not approach zero). Hence the series does not converge for all z such that $|z| \le 1$, and so cannot possibly be uniformly convergent in this region.

The series does not converge for $y = 0$, i.e., if z is real. In this case,

$$\left|\frac{\cos nz}{n^3}\right| \left|\frac{\cos nx}{n^3}\right| \le \frac{1}{n^3} = M_n \text{ (say)} .$$

Now $\sum M_n$ is convergent (as hyperharmonic series with $p = 3 > 1$) by Weierstrass M-test. The given series converges uniformly in any interval on the real axis.

Theorem 2.27. *Let $\langle f_n(z)\rangle$ be a sequence of analytic functions on a domain $\mathcal{D} \subseteq \mathbb{C}$. If $f_n(z) \to f(z)$ uniformly on $\mathcal{D}$, then $f(z)$ is analytic on $\mathcal{D}$.*

Proof: Let $z_0 \in \mathcal{D}$. Since $\mathcal{D}$ is open, there exists an $r > 0$ such that the open disc $\mathcal{S}(z_0, r) \subseteq \mathcal{D}$. Let Γ be a closed contour in $\mathcal{S}(z_0, r)$. Since analytic functions are continuous and the limit of a uniformly convergent sequence of continuous functions is continuous, $f(z)$ is continuous. Since, for each n, $f_n(z)$ is analytic in $\mathcal{D}$, by Cauchy's theorem $\displaystyle\int_\Gamma f_n(z)dz = 0$. Then, by Eq. (2.19), we have

$$\int_\Gamma f(z)dz = \lim_{n\to\infty} \int_\Gamma f_n(z)dz = 0.$$

By the Corollary 2.6.1, $f(z)$ is continuous on $\mathcal{D}$. Therefore, by Morera's theorem 5.14, $f(z)$ is analytic in the open disc $\mathcal{S}(z_0, r)$. Since $z_0 \in \mathcal{D}$ is arbitrary, $f(z)$ is analytic in $\mathcal{D}$. □

Corollary 2.6.4. Let $\langle u_n(z)\rangle$ be a sequence of analytic functions on a domain $\mathcal{D} \subseteq \mathbb{C}$. If $\sum\limits_{n=1}^{\infty} u_n(z)$ converges to $S(z)$ uniformly on $\mathcal{D}$, then $S(z)$ is analytic on $\mathcal{D}$.

Exercises

Multiple-choice Questions

1. Let $S_n = \sum\limits_{k=1}^{n} \frac{1}{k}$. Which of the following is true?

   (a) $S_{2^n} \ge \dfrac{n}{2}$ for every $n \ge 1$ (b) S_n is a bounded sequence

   (c) $|S_{2^n} - S_{2^{n-1}}| \to 0$ as $n \to \infty$ (d) $\dfrac{1}{n} S_n \to 1$ as $n \to \infty$

2. Consider the following power series in the complex variable z: $f(z) = \sum_{n=1}^{\infty} n \log n z^n$, $g(z) = \sum_{n=1}^{\infty} \frac{e^{n^2}}{n} z^n$. If r and R are the radii of convergence of f and g respectively, then

   (a) $r = 0, R = 1$ (b) $r = 1, R = 0$ (c) $r = 1, R = \infty$ (d) $r = \infty, R = 1$

3. Let $\langle z_n\rangle$ be a sequence of distinct points in $\mathcal{S}(0,1) = \{z \in \mathbb{C} : |z| < 1\}$ with $\lim\limits_{n\to\infty} z_n = 0$. Consider the following statements P and Q:

   P: There exists a unique analytic function f on $\mathcal{S}(0,1)$ such that $f(z_n) = \sin(z_n)$ for all n.

   Q: There exists a unique analytic function f on $\mathcal{S}(0,1)$ such that $f(z_n) = 0$ if n is even and $f(z_n) = 1$ if n is odd.

   Which of the above statements hold true?

   (a) Both P and Q (b) Only P (c) Only Q (d) Neither P nor Q

4. Analytic function has
   (a) A power series expansion about each point of its domain
   (b) A power series expansion about some point of its domain
   (c) A power series expansion about no point of its domain
   (d) None of these

5. Let $f : \mathbb{C} \to \mathbb{C}$ be an arbitrary analytic function satisfying $f(0) = 0$ and $f(1) = 2$. Then:
   (a) There exists a sequence $\{z_n\}$ such that $|z_n| > n$ and $|f(z_n)| > n$
   (b) There exists a sequence $\{z_n\}$ such that $|z_n| > n$ and $|f(z_n)| < n$
   (c) There exists a bounded sequence $\{z_n\}$ such that $|f(z_n)| > n$
   (d) There exists a sequence $\{z_n\}$ such that $|z_n| \to 0$ and $|f(z_n)| \to 2$

6. The radius of convergence of the power series $\sum_{n=0}^{\infty} 4^{(-1)^n n} z^{2n}$ is
   (a) 0 2 (c) 0.5 (d) ∞

7. If the power series $\sum_{n=0}^{\infty} a_n(z+3-i)^n$ converges at $5i$ and diverges at $-3i$, then the power series
   (a) converges at $-2+5i$ and diverges at $2-3i$
   (b) converges at $2-3i$ and diverges at $-2+5i$
   (c) converges at both $2-3i$ and $-2+5i$
   (d) diverges at both $2-3i$ and $-2+5i$.

8. The radius of convergence of $\sum_{n=0}^{\infty} \frac{1}{n^3}\left(1+\frac{1}{n}\right)^{n^2} - z^n$ is
   (a) e (b) $1/e$ (c) 1 (d) ∞

9. The power series $\sum_{n=0}^{\infty} 2^{-n} z^{2n}$ converges if
   (a) $|z| < 2$ (b) $|z| \le 2$ (c) $|z| < \sqrt{2}$ (d) $|z| \ge \sqrt{2}$

10. Let $S = \{0\} \cup \left\{\frac{1}{4n+7}; n = 1, 2, \cdots\right\}$. Then, the number of analytic functions which vanish only on S is
    (a) infinite (b) 0 (c) 1 (d) 2

11. If the radius of convergence of the series $f(z) = \sum_{n=0}^{\infty} a_n(z-z_0)^n$ is finite, then $f(z)$ has at least one singularity
    (a) on the circle of convergence (b) inside the circle of convergence
    (c) outside the circle of convergence (d) anywhere

12. The radius of convergence of the series $\sum_{n=1}^{\infty} \frac{z^n}{n^a}, a \in \mathbb{R}$
    (a) is 1 (b) cannot say unless the value of a is specified
    (c) diverges of all value of a (d) diverges only for specified values of a

13. Let $f(z) = \sum_{n=0}^{\infty} a_n(z-z_0)^n$ for $|z-z_0| < R$ and this series has radius of convergence $R > 0$, then on $S(0, R)$, the function f is
    (a) differentiable (b) infinitely differentiable (c) integrable (d) constant

14. If the series $P = \sum_{n=0}^{\infty} a_n(z-z_0)^n$ has radius of convergence $R > 0$, then on $S(0, R)$ the power series P is
    (a) constant (b) analytic (c) not analytic (d) divergent

15. If $f(z) = \sum_{n=0}^{\infty} a_n(z - z_0)^n$ for $|z - z_0| < R$ and this series has radius of convergence $R > 0$, then

(a) for $n \geq 0$, $a_n = f^{(n)}(a)$ (b) $n \geq 0$, $a_n = 1/f^{(n)}(a)$

(c) for $n > 0$, $a_n = n!f^{(n)}(a)$ (d) $n \geq 0$, $a_n = \frac{1}{n!}f^{(n)}(a)$

16. If $f(z) = \sum_{n=0}^{\infty} a_n(z - z_0)^n$ for $|z - z_0| < R$ and this series has radius of convergence $R > 0$, then

(a) $f^{(k)} = \sum_{n=k}^{\infty} n(n-1)\cdots(n-k+1)a_n(z-z_0)^{n-k}$ for each $k < 1$ and $|z - z_0| = R$

(b) $f^{(k)} = \sum_{n=k}^{\infty} n(n-1)\cdots(n-k+1)a_n(z-z_0)^{n-k}$ for $|z - z_0| > R$

(c) $f^{(k)} = \sum_{n=k}^{\infty} n(n-1)\cdots(n-k+1)a_n(z-z_0)^{n-k}$ for each $k \geq 1$ and $|z - z_0| < R$

(d) None of these

17. Let $f(z) = \sum_{n=0}^{\infty} a_n(z - z_0)^n$ for $|z - z_0| < R$ and this series have radius of convergence $R > 0$. Consider the series $S = \sum\_n = k^{\infty} n(n-1)\cdots(n-k+1)a_n(z-z_0)^{n-k}$. Then

(a) For each $k \geq 1$, S has radius of convergence $R > 0$

(b) For each $k < 0$, S has radius of convergence $R < 0$

(c) For each $k \geq 1$, S has no radius of convergence

(d) For each $k < 0$, S has no radius of convergence

18. Let k be a positive integer. The radius of convergence of the series $\sum_{n=0}^{\infty} \frac{(n!)^k}{(kn)!} z^n$ is:

(a) k (b) k^k (c) k^{-k} (d) ∞

19. Let $P(x)$ be a polynomial of degree $d \geq 2$. The radius of convergence of the power series $\sum_{n=0}^{\infty} P(n)z^n$ is:

(a) 0 (b) 1 (c) ∞ (d) Dependent on d

20. Let $\sum_{n=0}^{\infty} a_n z^n$ be a convergent power series such that $\lim_{n\to\infty} \frac{a_{n+1}}{a_n} = R > 0$. Let $P(x)$ be a polynomial of degree d. Then the radius of convergence of $\sum_{n=0}^{\infty} P(n)z^n$ is:

(a) R (b) d (c) Rd (d) $R + d$

21. Consider the power series $\sum_{n\geq 1} a_n z^n$, where a_n = number of divisors of n^{50}. Then the radius of convergence of $\sum_{n\geq 1} a_n z^n$ is

(a) 1 (b) 50 (c) 1/50 (d) 0

22. The domain of convergence of the power series $\sum_{n=1}^{\infty} \frac{n^4}{(n+1)(n+2)} z^n$ is:

(a) $\{z : |z| < 1\}$ (b) $\{z : |z| \leq 1\}$ (c) $\{0\}$ (d) $\{z : |z| < e\}$

23. The radius of convergence of the series $\sum_{n=1}^{\infty} z^{n^2}$ is:

(a) 0 (b) ∞ (c) 1 (d) 2

24. The radius of convergence of the power series $\sum_{n=1}^{\infty} z^{n!}$ is:

(a) 0 (b) ∞ (c) 1 (d) a real number greater than 1

Answers

1(b); 2(c); 3(b); 4(a); 5(c); 6(c); 7(a); 8(b); 9(b); 10(b); 11(a); 12(d); 13(b); 14(b); 15(d); 16(c); 17(a); 18(b); 19(b); 20(a); 21(d); 22(b); 23(b); 24(b)

Review Questions [Short-answer Type]

1. Find the sum of the following convergent series:

   (a) $\sum_{k=0}^{\infty}\left(\frac{k}{3}\right)^k$ (b) $\sum_{k=14}^{\infty}\left(\frac{1}{2k}\right)^k$ (c) $\sum_{k=0}^{\infty}\left(\frac{1}{3}\right)^{2k}$ (d) $\sum_{k=0}^{\infty}(-1)^k\left(\frac{2}{3}\right)^k$

   (e) $\sum_{k=0}^{\infty}\frac{3}{(1+i)^k}$ (f) $\sum_{k=0}^{\infty}\left[\frac{1}{k+2}-\frac{1}{k+1}\right]$

2. Show that if $\lim_{n\to\infty} z_n = z_0$, then (i) $\lim_{n\to\infty}|z_n| = |z_0|$; (ii) $\lim_{n\to\infty}\overline{z}_n = \overline{z}_0$.

3. Let the sequence $\{z_n\}$ converge to a number z_0. Then show that there exists $M > 0$ such that $|z_n| \le M \;\forall\, n \in \mathbb{N}$.

4. Let f be a function defined in a neighbourhood of z_0. Prove that the function f is continuous at z_0 if and only if for any sequence $\{z_n\}$ converging to z_0 the condition $\lim_{n\to\infty} f(z_n) = f(z_0)$ holds.

5. Find the limiting points of the following sets:

   (i) $\sqrt[n]{n}+i^n$ (ii) $(-1)^n(1+i)\dfrac{n-1}{n}$ (iii) i^n iv) $\left(\dfrac{n-1}{n+1}\right)\cos\dfrac{n\pi}{3}+i\left(\dfrac{n+1}{n}\right)\sin\dfrac{n\pi}{6}$

6. Using the ratio test, show that the following series converge:

   (a) $\sum_{k=1}^{\infty}\frac{1}{k!}$ (b) $\sum_{k=1}^{\infty}\frac{(3+i)^k}{k!}$ (c) $\sum_{k=1}^{\infty}\frac{k^2}{4^k}$ (d) $\sum_{k=1}^{\infty}\frac{k!}{k^k}$

7. Check whether the following sequences are convergent or divergent:

   (i) $\left\{\dfrac{1}{n}\tan in\right\}$ (ii) $\left\{\dfrac{n^2-3in+2}{3n^2+4in-2}\right\}$ (iii) $\left\{\left(\dfrac{1-i\alpha}{1+i\alpha}\right)^n\right\},\ \alpha\neq i$

   (iv) $\left\{\dfrac{1}{n!}(1+i)^n\right\}$ (v) $\left\{\left(\dfrac{1+i}{\sqrt{3}+i}\right)^n\right\}$ (vi) $\left\{\sqrt[n]{n}+\dfrac{n}{2^n}i\right\}$

   (vii) $\left\{\arg(1+in^2)\right\}$ (viii) $\left\{\dfrac{1}{2^n}+\dfrac{n}{n(n+1)}i\right\}$

   If convergent, find the limit of the complex sequence.

8. Using the ratio test, find a domain in which convergence holds for each of the following series of functions:

   (a) $\sum_{k=1}^{\infty}kz^k$ (b) $\sum_{k=0}^{\infty}\frac{(z-i)^k}{2^k}$ (c) $\sum_{k=0}^{\infty}\frac{z^k}{k!}$ (d) $\sum_{k=0}^{\infty}(z+5i)^{2k}(k+1)^2$

9. Determine the limiting points of the sequence $\{z_n\}$ of complex numbers, where

$$z_n = \left(1+\frac{1}{n}\right)\left(\cos\frac{n\pi}{\alpha}+i\sin\frac{n\pi}{\alpha}\right),$$

   in which α is (i) real and rational (ii) real and irrational.

10. If S is compact and $f : S \to \mathbb{C}$ is continuous on S, prove that it is uniformly continuous.

11. Test the convergence (C) or divergence (D) of the series $\sum_{n=1}^{\infty}a_n$, where $a_n =$

   (i) $\dfrac{3^n}{n+1}$ (ii) $\dfrac{n^2+i}{n^3}$ (iii) $\dfrac{n-2i}{3n^3}$ (iv) $\dfrac{3n+2i+5}{i^n}$

   (v) $\left(\dfrac{n}{n+1}\right)^{4n^2} i^n$ (vi) $\dfrac{\text{Log}n}{n!}\,i^n$ (vii) $\dfrac{1}{2^n+1}$ (viii) $\dfrac{n}{3^n-1}$

   (ix) $\dfrac{n+3}{3n^2-n+2}$ (x) $\dfrac{(-1)^n}{4n+3}$ (xi) $\dfrac{2n-1}{\sqrt{n^3+n+2}}$ (xii) $\dfrac{ne^{n\pi i/4}}{e^n-1}$

12. Prove that the complex power series $\sum_{k=1}^{\infty}\left(k+ik^2\right)^{-1}$ converges absolutely.
13. Check for the pointwise (P) and uniform convergence (U) of the series $\sum_{n=0}^{\infty} u_n(z)$ on the given domain, where $u_n(z)$ is

(i) $\dfrac{z^n}{(n+1)(n+2)};|z|\le 1$ (ii) $\dfrac{1}{(z+n+1)^2};\mathrm{Re}\,z>0$ (iii) $\dfrac{z^n}{n^2};|z|<1$

(iv) $\dfrac{1}{(1+n^2+z^n)};1<|z|<2$ (v) $\dfrac{1}{(1+n^2z^2)};|z|\le 1$ (vi) $\dfrac{z^n}{n};|z|<1$

14. Does the sequence of functions

$$f_n(z)=\begin{cases} n|z| & \text{if } |z|\le \frac{1}{n}\\ 1 & \text{if } \frac{1}{n}\le |z|\le 1\end{cases}$$

converge uniformly on the unit disc $\mathcal{S}[0,1]$?

15. Show that the sequence $\{f_n(z)\}$ of complex functions, where
    (a) $f_n(z)=1/nz$ is not uniformly convergent in the region $0<|z|<1$ but is uniformly convergent in the region $\varepsilon_0<|z|<1, \varepsilon_0>0$.
    (b) $f_n(z)=1/(1+nz)$ is uniformly convergent to zero for all z such that $|z|\ge 2$.
16. Show that the power series $\sum_{n=0}^{\infty} a_n z^n$ and $\sum_{n=k}^{\infty} a_n z^{n-k}$ both have the same radius of convergence.
17. Let the radius of convergence of $\sum_{n=0}^{\infty} a_n z^n$ be R. Then show that the power series $\sum_{n=0}^{\infty} a_n R^n z^n$ has unit radius of convergence.
18. Find the radius of convergence of each of the following power series:

(a) $\sum_{n=0}^{\infty} nz^n$ (b) $\sum_{n=0}^{\infty}\frac{z^n}{3^n}$ (c) $\sum_{n=0}^{\infty}\frac{z^n}{1+2^n}$ (d) $\sum_{n=0}^{\infty}\frac{n!}{n^n}z^n$

(e) $\sum_{n=0}^{\infty} n^p z^n$ (f) $\sum_{n=0}^{\infty}\frac{z^n}{n^n}$ (g) $\sum_{n=0}^{\infty} q^{n^2}z^n; |q|<1$ (h) $\sum_{n=0}^{\infty}\left(\frac{z}{2+(-1)^n}\right)^n$

(i) $\sum_{n=0}^{\infty}\frac{2^n z^n}{3^n+4^n}$ (j) $\sum_{k=0}^{\infty}\left[2+(-1)^k\right]^k z^k$ (k) $\sum_{n=1}^{\infty}\frac{n!}{n^n}z^n$ (l) $\sum_{n=0}^{\infty} z^{n!}$ (m) $\sum_{n=0}^{\infty} 2^n z^{n^2}$

(n) $\sum_{n=1}^{\infty}\frac{2}{3n}z^{2n}$

19. If the radius of convergence for the series $\sum_{n=0}^{\infty} a_n z^n$ is R, find the radius of convergence of the following:

(a) $\sum_{n=0}^{\infty} n^3 a_k z^k$ (b) $\sum_{n=0}^{\infty} a_n^4 z^k$ (c) $\sum_{n=0}^{\infty} a_n z^{2n}$ (d) $\sum_{n=0}^{\infty} a_n z^{n+7}$ (e) $\sum_{n=1}^{\infty} n^{-n} a_n z^n$

20. Determine the region of convergence of the following power series:

(a) $\sum_{n=0}^{\infty}\frac{(-1)^n}{z+n}$ (b) $\sum_{n=0}^{\infty}\frac{z^n}{1-z^n}$ (c) $\sum_{n=0}^{\infty}\frac{z^n}{1+z^{2n}}$

21. Show that the sequence of functions $f_n(z)=\dfrac{z^n}{z^n-3^n}; n\in\mathbb{N}$ converges to zero for $|z|<3$ and to 1 for $|z|>3$

Answers

5. (i) $1\pm i, 0, 2$; (ii) $\pm(1+i)$; (iii) None; (iv) $1, -1\pm i, \frac{1}{2}+\frac{1}{2}i, -\frac{1}{2}\pm\frac{\sqrt{3}}{2}i$
11. (i) D; (ii) D; (iii) C; (iv) D; (v) C; (vi) C; (vii) C; (viii) C; (ix) C; (x) D; (xi) D; (xii) C
13. (i) U; (ii) NU (Not uniform); (iii) U; (iv) U; (v) P; (vi) NU

Review Questions [Long-answer Type]

1. Prove that a series converges uniformly if and only is the series of real parts and the series of imaginary parts of the terms occuring in the original series converge uniformly.
2. Let $f : \mathcal{D} \to \mathbb{C}$, where $\mathcal{D} \subset \mathbb{C}$, let $z_0 \in D$, and let $\{z_n\}$ be a sequence in D such that $z_n \to z_0$ as $n \to \infty$. Prove that f is continuous at z_0, then $f(z_n) \to f(z_0)$ as $n \to \infty$.
3. Show that the series $\sum_{n=1}^{\infty} \frac{1}{n+i}$ diverges but $\sum_{n=1}^{\infty} \frac{i^n}{n^2+1}$ converges. Also, show that the series $\sum_{n=1}^{\infty} \frac{(-1)^n}{n}(3+4i)$ is conditionally convergent.
4. Verify whether the two series $\sum_{n=1}^{\infty} \frac{1}{2^n}\left(\frac{1+i}{1-i}\right)^n$ and $\sum_{n=1}^{\infty} \frac{i^n}{n(n+1)}$ are convergent or not? Find also the range for which the series $\sum_{n=0}^{\infty} \left(\frac{z-i}{z+i}\right)^n$ is convergent.
5. Prove that the limit function of a uniformly convergent sequence of complex function is itself continuous.
6. Prove that the series of complex numbers (a) $1 + \left(\frac{i}{3}\right) + \left(\frac{i}{3}\right) + \cdots$ converges and converge to $\frac{1}{10}(9+3i)$; (b) $\sum_{n=1}^{\infty} \frac{\omega^n}{5^{n/2}}$; $\omega = \sqrt{3} + i$ is convergent.
7. Show that the following series are convergent:

   (a) $\sum_{k=1}^{\infty} \frac{1}{k(k+1)}$ (b) $\sum_{k=1}^{\infty} \frac{\sin k^2}{k^{3/2}}$ (c) $\sum_{k=1}^{\infty} \frac{k^2 i^k}{k^4+1}$ (d) $\sum_{k=2}^{\infty} (-1)^k \left(\frac{5k+8}{k^3-1}\right)$
8. Find the circle of convergence of each of the following power series:

   (a) $\sum_{k=0}^{\infty} k^3 z^k$ (b) $\sum_{k=0}^{\infty} 2^k (z-1)^k$ (c) $\sum_{k=0}^{\infty} k! z^k$ (d) $\sum_{k=0}^{\infty} \frac{z^{2k}}{4^k}$

   (e) $\sum_{k=0}^{\infty} \frac{(-1)^k k}{3^k}(z-i)^k$ (f) $\sum_{k=1}^{\infty} \frac{(3-i)^k}{k^2}(z+2)^k$
9. Find the region of convergence of $\sum a_n(z)$, where $a_n(z) =$

   (i) $\dfrac{(-1)^n}{n!} z^n$ (ii) $\dfrac{1}{n^2.3^n}\left(\dfrac{z+1}{z-1}\right)^n$ (iii) $\dfrac{(z+i)^n}{(n+1)(n+2)}$

   (iv) $\dfrac{(iz-1)^n}{2+i}$ (v) $\dfrac{e^{2\pi i n z}}{(n+1)^{\frac{3}{2}}}$ (vi) $\dfrac{(-1)^n.n}{4^n(n^2+1)^{\frac{5}{2}}}(z-i)^n$

10. For each of the following, determine if the given series converges or diverges:

    (a) $\sum_{k=0}^{\infty} \left(\frac{1+2i}{1-i}\right)^k$ (b) $\sum_{k=1}^{\infty} \left(i^k - \frac{1}{k^2}\right)$ (c) $\sum_{k=1}^{\infty} \frac{k!}{5^k}$

    (d) $\sum_{k=1}^{\infty} \frac{1}{k^2 3^k}$ (e) $\sum_{k=1}^{\infty} \frac{k i^k}{2k+1}$ (f) $\sum_{k=1}^{\infty} \frac{(-1)^k k^3}{(1+i)^k}$
11. Find the radius of convergence for each of the following power series (where n varies from 1 to ∞):

    (i) $\sum n^n z^n$ (ii) $\sum \frac{2+in}{2^n} z^n$ (iii) $\sum n\left(n^{\text{Log} n}\right) z^{n-1}$

    (iv) $\sum \frac{(n+1)z^n}{(n+2)(n+3)}$ (v) $\sum \frac{z^n}{n^3}$ (*vi*) $\sum \frac{(z-1)^n}{2^n}$

    (vii) $\sum 3^n (z+1)^n$ (viii) $\sum \left(\frac{2-n}{1+in^2}\right) z^n$ (ix) $\sum n!\, z^n$

    (x) $\sum n^{\text{Log} n} z^n$ (xi) $\sum \frac{(-1)^n}{n}(z+2i)^n$ (*xii*) $\sum \frac{(-1)^n}{n}(z-2i)^n$

    (xiii) $\sum_{n=2}^{\infty} \frac{z^n}{\log n}$ (xiv) $\sum \frac{n(n!)}{z^{1-n}}$ (xv) $\sum \left(1+\frac{1}{n}\right)^{n^2} z^n$

(xvi) $\sum \left(1-\frac{1}{n}\right)^{n^2} z^n$ (xvii) $\sum \frac{z^n}{n}$ (xviii) $\sum \frac{nz^{n+1}}{n+1}$

(xviv) $\sum_{n=1}^{\infty} (\log n)^n z^n$ (xx) $\sum \left(\frac{n\sqrt{2+i}}{1+2in}\right) z^n$ (xxi) $\sum nz^{n-1}$

(xxii) $\sum \frac{z^n}{1+in^a}$ (xxiii) $\sum \left(\frac{n}{n+1}\right)^{n^2} z^n$ (xxiv) $\sum \left(\frac{n\sqrt{2+i}}{1+2in}\right) z^n$

12. Show that the power series

 (a) $1+\frac{a\cdot b}{1\cdot c}z+\frac{a(a+1)b(b+1)}{1\cdot 2\cdot c(c+1)}z^2+\cdots$ has unit radius of convergence.

 (b) $\frac{z}{2}+\frac{1\cdot 3}{2\cdot 5}z^2+\frac{1\cdot 3\cdot 5}{2\cdot 5\cdot 8}z^3+\cdots$ has radius of convergence $3/2$.

 (c) $1+z+\frac{z^2}{2^2}+\frac{z^3}{3!}+\frac{z^4}{2^4}+\frac{z^5}{5!}+\frac{z^6}{2^6}+\cdots$ has radius of convergence $\sqrt{2}$.

 (d) $\sum a_n z^n$ such that $a_0=0, a_1=1$ and $a_n=a_{n-1}+a_{n-2}, n\geq 2$, has the radius of convergence $1+2/\sqrt{5}$.

 (e) $\sum_{n=0}^{\infty} \frac{(n!)^2}{(2n)!} z^n$ does not converge for any point on the circle of convergence.

 (f) $\sum_{n=1}^{\infty} (1)^{n-1} \frac{z^{2n-1}}{(2n-1)!}$ is converge for every finite value of z.

13. Use power series $f(z)=\sum_{n=1}^{\infty} a_n z^n$ to solve the functional equation $f(z)=z+f(z^2)$.

14. Let the radius of convergence of $\sum_{n=0}^{\infty} a_n z^n$ be $R(0<R<\infty)$. Show that the radius of convergence of

 (i) $\sum_{n=0}^{\infty} n^n a_n z^n$ is 0 (ii) $\sum_{n=0}^{\infty} \frac{a_n}{n!} z^n$is ∞ (iii) $\sum_{n=0}^{\infty} n^k a_n z^n; k=0,1,\cdots$

 is R.

15. For the power series $f(z)=\sum_{n=0}^{\infty} a_n z^n$, let the radius of convergence be R.

 (a) Show that $\overline{f(\bar{z})}$ is a power series in z with the same radius of convergence R.

 (b) Suppose $R>1$ and $|f(e^{i\theta})|\leq 4$ for $0\leq\theta\leq\pi$ and $|f(e^{i\theta})|\leq 9$ for $\pi\leq\theta\leq 2\pi$. Show that $|f(0)|\leq 6$.

16. Examine the behaviour of the following power series on the circle of convergence (CR).

 (a) $\sum_{n=1}^{\infty} (-1)^n \frac{z^n}{n}$ (b) $\sum_{n=2}^{\infty} \frac{z^n}{n(\log n)^2}$ (c) $\sum_{n=0}^{\infty} \frac{z^{4n}}{1+4n}$

 (d) $\sum_{n=0}^{\infty} (-1)^n \frac{z^p}{n\ \log n}$ (e) $\sum_{n=2}^{\infty} \frac{z^{2n}}{n-\sqrt{n}}$

17. Determine the value(s) of the constant k that makes the radius of convergence of the power series $\sum_{n=2}^{\infty} \frac{k^n}{\ln n} z^n$ equal to 5.

18. Determine the domain of convergence of the following power series:

 (i) $\sum_{n=0}^{\infty} \left(\frac{iz-1}{2+i}\right)^n$ (ii) $\sum_{n=0}^{\infty} n!\ z^n$ (iii) $\sum_{n=0}^{\infty} \left(\frac{2i}{z+i+1}\right)^n$

19. (a) **Abel's theorem:** Let $\sum a_n (z-z_0)^n$ have radius of convergence 1 and suppose that $\sum a_n$ converges to A. Prove that $\lim\limits_{r\to 1_-} \sum a_n r^n = A$.

 (b) Use Abel's theorem to prove that $\log 2 = 1-\dfrac{1}{2}+\dfrac{1}{3}-\cdots$

20. Find the series expansion of $\frac{1}{z}(e^z-1)$ about 0 and determine its radius of convergence.

21. Determine regions in which each of the following series $\sum u_n(z)$ is uniformly convergent:

$$\text{(i)}\quad u_n(z) = \frac{z^n}{3^n+1}; |z| \leq R < 3 \qquad \text{(ii)}\quad u_n(z) = \frac{1}{n^2}(z-i)^{2n}; |z-i| \leq 1$$
$$\text{(iii)}\quad u_n(z) = \frac{1}{z^n(n+1)}; |z| \geq R > 1 \quad \text{(iv)}\quad u_n(z) = \frac{\sqrt{n+1}}{n^2+|z|^2}; \forall z \in \mathbb{C}$$

22. Determine whether the sequence $\{u_n(z)\}$ of functions given by

(a) $u_n(z) = 3z + \frac{4z^2}{n}$ converges uniformly to $3z$ for all z in $\mathcal{S}[0, 1]$.
(b) $u_n(z) = \frac{nz}{n^2+z^2}$ converges uniformly to zero for all z inside $\mathcal{S}(0, 3)$.

23. Show that the series $\sum_{n=1}^{\infty} n^{-z}$ converges absolutely and uniformly in any bounded closed region in which Re $(z) > 1$.
24. Use power series to solve the fundamental equation $f(z)z + f(z^2)$ for the analytic function f.
25. Find the limit points of the $\{z_n\}$, where z_n is given by
(i) $z_n = \dfrac{i^n}{n}; n = 1, 2, \ldots$ (ii) $z_n = (-1)^n(1+i)\dfrac{n-1}{n}; n = 1, 2, \ldots$
26. Prove that if R_1 and R_2 are the radii of convergence of the series $\sum a_n z^n$ and $\sum b_n z^n$, then the series $\sum a_n b_n z^n$ has a radius of convergence $R \geq R_1 R_2$.
27. Let $f(z) = \sum_{k=1}^{\infty} a_k z^k = 1 + z + 2z^2 + 3z^3 + 5z^4 + 8z^5 + 13z^6 + 21z^7 + \cdots$, where the coefficients a_k are the Fibonacci numbers defined by $a_0 = 1, a_1 = 1$, $a_k = a_{k-1} + a_{k-2}; k \geq 2$. Show that $f(z) = \frac{1}{1-z-z^2}$; $z \in \mathcal{S}(0, R)$, where $R = \frac{\sqrt{5}-1}{2}$.
28. Let $f(z) = \sum_{k=1}^{\infty} a_k z^k$, where the coefficients a_k are the Lucas numbers defined by $a_0 = 1, a_1 = 3$, $a_k = a_{k-1} + a_{k-2}; k \geq 2$. Show that $f(z) = \frac{2z+1}{1-z-z^2}$; $z \in \mathcal{S}(0, R)$, where $R = \frac{\sqrt{5}-1}{2}$.
29. Let the power series $f(z) = \sum_{n=0}^{\infty} a_n z^n$ have radius of convergence $R > 0$. Substitute $z = z_0 + (z - z_0)$ and show that the radius of convergence of the resulting power series in $z - z_0$ is not less than $R - |z_0|$.
30. Prove that set E is a discrete subset of an open set U if and only if no sequence of distinct points of E converges to a point in U.
31. Let G be a region and suppose $f_n : G \to \mathbb{C}$ is analytic for each $n \geq 1$. Suppose that $\langle f_n \rangle$ converges uniformly to a function $f : G \to \mathbb{C}$. Show that f is analytic.
32. Show that if $f : \mathbb{C} \to \mathbb{C}$ is a continuous function such that f is analytic in $[-1, 1]$, then f is an entire function.

Answers

9. (i) $|z| < \infty$; (ii) $\left|\frac{z+1}{z-1}\right| \leq 3$; (iii) $|z+i| \leq 1$; (iv) $|z+1| < \sqrt{5}$; (v) Im $(z) \geq 0$; (vi) $|z-i| \leq 4$
11. (i) 0; (ii) 2; (iii) 1; (iv) 1; (v) 1; (vi) 2; (vii) 1/3; (viii) 4; (ix) 0; (x) 1; (xi) 1; (xii) 1; (xiii) 1; (xiv) 0; (xv) e^{-1}; (xvi) e; (xvii) 1; (xviii) 1; (xviv) 0; (xx) 1; (xxi) 1; (xxii) 1; (xxiii) e; (xxiv) 1
16. (a) Convergent $\forall z (\neq -1)$ on the CR; (b) Absolutely convergent $\forall z$ on the CR; (c) Convergent $\forall z (\neq \pm 1, \pm i)$ on the CR; (d) Convergent $\forall z (\neq -1)$ on the CR; (e) Convergent $\forall z (\neq \pm 1)$ on the CR
17. $\pm 1/5$ 18. (i) $|z+i| < \sqrt{5}$; (ii) Only at $z = 0$; (iii) $|z+i+1| > 2$

3 Complex Differentiation

In this chapter we study in detail the concepts of limit and continuity for functions of a complex variable. We also introduce the notion of differentiability for functions of a complex variable and see how the complex differentiation of one complex variable sometimes behaves like the derivative of a real function of one variable and at other times is comparable to the partial derivatives of a real function of two variables. In the process, we are led to the notion of analytic functions which play a very important role in the study of complex analysis. In the last part of the chapter we will discuss harmonic functions and their relationship with analytic functions.

3.1 Functions of Complex Variables

Let $\mathcal{D} \subset \mathbb{C}$ be a domain.

Definition 3.1.1. A *function* $f : \mathcal{D} \to \mathbb{C}$ is defined as a rule which assigns each complex number $z = x + iy \in \mathcal{D}$ to a complex number $w \in \mathbb{C}$ (Fig. 3.1).

The number w is a called the *value of the function* f at z and we write $w = f(z)$. The variable z is sometimes called an *independent variable*, while w is called a *dependant variable* and f the complex function of a complex variable in $\mathcal{D}$. The set $\mathcal{D}$, a region of the complex plane, is called the *domain of the definition of* f. The function is said to be defined on $\mathcal{D}$.

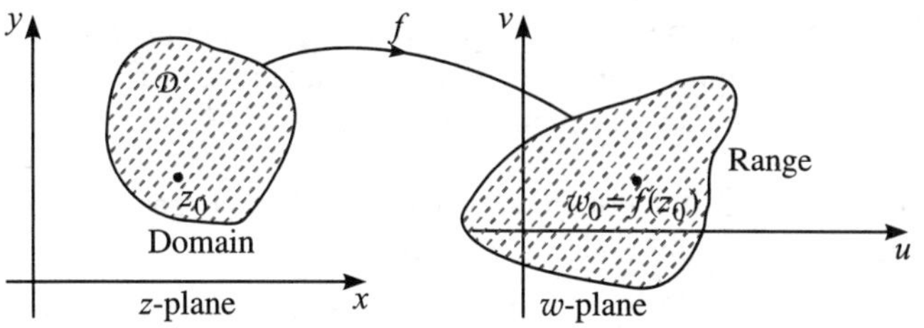

Figure 3.1 *Geometrical representation of* $w = f(z)$

Let $w = u + iv$ be the value of f at $z = x + iy$. Then

$$w = f(z) = f(x + iy) = u(x, y) + iv(x, y). \tag{3.1}$$

$u(x, y)$ is called the *real part* and $v(x, y)$ is called the *imaginary part* of the function $f(z)$ and are respectively denoted by $\text{Re}\, f(z)$ and $\text{Im}\, f(z)$, i.e.,

$$\text{Re}\, f(z) = u(x, y) \text{ and } \text{Im}\, f(z) = v(x, y). \tag{3.2}$$

To have a geometric representation of $w = f(z)$, (Fig. 3.1) it is convenient to draw separate complex planes for the variables z and w so that corresponding to each point $z = x + iy$ of the z-plane there is a point $w = u + iv$ in the w-plane.

(i) If $v(x, y) = 0$, then the function f is a real-valued function of the complex variable.

(ii) The value of the function $w = f(z)$ at $z_0 = x_0 + iy_0$ is often written as

$$f(z_0) = f(x_0 + iy_0) = u(x_0, y_0) + iv(x_0, y_0). \tag{3.3}$$

(iii) In polar coordinates, $z = re^{i\theta}$ So $w = f(z)$ can also be expressed as

$$f(z) = f(re^{i\theta}) = u(r, \theta) + iv(r, \theta). \tag{3.4}$$

Example 3.1.1. *Write $f(x) = 1/z$ in the form of $u + iv$. Find also the domain of definition of the function $f(z) = 1/z$.*

Solution. Since $z = x + iy$,

$$w = u + iv = \frac{1}{z} = \frac{1}{x + iy} = \frac{x - iy}{x^2 + y^2} = \frac{x}{x^2 + y^2} - i\frac{y}{x^2 + y^2}$$

$$u(x, y) = \frac{x}{x^2 + y^2} \text{ and } v(x, y) = -\frac{y}{x^2 + y^2} \quad \forall z \neq 0.$$

Since $f(z)$ is defined for all $z \in \mathbb{C}$, $z \neq 0$, the domain of definition of $f(z)$ is the whole complex plane excluding the point $z = 0$, i.e., the domain of definition is $\mathcal{D} = \mathbb{C} - \{0\}$.

Definition 3.1.2. [Functions defined on open sets]: We take open sets because when we study the properties of a function, epecially the limit of the function at a point, then it is possible to approach that point from all directions and so there must be a disc surrounding that point where the function is defined. So the function at a point in its neighbourhoods is defined.

Definition 3.1.3. [Functions defined in connected sets]: If a set is not connected, then it falls into two pieces and essentially a functions on such a set is a different function in each piece. So, we can reduce the study of function to just a study of the function in a single piece. That is why we always study a function defined on open connected sets which are domains.

Definition 3.1.4. [Single-valued functions]: A function f that assigns only one value of w to each value of z in the domain of the definition is called a *single-valued function.*

For example, $f(z) = z$ is single valued.

Definition 3.1.5. [Open mapping]: A function defined on an open set $\mathcal{D}$ is said to be an *open mapping* if for every open subset U of $\mathcal{D}$, the image $f(U)$ is open.

3.1.1 Univalent and Inverse Functions

Let A and B be two non-empty subsets of $\mathbb{C}$. If the range R of f is contained in B, then the function f on A to B is called a *mapping of A into B* and if $R = B$, then the function is called a *mapping of A onto R* since every element of $w \in R$ is an image of at least one point in A.

Definition 3.1.6. If f is a function defined on a set A, then the mapping $w = f(z)$ is called *one-to-one* if

$$z_1 \neq z_2 \Rightarrow f(z_1) \neq f(z_2) \ \forall z_1, z_2 \in A. \tag{3.5}$$

If the mapping $w = f(z)$ is one-to-one, then the function f is called *univalent*. A function f is univalent at a point z_0 if it is univalent in the neighbourhood of z_0.

Suppose f maps A in a one-to-one fashion onto B. Then there exists an inverse mapping on B onto A. This mapping is called *inverse function* of f, which is denoted by f^{-1} and written as $z = f^{-1}(w)$ if $w = f(z)$. Clearly, if $f : A \to B$ is univalent on A, then f^{-1} is defined on B and is univalent therein.

Example 3.1.2. *Show that $f(z) = 2z + z^2$ is univalent in the domain $|z| < 1$ but not univalent in the whole complex plane.*

Solution. Let z_1 and z_2 be two points in the domain $|z| < 1$. Then for $|z| < 1$,

$$\text{Re}[z_1 + z_2 + 2] = \text{Re}(z_1) + \text{Re}(z_2) + 2 > -1 - 1 + 2 = 0.$$

Thus $z_1 + z_2 + 2 \neq 0$ in $|z| < 1$. Then

$$\begin{aligned} f(z_1) = f(z_2) \quad &\Rightarrow \quad (z_1 - z_2)(z_1 + z_2 + 2) = 0 \\ &\Rightarrow \quad z_1 - z_2 = 0 \Rightarrow z_1 = z_2 \\ &\Rightarrow \quad f \text{ is one-to-one.} \end{aligned}$$

Thus f is univalent in the domain $|z| < 1$. If we choose $z_1 = 1$ and $z_2 = -3$, then $z_1 + z_2 + 2 = 0$.

Thus $f(z_1) = f(z_2) \Rightarrow z_1 \neq z_2$.

Thus, f is not univalent in the whole complex plane.

3.1.2 Conjugation and Composition of Functions

For $f(z) = u(x, y) + iv(x, y)$, where $u(x, y)$ and $v(x, y)$ are real-valued functions. We define

$$\overline{f(z)} = u(x, y) - iv(x, y) \text{ and } f(\bar{z}) = u(x, -y) + iv(x, -y). \tag{3.6}$$

Observe that $\overline{f(z)}$ and $f(\bar{z})$ are different functions.

Definition 3.1.7. Let a function f be defined on domain $\mathcal{D}_1$ and another function g be defined on the domain $\mathcal{D}_2$. Also, let, for every $z \in \mathcal{D}_1$, $f(z) \in \mathcal{D}_2$. Then, for every $z \in \mathcal{D}_1$, the *composition* f with g or the *superposition* of g on f is the association $g \circ f$ defined by

$$(g \circ f)(z) = g(f(z)). \tag{3.7}$$

The function $g(f(z))$ is also called a *function of a function* or *composite function.*

3.2 Limits

Let $w = f(z)$ be any single-valued function of z defined in a bounded a closed, deleted neighbourhood of $z = z_0$.

Definition 3.2.1. Let $\mathcal{D} \subset \mathbb{C}$ be a domain, $z_0 \in \mathcal{D}$, $l \in \mathbb{C}$ and $f : \mathcal{D} \to \mathbb{C}$. We say that $f(z) \to l$ as $z \to z_0$ if given $\varepsilon > 0$, $\exists\, \delta = \delta(\varepsilon) > 0$ such that

$$\left|f(z) - l\right| < \varepsilon \text{ whenever } 0 < |z - z_0| < \delta. \tag{3.8}$$

Symbolically, we write $\lim\limits_{z \to z_0} f(z) = l$.

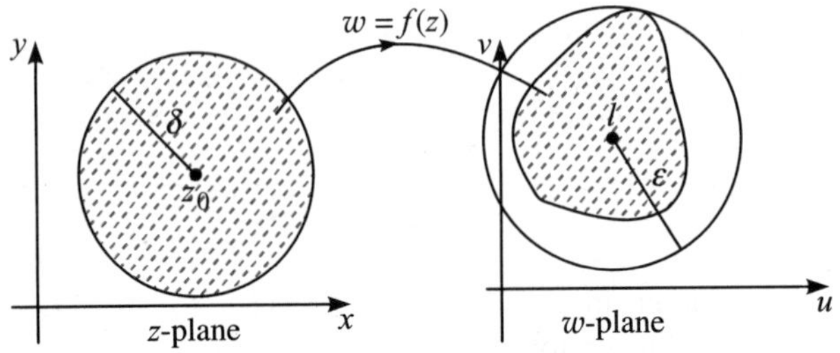

Figure 3.2 *Geometrical representation of Eq. (3.8)*

Geometrically (Fig. 3.2), the definition (3.8) states that given any open disc $\mathcal{S}(l, \varepsilon)$, there exists an open disc $\mathcal{S}(z_0, \delta)$ such that for every point $z(\neq z_0)$ in the disc $|z - z_0| < \delta$ the image $w = f(z)$ lies on the disc $|f(z) - l| < \varepsilon$. Practically, if we get two limits as $z \to z_0$ along two different paths, then we say that the limit does not exist.

Example 3.2.1. *Using the definition of limit, prove that* $\lim\limits_{z \to i} f(z) = -1$, *where*

$$f(z) = \begin{cases} z^2 & \text{if } z \neq i \\ 0 & \text{if } z = i. \end{cases}$$

Solution. We observe that, the function f is defined for all $z \in \mathbb{C}$. We must show that given any $\varepsilon > 0$, we can find $\delta = \delta(\varepsilon)$ such that

$$|z^2 + 1| < \varepsilon \text{ whenever } 0 < |z - i| < \delta.$$

Note that if we can find $\delta > 0$ satisfying the requirements of the definition, then we can choose another $\delta \leq 1$ satisfying the requirements of the definition. Now $0 < |z - i| < 1$ gives

$$\begin{aligned} & |z + i| = |z - i + 2i| \leq |z - i| + |2i| < 1 + 2 = 3 \\ \Rightarrow \quad & |z^2 + 1| = |(z + i)(z - i)| < 3|z - i|. \end{aligned}$$

Hence, if we choose $\delta = \min\{1, \varepsilon/3\}$, we get

$$|z^2 + 1| < 3|z - i| < \varepsilon \text{ whenever } 0 < |z - i| < \delta.$$

Hence by definition, $\lim\limits_{z \to i} f(z) = -1$.

Theorem 3.1. *When the limit of a function $f(z)$ exists as $z \to z_0$, then the limit has unique value.*

Proof: Suppose that $\lim\limits_{z \to z_0} f(z)$ has two values, l_1 and l_2. Then corresponding to an $\varepsilon > 0$, $\exists$ positive numbers δ_1, δ_2 (usually depending on ε) such that

$$\begin{aligned} |f(z) - l_1| &< \varepsilon/2 \text{ whenever } 0 < |z - z_0| < \delta_1 \\ |f(z) - l_2| &< \varepsilon/2 \text{ whenever } 0 < |z - z_0| < \delta_2. \end{aligned}$$

Let us choose $\delta = \min\{\delta_1, \delta_2\}$. Then, if $0 < |z - z_0| < \delta$, we have

$$\begin{aligned} |l_1 - l_2| &= |l_1 - f(z) + f(z) - l_2| \\ &\leq |l_1 - f(z)| + |f(z) - l_2| = |f(z) - l_1| + |f(z) - l_2| \\ &< \frac{\varepsilon}{2} + \frac{\varepsilon}{2} = \varepsilon. \end{aligned}$$

Since $\varepsilon > 0$ is arbitrary, $|l_1 - l_2| = 0$ so that $l_1 = l_2$. Thus, when the limit of a function exists it is unique. If $f(z)$ is multiple-valued, the limit as $z \to z_0$ may depend on the particular branch. □

Example 3.2.2. *Prove that* $\{f(z) - f(0)\}/z \to 0$ *as* $z \to 0$ *along any radius vector, but not as* $z \to 0$ *in any manner, if*

$$f(z) = \begin{cases} \dfrac{x^3 y(y - ix)}{x^6 + y^2} & z \neq 0 \\ 0 & z = 0. \end{cases}$$

Solution. Let $z \to 0$ along the radius vector $y = mx$. Then

$$\begin{aligned} \lim_{z\to 0} \frac{f(z) - f(0)}{z} &= \lim_{z\to 0} \left[\frac{x^3 y(y - ix)}{x^6 + y^2} \cdot \frac{1}{z}\right] \\ &= \lim_{z\to 0} \left[\frac{x^3 y(y - ix)}{(x^6 + y^2)(x + iy)}\right] = \lim_{x\to 0} \left[\frac{x^3 mx(mx - ix)}{(x^6 + m^2x^2)(x + imx)}\right] \\ &= \frac{m(m - i)}{1 + im} \lim_{x\to 0} \frac{x^2}{(m^2 + x^4)} = \frac{m(m - i)}{1 + im} \cdot \frac{0}{m^2} = 0, \end{aligned}$$

which is independent of m. Therefore, $\lim_{z\to 0}\{f(z) - f(0)\}/z$ exists and is equal to 0 along the radius vector $y = mx$. Now, let us consider the path $y = mx^3$ along which $z \to 0$. Then,

$$\begin{aligned} \lim_{z\to 0} \frac{f(z) - f(0)}{z} &= \lim_{z\to 0} \left[\frac{x^3 y(y - ix)}{x^6 + y^2} \cdot \frac{1}{z}\right] \\ &= \lim_{z\to 0} \left[\frac{x^3 y(y - ix)}{(x^6 + y^2)(x + iy)}\right] = \lim_{x\to 0} \left[\frac{x^3 mx^3(mx^3 - ix)}{(x^6 + m^2x^6)(x + imx^3)}\right] \\ &= \lim_{x\to 0} \frac{mx^7(mx^2 - i)}{x^7(1 + m^2)} = \lim_{x\to 0} \frac{m(mx^2 - i)}{1 + m^2} = -\frac{im}{1 + m^2}, \end{aligned}$$

which depends on the values of m. Consequently, $\lim_{z\to 0}\{f(z) - f(0)\}/z$ does not exist.

3.2.1 Limit Involving Point of Infinity

The extended complex plane $\mathbb{C}_\infty$ helps in defining the limits involving point at infinity. By means of the transformation $w = 1/z$, the point $z = 0$ is mapped into $w = \infty$, called the *point of infinity* in the w-plane. Similarly, we denote by $z = \infty$, the *point of infinity* in the z-plane. To consider the behaviour of $f(z)$ at $z = \infty$, we examine the behaviour of $f(1/w)$ at $w = 0$.

Definition 3.2.2. [Limit at infinity]: Let f be a function defined on an unbounded set $\mathcal{D}$ in the complex plane. Then the function $f(z)$ has a limit l as $z \to \infty$ if for a given $\varepsilon > 0$, $\exists\, M = M(\varepsilon) > 0$ such that

$$|f(z) - l| < \varepsilon \text{ whenever } z \in \mathcal{D} \text{ and } |z| > M. \tag{3.9}$$

Symbolically, $\lim_{z\to\infty} f(z) = l$ or $\lim_{|z|\to\infty} f(z) = l$.

Definition 3.2.3. [Infinite limit]: Let f be defined in some neighbourhood of a point z_0. Then the function $f(z)$ has an *infinite limit* as $z \to z_0$ if for a given $L > 0$, we can find δ such that

$$|f(z) - w_0| > L \text{ whenever } 0 < |z - z_0| < \delta. \tag{3.10}$$

Symbolically, $\lim_{z \to z_0} f(z) = \infty$ or $\lim_{z \to z_0} |f(z)| = \infty$.

Definition 3.2.4. Let f be defined on an unbounded set $\mathcal{D}$ in $\mathbb{C}$. Then f has an *infinite limit* as $z \to \infty$, if for a given $L > 0$, $\exists\, M > 0$ such that

$$|f(z)| > L \text{ whenever } z \in \mathcal{D} \text{ and } |z| > M. \tag{3.11}$$

Symbolically, $\lim_{z \to \infty} f(z) = \infty$ or $\lim_{|z| \to \infty} |f(z)| = \infty$.

Example 3.2.3. *Evaluate* $\lim_{z \to \infty} \dfrac{iz^3 + iz - 1}{(2z + 3)(z - i)^2}$.

Solution. Let $w = 1/z$, so that as $z \to \infty, w \to 0$. Therefore,

$$\lim_{z \to \infty} \frac{iz^3 + iz - 1}{(2z + 3)(z - i)^2} = \lim_{w \to 0} \frac{(i/w^3) + (i/w) - 1}{(2/w + 3)(1/w - i)^2} = \frac{i + iw^2 - w^3}{(2 + 3wi)(1 - iw)^2} = \frac{i}{2}.$$

Theorem 3.2. [Limit in terms of real and imaginary parts] : *Let a complex valued function be* $f(z) = u(x, y) + iv(x, y)$, *where* $z = x + iy$ *and* $z_0 = x_0 + iy_0$. *Let the function* f *be defined in a domain* $\mathcal{D}$, *except possibly at the point* z_0 *in* $\mathcal{D}$. *Then*

$$\lim_{z \to z_0} f(z) = w_0 = u_0 + iv_0$$

if and only if $\lim_{(x,y) \to (x_0,y_0)} u(x, y) = u_0$ *and* $\lim_{(x,y) \to (x_0,y_0)} v(x, y) = v_0$.

Proof: First, we suppose that $\lim_{z \to z_0} f(z)$ exists. Hence by definition, corresponding to an $\varepsilon > 0$, $\exists$ a $\delta(\varepsilon) > 0$ such that

$$|f(z) - w_0| < \varepsilon \text{ whenever } 0 < |z - z_0| < \delta$$

$$\text{or,} \quad \left|\left\{u(x, y) + iv(x, y)\right\} - \{u_0 + iv_0\}\right| < \varepsilon \text{ whenever } 0 < |z - z_0| < \delta$$

$$\text{or,} \quad \left|\left\{u(x, y) - u_0\right\} + i\left\{v(x, y) - v_0\right\}\right| < \varepsilon \text{ whenever } 0 < |z - z_0| < \delta. \tag{3.12}$$

We know that $|\text{Re}\,(z)| \leq |z|$ and $|\text{Im}\,(z)| \leq |z|$ for any complex number z. So from Eq. (3.12) we get,

$$\left|u(x, y) - u_0\right| \leq \left|\left\{u(x, y) - u_0\right\} + i\left\{v(x, y) - v_0\right\}\right| < \varepsilon,$$

$$\text{and} \quad \left|v(x, y) - v_0\right| \leq \left|\left\{u(x, y) - u_0\right\} + i\left\{v(x, y) - v_0\right\}\right| < \varepsilon$$

whenever $0 < |(x - x_0) + i(y - y_0)| < \delta$, i.e., $0 < |x - x_0| < \delta$ and $0 < |y - y_0| < \delta$. Therefore,

$$\lim_{(x,y) \to (x_0,y_0)} u(x, y) = u_0 \text{ and } \lim_{(x,y) \to (x_0,y_0)} v(x, y) = v_0.$$

Conversely, let

$$\lim_{(x,y) \to (x_0,y_0)} u(x, y) = u_0 \text{ and } \lim_{(x,y) \to (x_0,y_0)} v(x, y) = v_0.$$

Then, corresponding to an $\varepsilon > 0$, $\exists$ a $\delta_1, \delta_2 > 0$ such that

$$\Big|u(x,y) - u_0\Big| < \frac{\varepsilon}{2} \text{ whenever } 0 < |z - z_0| < \delta_1$$

$$\Big|v(x,y) - v_0\Big| < \frac{\varepsilon}{2} \text{ whenever } 0 < |z - z_0| < \delta_2.$$

Let $\delta = \{\delta_1, \delta_2\}$, then

$$\begin{aligned} |f(z) - w_0| &= \Big|\Big\{u(x,y) - u_0\Big\} + i\Big\{v(x,y) - v_0\Big\}\Big| \\ &\leq \Big|u(x,y) - u_0\Big| + \Big|v(x,y) - v_0\Big| \\ &\leq \frac{\varepsilon}{2} + \frac{\varepsilon}{2} = \varepsilon \text{ whenever } 0 < |z - z_0| < \delta. \end{aligned}$$

Hence, $\lim\limits_{z \to z_0} f(z) = w_0 = u_0 + iv_0$. □

Theorem 3.3. [Algebraic operations with limits]: *Let f and g be two functions whose limits at z_0 exist. Let $\lim\limits_{z \to z_0} f(z) = w_0$ and $\lim\limits_{z \to z_0} g(z) = w_1$. Then*

(i) $\lim\limits_{z \to z_0} \Big[f(z) \pm g(z)\Big] = w_0 \pm w_1$; (ii) $\lim\limits_{z \to z_0} \Big[f(z) \cdot g(z)\Big] = w_0 \cdot w_1$

(iii) $\lim\limits_{z \to z_0} \left[\dfrac{f(z)}{g(z)}\right] = \dfrac{w_0}{w_1}$ *if* $w_1 \neq 0$.

Proof: (i) Since $\lim\limits_{z \to z_0} f(z) = w_0$ and $\lim\limits_{z \to z_0} g(z) = w_1$, corresponding to an $\varepsilon > 0$, $\exists\ \delta_1, \delta_2 > 0$ such that

$$|f(z) - w_0| < \frac{\varepsilon}{2} \text{ whenever } 0 < |z - z_0| < \delta_1$$

$$|g(z) - w_1| < \frac{\varepsilon}{2} \text{ whenever } 0 < |z - z_0| < \delta_2.$$

Let $\delta = \min\{\delta_1, \delta_2\}$. Now,

$$\begin{aligned} |\{f(z) \pm g(z)\} - \{w_0 \pm w_1\}| &= |\{f(z) - w_0\} \pm \{g(z) - w_1\}| \\ &\leq |\{f(z) - w_0\}| + |\{g(z) - w_1\}| \\ &< \frac{\varepsilon}{2} + \frac{\varepsilon}{2} = \varepsilon \text{ whenever } 0 < |z - z_0| < \delta. \end{aligned}$$

Therefore, by definition $\lim\limits_{z \to z_0} [f(z) \pm g(z)] = w_0 \pm w_1$.

(ii) Since $\lim\limits_{z \to z_0} f(z) = w_0$, corresponding to an $\varepsilon_1 > 0$, $\exists\ \delta_1 = \delta_1(\varepsilon_1) > 0$ such that

$$|f(z) - w_0| < \varepsilon_1 \text{ whenever } 0 < |z - z_0| < \delta_1.$$

Now, for $0 < |z - z_0| < \delta_1$, we have

$$\begin{aligned} |f(z)| &= |f(z) - w_0 + w_0| \\ &\leq |f(z) - w_0| + |w_0| < \varepsilon_1 + |w_0| = p_1 (\neq 0), \quad \text{say}. \end{aligned}$$

Since $\lim_{z\to z_0} f(z) = w_0$ and $\lim_{z\to z_0} g(z) = w_1$ corresponding to an $\varepsilon > 0$, $\exists\, \delta_2, \delta_3 > 0$ such that

$$|f(z) - w_0| < \frac{\varepsilon}{2\{|w_1| + 1\}} \text{ whenever } 0 < |z - z_0| < \delta_2$$

$$|g(z) - w_1| < \frac{\varepsilon}{2p_1} \text{ whenever } 0 < |z - z_0| < \delta_3.$$

Let $\delta = \min\{\delta_1, \delta_2, \delta_3\}$. Now,

$$\begin{aligned} |f(z)g(z) - w_0w_1| &= |f(z)\{g(z) - w_1\} + w_1\{f(z) - w_0\}| \\ &\leq |f(z)||g(z) - w_1| + |w_1||f(z) - w_0| \\ &< p_1|g(z) - w_1| + \{|w_1| + 1\}|f(z) - w_0| \\ &< p_1 \cdot \frac{\varepsilon}{2p_1} + \{|w_1| + 1\}\frac{\varepsilon}{2\{|w_1| + 1\}} \\ &< \frac{\varepsilon}{2} + \frac{\varepsilon}{2} = \varepsilon \text{ whenever } 0 < |z - z_0| < \delta. \end{aligned}$$

Therefore, by definition $\lim_{z\to z_0} f(z)g(z) = w_0w_1$.

(iii) Since $\lim_{z\to z_0} g(z) = w_1$, corresponding to an $\varepsilon_1 = |w_1|/2 > 0$, $\exists\, \delta_1 = \delta_1(\varepsilon_1) > 0$ such that

$$|g(z) - w_1| < \varepsilon_1 \text{ whenever } 0 < |z - z_0| < \delta_1.$$

Now, for $0 < |z - z_0| < \delta_1$, we have

$$\begin{aligned} |w_1| &= |w_1 - g(z) + g(z)| \\ &\leq |w_1 - g(z)| + |g(z)| < \varepsilon_1 + |g(z)| \\ &\Rightarrow |g(z)| > |w_1| - \varepsilon_1 = \frac{1}{2}|w_1|. \end{aligned}$$

Since $\lim_{z\to z_0} f(z) = w_0$ and $\lim_{z\to z_0} g(z) = w_1$, corresponding to an $\varepsilon > 0$, $\exists\, \delta_2, \delta_3 > 0$ such that

$$|f(z) - w_0| < \frac{\varepsilon|w_1|}{4} \text{ whenever } 0 < |z - z_0| < \delta_2,$$

$$|g(z) - w_1| < \frac{\varepsilon|w_1|^2}{4\{|w_0| + 1\}} \text{ whenever } 0 < |z - z_0| < \delta_3.$$

Let $\delta = \min\{\delta_1, \delta_2, \delta_3\}$. Now,

$$\begin{aligned} \left|\frac{f(z)}{g(z)} - w_0w_1\right| &= \left|\frac{w_1f(z) - w_0g(z)}{w_1g(z)}\right| = \left|\frac{w_1\{f(z) - w_0\} - w_0\{g(z) - w_1\}}{w_1g(z)}\right| \\ &\leq \left|\frac{w_1\{f(z) - w_0\} - w_0\{g(z) - w_1\}}{w_1\, g(z)}\right| \\ &\leq \frac{1}{|g(z)|}|f(z) - w_0| + \frac{|w_0|}{|w_1|\,|g(z)|}|g(z) - w_1| \\ &< \frac{2}{|w_1|}|f(z) - w_0| + \frac{2\{|w_0| + 1\}}{|w_1|^2}|g(z) - w_1| \\ &< \frac{2}{|w_1|}\,\frac{\varepsilon|w_1|}{4} + \frac{2\{|w_0| + 1\}}{|w_1|^2}\,\frac{\varepsilon|w_1|^2}{4\{|w_0| + 1\}} \\ &< \frac{\varepsilon}{2} + \frac{\varepsilon}{2} = \varepsilon \text{ whenever } 0 < |z - z_0| < \delta. \end{aligned}$$

Therefore, by definition $\lim\limits_{z\to z_0} f(z)/g(z) = w_0/w_1$. □

Example 3.2.4. *Evaluate* $\lim\limits_{z\to 2i} \dfrac{(2z+3)(z-1)}{z^2-2z+4}$.

Solution. We see that the limits of numerator and denominator are each non-zero. So the theorems on limits to apply. Using the rules of limit, we get

$$\lim_{z\to 2i} \frac{(2z+3)(z-1)}{z^2-2z+4} = \frac{\lim\limits_{z\to 2i}(2z+3)\lim\limits_{z\to 2i}(z-1)}{\lim\limits_{z\to 2i}(z^2-2z+4)}$$

$$= \frac{(3-4i)(-2i-1)}{4i} = -\frac{1}{2} + \frac{11}{4}i.$$

Theorem 3.4. *Let at* z_0 *the limit of* f *exist. Now,* $\lim\limits_{z\to z_0} f(z) = l$,

(i) *then* $\lim\limits_{z\to z_0} \overline{f(z)} = \bar{l}$; (ii) *then* $\lim\limits_{z\to z_0} |f(z)| = |l|$;

(iii) *if and only if* $\lim\limits_{z\to z_0} \operatorname{Re} f(z) = \operatorname{Re} l$ *and* $\lim\limits_{z\to z_0} \operatorname{Im} f(z) = \operatorname{Im} l$.

Proof: (i) Let $\varepsilon > 0$ be chosen arbitrary. Since $\lim\limits_{z\to z_0} f(z) = l$, corresponding to that $\varepsilon > 0$, $\exists$ a positive number δ such that

$$|f(z) - l| < \varepsilon \text{ whenever } 0 < |z - z_0| < \delta.$$

Thus for $0 < |z - z_0| < \delta$, we have

$$\left|\overline{f(z)} - \bar{l}\right| = \left|\overline{f(z) - l}\right| = \left|f(z) - l\right| < \varepsilon.$$

Hence by definition $\lim\limits_{z\to z_0} f(\bar{z}) = \bar{l}$.

(ii) Since $\lim\limits_{z\to z_0} f(z) = l$, so, given any $\varepsilon > 0$, $\exists$ a positive number $\delta(\varepsilon)$ such that

$$|f(z) - l| < \varepsilon \text{ whenever } 0 < |z - z_0| < \delta.$$

Hence for $0 < |z - z_0| < \delta$, we have

$$\left||f(z)| - |l|\right| \leq \left|f(z) - l\right| < \varepsilon.$$

Hence by definition $\lim\limits_{z\to z_0} |f(z)| = |l|$.

(iii) Let $\lim\limits_{z\to z_0} f(z) = l$. Since $\operatorname{Re} f(z) = \frac{1}{2}[f(z) + \overline{f(z)}]$, we have

$$\lim_{z\to z_0} \operatorname{Re} f(z) = \frac{1}{2}[\lim_{z\to z_0} f(z) + \lim_{z\to z_0} \overline{f(z)}]$$

$$= \frac{1}{2}[l + \bar{l}] = \operatorname{Re}\ l.$$

Similarly, $\lim\limits_{z\to z_0} \operatorname{Im} f(z) = \operatorname{Im}\ l$. Conversely, let $\lim\limits_{z\to z_0} \operatorname{Re} f(z) = \operatorname{Re}\ l$ and $\lim\limits_{z\to z_0} \operatorname{Im} f(z) = \operatorname{Im}\ l$. Since $f(z) = \operatorname{Re} f(z) + i\operatorname{Im} f(z)$, it follows that $\lim\limits_{z\to z_0} f(z) = \operatorname{Re}\ l + \operatorname{Im}\ l = l$. □

3.2.2 Limit of Polynomial Functions

A polynomial function of z is an expression of the form

$$P(z) = a_0 z^n + a_1 z^{n-1} + \cdots + a_{n-1} z + a_n,$$

where $a_0, a_1, \cdots, a_n$ are complex constants, $a_0 \neq 0$ and n is a positive integer called the degree of the polynomial $P(z)$.
The quotients of the polynomials are called rational functions and are defined as

$$R(z) = \frac{a_0 z^n + a_1 z^{n-1} + \cdots + a_{n-1} z + a_n}{b_0 z^m + b_1 z^{m-1} + \cdots + b_{m-1} z + b_m}.$$

Here, the degree of the numerator is n and that of denominator is m provided $a_0 \neq 0$ and $b_0 \neq 0$. The rational functions are defined at each point z where the denominator does not vanish.

The limit of any polynomial $P(z)$ as $z \to z_0$ is the value of the polynomial at that point, i.e., $\lim\limits_{z \to z_0} P(z) = P(z_0)$.

Example 3.2.5. *Evaluate* $\lim\limits_{z \to -e^{i\pi/3}} (z + e^{i\pi/3}) \dfrac{z}{z^3 - 1}$.

Solution. Since the limits of numerator and denominator are each zero, the terms on limits fail to apply. Therefore,

$$\begin{aligned}
\lim_{z \to -e^{i\pi/3}} \frac{z(z + e^{i\pi/3})}{z^3 - 1} &= \lim_{z \to -e^{i\pi/3}} \frac{z(z + e^{i\pi/3})}{(z-1)(z^2 + z + 1)} \qquad \left[\text{form } \frac{0}{0}\right] \\
&= \lim_{z \to -e^{i\pi/3}} \frac{z(z + e^{i\pi/3})}{(z-1)\{(z + \frac{1}{2})^2 + \frac{3}{4}\}} \\
&= \lim_{z \to -e^{i\pi/3}} \frac{z(z + e^{i\pi/3})}{(z-1)(z + \frac{1}{2} + i\frac{\sqrt{3}}{2})(z + \frac{1}{2} - i\frac{\sqrt{3}}{2})} \\
&= \lim_{z \to -e^{i\pi/3}} \frac{z(z + e^{i\pi/3})}{(z-1)(z + e^{i\pi/3})(z + e^{-i\pi/3})} \\
&= \lim_{z \to -e^{i\pi/3}} \frac{z}{(z-1)(z + e^{-i\pi/3})} \quad \text{as } z \neq -e^{i\pi/3} \\
&= \frac{-e^{i\pi/3}}{(-e^{i\pi/3} - 1)(-e^{i\pi/3} + e^{-i\pi/3})} = -\frac{1}{6} + \frac{\sqrt{3}}{6} i.
\end{aligned}$$

Example 3.2.6. *Evaluate* $\lim\limits_{z \to 2e^{i\pi/3}} \dfrac{z^3 + 8}{z^4 + 4z^2 + 16}$.

Solution. Since the limits of numerator and denominator are each zero, the terms on limits fail to apply. Thus,

$$\begin{aligned}
\lim_{z \to 2e^{i\pi/3}} \frac{z^3 + 8}{z^4 + 4z^2 + 16} &= \lim_{z \to 2e^{i\pi/3}} \frac{(z+2)(z^2 - 2z + 4)}{(z^2 + 2)^2 + 12} \qquad \left[\text{form } \frac{0}{0}\right] \\
&= \lim_{z \to 2e^{i\pi/3}} \frac{(z+2)\{(z-1)^2 - (i\sqrt{3})^2}{(z^2 + 2)^2 - (2\sqrt{3}i)^2} \\
&= \lim_{z \to 2e^{i\pi/3}} \frac{(z+2)(z - 2e^{i\pi/3})(z - 2e^{5i\pi/3})}{(z - 2e^{i\pi/3})(z - 2e^{2i\pi/3})(z - 2e^{4i\pi/3})(z - 2e^{5i\pi/3})}
\end{aligned}$$

$$= \lim_{z\to 2e^{i\pi/3}} \frac{(z+2)}{(z-2e^{2i\pi/3})(z-2e^{4i\pi/3})} \quad \text{as } z \neq 2e^{i\pi/3}$$

$$= \frac{2e^{i\pi/3}+2}{(2e^{i\pi/3}-2e^{2i\pi/3})(2e^{i\pi/3}-2e^{4i\pi/3})} = \frac{3}{8} - \frac{\sqrt{3}}{8}i.$$

3.3 Continuity

Let a function $w = f(z)$ of a complex variable z be defined in a bounded closed region $\mathcal{D}$ of the complex plane.

Definition 3.3.1. This function $f(z)$ is said to be continuous at a point $z_0 \in \mathcal{D}$ if corresponding to an $\varepsilon > 0$, there exists a positive number $\delta(\varepsilon)$ such that

$$|f(z) - f(z_0)| < \varepsilon \text{ whenever } 0 < |z - z_0| < \delta, \tag{3.13}$$

which is also written as $\lim_{z\to z_0} f(z) = f(z_0)$.

A useful strategy for showing that $f(z)$ is continuous at z_0 is to obtain an estimate of the form

$$|f(z) - f(z_0)| \leq K|z - z_0|; z \in \mathcal{N}(z_0).$$

A function is said to be continuous in a domain $\mathcal{D}$ if it is continuous at every point of $\mathcal{D}$. Now,

(i) If the function $f(z)$ is continuous at a point z_0, then we can write $\lim_{z\to z_0} f(z) = f(\lim_{z\to z_0} z)$.

(ii) The estimates $|\text{Re}\,(z - z_0)| \leq |z - z_0|$, $|\text{Im}\,(z - z_0)| \leq |z - z_0|$, and $||z| - |z_0|| \leq |z - z_0|$ show respectively that the function $\text{Re}\, z$, $\text{Im}\, z$ and $|z|$ are continuous at z_0.

(iii) If the function $f(z)$ is continuous at a point z_0, then $|f(z)|, f(\bar{z}), \overline{f(z)}$ are also continuous thereat.

(iv) If f and g are continuous at a point z_0, then $f + g, f \cdot g$ and $\bar{f}$ are continuous at z_0 and f/g is continuous at z_0 if $g(z_0) \neq 0$.

(v) f is continuous at z_0 iff $\text{Re}\, f$ and $\text{Im}\, f$ are continuous at z_0.

(vi) Any polynomial $P(z)$ is continuous at each point of the complex plane and any rational function $P(z)/Q(z)$ is continuous at all points where $Q(z) \neq 0$.

If the function $f(z)$ is not continuous at the point z_0, then $f(z)$ is said to be discontinuous at z_0 and the point z_0 is known as the *point of discontinuity*.

Example 3.3.1. *Prove that the function $f(z)$ given by*

$$f(z) = \begin{cases} \text{Im}\, z/|z| & \text{if } z \neq 0 \\ 0 & \text{if } z = 0 \end{cases}$$

is not continuous at $z = 0$.

Solution. Here the function $f(z)$ is defined in the neighbourhood of $z = 0$. If $f(z)$ is continuous at the origin, then

$$f(0) = \lim_{z\to 0} f(z) = \lim_{z\to 0} \frac{\text{Im}\, z}{|z|} = \lim_{z\to 0} \frac{y}{\sqrt{x^2 + m^2x^2}}$$

exists. Let us choose the path $y = mx$ along which $z \to 0$. Then

$$\lim_{z\to 0} \frac{\text{Im } z}{|z|} = \lim_{z\to 0} \frac{y}{\sqrt{x^2+y^2}} = \lim_{x\to 0} \frac{mx}{\sqrt{x^2+m^2x^2}} = \frac{m}{\sqrt{1+m^2}},$$

which is different for different values of m. Consequently, the limit does not exist. Therefore, $\lim\limits_{z\to 0} f(z)$ does not exist and hence the function is not continuous at the origin.

Theorem 3.5. [Neighbourhood property]: *If $f(z)$ be a continuous at $z = z_0$ and $f(z_0) \neq 0$, then $f(z) \neq 0, \forall z$ in $|z - z_0| < \delta$.*

Proof: It is given that $f(z)$ is non-zero at z_0, i.e., $f(z_0) \neq 0 \Rightarrow |f(z_0)| \neq 0$. Since $f(z)$ is continuous at z_0, corresponding to an $\varepsilon > 0$ there exists $\delta = \delta(\varepsilon) > 0$ such that,

$$\Big|f(z) - f(z_0)\Big| < \varepsilon \text{ whenever } 0 < |z - z_0| < \delta$$

$$\Rightarrow \quad \Big|f(z) - f(z_0)\Big| < \frac{|f(z_0)|}{2} \text{ whenever } 0 < |z - z_0| < \delta,$$

where we choose $\varepsilon = |f(z_0)|/2$, as $f(z_0) \neq 0$. Now, if $f(z) = 0$ for some point z in the neighbourhood $|z - z_0| < \delta$, then $|f(z_0)| < |f(z_0)/2$, which is not possible. Hence $f(z) \neq 0 \, \forall z$ in the neighbourhood of $|z - z_0| < \delta$. Thus, if a function $f(z)$ is continuous and non-zero at z_0, then there exists some neighbourhood of z_0 in which the function $f(z)$ is non-zero. □

Lemma 3.3.1. Suppose f is continuous and assumes only integer values on a domain $\mathcal{D}$. The f is a constant on $\mathcal{D}$.

Proof: Since $\mathcal{D}$ is a domain, it is open and connected. For each integer n, let $\mathcal{D}_n \subset \mathcal{D}$ be such that for $z \in \mathcal{D}_n$, $f(z) = n$.

(i) The sets $\mathcal{D}_n$ are disjoint and their union is $\mathcal{D}$.

(ii) Let $z_0 \in \mathcal{D}_n$ so that $f(z_0) = n$. Since $f(z)$ is continuous at z_0, there exists an open disc $\mathcal{S}(z_0, r)$ such that

$$|f(z) - n| < 1/2 \forall z \in \mathcal{S}(z_0, r).$$

However, since $f(z)$ is integer-valued, this implies that $f(z) = n$ for all $z \in \mathcal{S}(z_0, r)$. Hence, each $\mathcal{D}_n$ is open.

Now, since a connected set cannot be the union of disjoint open sets, only one of the $\mathcal{D}_n$ can be non-empty, say $\mathcal{D}_{n0}$, and thus $f(z) = n_0$ for all z in $\mathcal{D}$. □

Example 3.3.2. *Show that the function Arg z is discontinuous at each point on the non-positive real axis.*

Solution. For a given $z(\neq 0) \in \mathbb{C}$, $\arg z = \{\Theta + 2n\pi; n \in \mathbb{Z}\}$, where, $\Theta = \tan^{-1}\left(\frac{\text{Im } z}{\text{Re } z}\right)$. Suppose there is a continuous function $\Theta : \mathbb{C}/\{0\} \to \mathbb{R}$ such that $\Theta(z) \in \arg z$. Then define

$$g(t) = \frac{1}{2\pi}\left[\Theta\left(e^{it}\right) + \Theta\left(e^{-it}\right)\right]; \quad t \in \mathbb{R}.$$

g is continuous and g is integer-valued as $\Theta\left(e^{it}\right) = t + 2n_0\pi$ and $\Theta\left(e^{-it}\right) = -t + 2n_1\pi$. g has to be constant. But

$$\begin{aligned} g(0) &= \frac{1}{2\pi}\Big[\Theta(e^0) + \Theta(e^0)\Big] = \frac{1}{2\pi}\Big[\Theta(1) + \Theta(1)\Big] \\ &= \frac{1}{2\pi} 2\Theta(1) = \frac{1}{2\pi} 2 \cdot 2n\pi = 2n = \text{ even integer} \end{aligned}$$

$$\begin{aligned} g(\pi) &= \frac{1}{2\pi}\Big[\Theta(e^{i\pi}) + \Theta(e^{-i\pi})\Big] = \frac{1}{2\pi}\Big[\Theta(-1) + \Theta(-1)\Big] \\ &= \frac{1}{2\pi}2\Theta(-1) = \frac{1}{2\pi}2\cdot(2n\pi + \pi) = 2n + 1 = \text{ odd integer} \end{aligned}$$

This is impossible. So no such function Θ exists.

3.3.1 Uniform Continuity

Let $f : \mathcal{D} \to \mathbb{C}$ be continuous in a region $\mathcal{D} \subset \mathbb{C}$. f is called *uniformly continuous* on $\mathcal{D}$ if for any $\varepsilon > 0$ we can find $\delta = \delta(\varepsilon) > 0$ such that

$$\Big|f(z_1) - f(z_2)\Big| < \varepsilon \text{ whenever } 0 < |z_1 - z_2| < \delta, \tag{3.14}$$

where z_1 and z_2 are any two points of the region $\mathcal{D}$.

Example 3.3.3. *Prove that $f(z) = z^2$ is uniformly continuous in the region $|z| < 1$.*

Solution. Let z_1 and z_2 be any two points in $|z| < 1$. Then

$$\begin{aligned} |f(z_1) - f(z_2)| &= |z_1^2 - z_2^2| = |(z_1 + z_2)(z_1 - z_2)| \\ &= |z_1 + z_2|\,|z_1 - z_2| \le \{|z_1| + |z_2|\}\,|z_1 - z_2| \\ &< 2|z_1 - z_2| < \varepsilon \text{ whenever } 0 < |z_1 - z_2| < \delta, \end{aligned}$$

where, $\delta = \varepsilon/2$ depends only on ε and not on z_0. Hence $f(z) = z^2$ is uniformly continuous in the region $|z| < 1$.

3.4 Derivability

Let $\mathcal{D}(\subset \mathbb{C})$ be an open set and $z_0 = x_0 + iy_0 \in \mathcal{D}$. Let a complex valued function $f : \mathcal{D} \to \mathbb{C}$ be defined in $\mathcal{D}$.

Definition 3.4.1. The function $f : \mathcal{D} \to \mathbb{C}$ is said to be *derivable at a point* $z_0 \in \mathcal{D}$ if

$$\begin{aligned} \lim_{\Delta z \to 0} \frac{\Delta w}{\Delta z} &= \lim_{\Delta z \to 0} \frac{f(z_0 + \Delta z) - f(z_0)}{\Delta z} \\ &= \lim_{z \to z_0} \frac{f(z) - f(z_0)}{z - z_0}; \quad \Delta z = z - z_0, \end{aligned} \tag{3.15}$$

provided the limit exists. This unique limit is called the *complex derivative of* $f(z)$ at z_0 and is denoted by $f'(z_0)$.

(i) If f is derivable at each point of $\mathcal{D}$, we say that f is derivable on $\mathcal{D}$. If f is derivable on $\mathcal{D}$, then $f'(z_0)$ defines a function $f' : \mathcal{D} \to \mathbb{C}$.

(ii) If $f' : \mathcal{D} \to \mathbb{C}$ is continuous, then we say that f is *continuously derivable*.

Definition 3.4.1 at once suggests that, corresponding to an $\varepsilon > 0$, $\exists\, \delta = \delta(\varepsilon) > 0$ such that

$$\left|\frac{f(z) - f(z_0)}{z - z_0} - f'(z_0)\right| < \varepsilon \text{ whenever } 0 < |z - z_0| < \delta \tag{3.16}$$

along the path where z approaches z_0 in the domain $\mathcal{D}$.

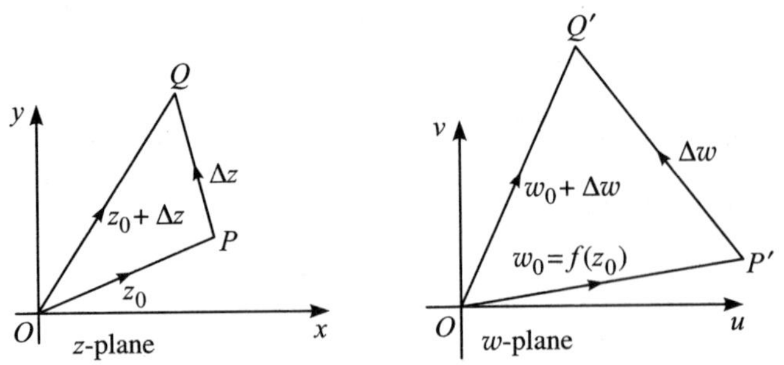

Figure 3.3 *Geometrical interpretation of the derivative*

Let a point P in the z-plane be represented by z_0 and its image P' in the w-plane be represented by w_0 under the transformation $w = f(z)$ (Fig. 3.3). Also, let $f(z)$ be a single-valued function so that the point z_0 maps only to one point w_0. On giving an increment of $\triangle z$ to z_0, we get the point Q. This point has the image Q' in the w-plane. Clearly, $P'Q'$ represents the complex number $\triangle w = f(z_0 + \triangle z) - f(z_0)$. Hence the derivative of $f(z)$ at z_0, if it exists, is given by

$$\lim_{\triangle z \to 0} \frac{f(z_0 + \triangle z) - f(z_0)}{\triangle z} = \lim_{Q \to P} \frac{Q'P'}{QP}, \tag{3.17}$$

i.e., the limit of ratio $Q'P'$ to QP as Q approaches point P.

Definition 3.4.2. [Differentiable]: A function $f : \mathcal{D} \to \mathbb{C}$ is said to be *differentiable at a point* $z_0 \in \mathcal{D}$ if the difference $f(z) - f(z_0)$; $z \in \mathcal{D}$, can be written in the form

$$\triangle f = f(z) - f(z_0) = A \triangle z + \varepsilon(z, z_0) \triangle z; \tag{3.18}$$

where A is a constant independent of $\Delta z = z - z_0$; z and $\varepsilon(z, z_0) \to 0$ as $\triangle z \to 0$. Further, if $f(z)$ is derivable at z_0, then $A = f'(z_0)$.

Theorem 3.6. *If $f : \mathcal{D} \to \mathbb{C}$ be differentiable at a point $z_0 \in \mathcal{D}$, then it is continuous there at.*

Proof: Let $f : \mathcal{D} \to \mathbb{C}$ be differentiable at a point $z_0 \in \mathcal{D}$, then by definition

$$f(z) - f(z_0) = A \triangle z + \varepsilon(z, z_0) \triangle z,$$

where A is a constant independent of $\Delta z = z - z_0$; z and $\varepsilon(z, z_0) \to 0$ as $\triangle z \to 0$. Therefore,

$$\lim_{z \to z_0} [f(z) - f(z_0)] = 0 \Rightarrow \lim_{z \to z_0} f(z) = f(z_0),$$

which shows that f is continuous. □

The converse of this theorem is not always true. For example, $f(z) = \bar{z}$, is continuous everywhere but is nowhere differentiable.

Example 3.4.1. *Prove that the function $f(z)$ given by*

$$f(z) = \begin{cases} z \operatorname{Re} z/|z| & \text{if } z \neq 0 \\ 0 & \text{if } z = 0 \end{cases}$$

is continuous at $z = 0$ but not differentiable there at.

Solution. First we shall prove that $\lim_{z\to 0} f(z) = f(0) = 0$. Now

$$|f(z) - f(0)| = |f(z) - 0| = \left|\frac{z \text{ Re } z}{|z|}\right| = |\text{Re } z| \leq |z|.$$

Therefore, for any given $\varepsilon > 0$ if we choose $\delta = \varepsilon$, we get

$$|z| = |z - 0| < \delta \Rightarrow |f(z) - f(0)| < \varepsilon.$$

Hence f is continuous at the origin. If $f(z)$ is differentiable at the origin, then

$$f'(0) = \lim_{z\to 0} \frac{f(z) - f(0)}{z - 0} = \lim_{z\to 0} \frac{z \text{ Re } z/|z| - 0}{z} = \lim_{z\to 0} \frac{\text{Re } z}{|z|}$$

exists. Let us choose the path $y = mx$ along which $z \to 0$. Therefore,

$$\lim_{z\to 0} \frac{\text{Re } z}{|z|} = \lim_{x\to 0} \frac{x}{\sqrt{x^2 + y^2}} = \lim_{x\to 0} \frac{x}{\sqrt{x^2 + m^2x^2}} = \frac{1}{\sqrt{1 + m^2}},$$

which is different for different values of m. Consequently, the limit does not exist. Therefore, $f'(0)$ does not exist and hence the function is not differentiable at the origin.

Definition 3.4.3. [Derivative along a parameterized curve]: Let $\gamma : \gamma(t) = x(t) + iy(t), 0 \leq t \leq 1$, be a smooth parameterized curve with $z_0 = \gamma(t_0)$ for $t_0 \in [0, 1]$. If $\gamma'(t_0) \neq 0$, then

$$\gamma'(t_0) = \lim_{t\to t_0} \frac{\gamma(t) - \gamma(t_0)}{t - t_0} = x'(t_0) + iy'(t_0) \tag{3.19}$$

is defined as the *tangent to the curve at* z_0. Moreover, $\arg \gamma'(t_0)$ represents the direction of the tangent to the curve γ at z_0.

Differentiation Rules

The complex derivative satisfies the usual rules for differentiating sums, products and quotients.

(i) If $n \in \mathbb{N}$, then $\frac{d}{dz}\, z^n = n\, z^{n-1}$ and when n is a negative integer, then this formula remains valid, provided that $z \neq 0$.

(ii) Let the derivatives of two functions $f(z)$ and $g(z)$ exist at a point z. Then

(a) $\frac{d}{dz}\Big[f(z) \pm g(z)\Big] = f'(z) \pm g'(z)$.

(b) $\frac{d}{dz}\Big[f(z) \cdot g(z)\Big] = f'(z) \cdot g(z) + f(z) \cdot g'(z)$.

(c) $\frac{d}{dz}\left[\frac{f(z)}{g(z)}\right] = \frac{f'(z) \cdot g(z) - f(z) \cdot g'(z)}{[g(z)]^2}, \quad g(z) \neq 0.$

(iii) **Linearity property** : The identity $\frac{d}{dz}[c_1 f(z) + c_2 g(z)] = c_1 f'(z) + c_2 g'(z)$ expresses the fact that the complex differential operator is linear.

(iv) **Chain rule** : Let $f(z)$ have a derivative at z_0 and $g(z)$ a derivative at the $f(z_0)$. Then the function $F(z) = g[f(z)]$ has a derivative at z_0 and

$$F'(z_0) = g'[f(z_0)]\; f'(z_0).$$

Example 3.4.2. *Using the rules of differentiation, find* $\frac{d}{dz}\left\{[\sin(iz-2)]^{\tan^{-1}(z+3i)}\right\}$.

Solution. Using the rules of differentiation, we get

$$\frac{d}{dz}\Big[\sin(iz-2)\Big]^{\tan^{-1}(z+3i)} = e^{\tan^{-1}(z+3i)\,\log\{\sin(iz-2)\}}\frac{d}{dz}\Big[\tan^{-1}(z+3i)\,\log\{\sin(iz-2)\}\Big]$$

$$= \Big[\sin(iz-2)\Big]^{\tan^{-1}(z+3i)}\Big[\frac{d}{dz}\{\tan^{-1}(z+3i)\}\,\log\{\sin(iz-2)\} + \tan^{-1}(z+3i)\,\frac{d}{dz}\{\log\sin(iz-2)\}\Big]$$

$$= \Big[\sin(iz-2)\Big]^{\tan^{-1}(z+3i)}\Big[\frac{\frac{d}{dz}(z+3i)}{1+(z+3i)^2}\,\log\{\sin(iz-2)\} + \tan^{-1}(z+3i)\,\frac{\cos(iz-2)}{\sin(iz-2)}\frac{d}{dz}(iz-2)\Big]$$

$$= \Big[\sin(iz-2)\Big]^{\tan^{-1}(z+3i)}\Big[\frac{\log\{\sin(iz-2)\}}{1+(z+3i)^2} + i\frac{\tan^{-1}(z+3i)}{\tan(iz-2)}\Big].$$

Example 3.4.3. *Using the rules of differentiation, find* $\frac{d^2}{dz^2}\Big[\cos^{-1}(\log z)\Big]$.

Solution. Using the rules of differentiation, we get

$$\frac{d}{dz}\Big[\cos^{-1}(\log z)\Big] = -\frac{1}{\sqrt{1-(\log z)^2}}\,\frac{d}{dz}(\log z)$$

$$= -\frac{1}{z\sqrt{1-(\log z)^2}} = -z^{-1}\Big\{1-(\log z)^2\Big\}^{-1/2}.$$

Again, using the rules of differentiation, we get

$$\frac{d^2}{dz^2}\Big[\cos^{-1}(\log z)\Big] = -\frac{d}{dz}\Big[z^{-1}\Big\{1-(\log z)^2\Big\}^{-1/2}\Big]$$

$$= \frac{1}{z^2}\Big\{1-(\log z)^2\Big\}^{-1/2} + \frac{1}{2z}\Big\{1-(\log z)^2\Big\}^{-3/2}\frac{d}{dz}\Big\{1-(\log z)^2\Big\}$$

$$= \frac{1}{z^2}\Big\{1-(\log z)^2\Big\}^{-1/2} + \frac{1}{2z}\Big\{1-(\log z)^2\Big\}^{-3/2}\,\frac{-2\log z}{z}$$

$$= \frac{1}{z^2\Big\{1-(\log z)^2\Big\}^{3/2}}\Big[1-(\log z)^2-\log z\Big].$$

Example 3.4.4. *If* $w^2-2w+\sin 2z=0$, *find* $(i)\,\frac{dw}{dz}$ *and* $(ii)\,\frac{d^2w}{dz^2}$.

Solution. (i) Differentiating with respect to z, considering w as an implicit function of z, we have

$$\frac{d}{dz}(w^2) - \frac{d}{dz}(2w) + \frac{d}{dz}\sin 2z = 0$$

$$\Rightarrow\quad 2w\frac{dw}{dz} - 2\frac{dw}{dz} + \cos 2z = 0 \Rightarrow \frac{dw}{dz} = \frac{\cos 2z}{2(1-w)}.$$

(ii) Again differentiating we get

$$\begin{aligned}
\frac{d^2w}{dz^2} &= \frac{d}{dz}\left(\frac{dw}{dz}\right) = \frac{d}{dz}\left[\frac{\cos 2z}{2(1-w)}\right] = \frac{-2(1-w)\sin 2z - \cos 2z\left(-\frac{dw}{dz}\right)}{(1-w)^2} \\
&= \frac{-2(1-w)\sin 2z + \cos 2z \cdot \frac{\cos 2z}{2(1-w)}}{(1-w)^2} \\
&= \frac{-2(1-w)^2 \sin 2z + \cos^2 z}{(1-w)^3}.
\end{aligned}$$

Theorem 3.7. [L'Hôpital's rule]: *If $f(z)$ and $g(z)$ are analytic at z_0, and $f(z_0) = g(z_0) = 0$ but $g'(z_0) \neq 0$, then*

$$\lim_{z\to z_0} \frac{f(z)}{g(z)} = \frac{f'(z_0)}{g'(z_0)}. \tag{3.20}$$

Proof: If $f(z)$ is analytic in a domain $\mathcal{D}$ and $z_0 \in \mathcal{D}$, then by definition

$$\begin{aligned}
& \lim_{z\to z_0} \frac{f(z) - f(z_0)}{z - z_0} = f'(z_0) \\
\Rightarrow \quad & \frac{f(z) - f(z_0)}{z - z_0} = f'(z_0) + \eta_1 \quad \text{where } \lim_{z\to z_0} \eta_1 = 0 \\
\Rightarrow \quad & f(z) = f(z_0) + (z - z_0)f'(z_0) + (z - z_0)\eta_1 \\
\Rightarrow \quad & f(z) = (z - z_0)f'(z_0) + (z - z_0)\eta_1 \quad \text{as } f(z_0) = 0.
\end{aligned}$$

Similarly, $g(z) = (z - z_0)g'(z_0) + (z - z_0)\eta_2$, where, $\eta_2 \to 0$ as $z \to z_0$. Now

$$\begin{aligned}
\lim_{z\to z_0} \frac{f(z)}{g(z)} &= \lim_{z\to z_0} \frac{(z - z_0)f'(z_0) + (z - z_0)\eta_1}{(z - z_0)g'(z_0) + (z - z_0)\eta_2} \\
&= \lim_{z\to z_0} \frac{f'(z_0) + \eta_1}{g'(z_0) + \eta_2} = \frac{f'(z_0)}{g'(z_0)}.
\end{aligned}$$

This is well celebrated *L'Hôpital's rule*. □

Example 3.4.5. *Evaluate* $\lim_{z\to 0}(\sin z/z)^{1/z^2}$.

Solution. Since $\lim_{z\to 0}(\sin z/z) = 1$ and $\lim_{z\to 0}(1/z^2) = \infty$, the required limit has the intermediate form 1^∞. Let

$$w = \left(\frac{\sin z}{z}\right)^{1/z^2} \Rightarrow \log w = \frac{\log \sin z - \log z}{z^2},$$

where we consider the principal branch of the logarithm. Using L'Hôpital's rule (3.20), we get

$$\begin{aligned}
\lim_{z\to 0} \log w &= \lim_{z\to 0} \frac{\log \sin z - \log z}{z^2} = \lim_{z\to 0} \frac{(\cos z/\sin z) - (1/z)}{2z} \\
&= \lim_{z\to 0} \frac{z\cos z - \sin z}{2z^2 \sin z}; \quad \text{form } \frac{0}{0} \\
&= \lim_{z\to 0} \frac{\cos z - z\sin z - \cos z}{4z\sin z + 2z^2\cos z}; \quad \text{form } \frac{0}{0}
\end{aligned}$$

$$= \lim_{z\to 0} \frac{-\sin z - z\cos z}{4\sin z + 4z\cos z + 4z\cos z - 2z^2\sin z}; \quad \text{form } \frac{0}{0}$$

$$= \lim_{z\to 0} \frac{-\cos z - \cos z + z\sin z}{4\cos z + 8\cos z - 8z\sin z - -4z\sin z - 2z^2\cos z}$$

$$= \frac{-1-1+0}{4+8-0-0-0} = -\frac{1}{6}.$$

But since the logarithm is a continuous function, we have

$$\lim_{z\to 0} \log w = \log\left(\lim_{z\to 0} w\right) = -\frac{1}{6} \Rightarrow \lim_{z\to 0} w = e^{-1/6}.$$

3.5 Analytic Functions

The differentiability of a function f is defined at a point, but analyticity is defined on an open set. In this section we shall define analytic functions with some examples. It is also shown that the Cauchy–Riemann equations hold for the real and imaginary parts of an analytic function.

Definition 3.5.1. A function $f : G \to \mathbb{C}$ is said to be *analytic* on the open set G if f is continuously differentiable on G. If G is not an open set, then we say f is analytic in G if f is analytic in an open set containing G.

The function defined and differentiable at each point z_0 of the domain $\mathcal{D}$ is said to be *analytic* in that domain.

The function being analytic at a point z_0 can also be described in several ways:

(i) It should be differentiable not only the point z_0, but in the neighbourhood of z_0; the first derivative exists in the neighbourhood of z_0.

(ii) The function is analytic at a point if the function can be expressed as a convergent power series centred at that point in a small disc surrounding that point and if it happens at every point.

One definition of the analytic function says that the function is differentiable once in the small neighbourhood of the point, while another definition says that it is represented by a convergent power series centred at that point. The relationship between the above two definitions is that they require great things of complex valued functions because complex power series is infinitely differentiable within its region of convergence (disc centred at the point z_0; then points on the boundary may or may not be included). For example:

(i) The function $f(z) = z^n, n \in \mathbb{N}$, has continuous complex derivative at all points and thus is analytic in $\mathbb{C}_\infty$.

(ii) For the rational function $f(z) = P(z)/Q(z)$, where $P(z)$ and $Q(z)$ are polynomials, let $\{z_1, z_2, \cdots, z_r\}$ be the roots of $Q(z)$. By the quotient rule, $f'(z)$ exists for all $z \in \mathcal{D} = \mathbb{C} - \{z_1, z_2, \cdots, z_r\}$. Since $\mathcal{D}$ is open, f is analytic in $\mathcal{D}$.

(iii) The function $f(z) = |z|^2$ is not analytic at any point as its derivative exists only $z = 0$ and not throughout any neighbourhood.

Definition 3.5.2. For a function to be *holomorphic* at a point, it needs to be complex differentiable at that point. If a function is *holomorphic* in the whole domain of f, then it is analytic at every point.

For complex-valued functions, *analyticity* and *holomorphic* or *regular* are equivalent.

RESULT **21.** For the real-valued function, homomorphic $\Rightarrow$ analyticity, but not the converse. The real functions of complex variables are nowhere analytic unless they are constant-valued. Thus functions like $|z|$, Re z, Im z are nowhere analytic.

Definition 3.5.3. An *analytic isomorphism* $f : U \to V$ from one open set to another is an analytic function if there exists another analytic function $g : V \to U$ satisfying

$$f \circ g = \text{id}_V \text{ and } g \circ f = \text{id}_U, \tag{3.21}$$

where id is the identity function. An analytic automorphism of U is an analytic isomorphism of U with itself.

Theorem 3.8. [The necessary condition for $f(z)$ to be analytic]: *If $w = f(z) = u(x, y) + iv(x, y)$ is analytic on a domain $\mathcal{D}$, then the partial derivatives u_x, u_y, v_x, v_y should exist and satisfy the equation*

$$u_x = v_y \text{ and } u_y = -v_x \tag{3.22}$$

at each point of the domain $\mathcal{D}$.

Proof: Let $z = x + iy$ so that $\Delta z = \Delta x + i\Delta y$. Since $w = f(z)$ is analytic, the limit

$$\begin{aligned} f'(z) &= \lim_{\Delta z\to 0} \frac{\Delta w}{\Delta z} = \lim_{\Delta z\to 0} \frac{f(z+\Delta z) - f(z)}{\Delta z} \\ &= \lim_{(\Delta x,\Delta y)\to(0,0)} \frac{\{u(x+\Delta x, y+\Delta y) + iv(x+\Delta x, y+\Delta y)\} - \{u(x,y) + iv(x,y)\}}{\Delta x + i\Delta y} \\ &= \lim_{(\Delta x,\Delta y)\to(0,0)} \frac{\{u(x+\Delta x, y+\Delta y) - u(x,y)\} + i\{v(x+\Delta x, y+\Delta y) - v(x,y)\}}{\Delta x + i\Delta y} \end{aligned} \tag{3.23}$$

must exist and can be computed by allowing $\Delta z \to 0$ from any covariant direction in the complex plane.

Case I: Choose that $\Delta z \to 0$ along the imaginary axis, so that $\Delta x = 0$ and $\Delta z = i\Delta y$, i.e., $(\Delta x, \Delta y) \to (0, 0)$ horizontally through the points $(0, \Delta y)$. In this case Eq. (3.23) becomes,

$$\begin{aligned} \lim_{\Delta z\to 0} \frac{\Delta w}{\Delta z} &= \lim_{\Delta y\to 0} \left[\frac{u(x, y+\Delta y) - u(x,y)}{i\Delta y} + \frac{v(x, y+\Delta y) - v(x,y)}{\Delta y}\right] \\ &= -i \lim_{\Delta y\to 0} \frac{u(x, y+\Delta y) - u(x,y)}{\Delta y} + \lim_{\Delta y\to 0} \frac{v(x, y+\Delta y) - v(x,y)}{\Delta y} \\ &= -i\frac{\partial u}{\partial y} + \frac{\partial v}{\partial y} = -iu_y + v_y, \end{aligned} \tag{3.24}$$

provided the partial derivatives u_y, v_y exist.

Case II: Choose $\Delta z \to 0$ along the real axis so that $\Delta y = 0$ and $\Delta z = \Delta x$, i.e., $(\Delta x, \Delta y) \to (0, 0)$ vertically through the points $(\Delta x, 0)$. In this case Eq. (3.23) becomes,

$$\begin{aligned} \lim_{\Delta z\to 0} \frac{\Delta w}{\Delta z} &= \lim_{\Delta x\to 0} \left[\frac{u(x+\Delta x, y) - u(x,y)}{\Delta x} + i\frac{v(x+\Delta x, y) - v(x,y)}{\Delta x}\right] \\ &= \lim_{\Delta x\to 0} \frac{u(x+\Delta x, y) - u(x,y)}{\Delta x} + \lim_{\Delta x\to 0} \frac{v(x+\Delta x, y) - v(x,y)}{\Delta x} \\ &= \frac{\partial u}{\partial x} + i\frac{\partial v}{\partial x} = u_x + iv_x, \end{aligned} \tag{3.25}$$

provided the partial derivatives u_x, v_x exist.

Now, $f(z)$ cannot possibly be analytic unless Eqs. (3.40) and (3.25) are identical. Since $\lim\limits_{\Delta z \to 0} \dfrac{\Delta w}{\Delta z}$ exists and is unique, so a necessary condition that $f(z)$ to be an analytic is

$$u_x + iv_x = -iu_y + v_y \Rightarrow u_x = v_y \text{ and } u_y = -v_x$$

on equating the real and imaginary parts. □

These two equations are called the *Cauchy–Riemann partial differential equations* for u and v. Therefore,

$$f'(z) = u_x + iv_x = -iu_y + v_y. \tag{3.26}$$

DEDUCTION **6.** From Eq. (3.26) we get

$$|f'(z)|^2 = u_x^2 + v_x^2 = u_y^2 + v_y^2 \text{ and } |f'(z)|^2 = u_x^2 + u_y^2 = v_x^2 + v_y^2.$$

Further

$$\begin{aligned}
|f'(z)|^2 &= f'(z) \cdot \overline{f'(z)} = \Big(u_x + iv_x\Big) \cdot \Big(iu_y + v_y\Big) \\
&= u_x v_y - u_y v_x + i(u_x u_y + v_x v_y) \\
&= u_x v_y - u_y v_x = \begin{vmatrix} u_x & u_y \\ v_x & v_y \end{vmatrix} = \frac{\partial(u, v)}{\partial(x, y)}.
\end{aligned} \tag{3.27}$$

RESULT **22.** The Cauchy–Riemann (C-R) equations (3.22) provide a necessary condition for differentiability at a point. Hence if the C-R equations are not satisfied for complex function at any point, then we can conclude that the function is not differentiable.

For example, consider $f(z) = \bar{z} = x - iy$, where $u(x, y) = x$ and $v(x, y) = -y$. Here $u_x = 1$ and $v_y = -1$. Since $u_x \neq v_y$, the C-R equations are not satisfied at any point z. Hence $f(z) = \bar{z}$ is nowhere differentiable.

RESULT **23.** The C-R equations (3.22) are not the sufficient condition for differentiability at a point. For example consider a function

$$f(z) = \begin{cases} \dfrac{xy}{\sqrt{x^2 + y^2}} & \text{if } z \neq 0 \\ 0 & \text{if } z = 0 \end{cases}$$

Let $f = u + iv$. Then $u(x, y) = xy/\sqrt{x^2 + y^2}$ and $v(x, y) = 0$. Now at the origin we have

$$\begin{aligned}
\frac{\partial u}{\partial x} &= \lim_{h \to 0} \frac{u(0 + h, 0) - u(0, 0)}{h} = \lim_{h \to 0} \frac{0 - 0}{h} = 0 \\
\frac{\partial u}{\partial y} &= \lim_{k \to 0} \frac{u(0, 0 + k) - u(0, 0)}{h} = \lim_{k \to 0} \frac{0 - 0}{k} = 0
\end{aligned}$$

Similarly, $\dfrac{\partial v}{\partial x} = \dfrac{\partial v}{\partial y} = 0$ at origin. Thus, $u_x = v_y$ and $u_y = -v_x$ at the origin and hence the C-R equations (3.22) are satisfied at the origin. Let us choose the path $y = mx$ along which $z \to 0$. Then

$$\lim_{z \to 0} f(z) = \lim_{x \to 0} \frac{x \cdot mx}{\sqrt{x^2 + m^2 x^2}} = \lim_{x \to 0} \frac{mx}{\sqrt{1 + m^2}} = 0 = f(0).$$

This shows that $f(z)$ is continuous at the origin. If $f(z)$ is analytic at the origin, then

$$f'(0) = \lim_{z\to 0} \frac{f(z) - f(0)}{z} = \lim_{z\to 0} \frac{[xy/\sqrt{x^2+y^2}] - 0}{x+iy}$$

exists. Let us choose the path $y = mx$ along which $z \to 0$. Therefore

$$\lim_{z\to 0} \frac{xy}{(x+iy)\sqrt{x^2+y^2}} = \lim_{x\to 0} \frac{x\cdot mx}{x(1+im)x\sqrt{1+m^2}} = \frac{m}{(1+im)\sqrt{1+m^2}},$$

which is different for different values of m. Consequently, the limit does not exist. Therefore, $f'(0)$ does not exist and hence the function is not analytic at the origin.

Example 3.5.1. *Show that the function $f(z)$, where*

$$f(z) = \begin{cases} \dfrac{x^3(1+i) - y^3(1-i)}{x^2+y^2} & \text{if } z \neq 0 \\ 0 & \text{if } z = 0 \end{cases}$$

is continuous at the origin, where the C-R equations are satisfied but $f'(0)$ does not exist.

Solution. Let us consider $f = u + iv$ so that

$$u + iv = \frac{x^3 - y^3}{x^2+y^2} + i\frac{x^3+y^3}{x^2+y^2} \Rightarrow u = \frac{x^3-y^3}{x^2+y^2}; v = \frac{x^3+y^3}{x^2+y^2}.$$

Let us choose the path $y = mx$ along which $z \to 0$. Then

$$\lim_{z\to 0} f(z) = \lim_{x\to 0} \left[\frac{x^3(1+i) - m^3x^3(1-i)}{x^2+m^2x^2}\right] = 0 = f(0).$$

This shows that $f(z)$ is continuous at the origin. Also, at the origin

$$\begin{aligned} \frac{\partial u}{\partial x} &= \lim_{h\to 0} \frac{u(0+h,0) - u(0,0)}{h} = \lim_{h\to 0} \frac{(h^3/h^2) - 0}{h} = 1 \\ \frac{\partial u}{\partial y} &= \lim_{k\to 0} \frac{u(0,0+k) - u(0,0)}{h} = \lim_{k\to 0} \frac{(-k^3/k^2) - 0}{k} = -1. \end{aligned}$$

Similarly, $\dfrac{\partial v}{\partial x} = \dfrac{\partial v}{\partial y} = 1$ at the origin. Thus, $u_x = v_y$ and $u_y = -v_x$ at the origin and hence the C-R equations (3.22) are satisfied at the origin. If $f(z)$ is analytic at the origin, then

$$f'(0) = \lim_{z\to 0} \frac{f(z) - f(0)}{z} = \lim_{z\to 0} \left[\frac{x^3-y^3}{x^2+y^2} + i\frac{x^3+y^3}{x^2+y^2}\right] \cdot \frac{1}{x+iy}$$

exists. Let us choose the path $y = mx$ along which $z \to 0$. Therefore

$$\begin{aligned} \lim_{z\to 0} \frac{f(z) - f(0)}{z} &= \lim_{x\to 0} \left[\frac{x^3(1-m^3)}{x^2(1+m^2)} + i\frac{x^3(1+m^3)}{x^2(1+m^2)}\right] \cdot \frac{1}{x(1+im)} \\ &= \left[\frac{1-m^3}{1+m^2} + i\frac{1+m^3}{1+m^2}\right] \cdot \frac{1}{1+im}, \end{aligned}$$

which is different for different values of m. Consequently, the limit does not exist. Therefore, $f'(0)$ does not exist and hence the function is not analytic at the origin.

Theorem 3.9. [The sufficient condition for $f(z)$ to be analytic]: *Let $\mathcal{D}$ be a domain in the plane and let $f(z) = u(z) + iv(z)$. Also let the functions u and v (i) have continuous first order derivatives in $\mathcal{D}$, (ii) satisfy the C-R equations (3.22) in $\mathcal{D}$. Then f is analytic in $\mathcal{D}$.*

Proof: Since u_x and u_y are supposed to be continuous,

$$\begin{aligned}\Delta u &= u(x+\Delta x, y+\Delta y) - u(x,y)\\ &= u(x+\Delta x, y+\Delta y) - u(x, y+\Delta y) + u(x, y+\Delta y) - u(x,y)\\ &= \Big(\frac{\partial u}{\partial x} + \varepsilon_1\Big)\Delta x + \Big(\frac{\partial u}{\partial y} + \eta_1\Big)\Delta y\\ &= \frac{\partial u}{\partial x}\Delta x + \frac{\partial u}{\partial y}\Delta y + \varepsilon_1 \Delta x + \eta_1 \Delta y, \qquad (3.28)\end{aligned}$$

where, ε_1 and η_1 are functions of $(\Delta x, \Delta y)$ such that $(\varepsilon_1, \eta_1) \to (0,0)$ as $(\Delta x, \Delta y) \to (0,0)$ and u_x and u_y are independent of Δx and Δy. Again, since v_x and v_y are supposed to be continuous,

$$\begin{aligned}\Delta v &= v(x+\Delta x, y+\Delta y) - v(x,y)\\ &= v(x+\Delta x, y+\Delta y) - v(x, y+\Delta y) + v(x, y+\Delta y) - v(x,y)\\ &= \Big(\frac{\partial v}{\partial x} + \varepsilon_2\Big)\Delta x + \Big(\frac{\partial v}{\partial y} + \eta_2\Big)\Delta y\\ &= \frac{\partial v}{\partial x}\Delta x + \frac{\partial v}{\partial y}\Delta y + \varepsilon_2 \Delta x + \eta_2 \Delta y, \qquad (3.29)\end{aligned}$$

where, ε_2 and η_2 are functions of $(\Delta x, \Delta y)$ such that $(\varepsilon_2, \eta_2) \to (0,0)$ as $(\Delta x, \Delta y) \to (0,0)$ and v_x and v_y are independent of Δx and Δy. Therefore,

$$\Delta w = \Delta u + i\Delta v = \Big(\frac{\partial u}{\partial x} + i\frac{\partial v}{\partial x}\Big)\Delta x + \Big(\frac{\partial u}{\partial y} + i\frac{\partial v}{\partial y}\Big)\Delta y + \varepsilon\Delta x + \eta\Delta y, \qquad (3.30)$$

where $\varepsilon = \varepsilon_1 + i\varepsilon_2$, $\eta = \eta_1 + i\eta_2$ are functions of $\Delta x, \Delta y$ such that $(\varepsilon, \eta) \to (0,0)$ as $(\Delta x, \Delta y) \to (0,0)$. Using C-R equations (3.22), Eq. (3.30) can be written as

$$\begin{aligned}\Delta w &= \Big(\frac{\partial u}{\partial x} + i\frac{\partial v}{\partial x}\Big)\Delta x + \Big(-\frac{\partial v}{\partial x} + i\frac{\partial u}{\partial x}\Big)\Delta y + \varepsilon\Delta x + \eta\Delta y\\ &= (\frac{\partial u}{\partial x} + i\frac{\partial v}{\partial x})(\Delta x + i\Delta y) + \varepsilon\Delta x + \eta\Delta y\\ &= (\frac{\partial u}{\partial x} + i\frac{\partial v}{\partial x})\Delta z + \varepsilon\Delta x + \eta\Delta y.\\ \therefore f'(z) &= \frac{dw}{dz} = \lim_{\Delta z \to 0}\frac{\Delta w}{\Delta z} = \frac{\partial u}{\partial x} + i\frac{\partial v}{\partial x}, \qquad (3.31)\end{aligned}$$

so that the derivative exists and is unique. Since u_x and v_x are continuous, f' is continuous and f is analytic in $\mathcal{D}$. □

Using Eqs. (3.40) and (3.25), we get

$$\frac{dw}{dz} = f'(z) = \frac{\partial w}{\partial x} = -i\frac{\partial w}{\partial y}. \qquad (3.32)$$

Example 3.5.2. *Show that $f(z) = \sin x \cosh y + i \cos x \sinh y$ is continuous as well as analytic.*

Solution. Let $f(z) = u + iv$. Then $u(x, y) = \sin x \cosh y$ and $v(x, y) = \cos x \sinh y$. Both are continuous everywhere and so, $f(z)$ is also continuous everywhere. Again,

$$u_x = \cos x \cosh y; \; u_y = \sin x \sinh y; \; v_x = -\sin x \sinh y; \; v_y = \cos x \cosh y.$$

Since they are C-R equations (3.22) are satisfied everywhere and the functions u_x, u_y, v_x and v_y are non-zero, they are everywhere continuous. Hence the conditions of the Theorem 3.9 are satisfied at all points in $\mathbb{C}$. Thus $f(z)$ is analytic everywhere.

DEDUCTION **7.** [Complex form of Cauchy–Riemann equations]: Since $z = x + iy \in \mathbb{C}$, we have $x = \frac{1}{2}(z + \bar{z})$ and $y = \frac{1}{2i}(z - \bar{z})$. Therefore,

$$\frac{\partial x}{\partial z} = \frac{1}{2}, \frac{\partial y}{\partial z} = \frac{1}{2i}, \frac{\partial x}{\partial \bar{z}} = \frac{1}{2} \text{ and } \frac{\partial y}{\partial \bar{z}} = -\frac{1}{2i}.$$

Let $f(z) = u(x, y) + iv(x, y)$. Then we have

$$f(z) = u\Big(\frac{z + \bar{z}}{2}, \frac{z - \bar{z}}{2i}\Big) + iv\Big(\frac{z + \bar{z}}{2}, \frac{z - \bar{z}}{2i}\Big).$$

Finally, treating $z, \bar{z}$ as independent variable, by chain rule of differentiation, we get

$$\begin{aligned}
\frac{\partial u}{\partial \bar{z}} &= \frac{\partial u}{\partial x} \cdot \frac{\partial x}{\partial \bar{z}} + \frac{\partial u}{\partial y} \cdot \frac{\partial y}{\partial \bar{z}} = \frac{1}{2} \cdot \frac{\partial u}{\partial x} - \frac{1}{2i} \cdot \frac{\partial u}{\partial y} \\
&= \frac{1}{2}\Big[\frac{\partial u}{\partial x} - i\frac{\partial u}{\partial y}\Big] = \frac{1}{2}(u_x + iu_y). \\
\frac{\partial v}{\partial \bar{z}} &= \frac{\partial v}{\partial x} \cdot \frac{\partial x}{\partial \bar{z}} + \frac{\partial v}{\partial y} \cdot \frac{\partial y}{\partial \bar{z}} = \frac{1}{2} \cdot \frac{\partial v}{\partial x} - \frac{1}{2i} \cdot \frac{\partial v}{\partial y} \\
&= \frac{1}{2}\Big[\frac{\partial v}{\partial x} + i\frac{\partial v}{\partial y}\Big] = \frac{1}{2}(v_x + iv_y).
\end{aligned}$$

Therefore, from $f(z) = u + iv$, we get

$$\begin{aligned}
\frac{\partial f}{\partial \bar{z}} &= \frac{\partial u}{\partial \bar{z}} + i\frac{\partial v}{\partial \bar{z}} = \frac{1}{2}\Big[(u_x + iu_y) + i(v_x + iv_y)\Big] \\
&= \frac{1}{2}\Big[(u_x - v_y) + i(u_x + v_x)\Big] = 0, \qquad (3.33)
\end{aligned}$$

which is known as the *complex form of the C-R equations*. From the above, it is clear that every analytic function is independent of $\bar{z}$.

DEDUCTION **8.** [Polar form of Cauchy-Riemann equations]: Let a function $f(z) = u(r, \theta) + iv(r, \theta)$ be defined in some neighbourhood of a non-zero point $z_0 = r_0 e^{i\theta_0}$ and the partial derivatives $u_r, u_\theta, v_r, v_\theta$ exist everywhere in the neighbourhood. Now,

$$x = r\cos\theta; y = r\sin\theta \Rightarrow r = \sqrt{x^2 + y^2} \text{ and } \theta = \tan^{-1}\frac{y}{x}.$$

$$\begin{aligned}
\Rightarrow \quad & \frac{\partial x}{\partial r} = \cos\theta; \; \frac{\partial y}{\partial r} = \sin\theta; \; \frac{\partial x}{\partial \theta} = -r\sin\theta; \; \frac{\partial y}{\partial \theta} = r\cos\theta \\
& \frac{\partial r}{\partial x} = \frac{x}{r} = \cos\theta; \; \frac{\partial \theta}{\partial x} = \frac{-(y/x^2)}{1 + (y/x)^2} = -\frac{\sin\theta}{r} \\
& \frac{\partial r}{\partial y} = \frac{y}{r} = \sin\theta; \; \frac{\partial \theta}{\partial y} = \frac{(1/x)}{1 + (y/x)^2} = \frac{\cos\theta}{r}.
\end{aligned}$$

For $u(x,y)$, we have

$$u_r = u_x \frac{\partial x}{\partial r} + u_y \frac{\partial y}{\partial r} \text{ and } u_\theta = u_x \frac{\partial x}{\partial \theta} + u_y \frac{\partial y}{\partial \theta}.$$

$$\Rightarrow \quad u_r = u_x \cos\theta + u_y \sin\theta \text{ and } u_\theta = -u_x r \sin\theta + u_y r \cos\theta. \tag{3.34}$$

Now, for $v(x,y)$, using C-R equations (3.22), we have

$$v_r = v_x \frac{\partial x}{\partial r} + v_y \frac{\partial y}{\partial r} \text{ and } v_\theta = v_x \frac{\partial x}{\partial \theta} + v_y \frac{\partial y}{\partial \theta}.$$

$$\Rightarrow \quad v_r = v_x \cos\theta + v_y \sin\theta \text{ and } v_\theta = -v_x r \sin\theta + v_y r \cos\theta.$$

$$\Rightarrow \quad v_r = -u_y \cos\theta + u_x \sin\theta \text{ and } v_\theta = u_y r \sin\theta + u_x r \cos\theta. \tag{3.35}$$

Comparing Eqs. (3.34) and (3.35) , we get

$$u_r = \frac{1}{r}\, v_\theta \text{ and } u_\theta = -r\, v_r, \tag{3.36}$$

which are the *C-R equations in polar form.*

Solving the equations (3.34) for u_x and equations (3.35) for v_x, we get

$$u_x = u_r \cos\theta - u_\theta \frac{\sin\theta}{r} \text{ and } v_x = v_r \cos\theta - v_\theta \frac{\sin\theta}{r}$$

$$\Rightarrow \quad u_x = u_r \cos\theta + v_r \sin\theta \text{ and } v_x = v_r \cos\theta - u_r \sin\theta,$$

where we use the equations (3.36). Therefore,

$$\begin{aligned} f'(z) &= u_x + iv_x = u_r \cos\theta + v_r \sin\theta + i(v_r \cos\theta - u_r \sin\theta) \\ &= (\cos\theta - i\sin\theta)(u_r + iv_r) = e^{-i\theta}(u_r + iv_r). \end{aligned} \tag{3.37}$$

DEDUCTION **9.** Here we calculate the derivative of w in the polar form. As $w_x = u_x + iv_x$, so,

$$\begin{aligned} \frac{dw}{dz} &= \frac{\partial w}{\partial x} = \frac{\partial w}{\partial r} \cdot \frac{\partial r}{\partial x} + \frac{\partial w}{\partial \theta} \cdot \frac{\partial \theta}{\partial x} \\ &= \frac{\partial w}{\partial r} \cos\theta - \frac{\partial w}{\partial \theta} \frac{\sin\theta}{r} = \frac{\partial w}{\partial r} \cos\theta - \left(\frac{\partial u}{\partial \theta} + \frac{\partial v}{\partial \theta}\right)\frac{\sin\theta}{r} \\ &= \frac{\partial w}{\partial r} \cos\theta - \left(-r\frac{\partial v}{\partial r} + ir\frac{\partial u}{\partial r}\right)\frac{\sin\theta}{r}; \quad \text{using Eq. (3.36)} \\ &= \frac{\partial w}{\partial r} \cos\theta - i\left(\frac{\partial u}{\partial r} + i\frac{\partial v}{\partial r}\right)\sin\theta \\ &= \frac{\partial w}{\partial r} \cos\theta - i\frac{\partial w}{\partial r}\sin\theta = (\cos\theta - i\sin\theta)\frac{\partial w}{\partial r} = e^{-i\theta}\frac{\partial w}{\partial r}. \end{aligned}$$

Similarly, we have

$$\frac{dw}{dz} = \frac{\partial w}{\partial y} = -\frac{i}{r}(\cos\theta - i\sin\theta)\frac{\partial w}{\partial \theta} = -\frac{i}{r}\, e^{-i\theta}\frac{\partial w}{\partial \theta}.$$

Example 3.5.3. *Find the analytic function $f(z) = u(r,\theta) + iv(r,\theta)$ such that $v(r,\theta) = r^2\cos 2\theta - r\cos\theta + 2$.*

Solution. Using the C-R equations (3.36), we get

$$\begin{aligned}
\frac{\partial u}{\partial r} &= \frac{1}{r}\frac{\partial v}{\partial \theta} = \frac{1}{r}[-2r^2 \sin 2\theta + r \sin\theta] = -2r\sin 2\theta + \sin\theta, \\
\frac{\partial u}{\partial \theta} &= -r\,\frac{\partial v}{\partial r} = -r[2r\cos 2\theta - \cos\theta] = -2r^2\cos 2\theta + r\cos\theta.
\end{aligned}$$

Integrating the first equation, we get

$$u = \int(-2r\sin 2\theta + \sin\theta)dr = -r^2\sin 2\theta + r\sin\theta + \phi(\theta),$$

where $\phi(\theta)$ is an arbitrary function of θ or constant. Using the second equation we get

$$\begin{aligned}
& \frac{\partial u}{\partial \theta} = -2r^2\cos 2\theta + r\cos\theta + \phi'(\theta) \\
\Rightarrow \quad & -2r^2\cos 2\theta + r\cos\theta = -2r^2\cos 2\theta + r\cos\theta + \phi'(\theta) \\
\Rightarrow \quad & \phi'(\theta) = 0 \Rightarrow \phi(\theta) = c; \quad c = \text{ arbitrary constant.}
\end{aligned}$$

Thus, $u(r,\theta) = -r^2\sin 2\theta + r\sin\theta + c$. Therefore

$$\begin{aligned}
f(z) &= \Big[-r^2\sin 2\theta + r\sin\theta + c\Big] + i\Big[r^2\cos 2\theta - r\cos\theta + 2\Big] \\
&= i\,r^2(\cos 2\theta + i\sin 2\theta) - ir(\cos\theta + i\sin\theta) + c + 2i \\
&= i\,r^2\,e^{2i\theta} - i\,r\,e^{i\theta} + c + 2i,
\end{aligned}$$

which is the required analytic function.

3.5.1 Inverse Function

Definition 3.5.4. Let $w = f(z)$ be an analytic function, where corresponding to any point z_0 there is a point w_0. Now, if the relation $z = F(w)$ is such that corresponding to the value w_0 of w there is a value z_0 of z, then $z = F(w)$ is called the *inverse function* of $w = f(z)$.

If $f'(z_0) \neq 0$, then w_0 is a *regular point of* $z = F(w)$, i.e., z is analytic in the neighbourhood of w_0. Also, $F'(w_0) = 1/f'(z_0)$. If $f'(z_0) \neq 0$, $F'(w_0)$ should be finite and $F(w)$ is analytic at w_0. Hence, the function $z = F(w)$ will *cease to be analytic* where

$$\frac{dw}{dz} = f'(z) = 0 = \frac{\partial w}{\partial x}; \quad \text{using Eq. (3.32).} \tag{3.38}$$

It follows that for $w = f(z)$, $z = F(w)$ will *cease to be analytic* where

$$\frac{dz}{dw} = F'(w) = 0 = \frac{\partial z}{\partial u}; \quad \text{using Eq. (3.32).} \tag{3.39}$$

Example 3.5.4. *For what values of z do the function $w = u + iv$ defined by $z = \sin u\cosh v + i\cos u\sinh v$ cease to be analytic?*

Solution. Given $z = \sin u\cosh v + i\cos u\sinh v$, where $w = u + iv$. Now,

$$\frac{\partial z}{\partial u} = \cos u\cosh v - i\sin u\sinh v = \frac{dz}{dw}; \quad \text{using Eq. (3.32).}$$

Squaring and adding we get

$$\begin{aligned}
z^2 + \left(\frac{dz}{dw}\right)^2 &= (\sin u \cosh v + i \cos u \sinh v)^2 + (\cos u \cosh v - i \sin u \sinh v)^2 \\
&= \sin^2 u \cosh^2 v - \cos^2 u \sinh^2 v + \cos^2 u \cosh^2 v - \sin^2 u \sinh^2 v \\
&= (\sin^2 u + \cos^2 u)(\cosh^2 v - \sinh^2 v) = 1. \\
&\Rightarrow \frac{dz}{dw} = \sqrt{1 - z^2}.
\end{aligned}$$

Hence, w ceases to be analytic when

$$\frac{dz}{dw} = 0 \Rightarrow \sqrt{1 - z^2} = 0 \Rightarrow z = \pm 1.$$

3.5.2 Orthogonal System

Definition 3.5.5. Two families of curves $u(x, y) = \alpha$ and $v(x, y) = \beta$ are said to form an *orthogonal system* if they intersect at right angles at each of their point of intersection.

Example 3.5.5. *If $f(z) = u + iv$ is an analytic function, regular in a domain $\mathcal{D}$, where $f'(z)$ does not vanish, then prove that the curves $u(x, y)$ = constant, $v(x, y)$ = constant form two orthogonal families.*

Solution. Given that $f(z) = u + iv$ is analytic; hence,

$$f'(z) = u_x + iv_x = v_y - iu_y \neq 0 \text{ in } \mathcal{D}.$$

Given two curves $u(x, y)$ = constant and $v(x, y)$ = constant; hence,

$$u_x\, dx + u_y\, dy = 0 \text{ and } v_x\, dx + v_y\, dy = 0.$$

Now, $u_x + iv_x \neq 0$ implies that both u_x and v_x are not simultaneously zero. Hence, if $u_x \neq 0$, by the C-R equations, we have $v_y = u_x \neq 0$.

Again $v_y - iu_y \neq 0$ implies that simultaneously v_y and u_y are not zeros. So, $v_y \neq 0, u_y \neq 0$. But $u_y \neq 0$ implies $v_x \neq 0$. Hence for u = constant, we have

$$\frac{dy}{dx} = -\frac{\partial u}{\partial x} \Big/ \frac{\partial u}{\partial y} = m_1 \quad \text{(say)}$$

and from v = constant, we have

$$\frac{dy}{dx} = -\frac{\partial v}{\partial x} \Big/ \frac{\partial v}{\partial y} = \frac{\partial u}{\partial y} \Big/ \frac{\partial u}{\partial x} = m_2 \quad \text{(say)},$$

using the C-R equations (3.22). Thus $m_1 \cdot m_2 = -1$ shows that u = constant, v = constant form two orthogonal families.

3.6 Harmonic Function

Definition 3.6.1. Any real-valued function $\phi = \phi(x, y)$ which possesses continuous second order partial derivatives and satisfies Laplace's equation

$$\nabla^2 \phi(x, y) = \frac{\partial^2 \phi}{\partial x^2} + \frac{\partial^2 \phi}{\partial y^2} = \phi_{xx} + \phi_{yy} = 0$$

is called a *harmonic function*.

For example, the temperatures $T(x, y)$ in their plates lying in the xy plane are harmonic.

Theorem 3.10. *If $f = u + iv$ be an analytic in a domain $\mathcal{D} \subset \mathbb{C}$, then u and v are both harmonic functions.*

Proof: Since f is analytic in $\mathcal{D}$, the C-R equations (3.22) are satisfied throughout $\mathcal{D}$. Differentiating both sides of equations (3.22), we get

$$u_{xx} = v_{xy}; \quad u_{xy} = -v_{xx}; \quad u_{yx} = v_{yy}; \quad u_{yy} = -v_{yx}.$$

Since f is analytic, u and v have continuous derivatives of all orders. Hence, $u_{xy} = u_{yx}$ and $v_{xy} = v_{yx}$. Thus it follows that

$$u_{xx} + u_{yy} = 0 \text{ and } v_{xx} + v_{yy} = 0,$$

i.e., u and v are harmonic in $\mathcal{D}$. □

(i) Let $u : \mathcal{D} \to \mathbb{R}$ be harmonic and a harmonic function $v : \mathcal{D} \to \mathbb{R}$ be such that $f = u + iv$ is analytic in $\mathcal{D}$, then v is called *harmonic conjugate* of u.

(ii) The harmonic conjugate v is unique up to the addition of a constant. If v_1 and v_2 are two harmonic conjugates of u, then $i(v_1 - v_2) = (u + iv_1) - (u + iv_2)$ is analytic on $\mathcal{D}$ and only takes on purely imaginary values. Thus $v_1 - v_2$ is a real-valued analytic function, and hence constant in $\mathcal{D}$.

RESULT **24.** Laplace's equation provides a necessary condition for the function to be the real or imaginary part of an analytic function. For example:

(i) Let $u(x, y) = x^2 + y$. Then $\frac{\partial^2 u}{\partial x^2} + \frac{\partial^2 u}{\partial y^2} = 2$. Thus u is not harmonic and hence it cannot be the real part of any analytic function.

(ii) Let $u(z) = \log |z|$ on $\mathcal{D}^* = \mathbb{C} - \{0\}$; then it is harmonic. Since it would not be possible to define an analytic branch of the logarithm on $\mathcal{D}^*$, so u has no conjugate harmonic.

Theorem 3.11. *Let $\mathcal{D}$ be either the whole plane $\mathbb{C}$ or some open disc. If $u : \mathcal{D} \to \mathbb{R}$ is a harmonic function, then u has a harmonic conjugate.*

Proof: Let $\mathcal{D} = \mathcal{S}(0, R)$, where $0 < R \leq \infty$. Here, we shall find a harmonic function v such that u and v satisfy the C-R equations. For this, define

$$v(x, y) = \int_0^y u_x(x, t)\, dt + \phi(x)$$

and determine ϕ so that $v_x = -u_y$. Differentiating both sides of this equation with respect to x gives (using Leibnitz's rule for differentiation)

$$\begin{aligned} v_x(x, y) &= \int_0^y u_{xx}(x, t)\, dt + \phi'(x) = -\int_0^y u_{yy}(x, t)\, dt + \phi'(x) \\ &= -u_y(x, y) + u_y(x, 0) + \phi'(x). \end{aligned}$$

Thus, ϕ is such that $\phi'(x) = -u_y(x, 0)$, so that

$$\phi(x) = -\int_0^x u_y(s, 0)\, ds.$$

Thus, $v(x, y)$ becomes

$$v(x, y) = \int_0^y u_y(x, t)\, dt - \int_0^x u_y(s, 0)\, ds,$$

where both u and v satisfy the C-R equations.

Example 3.6.1. *Prove that* $u(x,y) = e^{-x}(x\sin y - y\cos y)$ *is harmonic. Hence find its harmonic conjugate* $v(x,y)$. *Hence find* $f(z)$.

Solution. For the function $u(x,y) = e^{-x}(x\sin y - y\cos y)$, we have

$$\begin{aligned} &u_x = e^{-x}(\sin y - x\sin y + y\cos y); \qquad u_y = e^{-x}(x\cos y + y\sin y - \cos y),\\ &u_{xx} = e^{-x}(-2\sin y + x\sin y - y\cos y); \; u_{yy} = e^{-x}(-x\sin y + 2\sin y + \cos y).\\ \therefore\quad &u_{xx} + u_{yy} = 0. \end{aligned}$$

Hence $u(x,y)$ is harmonic. Let $v(x,y)$ be the conjugate harmonic function. Using the total differential formula, we have,

$$\begin{aligned} dv &= \frac{\partial v}{\partial x}\,dx + \frac{\partial v}{\partial y}\,dy = -\frac{\partial u}{\partial y}\,dx + \frac{\partial u}{\partial x}\,dy, \quad \text{by Eq. (3.22)}\\ &= -e^{-x}(x\cos y + y\sin y - \cos y)\,dx + e^{-x}(\sin y - x\sin y + y\cos y)\,dy\\ &= d\Big\{e^{-x}(y\sin y + x\cos y)\Big\}.\\ \therefore v(x,y) &= e^{-x}(y\sin y + x\cos y + c, \end{aligned}$$

where, $c \in \mathbb{R}$, a constant. The corresponding analytic function f is given by

$$\begin{aligned} f(z) &= u + iv = e^{-x}[x\sin y - y\cos y + i\{y\sin y + x\cos y\}] + ic\\ &= e^{-x}\Big[(x+iy)\Big(\frac{e^{iy} - e^{-iy}}{2i}\Big) + (ix - y)\Big(\frac{e^{iy} + e^{-iy}}{2}\Big)\Big] + ic\\ &= i(x+iy)e^{-(x+iy)} = ize^{-z} + ic. \end{aligned}$$

Example 3.6.2. *Prove that* $u(x,y) = \frac{1}{2}\log(x^2+y^2)$ *is harmonic on* $\mathcal{D}^* = \mathbb{C} - \{0\}$ *and find its harmonic conjugate. Hence find the analytic function* $f(z)$.

Solution. Since $u(x,y) = \frac{1}{2}\log(x^2+y^2)$, we have

$$\frac{\partial u}{\partial x} = \frac{x}{x^2+y^2}; \quad \frac{\partial^2 u}{\partial x^2} = \frac{x^2+y^2-2x^2}{(x^2+y^2)^2} = \frac{y^2-x^2}{(x^2+y^2)^2},$$

$$\frac{\partial u}{\partial y} = \frac{y}{x^2+y^2}; \quad \frac{\partial^2 u}{\partial y^2} = \frac{x^2+y^2-2y^2}{(x^2+y^2)^2} = \frac{x^2-y^2}{(x^2+y^2)^2},$$

so that $u_{xx} + u_{yy} = 0$. Hence $u : \mathcal{D}^* \to \mathbb{R}$ is harmonic on $\mathcal{D}^*$. Let $v : \mathcal{D}^* \to \mathbb{R}$ be the harmonic conjugate of u. Using the C-R equations (3.22), we have

$$\begin{aligned} dv &= \frac{\partial v}{\partial x}\,dx + \frac{\partial v}{\partial y}\,dy = -\frac{\partial u}{\partial y}\,dx + \frac{\partial u}{\partial x}\,dy\\ &= -\frac{y}{x^2+y^2}\,dx + \frac{x}{x^2+y^2}\,dy = \frac{x\,dy - y\,dx}{x^2+y^2}.\\ \therefore v(x,y) &= \tan^{-1}(y/x) + c, \text{ where } c \in \mathbb{R}. \end{aligned}$$

The corresponding analytic function $f(z) = u + iv$ is

$$\begin{aligned} f(z) &= \frac{1}{2}\log(x^2+y^2) + i\Big[\tan^{-1}(y/x) + c\Big]\\ &= \frac{1}{2}\log(z\cdot\bar{z}) + i\tan^{-1}\Big(\frac{1}{i}\frac{z-\bar{z}}{z+\bar{z}}\Big) + ic. \end{aligned}$$

Thus u is the real part of the analytic function $f(z)$ defined in the neighbourhood of each point of $\mathcal{D}^*$.

Example 3.6.3. *If $f(z) = u + iv$ and $u - v = e^x(\cos y - \sin y)$, find w in terms of z.*

Solution. From the given relation $u - v = e^x(\cos y - \sin y)$ we get

$$u_x - v_x = e^x(\cos y - \sin y); u_y - v_y = e^x(-\sin y - \cos y) = -v_x - u_x,$$

where, we use the C-R equations (3.22). Solving, we get

$$u_x = e^x \cos y = v_y \text{ and } v_x = e^x \sin y = -u_y.$$

Using the formula $f'(z) = u_x + iv_x$, we get

$$\begin{aligned} f'(z) &= e^x \cos y + ie^x \sin y = e^x(\cos y + i \sin y) = e^x \cdot e^{iy} = e^{x+iy} = e^z. \\ \therefore f(z) &= e^z + c; \text{ where } c = \text{ constant.} \end{aligned}$$

This is the required analytic function, where $c \in \mathbb{C}$ is a constant.

Example 3.6.4. *Show that a harmonic function $u(x, y)$ satisfies $\frac{\partial^2 u}{\partial z\, \partial \bar{z}} = 0$.*

Solution. Let $u(x, y)$ be a harmonic function. Then $u_{xx} + u_{yy} = 0$ and have continuous second order partial derivatives, i.e., $u_{xy} = u_{yx}$. If $z = x + iy \in \mathbb{C}$, then

$$x = \tfrac{1}{2}(z + \bar{z}) \text{ and } y = \tfrac{1}{2i}(z - \bar{z})$$

$$\tfrac{\partial x}{\partial z} = \tfrac{1}{2}, \tfrac{\partial y}{\partial z} = \tfrac{1}{2i}, \tfrac{\partial x}{\partial \bar{z}} = \tfrac{1}{2} \text{ and } \tfrac{\partial y}{\partial \bar{z}} = -\tfrac{1}{2i}.$$

Thus the given function takes the form

$$u = u(x, y) = u\Big(\frac{z + \bar{z}}{2}, \frac{z - \bar{z}}{2i}\Big).$$

Finally, treating $z, \bar{z}$ as independent variables, by chain rule of differentiation, we get

$$\begin{aligned} \frac{\partial u}{\partial \bar{z}} &= \frac{\partial u}{\partial x} \cdot \frac{\partial x}{\partial \bar{z}} + \frac{\partial u}{\partial y} \cdot \frac{\partial y}{\partial \bar{z}} = \frac{1}{2} \cdot \frac{\partial u}{\partial x} - \frac{1}{2i} \cdot \frac{\partial u}{\partial y} \\ &= \frac{1}{2}\Big[\frac{\partial u}{\partial x} - i\frac{\partial u}{\partial y}\Big] = \frac{1}{2}(u_x - iu_y). \end{aligned}$$

Again, differentiating with respect to z, we get

$$\begin{aligned} \frac{\partial^2 u}{\partial z\, \partial \bar{z}} &= \frac{\partial}{\partial z}\Big(\frac{\partial u}{\partial \bar{z}}\Big) = \frac{1}{2}\frac{\partial}{\partial z}(u_x - iu_y) \\ &= \frac{1}{2}\Big[\frac{\partial}{\partial x}(u_x - iu_y)\frac{\partial x}{\partial z} + \frac{\partial}{\partial y}(u_x - iu_y)\frac{\partial y}{\partial \bar{z}}\Big] \\ &= \frac{1}{2}\Big[(u_{xx} - iu_{xy}) \cdot \frac{1}{2} - (u_{yx} - iu_{yy}) \cdot \frac{1}{2i}\Big] \\ &= \frac{1}{4}\Big[(u_{xx} + u_{yy}) + i(u_{yx} - u_{xy})\Big] = 0, \end{aligned}$$

which is known as the *complex form of Laplace equation.*

Example 3.6.5. *Let $f(z) = u + iv$ be an analytic function in a region $\mathcal{D}$. Show that both u and v satisfy the equation*

$$\frac{\partial^2 \psi}{\partial r^2} + \frac{1}{r} \cdot \frac{\partial \psi}{\partial r} + \frac{1}{r}\frac{\partial^2 \psi}{\partial \theta^2} = 0.$$

Solution. Differentiate the first equation partially with respect to r and the second partially with respect to θ of Eq. (3.36). Then,

$$\begin{aligned} v_{r\theta} &= \frac{\partial^2 v}{\partial r\,\partial\theta} = \frac{\partial}{\partial r}\Big(\frac{\partial v}{\partial \theta}\Big) = \frac{\partial}{\partial r}\Big(r\,\frac{\partial u}{\partial r}\Big) = r\frac{\partial^2 u}{\partial r^2} + \frac{\partial u}{\partial r} \\ v_{\theta r} &= \frac{\partial^2 v}{\partial\theta\partial r} = \frac{\partial}{\partial\theta}\Big(\frac{\partial v}{\partial r}\Big) = \frac{\partial}{\partial\theta}\Big(-\frac{1}{r}\frac{\partial u}{\partial\theta}\Big) = -\frac{1}{r^2}\frac{\partial^2 u}{\partial\theta^2}. \end{aligned}$$

Assuming the continuity of the second partial derivative, i.e., $v_{r\theta} = v_{\theta r}$, we get

$$r\frac{\partial^2 u}{\partial r^2} + \frac{\partial u}{\partial r} = -\frac{1}{r^2}\frac{\partial^2 u}{\partial\theta^2} \Rightarrow \frac{\partial^2 u}{\partial r^2} + \frac{1}{r}\cdot\frac{\partial u}{\partial r} + \frac{1}{r}\frac{\partial^2 u}{\partial\theta^2} = 0.$$

Similarly, eliminating u, we find that

$$\frac{\partial^2 v}{\partial r^2} + \frac{1}{r}\cdot\frac{\partial v}{\partial r} + \frac{1}{r}\frac{\partial^2 v}{\partial\theta^2} = 0.$$

This equation is called the *polar form of Poisson equation.*

RESULT **25.** [Milne–Thomson's method of constructing regular function]:
(a) Given $f(z) = u(x, y) + iv(x, y)$ and $x = \frac{1}{2}(z + \bar{z})$, $y = \frac{1}{2i}(z - \bar{z})$, we may write

$$f(z) = u\Big[\frac{1}{2}(z+\bar{z}), \frac{1}{2i}(z-\bar{z})\Big] + i\,v\Big[\frac{1}{2}(z+\bar{z}), \frac{1}{2i}(z-\bar{z})\Big]. \tag{3.40}$$

We may regard this as a formal identity in two independent variables, z and $\bar{z}$.
(b) Let $\dfrac{\partial u}{\partial x} = \psi_1(x, y)$ and $\dfrac{\partial u}{\partial y} = \psi_2(x, y)$. Then

$$\begin{aligned} f'(z) &= \frac{\partial u}{\partial x} - i\frac{\partial u}{\partial y} = \psi_1(x, y) - i\psi_2(x, y) \\ \Rightarrow f'(x) &= \psi_1(x, 0) - i\psi_2(x, 0) \quad \text{putting } y = 0 \\ \Rightarrow f'(z) &= \psi_1(z, 0) - i\psi_2(z, 0) \quad \text{replacing } x \text{ by } z \\ \Rightarrow f(z) &= \int \Big[\psi_1(z, 0) - i\psi_2(z, 0)\Big]\, dz + c_1. \end{aligned} \tag{3.41}$$

(c) Let $\dfrac{\partial v}{\partial y} = \phi_1(x, y)$ and $\dfrac{\partial v}{\partial x} = \phi_2(x, y)$. Then

$$\begin{aligned} f'(z) &= \frac{\partial v}{\partial y} + i\frac{\partial v}{\partial x} = \phi_1(x, y) + i\phi_2(x, y) \\ \Rightarrow f'(x) &= \phi_1(x, 0) + i\phi_2(x, 0) \quad \text{putting } y = 0 \\ \Rightarrow f'(z) &= \phi_1(z, 0) + i\phi_2(z, 0) \quad \text{replacing } x \text{ by } z \\ \Rightarrow f(z) &= \int \Big[\phi_1(z, 0) + i\phi_2(z, 0)\Big]\, dz + c_2. \end{aligned} \tag{3.42}$$

The above method is known as the *Milne–Thomson's method.*

Example 3.6.6. *Construct the analytic function $f = u + iv$, where $u(x, y) = x^3 - 3xy^2 + 3x + 1$, using Milne–Thomson's method.*

Solution. Let $u_x = 3x^2 - 3y^2 + 3 = \psi_1(x, y)$ and $u_y = -6xy = \psi_2(x, y)$. Then using Eq. (3.41), we get

$$\begin{aligned} f(z) &= \int \Big[\psi_1(z, 0) - i\psi_2(z, 0)\Big]\, dz + c_1 \\ &= \int \Big[3z^2 + 3 - i\,0\Big]\, dz + c_1 = z^3 + 3z + c_1. \end{aligned}$$

Example 3.6.7. *Show that $u(x, y) = \sin x \cosh y + 2\cos x \sinh y + x^2 - y^2 + 4xy$ is harmonic. Find the analytic function f with the given u for its real part.*

Solution. Since $u(x, y) = \sin x \cosh y + 2\cos x \sinh y + x^2 - y^2 + 4xy$,

$$\begin{aligned} u_x &= \cos x \cosh y - 2\sin x \sinh y + 2x + 4x = \psi_1(x, y) \\ u_y &= \sin x \sinh y + 2\cos x \cosh y - 2y + 4x = \psi_2(x, y) \\ u_{xx} &= -\sin x \cosh y - 2\cos x \sinh y + 2 \\ u_{yy} &= \sin x \cosh y + 2\cos x \sinh y - 2, \end{aligned}$$

so that $u_{xx} + u_{yy} = 0$. Hence $u(x, y)$ is a harmonic function. Using Milne–Thompson formula (3.41),

$$\begin{aligned} f(z) &= \int \Big[\psi_1(z, 0) - i\psi_2(z, 0)\Big]\, dz + c \\ &= \int \Big[\cos z + 2z - i(2\cos z + 4z)\Big] dz \\ &= \sin z + z^2 - 2i \sin z - 2iz^2 + c, \end{aligned}$$

where c is a constant. This is the required analytic function.

Example 3.6.8. *Let $f(z) = u + iv$ be an analytic function in a domain $\mathcal{D}$. Then prove that $f(z)$ is constant in $\mathcal{D}$ if any one of the following conditions hold:*

(i)	$f'(z) = 0, \forall z \in \mathcal{D}$	(ii)	$\mathrm{Re}\,\{f(z)\} = u(x, y) = constant$
(iii)	$\mathrm{Im}\,\{f(z)\} = v(x, y) = constant$	(iv)	$\lvert f(z)\rvert = \sqrt{u^2 + v^2} = constant$
(v)	$\arg f(z) = constant$		

Solution. (i) In view of the C-R equations (3.22), we have the expression for $f'(z)$ as in Eq.(3.31). Thus if $f'(z) = 0$, then

$$u_x + iv_x = 0 \text{ and } v_y - iu_y = 0.$$

Equating the real and imaginary parts to 0, we get $u_x = 0$; $v_x = 0$; $v_y = 0$ and $u_y = 0$ in $\mathcal{D}$. Now, as $\mathcal{D}$ is connected, we have $u =$ constant and $v =$ constant. Consequently, $f = u + iv$ is also constant in $\mathcal{D}$. The above result is not true if the domain of $f(z)$ is not connected. For example, let $\mathcal{D} = \{z : |z| < 1\} \cup \{z : |z| > 2\}$. Clearly, $\mathcal{D}$ is not a connected subset of $\mathbb{C}$ so that $\mathcal{D}$ is not a region. Let $f : \mathcal{D} \to \mathbb{C}$ be defined by

$$f(z) = \begin{cases} 1 & \text{if } |z| < 1 \\ 2 & \text{if } |z| > 2. \end{cases}$$

Clearly, f is analytic and $f'(z) = 0 \forall z \in \mathcal{D}$ but f is not a constant in $\mathcal{D}$.
(ii) If Re$\{f(z)\} = u =$ constant, then $u_x = 0 = u_y$. In view of the C-R equations (3.22), we get

$$f'(z) = u_x + iv_x = u_x - iu_y = 0 \quad \text{as } u_x = 0 = u_y.$$

Hence $f(z)$ is a constant.
(iii) If Im$\{f(z)\} = v =$ constant, then $v_x = 0 = v_y$. In view of the C-R equations (3.22), we get

$$f'(z) = u_x + iv_x = v_y + iv_x = 0; \quad \text{as } v_x = 0 = v_y.$$

Hence $f(z)$ is a constant.
(iv) As usual, let $f(z) = u(x,y) + iv(x,y)$. If $|f| = 0$, then $f = 0$. Otherwise,

$$|f(z)|^2 = u^2 + v^2 = c; \text{ where } c \in \mathbb{R}^+.$$

Taking the partial derivatives with respect to x and y, we have

$$2u\frac{\partial u}{\partial x} + 2v\frac{\partial v}{\partial x} = 0 \text{ and } 2u\frac{\partial u}{\partial y} + 2v\frac{\partial v}{\partial y} = 0$$

$$\Rightarrow \quad u\frac{\partial u}{\partial x} - v\frac{\partial u}{\partial y} = 0 \text{ and } u\frac{\partial u}{\partial y} + v\frac{\partial u}{\partial x} = 0; \quad \text{using Eq. (3.22)}$$

$$\Rightarrow \quad (u^2 + v^2)u_x = 0 \text{ and } (u^2 + v^2)u_y = 0$$

$$\Rightarrow \quad u_x = u_y = 0; \text{ as } (u^2 + v^2) = c \neq 0.$$

Similarly, $v_x = v_y = 0$. Therefore,

$$f'(z) = u_x + iv_x = u_y + iv_y = 0 \text{ for } z \in \mathcal{D}.$$

Thus f is a constant. Thus the analytic function cannot have a constant absolute value without reducing the constant.
(v) As $f(z) = u(x,y) + iv(x,y)$, so $\arg f(z) = \tan^{-1}\frac{v}{u}$. Since $\arg f(z) =$ constant, we have $\arg f(z) = c.$ (say), where c is a constant. Therefore

$$\tan^{-1}\frac{v}{u} = c \Rightarrow \frac{v}{u} = \tan c.$$

$$\therefore \quad (1 - i\tan c)f(z) = \left(1 - i\frac{v}{u}\right)(u + iv) = \frac{(u+iv)(u-iv)}{u} = \frac{1}{u}|f(z)|^2.$$

Now $(1 - i\tan c)f(z)$ is analytic and is a real-valued function.

$$\therefore \text{Im}\Big[(1 - i\tan c)\ f(z)\Big] = 0$$

shows that $(1 - i\tan c)\ f(z)$ is constant. Since $(1 - i\tan c)$ is a constant, it follows that any analytic function $f(z)$ with $\arg f(z)$ constant is itself a constant function.

Example 3.6.9. *If $f(z)$ and $\overline{f(z)}$ are analytic in a domain $\mathcal{D}$, show that $f(z)$ is constant in $\mathcal{D}$.*

Solution. Let $f(z) = u(x,y) + iv(x,y)$. Then

$$\overline{f(z)} = u(x,y) - iv(x,y) = u(x,y) + i\{-v(x,y)\}.$$

Given $f(z)$ is analytic in a domain $\mathcal{D}$, it satisfies C-R equations (3.22). Therefore

$$2u_x = 0 \text{ and } 2u_y = 0.$$

$$\Rightarrow \quad f'(z) = u_x + iv_x = 0 \Rightarrow f(z) \text{ is constant in } \mathcal{D}.$$

Example 3.6.10. *If $u+v=\sin 2x/(\cosh 2y-\cos 2x)$, and $f(z)=u+iv$ is an analytic function of $z\in\mathbb{C}$, find $f(z)$ in terms of z.*

Solution. Here $u+iv=f(z)$, so that $iu-v=if(z)$. Adding, we get

$$(1+i)f(z)=(u-v)+i(u+v)=U+iV;\text{ where, } U=u-v, V=u+v.$$

Since $f(z)=u+iv$ is analytic, so $F(z)=(1+i)f(z)=U+iV$ is also an analytic function. Now,

$$\frac{\partial V}{\partial y}=u_y+v_y=-\frac{\sin 2x\cdot 2\sinh 2y}{(\cosh 2y-\cos 2x)^2}=\phi_1(x,y)$$

$$\frac{\partial V}{\partial x}=-\frac{2\cos 2x(\cosh 2y-\cos 2x)+2\sin^2 2x}{(\cosh 2y-\cos 2x)^2}=\phi_2(x,y).$$

By using Milne–Thompson formula (3.42), the corresponding analytic function f is given by

$$\begin{aligned}F(z) &= (1+i)f(z)=\int\Big[\phi_1(z,0)+i\phi_2(z,0)\Big]\,dz+c\\ &= 0+i\int\frac{\cos 2z(1-\cos 2z)-2\sin^2 2z}{(1-\cos 2z)^2}\,dz+c\\ &= -2i\int\frac{dz}{1-\cos 2z}+c=-i\int\operatorname{cosec}^2 z\,dz+c=i\cot z+c\\ f(z) &= \frac{i}{i+1}\cot z+\frac{c}{1+i}=\frac{1+i}{2}\cot z+c_1,\end{aligned}$$

where c_1 is a constant.

Example 3.6.11. *If $u-v=(\cos x+\sin x-e^{-y})/(2\cos x-e^{y}-e^{-y})$, and $f(z)=u+iv$ is an analytic function of z, find $f(z)$ subject to the condition $f(\pi/2)=0$.*

Solution. Here $u+iv=f(z)$ so that $iu-v=if(z)$. Adding, we get

$$(1+i)f(z)=(u-v)+i(u+v)=U+iV;\text{ where, } U=u-v, V=u+v.$$

Since $f(z)$ is analytic, so $F(z)=(1+i)f(z)=U+iV$ is also an analytic function. Now,

$$U=u-v=\frac{\cos x+\sin x-e^{-y}}{2\cos x-2\cosh x}=\frac{1}{2}\left[1+\frac{\sin x+\sinh y}{\cos x-\cosh y}\right].$$

Differentiating partially with respect to x and y, we get

$$U_x=\frac{1}{2}\cdot\frac{1+\sin x+\sinh y-\cos x\cosh y}{(\cos x-\cosh y)^2}=\psi_1(x,y);$$

$$U_y=\frac{1}{2}\cdot\frac{-1+\sin x+\sinh y+\cos x\cosh y}{(\cos x-\cosh y)^2}=\psi_2(x,y).$$

Using Milne–Thomson formula (3.41), the corresponding analytic function $f(z)$ is given by

$$\begin{aligned}F(z) &= (1+i)f(z)=\int\Big[\psi_1(z,0)-i\psi_2(z,0)\Big]\,dz+c\\ &= \frac{1}{2}(1+i)\int\frac{1}{1-\cos z}\,dz=\frac{1}{4}(1+i)\int\operatorname{cosec}^2(z/2)\,dz+c\\ &= -\frac{1}{2}(1+i)\cot(z/2)+c.\\ f(z) &= -\frac{1}{2}\cot(z/2)+\frac{c}{1+i}=-\frac{1}{2}\cot(z/2)+c_1,\end{aligned}$$

where c_1 is a constant. Given that $f(\pi/2) = 0$, $c_1 = \frac{1}{2}$. Therefore, $f(z) = \frac{1}{2}(1 - \cot(z/2))$.

Example 3.6.12. *If* $\operatorname{Im}\{f'(z)\} = 6x(2y-1)$ *and* $f(0) = 3 - 2i$, $f(1) = 6 - 5i$, *find* $f(1+i)$.

Solution. Let $f(z) = u + iv$. Then according to formula (3.26), $f'(z) = u_x + iv_x$. Given

$$\begin{aligned}
&\operatorname{Im}\{f'(z)\} = 6x(2y-1) = v_x \\
\Rightarrow\quad & v(x,y) = 3x^2(2y-1) + F(y); \quad F(y) = \text{ arbitrary function} \\
\Rightarrow\quad & v_y = 6x^2 + F'(y) = u_x; \quad \text{using Eq. (3.22)} \\
\Rightarrow\quad & u(x,y) = 2x^3 + x\,F'(y).
\end{aligned}$$

Now, using the C-R equations (3.22), we get

$$\begin{aligned}
& v_x = 6x(2y-1) = -u_y = -xF''(y) \\
\Rightarrow\quad & F''(y) = -6(2y-1) \Rightarrow F(y) = 3y^2 - 2y^3 + c_1 y + c_2.
\end{aligned}$$

Therefore, $u(x,y) = 2x^3 + x(6y - 6y^2 + c_1)$ and $v(x,y) = 3x^2(2y-1) + 3y^2 - 2y^3 + c_1 y + c_2$. Let $v_y = \phi_1(x,y)$ and $v_x = \phi_2(x,y)$. Then using Milne–Thomson's formula (3.42), we get

$$\begin{aligned}
f(z) &= \int \Big[\phi_1(z,0) + i\phi_2(z,0)\Big]\,dz + c_3 \\
&= \int [6z^2 + c_1 - i\,6z]dz + c_3 = 2z^3 + c_1 z - 3iz^2 + c_3.
\end{aligned}$$

Given $f(0) = 3 - 2i$, so, $c_3 = 3 - 2i$; and given $f(1) = 6 - 5i$, so,

$$6 - 5i = 2 + c_1 - 3i + c_3 \Rightarrow c_1 = 1.$$

Thus the analytic function is $f(z) = 2z^3 + z - 3iz^2 + 3 - 2i$. Accordingly,

$$f(1+i) \;=\; 2(1+i)^3 + (1+i) - 3i(1+i)^2 + 3 - 2i = 6 + 3i.$$

Example 3.6.13. *If* $f(z) = u + iv$ *is an analytic function of* $z = x + iy$ *and* ψ *any function of* x *and* y *with differential coefficient of the first and second orders, then prove that*

(i) $\left(\dfrac{\partial\psi}{\partial x}\right)^2 + \left(\dfrac{\partial\psi}{\partial y}\right)^2 = \left[\left(\dfrac{\partial\psi}{\partial u}\right)^2 + \left(\dfrac{\partial\psi}{\partial v}\right)^2\right]\Big|f'(z)\Big|^2$;

(ii) $\dfrac{\partial^2\psi}{\partial x^2} + \dfrac{\partial^2\psi}{\partial y^2} = \left[\dfrac{\partial^2\psi}{\partial u^2} + \dfrac{\partial^2\psi}{\partial v^2}\right]\Big|f'(z)\Big|^2$; (iii) $\left[\dfrac{\partial^2}{\partial x^2} + \dfrac{\partial^2}{\partial y^2}\right]\Big|f(z)\Big|^2 = 4\Big|f'(z)\Big|^2$.

Solution. Since $f(z) = u(x,y) + iv(x,y)$ is an analytic function of z, we have

$$|f'(z)| = \frac{\partial u}{\partial x} + i\frac{\partial v}{\partial x} \Rightarrow |f'(z)|^2 = \left(\frac{\partial u}{\partial x}\right)^2 + \left(\frac{\partial v}{\partial x}\right)^2.$$

(i) Since ψ is any function of x, y, we have, using chain rule of differentiation

$$\begin{aligned}
\frac{\partial\psi}{\partial x} &= \frac{\partial\psi}{\partial u}\cdot\frac{\partial u}{\partial x} + \frac{\partial\psi}{\partial v}\cdot\frac{\partial v}{\partial x} \\
\frac{\partial\psi}{\partial y} &= \frac{\partial\psi}{\partial u}\cdot\frac{\partial u}{\partial y} + \frac{\partial\psi}{\partial v}\cdot\frac{\partial v}{\partial y} = -\frac{\partial\psi}{\partial u}\cdot\frac{\partial v}{\partial x} + \frac{\partial\psi}{\partial v}\cdot\frac{\partial u}{\partial x}
\end{aligned}$$

in view of the C-R equations (3.22). Squaring and adding, we get

$$\begin{aligned}\left(\frac{\partial\psi}{\partial x}\right)^2+\left(\frac{\partial\psi}{\partial y}\right)^2 &= \left[\frac{\partial\psi}{\partial u}\cdot\frac{\partial u}{\partial x}+\frac{\partial\psi}{\partial v}\cdot\frac{\partial v}{\partial x}\right]^2+\left[-\frac{\partial\psi}{\partial u}\cdot\frac{\partial v}{\partial x}+\frac{\partial\psi}{\partial v}\cdot\frac{\partial u}{\partial x}\right]^2\\ &= \left(\frac{\partial\psi}{\partial u}\right)^2\left(\frac{\partial u}{\partial x}\right)^2+\left(\frac{\partial\psi}{\partial v}\right)^2\left(\frac{\partial v}{\partial x}\right)^2+\left(\frac{\partial\psi}{\partial u}\right)^2\left(\frac{\partial v}{\partial x}\right)^2+\left(\frac{\partial\psi}{\partial v}\right)^2\left(\frac{\partial u}{\partial x}\right)^2\\ &= \left[\left(\frac{\partial\psi}{\partial u}\right)^2+\left(\frac{\partial\psi}{\partial v}\right)^2\right]\left[\left(\frac{\partial u}{\partial x}\right)^2+\left(\frac{\partial v}{\partial x}\right)^2\right]\\ &= \left[\left(\frac{\partial\psi}{\partial u}\right)^2+\left(\frac{\partial\psi}{\partial v}\right)^2\right]\left|f'(z)\right|^2.\end{aligned}$$

(ii) Since $w=f(z)=u+iv$, and $\bar{w}=f(\bar{z})=u-iv$, we have $u=(w+\bar{w})/2$; $v=(w-\bar{w})/2i$, where z and $\bar{z}$ are independent variables. Now,

$$\frac{\partial}{\partial w}=\frac{\partial u}{\partial w}\frac{\partial}{\partial u}+\frac{\partial v}{\partial w}\frac{\partial}{\partial v}=\frac{1}{2}\left(\frac{\partial}{\partial u}-i\frac{\partial}{\partial v}\right)$$

$$\frac{\partial}{\partial \bar{w}}=\frac{\partial u}{\partial \bar{w}}\frac{\partial}{\partial u}+\frac{\partial v}{\partial \bar{w}}\frac{\partial}{\partial v}=\frac{1}{2}\left(\frac{\partial}{\partial u}+i\frac{\partial}{\partial v}\right)$$

or, $$\frac{\partial}{\partial w}\frac{\partial}{\partial \bar{w}}=\frac{1}{4}\left(\frac{\partial}{\partial u}-i\frac{\partial}{\partial v}\right)\left(\frac{\partial}{\partial u}+i\frac{\partial}{\partial v}\right)$$

or, $$4\,\frac{\partial^2}{\partial w\partial\bar{w}}=\frac{\partial^2}{\partial u^2}+\frac{\partial^2}{\partial v^2}\Rightarrow\frac{\partial^2\psi}{\partial u^2}+\frac{\partial^2\psi}{\partial v^2}=4\,\frac{\partial^2\psi}{\partial w\partial\bar{w}}.$$

Similarly, as $z=x+iy$ and $\bar{z}=x-iy$, we have

$$\frac{\partial^2\psi}{\partial x^2}+\frac{\partial^2\psi}{\partial y^2}=4\,\frac{\partial^2\psi}{\partial z\partial\bar{z}}.$$

Therefore, from the previous result, we get

$$\begin{aligned}\frac{\partial^2\psi}{\partial u^2}+\frac{\partial^2\psi}{\partial v^2} &= 4\,\frac{\partial^2\psi}{\partial w\partial\bar{w}}=4\left(\frac{\partial}{\partial z}\frac{\partial z}{\partial w}\right)\left(\frac{\partial}{\partial\bar{z}}\frac{\partial\bar{z}}{\partial\bar{w}}\right)\psi\\ &= 4\left[\frac{1}{f'(z)}\frac{\partial}{\partial z}\right]\left[\frac{1}{f'(\bar{z})}\frac{\partial}{\partial\bar{z}}\right] \quad \text{as } z,\bar{z} \text{ are independent.}\\ &= \frac{4}{f'(z)\,f'(\bar{z})}\frac{\partial^2\psi}{\partial z\partial\bar{z}}=\frac{1}{|f'(z)|^2}\left[\frac{\partial^2\psi}{\partial x^2}+\frac{\partial^2\psi}{\partial y^2}\right].\end{aligned}$$

Hence the result follows.
(iii) Proceding as in (ii), we have

$$\begin{aligned}\left[\frac{\partial^2}{\partial x^2}+\frac{\partial^2}{\partial y^2}\right]|f(z)|^2 &= 4\,\frac{\partial^2}{\partial z\partial\bar{z}}\left[f(z)\,f(\bar{z})\right]\\ &= 4\,\frac{\partial}{\partial z}\left[f(z)\,f'(\bar{z})\right]=4f'(z)\,f'(\bar{z})=4|f'(z)|^2.\end{aligned}$$

Example 3.6.14. *If $f(z)$ is an analytic function of z, prove that*

$$\left[\frac{\partial^2}{\partial x^2}+\frac{\partial^2}{\partial y^2}\right]|\mathrm{Re}\,\{f(z)\}|^2=2|f'(z)|^2.$$

Solution. As $f(z) = u + iv$, Re $f(z) = u$. Given $f(z)$ is analytic, $f'(z) = u_x + iv_x$.

$$\therefore \quad \frac{\partial}{\partial x}u^2 = 2u\frac{\partial u}{\partial x}; \quad \frac{\partial^2}{\partial x^2}u^2 = 2\left(\frac{\partial u}{\partial x}\right)^2 + 2u\frac{\partial^2 u}{\partial x^2}$$

$$\frac{\partial}{\partial y}u^2 = 2u\frac{\partial u}{\partial y}; \quad \frac{\partial^2}{\partial y^2}u^2 = 2\left(\frac{\partial u}{\partial y}\right)^2 + 2u\frac{\partial^2 u}{\partial y^2}.$$

Since u is a harmonic function $u_{xx} + u_{yy} = 0$, using the Cauchy–Riemann equations (3.22) we get

$$\begin{aligned}\left[\frac{\partial^2}{\partial x^2} + \frac{\partial^2}{\partial y^2}\right]u^2 &= 2\left[\left(\frac{\partial u}{\partial x}\right)^2 + \left(\frac{\partial u}{\partial y}\right)^2\right] + 2u\left[\frac{\partial^2 u}{\partial x^2} + \frac{\partial^2 u}{\partial y^2}\right] \\ &= 2\left[\left(\frac{\partial u}{\partial x}\right)^2 + \left(\frac{\partial u}{\partial y}\right)^2\right] = 2\left[\left(\frac{\partial u}{\partial x}\right)^2 + \left(\frac{\partial v}{\partial x}\right)^2\right] = 2|f'(z)|^2.\end{aligned}$$

3.7 Singular Points

A function $f(z)$ may be analytic in a domain except possibly for a finite number of points. These points are called *singular points* or *singularities* of $f(z)$ in the domain.

Let $z_0 \in \mathbb{C}$ be a complex number and the domain $\mathcal{D} = \mathcal{S}(z_0, \delta) = \{z \in \mathbb{C} : d(z, z_0) < \delta\}$ be an open disc centred at z_0. Let $\mathcal{D}^* = \mathcal{D} - \{z_0\} = \{z \in \mathbb{C} : 0 < |z - z_0| < \delta\}$ be an open set.

Definition 3.7.1. A point z_0 is called a *singular point* or *singularity* of a function f if

(i) $f(z)$ is not analytic at z_0 and

(ii) f is analytic at some point of every disc $\mathcal{D}$.

Thus at a singular point z_0 the function ceases to be analytic but is analytic at some point in every neighbourhood of the point z_0. For example,

(i) Consider, the function $f(z) = 1/z$. Then $f'(z) = -1/z^2$ for all $z \neq 0$. Thus $f(z)$ is analytic except at $z = 0$. Therefore, $z = 0$ is evidently a singular point of the function $f(z) = 1/z$.

(ii) Let $f(z) = |z|^2$. Since it is nowhere analytic, it has no singular points.

Singular points are of two types: (i) isolated singular points and (ii) non-isolated singular points.

If z_0 is not a singular point and we can find $\delta > 0$ such that $|z - z_0| = \delta$ encloses no singular point, then we call z_0 an *ordinary point* of the function $f(z)$.

3.7.1 Isolated Singularity

Definition 3.7.2. A point z_0 is called an *isolated singularity* or *isolated singular point* of a complex valued function $f(z)$ if

(i) it is not analytic at z_0 and

(ii) there exists $\delta > 0$ such that $f(z)$ is analytic in $\mathcal{D}^*$.

Observe that f must be discontinuous at isolated singular points. For example, let $f(z) = z/(z^2 - 1)$. Then $f(z)$ is analytic everywhere except at $z = -1, 1$. Therefore, $z = -1, 1$ are the isolated singular points.

If no such δ can be found, we call z_0 a *non-isolated singularity.* For example, if $f(z) = \log z$, then $\{z \in \mathbb{C} : y = 0, x \leq 0\}$ are the non-isolated singular points.

3.7.2 Pole

Definition 3.7.3. If $z = z_0$ is an isolated singularity of f, then z_0 is a *pole* of f if

$$\lim_{z\to z_0} |f(z)| = \infty.$$

That is, for any $M > 0$, there is a number $\delta > 0$ such that $|f(z)| \geq M$ whenever $z \in \mathcal{D}^*$.

Thus, if f has a pole of order $m \in \mathbb{N}$ at z_0, then m is the smallest integer such that

$$\lim_{z\to z_0} (z - z_0)^m f(z) = c \neq 0.$$

A pole of order 1 ($m = 1$) is called a *simple pole* and a pole of order 2 ($m = 2$) is called a *double pole.* For other finite values of m, the corresponding pole is called *a multiple pole of order* m. For example:

(i) Let $f(z) = 1/(z-1)^4$, the isolated singularity in the finite z-plane be located at $z = 1$. Now

$$\lim_{z\to 1}(z-1)^4 f(z) = \lim_{z\to 1}(z-1)^4 \frac{1}{(z-1)^4} = 1 \neq 0.$$

Therefore $f(z)$ has a pole of order 4 at $x = 1$.

(ii) Let $f(z) = (z^2 + 3z - 10)/(z-2)$. The isolated singularity in the finite z-plane be located at $z = 2$. Now,

$$\lim_{z\to 2}(z-2)f(z) = \lim_{z\to 2}(z-2)\frac{z^2+3z-10}{z-2} = 0.$$

Therefore, $z = 2$ is not a pole of $f(z)$.

(iii) Let $f(z) = (z^2 + 2z - 3)/(z-1)^2$. The isolated singularity in the finite z-plane is located at $z = 1$. Now,

$$\begin{aligned}\lim_{z\to 1}(z-1)^2 f(z) &= \lim_{z\to 1}(z-1)^2\frac{z^2+2z-3}{(z-1)^2} = \lim_{z\to 1}(z^2+2z-3) = 0\\ \lim_{z\to 1}(z-1)f(z) &= \lim_{z\to 1}(z-1)\frac{z^2+2z-3}{(z-1)^2} = \lim_{z\to 1}(z+3) = 4 \neq 0.\end{aligned}$$

Therefore $f(z)$ has a pole of order 1 at $z = 1$.

(iv) Let $f(z) = (3z-2)/[(z-1)^2(z+1)(z-4)]$. The isolated singularities in the finite z-plane are located at $z = 1, -1$ and $z = 4$. Now,

$$\begin{aligned}\lim_{z\to 1}(z-1)^2 f(z) &= \lim_{z\to 1}\frac{3z-2}{(z+1)(z-4)} = -\frac{1}{6} \neq 0.\\ \lim_{z\to -1}(z+1)f(z) &= \lim_{z\to -1}\frac{3z-2}{(z-1)^2(z-4)} = \frac{1}{4} \neq 0.\\ \lim_{z\to 4}(z-4)f(z) &= \lim_{z\to 4}\frac{3z-2}{(z-1)^2(z+1)} = \frac{2}{9} \neq 0.\end{aligned}$$

Therefore $f(z)$ has a double pole at $z = 1$ and simple poles at $z = -1, 4$.

(v) Let $f(z) = \sinh z/z^4$. The isolated singularity in the finite z-plane is located at $z = 0$. Now,

$$\lim_{z\to 0} z^4 f(z) = \lim_{z\to 0} z^4 \frac{\sinh z}{z^4} = \lim_{z\to 0} \sinh z = 0$$

$$\lim_{z\to 0} z^3 f(z) = \lim_{z\to 0} z^3 \frac{\sinh z}{z^4} = \lim_{z\to 0} \frac{\sinh z}{z} = 1 \neq 0.$$

Therefore $f(z)$ has a pole of order 3 at $z = 0$.

(vi) The function $f(z) = z^4 \ \sin \frac{1}{z+1}$ has a pole of order five at the point infinity.

Definition 3.7.4. Let m be a positive integer. If

$$f(z) = (z - z_0)^m g(z), \quad \text{where } g(z_0) \neq 0,$$

then z_0 is called a *zero* of order m of $f(z)$. If $m = 1$, z_0 is called a *simple zero*. In such a case z_0 is a pole of order m of the function $1/f(z)$.

Theorem 3.12. *Zeros of an analytic function are isolated.*

Proof: Let $f(z)$ be analytic in the neighbourhood of z_0 and $f(z_0) = 0$. So if the zero z_0 of $f(z)$ is of some finite order, say r, then,

$$f(z) = (z - z_0)^r \ g(z); \ \ g(z_0) \neq 0,$$

where $g(z)$ is analytic at z_0. Also, $g(z)$ is clearly continuous there. By neighbourhood property of continuous function, $g(z) \neq 0$ in the deleted neighbourhood $\mathcal{N}'(z_0, \delta)$ of z_0. So, in $\mathcal{N}'(z_0, \delta)$, the analytic function $f(z) \neq 0$. □

Therefore, given a zero of an analytic function, there is always a small discs surrounding that zero where there are no other zeros. So, if we have different zeros, they are separated from each other by small open discs centred at the zeros.

RESULT **26.** [Identity theorem]: If an analytic function has zeros on every point of a set which has an accumulation point in the domain of analyticity, then the analytic function is identically zero. In other words, if we have a sequence of zeros of an analytic function which converges to a point where the analytic function is defined and analytic, then, due to continuity, that must be zero. Since the neighbourhood of that point contains zero, so the said point is not isolated.

3.7.3 Removable Singularity

Definition 3.7.5. Let f be an analytic function on $\mathcal{D}-\{z_0\}$. Then, the point z_0 is said to be a *removable singularity* of f if $\lim\limits_{z\to z_0} f(z)$ exists finitely in $\mathbb{C}$.

Here, in fact, $f(z)$ is not defined at z_0 but $\lim\limits_{z\to z_0} f(z)$ exists.

For example:

(i) The function $f(z) = \dfrac{z^2 - z_0^2}{z - z_0}$; $\quad z \neq z_0$, has removal singularity at $z = z_0$ because, $f(z_0)$ is not defined but $\lim\limits_{z\to z_0} f(z) = 2z_0$.

(ii) Let $f(z) = (z - \sin z)/z^3$. Clearly, $z = 0$ is an isolated singular point for $f(z)$ in the finite z-plane. Also,

$$\lim_{z\to 0} f(z) = \lim_{z\to 0} \frac{z - \sin z}{z^3} = \lim_{z\to 0} \frac{1 - \cos z}{3z^2} = \lim_{z\to 0} \frac{\sin z}{6z} = \frac{1}{6}.$$

Hence $z = 0$ is a removable singularity. The singularity can be removed by defining $f(0) = 1/6$.

RESULT **27.** The isolated singularity z_0 of f is the removable singularity if

(i) f is bounded in a deleted neighbourhood of z_0;

(ii) $\lim_{z \to z_0} f(z)$ exists; and (iii) $\lim_{z \to z_0} (z - z_0) f(z) = 0$.

In other words, f can be extended to z_0 too so that f is analytic in the domain. This is known as *Riemann's removable singularity theorem.*

(i) Let f and g are analytic and both have a zero of order n at z_0. Then they can be written as

$$f(z) = (z - z_0)^n f_0(z), \quad g(z) = (z - z_0)^n g_0(z),$$

where f_0 and g_0 are both analytic and non-zero at z_0. Thus

$$\lim_{z \to z_0} \frac{f(z)}{g(z)} = \frac{f_0(z_0)}{g_0(z_0)}$$

exists. This means that $f(z)/g(z)$ has a removable singularity at z_0. For example, the functions $f(z) = \dfrac{e^z - 1}{z}$ and $f(z) = \dfrac{z - \sin z}{z \sin z}$ has a removable singularity at $z = 0$.

(ii) Let f be analytic in a deleted neighbourhood of z_0. Then f has a pole of order n if and only if

(a) $(z - z_0)^n f$ has a removable singularity at z_0, i.e., $\lim_{z \to z_0} (z - z_0)^{n+1} f(z) = 0$;

(b) $\lim_{z \to z_0} (z - z_0)^n f(z) \neq 0$ and $(z - z_0)^n f(z)$ is bounded.

3.7.4 Essential Singularity

Definition 3.7.6. Let z_0 be an isolated singularity of $f(z)$. The singular point $z = z_0$ is an *essential singularity* if we cannot find any $m \in \mathbb{N}$ such that

$$\lim_{z \to z_0} (z - z_0)^m f(z) = c \neq 0.$$

Thus, any singularity z_0 which is not a pole, branch point or removable singularity for $f(z)$ is called an *essential singularity*. For example:

(i) Let $f(z) = e^{1/z}$. Obviously, $z = 0$ is an isolated singularity for $f(z)$. Now,

$$\lim_{z \to 0} z^m f(z) = \lim_{z \to 0} z^m \left[1 + \frac{1}{z} + \frac{1}{2! z^2} + \cdots\right] \neq 0 \text{ for no } m \in \mathbb{N}.$$

Hence $f(z)$ has an essential singularity at $z = 0$.

(ii) Let $f(z) = z^2 \sin(1/z)$. Obviously, $z = 0$ is an isolated singularity for $f(z)$. Now,

$$\begin{aligned} \lim_{z \to 0} z^m f(z) &= \lim_{z \to 0} z^{m+2} \sin(1/z); \quad \frac{1}{z} = t \\ &= \lim_{t \to \infty} \frac{\sin t}{t^{m+2}} \neq 0; \text{ for no } m \in \mathbb{N}. \end{aligned}$$

Hence $f(z)$ has an essential singularity at $z = 0$.

(iii) The function $f(z) = ze^{1/z^2}$, $f(z) = e^{-1/z}$, $f(z) = \sin\frac{1}{z}$, $f(z) = \cos\frac{1}{z}$, etc., has an essential singularities at $z = 0$.

(iv) The functions $\sin z$ and $\cos z$ are all non-analytic at the point infinity because $\sin(1/z), \cos(1/z)$ have an isolated essential singularity at $z = 0$.

If a function is single-valued and has a singularity, then the singularity is either a pole or an essential singularity. For this reason a pole is sometimes called a *non-essential singularity.*

Theorem 3.13. *If $f(z)$ has an isolated singularity at z_0, then $f(z)$ has a pole at $z = z_0$ if and only if* $\lim_{z\to z_0} f(z) = \infty$.

Proof: Let f have a pole of order m at z_0 $f(z)$ can be written as

$$f(z) = (z - z_0)^{-m}\,\phi(z), \quad \phi(z_0) \neq 0,$$

and $\phi(z)$ is analytic and non-zero at z_0. Now

$$\begin{aligned} &\lim_{z\to z_0}\frac{1}{f(z)} = \lim_{z\to z_0}\frac{(z-z_0)^m}{\phi(z)} = \frac{0}{\phi(z_0)} = 0 \\ \Rightarrow\quad &\lim_{z\to z_0} f(z) = \infty. \end{aligned}$$

Conversely, let z_0 be a pole but $\lim_{z\to z_0} f(z) \neq \infty$. Then, $\lim_{z\to z_0} |f(z)|$ is either finite or $\lim_{z\to z_0} f(z)$ does not exist.

(i) In the first case, by the Riemann's removability theorem, z_0 is a removable singular point.
(ii) In the second case, z_0 is an essential singular point.

This is a contradiction to the hypothesis that z_0 is a pole. □

3.8 Meromorphic Functions

The term meromorphic comes from the Ancient Greek word 'meros', meaning 'part'. A meromorphic function is analytic on a domain $\mathcal{D}$ except at a set of isolated points which are poles for the function. It is clear that the class of meromorphic functions includes analytic and rational functions.

Definition 3.8.1. Let $\mathcal{D} \subset \mathbb{C}$ be an open set. If the function $f : \mathcal{D} \to \mathbb{C}$ is defined and analytic in $\mathcal{D}$ except possibly at a discrete set of singular points Ω of f, which are isolated poles, then we say that f is *meromorphic on* $\mathcal{D}$.

If z_0 is such a point, then there exists an integer m such that $(z - z_0)^m f(z)$ is holomorphic in the neighbourhood of z_0. Thus, f is the quotient of two holomorphic functions in the neighbourhood of a point. We say that f is meromorphic at a point z_0 if f is meromorphic on some open set Ω containing z_0. For example,

(i) Let $f(z) = z/(z-1)(z+3)^2$. f is analytic everywhere in the finite complex plane $\mathbb{C}/\{1, -3\}$ and the singularities are the poles $z = 1, -3$. Thus f is a meromorphic function.

(ii) Let $f(z) = 1/\sinh z$. The singularities of f are poles at $z = n\pi i$, $n \in \mathbb{Z}$ and ∞ is the limit point of the poles. Thus, f is meromorphic in $\mathbb{C}$.

(iii) Let $f(z) = p(z)/q(z)$ be a rational polynomial function where p and q are polynomials with no common zeros. If the degree of p is less than or equal to the degree of q, then f has only a finite number of poles and the point infinity is not a pole. On the other hand, if the degree of p is greater than the degree of q, then, say,

$$\begin{aligned} f(z) &= \frac{a_0 z^m + a_1 z^{m-1} + \cdots + a_{m-1}z + a_m}{b_0 z^n + b_1 z^{n-1} + \cdots + b_{n-1}z + b_n}; m > n \\ &= c_0 z^{m-n} + c_1 z^{m-n-1} + \cdots + c_{m-n-1}z + c_{m-n} + \frac{r(z)}{q(z)}, \end{aligned}$$

where the degree of $r(z) \le n-1$. This shows that the point infinity is a pole of order $(m-n)$ and there lies a finite number of poles in the unextended plane $\mathbb{C}$. Thus f is meromorphic in $\mathbb{C}$.

(iv) The quotient of meromorphic functions are meromorphic, provided the denominator is not identically zero.

(v) Every analytic function in a domain is obviously meromorphic function. Consequently, sums and products of meromorphic functions are also meromorphic.

Definition 3.8.2. A function f is said to be *meromorphic at* $z = \infty$ if $g(z) = f(1/z)$ is meromorphic in the neighbourhood of $z = 0$.

Suppose that f is a meromorphic function on $\mathcal{D}$. Define $f : \mathcal{D} \to \mathbb{C}_\infty$ by $f(z) = \infty$ whenever z is a pole of f. Then f is continuous from $\mathcal{D}$ into the extended complex plane $\mathbb{C}_\infty$.

Definition 3.8.3. The *order of a meromorphic function* f at a point z_0 is the order of the zero if $f(z_0) = 0$, and it is minus the order of the pole if f has a pole at z_0. If z_0 is neither a pole nor a zero of f, the *order of* f at z_0 is defined to be *zero*.

RESULT **28.** A meromorphic function does not have any essential singularity in the finite complex plane. It follows that a meromorphic function can have an essential singularity at ∞ only in $\mathbb{C}_\infty$.

This observation implies the following results in the particular cases below taken as examples:

(i) The functions $f_1(z) = \dfrac{e^{1/z}}{z^2+4}$ and $f_2(z) = \dfrac{1}{z^2-4} \sin \dfrac{1}{z}$ has an essential singularity at $z = 0$. So the functions f_1 and f_2 are meromorphic in $\mathbb{C}/\{0\}$, but not in $\mathbb{C}$.

(ii) $f(z) = \cot z$ has simple poles $z = n\pi, n \in \mathbb{N}$ as singularities. Thus f is meromorphic in $\mathbb{C}$. The limit point of these poles is ∞; so, f is not meromorphic in the extended complex plane $\mathbb{C}_\infty$. Similarly, the functions $\tan z, \cot z, \operatorname{cosec} z$, $\operatorname{cosech} z$, $1/\cos z$, $1/\cosh z$, etc., are all meromorphic functions in $\mathbb{C}$ but not in $\mathbb{C}_\infty$.

RESULT **29.** Let $\mathcal{D}$ be a bounded subset in $\mathbb{C}$. A meromorphic function f in $\mathbb{C}$ have only a finite number of poles in $\mathcal{D}$.

This observation implies the following results in the particular cases given below as examples:

(i) Let $f_1(z) = \dfrac{1}{\sin(\pi/z)}$. The set of all poles of f_1 is $\Omega = \left\{\dfrac{1}{n} : n \in \mathbb{Z}\right\}$, which is a bounded set. The limit point of Ω is zero, which is a non-isolated essential singularity of f_1. Thus f_1 is not a meromorphic function in $\mathbb{C}$, but it is in $\mathbb{C}/\{0\}$.

(ii) Let $f_2(z) = \dfrac{1}{e^z+1}$. Then f_2 is a meromorphic function in $\mathbb{C}$.

RESULT **30.** Every meromorphic function ψ in $\mathbb{C}$ that omits three distinct values $z_0, z_1, z_2 \in \mathbb{C}$ is necessarily constant. A value that a meromorphic function does not assume is known as *Picard's exceptional value.*

Theorem 3.14. *A complex function $f(z)$ is rational function if and only if it is meromorphic in $\mathbb{C}_\infty$.*

Proof [Necessary condition]: Let $f(z)$ be a rational function, i.e., $f(z) = p(z)/q(z)$, where $p(z)$ and $q(z)$ are polynomials. By Theorem 5.32, the zeros of $q(z)$ are the poles of $f(z)$. Hence, $f(z)$ is a meromorphic function.

[Sufficient condition]: Let $f(z)$ is meromorphic function in the extended complex plane. Then $f(z)$ is analytic and it has only poles as singularities in $\mathbb{C}_\infty$. We note that $f(z)$ can only have finite number of poles; otherwise the poles of $f(z)$ would have a limit point z_0 (possibly at infinity) which would be a pole and non-isolated, which is a contradiction to the definition of a pole.

Let $z_1, z_2, z_3, \cdots, z_k$ be the poles of $f(z)$ with multiplicities $\alpha_1, \alpha_2, \alpha_3, \cdots, \alpha_k$. Let $g(z)$ be a function defined as

$$g(z) = f(z)(z - z_1)^{\alpha_1}(z - z_2)^{\alpha_2} \cdots (z - \alpha_k)^{\alpha_k}.$$

Then $g(z)$ is analytic in $\mathbb{C}$ and so $g(z) = \sum_{n=0}^{\infty} a_n z^n$; $|z| < \infty$.

Case I: If $g(z)$ is analytic at $z = \infty$, then using Theorem 5.37, $g(z)$ reduces to a constant function.

Case II: If $g(z)$ has a pole at $z = \infty$, then using Theorem 5.37, $g(z)$ is a polynomial. Thus, in both cases $f(z)$ is a rational function. □

Note: Simple poles are the only singularity of $\cot z$ even though it is not a rational function because $z = \infty$ is a non-isolated essential singularity.

3.9 Entire Function

An *entire function* is a complex function $f(z)$ that is analytic at each point in the entire complex plane.

(i) We know that the derivative of any polynomial function

$$P(z) = a_0 z^n + a_1 z^{n-1} + \cdots + a_{n-1} z + a_n; \;\; a_0 \neq 0$$

exists everywhere. It follows that every polynomial is an entire function.

(ii) Entire functions which are not polynomials (e.g. $e^z, \sin z, \cos z, \cosh z, \sinh z, e^{\sin z}$, etc.) are called *entire transcendental functions.*

The followings are the important results of the entire function:

(i) The range of a bounded entire function is a singleton set.

(ii) An entire function $f : \mathcal{D} \longrightarrow \mathbb{C}$ is said to *omit a value* z_0 if $z_0 \in \mathbb{C} - f(\mathcal{D})$. For example, the entire function e^z omits 0.

(iii) A non-zero constant function is the only (rational) function having no zeros or poles.

(iv) If f be entire and $z_0 \in \mathbb{C}$ is arbitrary, then f admits a Taylors's series expansion:

$$f(z) = \sum_{n=0}^{\infty} a_n (z - z_0)^n; \;\; a_n = \frac{f^{(n)}(z_0)}{n!}.$$

If $a_n = 0$ for $n \geq 1$, then $f(z) = f(z_0) = a_0$, i.e., f is a constant function. Note that entire functions are just Taylor's series about any point z_0 with infinite radius of convergence.

(a) If $a_k \neq 0$ for some integer $k \geq 1$ but $a_n = 0$ for all $n > k$, then f is a polynomial of degree $k \geq 1$.

(b) If $a_k \neq 0$ for for an infinite number of values of k, then f becomes an entire transcendental function.

(v) A non-constant entire function is either a polynomial of degree n, so that it has a pole of order n at infinity; else it is an entire transcendental function in which case it will have an essential singularity at infinity.

(vi) An entire function is always an analytic, differentiable and continuous function. But the converse is not true.

Example 3.9.1. *Let $f(z)$ be an analytic function such that $|f(z)| < k|z|\ \forall z \in \mathbb{C}$, for some $k > 0$. If $f(1) = i$, find the value of $f(i)$.*

Solution. Here, $f(z)$ be an analytic function such that $|f(z)| < k|z|\ \forall z \in \mathbb{C}$, and $f(1) = i$. Let $f(z) = kz$, as $f(1) = i = k \Rightarrow k = i$ and $f(z) = iz$.

$$\therefore f(z) = iz, \text{ i.e., } f(i) = i \times i = -1.$$

Example 3.9.2. *Let $f(z)$ be an entire function such that for some constant α, $|f(z)| \leq \alpha\,|z|^3$ for $|z| \geq 1$ and $f(z) = f(iz)\ \forall z \in \mathbb{C}$. Prove that $f(z)$ is constant.*

Solution. Let $f(z) = a_0 + a_1 z + a_2 z^2 + a_3 z^3$. Since $f(z) = f(iz)$, we get

$$\begin{aligned} & a_0 + a_1 z + a_2 z^2 + a_3 z^3 = a_0 + a_1(iz) - a_2 z^2 - i a_3 z^3 \\ \Rightarrow\quad & a_1(1-i)z + 2a_2 z^2 + (1+i)a_3 z^3 = 0 \\ \Rightarrow\quad & a_1 = 0,\ \ a_3 = 0 \text{ and } a_2 = 0 \text{ as } 1 \neq i \\ \Rightarrow\quad & f(z) = a_0, \quad \text{i.e., a constant.} \end{aligned}$$

RESULT **31.** Let a function $f : \mathbb{C} \to \mathbb{C}$ be entire. Then

(i) f is constant;

(ii) the set of values that $f(z)$ assumes is either the whole complex plane $\mathbb{C}$, i.e., f is surjective, or;

(iii) f is onto $\mathbb{C} - \{z_0\}$ for some point $z_0 \in \mathbb{C}$.

Stated in another way, an entire function which misses 2 points in the codomain is actually a constant. This is the well known *Picard's little theorem*.

RESULT **32.** A function with an essential singularity assumes every complex value, except possibly one, infinitely often in every neighbourhood of this singularity.

This observation implies the following results in the particular cases below taken an examples:

(i) Consider the function $f_1(z) = e^{1/z}$. Then f_1 has an essential singularity at $z = 0$. If we choose the sequence $\langle z_n \rangle = \langle 1/n \rangle$, then $z_n \to 0$ and $f_1(z_n) = e^n \to \infty$ as $n \to \infty$. If we choose the sequence $\langle z_n \rangle = \langle -1/n \rangle$, then $z_n \to 0$ and $f_1(z_n) = e^{-n} \to 0$ as $n \to \infty$. If we choose the sequence $\langle z_n \rangle = \langle 1/(\log|w| + i\,\text{Arg}\,w + 2n\pi i) \rangle$, where $w \neq 0$, then $z_n \to 0$ and $f_1(z_n) = e^{\log|w| + i\text{Arg}\,w} = w$. Here, since for each n, $f_1(z_n) = w$, $e^{1/z}$ is never zero.

(ii) Consider the function $f_2(z) = \sin\frac{1}{z}$. Then f_2 has an essential singularity at $z = 0$. If we choose the sequence $\langle z_n\rangle = \langle i/n\rangle$, then $z_n \to 0$ and $f_2(z_n) = \sin\frac{n}{i} = -\sin(in) = -i\sinh n \to \infty$ as $n \to \infty$. If we choose the sequence $\langle z_n\rangle = \langle i/[\log(iw + \sqrt{1-w^2}) + 2n\pi i]\rangle$, where $w \neq \infty$, then $z_n \to 0$ and $f_2(z_n) = \sin -i[\log(iw + \sqrt{1-w^2}) + 2n\pi i] = \sin(\sin^{-1} w)w$.

(iii) Let $f_3(z) = e^{z/\sin z}$. The only possible singularities are $z = k\pi, k = 0, 1, 2, \cdots$. Since $\lim\limits_{z\to 0} f(z) = e$, so $f_3(z)$ has a removable singularity at $z = 0$. We claim that $z = k\pi$, $k \neq 0$ are essential singularities of $f_3(z)$. For this, let k be even. Then, as $\lim\limits_{z\to k\pi_+} \frac{z}{\sin z} = \infty$ and $\lim\limits_{z\to k\pi_-} \frac{z}{\sin z} = -\infty$, it follows that $\lim\limits_{z\to k\pi_+} f(z) = \infty$ and $\lim\limits_{z\to k\pi_-} f(z) = 0$. From the first limit we find that $z = k\pi$ cannot be a removable singularity, whereas from the second limit it is clear that $z = k\pi$ cannot be a pole. Thus, the only possibility left is that $z = k\pi$ is an essential singularity. A similar argument holds when k is odd.

RESULT **33.** An entire function $f(z)$ has a

(i) removable singularity at $z = \infty$ if and only if it is a constant;

(ii) pole of order m at $z = \infty$ if and only if it is a polynomial of degree m.

Thus it follows that $f(z)$ has an essential singularity at $z = \infty$ if and only if $f(z)$ is not a polynomial of finite degree, i.e., it is an entire transcendental function, such as e^z, $\sin z$.

RESULT **34.** An entire function $f(z)$ is rational if and only if it has at most poles on the extended complex plane $\mathbb{C}_\infty$.

3.10 Multi-valued Functions

In this section, we shall study multi-valued functions, in particular, multi-valued logarithm function, the power function z^α, where both z, and α are complex numbers; $z \neq 0$. The ideas of branch, branch points, branch cut, and branch point at infinity will be explained with several examples. Also, Riemann surfaces for different multi-valued functions will be constructed.

Definition 3.10.1. A complex function f that assigns more than one value of w to some or all values of z in the domain of definition $\mathcal{D}$ is called a *multi-valued function.*

So far we have considered single-valued functions, i.e., one-to-one mapping, or many-to-one mapping. In the latter case, under certain restrictions, inverse mappings give rise to multi-valued functions, i.e., one-to-many. For example,

$$z = e^w;\ z = w^2;\ z = \sin w;\ z = \cos w$$

For each of these functions, a given value of z corresponds to more than one value of w. $w = f^{-1}(z)$ is multi-valued and $z = f(w)$ is single-valued; given w, there is a unique value of z. Here our aim is to:

(i) determine all possible values of the inverse function w and

(ii) construct an inverse function which is single-valued in some region $\mathcal{D}$ in the complex plane.

3.10.1 Branch

Whenever a complex-valued function f is multi-valued, we have to introduce branch cuts in the complex plane $\mathbb{C}$ in order to allow analyticity.

Definition 3.10.2. Suppose f is a multiple-valued function defined in $\mathcal{D}$. A *branch of* f is a single -valued analytic function g in some domain $\mathcal{D}$ obtained from f in such a way that

(i) g is continuous on $\mathcal{D}$ and

(ii) for each $z \in \mathcal{D}$, the value of $g(z)$ is one of the values of f.

A multi-valued function can be considered as a collection of single-valued functions, each member of which is called a *branch* of the function. If we consider one particular member as a *principal branch* of the multi-valued function, the value of the function corresponding to this branch is the *principal value*. For example, let $f(z) = z^{1/2} : z \neq 0$ be a square root function. Let $z = re^{i\theta}$, where $r = |z|$ and $\theta =$ Arg $(z) \in (-\pi, \pi]$. Then

$$f_k(z) = \sqrt{r}\, e^{i\left(\theta + 2k\pi\right)/2}; k = 0, 1.$$

The principal branch of f is

$$f_0(z) = \sqrt{r}\, e^{\frac{i\theta}{2}} = \sqrt{r}\left[\cos\frac{\theta}{2} + i\sin\frac{\theta}{2}\right].$$

Further, if $u(r,\theta) = \sqrt{r}\cos\frac{\theta}{2}$ and $v(r,\theta) = \sqrt{r}\sin\frac{\theta}{2}$, then

$$u_r = \frac{1}{2\sqrt{r}}\cos\frac{\theta}{2};\ u_\theta = -\frac{\sqrt{r}}{2}\sin\frac{\theta}{2};\ v_r = \frac{1}{2\sqrt{r}}\sin\frac{\theta}{2};\ u_\theta = -\frac{\sqrt{r}}{2}\cos\frac{\theta}{2}$$

$$\Rightarrow \quad r\, u_r = v_\theta;\ \text{and}\ u_\theta = -r\, v_r.$$

This shows that the C-R equations are satisfied and so f_0 is analytic. The other branch of the function f is

$$\begin{aligned} f_1(z) &= \sqrt{r}\, e^{i\left(\theta + 2\pi\right)/2} = \sqrt{r}\left[\cos\left(\frac{\theta + 2\pi}{2}\right) + i\sin\left(\frac{\theta + 2\pi}{2}\right)\right] \\ &= \sqrt{r}\, e^{\frac{i\theta}{2}}\left[\cos\pi + i\sin\pi\right] = -f_0(z). \end{aligned}$$

Thus, f_0 and f_1 can be taken as the two branches of the multi-valued square root function f as depicted in the Fig. 3.4.

Definition 3.10.3. [Branch cut]: A *branch cut* or *slit* is a line or curve (with ends possibly open, closed, or half-open) in the complex plane across which an analytic multi-valued function is discontinuous. Thus a *branch cut for a branch* g of a multiple-valued function f is a portion of the curve that is excluded from the domain of f so that g is continuous on the remaining points.

For convenience, branch cuts are often taken as lines or line segments. Branch cuts (even those consisting of curves) are also known as *cut lines* or *branch lines*. For example, let $f(z) = z^{1/2}$, and consider the line $L = \{z = x + iy : \text{Re } z < 0, \text{Im } z > 0\}$ as the negative real axis. Let $z_0 = r_0 e^{i\pi}$ be any point in L. Now

$$\lim_{(r,\theta)\to(r_0,\pi)} f_0(re^{i\theta}) = \lim_{(r,\theta)\to(r_0,\pi)} \sqrt{r}\left[\cos\frac{\theta}{2} + i\sin\frac{\theta}{2}\right] = i\sqrt{r_0},$$

$$\lim_{(r,\theta)\to(r_0,-\pi)} f_0(re^{i\theta}) = \lim_{(r,\theta)\to(r_0,-\pi)} \sqrt{r}\left[\cos\frac{\theta}{2} + i\sin\frac{\theta}{2}\right] = -i\sqrt{r_0}.$$

The two limits are different as z approaches to z_0 in two different paths and consequently f_0 is not continuous at z_0. But z_0 is an arbitrary point in L, so, f_0 and similarly f_1 is discontinuous in L.

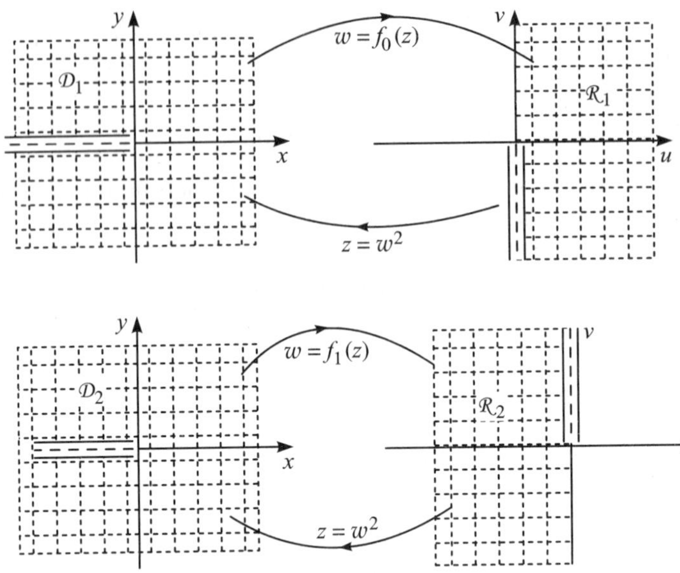

Figure 3.4 *Branches of $z^{1/2}$*

The line L is called a branch cut for the functions f_0 and f_1. The non-injective function $w = z^{1/2}$ has two branches, represented by $f_0(z)$ and $f_1(z)$ in Fig. 3.4.

The respective domains, $\mathcal{D}_1$ and $\mathcal{D}_2$ of these functions, are obtained by cutting the z-plane along the negative x-axis. The range $\mathcal{R}_1$ for $g_1(z)$ consists of the right half-plane and the positive v-axis; the range $\mathcal{R}_2$ for $g_2(z)$ consists of the left half-plane and the negative v-axis. The sets $\mathcal{R}_1$ and $\mathcal{R}_2$ are glued together along the positive and negative v-axes to form the w-plane with the origin deleted.

Definition 3.10.4. [Branch point]: A point with the property that it is on the branch cut of every branch is called a *branch point of a multiple valued function f*. Alternatively, a *branch point* is a point z_0 with the property: If we traverse any circle centered at z_0 with sufficiently small radius starting at a point z_1, then the values of any branch do not return to the value z_1.

It should be noted, however, that the endpoints of branch cuts are not necessarily branch points. Thus a branch point is one for which if we enclose it with a curve, the function changes discontinuously as the variable makes a complete loop over the curve. For example, let $f(z) = z^{1/2}$. Let P be a point on the z-plane, where $w_1 = \sqrt{z_1}$ and $\operatorname{Arg} z_1 = \theta_1, \theta_1 \in (0, 2\pi)$. If $z_1 = r_1\, e^{i\theta_1}$ then at P, $w_1 = \sqrt{r_1}\, e^{i\theta_1/2}$. We now encircle the regions along closed curve Γ_1 through P. Then, upon travelling anticlockwise once, we have $\theta = \theta_1 + 2\pi$ and

$$w = \sqrt{r_1}\, e^{i(\theta_1+2\pi)/2} = -\sqrt{r_1}\, e^{i\theta_1/2} = -w_1 \text{ at } P.$$

This shows that w has changed discontinuously after performing a loop about $z = 0$ (Fig. 3.5 (a)). This establishes $z = 0$ as a branch point.

Now, we consider another loop, a closed curve Γ_2 around some point z^* which does not enclose the origin (Fig. 3.5 (b)). As before, $z_1 = r_1 e^{i\theta_1}$ and $w_1 = \sqrt{r_1} e^{i\theta_1/2}$. Upon returning to P, traveling anticlockwise, $\theta = \theta_1$ again. Hence w is continuous after performing the loop. So $z = z^*$ is not a branch point of $f(z)$.

Example 3.10.1. *Discuss the multivaluedness of the composite function $f(z) = (z^2 - 1)^{1/2}$ and introduce cuts to obtain single-valued branches.*

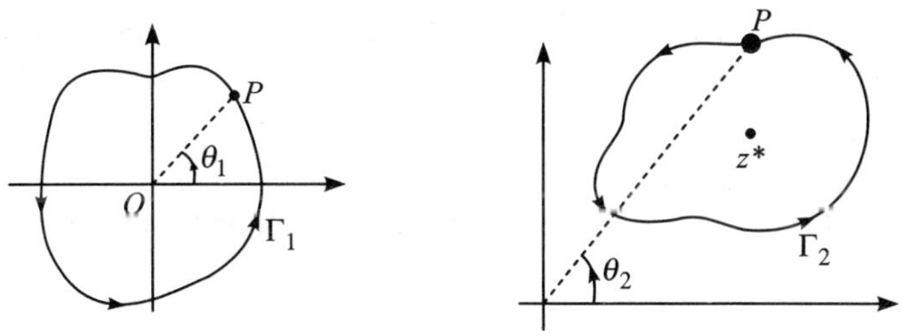

(a) (b)

Figure 3.5 *Branch point*

Solution. Let $z-1=r_1\, e^{i\theta_1}$ and $z+1=r_2\, e^{i\theta_2}$. Then $f(z)=\sqrt{r_1 r_2}\; e^{i(\theta_1+\theta_2)/2}$.
If now z traverses from the point z_0 and forms a simple closed contour Γ_1 enclosing the point where the point $z=-1$ lies outside Γ_1, the value of f at z_0 changes to

$$\begin{aligned} &= \sqrt{r_1 r_2}\; e^{i(\theta_1+\theta_2+2\pi)/2} = \sqrt{r_1 r_2}\; e^{i(\theta_1+\theta_2)/2} e^{i\pi} \\ &= \sqrt{r_1 r_2}\; e^{i(\theta_1+\theta_2)/2}(\cos\theta + i\sin\theta) = -f_0. \end{aligned}$$

Thus the branch points are $z=-1,1$ (Fig. 3.6). In order to obtain single-valued branches we introduce two different set of branch cuts:

(i) A branch cut $L_1=\{x+iy : -1<\text{Re } z<1, \text{Im } z=0\}$ on the real axis (Fig. 3.7):

In this case, consider the closed counter Γ_3 enclosing the branch points -1 and 1. Here $f(z)$ returns to the value

$$\sqrt{r_1 r_2} e^{i(\theta_1+2\pi+\theta_2+2\pi)/2} = \sqrt{r_1 r_2} e^{i(\theta_1+\theta_2)/2} = f_0.$$

Thus it is a single-valued branch.

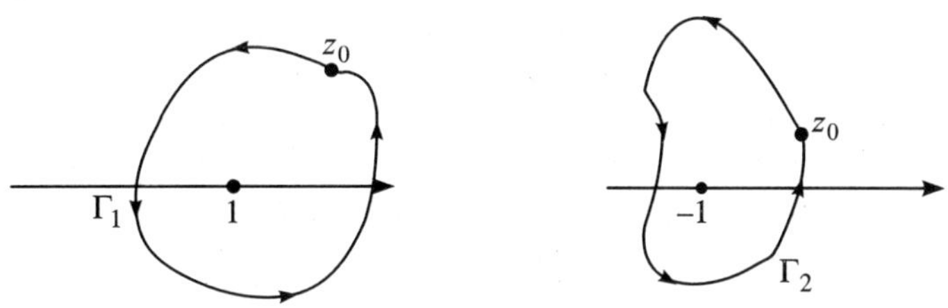

(a) (b)

Figure 3.6 *Branch points of* $(z^2-1)^{1/2}$

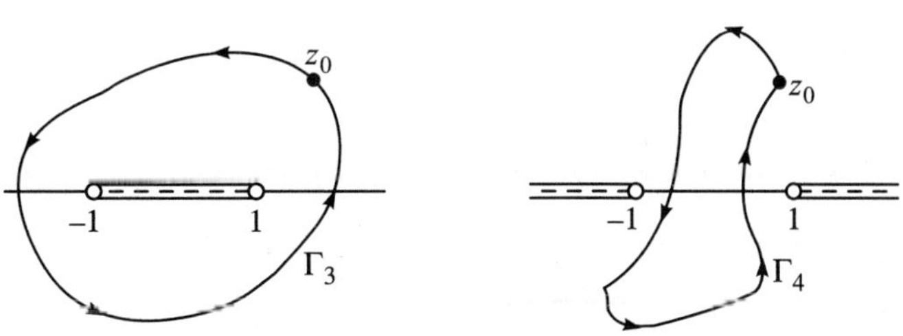

Figure 3.7 *Two possible branch cuts of* $(z^2-1)^{1/2}$

(ii) Two branch cuts on the real axis, $(-\infty, -1)$ and $(1, \infty)$: Here the contour Γ_4 does not enclose any of branch points; so $f(z)$ remains single-valued as z makes a complete loop through Γ_4 initiating from z_0.

Definition 3.10.5. [Branch points at infinity]: The *branch point at infinity* is defined at the point $\zeta = 0$ by using the map from the point ∞ to the origin with the transformation $\zeta = z^{-1}$.

For example:

(i) Consider $f(z) = z^{1/2}$. Making the transformation $\zeta = z^{-1}$, we get

$$f(\zeta) = \zeta^{-1/2} = \frac{1}{|\zeta|} e^{-i\text{Arg }(\zeta)/2}.$$

Thus $\zeta^{-1/2}$ has a branch point at zero; so $z^{1/2}$ has a branch point at infinity.

(ii) Let $f(z) = \log z$. Then the transformation $\zeta = z^{-1}$ gives

$$f(\zeta) = \log(1/\zeta) = -\log\zeta.$$

But $\log\zeta$ has a branch point at $\zeta = 0$. Thus $\log z$ has a branch point at infinity.

(iii) Let $f(z) = (z^2 - 1)^{1/2}$. Then the transformation $\zeta = z^{-1}$ gives

$$f(\zeta) = \left(\frac{1}{\zeta^2} - 1\right)^{1/2} = \pm\frac{1}{\zeta}\left(1 - \zeta^2\right)^{1/2},$$

which shows that $f(\zeta)$ does not have a branch point at $\zeta = 0$. Thus $f(z)$ does not have a branch point at infinity.

Detection of branch points

We have the definition of a branch point, but we do not have a convenient criterion for determining if a particular function has a branch point. We will see later on that $\log z$ and z^k for non-integer k have branch points at zero and infinity. The inverse trigonometric functions like $\sin^{-1} z, \cos^{-1} z$, etc., also have branch points, but they can be written in terms of the logarithm and the square root. In fact, all the elementary functions with branch points can be written in terms of $\log z$ and z^k. Furthermore, note that the multi-valuedness of z^k comes from the logarithm, $z^k = e^{k\log z}$. This gives us a way of determining branch points of a function if there are any.

RESULT **35.** Let $f(z)$ be a single-valued function. Then $\log f(z)$ and $(f(z))^k$ may have branch points only where $f(z)$ is zero or singular.

For example:

(i) Consider the function $f_1(z) = \left(z^2\right)^{1/2}$. Now,

$$f_1(z) = \left(z^2\right)^{1/2} = \pm\sqrt{z^2} = \pm z.$$

Due to $(\cdot)^{1/2}$, f_1 is multi-valued. The possible branch points of $f_1(z)$ are 0 and ∞ only. Now, as $\left\{(e^{i0})^2\right\}^{1/2} = 1$, then as

$$\left\{\left(e^{2\pi i}\right)^2\right\}^{1/2} = \left(e^{4\pi i}\right)^{1/2} = e^{2\pi i} = 1,$$

the function f_1 does not change value when we walk around the origin. We can also consider this to be a path around infinity. Thus f_1 is multi-valued but has no branch points.

(ii) Consider the function $f_2(z) = \left(z^{1/2}\right)^2$. As

$$\left(z^{1/2}\right)^2 = \left(\pm\sqrt{z}\right)^2 = z,$$

so, f_2 is single-valued and there is no question of branch points.

(iii) Consider the function $f_3(z) = \left(z^{1/2}\right)^3$. As

$$\left(z^{1/2}\right)^3 = \left(\pm\sqrt{z}\right)^3 = \pm\left(\sqrt{z}\right)^3,$$

so, f_3 is a multi-valued function. The possible branch points of $f_3(z)$ are 0 and ∞ only. Now, as $\left\{(e^{i0})^{1/2}\right\}^3 = 1$, then as

$$\left\{\left(e^{2\pi i}\right)^{1/2}\right\}^3 = \left(e^{\pi i}\right)^3 = e^{3\pi i} = -1,$$

the function changes value when we walk around the origin. Thus $z = 0$ and singularity $z = \infty$ are the branch points.

(iv) Consider the complex-valued function $f_4(z) = \log(z^2 + z - 2)$. The singularity (pole) of $f(z)$ are

$$z^2 + z - 1 = 0 \Rightarrow z = -2, 1.$$

Thus the possible branch points of f_4 are $-2, 1$ and ∞.

RESULT **36.** (i) Let $f(z)$ be single-valued and have either a zero or a singularity at $z = z_0$. Then $\{f(z)\}^k$ may have a branch point at $z = z_0$. If $f(z)$ is not a power of k, then we are not sure whether $\{f(z)\}^k$ changes value when we walk around z_0.

(ii) Let $f(z) = p(z) \cdot q(z)$, where $p(z)$ is finite and non-zero at z_0. Then, from $q(z)$ we know how fast $f(z)$ vanishes or tends to infinity. Again,

$$\left\{f(z)\right\}^k = \left\{p(z)\right\}^k \cdot \left\{q(z)\right\}^k$$

and $\{p(z)\}^k$ does not have a branch point at z_0, so that $\{f(z)\}^k$ has a branch point at z_0 if and only if $\{q(z)\}^k$ has a branch point there.

(iii) Since $f(z) = p(z) \cdot q(z)$,

$$\log f(z) = \log p(z) + \log q(z)$$

We notice that $\log f(z)$ has a branch point at z_0 if and only if $\log q(z)$ has a branch point there.
For example:

(i) Consider $f_1(z) = \sin z^{1/2}$. As

$$\sin z^{1/2} = \sin(\pm\sqrt{z}) = \pm\sin(\sqrt{z}),$$

so it is multi-valued. It has two branches and and the possible branch points are zero and infinity. Consider the unit circle $|z| = 1$ which is a path around the origin and infinity. If $\sin(e^{0i})^{1/2} = \sin 1$, then as

$$\sin\left\{\left(e^{2\pi i}\right)^{1/2}\right\} = \sin\left(e^{\pi i}\right) = \sin(-1) = -\sin 1,$$

there are branch points at the origin and infinity.

(ii) Consider $f_2(z) = (\sin z)^{1/2} = \pm\sqrt{\sin z}$. So the function f_2 is multi-valued and has two branches. The sine function vanishes at $z = n\pi$ and is singular at infinity. These may be the branch points of the function. Consider the point $z = n\pi$. Then we can express $\sin z = (z - n\pi)\dfrac{\sin z}{z - n\pi}; n \in \mathbb{Z}$. But,

$$\lim_{z\to n\pi} \frac{\sin z}{z - n\pi} = \lim_{z\to n\pi} \frac{\cos z}{1} = (-1)^n.$$

So, $f_2(z)$ has branch points at $z = n\pi$, as $(z - n\pi)^{1/2}$ has a branch point at $z = n\pi$. Here the branch points are $z = n\pi$; $n = 0, \pm1, \cdots$ and they go to infinity. So, it is not possible to make a path that encloses infinity and no other singularities. The point at infinity is a non-isolated singularity. A point can be a branch point only if it is an isolated singularity.

(iii) Consider $f_3(z) = z^{1/2}\cos z^{1/2}$. Then it can be written as

$$z^{1/2}\cos z^{1/2} = \pm\sqrt{z}\,\cos(\pm\sqrt{z}) = \pm\sqrt{z}\,\cos(\sqrt{z}).$$

The function f_3 is multi-valued. If $(e^{i0})^{1/2}\cos(e^{i0})^{1/2} = \cos(1)$,

$$\left(e^{i2\pi}\right)^{1/2}\cos\left[\left(e^{i2\pi}\right)^{1/2}\right] = (-1)\cos(e^{i\pi}) = -\cos(-1) = -\cos 1.$$

There are branch points at $z = 0$ and $z = \infty$.

(iv) Consider $f_4(z) = (\sin z^2)^{1/2} = \pm\sqrt{\sin z^2}$; f_4 is multi-valued. Now, since $\sin z^2 = 0$ at $z = \sqrt{n\pi}$, there may be branch points there. Now, f_4 can be written as $\sin z^2 = z^2\dfrac{\sin z^2}{z^2}$, but

$$\lim_{z\to 0} \frac{\sin z^2}{z^2} = \lim_{z\to 0} \frac{2z\cos z^2}{2z} = 1.$$

So, $f_4(z)$ does not have a branch point at $z = 0$ as $(z^2)^{1/2}$ does not have a branch point there. Next, consider the point $z = \sqrt{n\pi}$. Now, f_4 can be written as $\sin z^2 = \left(z - \sqrt{n\pi}\right)\dfrac{\sin z^2}{z - \sqrt{n\pi}}$, but

$$\lim_{z\to\sqrt{n\pi}} \frac{\sin z^2}{z - \sqrt{n\pi}} = \lim_{z\to\sqrt{n\pi}} \frac{2z\cos z^2}{1} = 2\sqrt{n\pi}\,(-1)^n.$$

Since $(z - \sqrt{n\pi})^{1/2}$ has a branch point at $z = \sqrt{n\pi}$, $(\sin z^2)^{1/2}$ too has a branch point there. Thus we see that $(\sin z^2)^{1/2}$ has branch points at $z = \sqrt{n\pi}$ for $n \in \mathbb{Z}/\{0\}$. This is the set of numbers $\left\{\pm\sqrt{\pi}, \pm\sqrt{2\pi}, \cdots, i\sqrt{\pi}, \pm i\sqrt{2\pi}, \cdots\right\}$. The point at infinity is a non-isolated singularity and hence it is not included in the set of branch points.

Example 3.10.2. *Find the branch points and the branch lines of the function $f(z) = (z^2+1)^{1/2}$ and show that a complete circuit around the branch points makes no change in the branches of the $f(z)$.*

Solution. Here $f(z) = (z^2+1)^{1/2} = \{(z-i)(z+i)\}^{1/2}$, so the possible branch points are $i, -i$. Let us take a closed curve Γ enclosing the point i but not the point $-i$. Then, as the point z goes around Γ_1 once in anticlockwise direction,

$$\text{change in } \arg(z-i) = 2\pi; \text{ change in } \arg(z+i) = 0 \Rightarrow \text{change in } \arg w = \pi.$$

Thus f does not return to its original value, i.e., a change in branch has occurred. As the complete circuit around $z = i$ changes the branches of the function, therefore, $z = i$ is a branch point. Similarly, if we take Γ_2 to be a curve enclosing $z = -i$ but not $z = i$, then $z = -i$ is another branch point of $f(z)$. We consider the point at infinity:

$$f(1/\zeta) = \left\{\left(\frac{1}{\zeta} - i\right)\left(\frac{1}{\zeta} + i\right)\right\}^{1/2} = \frac{1}{\zeta}\left\{(1-i\zeta)(1+i\zeta)\right\}^{1/2}$$

Since $f(1/\zeta)$ does not have a branch point at $\zeta = 0$, $f(z)$ does not have a branch point at infinity.

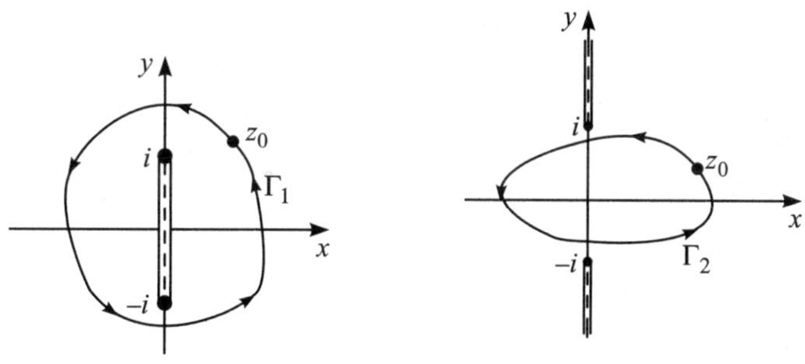

(a) (b)

Figure 3.8 *Branch cut and points of* $(z^2+1)^{1/2}$

Here, we give two possible branch cuts:

(i) The line $L_1 = \{x + iy : x = 0, -1 \le y \le 1\}$ as depicted in the Fig. 3.8(a): If Γ_1 encloses both the branch points, $z = \pm i$, and the point z goes around Γ_1 in anticlockwise direction, then,

$$\text{change in } \arg(z-i) = 2\pi; \text{ change in } \arg(z+i) = 2\pi \Rightarrow \text{change in } \arg w = 2\pi.$$

Hence the complete circuit around the branch points makes no change in the branches of the function $f(z) = (z^2+1)^{1/2}$. The line $-1 \le y \le 1$ continues the branch line.

(ii) Two branch cuts on the imaginary axis, $(-\infty, -i)$ and (i, ∞) as depicted in the Fig. 3.8(b). Here the contour Γ_2 does not enclose any of the branch points, and so $f(z)$ remains single-valued as z makes a complete round through Γ_2 initiating from z_0.

3.10.2 Riemann Surfaces

In the previous section we defined branches of multiple-valued functions by ommiting all but one of its images at every point. However, G. F. B. Riemann introduced an alternating concept to branch cuts, called *elementary surfaces* corresponding to every multi-valued function. The idea is ingenious—a geometric construction that permits surfaces to be the domain or range of a multi-valued functions.

We treat all branches of a multi-valued function as a bijective single-valued function on a domain that consists of many sheets of the complex plane. These sheets are then glued together so that in moving from one sheet to another we pass continuously from one branch of the multi-valued function to another. This glued structure of sheets is called a *Riemann surface for the multi-valued function.*

The concept of Riemann surfaces has the advantage that the various values of multiple-valued functions are obtained in a continuous fashion.

Riemann surface for $f(z) = z^{1/2}$

We have seen that $f(z) = z^{1/2}$ possesses two branch points $z = 0$ and $z = \infty$. To utilise the developments, we introduce a branch cut along the negative real axis. The given function has two values for any $z \neq 0$:

$$f_1(z) = r^{1/2}e^{i\theta/2}; -\pi < \theta \leq \pi \text{ and } f_2(z) = r^{1/2}e^{i\theta/2}; \pi < \theta \leq 3\pi.$$

Each function f_1 and f_2 is single-valued on the domain formed by cutting the z-plane along the negative real axis.

Let $\mathcal{D}_1$ and $\mathcal{D}_2$ be the domains of f_1 and f_2 respectively. The range set for f_1 is the set $\mathcal{R}_1$ consisting of the right-half plane and the positive imaginary axis of w-plane; the range set for f_2 is the set $\mathcal{R}_2$ consisting of the right-half plane and the negative imaginary axis of w-plane.

The sets $\mathcal{R}_1$ and $\mathcal{R}_2$ are glued together along the positive imaginary axis and the negative imaginary axis to form the w-plane with the origin deleted. We stuck $\mathcal{D}_1$ directly above $\mathcal{D}_2$ so that the upper sheet $\mathcal{D}_1$ maps to the right w half-plane $\mathcal{R}_1$ (for portions of $\mathcal{D}_1$ and $\mathcal{D}_2$ see Figure 3.9) and the lower sheet $\mathcal{D}_2$ maps to the left w half-plane $\mathcal{R}_2$. The dashed edge of $\mathcal{D}_1$ in the upper-half plane is joined to the black edge of $\mathcal{D}_2$ in the lower-half plane, and the black edge of $\mathcal{D}_1$ in the lower-half plane is joined to the dashed edge of $\mathcal{D}_2$ in the upper-half plane (it is needless to mention that the line of joining is the negative real-axis). When these domains are glued together in this manner, they form a Riemann surface domain for the mapping $w = f(z) = z^{1/2}$ shown in Fig. 3.9 for some finite r.

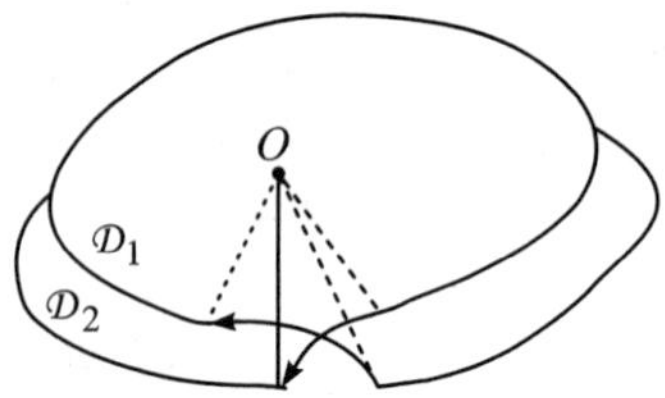

Figure 3.9 *Riemann surface of $z^{1/2}$.*

The origin on a Riemann surface which is common to both sheets, and a curve around the origin on the surface must wind around it twice in order to be a closed surface is called a *branch point.*

Therefore, we have shown geometrically that the function $z \mapsto z^{1/2}$ is a continuous bijection from $\mathbb{C}$ onto the surface.

Example 3.10.3. *Form a Riemann surface for $f(z) = (z-1)^{1/3}$ taking a branch cut along the line $y = 0, x \geq 1$. Detect the point where the function takes the value $\sqrt{2}(i-1)$.*

Solution. Let $r = |z - 1$ and $\theta = \arg(z-1)$, where $0 \leq \theta \leq 2\pi$. Here the Riemann surface consists of three domains:

$$\begin{aligned} \mathcal{D}_1: &\quad f_1(z) = r^{1/3}\, e^{i\theta/3};\ 0 \leq \theta < 2\pi \\ \mathcal{D}_2: &\quad f_2(z) = r^{1/3}\, e^{i\theta/3};\ 2\pi \leq \theta < 4\pi \\ \mathcal{D}_3: &\quad f_3(z) = r^{1/3}\, e^{i\theta/3};\ 4\pi \leq \theta < 6\pi \end{aligned}$$

Each function f_1, f_2 and f_3 is single-valued on the domain formed by cutting the z-plane along the line $y = 0, x \geq 1$.

We place $\mathcal{D}_1$ on the top, followed by $\mathcal{D}_2$ and $\mathcal{D}_3$. The edge of $\mathcal{D}_1$ in the upper-half plane is joined to the edge of $\mathcal{D}_2$ in the lower-half plane and the edge of $\mathcal{D}_2$ in the upper-half plane is joined to the edge of $\mathcal{D}_3$ in the lower-half plane and finally, the edge of D_3 in the upper-half plane is joined to the edge of $\mathcal{D}_1$ in the lower-half plane.

Let $\zeta - \sqrt{2}(i-1)$. Then

$$f(\zeta) = -2\Big(\frac{1}{\sqrt{2}} - \frac{i}{\sqrt{2}}\Big) = 2e^{i\pi}e^{-i\pi/4} = 2e^{i\left(\frac{\pi}{4}+2\pi\right)/3}.$$

So, $\zeta - 1 = 2^3 e^{i\pi/4}$, i.e., $\zeta = 1 + 8e^{i\pi/4}$ lying in the domain $\mathcal{D}_2$.

Exercises

Multiple-choice Questions

1. Let $f(x,y) = \sqrt{|xy|}$. Then

   (a) f_x, f_y do not exist at $(0,0)$ (b) $f_x(0,0) = 1$
   (c) $f_y(0,0) = 0$ (d) f is differentiable at $(0,0)$

2. Let $f : \mathbb{C} \to \mathbb{C}$ such that $f(z) = e^z$. Then

   (a) $f(z)$ is univalent but not onto
   (b) $f(z)$ is univalent in $|z| < a$ for $a > \pi$ but not onto
   (c) $f(z)$ is univalent in $|z| < \pi$ and onto on $\mathbb{C}$
   (d) $f(z)$ is univalent in $|z| < \pi$ but not onto therein

3. For the function $f(z) = e^{-1/z^4}$ at $z = 0$:

   (a) Limit does not exist (b) Limit exists but $f(z)$ is discontinuous
   (c) $f(z)$ is continuous but not differentiable (d) $f(z)$ is differentiable

4. The function $f(z) = |z|^2$ is:

   (a) Differentiable everywhere (b) Differentiable only at the origin
   (c) Not differentiable anywhere (d) Differentiable on real axis

5. Let $f(z) = \frac{|z|}{\mathrm{Re}\,(z)}$ or 0 according as $\mathrm{Re}\,(z) \neq 0$ or $\mathrm{Re}\,(z) = 0$. Then at $z = 0$, the function $f(z)$ is

   (a) Having non-zero limit (b) Differentiable
   (c) Continuous but not differentiable (d) Neither continuous nor differentiable.

6. Which of the following statement(s) is correct about $f(z) = |z|$?
   (P): It is everywhere continuous; (Q): It is nowhere differentiable except zero; (R): It has no singularity.

   (a) Only P (b) Only P and R (c) Only Q and R (d) P, Q, R

7. The function $f(z) = \begin{cases} \bar{z}^2/z^2 & \text{if } z \neq 0 \\ 0 & \text{if } z = 0 \end{cases}$

   (a) Satisfies C-R equation at $z = 0$ (b) Is not continuous at $z = 0$
   (c) Is differentiable (d) Is analytic at $z = 1$

8. An analytic function $f(z)$ is such that $\mathrm{Re}\,\{f'(z)\} = 2y$ and $f(1+i) = 2$. Then the imaginary part of $f(z)$ is:

   (a) $-2xy$ (b) $x^2 - y^2$ (c) $2xy$ (d) $y^2 - x^2$

9. If $f(z) = u + iv$ is an analytic function of $z \neq 0$ and $v(x, y) = (x - y)/(x^2 + y^2)$, $f(z)$ is equal to

   (a) $\frac{1+i}{z} + c$ (b) $\frac{1-i}{z} + c$ (c) $\frac{2-i}{z} + c$ (d) $\frac{2+i}{z} + c$

10. Let $f(z)$ be analytic on $\mathcal{D} = \{z \in \mathbb{C} : |z - 1| < 1\}$ such that $f(1) = 1$. If $f(z) = f(z^2) \forall z \in \mathcal{D}$, then which of the following statements is not correct?

    (a) $f(z) = [f(z)]^2 \ \forall z \in \mathcal{D}$ (b) $f(z/2) = \frac{1}{2}f(z) \ \forall z \in \mathcal{D}$
    (c) $f(z^3) = [f(z)]^3 \ \forall z \in \mathcal{D}$ (d) $f'(1) = 0$.

11. Let $f : \mathbb{R} \to \mathbb{R}$ be a twice continuously differentiable function, with $f(0) = f(1) = f'(0) = 0$. Then

    (a) f'' is a zero function (b) $f''(0)$ is zero
    (c) $f''(x) = 0$ for some $x \in (0, 1)$ (d) f'' never vanishes

12. If G be an open set in $\mathbb{C}$ and $f : G \to \mathbb{C}$, then f is differentiable at $a \in G$ if

    (a) $\lim\limits_{h\to 0}[f(a + h) - f(a)]$ exists (b) $\lim\limits_{h\to 0}[f(a + h) - f(a)]$ does not exist

    (c) $\lim\limits_{h\to 0} \frac{f(a+h)-f(a)}{h}$ exists (d) $\lim\limits_{h\to 0} \frac{f(a+h)-f(a)}{h}$ does not exist

13. Let G be an open set in $\mathbb{C}$ and $f : G \to \mathbb{C}$. f is an analytic function if

    (a) f is continuously differentiable on G (b) f is continuously integrable on G
    (c) f is continuous on G (d) f is differentiable on G

14. The function $f(z) = |z|^2 + i\bar{z} + 1$ is differentiable at

    (a) i (b) 1 (c) $-i$ (d) No point in $\mathbb{C}$

15. Let G be an open connected set in $\mathbb{C}$ and $f : G \to \mathbb{C}$ is analytic function. Let $u(x, y) = \operatorname{Re} f(x + iy)$, $v(x, y) = \operatorname{Im} f(x + iy)$, where $z = x + iy \in G$. Then the C-R equations are:

    (a) $\frac{\partial u}{\partial x} = \frac{\partial v}{\partial y}, \frac{\partial u}{\partial y} = \frac{\partial v}{\partial x}$ (b) $\frac{\partial u}{\partial x} = -\frac{\partial v}{\partial y}, \frac{\partial u}{\partial y} = -\frac{\partial v}{\partial x}$

    (c) $\frac{\partial u}{\partial x} = \frac{\partial v}{\partial y}, \frac{\partial u}{\partial y} = -\frac{\partial v}{\partial x}$ (d) $\frac{\partial u}{\partial x} = \frac{\partial v}{\partial y} = \frac{\partial u}{\partial y} = \frac{\partial v}{\partial x}$

16. Let G be an open set in $\mathbb{C}$. Then the real-valued function $u(x, y)$, where $(x, y) \in G$, is a harmonic function if:

    (a) $u(x, y)$ has first partial derivative
    (b) $u(x, y)$ has second partial derivative
    (c) $u(x, y)$ has continuous first partial derivative
    (d) $u(x, y)$ has continuous second partial derivatives

17. Let G be an open set in $\mathbb{C}$. Then the real-valued function $u(x, y)$, where $(x, y) \in G$, is a harmonic function if it satisfies

    (a) Laplace equation (b) C-R equations (c) Heat equation (d) None of these

18. Consider the functions $f(z) = x^2 + iy^2$ and $g(z) = x^2 + y^2 + ixy$. At $z = 0$

    (a) f is analytic but not g (b) g is analytic but not f
    (c) Both f and g are analytic (d) Neither f nor g is analytic

19. Consider a function $f(z) = u + iv$ defined on $|z - i| < 1$. Then, $f(z)$ is analytic for u equals to

    (a) $x^2 + y^2$ (b) $\ln(x^2 + y^2)$ (c) e^{xy} (d) $e^{x^2-y^2}$

20. Let $u(x, y) = x^3 + ax^2y + bxy^2 + 2y^3$ be a harmonic function and $v(x, y)$ its harmonic conjugate. If $v(0, 0) = 1$, then $|a + b + v(1, 1)|$ is equal to

    (a) 10 (b) 20 (c) 30 (d) 40

21. At $z = 0$, the function $f(z) = z^2\overline{z}$

    (a) Does not satisfy C-R equations (b) Satisfies CR equations but is not differentiable
    (c) Is differentiable (d) Is analytic

22. If G is open and connected and $f : G \to \mathbb{C}$ is differentiable with $f'(z) = 0$ for all $z \in G$, then

    (a) f is a constant function (b) f is monotone function
    (c) f is decreasing function (d) f is increasing function

23. Let G and Ω be open subsets of $\mathbb{C}$. Let $f : G \to \mathbb{C}$ and $g : \Omega \to \mathbb{C}$ be continuous functions such that $f(G) \subset \Omega$ and $g(f(z)) = z$ for all $z \in G$. If g is differentiable, $g'(z) \neq 0$ and also f is differentiable and $f(z) = 1/g'(f(z))$ implies:

    (a) If f is analytic then g is analytic (b) If g is analytic then f is analytic
    (c) Both f and g are analytic (d) Neither f nor g are analytic

24. Let $u(x, y)$ be the real part of an entire function $f(z) = u + iv$. If C is the positively oriented boundary of the rectangular region $\mathcal{R}$ in $\mathbb{R}^2$, then $\oint_C \left[\frac{\partial u}{\partial y}\, dx - \frac{\partial u}{\partial x}\, dy\right]$ is equal to

    (a) 1 (b) 0 (c) 2π (d) π **[Hint:** $\int\int_{\mathcal{R}} \left(\frac{\partial^2 u}{\partial x \partial y} - \frac{\partial^2 u}{\partial y \partial x}\right) dx\, dy$]

25. Let S be a disc $|z| < 3$ in the complex plane and let $f : S \to \mathbb{C}$ be an analytic function such that $f\left(1 + \frac{\sqrt{2}}{n} i\right) = -\frac{2}{n^2}$ for each $n \in \mathbb{N}$. Then $f(\sqrt{2})$ is equal to

    (a) $3 - 2\sqrt{2}$ (b) $3 + 2\sqrt{2}$ (c) $2 - 3\sqrt{2}$ (d) $2 + 3\sqrt{2}$

26. Let $f(z) = \cos z - \frac{\sin z}{z}$; $z \in \mathbb{C} - \{0\}$ and $f(0) = 0$. Then, $f(z)$ has a zero at $z = 0$ of the order

    (a) 0 (b) 1 (c) 2 (d) Greater than 2

27. Let $f(z) = \cos z - \dfrac{\sin z}{z}$; $z \in \mathbb{C} - \{0\}$ and $f(0) = 0$. Also let $g(z) = \sinh z$ for $z \in \mathbb{C}$. Then, $f(z)/zg(z)$ has a zero at $z = 0$ of the order

    (a) 1 (b) 2 (c) 3 (d) greater than 3

28. Define $f : \mathbb{C} \to \mathbb{C}$ by $f(z) = \begin{cases} 0; & \text{if Re}\,(z) = 0 \text{ or Im}\,(z) = 0 \\ z; & \text{otherwise.} \end{cases}$

    Then the set of points, where f is analytic is

    (a) $\{z : \text{Re}\,(z) \neq 0 \text{ and Im}\,(z) \neq 0\}$ (b) $\{z : \text{Re}\,(z) \neq 0\}$
    (c) $\{z : \text{Re}\,(z) \neq 0 \text{ or Im}\,(z) \neq 0\}$ (d) $\{z : \text{Im}\,(z) \neq 0\}$

29. Which of the following is not the real part of an analytic function?

    (a) $x^2 - y^2$ (b) $1/(1 + x^2 + y^2)$ (c) 0 (d) 1

30. In order for the function $f(z) = |z|^2/z, z \neq 0$ to be continuous at $z = 0$, $f(0)$ should be equal to

    (a) 2 (b) -1 (c) 0 (d) 1.

31. Let G be an open connected set in $\mathbb{C}$ and $f : G \to \mathbb{C}$ an analytic function. Also, let $z = x+iy \in G$. Then the harmonic conjugate of $u(x, y) = \text{Re}\, f(z)$ is

    (a) $-\text{Im}\, f(z)$ (b) $1/\text{Im}\, f(z)$ (c) $\text{Im}\, f(z)$ (d) $f(z)$

32. Let G be an open connected set in $\mathbb{C}$ and $f : G \to \mathbb{C}$ an analytic function. Also, let $z = x+iy \in G$. Then the harmonic conjugate of $v(x, y) = \text{Im } f(z)$ is

    (a) $-\text{Re } f(z)$ (b) $1/\text{Re } f(z)$ (c) $\text{Re } f(z)$ (d) $f(z)$

33. If $f(z)$ is analytic and equal $u(x, y) + iv(x, y)$, then $f'(z) = ?$

    (a) $u_x - iv_y$ (b) $u_x - iv_x$ (c) $u_y - iv_x$ (d) None of these

34. An analytic function is

    (a) Single-valued (b) Bounded (c) Differentiable (d) All of these

35. The value of k so that $2x - x^2 + ky^2$ may be harmonic is

    (a) 0 (b) 1 (c) 2 (d) 3

36. If $|f(z)|$ is constant, then $f(z)$ is

    (a) Variable (b) Partially variable and constant (c) Constant (d) None of the above

37. If $u = x^2 - y^2$, then the corresponding analytic function is (c = constant)

    (a) $z + c$ (b) $z^2 + c$ (c) $z^3 + c$ (d) $z^4 + c$

38. Let $u(x, y) = 2x(1 - y) \; \forall \; x, y \in \mathbb{R}$. Then, a function v such that $f = u + iv$ is analytic is

    (a) $x^2 - (y - 1)^2$ (b) $(x - 1)^2 - y^2$ (c) $(x - 1)^2 + y^2$ (d) $x^2 + (y - 1)^2$

39. The function $f(z) = x^2 + iy^3$

    (a) Does not satisfy C-R equations at $z = 0$ (b) Is differentiable
    (c) Is analytic (d) Is not analytic anywhere

40. At $z = 0$, the function $f(z) = |z|^4$ is

    (a) Analytic (b) Differentiable but not analytic
    (c) Differentiable (d) Analytic but not differentiable

41. The function $f(z) = e^{y+ix}$ is

    (a) Analytic (b) Nowhere analytic
    (c) Differentiable (d) Analytic but not differentiable

42. At $z = 0$, the function $f(z) = \sqrt{|xy|}$

    (a) Does not satisfy C-R equations (b) Satisfies C-R equations but not analytic
    (c) Satisfies C-R equations but differentiable (d) Satisfies C-R equations but not differentiable

43. Consider $u(x, y) = x^2 - y^2$ and $v(x, y) = -y/(x^2 + y^2)$. Then

    (a) Both are harmonic and $u + iv$ is an analytic.
    (b) Both are not harmonic and $u + iv$ is an analytic.
    (c) Both are harmonic and $u + iv$ is an not analytic.
    (d) Both are not harmonic and $u + iv$ is not an analytic.

44. Let $f : \mathbb{C} \to \mathbb{C}$ be a complex valued function given by $f(z) = u(x, y) + iv(x, y)$. Suppose $v(x, y) = 3xy^2$. Then

    (a) f cannot be holomorphic on $\mathbb{C}$ for any choice of u.
    (b) f is holomorphic on $\mathbb{C}$ for a suitable choice of u.
    (c) f is holomorphic on $\mathbb{C}$ for all choices of u.
    (d) v is not differentiable as a function of x and y.

45. Let f be an analytic function on $\overline{\mathcal{D}} = \{z \in \mathbb{C} : |z| \leq 1\}$. Assume that $|f(z)| \leq 1$ for each $z \in \overline{\mathcal{D}}$. Then, which of the following is not a possible value of $(e^f)''(0)$?

(a) 2 (b) 6 (c) $7e^{1/9}/9$ (d) $\sqrt{2} + i\sqrt{2}$

46. If $f(z) = u + iv$ is an analytic function of $z = x + iy$ and $u - v = (x - y)(x^2 + 4xy + y^2)$. Then $f(z)$ =? (c – constant)

(a) $iz + c$ (b) $iz^2 + c$ (c) $iz^3 + c$ (d) $iz^4 + c$

47. Let Im $\{f'(z)\} = 6x(2y - 1)$ and $f(0) = 3 - 2i$, $f(1) = 6 - 5i$. Then $f(1 + i)$ =?

(a) $3 + 6i$ (b) $3 - 6i$ (c) $6 + 3i$ (d) $6 - 3i$

48. Let f be an entire function on $\mathbb{C}$ such that $|f(z)| \leq 100 \log |z|$ for each z with $|z| \geq 2$. If $f(i) = 2i$, then $f(1)$ must be

(a) 2 (b) $2i$ (c) i (d) Cannot be determined

49. Let $f(z)$ be an entire function such that $|f(z)| < k|z| \forall z \in \mathbb{C}$ for some $k > 0$. If $f(1) = i$, the value of $f(i)$ is

(a) 1 (b) -1 (c) i (d) $-i$

50. Let $f(z)$ be an entire function such that for some constant k, $|f(z)| \leq k|z|^3$ for $|z| \geq 1$ and $f(z) = f(iz), \forall z \in \mathbb{C}$. Then

(a) $f(z) = kz^3, \forall z \in \mathbb{C}$ (b) $f(z)$ is a constant
(c) $f(z)$ is a quadratic polynomial (d) no such $f(z)$ exists

51. Let $f(z)$ be an entire function. If f satisfies the following two equations $f(z + 1) = f(z)$, $f(z + i) = f(z)$, for every $z \in \mathbb{C}$, then

(a) $f'(z) = f(z)$ (b) $f(z) \in \mathbb{R}$, for all $z \in \mathbb{C}$
(c) f = constant (d) f is a nonconstant polynomial

52. For the function $f(z) = (z - \sin z)/z^3$ the point $z = 0$ is

(a) a pole of order 3 (b) a pole of order 2
(c) an essential singularity (d) a removable singularity

53. For the function $f(z) = (1 - e^{-z})/z$, the point $z = 0$ is

(a) an essential singularity (b) a pole of order 0
(c) a pole of order one (d) a removable singularity

Answers:

1(c); 2(d); 3(a); 4(b); 5(d); 6(b); 7(b); 8(d); 9(a); 10(a); 11(c); 12(c); 13(a); 14(c); 15(c); 16(d); 17(a); 18(d); 19(b); 20(a); 21(a); 22(a); 23(b); 24(b); 25(b); 26(c); 27(b); 28(a); 29(b); 30(c); 31(c); 32(c); 33(c); 34(d); 35(b); 36(c); 37(b); 38(a); 39(d); 40(b); 41(b); 42(b); 43(c); 44(b); 45(b); 46(c); 47(c); 48(b); 49(b); 50(b); 51(c); 52(d); 53(d)

Review Questions [Short-answer Type]

1. If $w = f(z) = z^2$, find the value of $w = u + iv$ which corresponds to

(a) $z = 1 - 3i$. (b) $z = -2 + i$.

Draw the graph of the corresponding values in the w and z planes.

2. If $w = f(z) = z(2 - z)$, find the value of $w = u + iv$ which corresponds to

(a) $z = 1 + i$. (b) $z = 2 - 2i$.

Draw the graph of the corresponding values in the w and z planes.

3. If $w = f(z) = (z+2)/(2z-1)$:
   (a) Find $f(0), f(i)$ and $f(1+i)$.
   (b) Find z such that $f(z) = i, f(z) = 2-3i$.
   (c) Show that z is single-valued.

4. Find the domain of definition for each of the following functions:
   (a) $f(z) = \frac{1}{1-|z|^2}$ (b) $f(z) = \frac{1}{z^2+1}$ (c) $f(z) = \frac{\bar{z}}{\bar{z}+z}$ (d) $f(z) = \frac{z}{z^2+3}$

5. State the domain and range of f where $f(z)$ is determined by the expression
   (a) $\bar{z}$ (b) $az^2 + bz + c$, where $a, b, c \in \mathbb{C}$ constants. (c) $\frac{z+3}{z^2+1}$
   (d) $\frac{5}{z^3-8}$ (e) $|z|$ (f) Re z (g) Im z

6. Prove that each of the following functions $f(z); z \neq 0$, does not have a limit as $z \to 0$.
   (a) $f(z) = \frac{\sqrt{xy}}{x+iy}$ (b) $f(z) = \frac{-x^3yi}{x^6+y^2}$ (c) $f(z) = \frac{xy}{x^2+y^2}$

7. Which of the following functions is univalent?
   (a) $f(z) = z^2 + 3z + 1;\ |z| < 1$ (b) $f(z) = z/(1-z^2);\ |z| < 1$

8. Using the definition of limit:
   (a) Find $\lim\limits_{z \to z_0} f(z)$ if $f(z) = \begin{cases} z^2 & \text{if } z \neq z_0 \\ 0 & \text{if } z = z_0. \end{cases}$
   (b) Prove that $\lim\limits_{z \to z_0} \bar{z} = \bar{z}_0$.
   (c) Prove that $\lim\limits_{z \to z_0} \text{Re } z = \text{Re } z_0$.
   (d) Prove that $\lim\limits_{z \to i} \frac{3z^4-2z^3+8z^2-2z+5}{z-i} = 4(1+i)$.

9. Show that the following functions are continuous for all $z \in \mathbb{C}$.
   (a) $|z|$ (b) $|z|^2$ (c) z^2 (d) Re z (e) Im z.
   (f) A polynomial $P(z)$ (g) $z^2 - 2z + 3$.

10. Prove that if the function $f(z)$ is continuous in a closed region $\mathcal{R}$, it is bounded in that region $\mathcal{R}$.

11. If the functions $f(z)$ and $g(z)$ are continuous at z_0, prove that $3f(z) - 4ig(z)$ is also continuous at z_0.

12. Prove that $f(z) = 1/z$ is not uniformly continuous in the region $|z| < 1$.

13. Prove that the function
   (a) $f(z) = \dfrac{z^3+1}{z^3+9}$ is continuous and bounded in the region $|z| \leq 2$.
   (b) $f(z) = \begin{cases} \frac{\text{Re } z^2}{|z|^2} & \text{if } z \neq 0 \\ 0 & \text{if } z = 0 \end{cases}$ is not continuous at $z = 0$.
   (c) $f(z) = \begin{cases} \frac{z\text{Re } z}{|z|} & \text{if } z \neq 0 \\ 0 & \text{if } z = 0 \end{cases}$ is continuous in the entire complex plane.
   (d) $f(z) = \begin{cases} \frac{z^2+4}{z-2i} & \text{if } z \neq 2i \\ 2+3i & \text{if } z = 2i \end{cases}$ is not continuous at $z = 2i$.
   (e) $f(z) = \begin{cases} \frac{z^3-iz^2+z-i}{z-i} & \text{if } z \neq i \\ 0 & \text{if } z = i \end{cases}$ is continuous at $z = i$.

14. Classify each of the following as a polynomial, a rational function or a composite function.

    (a) $f(z) = (3 + 2i)z^2 + 5z + 7i$. (b) $f(z) = \frac{3z^4 - 2z^3 + 8z^2 - 2z + 5}{z - i}$.

    (c) $f(z) = \frac{iz^2 + 7}{3iz + 2}$ (d) $f(z) = \left| \frac{z^2 - 2z + 4}{z^4 - 4} \right|$

    For each function, indicate all the values of z at which the function is continuous. Justify your answer.

15. Find all points of discontinuity for the following functions:

    (i) $\frac{3z^2+4}{z^4-16}$ (ii) $\frac{1}{z} - \sec z$ (iii) $\cot z$ (iv) $\frac{\tanh z}{z^2+1}$

16. Show that

    (a) $f(z) = 3z - 2$ is uniformly continuous (UC) in the region $|z| \leq 10$.
    (b) $f(z) = 1/z^2$ is not UC in $|z| \leq 1$ but is UC in $1/2 \leq |z| \leq 1$.
    (c) $f : \mathbb{C} - \{0\} \to \mathbb{C}$, defined by $z \to 1/z$, is not UC in $|z| < 1$.
    (d) $f : \mathbb{C} \to \mathbb{C}$, defined by $z \to x^2 - iy^2$, is not UC.

17. Using the definition, show that $f'(z)$ does not exist at any point z, when $f(z) =$

    (a) Re z (b) Im z (c) $\bar{z}^2$ (d) $z^2\bar{z}$

18. Using the differentiation formulas, find $f'(z)$ for the following functions:

    (a) $f(z) = (iz - 1)^{-3}$ (b) $f(z) = (z^2 - 2z)/(z^4 + 1)$ (c) $f(z) = (2z - i)/(z + 2i)$

19. Find the derivatives of the following:

    (i) $\{\sin^{-1}(2z - 1)\}^2$ (ii) $z \tan^{-1}(\log z)$ (iii) $\cos^{-1}(\sin z - \cos z)$
    (iv) $\log(\cot^{-1} z^2)$ (v) $\left\{ \tanh^{-1}(iz + 2) \right\}^{-1}$ (vi) $\coth^{-1}(z \operatorname{cosec} 2z)$
    (vii) $(z - 3i)^{4z+2}$

20. Find $f'(z)$, where it exists, for the following functions, where $f(z) =$

    (i) $y^2 - 3ix^2$ (ii) $e^{\bar{z}}$ (iii) $e^z \sin z$ (iv) $z^2 e^{z^2}$ (v) $z^2 e^{|z|}$

21. For each of the following functions, determine the set of points at which it is (i) differentiable and (ii) analytic.

    (a) $f(z) = \frac{2z^2+6}{z(z^2+4)}$ (b) $f(z) = 6\bar{z}^2 - 2\bar{z} - 4i|z|^2$
    (c) $f(z) = (x^3 + 3xy^2 - 3x) + i(y^3 + 3x^2y - 3y)$
    (d) $f(z) = (3x^2 + 2x - 3y^2 - 1) + i(6xy + 2y)$
    (e) $f(z) = e^{y^2 - x^2} \left[\cos(2xy) - i \sin(2xy) \right]$

    Find the derivative where it exists.

22. Show that

    (a) $f(z) = xy + iy$ is everywhere continuous but not analytic.
    (b) $f(z) = z + 2\bar{z}$ is not analytic anywhere in $\mathbb{C}$.
    (c) $f(z) =$ Arg z is discontinuous at each point on the non-positive real axis.

23. Verify the C-R equations and test the analyticity of each of the following functions:

(i) $\sin 2z$ (ii) $\cosh z$ (iii) $z^2 + 5iz + 3 - i$ (iv) $2xy + i(x^2 - y^2)$

(v) $z^3 + z$ (vi) $|z|^4$ (vii) $e^x(\cos y + i \sin y)$ (viii) $e^x(\cos y - i \sin y)$

(ix) $\sinh z$ (x) $z\, e^{-z}$ (xi) $z\,|z|$ (xii) $e^{-x}(\cos y - i \sin y)$

If analytic, find their derivatives.

24. Find the constant k such that the function

(a) $u(x, y) = kx^2 - y^2 + xy$ is harmonic. Hence find the conjugate harmonic and the analytic function.

(b) $f(z) = \frac{1}{2}\log(x^2 + y^2) + i \tan^{-1} \frac{kx}{y}$ is an analytic function.

(c) $f(z)$ expressed in polar coordinates as $f(z) = r^2 \cos 2\theta + ir^2 \sin k\theta$ is analytic.

25. Map $G = \mathbb{C}/\{z \in \mathbb{C} : -1 \leq z \leq 1\}$ onto the open disc on an analytic function f. Can f be one-to-one?

26. Determine the constants

(i) a and b in order that the function $f(z) = (x^2 + ay^2 - 2xy) + i(bx^2 - y^2 + 2xy)$ is analytic. Find $f'(z)$.

(ii) a, b and c so that the function $f(z) = (ay^3 + ix^3) + xy(bx + icy)$ is analytic. Express $f'(z)$ in the form $g(z)$.

27. If $f(x + iy) = re^{i\theta}$ is analytic, show that $\frac{\partial r}{\partial x} = r\frac{\partial \theta}{\partial y}$.

28. For what value of z does the function $f(z) = u + iv$ defined by (i) $z = e^{-v}(-\cos u + i \sin u)$ and (ii) $z = \sin u\, \cosh v + i \cos u\, \sinh v$, cease to be analytic?

29. Show that:

(a) If v and V are harmonic conjugates of $u(x, y)$ in a domain $\mathcal{D}$, then $v(x, y)$ and $V(x, y)$ can differ at most by an additive constant.

(b) If v is harmonic conjugate for u, then $-u$ is the harmonic conjugate to v.

30. Show that each of the following functions are differentiable in the given domain of definition and find $f'(z)$.

(a) $f(z) = \sqrt{r}\, e^{i\theta/2};\ r > 0, -\pi < \theta < \pi$.

(b) $f(z) = e^{-\theta} \cos(\log r) + ie^{-\theta} \sin(\log r),\ \ r > 0, 0 < \theta < 2\pi$.

31. Let $f(z)$ be an analytic function in domain $\mathcal{D}$. Show that another function $g(z)$ defined on $\mathcal{D}_1 = \{z : \bar{z} \in \mathcal{D}\}$ by $g(z) = f(\bar{z})$ is analytic as a function of z in $\mathcal{D}_1$.

32. Show that if f is analytic on $\mathcal{D}$, then $g(z) = \overline{f(\bar{z})}$ is analytic on the reflected domain $\mathcal{D}_1 = \{\bar{z} : z \in \mathcal{D}\}$, and $g'(z) = \overline{f'(\bar{z})}$.

33. Show that $u + iv = (x - iy)/(x - iy + a)$, where $a \neq 0$, is not an analytic function of $z = x + iy$, whereas $u - iv$ is such a function.

34. Let $f(z)$ be an analytic function in a domain $\mathcal{D}$. Prove that $f(z)$ must be constant in $\mathcal{D}$, if

(a) $f(z)$ is real-valued for all z in $\mathcal{D}$. (b) Re $f(z) = [\text{Im } f(z)]^2$ for all $z \in \mathcal{D}$.

(c) f and $\bar{f}$ are both analytic on the domain $\mathcal{D}$. (d) $|f|$ is constant. **[Hint:** $\bar{f} - |f|^2/f$**]**

35. Let f and g be two analytic functions on region $\mathcal{D}$ such that Re (f) = Re (g) $\forall z \in \mathcal{D}$. Show that $f(z)$ and $g(z)$ can differ at most by an additive constant.

36. Prove the following:

(a) $\frac{\partial}{\partial x} \equiv \frac{\partial}{\partial z} + \frac{\partial}{\partial \bar{z}}$ (b) $\frac{\partial}{\partial y} \equiv i(\frac{\partial}{\partial z} + \frac{\partial}{\partial \bar{z}})$ (c) $4\frac{\partial}{\partial z\, \partial \bar{z}} \equiv \frac{\partial^2}{\partial x^2} + \frac{\partial^2}{\partial y^2}$

37. Find an analytic function in the disc $|z-1| < 1$ whose real part is $\log \sqrt{x^2+y^2}$.

38. Find the analytic function $f(z)$ such that $\text{Re}\{f'(z)\} = 3x^2 - 4y - 3y^2$ and $f(1+i) = 0$.

39. Suppose that $f(z)$ has a pole of order m at z_0. Show that $f'(z)$ has a pole of order $(m+1)$ at z_0.

40. Prove that the functions (i) $u(z)$ and $u(\bar{z})$; and (ii) $u(x,y)$ and $u(x^2-y^2, 2xy)$ are simultaneously harmonic.

41. Locate and determine the order of zeros of the following functions:

(a) $e^{2z} - e^z$ (b) $z^2 \sinh z$ (c) $z^4 \cos^2 z$ (d) $z^3 \cos z^2$

(e) $\frac{5z+3}{(1-z)^3 \sin^2 z}$ (f) $\sin(1 - \frac{1}{z})$ (g) $\frac{\tan z}{z}$ (h) $\tanh z$

42. Find and classify the singularities of the following functions:

(i) $\sqrt{z(z^2+1)}$ (ii) $\frac{z^2+1}{z^{3/2}}$ (iii) $\sin^{-1}(1/z)$ (iv) $\frac{\log(z+3i)}{z^2}$

(v) $\frac{z^2-3z}{z^2+2z+2}$ (vi) $\frac{z^2-2z+3}{z-2}$ (vii) $\frac{z+3}{z^2-1}$ $(viii)$ $\frac{\cos z}{(z+i)^2}$

(ix) $(z-i)\sin(\frac{1}{z+2i})$

43. Prove that if f has a simple pole at z_0, then there exists a punctured neighbourhood of z_0 on which f is one-to-one.

44. Show that $z = 0$ is the removable singularity of the following functions:

(a) $\frac{e^{z^2}-1}{z}$ (b) $\frac{\sin z - z}{z^2}$ (c) $\frac{1-\frac{z^2}{2}-\cos z}{\sin z^2}$

Furthermore, define $f(0)$ such that these functions are analytic at $z = 0$.

45. Find and classify the isolated singularities(and order) of the following functions:

(a) $\frac{z^3+1}{z^2(z-1)}$ (b) $z^2 e^{1/z}$ (c) $\frac{z}{\cos z}$ (d) $\frac{\sin 3z}{z^2} - \frac{3}{z}$ (e) $\frac{z(1+z)}{1-\cos z}$

(f) $\cos\frac{1}{z}\frac{\sin(z-1)}{z^2+1}$ (g) $\frac{(z^2-1)\cos \pi z}{(z+2)(2z-1)(z^2+1)^2 \sin^2 \pi z}$ (h) $\frac{\cos z}{z^2+1} + 4z$

(i) $\cos(1-\frac{1}{z})$ (j) $\tan z$ (k) $\cot\frac{1}{z}$ (l) $\frac{1}{e^z-1}$

46. What is the order of the pole of $f(z) = \frac{1}{(2\cos z - 2 + z^2)^2}$ at $z = 0$?

47. Show that $f(z) = \frac{(z+3i)^5}{(z^2-2z+5)^2}$ has a double pole at $z = 1 \pm 2i$ and a simple pole at infinity.

48. Prove that if $f(z)$ has an essential singularity at z_0, then so does the function $e^{f(z)}$.

49. State why the function $f(z) = 2z^2 - 3 - ze^z + e^{-z}$ is entire.

50. Verify Result 32 for the functions $\cos\frac{1}{z}$ and $\cosh\frac{1}{z}$.

51. Show in two ways that the function $f(z) = e^{z^2}$ is entire. What is its derivative?

52. If $f(z)$ is an entire function, show that $f(e^z)$ and $f(z^2)$ are entire functions.

53. Show that the following functions are entire:

(a) $\begin{cases} \frac{\cos z}{z^2 - \frac{\pi^2}{4}} & \text{when } z \neq \pm\frac{\pi}{2} \\ -\frac{1}{\pi} & \text{when } z = \pm\frac{\pi}{2} \end{cases}$ (b) $\begin{cases} \frac{e^z - e^{-z}}{2z} & \text{when } z \neq 0 \\ 1 & \text{when } z = 0 \end{cases}$

54. Which of the following functions are meromorphic in the whole plane?

(a) $\frac{e^z-1-z}{z^4}$ (b) $\frac{z}{\sin^2 z}$ (c) $2z+z^3$ (d) $\operatorname{Log} z$

(e) $\frac{\sin z}{z^3+1}$ (f) $\frac{2i}{(z-3)^2}+\cos z$ (g) $\tan z$ (h) $e^{1/z}$

55. Determine the branch points of the function $w =$

(i) $\{(z^2-z)(z+2)\}^{1/2}$ (ii) $\cos(z^{1/2})$ (iii) $(z+i)^{-z}$ (iv) $\log e^z$

56. Construct a branch of $f(z) = \log(\frac{z-1}{z+1})$ which is analytic at the origin and takes the values $5\pi i$ there. [**Hint:** Let $z-1 = re^{i\theta}$ and $z+1 = \rho e^{i\psi}$. Then the answer is $4\pi \leq \theta < 6\pi$ and $\pi \leq \psi < 3\pi$.]

57. Construct a Riemann surface for the following functions:

(i) $w = z^{1/3}$ (ii) $w = (z^2+1)^{1/2}$ (iii) $w = \log\frac{z+1}{z-1}$ (iv) $w = \sin^{-1} z$

Answers

1. (a) $-8-6i$; (b) $3-4i$ 2. (a) 2; (b) $4+4i$

3. (a) $2, -i, 1-i$; (b) $-i, (2+i)/3$ 7. (a) No; (b) Yes

15. (i) $\pm 2, \pm 2i$; (ii) $(2n+1)\pi/2, n \in \mathbb{Z}$; (iii) $n\pi, n \in \mathbb{Z}$; (iv) $\pm i, (2n+1)i\pi/2, n \in \mathbb{Z}$

18. (a) $-3i(iz-1)^{-4}$; (b) $(-2z^5+6z^4+2z-2)/(z^4+1)^2$; (c) $5i/(z+2i)^2$

19. (i) $2\sin^{-1}(2z-1)/(z-z^2)^{1/2}$; (ii) $\frac{1}{1+(\log z)^2}+\tan^{-1}(\log z)$; (iii) $-(\sin z+\cos z)/\sqrt{\sin 2z}$;
(iv) $-2z/(1+z^4)\cot^{-1} z^2$; (v) $\frac{-i}{1-(iz+2)^2}\{\tanh^{-1}(iz+2)\}^{-2}$;
(vi) $\operatorname{cosec} 2z(1-2z\cot 2z)/(1-z^2\operatorname{cosec}^2 2z)$; (vii) $(z-3i)^{4z+2}[(4z+2)+4\log(z-3i)]$

20. (i) $-6ix$ on $y=3x$; (ii) does not exist anywhere; (iii) $e^z(\sin z+\cos z)$; (iv) $2z(1+z^2)e^{z^2}$; (v) nowhere differentiable

23. (i) Analytic(A); (ii) A; (iii) A; (iv) Not analytic (NA); (v) A; (vi) Differentiable but NA at $z=0$; (vii) A; (viii) NA; (ix) A; (x) A; (xi) A; (xii) A

24. (a) $1, f(z) = z^2 - \frac{1}{2}z^2$; (b) -1; (c) 2 26. (i) $a=-1, b=1$; (ii) $1, -3, -3; 3iz^2$

28. (i) $z=0$; (ii) $z=\pm 1$ 38. z^3+2iz^2+6-2i

42. (i) $0, \pm i$, branch points; (ii) $0, \infty$, branch points; (iii) 0, essential singularity; (iv) $-3i$, branch point; 0, simple pole; (v) $-1\pm i$, simple poles; (vi) 2, simple pole; (vii) $\pm 1, \infty$, simple poles; (viii) $-i$, pole of order 3; (ix) $-2i$, essential singularity

Review Questions [Long-answer Type]

1. Express each of the following functions in the form $f(z) = u(x,y) + iv(x,y)$.

(a) $w = 2z^2 - 3iz$. (b) $w = 3z^2+5z+i+1$ (c) $w = z^3+z+1$

(d) $w = \frac{z}{1+z}$. (e) $w = z + \frac{1}{z}$.

(f) $w = \frac{1-z}{1+z}$. (g) $w = \frac{2z^2+3}{|z-1|}$

2. Show that when $T(z) = \frac{az+b}{cz+d}, ad-bc \neq 0$: (a) $\lim_{z\to\infty} T(z) = \infty$, if $c=0$; (b) $\lim_{z\to\infty} T(z) = a/c$ and $\lim_{z\to -d/c} T(z) = \infty$ if $c \neq 0$.

3. Discuss $\lim_{z\to 0}\{f(z)-f(0)\}/z$ along any radius vector, where

(a) $f(z)=\begin{cases}\frac{x^2y^5(x+iy)}{x^4+y^{10}}; & z\neq 0\\ 0 & z=0\end{cases}$ (b) $f(z)=\begin{cases}\frac{xy^3(x+iy)}{x^2+y^6} & z\neq 0\\ 0 & z=0\end{cases}$

(c) $f(z)=\begin{cases}\frac{xy^2(x+iy)}{x^2+y^4}; & z\neq 0\\ 0 & z=0\end{cases}$

Discuss whether $\lim_{z\to 0}\{f(z)-f(0)\}/z$ exists or not. Verify if they are analytic or not.

4. If $P(z)=a_0z^n+a_1z^{n-1}+\cdots+a_{n-1}z+a_n, a_0\neq 0$ and $Q(z)=b_0z^m+b_1z^{m-1}+\cdots+b_{m-1}z+b_m, b_0\neq 0$ are polynomials of degree n and m respectively, then prove that

(a) $\lim_{z\to 0}\frac{P(z)}{Q(z)}=\frac{a_n}{b_m}$. (b) $\lim_{z\to\infty}\frac{P(z)}{Q(z)}=\begin{cases}0 & \text{if } m>n\\ \text{does not exist} & \text{if } m<n\\ \frac{a_0}{b_0} & \text{if } m=n.\end{cases}$

5. Suppose $f(z)=az^2+bz\bar{z}+c\bar{z}^2$, where a,b and c are fixed complex numbers. Show that $f(z)$ is complex differentiable at z if and only if $bz+2c\bar{z}=0$. Where is $f(z)$ analytic?

6. Prove that the complex function is continuous at the point $z_0=x_0+iy_0$ if and only if the real and imaginary parts of $f(z)$, regarded as functions of two variables x and y, are continuous at the point (x_0,y_0).

7. Prove that a function f with open domain $\mathcal{D}$ is continuous at a point $z_0\in\mathcal{D}$ if and only if whenever $\langle z_n\rangle\subset\mathcal{D}$ is a sequence converging to z_0, the sequence $\langle f(z_n)\rangle$ converges to $f(z_0)$.

8. Prove that if the function $f(z)$ is continuous in a closed region $\mathcal{D}$, it is uniformly convergent in $\mathcal{D}$.

9. Let $\mathcal{D}_1$ and $\mathcal{D}_2$ be domains, and let $f:\mathcal{D}_1\to\mathbb{C}$ and $g:\mathcal{D}_2\to\mathbb{C}$ be analytic functions such that $f(\mathcal{D}_1)\subset\mathcal{D}_2$. Show that $g\circ f:\mathcal{D}_1\to\mathbb{C}$ is analytic.

10. Suppose $f=u+iv$ is analytic in a rectangle with sides parallel to the coordinate axes and satisfies the relation $u_x+v_y=0$ for all x and y. Show that there exist a real constant c and a complex constant d such that $f(z)=-cz+id$.

11. Show that the Cauchy–Riemann conditions are equivalent to $w_{\bar{z}}=0$. Hence, deduce that the function $w=f(z)=\bar{z}e^{-|z|^2}$ is not analytic.

12. For each of the following functions determine the points at which the function is not analytic (singular points).

$(i)\ \frac{az+b}{cz+d}\quad (ii)\ \frac{(2+z^3)^4}{z-1}\quad (iii)\ \frac{3z-2}{z^2+2z+5}\quad (iv)\ \frac{z^3-1}{z(z^2-1)}\quad (v)\ \frac{z^2+1}{(z+2)(z^2+2z+2)}$

Determine the derivatives at all other points.

13. If $f(z)$ is the given function, state where $f'(z)$ exists and find $f'(z)$ at these points.

(i) $ax+by+i(ay-bx); a,b\in\mathbb{R}$ (ii) $e^x(\cos y+i\sin y)$
(iii) $\sin x\cosh y+i\cos x\sinh y$ (iv) $(y-2)^3+ix^3$ (v) $z\operatorname{Re} z$

State also where each function is analytic. Which functions of the above are entire?

14. Let $k\in\mathbb{R}$ be fixed. Find the set of $z\in\mathbb{C}$ for which $\sum_{n=1}^{\infty}n^{i(z^2+k)}$ represents an analytic function.

15. Evaluate the following limits by using L'Hôpital's rule (3.20).

(i) $\lim_{z\to i}\frac{z^{10}+1}{z^6+1}$ (ii) $\lim_{z\to i}\frac{z^2-2iz-1}{z^4+2z^2+1}$ (iii) $\lim_{z\to 0}\frac{1-\cos z}{z^2}$

(iv) $\lim_{z\to 0}\frac{z-\sin z}{z^3}$ (v) $\lim_{z\to 0}\frac{1-\cos z}{\sin z^2}$ (vi) $\lim_{z\to 0}(\cos z)^{1/z^2}$

16. Consider the two annulus domains $\mathcal{D}_a = \{z \in \mathbb{C} : a < |z| < 1\}$ and $\mathcal{D}_b = \{z \in \mathbb{C} : b < |z| < 1\}$. Define $f : \mathcal{D}_a \to \mathcal{D}_b$ by

$$f\left(re^{i\theta}\right) = \left[\left(\frac{1-b}{1-a}\right)r + \left(\frac{b-a}{1-a}\right)\right]e^{i\theta}$$

Show that (i) f is bijective and (ii) f is analytic if and only if $a = b$.

17. Show that $f'(0)$ does not exist and yet the C-R equations are satisfied at $z = 0$ if

$$f(z) = \begin{cases} \frac{ax^3-by^3}{ax^2+by^2} + i\frac{ax^3+by^3}{ax^2+by^2} & \text{if } z \neq 0 \\ 0 & \text{if } z = 0;\ \ ab > 0. \end{cases}$$

18. Express the C-R equations in terms of the curvilinear coordinates (ξ, η) where $x = e^{\xi}\cosh\eta$ and $y = e^{\xi}\sinh\eta$.

19. Show that the given functions $u(x, y)$ are harmonic. Hence find the conjugate harmonic.

(i) $2xy + 3y$ (ii) $3x^2y + 2x^2 - y^3 - 2y^2$ (iii) $2x(1-y)$
(iv) $2x - x^3 + 3xy^2$ (v) $e^x\ \cos y$ (vi) $y/(x^2+y^2)$
(vii) $\sinh x \sin y$

Also find the analytic function $f(z)$.

20. Show that a necessary and sufficient condition for a function $f(z) = u(x, y) + iv(x, y)$ to be analytic in a domain $\mathcal{D}$ is that its real part u and imaginary part v be conjugate harmonic functions in $\mathcal{D}$.

21. Construct the analytic function $f(z) = u + iv$, where $u(x, y) =$

(i) $e^x\ \cos y$ (ii) $x^3 - 3xy^2$ (iii) $x^3 - 3xy^2 + 3x^2 - 3y^2 + 1$

(iv) $y^3 - 3x^2y$ (v) $x^3 - 3xy^2 + 3x + 1$ (vi) $e^{-x}\{(x^2-y^2)\cos y + 2xy\ \sin y\}$

(vii) $\cos x\cosh y$ (viii) $e^x(x\cos y - y\sin y)$ (ix) $\frac{2\cos x\cosh y}{\cos 2x+\cosh 2y}$

(x) $\sin x\ \cosh y$ (xi) $\frac{e^{2x}}{x\cos 2y - y\sin 2y}$ (xii) $\frac{2\sin 2x}{e^{2y}+e^{-2y}-2\cos 2x}$

22. Construct the analytic functions whose imaginary part $v(x, y)$ is

(i) $3x^2y - y^3$ (ii) $x^3 - 3xy^2 + 3x^2 - 3y^2 + 1$ (iii) $e^x\ \sin y$
(iv) $(x-y)/(x^2+y^2)$ (v) $\cos x\ \cosh y$

23. Prove that $u(x, y) = x^2 - y^2$ and $v(x, y) = -y/(x^2+y^2)$ are both harmonic but $u + iv$ is not analytic.

24. Show using Laplace's equation in polar coordinates that $u(re^{i\theta}) = \theta\log r$ is harmonic. Use the polar form of the CR equations to find a harmonic conjugate v of u. What is the analytic function $u + iv$?

25. If $k \in \mathbb{R}$, show that $u(x,y) = e^{-2kxy}\cos k(x^2 - y^2)$ is harmonic in $\mathbb{R}^2$. Find all its harmonic conjugates $v(x,y)$ in $\mathbb{R}^2$. Write $f = u + iv$ as function of z with $f(0) = 1$.

26. Use the Cauchy–Riemann equations to prove that if f is analytic on an open set U, then the function g defined by $g(z) = \overline{f(\bar{z})}$ is analytic on the open set $\{\bar{z} : z \in U\}$.

27. If $f(z) = u + iv$ is an analytic function of $z = x + iy \in \mathbb{C}$, find $f(z)$ in terms of z if

    (i) $u - v = e^x(\cos y - \sin y)$ (ii) $u - v = (x - y)(x^2 + 4xy + y^2)$

    (iii) $u + v = x/(x^2 + y^2)$ given $f(1) = 1$

28. If $f(z) = u + iv$ is a regular function of z in any domain, prove that

    (a) $(\frac{\partial}{\partial x}|f(z)|)^2 + (\frac{\partial}{\partial y}|f(z)|)^2 = |f'(z)|^2$.

    (b) $(\frac{\partial^2}{\partial x^2} + \frac{\partial^2}{\partial y^2})|f(z)|^p = p^2|f(z)|^{p-2}|f'(z)|^2$.

    (c) $(\frac{\partial^2}{\partial x^2} + \frac{\partial^2}{\partial y^2})|u|^p = p(p-1)|u|^{p-2}|f'(z)|^2$.

29. If $f(z) = u + iv$ is analytic and $f(z) \neq 0$, prove that

    (a) $(\frac{\partial^2}{\partial x^2} + \frac{\partial^2}{\partial y^2})\log|f(z)| = 0$. (b) $(\frac{\partial^2}{\partial x^2} + \frac{\partial^2}{\partial y^2})\arg f(z) = 0$.

30. If $u - v = (e^{-y} - \cos x + \sin x)/(\cosh y - \cos x)$, and $f(z) = u + iv$ is an analytic function of z, find $f(z)$ subject to $f(\pi/2) = (3 - i)/2$.

31. An entire function $f = u + iv$ has the feature that $u_x v_y - u_y v_x = 1$ in $\mathbb{C}$. Demonstrate that f has the form $f(z) = az + b$, where a and b are some constants with $|a| = 1$.

32. Suppose that $f = u + iv$ is entire such that $u_x + v_y = 0$ in $\mathbb{C}$. Demonstrate that f has the form $f(z) = az + b$, where a, b are some constants with $\operatorname{Re} a = 0$.

33. Explain why the composition of two entire functions $f(z)$ and $g(z)$ is also entire. Also, explain why the linear combination $c_1 f(z) + c_2 g(z)$, where c_1 and c_2 are complex constants, is entire.

34. For the function $g(z)$ defined as $g(z) = \begin{cases} \sin z/z & \text{if } z \neq 0 \\ 0 & \text{if } z = 0, \end{cases}$ show that $g(z) = \sum\limits_{k=1}^{\infty}(-1)^k \frac{z^{2k}}{(2k+1)!}$; $z \in \mathbb{C}$. Hence deduce that the function $g(z)$ is entire.

35. Let $f(z)$ be an entire function such that $\lim\limits_{z\to\infty} \frac{f(z)}{z} = 0$. Show that $f(z)$ is a constant.

36. Suppose that $f(z)$ is entire and $\operatorname{Im} z \neq 0$ whenever $|z| \neq 1$. Prove that $f(z)$ is constant.

    [**Hint:** Picard's little theorem]

37. Find the branch points of the function $f(z) = (z^3 - z)^{1/3}$ and introduce the branch cuts. If $f(3) = 2\sqrt[3]{3}$, show that $f(-3) = -2\sqrt[3]{3}$.

38. Determine the branch points of the function $f(z) = (z^3 - 1)^{1/2}$.

39. Find a branch of each of the following multiple-valued functions that is analytic in the given domain:

    (a) $(z^2 - 1)^{1/2}$ in the interior of the unit disc $|z| < 1$

    (b) $(z^2 - 1)^{1/2}$ in the exterior of the unit circle $|z| > 1$

    (c) $(4 + z^2)^{1/2}$ in the complex plane slit along the imaginary axis from $-2i$ to $2i$

(d) $(z^4-1)^{1/2}$ in the exterior of the unit circle $|z|>1$
(e) $(z^3-1)^{1/3}$ in the exterior of the unit circle $|z|>1$

40. Identify the branch points of the function $f(z)=(z^3+z^2-6z)^{1/2}$ in the extended complex plane $\mathbb{C}_\infty$. Specify the branch cuts and select a branch that gives a single-valued function where it is continuous at $z=-1$ with $f(-1)=-\sqrt{6}$.

41. Find the branch points of $(z^{1/2}-1)^{1/2}$ in the finite complex plane. Introduce branch cuts to make the function single-valued.

42. Let $\mathcal{D}$ be the complex z-plane with a cut along the segment $[-1,1]$. Determine the regular branches of the function $f(z)=\{(1-z)/(1+z)\}^{1/2}$.

43. Split the function $f(z)=\sqrt{(z^2-4)(z^2-9)}$ into two regular branches in the domain $\mathcal{D}$: $\mathbb{C}/\{[-3,-2],[2,3]\}$.

44. Form the Riemann surface for $w=z^{1/n}$.
[**Hint:** Recall that this function has n branches, represented by the single-valued functions

$$f_k(z)=\left|z\right|^{1/n}\left[\cos\frac{\theta+2k\pi}{n}+i\sin\frac{\theta+2k\pi}{n}\right];k=0,1,\cdots,n-1.$$

The respective domains $S_0,S_1,\cdots,S_{n-1}$ of these functions are obtained by cutting the z-plane along the negative x-axis. Now we place these n copies of the cut z-plane directly on each other as depicted in the Fig. 3.10(a) for $n=4$.]

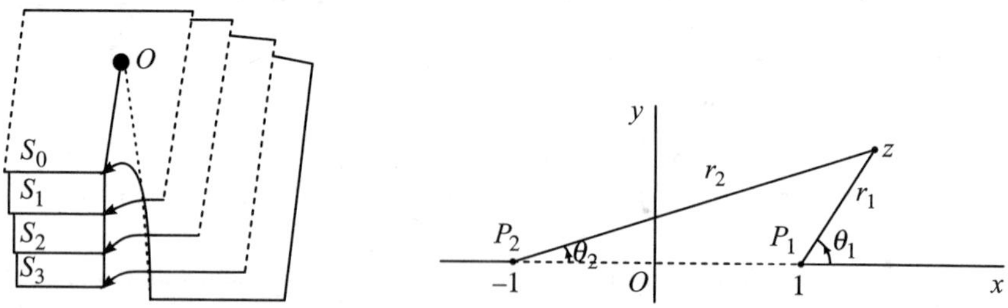

(a) (b)

Figure 3.10 *(a) Riemann surface of $z^{1/n}$; (b) Riemann surface of $(z^2-1)^{1/2}$*

45. Form the Riemann surface for the function $w=\sqrt{z^2-1}$.
[**Hint :** Branch points $-1,1$ but not at ∞. Let $z-1=r_1e^{i\theta_1}$ and $z+1=r_2e^{i\theta_2}$. Use Fig. 3.10(b).]

46. (a) Prove there does not exist a (single-valued) branch of $(z^2-1)^{1/3}$ that is analytic in $|z|>1$.
(b) Prove that there does exist a (single-valued) branch of $[(z-1)(z-2)(z-3)]^{1/3}$ that is analytic in $|z|>3$.

Answers

1. (a) $(2x^2-2y^2+3y)+i(4xy-3x)$; (b) $(3x^2-3y^2+5x+1)+(6xy+5y+1)i$;
(c) $(x^3-3xy^2+x+1)+(3x^2y-y^3+y)i$; (d) $\frac{x^2+x+y^2}{(x+1)^2+y^2}+\frac{y}{(x+1)^2+y^2}i$;
(e) $\frac{x(x^2+x+y^2+1)}{x^2+y^2}+\frac{y(x^2+x+y^2-1)}{x^2+y^2}i$; (f) $\frac{1-x^2-y^2}{(1+x)^2+y^2}+\frac{-2y}{(1+x)^2+y^2}i$; (g) $\frac{(2x^2-2y^2+3)+4xyi}{\sqrt{(x-1)^2+y^2}}$

12. (i) $-d/c$; (ii) 1; (iii) $-1 \pm 2i$; (iv) $0, -1$; (v) $-2, -1 \pm i$
13. (i) $\mathbb{C}$; (ii) $\mathbb{C}$; (iii) $\forall z \in \mathbb{C}$; (iv) $f'(2i) = 0$; (v) $f'(0) = 0$
15. (i) $5/3$; (ii) $-1/4$; (iii) $1/2$; (iv) $1/6$; (v) $1/2$; (vi) $e^{-1/2}$
19. (i) $e^x \sin y + c$; (ii) $4xy - x^3 + 3xy^2$ (iii) $x^2 - y^2 + 2y$; (iv) $xy - 3x^2y + y^3 + c$; (v) $e^x \sin y$; (iv) $x/(x^2 + y^2) + c$; (vii) $-\cosh x \cos y$
21. (i) $e^z + c$; (ii) $z^3 + c$; (iii) $z^3 + 3z^2 + c$; (iv) $i(z^3 + c)$; (v) $z^3 + 3z + c$; (vi) $z^2\, e^{-z} + c$; (vii) $\cos z + c$; (viii) $z\, e^z + c$; (ix) $\sec z + c$; (x) $\sin z + c$; (xi) $z\, e^{2z} + c$; (xii) $\cot z + c$
22. (i) $z^3 + c$; (ii) $i(z^3 + 3z) + c$; (iii) $e^z + c$; (iv) $(1 + i)/z + c$; (v) $i \cos z + c$
27. (i) $e^z + c$; (ii) $-iz^3 + c$; (iii) $\frac{1+i}{2z} + \frac{1-i}{2}$
30. $(1 - i)/2 + \cot(z/2)$

4 Elementary Transcendental Functions

We have considered earlier various elementary functions as polynomials and rational functions. In this section we study the elementary transcendental functions such as the exponential, logarithmic, trigonometric, and hyperbolic functions of a complex variable. We start by defining the complex exponential function and then use it to develop the others. Further, with the help of multivalued functions, we will explain the branches of logarithmic functions.

4.1 Complex Exponential Function

Consider the complex power series

$$\sum_{n=0}^{\infty} \frac{z^n}{n!} = 1 + \frac{z}{1!} + \frac{z^2}{2!} + \cdots + \frac{z^n}{n!} + \cdots, \tag{4.1}$$

which is of the form $\sum a_n z^n$, so that $a_n = \frac{1}{n!}$. Since

$$R = \lim_{n\to\infty} |\frac{a_n}{a_{n+1}}| = \lim_{n\to\infty} \frac{(n+1)!}{n!} = \lim_{n\to\infty} (n+1) = \infty,$$

so the complex power series (4.1) converges absolutely in $\mathbb{C}_\infty$ by D'Alembert's ratio test and the convergence is uniform on each compact subset $\mathbb{C}$. All of its derivatives are also analytic in the entire complex plane.

Definition 4.1.1. The sum of the complex power series (4.1) denoted by e^z or $\exp z$, given by

$$e^z = 1 + \frac{z}{1!} + \frac{z^2}{2!} + \cdots + \frac{z^n}{n!} + \cdots = \sum_{n=0}^{\infty} \frac{z^n}{n!} \tag{4.2}$$

is known as *exponential series* or *complex exponential function.* It is an analytic function which has no singularities.

The function e^z provides a basic tool for the application of complex variables to electrical circuits, control systems, wave propagation, and time-invariant physical systems.

Property 4.1.1. If $z = x + iy$, then the *complex exponential form* is defined by

$$e^z = e^{x+iy} = e^x \Big[\cos y + i \sin y\Big]. \tag{4.3}$$

$$\therefore \quad \text{Re}\,(e^z) = e^x \cos y \text{ and } \text{Im}\,(e^z) = e^x \sin y$$

which are defined for all $(x, y) \in \mathbb{R}^2$. Thus, for the function e^z, the domain is $\mathbb{C}$.

Property 4.1.2. Using definition (4.1) we get,

$$\frac{d}{dz}(e^z) = \frac{d}{dz}\sum \frac{z^n}{n!} = \sum_{n=0}^{\infty} \frac{nz^{n-1}}{n!} = \sum_{n=1}^{\infty} \frac{z^{n-1}}{(n-1)!} = e^z \forall z \in \mathbb{C}.$$

This shows that e^z is an entire function and $\frac{d}{dz}(e^z) = e^z$ on $\mathbb{C}$. From this we conclude that the *exponential function* $f(z)$ of complex variable z is the solution of the differential equation $f'(z) = f(z)$, with the initial condition: $f(0) = 1$.

Property 4.1.3. [The addition theorem of exponential function]: If z_1 and z_2 are complex numbers, then

$$\exp(z_1 + z_2) = e^{z_1+z_2} = e^{z_1} \cdot e^{z_2} = \exp(z_1) \cdot \exp(z_2) \tag{4.4}$$

which is the same form as the law of indices in algebra.

Proof: Let z and c be complex numbers in $\mathbb{C}$.

$$\begin{aligned} & \frac{d}{dz}(e^z e^{c-z}) = e^z\, e^{c-z} - e^z\, e^{c-z} = 0 \,\forall \in \mathbb{C} \\ \Rightarrow \quad & e^z\, e^{c-z} = k(\text{ constant}) \,\forall z \in \mathbb{C}. \end{aligned}$$

By putting $z = 0$, we obtain $e^c = k$. Hence

$$e^z\, e^{c-z} = e^c \,\forall z \in \mathbb{C}.$$

Now, putting $z = z_1$ and $c = z_1 + z_2$, we get Eq. (4.4). □

We see by induction that

$$e^{z_1+z_2+\cdots+z_n} = e^{z_1} \cdot e^{z_2} \cdots e^{z_n}. \tag{4.5}$$

Since $e^z\, e^{-z} = e^0 = 1$, so e^z is never zero.

Property 4.1.4. [Algebraic properties]: If z_1 and z_2 be complex numbers, then we have the following algebraic properties:

$$(i)\quad e^0 = 1 \quad (ii)\quad e^{z_1}e^{z_2} = e^{z_1+z_2} \quad (iii)\quad \frac{e^{z_1}}{e^{z_2}} = e^{z_1-z_2} \quad (iv)\quad (e^{z_1})^n = e^{nz_1}, n \in \mathbb{Z}$$

Property 4.1.5. [Modulus, Argument and Conjugate]: The complex number $w = e^z$ can be written in the polar form as:

$$\begin{aligned} & w = e^x\,[\cos y + i \sin y] = r\,[\cos\theta + i\sin\theta] \\ \Rightarrow \quad & r = |w| = e^x \text{ and } \theta = \arg w = y + 2n\pi, \text{ for } n \in \mathbb{Z}. \end{aligned} \tag{4.6}$$

From calculus we know that $e^x > 0, \forall x \in \mathbb{R}$, and so from Eq.(4.6) it follows that $|e^z| > 0$. This implies that $e^z \neq 0$ for all complex z.

A formula for the *conjugate of the complex exponential* e^z is given by

$$\begin{aligned} \overline{e^z} &= e^x \cos y - ie^x \sin y = e^x \cos(-y) + ie^x \sin(-y) \\ &= e^{x-iy} = e^{\overline{z}}, \quad \forall z \in \mathbb{C}. \end{aligned} \tag{4.7}$$

Example 4.1.1. *Find* $z = x + iy$ *such that* $e^z = 1 + i$.

Solution. Since $z = x + iy$, and $e^z = 1 + i$, we can write

$$e^{x+iy} = \sqrt{2}\Big(\frac{1}{\sqrt{2}} + i\frac{1}{\sqrt{2}}\Big) \;\Rightarrow\; e^x e^{iy} = \sqrt{2}\Big(\cos\frac{\pi}{4} + i\sin\frac{\pi}{4}\Big) = \sqrt{2}e^{\frac{i\pi}{4}}.$$

Equating the two non-zero complex numbers is exponential form,

$$e^x = \sqrt{2} \text{ and } y = \frac{\pi}{4} + 2n\pi; n = 0, \pm1, \pm2, \cdots$$

$$\Rightarrow \quad x = \sqrt{2} = \frac{1}{2}\ln 2 \text{ and } y = \Big(2n + \frac{1}{4}\Big)\pi; n = 0, \pm1, \pm2, \cdots$$

$$\Rightarrow \quad z = \frac{1}{2}\ln 2 + \Big(2n + \frac{1}{4}\Big)\pi; n = 0, \pm1, \pm2, \cdots$$

4.2 Complex Trigonometric Functions

We now define the trigonometric functions of a complex variable z by considering the following two complex power series:

$$\sum_{n=0}^{\infty}(-1)^n\frac{z^{2n}}{(2n)!} \text{ and } \sum_{n=0}^{\infty}(-1)^n\frac{z^{2n+1}}{(2n+1)!}. \tag{4.8}$$

If $z \in \mathbb{C}$ such that $z \neq 0$, then from the first series of (4.8), we have

$$\lim_{n\to\infty}\left|\frac{a_n}{a_{n+1}}\right| = \lim_{n\to\infty}\left|\frac{z^{2n}}{(2n)!}\frac{(2n+2)!}{z^{2n+2}}\right| = \lim_{n\to\infty}\frac{(2n+2)(2n+1)}{|z^2|} = \infty$$

and from the second series of (4.8) we have

$$\lim_{n\to\infty}\left|\frac{a_n}{a_{n+1}}\right| = \lim_{n\to\infty}\left|\frac{z^{2n+1}}{(2n+1)!}\frac{(2n+3)!}{z^{2n+3}}\right| = \lim_{n\to\infty}\frac{(2n+3)(2n+2)}{|z^2|} = \infty.$$

Thus the radii of convergence $R = \infty$ for both the power series, and by the ratio test the two series converge for each $z \neq 0$ in $\mathbb{C}$. But, clearly, the series converges for $z = 0$.

The two power series in Eq.(4.8), with all of their derivatives, define analytic functions on $\mathbb{C}$.

Definition 4.2.1. The sum of the powers series

$$\cos z = \sum_{n=0}^{\infty}(-1)^n\frac{z^{2n}}{(2n)!} \text{ and } \sin z = \sum_{n=0}^{\infty}(-1)^n\frac{z^{2n+1}}{(2n+1)!} \tag{4.9}$$

are defined as the *cosine* and *sine functions* respectively for each $z \in \mathbb{C}$.

Since the radii of convergence $R = \infty$, $\sin z$ and $\cos z$ are analytic functions of z, regular in every bounded domain of the Argand plane.

Property 4.2.1. By Eq. (4.9), for each $z \in \mathbb{C}$

$$\cos(-z) = \sum_{n=0}^{\infty}(-1)^n\frac{(-z)^{2n}}{(2n)!} = \sum_{n=0}^{\infty}(-1)^n(-1)^{2n}\frac{z^{2n}}{(2n)!} = \cos z.$$

This shows that $\cos z$ is an *even function.* Also, for each $z \in \mathbb{C}$,

$$\sin(-z) = \sum_{n=0}^{\infty}(-1)^n \frac{(-z)^{2n+1}}{(2n+1)!} = \sum_{n=0}^{\infty}(-1)^n(-1)^{2n+1}\frac{z^{2n+1}}{(2n+1)!} = -\sin z,$$

showing that $\sin z$ is an *odd function.*

Property 4.2.2. [Euler's formula]: Using the definition of (4.2), we have

$$\begin{aligned}
e^{iz} &= \sum \frac{(iz)^n}{n!} = 1 + iz + \frac{(iz)^2}{2!} + \frac{(iz)^3}{3!} + \cdots \infty \\
&= (1+iz) + \left[\frac{(iz)^2}{2!} + \frac{(iz)^3}{3!}\right] + \left[\frac{(iz)^4}{4!} + \frac{(iz)^5}{5!}\right] + \cdots \infty \\
&= \sum_{n=0}^{\infty} \frac{(iz)^{2n}}{(2n)!} + \sum_{n=0}^{\infty} \frac{(iz)^{2n+1}}{(2n+1)!} \\
&= \sum_{n=0}^{\infty}(i^2)^n \frac{z^{2n}}{(2n)!} + \sum_{n=0}^{\infty} i(i^2)^n \frac{z^{2n+1}}{(2n+1)!} \\
&= \sum_{n=0}^{\infty}(-1)^n \frac{z^{2n}}{(2n)!} + i\sum_{n=0}^{\infty}(-1)^n \frac{z^{2n+1}}{(2n+1)!} \\
&= \cos z + i \sin z \ \forall z \in \mathbb{C}. \qquad (4.10)
\end{aligned}$$

The formula (4.10) is called *Euler's formula.* Again from Eq. (4.10) we get

$$e^{-iz} = \cos(-z) + i\sin(-z) = \cos z - i \sin z. \qquad (4.11)$$

So by equations (4.10) and (4.11), we get, conversely,

$$\cos z = \frac{1}{2}(e^{iz} + e^{-iz}) \text{ and } \sin z = \frac{1}{2i}(e^{iz} - e^{-iz}) \qquad (4.12)$$

These formulae are called *Euler's exponential values* of $\sin z$ and $\cos z$.

The other trigonometric functions are $\tan z$, $\cot z$, $\sec z$ and cosec z can also be similarly defined.

(i) Using equations (4.10) and (4.11), we get

$$e^{2n\pi} = 1, e^{\pi i} = -1, e^{i\pi/2} = i, e^{-i\pi/2} = -i.$$

(ii) Let $z = x + iy \in \mathbb{C}$, where $x, y \in \mathbb{R}$. Then the Euler's formula (4.10) can be written as

$$\begin{aligned}
& e^z = e^{x+iy} = e^x \cdot e^{iy} = e^x(\cos y + i \sin y) \\
\Rightarrow \quad & \text{Re}\,(e^z) = e^x \cos y; \text{Im}\,(e^z) = e^x \sin y. \qquad (4.13)
\end{aligned}$$

(iii) If we take $z = x$ in (4.10), where $x \in \mathbb{R}$, we obtain the result

$$e^{ix} = \cos x + i \sin x = \text{cis}x. \qquad (4.14)$$

$$\therefore \quad |e^{ix}| = |\cos x + i \sin x| = \sqrt{\cos^2 x + \sin^2 x} = 1; x \in \mathbb{R}. \qquad (4.15)$$

(iv) For each $z = x + iy \in \mathbb{C}$, where $x, y \in \mathbb{R}$, we have

$$\begin{aligned}
|e^z| = |e^{x+iy}| &= |e^x.e^{iy}| = |e^x| \cdot |e^{iy}| \text{ by Eq. (4.15).} \\
&= |e^x| \cdot 1 = e^x = e^{\text{Re}\,(z)}.
\end{aligned}$$

Property 4.2.3. [The periodicity of e^z]: A fundamental property of the *complex exponential function* e^z is that it is periodic. Since, for all z

$$e^{z+2\pi i} = e^z \cdot e^{2\pi i} = e^z(\cos 2\pi + i \sin 2\pi) = e^z, \tag{4.16}$$

we see that e^z is periodic with a pure imaginary period $2\pi i$. Because Eq. (4.16) holds for all values of z, we also have

$$e^{z+2n\pi i} \;=\; e^z e^{2n\pi i} = e^z(\cos 2n\pi + i \sin 2n\pi) = e^z;\;\; n \in \mathbb{Z}. \tag{4.17}$$

This means that e^z is a periodic function with a pure imaginary period $2n\pi i$, where $n \in \mathbb{Z}$. Consequently, if we divide up the Argand z-plane into the infinite horizontal strips,

$$S_n = \{x + iy : -\infty < x < \infty, (2n-1)\pi < y \leq (2n+1)\pi; n \in \mathbb{Z}\},$$

as depicted in the Fig. 4.1, then e^z behaves in the same manner on each strip.

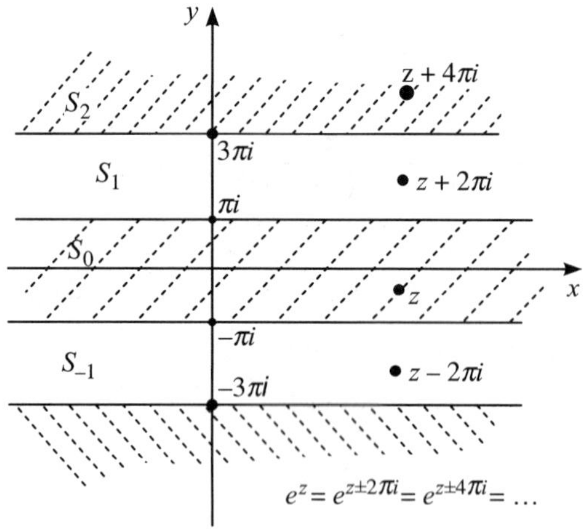

Figure 4.1 *The fundamental region of e^z*

Thus the complex exponential function e^z is one-to-one in any infinite horizontal strip S_n of width 2π in the Argand plane. If the point z is in the infinite horizontal strip $-\infty < x < \infty$, $-\pi < y \leq \pi$, shown in the Fig. 4.1, then the values

$$f(z) = e^z, f(z + 2\pi i) = e^{z+2\pi i}, f(z - 2\pi i) = e^{z-2\pi i},$$

and so on are the same. For these reasons any one of these strips is called a *fundamental region* of e^z.

Property 4.2.4. [The periodicity of $\sin z$ and $\cos z$]: Since e^z is a periodic function with period $2n\pi i$; $n \in \mathbb{Z}$ and e^{iz} is a periodic function with period 2π, hence it follows from Eq.(4.12) that $\sin z$ and $\cos z$ are also periodic functions of period 2π.

Property 4.2.5. Let us suppose that $e^{z_1} = e^{z_2}$.

$$\therefore 1 = e^0 = e^{z_1 - z_2} = e^{x_2 - x_1}\Big[\cos(y_2 - y_1) + i \sin(y_2 - y_1)\Big].$$

Thus $x_2 - x_1 = 0$ and $y_2 - y_1 = 2k\pi$ for some integer k. Hence

$$z_2 - z_1 = (x_2 - x_1) + i(y_2 - y_1) = 0 + 2k\pi i = 2k\pi i.$$

$$\therefore \quad e^{z_1} = e^{z_2} \Rightarrow z_2 = z_1 + 2k\pi i;\; k \in \mathbb{Z}. \tag{4.18}$$

Property 4.2.6. [Analyticity]: The derivative of $\cos z$ exists for each $z \in \mathbb{C}$ and is given by

$$\begin{aligned}\frac{d}{dz}\cos z &= \frac{d}{dz}\Big[\sum_{n=0}^{\infty}(-1)^n\frac{z^{2n}}{(2n)!}\Big] = \sum_{n=0}^{\infty}(-1)^n\frac{2nz^{2n-1}}{(2n)!}\\ &= \sum_{n=0}^{\infty}(-1)^n\frac{z^{2n-1}}{(2n-1)!} = -\sum_{n=0}^{\infty}(-1)^n\frac{z^{2n+1}}{(2n+1)!} = -\sin z\ \forall z \in \mathbb{C}.\end{aligned}$$

Likewise, the derivative of $\sin z$ exists for each $z \in \mathbb{C}$ and is given by

$$\begin{aligned}\frac{d}{dz}\sin z &= \frac{d}{dz}\Big[\sum_{n=0}^{\infty}(-1)^n\frac{z^{2n+1}}{(2n+1)!}\Big] = \sum_{n=0}^{\infty}(-1)^n\frac{(2n+1)z^{2n}}{(2n+1)!}\\ &= \sum_{n=0}^{\infty}(-1)^n\frac{z^{2n}}{(2n)!} = \cos z\ \forall z \in \mathbb{C}.\end{aligned}$$

Thus the functions $\cos z$ and $\sin z$ are entire functions.

We know $\tan z$ and $\sec z$ are analytic functions, regular in any domain in which $\cos z$ never vanishes, their derivatives being $\sec^2 z$ and $\sec z \tan z$ respectively. Similarly, $\cot z$ and $\operatorname{cosec} z$ are analytic functions, regular in any domain in which $\sin z$ never vanishes, their derivatives being $-\operatorname{cosec}^2 z$ and $-\cot z \operatorname{cosec} z$ respectively.

Property 4.2.7. [Addition theorem]: From the Eq.(4.12) and Eq.(4.4), we obtain the following identities:

$$\left.\begin{aligned}\cos(z_1 \pm z_2) &= \cos z_1 \cos z_2 \mp \sin z_1 \sin z_2\\ \sin(z_1 \pm z_2) &= \sin z_1 \cos z_2 \pm \cos z_1 \sin z_2\end{aligned}\right\} \tag{4.19}$$

Property 4.2.8. [Modulus]: Since $z = x + iy \in \mathbb{C}$, we have

$$\begin{aligned}\sin z &= \sin(x+iy) = \sin x \cos(iy) + \cos x \sin(iy)\\ &= \sin x\ \cosh y + i\cos x\ \sinh y\\ \Rightarrow |\sin z| &= \sqrt{\sin^2 x + \sinh^2 y}.\end{aligned}$$

Similarly, $|\cos z| = \sqrt{\cos^2 x + \sinh^2 y}$. The modulus of a complex trigonometric function can also be helpful in solving trigonometric equations.

Property 4.2.9. [Zeros]: The zeros of the complex sine and cosine functions are given by

$$\sin z = 0 \Rightarrow z = n\pi \text{ and } \cos z = 0 \Rightarrow z = (2n+1)\frac{\pi}{2},\quad n \in \mathbb{Z}. \tag{4.20}$$

Thus, all of the zeros of $\sin z$ and $\cos z$ are real.

Example 4.2.1. *Find all the zeros and periods of the function* $\cos(2iz + 13)$.

Solution. To find the zero of $\cos(2iz+13)$, we have

$$\cos(2iz+13) = 0 \Leftrightarrow 2iz + 13 = (n + \frac{1}{2})\pi, n \in \mathbb{N}$$

$$\therefore \quad z = \frac{1}{4}\Big\{(2n-1)\pi + 26\Big\}i, \quad \text{where } n \in \mathbb{N}$$

which are the required zeros. Let the period of the given function $\cos(2iz+13)$ be p. As function $\cos z$ has a period of 2π, we have

$$\cos[2i(z+p)+13] = \cos(2iz+13+2ip).$$

$$\therefore \quad 2ip = 2\pi \Rightarrow p = -\pi i.$$

Hence, πi is the period of the given function.

4.3 Complex Logarithmic Function

In real analysis, the natural logarithm function $\ln x$ is defined as an inverse function of the real exponential function e^x. The situation is very different in complex analysis because the complex exponential function e^z is not a one-to-one function on its domain $\mathbb{C}$. In this section we shall introduce the complex logarithmic function, study some of its properties, and then use it to define complex powers and inverse trigonometric functions.

Definition 4.3.1. Let $z \in \mathbb{C} - \{0\}$. If $w \in \mathbb{C}$ such that $e^w = z$, then w is called the *logarithm of* z and we write $w = \log z$.

Let $z = re^{i\theta} \in \mathbb{C} - \{0\}$, where $\theta \in (-\pi, \pi]$, and $w = u + iv$. Then,

$$e^z = w \Rightarrow e^{u+iv} = e^u \cdot e^{iv} = re^{i\theta}$$

$$\Rightarrow \quad e^u = r \text{ and } v = 2k\pi + \theta; \quad k \in \mathbb{Z}.$$

Now, $e^u = r$ implies $u = \log r = \ln r$, the natural logarithm of a positive real number r. It follows that the solutions of $e^w = z$ are given by

$$w = \log r + i(2k\pi + \theta); \quad k = 0, \pm 1, \pm 2, \cdots.$$

Then we define $w = \log z$ to be any of the infinitely many values as

$$w = \log z = \log r + i(2k\pi + \theta); \quad k \in \mathbb{Z}. \tag{4.21}$$

Therefore, equation $e^w = z$ tells us that, for $z \neq 0$

$$e^{\log z} = e^{\log r + i(2k\pi + \theta)} = e^w = z. \tag{4.22}$$

Equation (4.22) serves as the motivation of defining the multi-valued logarithmic function of a non-zero complex variable $z = re^{i\theta}$.

Definition 4.3.2. The *principal value logarithm* or *principal branch of the logarithm* of z (denoted by $\text{Log } z$) is the unique number given by

$$\text{Log } z = \log|z| + i\text{Arg}z, \quad \text{Arg } z \in (-\pi, \pi], z \in \mathbb{C} - \{0\}. \tag{4.23}$$

The function $\text{Log } z$ is defined by the Eq.(4.23) is a single-valued function and it coincides with natural logarithm when z is a positive real number.

Since $\arg z = \text{Arg } z + 2k\pi$, Eq. (4.21) can be rewritten as

$$\log z = \log r + i \arg z = \log r + i\theta; \theta \in (-\pi, \pi], \tag{4.24}$$

where $z \neq 0$. From Eq.(4.21) we see that all values of $\log z$ are given by

$$\begin{aligned} \log z &= \text{Log } |z| + i(2k\pi + \text{Arg } z) \\ &= \text{Log } z + 2k\pi i; \quad n \in \mathbb{Z} \text{ and } z \neq 0. \end{aligned} \tag{4.25}$$

Note that if z is real and positive $\text{Arg}z = 0$ and hence $\text{Log}|z|$ represents the *principal value of* $\log|z|$.

Property 4.3.1. We know, Arg $z \in (-\pi, \pi]$. The value of Arg z jumps by i as z crosses the negative real axis. Therefore, Log z is not continuous at any point on the non-positive real axis. However, at all points of the non-positive real axis, Log z is continuous.

Theorem 4.1. *Let z_1 and z_2 be any two nonzero complex numbers. Then* $\text{Log}(z_1 z_2) = \text{Log}\, z_1 + \text{Log}\, z_2 + 2k(z_1, z_2)\pi$, *where $k(z_1, z_2)$ is given by*

$$k(z_1, z_2) = \begin{cases} 0 & \text{if } -\pi < \arg z_1 + \arg z_2 \leq \pi; \\ +1 & \text{if } -2\pi < \arg z_1 + \arg z_2 \leq -\pi; \\ -1 & \text{if } \pi < \arg z_1 + \arg z_2 \leq 2\pi. \end{cases}$$

Proof: Using definition (4.23), we have

$$\begin{aligned} \text{Log}\,(z_1 z_2) &= \log|z_1 z_2| + i\,\text{Arg}\,(z_1 z_2) \\ &= \log|z_1| + \log|z_2| + i[\arg z_1 + \arg z_2 + 2n\pi(z_1, z_2)] \\ &= \log|z_1| + i\arg z_1 + \log|z_2| + i\arg z_2 + 2n\pi(z_1, z_2) \\ &= \text{Log}\, z_1 + \text{Log}\, z_2 + 2n\pi(z_1, z_2). \end{aligned}$$

This is the product rule of logarithms. □

Property 4.3.2. [Analyticity of logarithmic function]: We make a cross cut along the real negative axis from $-\infty$ to 0 (Fig. 4.2). Let

$$\mathcal{D} = \mathbb{C} - \{z = x + iy \in \mathbb{C} : y = 0, x \leq 0\} \tag{4.26}$$

be any bounded domain in this cut plane, so that no point of the cut belongs of $\mathcal{D}$; Log z is one-valued and continuous in $\mathcal{D}$. Let $f : \mathcal{D} \to \mathbb{C}$ defined by $z \to \text{Log}\, z$ and $g : \mathbb{C} \to \mathbb{C}$ be defined by $z \to e^z$. Then $f(\mathcal{D}) \subset \mathbb{C}$ and $g(f(z)) = z$ for every $z \in \mathcal{D}$.

Let $w = \text{Log}\, z$ and $w_1 = \text{Log}\, z_1$, so that $z = e^w$ and $w_1 = e^{z_1}$. Then $w_1 \to w$ as $z_1 \to z$ along any path in $\mathcal{D}$. Hence

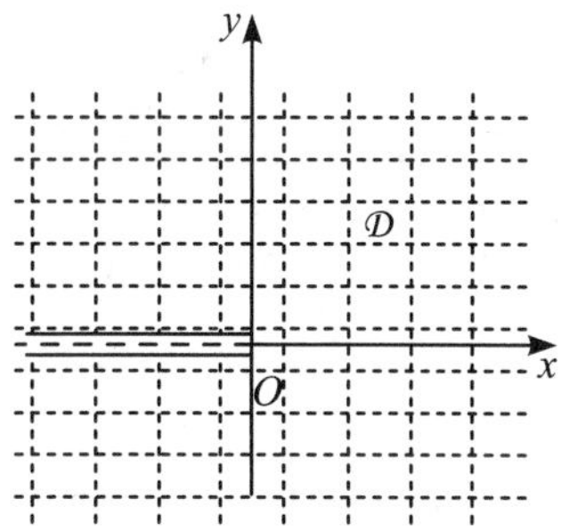

Figure 4.2 $\mathcal{D} = \mathbb{C} - \{z = x + iy \in \mathbb{C} : y = 0, x \leq 0\}$

$$\begin{aligned} \frac{\text{Log}\, z - \text{Log}\, z_1}{z - z_1} &= \frac{w - w_1}{e^w - e^{w_1}} = \frac{w - w_1}{(w - w_1) + \frac{1}{2!}(w^2 - w_1^2) + \cdots} \\ &= \frac{1}{1 + \frac{1}{2!}(w + w_1) + \frac{1}{3!}(w^2 + ww_1 + w_1^2) + \cdots} \\ &= \frac{1}{1 + \frac{w}{1!} + \frac{w^2}{2!} + \cdots} \quad \text{as } w \to w_1. \end{aligned}$$

Therefore,

$$\lim_{z \to z_1} \frac{w - w_1}{z - z_1} = \frac{1}{e^w} = \frac{1}{z} \text{ for all } z \in \mathcal{D}.$$

Thus, a branch of logarithm is analytic and $\dfrac{d}{dz}(\text{Log}\, z) = \dfrac{1}{z}$ on $\mathcal{D}$.

A line used to create a domain of analyticity is called a *branch line* or *branch cut*. Any point that must lie on a branch cut-no matter what branch is used is called a *branch point of a multi-valued function*. For example, the non-positive real axis shown in Figure 4.2 is a branch cut for Log z, and the point $z = 0$ is a *branch point*.

Property 4.3.3. [Branch of a logarithm]: From Eq. (4.25), the logarithm of z is given by

$$\begin{aligned} f(z) &= \log z = \operatorname{Log} z + 2k\pi i = \log r + i(\theta + 2k\pi); \quad k \in \mathbb{Z} \\ &= u(r,\theta) + iv(r,\theta), \text{ say }, r \geq 0, -\infty < \theta < \infty, \end{aligned} \tag{4.27}$$

where, $u(r,\theta) = \log r$, $v(r,\theta) = 2k\pi + \theta$, $r \geq 0, -\infty < \theta < \infty$. $f(z)$ has a singular point at $z = 0$, i.e., $r = 0$. So, to define $f(z)$ we have to restrict $r \neq 0$, i.e., $z \neq 0$. Further, to make it single-valued, we have to restrict θ to either

$$(i) \quad 0 \leq \theta < 2\pi \text{ or } \quad (ii) \quad -\pi \leq \theta < \pi \text{ or } \quad (iii) \quad \alpha \leq \theta < 2\pi + \alpha, \alpha \in \mathbb{R}.$$

With these restrictions, $f(z)$ is a single-valued function that exists everywhere in the restricted complex z-plane if $v(r,\theta) = \theta$. Consider the single-valued function $g(z)$, as in Eq. (4.24):

$$g(z) = \log z = \ln r + i\theta; \quad r \geq 0, \quad 0 \leq \theta < 2\pi. \tag{4.28}$$

Further, if $u(r,\theta) = \ln r$ and $v(r,\theta) = \theta$, we have

$$u_r(r,\theta) = \frac{1}{r}, u_\theta(r,\theta) = 0; v_r(r,\theta) = 0, v_\theta(r,\theta) = 1$$

$$\Rightarrow \qquad u_r = \frac{1}{r}v_\theta \qquad \text{and} \qquad \frac{1}{r}u_\theta = -v_r,$$

which shows that the Cauchy–Riemann equations are satisfied with the continuity conditions on the partial derivatives. Hence g is analytic.

But in $0 \leq \theta < 2\pi$, if we choose any point (r,θ), we get

$$\lim_{\theta \to 0_+} v(r,\theta) = 0 \text{ and } \lim_{\theta \to 2\pi_{-0}} v(r,\theta) = 2\pi,$$

implying that $v(r,\theta)$ is not continuous at any point on the ray comprising of the real z-axis including $z = 0$. Thus, every point on the real z-axis is a singular point of $\log z$. Therefore, for analyticity of $f(z) = \log z$, we have to restrict r and θ such that

$$g(z) = \log z = \ln r + i\theta \text{ in } \mathcal{D} = \{|z| = r > 0, \; 0 < \theta < 2\pi\}. \tag{4.29}$$

This $u(r,\theta)$ and $v(r,\theta)$ are both continuous on $\mathcal{D}$. We have that if $\ln r$ on $\mathcal{D}$ is a branch of logarithm, it is called the *principal branch of logarithm*. This means that the branch point of $f(z) = \log z$ is $z = 0$, and the branch cut for the principal branch is the positive real axis $\theta = 0$ as shown in Fig. 4.3.

If instead, we choose $-\pi < \theta < \pi$, restricting $\theta = -\pi$, the branch cut is $\theta = -\pi$ and if we choose $\alpha < \theta < 2\pi + \alpha$, the branch cut is $\theta = \alpha$ and $r = 0$, implying that $z = 0$ is the common branch point for all the cuts.

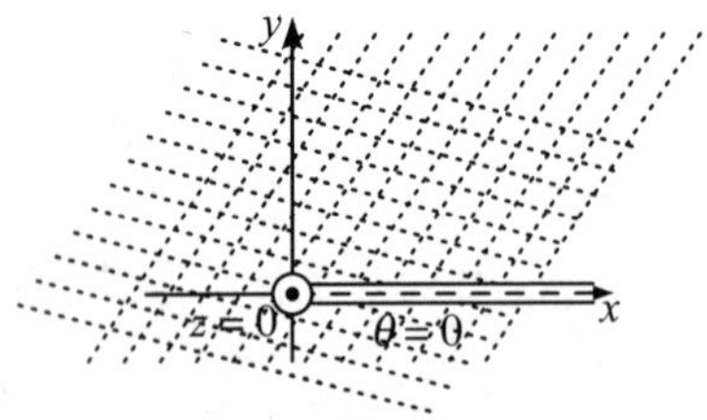

Figure 4.3 *The domain $\mathcal{D}$ of the branch g*

Property 4.3.4. Suppose $f : \mathcal{D} \to \mathbb{C}$ is a principle branch of logarithm, then $g : \mathcal{D} \to \mathbb{C}$ another branch of logarithm iff $g(z) = f(z) + 2k\pi i$ for some integer k.

Proof: Let, for some integer k, $g(z) = f(z) + 2k\pi i$. Then

$$e^{g(z)} = e^{f(z)+2k\pi i} = e^{f(z)}[\cos 2k\pi + i \sin 2k\pi] = e^{f(z)} = z.$$

Thus g is continuous as well and hence g is also a branch of logarithm.

Conversely, if f and g are both branches of logarithm, then by definition

$$e^{f(z)} = e^{g(z)} = z \text{ for each } z \in \mathcal{D}$$

$$\Rightarrow \quad f(z) = g(z) + 2k\pi i; \quad k \in \mathbb{Z}.$$

This proves the property. □

RESULT **37.** Thus we have if $\mathcal{D} \subset \mathbb{C}$ is a region and f is a branch of logarithm, then the totality of branches of logarithm are the functions $f(z) + 2k\pi i; \quad k \in \mathbb{Z}$.

Property 4.3.5. [Riemann surface for $\log z$] : Corresponding to each $z \neq 0$, the multi-valued function

$$\log z = \text{ Log } |z| + i \arg z = \text{ Log } |z| + i \text{ Arg } z + 2k\pi i; k = 0, 1, 2, \cdots$$

has a countably many values. To describe $\log z$ as a single-valued function, we replace the z-plane with the origin deleted by a surface on which a new point is located whenever the argument of the number z is increased by 2π, or an interval multiple of 2π. Thus the Riemann surface of $\log z$ requires infinitely many copies S_k for $k = \cdots, -n, \cdots, -1, 0, 1, \cdots, n, \cdots$ of the z-plane cut along the negative x-axis (Fig. 4.4). Now, we stick these cut planes directly on each other so that the corresponding points have the same position. We join the sheet S_k to S_{k+1} as follows.

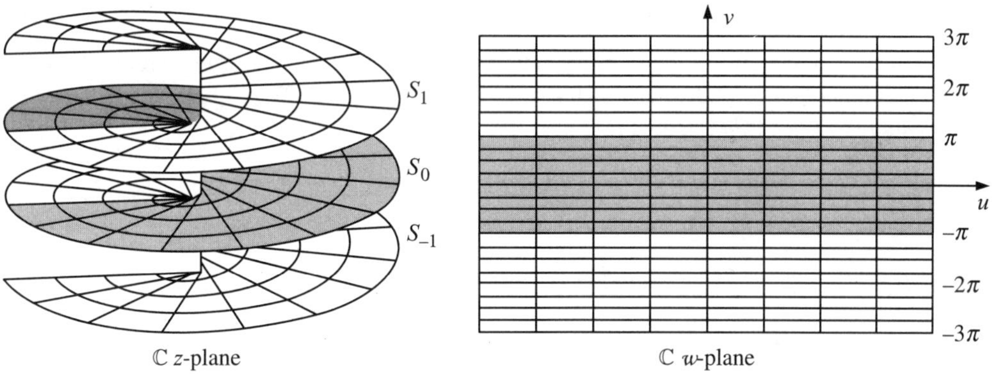

Figure 4.4 *Riemann surface of $w = \log z$*

For each integer k, the edge of the sheet S_k in the upper half-plane is joined to the edge of the sheet S_{k+1} in the lower-half plane. The Riemann surface for the domain of the function $w = \log z$ looks like a spiral staircase that extends upward on the sheets $S_1, S_2, \cdots$ and downward on the sheets $S_{-1}, S_{-2}, \cdots$ as depicted in the Fig. 4.4. For S_k, we use

$$z = re^{i\theta}; \ r = |z| \text{ and } (2k-1)\pi < \theta \leq (2k+1)\pi.$$

The coordinates r and θ of a point on any sheet can be considered as the polar coordinates of the projection of the point onto the original z-plane, the angular coordinate θ being restricted to a definite range of 2π radians on each sheet.

Consider any continuous curve on this connected surface of infinitely many sheets. As a point z describes that curve, the values of $\log z$ vary continuously since θ, in addition to r, varies continuously; $\log z$ now assumes just one value corresponding to each point on the curve.

Again, for S_k, the correct branch of $\log z$ on each sheet is

$$\log z = \log r + i\theta, \text{ where } r = |z| \text{ and } (2k-1)\pi < \theta \leq (2k+1)\pi.$$

The surface described here is a Riemann surface for $\log z$. It is a connected surface of infinitely many sheets, arranged so that $\log z$ is a single-valued function of points on it.

4.3.1 The general Power Function z^α

Definition 4.3.3. If z and α are given complex numbers such that $z \neq 0$, then z^α is defined by the equation

$$z^\alpha = e^{\alpha \log z} = |z|^\alpha e^{i\alpha \arg z}, \tag{4.30}$$

where $\arg z = \text{ Arg}z + 2n\pi; n \in \mathbb{Z}$. Powers of z are, in general, multi-valued.

If a denotes the principal value of $\arg z$, then

$$z^\alpha = e^{\alpha \log |z| + i\alpha a + 2n\alpha\pi i} = e^{\alpha \log |z|} \cdot e^{i\alpha a} \cdot e^{2n\alpha\pi i}. \tag{4.31}$$

Thus, in general, z^α is infinite-valued and the different values of z^α are obtained by giving different $n \in \mathbb{Z}$ in Eq. (4.31). The principal value of z^α is defined to be a unique number $e^{\alpha \text{Log} z}$. For example

$$i^i = e^{i\text{Log} i} = e^{i[\log |i| + i \arg i]} = e^{-\frac{\pi}{2}}.$$

If $z = re^{i\theta}$ and θ_0 is a real constant, the function

$$\log z = \log r + i\theta; \quad r > 0, \theta_0 < \theta < \theta_0 + 2\pi \tag{4.32}$$

is single-valued and analytic in the domain indicated, as is the composite function $e^{\log z}$. Thus the function z^α defined by Eq. (4.30), in which $\log z$ is given by Eq. (4.32), is single-valued and analytic in the domain $r > 0, \theta_0 < \theta < \theta_0 + 2\pi$.

Theorem 4.2. *If z, α_1 and α_2 are complex numbers and $z \neq 0$, then*

$$z^{\alpha_1} \cdot z^{\alpha_2} = z^{\alpha_1 + \alpha_2}.$$

If z_1, z_2 and α are complex numbers and $z_1 z_2 \neq 0$, then

$$(z_1\, z_2)^\alpha = z_1^\alpha\, z_2^\alpha\, e^{2n\pi i\alpha} \textit{ for some } n \in \mathbb{Z}.$$

Proof: Using Eq. (4.30), we get

$$z^{\alpha_1 + \alpha_2} = e^{(\alpha_1 + \alpha_2)\text{Log} z} = e^{\alpha_1 \text{Log} z} + e^{\alpha_2 \text{Log} z} = z^{\alpha_1} + z^{\alpha_2}$$

Also, using Eq. (4.30), we get

$$\begin{aligned}(z_1\, z_2)^\alpha &= e^{\alpha \text{Log}(z_1\, z_2)} = e^{\alpha \text{Log}(z_1 + \text{Log} z_2 + 2n\pi i)} \\ &= e^{\alpha \text{Log} z_1}\, e^{\text{Log} z_2}\, e^{2n\pi i} = z_1^\alpha\, z_2^\alpha\, e^{2n\pi i\alpha}.\end{aligned}$$

□

Property 4.3.6. [Derivative of z^α]: If $f : \mathcal{D} \to \mathbb{C}$; where $\mathcal{D}$ is a principal branch of logarithm (Eq. 4.26), then for $g(z) = z^\alpha$ the principal branch is taken so that $g : \mathcal{D} \to \mathbb{C}$. The derivative of that branch of the multiple-valued power function z^α can be written as

$$\begin{aligned}\frac{d}{dz}(z^\alpha) &= \frac{d}{dz}\left(e^{\alpha \log z}\right) = e^{\alpha \log z} \cdot \frac{\alpha}{z} \\ &= \alpha \frac{e^{\alpha \log z}}{e^{\log z}} = \alpha e^{(\alpha - 1)\log z} = \alpha\, z^{\alpha - 1}.\end{aligned} \tag{4.33}$$

In particular, when $\theta_0 = -\pi$, so that $-\pi < \theta < \pi$, the function $z^\alpha = e^{\alpha \log z}$; $z \neq 0$, is called the *principal branch* of the multiple-valued power function given by Eq. (4.30). It is single-valued and analytic in domain $r > 0, -\pi < \theta < \pi$.

RESULT **38.** [Phase change lemma]: Let $g(z)$ be a single-valued function that is defined and continuous near z_0. For any continuously-varying branch of $(z-z_0)^{\alpha}$, the function $f(z) = (z-z_0)^{\alpha} g(z)$ is multiplied by the phase factor $e^{2\pi i\alpha}$ when z traverses a complete circle about z_0 in the positive direction. For example,

(i) If α is an integer, the phase factor of z^{α} at 0 is $e^{2\pi i\alpha} = 1$, in accordance with the fact that z^{α} is single-valued.

(ii) The phase factor of $\sqrt{z-z_0}$ at z_0 is $e^{\pi i} = -1$. As z traverses a circle about z_0, the values of $f(z) = \sqrt{z-z_0}$ returns to $-f(z)$. The phase factor of $1/\sqrt{z-z_0} = i/\sqrt{z-z_0}$ at z_0 is also -1.

Property 4.3.7. [Riemann surface associated with cos]: Consider $0 + iy \in \mathbb{C}$. Then

$$\cos(0+iy) = \cos 0 \cosh y + \sin 0 \sinh y = \cosh y \geq 1.$$

Thus, the non-negative imaginary axis $\{(0,y) : y \geq 0\}$ is mapped into $\{(u,0) : u \geq 1\}$ and

$$\cos : \{(0,y) : y \geq 0\} \longrightarrow \{(u,0) : u \geq 1\}$$

is onto. Indeed, for a given $u > 1$, using $\cosh(y) \to \infty$ as $y \to \infty$, we find $y > 0$ such that $\cosh(y) > u$. Therefore, applying the intermediate theorem, we can find $y_0 \in (0,y)$ such that $\cosh(y_0) = u$. cosh being an even function, we also observe that $\{(0,y) : y \leq 0\}$ is mapped onto $\{(u,0) : u \geq 1\}$. By a similar argument, one can verify that cos maps:

(i) both of $\{(\pi,y) : y \geq 0\}$ and $\{(\pi,y) : y \leq 0\}$ onto $\{(u,0) : u \leq 1\}$,

(ii) $\{(x,y) : 0 < x < \pi, y > 0\}$ onto the lower half-plane $\{(u,v) : v < 0\}$,

(iii) $\{(x,y) : 0 < x < \pi, y < 0\}$ onto the upper half-plane $\{(u,v) : v > 0\}$,

(iv) $\{(x,y) : 0 < x < \pi, y = 0\}$ onto the line segment $\{(u,0) : |u| < 1\}$.

Now, we claim that

$$\{(x,y) : 0 < x < \pi\} \to \mathbb{C}/\{(u,v) \in \mathbb{C} : |u| \geq 0, v = 0\}$$

is a bijection. As the injectivity of cos on $\{(x,y) : 0 < x < \pi\}$ is obvious, we only verify the surjectivity of cos. Let

$$w \in \mathbb{C}/\{(u,v) \in \mathbb{C} : |u| \geq 0, v = 0\}$$

be arbitrary and let $\theta = (w \pm \sqrt{w^2-1})$. If $\theta \in (0,\pi)$, then we take $z = \theta - i\log|w \pm \sqrt{w^2-1}$ so that $z \in \{(x,y) : 0 < x < \pi\}$ and $\cos z = w$. Otherwise, $\theta \in (\pi, 2\pi)$. In this case, we choose the required pre-image by

$$z = 2\pi - \theta - i\log|w \pm \sqrt{w^2-1}|.$$

By the similar argument, we obtain

$$\cos : \{(x,y) : (k-1)\pi < z < k\pi\} \to \mathbb{C}/\{(u,v) \in \mathbb{C} : |u| \geq 0, v = 0\}$$

is a bijection for all $k \in \mathbb{N}$, with the additional information that

- the upper and lower half-strips are mapped onto the lower and upper half-planes respectively when k is odd.
- the upper and lower half-strips are mapped onto themselves when k is even.

Now we are ready to construct the Riemann surface associated with cos.

- We take countably infinite number of complex planes with two slits from 1 to ∞ and -1 to ∞ and denote them by

$$\{w \in \mathbb{C}/\{0\} : (2k+1)\pi \neq \arg w \in (2k\pi.(2k+2)\pi)\} \cup \{(u,0) \in \mathbb{C} : |u| \leq 1\}$$

  called *sheet-k*, $k \in \mathbb{Z}$.
- Paste the upper edge on the slit on the left side of sheet-k with the lower edge on the slit on the left side of sheet-$(k+1)$ and the lower edge on the slit on the left side of sheet-k with the upper edge on the slit on the left side of sheet-$(k+1)$.
- Paste the upper edge on the slit on the right side of the sheet-k with lower edge on the slit on the right side of sheet-$(k+1)$ and the lower edge on the slit on the right side of sheet-k with the upper edge on the slit on the right side of sheet-$(k+1)$.
- By gluing the line segment $\{(u,0) : |u| \leq 1\}$ in every sheet together, identify it as a single line segment on the surface.

Then, cos becomes a continuous bijection from $\mathbb{C}$ onto the surface.

4.4 Complex Hyperbolic Functions

Definition 4.4.1. The *hyperbolic cosine* and *sine* functions of complex variable $z \in \mathbb{C}$ are defined as

$$\cosh z = \frac{1}{2}\Big(e^z + e^{-z}\Big) = \sum_{k=0}^{\infty} \frac{z^{2k}}{(2k)!} \tag{4.34}$$

$$\sinh z = \frac{1}{2}\Big(e^z - e^{-z}\Big) = \sum_{k=0}^{\infty} \frac{z^{2k+1}}{(2k+1)!}. \tag{4.35}$$

The other hyperbolic functions $\tanh z$, $\operatorname{sech} z$, $\operatorname{cosech} z$, and $\coth z$ can be defined in terms of $\cosh z$ and $\sinh z$ in the customary manner.

Property 4.4.1. In the definition of $\cos z$ and $\sin z$, taking $z = iy$, we obtain

$$\begin{aligned} \cos(iy) &= \frac{1}{2}\Big(e^y + e^{-y}\Big) = \cosh y, \\ \sin(iy) &= \frac{1}{2i}\Big(e^{-y} - e^y\Big) = \frac{i}{2}\Big(e^y - e^{-y}\Big) = i \sinh y. \end{aligned} \tag{4.36}$$

Equations (4.36) are of great importance as they enable us to deduce the properties of the hyperbolic functions from the corresponding properties of the trigonometrical functions.

Property 4.4.2. Using the property 4.4.1, we get, for each $z = x + iy \in \mathbb{C}$,

$$\begin{aligned} \cos z &= \cos(x+iy) = \cos x \cos(iy) - \sin x \sin(iy) \\ &= \cos x \cosh y - i \sin x \sinh y. \end{aligned} \tag{4.37}$$

$$\begin{aligned} \sin z &= \sin(x+iy) = \sin x \cos(iy) + \cos x \sin(iy) \\ &= \sin x \cosh y + i \cos x \sinh y. \end{aligned} \tag{4.38}$$

These formula separate the real and imaginary parts of $\cos z$ and $\sin z$ so that they may be evaluated by tables of trigonometric and hyperbolic functions.

Property 4.4.3. Since e^z and e^{-z} are analytic in the entire complex plane $\mathbb{C}_\infty$, it follows that $\sinh z$, $\cosh z$ are also analytic in $\mathbb{C}_\infty$. Thus the functions $\cosh z$ and $\sinh z$ are entire and so are analytic functions. Further,

$$\frac{d}{dz}\sinh z = \frac{d}{dz}\left[\frac{1}{2}(e^z - e^{-z})\right] = \frac{1}{2}\left(e^z + e^{-z}\right) = \cosh z \quad \text{on } \mathbb{C}.$$

Similarly, $\frac{d}{dz}\cosh z = \sinh z$ on $\mathbb{C}$.

It should be observed that $\sinh z$ and $\cosh z$ are regular in every bounded domain. On the other hand, $\tanh z$ and $\mathrm{sech} z$ have singularities at the points where $\cosh z$ vanishes and $\coth z$ and $\mathrm{cosech} z$ at the points where $\sinh z$ vanishes.

Property 4.4.4. [Zeros]: The *zeros* of the complex sine and cosine hyperbolic functions are given by

$$\sinh z = 0 \Leftrightarrow -i\sin iz = 0 \Leftrightarrow iz = n\pi \Leftrightarrow z = n\pi i; n \in \mathbb{N},$$
$$\cosh z = 0 \Leftrightarrow z = (2n+1)\frac{\pi}{2}i; n \in \mathbb{N}.$$

The singularities of $\tanh z$ and $\mathrm{sech} z$ are the zeros of $\cosh z$, and hence they are analytic everywhere except at $z = (2n+1)\pi i/2$, where $n \in \mathbb{N}$. Similarly, the singularities of $\coth z$ and $\mathrm{cosech} z$ are the zeros of $\sinh z$, and hence they are analytic everywhere except at $z = n\pi i$, where $n \in \mathbb{N}$.

4.4.1 Inverse Trigonometric Functions

Here we will find inverse of the trigonometric functions which are multiple-valued and can be expressed in terms of logarithms. They made single-valued by choosing a particular branch of the log function (and the particular branch of the square root function if necessary).

Definition 4.4.2. The *inverse sine of* z or arcsin of z, $w = \sin^{-1} z$, is defined by the equation $z = \sin w$. For each $z \in \mathbb{C}$, there are infinitely many values of w such that $z = \sin w$.

To find these values of w (for a given z), we write

$$\begin{aligned} z &= \sin w = \frac{e^{iw} - e^{-iw}}{2i} = \frac{1}{2ie^{iw}}(e^{2iw} - 1) = \frac{1}{2ip}(p^2 - 1) \\ &\Rightarrow p^2 - 2izp - 1 = 0: \ p = e^{iw} \\ &\Rightarrow p = e^{iw} = iz \pm \sqrt{1 - z^2}, \end{aligned} \tag{4.39}$$

where $\sqrt{1-z^2}$ is a double-valued function. Thus for each $z \in \mathbb{C}$,

$$w = \sin^{-1} z = -i\log\left[iz + (1 - z^2)^{1/2}\right]. \tag{4.40}$$

Since the logarithm of the right-hand side of Eq. (4.40) is a multi-valued function, $\sin^{-1} z$ is also a multi-valued function. For a fixed number z, each number w given by the formula (4.40) is denoted by $\sin^{-1} z$ or $\arcsin z$.

When specific branches of the square root and the logarithm are used, the function $\sin^{-1} z$ becomes single-valued and analytic because it is a composite of analytic functions. Let the *principal arcsin of* z be denote by $\mathrm{Sin}^{-1} z$. Then

$$\mathrm{Sin}^{-1} z = -i\mathrm{Log}\left(iz + \sqrt{1-z^2}\right) = \frac{\pi}{2} + i\mathrm{Log}\left(z + i\sqrt{1-z^2}\right); \tag{4.41}$$

where,

$$\left.\begin{aligned}\sqrt{z^2-1} &= \sqrt{|z^2-1|}e^{i(\theta+\phi)/2}\\ \theta &= \arg(z-1) \in [0, 2\pi)\\ \phi &= \arg(z+1) \in (-\pi, \pi]\end{aligned}\right\}. \tag{4.42}$$

Let us consider $-1 < z = x < 1$. Then

$$\begin{aligned}\sqrt{x^2-1} &= \sqrt{|x^2-1|}e^{i(\theta+\phi)/2} = \sqrt{1-x^2}e^{i(\pi+0)/2}\\ &= \sqrt{1-x^2}\, e^{i\pi/2} = i\sqrt{1-x^2}.\end{aligned}$$

By using Eq. (4.41) we get

$$\begin{aligned}\sin^{-1} x &= \frac{\pi}{2} + i\text{Log}(x + i\sqrt{1-x^2})\\ &= \frac{\pi}{2} + i\Big[\text{Log}1 + i\text{Arg}(x + i\sqrt{1-x^2})\Big]\\ &= \frac{\pi}{2} - \text{Arg}(x + i\sqrt{1-x^2}).\end{aligned} \tag{4.43}$$

Now, $0 < \text{Arg}(x + i\sqrt{1-x^2}) < \pi$ where $-1 < x < 1$, and hence from Eq. (4.43) we get

$$-\frac{\pi}{2} < \sin^{-1} x < \frac{\pi}{2}.$$

We observe that $\text{Sin}^{-1}0 = 0$ and that $\text{Sin}^{-1}(\pm 1) = \pm\frac{\pi}{2}$. Thus if $-1 \le z = x \le 1$, then the value of Arcsin z given in Eq. (4.41) agrees with the value of Arcsin x used in the calculus. In fact, we choose our branch of $(z^2-1)^{1/2}$ and the principal branch of the logarithm so that Eq. (4.41) will hold.

Property 4.4.5. [Branch cut]: Let L_1 and L_{-1} be the rays from 1 and -1 respectively, given by

$$L_1 = \{z : \theta = \arg(z-1)\} \text{ and } L_{-1} = \{z : \phi = \arg(z+1)\}. \tag{4.44}$$

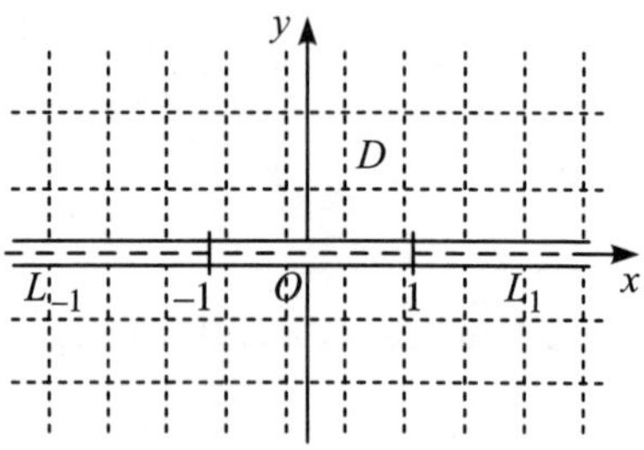

Figure 4.5 $\mathcal{D} = \mathbb{C} - (L_1 \cup L_{-1})$

The branch cut for $\sqrt{(z-1)(z+1)}$ is $L_1 \cup L_{-1}$, where L_1 is the closed ray $\theta = \arg(z-1) = 0$ and L_{-1} is the closed ray $\phi = \arg(z+1) = \pi$. Let $\mathcal{D}_1 = \mathbb{C} - L_1$ and $\mathcal{D}_{-1} = \mathbb{C} - L_{-1}$. Then

$$\mathcal{D} = \mathcal{D}_1 \cap \mathcal{D}_{-1} = \mathbb{C} - (L_1 \cup L_{-1}),$$

as depicted in Fig. 4.5. Using the branch of $(z^2-1)^{1/2}$ as in Eq. (4.42), we get $g(z) = z + \sqrt{z^2-1}$. Then g is analytic in $\mathcal{D}$. Also, using $K = \{(x, 0) : x \le 0\}$ for the non-positive real axis, we notice that

Log z is analytic in $(\mathbb{C}-K)$. Now $g(D)\subset\mathbb{C}-K$, so Log z is analytic in $\mathcal{D}$. Hence $\text{Sin}^{-1}z$ is analytic in $\mathcal{D}$ with

$$\frac{d}{dz}\text{Sin}^{-1}z=\frac{1}{(1-z^2)^{\frac{1}{2}}}.$$

Therefore, $L_1\cup L_{-1}$ is the branch cut for $\text{Sin}^{-1}z$.

Similarly, we can define

$$\text{Cosec}^{-1}z=\frac{1}{i}\ln\left(\frac{i+\sqrt{z^2-1}}{z}\right)+2k\pi i;\quad k\in\mathbb{Z}.$$

Similarly, we define other inverse trigonometric or circular functions, $\cos^{-1}z$, $\tan^{-1}z$, etc., which are multiple valued and can be expressed in terms of natural logarithm.

$$\text{cosec}^{-1}z=\frac{1}{i}\ln\left(\frac{i+\sqrt{z^2-1}}{z}\right)\quad\text{and}\quad\sec^{-1}z=\frac{1}{i}\ln\left(\frac{i+\sqrt{1-z^2}}{z}\right)$$
$$\tan^{-1}z=\frac{1}{2i}\ln\left(\frac{1+iz}{1-iz}\right)\quad\text{and}\quad\cot^{-1}z=\frac{1}{2i}\ln\left(\frac{z+i}{z-i}\right).$$

In all cases we omit an additive constant $2k\pi i$; $k=0,\pm1,\pm2,\cdots$ in the logarithm.

Example 4.4.1. *Construct a Riemann surface for the function* $\cos^{-1}z$.

Solution. The function $w=\cos^{-1}z=-i\log\left[z+\left(z^2-1\right)^{1/2}\right]$ has two sources of multi-valuedness; one due to the square root function $\left(z^2-1\right)^{1/2}$ and the other due to the logarithm, if any.

First we construct the branch of the square root

$$(z^2-1)^{1/2}=(z+1)^{1/2}(z-1)^{1/2}.$$

We see that there are branch points at $z=-1$ and $z=1$. In particular, we want $\cos^{-1}z$ to be defined for $z=x, x\in[-1,1]$. Hence we introduce the branch cuts on the lines $(-\infty,-1]$ and $[1,\infty)$. Let

$$z+1=re^{i\theta}\text{ and }z-1=\rho e^{i\psi}.$$

With the given branch cuts, the angles have possible ranges $-\pi\le\theta<\pi, 0\le\psi<2\pi$. Now we must determine if the logarithm introduces additional branch points. The only possibilities for branch points are where the argument of the logarithm is zero.

$$\begin{aligned}&z+(z^2-1)^{1/2}=0\\ \text{or,}\quad&z^2=z^2-1\Rightarrow0=-1.\end{aligned}$$

We see that the argument of the logarithm cannot be zero and thus there are no additional branch points. Here the Riemann surface consists of two sheets, $\mathcal{S}_0$ and $\mathcal{S}_1$, joined on the real axis $(-\infty,-1]\cup[1,\infty)$:

$$\mathcal{S}_0\begin{cases}0\le\psi<2\pi\\-\pi\le\theta<\pi\end{cases}\qquad\mathcal{S}_1\begin{cases}2\pi\le\psi<4\pi\\ \pi\le\theta<3\pi\end{cases}$$

A particular branch of this function can be obtained by taking

$$z+1=re^{i\theta},-\pi\le\theta<\pi;\quad z-1=\rho e^{i\psi},0\le\psi<2\pi.$$

Then adding these relations, we find $z = \frac{1}{2}\left(re^{i\theta} + \rho e^{i\psi}\right)$. Thus the function $z + (z^2-1)^{1/2}$ reduces to

$$\begin{aligned} z + \left(z^2-1\right)^{1/2} &= \frac{1}{2}\left(re^{i\theta} + \rho e^{i\psi}\right) + \left(r\phi\right)^{1/2} e^{i(\theta+\psi)/2} \\ &= \frac{re^{i\theta}}{2}\left[1 + \frac{\rho}{r}\, e^{i(\psi-\theta)} + 2\sqrt{\frac{\rho}{r}}\, e^{i(\psi-\theta)/2}\right] \\ &= \frac{re^{i\theta}}{2}\left[1 + \sqrt{\frac{\rho}{r}}\, e^{i(\psi-\theta)/2}\right]^2. \\ \therefore \log\left[z + \left(z^2-1\right)^{1/2}\right] &= \log\left(\frac{re^{i\theta}}{2}\right) + \log\left(1 + \sqrt{\frac{\rho}{r}}\, e^{i(\psi-\theta)/2}\right)^2. \end{aligned}$$

Then on $\mathcal{S}_0$,

$$\cos^{-1} z = -i\left[\log\left(\frac{re^{i\theta}}{2}\right) + \log\left(1 + \sqrt{\frac{\rho}{r}}\, e^{i(\psi-\theta)/2}\right)^2\right].$$

If a point lying on the sheet $\mathcal{S}_0$ is allowed to travel a path making a complete round of only the branch point $z = 1$, it enters sheet $\mathcal{S}_1$ from sheet $\mathcal{S}_0$. In this case the value ψ changes by 2π, while the value of θ remains unchanged. The change in $(\psi - \theta)/2$ is π. So in this case,

$$\cos^{-1} z = -i\left[\log\left(\frac{re^{i\theta}}{2}\right) + \log\left(1 - \sqrt{\frac{\rho}{r}}\, e^{i(\psi-\theta)/2}\right)^2\right]$$

on $\mathcal{S}_1$. Similarly we can analyse the case when the point on $\mathcal{S}_0$ encloses only the branch point $z = -1$ while travelling a complete round.

4.4.2 Inverse Hyperbolic Functions

Here we will find inverse of the hyperbolic functions, which are multiple-valued and can be expressed in terms of the natural logarithms. For each $z \in \mathbb{C}$, there are infinitely many values of w such that $z = \sinh w$. If $z = \sinh w$, then $w = \sinh^{-1} z$ is called the inverse hyperbolic sine of z.

In view of the relation $\cosh iw = \cos w$, we let

$$\begin{aligned} & z = \cosh iw = \cos w \\ \Rightarrow\quad & w = \cos^{-1} z \text{ and } iw = \cosh^{-1} z. \end{aligned}$$

For each $z \in \mathbb{C}$,

$$\begin{aligned} \cosh^{-1} z &= i\cos^{-1} z = i\Big[-i\log\left\{z + (z^2-1)^{\frac{1}{2}}\right\}\Big] \\ &= \log\left[z + (z^2-1)^{\frac{1}{2}}\right]. \end{aligned}$$

Similarly, in view of the relation $\sinh iw = i\sin w$, we let

$$\begin{aligned} z &= \sinh iw = i\sin w \\ \Rightarrow w &= \sin^{-1}(-iz) \text{ and } iw = \sinh^{-1} z \end{aligned}$$

Hence for each $z \in \mathbb{C}$

$$\begin{aligned} \sinh^{-1} z &= i\sin(-iz) = i\Big[-\imath\log i\{-iz + (-z^2-1)^{\frac{1}{2}}\}\Big] \\ &= \log\left[z + (z^2+1)^{\frac{1}{2}}\right]. \end{aligned}$$

Similarly, we define other inverse hyperbolic functions $\tanh^{-1}$, $\cosh^{-1}, \cdots$ etc., which are multiple valued and can be expressed in terms natural logarithm as

$$\operatorname{cosech}^{-1} z = \ln\left(\frac{1+\sqrt{1+z^2}}{z}\right), \qquad \operatorname{sech}^{-1} z = \ln\left(\frac{1+\sqrt{1-z^2}}{z}\right)$$

$$\tan^{-1} z = \ln\left(\frac{1+z}{1-z}\right), \qquad \cosh^{-1} z = \ln\left(\frac{z+1}{z-1}\right).$$

Exercises

Multiple-choice Questions

1. The principal value of $\log(i^{3/4})$ is

   (a) $i\pi$ (b) $i\pi/2$ (c) $i\pi/4$ (d) $i\pi/8$

2. Let G be an open-connected set in $\mathbb{C}$ and $f : G \to \mathbb{C}$. f is a branch of logarithm if

   (a) f is a continuous function
   (b) f is continuous function such that $z = e^{f(z)}$ for all $z \in G$
   (c) f is constant function such that $z = e^{f(z)}$ for all $z \in G$
   (d) f is decreasing function such that $z = e^{f(z)}$ for all $z \in G$

3. A branch of logarithm function is

   (a) analytic and its derivative is $1/z$ (b) not analytic and its derivative is $1/z$
   (c) analytic and its derivative is z (d) not analytic and its derivative is z

4. Let G be an open-connected set in $\mathbb{C}$ and $f : G \to \mathbb{C}$ a branch of the logarithm on $z \in G$. Then the totality of branch of $\log z$ are the functions:

   (a) $f(z) + i$ (b) $f(z) + 2\pi$ (c) $f(z) + 2k\pi i; k \in \mathbb{Z}$ (d) $f(z)/2k\pi i; k \in \mathbb{Z}$

5. The principal value of $\log(-2+3i)$ is

   (a) $\log\sqrt{13} - i\tan^{-1}(3/2)$ (b) $\log\sqrt{13} + i\tan^{-1}(3/2)$
   (c) $\log\sqrt{13} + i(\pi - \tan^{-1}(3/2))$ (d) $\log\sqrt{13} - i(\pi - \tan^{-1}(3/2))$

6. Which of the following is not multi-valued?

   (a) $\sin^{-1} z$ (b) $\tan^{-1} z$ (c) $\cos^{-1} z$ (d) None of these

7. The function $f(z) = \log(z^2 + z - 2)$ has branch points at

   (a) $1, -1$ (b) $2, -2$ (c) $1, -2$ (d) None of these

8. The function $\log(\cos \pi z)$ has:

   (a) A simple pole at $z = 1/2$ (b) A removable singularity at $z = 1/2$
   (c) An essential singularity at $z = 1/2$ (d) A branch point at $z = 1/2$

Answers: 1(d); 2(b); 3(a); 4(c); 5(c); 6(d); 7(c); 8(d)

Review Questions [Short-answer Type]

1. Separate the following equations into their real and imaginary parts: (i) $e^{\bar{z}}$ (ii) e^{z^2} (iii) e^{e^z}
2. Using the power series for e^z, prove that $\overline{e^z} = e^{\bar{z}}$ for each $z \in \mathbb{C}$.
3. Calculate the value of the function e^z at the points $z = -i\pi/2, 3i\pi/4, 1-2i$.

4. At which points z does e^z assume the values of $2, -1, i, -i/2, -1-i, 1+2i$?
5. How does e^z change when the point z goes off to infinity along a ray from the origin?
6. Find the power series expansion of the principal branch of the log function about the point $z = i$.
7. Show that

   (i) $\left|e^{z^2}\right| \le e^{|z|^2}$ (ii) $e^{z+\pi i} = -e^z$ (iii) $e^{(2+\pi i)/4} = (1+i)\sqrt{e/2}$

   (iv) $\cos i\bar{z} = \overline{\cos iz}$ (v) $e^{\bar{z}} = \overline{e^z}$ (vi) $(e^z)^n = e^{nz}, n \in \mathbb{Z}$

   (vii) $\cos \bar{z} = \overline{\cos z}$ (viii) $\sin \bar{z} = \overline{\sin z}$ (ix) $\tan \bar{z} = \overline{\tan z}$

   (x) $\sinh \bar{z} = \overline{\sinh z}$ (xi) $\cosh \bar{z} = \overline{\cosh z}$ (xii) $|e^{-2z}| < 1$ iff $\mathrm{Re}\,(z) > 0$

   (xiii) $\sin i\bar{z} = \overline{\sin iz} \Leftrightarrow z = n\pi i, n \in \mathbb{N}$ (xiv) $\tanh \bar{z} = \overline{\tanh z}$ at points where $\cosh z \neq 0$
8. Prove that every complex number of modulus 1 has the form $\cos\theta + i\sin\theta$ for some angle θ
9. Show that the function

   (a) $f(z) = e^z$ and $f(z) = e^{z^2}$ are entire functions.

   (b) $f(z) = e^{\bar{z}}$ is not analytic anywhere.
10. Determine the different values of the following expressions:
    $(-1)^{\sqrt{2}}, 2^{1-i}, (1+i)^{\sqrt{3}}, i^i$
11. Determine the values of the following:
    $\sin i, \cos i, \cos\left(\frac{\pi}{4} - i\right), \tan(1+i)$
12. Find the period and all the zeros of the functions: (i) $\sin(iz+2)$ (ii) $\cos(z-2i)$ (iii) $\tanh(z+\sqrt{3})$ (iv) $|\sinh z|$
13. Determine the different values of the following logarithms:
    $\log 1, \log(-1/2), \log(-i), \log(-1+i), \log(3+4i), \log\dfrac{a-ib}{a+ib}$
14. Find and plot $\log z$ for the following complex numbers z:

    (a) 2 (b) i (c) $1+i$ (d) $(1+i\sqrt{3})/2$

    Specify the principal value.
15. Which branches of the function $z^{3/4}$ assume the value $-1+i$, and at what points?
16. For the principal branch of the log function, find $\log(1-i)$.
17. Find $\log(1-i)$ for the branch of the log function determined ny the interval $[0, 2\pi)$.
18. Show that the principal branch of the log function can be described by the formula $\log z = \int_1^z \frac{1}{\zeta}d\zeta$ for $z \notin (-\infty, 0]$.

Answers

12. (i) $2n\pi i, n\pi - 2, n \in \mathbb{N}$; (ii) $2\pi, (n+\frac{1}{2})\pi + 2i, n \in \mathbb{N}$; (iii) $i\pi, -\sqrt{3} + (n+\frac{1}{2})\pi i, n \in \mathbb{N}$; (iv) $i\pi, n\pi i, n \in \mathbb{N}$

Review Questions [Long-answer Type]

1. Show that the real and imaginary parts of the function e^z satisfy both the Cauchy–Riemann equations and Laplace equation.
2. Construct the Riemann surface corresponding to e^z.
3. Investigate how the branches of the function $\log[(z-a)/(z-b)]$ pass from one to another as z moves about the plane.
4. Prove that the conjugates of the complex numbers $\sin z$ and $\cos z$ are $\sin \overline{z}$ and $\cos \overline{z}$ respectively. Hence show that $|\sin z|^2 = \frac{1}{2}(\cosh 2y - \cos 2x)$; $|\cos z|^2 = \frac{1}{2}(\cosh 2y + \cos 2x)$.
5. Determine all the roots of the following equations: $\sin z = i$, $\cos z = 2$, $\cot z = 1 + i$. Which of these roots belong to the fundamental domains of the functions, as given in the text?
6. Show that the real and imaginary parts of the functions $\tan(x + iy)$ and $\cot(x + iy)$ satisfy the Cauchy–Riemann equations.
7. Along what path in the plane must the point z move if: (a) Arc $\cos z$ is to go into the branch $-$ Arc $\cos z - 2\pi$, (b) Arc $\sin z$ is to go into the branch Arc $\sin z + 4\pi$ (c) Arc $\tan z$ is to go into the branch Arc $\tan z - 3\pi$?
8. If $\sqrt{z}$ is defined using the principal branch of the log function, analyze the behaviour of $\sqrt{z}$ near the cut line for this branch.
9. Assuming each branch of the log function is analytic, use the chain rule to prove that each such function has derivative $1/z$.
10. Find the power series expansion of (a) $\log z$ about $z = i$ and (b) $\sqrt{z}$ about $z = 1$ and find their radius of convergence.
11. Define explicitly a continuous branch of $\log z$ in the complex plane slit along the negative imaginary axis.
12. How would you make a branch cut to define a single-valued branch of the function $\log(z + i - 1)$? How about $\log(z - z_0)$?
13. Show that a branch of the multiple-valued function $z^{\frac{1}{n}}$ is $r^{\frac{n}{2}} e^{\frac{i\theta}{2}}$; $0 \leq \theta \leq n\pi$.
14. Determine the branches and branch points of the multi-valued functions:
    (i) $f(z) = (z^2 + 1)^{\frac{1}{2}}$ (ii) $f(z) = (1 - z^2)^{\frac{1}{2}}$.
15. Find all the values of z such that:
    (a) $e^z = 1 + \sqrt{3}\, i$ (b) $e^{4z} = i$ (c) $e^{2z-1} = 1$ (d) $e^{2z-1} = 1 + i$
    (e) $\sin z = \cosh 4$ (f) $\tan z = 2 - \cot z$ (g) $\sinh z = -i$ (h) $\cosh z = \frac{1}{2}$
16. Suppose the function $\arccos z$ has at the point $z = 2$ its principal value. Let z make one circuit about the point $z = 1$ and one circuit about the points $z = 1$ and $z = -1$ in a positive direction. What value does the function assume when z returns to the point $z = 2$?
17. Show that the functions $\sinh z$ and $\cosh z$ are unbounded.
18. Find all the zeros and periods of $\cosh z$ and $\sinh z$.

Answers

15. (a) $z = \ln 2 + (2n + \frac{1}{3})\pi i,\ n \in \mathbb{Z}$; (b) $z = \frac{1}{8}\pi i + \frac{1}{2} n\pi i,\ n \in \mathbb{Z}$; (c) $z = \frac{1}{2} + n\pi i,\ n \in \mathbb{Z}$; (d) $z = \frac{1}{2} + \frac{1}{4}\ln 2 + (\frac{1}{8} + n)\pi i,\ n \in \mathbb{Z}$; (e) $z = (2n + \frac{1}{2})\pi \pm 4i,\ n \in \mathbb{N}$; (f) $z = (n + \frac{1}{4})\pi,\ n \in \mathbb{N}$; (g) $z = (2n - \frac{1}{2})\pi i,\ n \in \mathbb{N}$; (h) $z = (2n \pm \frac{1}{3})\pi i,\ n \in \mathbb{N}$

5 Complex Integration

The properties of analytic functions are very much related to complex integration. So, complex integration brings out the properties of analytic functions and plays a very important role in various areas of science and engineering. We begin this chapter by defining terms such as simply-connected contour, multiply-connected contour, line integral and contour integral, along with examples. Using integration, we will prove an important result on analytic functions. Towards this, many important theorems such as Cauchy–Goursat theorem, Cauchy's integral formula, the derivative of analytic functions in terms of contour integrals, Moreras theorem, Liouvilles theorem and some related ones, and maximum modulus principle and their applications are covered.

5.1 Integrals of a Function

Definition 5.1.1. [Definite integrals of functions]: Let $f : [a, b] \to \mathbb{C}$ be a continuous function. Let $f(t) = u(t) + iv(t)$ for $t \in [a, b]$, where $u(t)$ and $v(t)$ are real-valued functions of real variable t. Then, the *definite integral* of $f(t)$ over $[a, b]$ is defined as

$$\int_a^b f(t)\, dt = \int_a^b u(t)\, dt + i \int_a^b v(t)\, dt, \tag{5.1}$$

provided the functions $u(t)$ and $v(t)$ are integrable in $[a, b]$. The integrals of $u(t)$ and $v(t)$ in Eq. (5.1) exist if these functions are piecewise continuous over $[a, b]$.

5.2 Contours

Unlike real functions, the complex-valued functions of a complex variable are integrated along curves in the complex plane, instead of intervals of real line. In the definition of a curve, we distinguish between one-dimensional geometric object in a plane (endowed with an orientation) and its parametrization, which is a mapping from a closed interval to $\mathbb{C}$ that is not uniquely determined. Some adequate classes of curves are given below.

5.2.1 Curve

Let $[a, b]; a < b$ be a bounded closed interval with $a, b \in \mathbb{R}$.

Definition 5.2.1. A *curve* is a continuous mapping $\gamma : [a, b] \to \mathbb{C}$ which is assumed to be of class C^1 (Fig.5.1). The curve has an induced direction due to the increasing value of the parameter in the parametric interval $[a, b]$.

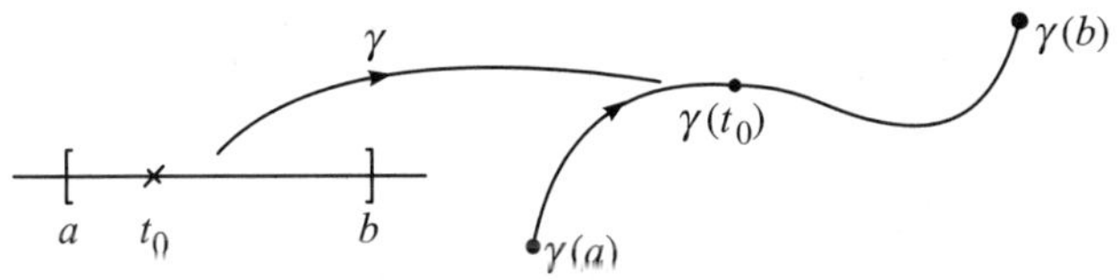

Figure 5.1 *Curve*

By a *point on the curve*, we mean a point z_0 such that $z_0 = \gamma(t_0)$ for some $t_0 \in [a, b]$. For a *curve* γ over a parametric interval $[a, b]$, the point $\gamma(a)$ is called the *initial point* and $\gamma(b)$ *the terminal point* of the curve. For instance:

(i) Consider $\gamma_1(t) = 3 + 2e^{it}; 0 \leq t \leq \pi$. Then γ_1 is the curve with parametric interval $[0, \pi]$ as depicted in the Fig. 5.2(a).

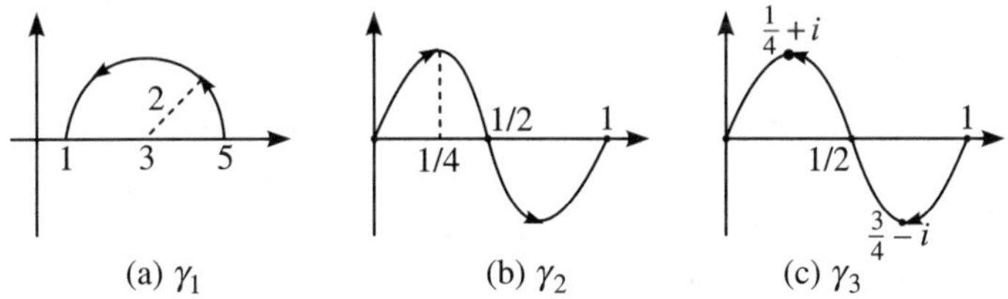

Figure 5.2 *Pictorial notation*

(ii) Consider $\gamma_2(t) = t + i \sin 2\pi t; 0 \leq t \leq 1$. Then γ_2 is the curve with parametric interval $[0, 1]$ as depicted in the Fig. 5 2 (b).

(iii) Consider $\gamma_3(t) = (1 - t) + i \sin(2\pi(1 - t)); 0 \leq t \leq 1$. Then γ_3 is the curve with parametric interval $[0, 1]$ as depicted in the Fig. 5.2 (c).

A parametric representation of a curve γ is given by $\gamma(t) = \gamma_1(t) + i\gamma_2(t)$, where $\gamma_1(t) = \text{Re}\{\gamma(t)\}$ and $\gamma_2(t) = \text{Im}\{\gamma(t)\}$ are continuous real-valued functions on $[a, b]$. The interval $[a, b]$ is called *parametric interval* of γ.

(i) A set $\gamma^* = \{\gamma(t) : a \leq t \leq b\}$ is called the *image* of a curve γ over a parametric interval $[a, b]$. We say that γ is *contained in a set* $\mathcal{D}(\subseteq \mathbb{C})$ when $\gamma^* \subseteq \mathcal{D}$.

(ii) A curve γ with parametric interval $[a, b]$ such that $\gamma(a) = \gamma(b)$ is called a *closed curve*. For example, $\gamma(t) = \cos t; t \in [-\pi, \pi]$ is the segment $[-1, 1]$ traversed twice from -1 to 1 and then from 1 to -1, and therefore considered as a closed curve.

(iii) A curve is *simple* if $\gamma(t)$ is one-to-one, i.e., for $t_1, t_2 \in [a, b]$ with $t_1 \neq t_2$, we have $\gamma(t_1) \neq \gamma(t_2)$, or in the case of a closed curve, $t_1 = a$ and $t_2 = b$. For example, circles, ellipses, rectangles, and triangles are simple closed curves.

(iv) A curve γ with a parameter interval $[a, b]$ is said to be *smooth* if γ has a continuous non-zero derivative on its parametric interval $[a, b]$. Thus a curve $\gamma(t) = \gamma_1(t) + i\gamma_2(t)$ is called *smooth* if $\gamma_1(t)$ and $\gamma_2(t)$ are smooth (i.e., they have finitely many derivatives). For example, $\gamma(t) = t^3 + it^2; t \in [-1, 1]$, being a continuously differentiable function of t on $[-1, 1]$, is a smooth curve.

(v) Given a curve $\gamma : [a, b] \to \mathbb{C}$, the opposite curve $-\gamma$ with parametric interval $[a, b]$ is defined by

$$(-\gamma)(t) = \gamma(a + b - t) : t \in [a, b].$$

In the above example, $\gamma_3 = -\gamma_2$. Note that $(-\gamma)(t) \neq -(\gamma(t))$. $-\gamma$ describes the same curve as γ, but in the reverse direction with the initial and terminal points interchanged.

(vi) Let γ be a curve in an open set G, and $f : G \to \mathbb{C}$ be a holomorphic function. Then the composite $f \circ \gamma$ is differentiable and

$$\Big(f \circ \gamma\Big)'(t) = f'(\gamma(t))\, \gamma'(t).$$

5.2.2 Path (Arc)

Definition 5.2.2. A *path* or *arc* is the set of all image points of a closed finite interval $\{\gamma(t) : t \in [a, b]\}$ under a continuous mapping $\gamma : [a, b] \to \mathbb{C}$.

(i) The path is *simple* if $\gamma(t_1) \neq \gamma(t_2)$ when $t_1 \neq t_2$.

(ii) The path is *closed* if it starts and ends at the same point, i.e., $\gamma(a) = \gamma(b)$.

(iii) The *trace* of the path γ is its image $\Gamma = \gamma([a, b]) = \{\gamma(t) : t \in [a, b]\}$ of γ, which is a subset of the plane. The trace of a path is always a compact set.

(iv) A differentiable arc is called *smooth arc* if $\gamma'(t) \neq 0$ for all $t \in (a, b)$. Actually, if a curve is *smooth*, it will have an infinite number of admissible parametrizations.

(v) If $\gamma(t)$ is one-to-one, that is, for $t_1, t_2 \in [a, b]$ with $t_1 \neq t_2$ we have $\gamma(t_1) \neq \gamma(t_2)$, we call the curve a *Jordan arc*. For example, $\gamma(t) = e^{it}; t \in [0, \pi]$, is a Jordan arc. A Jordan arc γ, such that $\gamma(a) = \gamma(b)$, is called a *Jordan curve*. A domain $\mathcal{D}$ bounded by a Jordan curve is called a *Jordan domain*.

(vi) A curve is an equivalence class of paths. The trace of a curve is the trace of any one of its members.

5.2.3 Contour (Piecewise Smooth Curve)

A piecewise smooth curve or a sectionally smooth arc is a finite number of smooth arcs joined end to end. Let $P = \{a = t_0 < t_1 < \cdots < t_n = b\}$ be a partition of the parametric interval $[a, b]$ of the curve γ.

Definition 5.2.3. A curve with a parametric equation $\gamma(t), t \in [a, b]$ is called *piecewise smooth curve* if $\gamma(t)$ is differentiably continuous for all but finite number of points of P so that the restriction of γ to each subinterval $[t_k, t_{k+1}]; 0 \leq k \leq n-1$ is a smooth curve. Furthermore, $\gamma(t)$ has both left limit and right limit at each $t_k; 1 \leq k \leq n$.

Definition 5.2.4. A *contour* is just a finite chain of continuous curves that is piecewise smooth.

For example, the boundaries of triangles, rectangles and squares are contours.

(i) A *contour* is defined to be *closed* if $\gamma(a) = \gamma(b)$ only. The boundaries of triangles, rectangles, circles are examples of closed contours.

(ii) A contour is called a *simple closed contour* or *Jordan arc* if it is a piecewise smooth curve and there is no self-intersection except that the initial point equals the final point.

Definition 5.2.5. Let γ_1 be a curve with parameter interval $[a, b]$ and γ_2 be a curve with parameter interval $[c, d]$. When the final point of γ_1 is the same as the initial point of γ_2 (i.e., $\gamma_1(b) = \gamma_2(c)$), then we define the *joint of* γ_1 *with* γ_2 as

$$\Big(\gamma_1 + \gamma_2\Big)(t) = \begin{cases} \gamma_1(t); & a \leq t \leq b \\ \gamma_2(t - b + c); & b \leq t \leq b - c + d. \end{cases} \tag{5.2}$$

Note that the join itself is a *curve* with parametric interval $[a, b - c + d]$.

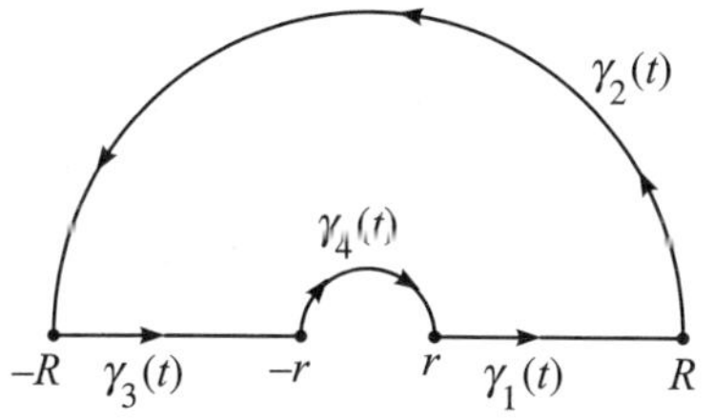

Figure 5.3

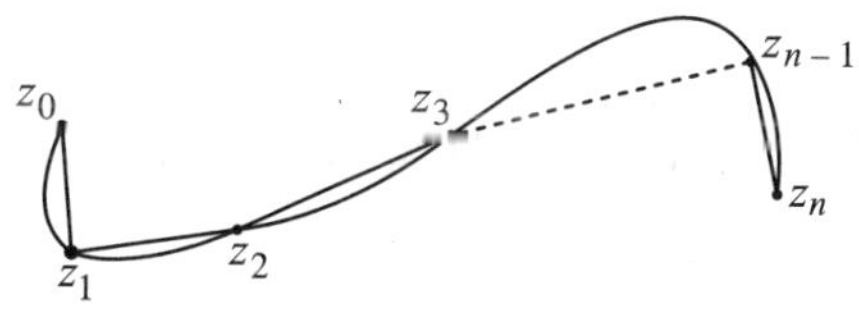

Figure 5.4 *Rectifiable arc*

We can likewise define the join of n curves, $\gamma_1, \gamma_2, \cdots, \gamma_n$, in that order as long as the final point of γ_{j-1} is the initial point $\gamma_j; 2 \leq j \leq n$. For example, we consider Fig. 5.3. The equations of the four curves in Fig. 5.3 are given by

$$\gamma_1(t) = r(1-t) + Rt; 0 \leq t \leq 1$$

$$\gamma_2(t) = Re^{it}; 0 \leq t \leq \pi$$

$$\gamma_3(t) = -R(1-t) - rt; 0 \leq t \leq 1$$

$$\gamma_4(t) = re^{-it}; -\pi \leq t \leq 0.$$

Thus the join of four curves is given by

$$\Big(\gamma_1 + \gamma_2 + \gamma_3 + \gamma_4\Big)(t) = \begin{cases} r(1-t) + Rt; & 0 \leq t \leq 1 \\ Re^{i(t-1)}; & 1 \leq t \leq \pi + 1 \\ -R(\pi + 2 - t) - r(t - \pi - 1); & \pi + 1 \leq t \leq \pi + 2 \\ re^{-i(t-\pi-2)}; & \pi + 2 \leq t \leq 2\pi + 2. \end{cases}$$

5.2.4 Rectifiable Curves

Let a curve be given in parametric form $z = z(t) = \gamma(t) = x(t) + iy(t)$, $t \in [a, b]$. If γ is piecewise smooth, then its *length* is given by

$$L = \int_\gamma |dz| = \int_a^b |\gamma'(t)| dt = \int_a^b \sqrt{[x'(t)]^2 + [y'(t)]^2}\ dt. \tag{5.3}$$

A curve of finite length i.e., $L < \infty$, is called a *rectifiable curve.*

Let P be an arbitrary partition of the parametric interval $[a, b]$ of the curve, whose parametric equation is $z = z(t) = \gamma(t); t \in [a, b]$. Corresponding to this partition P, Γ is divided into n subarcs (Fig. 5.4) $\Gamma_k = \text{arc } z_{k-1}z_k$; $k = 1, 2, \cdots, n$, where $z_k = \gamma(t_k)$; $k = 0, 1, 2, \cdots, n$. We join each of the points $z_0, z_1, \cdots, z_n$ to the next point by straight lines, so we obtain a polygonal curve. The length of this polygonal curve is evidently $\sum_{k=1}^n |z_k - z_{k-1}|$, which is the bounded variation, denoted by $V(\gamma; P)$. γ is a *rectifiable path* if γ is a function of bounded variation.

The *length of the curve* γ is defined by a non-negative real number

$$L(\gamma) = \sup_P \sum_{k=1}^n |z_k - z_{k-1}| = \sup_P \sum_{k=1}^n \Big|\gamma(t_k) - \gamma(t_{k-1})\Big|. \tag{5.4}$$

We say that curve γ is *rectifiable* if $L(\gamma)$ is finite in length.

Definition 5.2.6. A region determined by a closed curve C is defined by a component of the complement of C in the extended complex plane. This definition can be understood by looking at Fig. 5.5. For a given closed curve, among the regions determined by C, these must be only one unbounded region determined by C. That region is treated as the region containing ∞ in $\mathbb{C}_\infty$. In Fig. 5.5, R_2 and S_4 are the unbounded regions determined by C_1 and C_2 respectively.

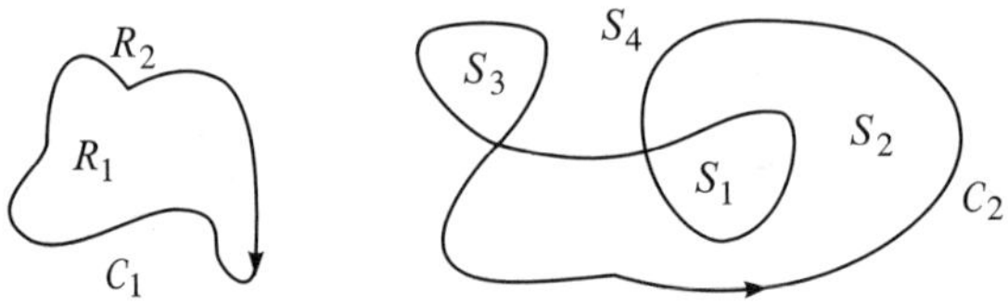

Figure 5.5 *Region determined by a closed curve*

Definition 5.2.7. [Connected region]: A region is said to be *connected* if any two points of the region can be joined by an arc that lies wholly within the region; that is, every point of the arc belongs to the region. Otherwise it is *disconnected.*

(i) **Simply-connected region** : A region $\mathcal{R}$ in $\mathbb{C}$ is said to be a *simply-connected region* (Fig. 5.6) if any closed curve lying entirely in $\mathcal{R}$ can be contracted continuously to a point without any portion of the curve passing out of $\mathcal{R}$ in that process of contraction.

  (a) Thus a region $\mathcal{R}$ in $\mathbb{C}$ is called simply-connected if the complement of $\mathcal{R}$ in $\mathbb{C}_\infty$ is also connected.

  (b) Geometrically, every closed curve in a simply-connected region $\mathcal{R}$ encloses only the points of $\mathcal{R}$.

For example, a circular region, a cubical region, a spherical region, etc., are examples of simply-connected region.

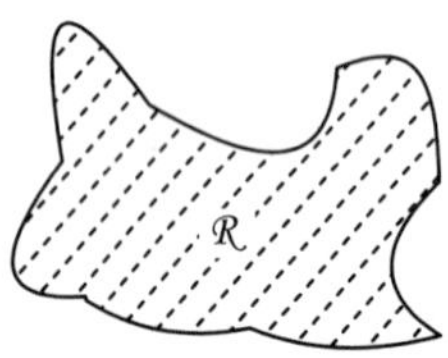

Figure 5.6 *A simply-connected region*

Figure 5.7 *A multiply-connected region*

(ii) **Multiply-connected region** : A region $\mathcal{R}$ in $\mathbb{C}$ which is not simply connected is called a *multiply-connected region*, Fig. 5.7. That is, the complement of $\mathcal{R}$ in $\mathbb{C}_\infty$ has more than two components or equivalently there exists a closed curve C in $\mathcal{R}$ which encloses a point of complement of $\mathcal{R}$. For example, the regions bounded by two concentric circles or spheres are multiply-connected regions.

Definition 5.2.8. [Orientation]: A simple closed contour is said to be *positively oriented* (anticlockwise) if the points in the bounded component fall to the left of the contour when it is traced; otherwise, it is *negatively oriented* (clockwise), as depicted in Fig. 5.8. With this convention it is clear that the positive direction of traversing a curve surrounding the point at infinity is the clockwise direction.

The bounded region for a simple closed contour will be called the *inside of the contour* when the contour is positively oriented. The unbounded component determined by a simple closed contour will be the inside of the contour when the contour is negatively oriented.

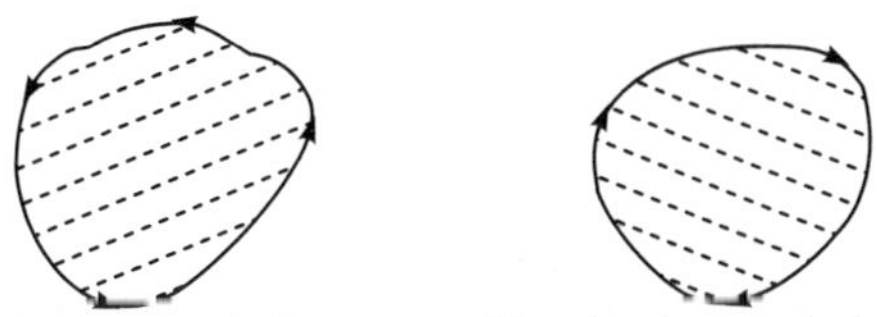

Positively oriented curve Negatively oriented curve

Figure 5.8 *Orientation of curves*

Definition 5.2.9. A simple open contour γ is said to be *positively oriented* if we traverse it from its initial point to its terminal point.

Now we shall state Jordan's curve theorem, which seems to be quite obvious; however, its proof is rather complicated and hence not stated.

Theorem 5.1. [Jordan's curve theorem]: *Any simple closed curve or simple closed contour γ divides the plane $\mathbb{C}$ in two domains: an 'interior' bounded domain and an 'exterior' unbounded domain. The curve is the common boundary of these domains.*

For example, consider the disjoint open discs $\mathcal{S}_1 = \{z \in \mathbb{C} : |z - z_0| < \rho\}$ and $\mathcal{S}_2 = \{z \in \mathbb{C} : |z - z_0| > \rho\}$ in the complex plane $\mathbb{C}$. Clearly, the circle $\gamma = \{z \in \mathbb{C} : |z - z_0| = \rho\}$ is a closed contour and the points on γ are the boundary points of $\mathcal{S}_1$ and $\mathcal{S}_2$. The interior of γ is $\mathcal{S}_1$, which is bounded, and the exterior of γ is $\mathcal{S}_2$, which is unbounded.

5.2.5 Homotopy of Curves

Let $\mathcal{D} \subset \mathbb{C}$ be a domain. For fixed $s \in [0,1]$ we have a closed curve $\gamma_s : [a,b] \to \mathcal{D}$ given by $\gamma_s(t) = H(t,s)$. The way to think of it is that these closed curves γ_s vary continuously from γ_0 to γ_1 (as s varies from 0 to 1).

Definition 5.2.10. Let $\gamma_0, \gamma_1 : [a,b] \to \mathcal{D}$ be two closed rectifiable curves in $\mathcal{D}$. The loop γ_0 is said to be *homotopic to* γ_1 in $\mathcal{D}$ if there exists a continuous map $H : [a,b] \times [0,1] \to \mathcal{D}$ satisfying the following:

(i) For each fixed t in $[a,b]$, the function $H(t,s)$ parametrizes a loop lying in $\mathcal{D}$ with $H(a,s) = H(b,s); \forall s \in [0,1]$.

(ii) The function $H(t,0)$ parametrizes the loop γ_0, i.e., $H(t,0) = \gamma_0(t); \forall t \in [a,b]$.

(iii) The function $H(t,1)$ parametrizes the loop γ_1, i.e., $H(t,1) = \gamma_1(t); \forall t \in [a,b]$.

Such H is called *deformation function* or a *homotopy from* γ_0 *to* γ_1 and is denoted by $\gamma_0 \simeq \gamma_1$. For example:

(i) Let $\mathcal{D} = \mathbb{C}/\{0\}$ and let $\gamma_0, \gamma_1 : [0,1] \to \mathcal{D}$ be $\gamma_0(t) = e^{2it}; 0 \le t \le 1; \gamma_1(t) = 2e^{2it}; 0 \le t \le 1$. Then a homotopy $H : [0,1] \times [0,1] \to \mathcal{D}$ can be given as $H(t,s) = (1+s)e^{2it}$.

(ii) If $\gamma_0, \gamma_1 : [a,b] \to \mathbb{C}$ are any pair of closed curves in $\mathbb{C}$ parametrised by the same interval $[a,b]$, then they are homotopic in $\mathbb{C}$ via $H(t,s) = (1-s)\gamma_0(t) + s\gamma_1(t)$.

(iii) If $\gamma_0, \gamma_1 : [a,b] \to G$ are any pair of closed curves in a convex set $G \subset \mathbb{C}$, then they are homotopic in G (by the same H as in the previous example).

Definition 5.2.11. A domain $\mathcal{D}$ is said to be *simply connected* if every closed curve in $\mathcal{D}$ is homotopic to a point in $\mathcal{D}$.

Definition 5.2.12. (i) If γ is a closed rectifiable curve in G, then γ is *homotopic to zero* ($\gamma \sim 0$) if it is homotopic to a constant curve.

(ii) If $\gamma : [a, b] \to G$ is a closed curve in a set $\mathcal{D} \subset \mathbb{C}$, we say that γ is *null homotopic* in $\mathcal{D}$ if γ is homotopic in $\mathcal{D}$ to a constant curve, such as the curve $\sigma : [a, b] \to \mathcal{D}$ with $\sigma(t) = \gamma(a)$.

5.3 Contour Integrals

An integral of a function f of a complex variable z that is defined on a contour C is denoted by $\int_C f(z)\, dz$ and is called a *complex integral.*

Definition 5.3.1. Let $\gamma : [a, b] \to \mathbb{C}$ be a smooth contour in $\mathbb{C}$ with parametric interval $[a, b]$. Let $f : \gamma \to \mathbb{C}$ be a continuous complex function. Then the *contour integral of* f over γ is defined as

$$\int_\gamma f(z)dz = \int_a^b f(\gamma(t)) \cdot \gamma'(t)\, dt. \tag{5.5}$$

Suppose γ is a piecewise smooth contour on the parametric interval $[a, b]$ and suppose that γ is smooth on $[a, t_1], [t_1, t_2], \cdots, [t_n, b]$, where $t_0 = a < t_1 < t_2 < \cdots << t_n < b = t_{n+1}$. Then

$$\int_\gamma f(z)\, dz = \sum_{k=0}^{n} \int_{t_k}^{t_{k+1}} f(\gamma(t)) \cdot \gamma'(t)\, dt.$$

For example, let $f(z) = z^n : n \in \mathbb{Z}$ ($z \neq 0$, if $n < 0$). Let $\gamma(t) = re^{it} : 0 \le t \le \pi$. Then

$$\begin{aligned}\int_\gamma f(z)\, dz &= \int_0^{2\pi} \left(re^{it}\right)^n \cdot rie^{it} dt = i\int_0^{2\pi} r^{n+1}\Big[\cos(n+1)t + i\sin(n+1)t\Big]dt \\ &= \int_0^{2\pi} -r^{n+1}\sin(n+1)t\, dt + i\int_0^{2\pi} r^{n+1}\cos(n+1)t\, dt.\end{aligned}$$

Consider the following two subcases:

(i) If $n \neq -1$,

$$\int_\gamma f(z)\, dz = r^{n+1}\frac{\cos(n+1)t}{n+1}\Big|_0^{2\pi} + ir^{n+1}\frac{\sin(n+1)t}{n+1}\Big|_0^{2\pi} = 0.$$

(ii) If $n = -1$,

$$\int_\gamma f(z)\, dz = \int_0^{2\pi} -r^{-1+1}(0)\, dt + i\int_0^{2\pi} r^{-1+1} \cdot 1\, dt = 2\pi i.$$

Therefore, the contour integral of f is given by

$$\int_\gamma z^n\, dz = \begin{cases} 0 & \text{if } n \neq -1 \\ 2\pi i & \text{if } n = -1. \end{cases}$$

This is known as *the fundamental integral.*

Definition 5.3.2. Let $\gamma : [a, b] \to \mathbb{C}$ be a rectifiable path and $P = \{t_0 < t_1 < \cdots < t_n\}$ be a partition of $[a, t]$. For $t \in [a, b]$, let $|\gamma|(t)$ be given by

$$|\gamma|(t) = V(\gamma; [a, t]) = \sup_P \sum_{k=1}^{n} \Big|\gamma(t_k) - \gamma(t_{k-1})\Big|.$$

This $|\gamma|(t)$ is increasing and so $|\gamma| : [a, b] \to \mathbb{C}$ is of bounded variation. If f is continuous on $\{\gamma\}$, define

$$\int_\gamma f|dz| = \int_a^b f(\gamma(t))\, d|\gamma|(t). \tag{5.6}$$

Property 5.3.1. Let γ be a contour with parameter interval $[a, b]$ and let $f, g : \gamma^* \to \mathbb{C}$ be continuous. Then

(i) $\displaystyle\int_\gamma \Big[f(z) \pm g(z)\Big] dz = \int_\gamma f(z)\, dz \pm \int_\gamma g(z)\, dz;$

(ii) $\displaystyle\int_\gamma c\, f(z)\, dz = c\int_\gamma f(z)\, dz;$ for any $c \in \mathbb{C}$;

(iii) $\displaystyle\int_{-\gamma} f(z)\, dz = -\int_\gamma f(z)\, dz.$

Property 5.3.2. Let γ be a contour with parametric interval $[a, b]$ and let $f : \gamma^* \to \mathbb{C}$ be continuous. Suppose $a < \tau < b$. Let γ_1 be the restriction of γ on $[a, \tau]$, i.e., $\gamma_1 = \gamma|_{[a,\tau]}$ and $\gamma_2 = \gamma|_{[\tau,b]}$. Then,

$$\int_\gamma f(z)\, dz = \int_{\gamma_1} f(z)\, dz + \int_{\gamma_2} f(z)\, dz.$$

Property 5.3.3. [Reparametrization]: Let $\widehat{\gamma}$ be another path with parameter interval $[c, d]$. Suppose that $\widehat{\gamma} = \gamma \circ \psi$, where ψ is a function which maps $[c, d]$ to $[a, b]$ and has positive continuous derivatives. Then

$$\int_\gamma f(z)\, dz = \int_{\widehat{\gamma}} f(z)\, dz.$$

Proof: Since $\widehat{\gamma}(t) = \gamma \circ \psi(t)$, by chain rule of differentiation, we have $\widehat{\gamma}'(t) = \gamma'(\psi(t)) \cdot \psi'(t)$. Thus

$$\begin{aligned}\int_{\widehat{\gamma}} f(z)\, dz &= \int_c^d f\Big(\widehat{\gamma}(t)\Big) \cdot \widehat{\gamma}'(t)\, dt \\ &= \int_c^d f\Big(\gamma(\psi(t))\Big) \cdot \gamma'(\psi(t)) \cdot \psi'(t)\, dt.\end{aligned}$$

Let $\tau = \psi(t)$ so that $d\tau = \psi'(t)dt$. When $t = c$, then $\tau = \psi(c) = a$, and $t = d$, $\tau = \psi(d) = b$. Therefore,

$$\int_{\widehat{\gamma}} f(z)\, dz = \int_a^b f\Big(\gamma(\tau)\Big)\gamma'(\tau)\, d\tau = \int_\gamma f(z)\, dz.$$

Hence the result. □

Property 5.3.4. Let γ_1 be a smooth curve with parameter interval $[a, b]$ and let γ_2 be a smooth curve with parameter interval $[c, d]$. Let $f : \gamma_1^* \cup \gamma_2^* \to \mathbb{C}$ be a continuous function where $\gamma_1(b) = \gamma_2(c)$. Then

$$\int_{\gamma_1+\gamma_2} f(z)\, dz = \int_{\gamma_1} f(z)\, dz + \int_{\gamma_2} f(z)\, dz.$$

Theorem 5.2. [Fundamental theorem of calculus (complex version): applying to contour integrals]: *Suppose that γ is a contour with parameter interval $[a, b]$ and f is defined on an open set containing γ^* and $f'(z)$ exists and is continuous at each point of γ^*. Then*

$$\int_\gamma f'(z)\, dz = \begin{cases} f\big(\gamma(b)\big) - f\big(\gamma(a)\big) & \\ 0 & \text{if } \gamma \text{ is closed.} \end{cases}$$

Proof: Assume that γ is smooth. So $f \circ \gamma : [a, b] \to \mathbb{C}$ is differentiable. Thus

$$(f \circ \gamma)'(t) = f'\big(\gamma(t)\big) \cdot \gamma'(t),$$

which is continuous. Therefore,

$$\begin{aligned} \int_\gamma f'(z)\, dz &= \int_a^b f'\big(\gamma(t)\big) \cdot \gamma'(t)\, dt = \int_a^b \big(f \circ \gamma\big)'(t)\, dt \\ &= \int_a^b \mathrm{Re}\Big(\big(f \circ \gamma\big)'(t)\Big)\, dt + i \int_a^b \mathrm{Im}\Big(\big(f \circ \gamma\big)'(t)\Big)\, dt \\ &= \mathrm{Re}\Big(\big(f \circ \gamma\big)'(t)\Big)\Big|_a^b + i\mathrm{Im}\Big(\big(f \circ \gamma\big)'(t)\Big)\Big|_a^b \\ &= f \circ \gamma(t)\Big|_a^b = f\big(\gamma(b)\big) - f\big(\gamma(a)\big). \end{aligned}$$

In particular, when γ is closed, $\gamma(b) = \gamma(a)$ and the theorem follows. □

When γ is piecewise smooth, then applying the previous theorem, we get

$$\int_\gamma f'(z)\, dz = \sum_{k=0}^{n-1} \int_{t_k}^{t_{k+1}} f'(\gamma(t))\, \gamma'(t)\, dt,$$

where $\gamma|_{[t_k, t_{k+1}]}$ is smooth.

Example 5.3.1. *Compute the integral $\int_\Gamma \cos z\, dz$, where Γ is the contour consisting of a semicircle γ_1 and a straight line segment γ_2 as shown in the Fig. 5.9.*

Solution. There is no need to parametrize Γ since the integrand $\cos z$ has the antiderivative $F(z) = \sin z$ for all $z \in \mathbb{C}$. Hence by Theorem 5.2, the value of the integral can be computed using only the endpoints of Γ:

$$\int_\Gamma \cos z\, dz = \sin z\Big|_{-3}^{6+3i} = \sin(6 + 3i) - \sin(-3).$$

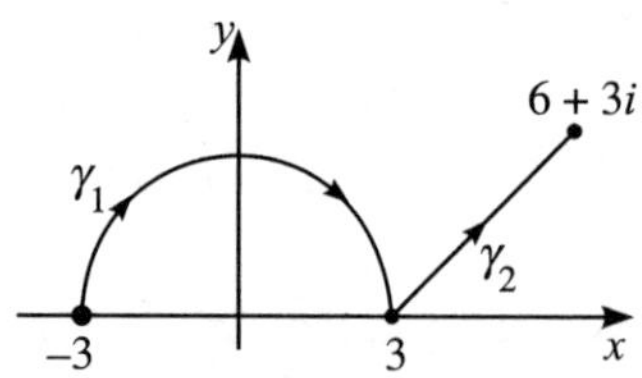

Figure 5.9 *Contour for Example 5.3.1*

Example 5.3.2. *Compute the integral $\int_\Gamma \frac{1}{z}\, dz$, where Γ is the contour as shown in the Fig. 5.10.*

Solution. At each point of the contour Γ_1 as depicted in the Fig. 5.10(a), the function z^{-1} is the derivative of the principal branch of $\log z$. Hence by Theorem 5.2, we have

$$\int_{\Gamma_1} \frac{1}{z}\, dz = \operatorname{Log} z\Big|_{-i}^{i} = \operatorname{Log}(i) - \operatorname{Log}(-i) = \frac{\pi i}{2} + \frac{\pi i}{2} = \pi i,$$

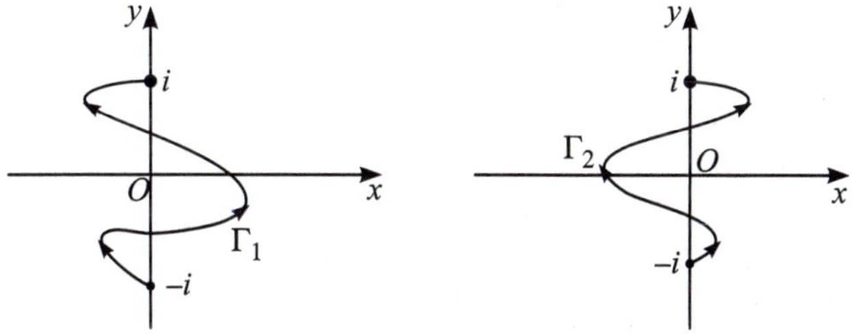

(a) (b)

Figure 5.10 *Contour for Example 5.3.2*

For the contour Γ_2 of the Fig. 5.10(b), we cannot employ the function $\operatorname{Log} z$, since its branch cut intersects Γ_2. We use instead Eq. (4.23): $\log z = \operatorname{Log}|z| + i\operatorname{Arg} z$, $\operatorname{Arg} z \in (0, 2\pi), z \in \mathbb{C} - \{0\}$, which is a branch of the logarithm without along the non-negative real axis. Hence by Theorem 5.2, we have

$$\int_{\Gamma_2} \frac{1}{z}\, dz = \log z\Big|_{-i}^{i} = \frac{\pi i}{2} - \frac{3\pi i}{2} = -\pi i.$$

Theorem 5.3. [Antiderivative theorem]: *Let f be a continuous function on a region $\mathcal{G}$. Then the following statements are equivalent:*

(i) f has an antiderivative F throughout $\mathcal{G}$, i.e., there is a function F defined on $\mathcal{G}$ such that F is analytic on $\mathcal{G}$ and $F'(z) = f(z)$, for all $z \in \mathcal{G}$.

(ii) $\int_\gamma f(z)\, dz = 0$ for every closed contour γ in $\mathcal{G}$.

(iii) Let $a, b \in \mathcal{G}$. Then $\int_\gamma f(z) dz$ is a constant for any γ lying in $\mathcal{G}$ with initial point a and terminal point b (contour integral is path independent).

Proof: (i) $\Rightarrow$ (ii) or (i) $\Rightarrow$ (iii) due to the fundamental theorem of calculus 5.2.
(ii) $\Rightarrow$ (iii): Let γ_1 and γ_2 be two contours with initial point a and final point b and let γ_1, γ_2 lie in $\mathcal{G}$. Then $\gamma_1 - \gamma_2$ is a closed contour γ with initial and final point a. By condition (ii),

$$\int_{\gamma_1 - \gamma_2} f(z)\, dz = 0 \Rightarrow \int_{\gamma_1} f(z)\, dz - \int_{\gamma_2} f(z)\, dz = 0.$$

So (iii) is true.
(iii) $\Rightarrow$ (i): Suppose $\int_\gamma f(z) dz$ depends only on initial and final points of γ for any γ that lies in $\mathcal{G}$. Define $F(z) = \int_\gamma f(z) dz$, where γ is any contour in G from z_0 to z. As we have assumed (iii), F is well defined and

$$F(z) = \int_\gamma f(z)\, dz = \int_{z_0}^{z} f(z)\, dz.$$

Let $z + h$ be any point in $\mathcal{G}$, near the point $z \in \mathcal{G}$ (Fig. 5.11). Thus

$$\begin{aligned}\left|\frac{F(z+h)-F(z)}{h} - f(z)\right| &= \left|\frac{1}{h}\Big[\int_{z_0}^{z+h} f(\zeta)\,d\zeta - \int_{z_0}^{z} f(\zeta)d\zeta\Big] - f(z)\right| \\ &= \left|\frac{1}{h}\int_{z}^{z+h} f(\zeta)\,d\zeta - f(z)\right|.\end{aligned}$$

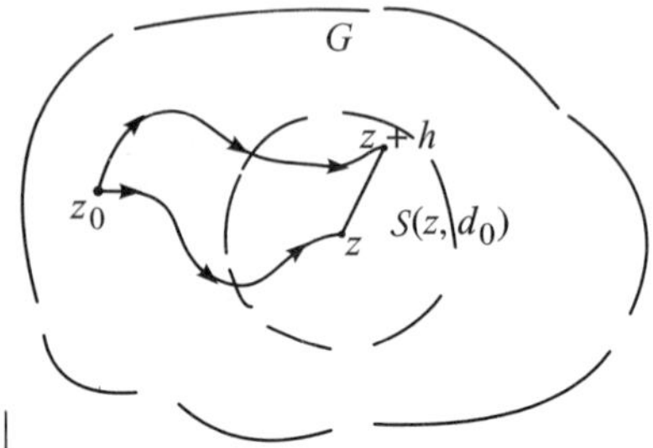

Figure 5.11 *Region $\mathcal{G}$*

Since $\mathcal{G}$ is open and $z \in \mathcal{G}$, there is a $\delta_0 > 0$ such that $\mathcal{S}(z;\delta_0) \subseteq G$.

When $|h| < \delta_0/2$, the straight line joining z and $z+h$ is contained in $\mathcal{G}$. So pick the straight line (contour) from z to $z+h$ for the above integral, when $|h| < \delta_0/2$. Then

$$\begin{aligned}\left|\frac{F(z+h)-F(z)}{h} - f(z)\right| &= \left|\frac{1}{h}\int_{[z,z+h]} \Big[f(\zeta) - f(z)\Big]\,d\zeta\right| \\ &\le \frac{1}{|h|}\int_{[z,z+h]} \Big|f(\zeta) - f(z)\Big|\,|d\zeta|.\end{aligned}$$

Now, given that, $f(z)$ is continuous in $\mathcal{G}$, so by definition, corresponding to an $\varepsilon > 0$, $\exists$ a $\delta_1 > 0$ such that

$$|f(\zeta) - f(z)| < \varepsilon \text{ whenever } |\zeta - z| < \delta_1. \tag{5.7}$$

We now choose δ such that $\delta = \min \delta_0, \delta_1$. Then the inequality (5.7) is satisfied for every point ζ on the line segment joining z to $z+h$. Hence for $|\zeta - z| < \delta$,

$$\begin{aligned}\left|\frac{F(z+h)-F(z)}{h} - f(z)\right| &\le \frac{1}{|h|}\int_{[z,z+h]} \Big|f(\zeta) - f(z)\Big|\,|d\zeta| \\ &< \frac{1}{|h|}\varepsilon\,|h| = \varepsilon.\end{aligned}$$

Therefore, F is differentiable at every point $z \in \mathcal{G}$ and $F'(z) = f(z)$. Thus (i) is true by (iii). □

Example 5.3.3. *Let Γ be the circle with centre a and radius r. Prove that*

$$\int_\Gamma \frac{dz}{(z-a)^n} = \begin{cases} 0 & \text{if } n \neq 1 \\ 2\pi i & \text{if } n = 1 \end{cases}, \quad n \in \mathbb{Z}.$$

Solution. The parametric equation of the circle Γ is given by $z - a = re^{it}$, $0 \le t \le 2\pi$. Then,

$$\begin{aligned}\int_\Gamma \frac{dz}{(z-a)^n} &= \int_0^{2\pi} \frac{i\,r\,e^{it}}{(re^{it})^n}\,dt = \frac{i}{r^{n-1}}\int_0^{2\pi} e^{i(1-n)t}\,dt \\ &= \frac{i}{r^{n-1}}\left[\frac{e^{i(1-n)t}}{i(1-n)}\right]_0^{2\pi}; \quad \text{provided } n \neq 1 \\ &= \frac{1}{(1-n)r^{n-1}}\Big[e^{i(1-n)2\pi} - 1\Big] \\ &= \frac{1}{(1-n)r^{n-1}}\Big[1 - 1\Big] = 0 \text{ if } n \neq 1.\end{aligned}$$

In particular, if $n = 1$, then

$$\int_\Gamma \frac{dz}{z-a} = \int_0^{2\pi} \frac{i\,r\,e^{it}}{re^{it}}\,dt = i\int_0^{2\pi} dt = 2\pi i.$$

Hence the result.

Example 5.3.4. *Let Γ be the semi circular path $\gamma(t) = 3e^{it}$ $(0 \le t \le \pi)$ and the branch of the function $z^{1/2}$ be $f(z) = z^{1/2} = e^{\log z/2}$ $(|z| > 0, 0 < \arg z < 2\pi)$. Then evaluate $I = \int_\Gamma z^{1/2}\,dz$.*

Solution. The parametric form of the contour is $\gamma(t) = 3e^{it}$ $(0 \le t \le \pi)$ (Fig. 5.12). Although the branch of the multi-valued function $z^{1/2}$ is not defined at the initial point $z = 3$ on Γ, the integral I nevertheless exists, as

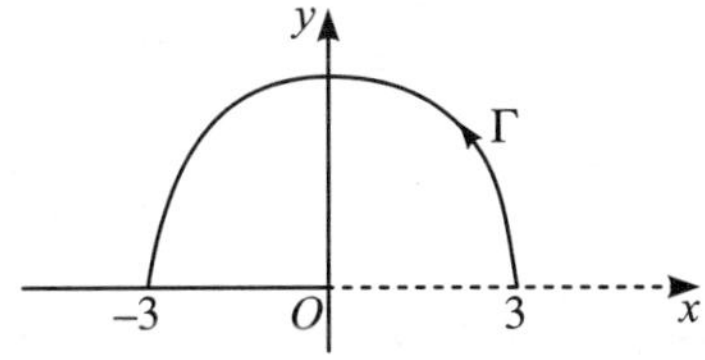

Figure 5.12 *Branch cut and contour integral*

$$f\Big(\gamma(t)\Big) = \exp\Big[\frac{1}{2}\Big(\ln 3 + it\Big)\Big] = \sqrt{3}\,e^{it/2},$$

which is piecewise continuous on $[0, \pi]$. Also, for continuity of $f(\gamma(t))\gamma'(t)$, we assign its value $i3\sqrt{3}$ at $t = 0$. Now, using result (5.5), we get

$$\begin{aligned}
I &= \int_0^\pi \Big(3e^{it}\Big)^{1/2} 3i\,e^{it}\,dt = 3\sqrt{3}\int_0^\pi ie^{i3t/2}\,dt \\
&= 3\sqrt{3}\Big[-\int_0^\pi \sin\frac{3t}{2}\,dt + i\int_0^\pi \cos\frac{3t}{2}\,dt\Big] \\
&= 3\sqrt{3}\frac{2}{3}\Big[\cos\frac{3t}{2}\,dt + i\sin\frac{3t}{2}\,dt\Big]_0^\pi = -2\sqrt{3}(1+i).
\end{aligned}$$

Example 5.3.5. *Evaluate $I = \int_C \bar{z}^2\,dz + z^2\,d\bar{z}$ along the curve C defined by $z^2 + 2z\bar{z} + \bar{z}^2 = (2-2i)z + (2+2i)\bar{z}$ from the point $z = 1$ to $z = 2+2i$.*

Solution. Let $z = x + iy$. The equation of the given curve can be written as

$$\begin{aligned}
& (z+\bar{z})^2 = 2(z+\bar{z}) - 2i(z-\bar{z}) \\
\Rightarrow\quad & (2x)^2 = 2\cdot 2x - 2i\cdot 2iy \Rightarrow y = x(x-1).
\end{aligned}$$

The parametric equation of the curve is $x = t, y = 2t^2 - t$, where $1 \le t \le 2$. Therefore,

$$\begin{aligned}
I &= \int_{z=1}^{2+2i} \Big[(x-iy)^2(dx+idy) + (x-iy)^2(dx-idy)\Big] \\
&= \int_{z=1}^{2+2i} \Big[\Big\{(x-iy)^2 + (x+iy)^2\Big\}dx + i\Big\{(x-iy)^2 - (x+iy)^2\Big\}dy\Big] \\
&= 2\int_{(1,0)}^{(2,2)} \Big[(x^2-y^2)dx + 2xy\,dy\Big] \\
&= 2\int_1^2 (3t^4 - 8t^3 + 5t^2)dt = 248/15.
\end{aligned}$$

RESULT **39.** Let $\gamma = \gamma_1 + \gamma_2$, where γ_1 and γ_2 respectively are given by $z_1(t) = it$ and $z_2(t) = t + i$; $t \in [0, 1]$ as depicted in the Fig. 5.13.

Furthermore, let $f(z) = (y - x) + 3ix^2$. It follows that

$$\begin{aligned}\int_\gamma f(z)\,dz &= \int_0^1 f(z_1(t))\cdot z_1'(t)\,dt + \int_0^1 f(z_2(t))\cdot z_2'(t)\,dt \\ &= \int_0^1 t\cdot i\,dt + \int_0^1 \Big[(1-t) + 3it^2\Big]\cdot 1\,dt \\ &= \frac{i}{2} + \frac{1}{2} + i = \frac{1}{2} + \frac{3i}{2}.\end{aligned}$$

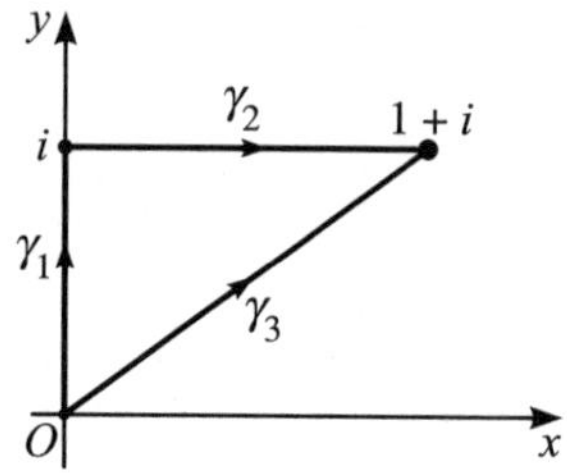

Figure 5.13 *Contour for Result 39*

Now, let γ_3 be given by $z_3(t) = (1+i)t; t \in [0,1]$. Then for the same function f, we have

$$\int_{\gamma_3} f(z)\,dz = \int_0^1 f(z_3(t))\cdot z_3'(t)\,dt = \int_0^1 \Big[(t-t) + 3it^2\Big]\cdot(1+i)\,dt = i(i+1) = i - 1.$$

Hence, although γ and γ_3 have the same initial and terminal points, $\int_\gamma f(z)\,dz \neq \int_{\gamma_3} f(z)\,dz$. Thus, from Theorem 5.2, this function f cannot have an antiderivative.

DEDUCTION **10** [Complex form of Green's theorem]: Let $f(z, \bar{z})$ be continuous and have continuous partial derivatives in a region $\mathcal{R}$ and on its boundary Γ. Let $f(z,\bar{z}) = u(x,y) + iv(x,y)$. Then,

$$\begin{aligned}\oint_\Gamma f(z,\bar{z})\,dz &= \oint_\Gamma (u+iv)(dx+idy) \\ &= \oint_\Gamma (u\,dx - v\,dy) + i\oint_\Gamma (v\,dx + u\,dy) \\ &= -\int\!\!\int_{\mathcal{R}} \Big(\frac{\partial v}{\partial x} + \frac{\partial u}{\partial y}\Big)dx\,dy + i\int\!\!\int_{\mathcal{R}}\Big(\frac{\partial u}{\partial x} - \frac{\partial v}{\partial y}\Big)dx\,dy \\ &= i\int\!\!\int_{\mathcal{R}}\Big[\Big(\frac{\partial u}{\partial x} - \frac{\partial v}{\partial y}\Big) + i\Big(\frac{\partial u}{\partial y} + \frac{\partial v}{\partial x}\Big)\Big]dx\,dy \\ &= 2i\int\!\!\int_{\mathcal{R}} \frac{\partial f}{\partial \bar{z}}\,dx\,dy = 2i\int\!\!\int_{\mathcal{R}}\frac{\partial f}{\partial \bar{z}}\,dA, \qquad (5.8)\end{aligned}$$

where $dA = dx\,dy$ represents the element of area $dxdy$. This represents the *complex form of Green's theorem.*

5.3.1 Estimation of Contour Integrals

The following two important theorems may be found useful in evaluating contour integrals.

Theorem 5.4. *Let* $f(t) = u(t) + iv(t)$ *be a continuous function of real variable* t *(defined on an interval* $[a,b]$*). Then,*

$$\Big|\int_a^b f(t)\,dt\Big| \leq \int_a^b |f(t)|\,dt. \qquad (5.9)$$

Proof: If $a = b$ or $\int_a^b f(t)\,dt = 0$, the above inequality holds. If $a < b$, and the value of the integral is a non-zero complex number $re^{i\theta}$. Therefore,

$$\int_a^b f(t)\,dt = r\,e^{i\theta} \Rightarrow r = e^{-i\theta}\int_a^b f(t)\,dt = \int_a^b e^{-i\theta} f(t)\,dt.$$

Also, $r = |re^{i\theta}| = |\int_a^b f(t)\, dt|$. Since r is a positive real number, so the integral on the right-hand side is also a positive real number. We know that the real part of a real number is the number itself. Therefore,

$$\begin{aligned} r &= \text{Re} \int_a^b e^{-i\theta} f(t)\, dt = \int_a^b \text{Re}\Big[e^{-i\theta} f(t)\Big]\, dt \\ &\le \int_a^b \Big|e^{-i\theta} f(t)\Big|\, dt, \quad \text{as Re } z \le |z| \\ &\le \int_a^b \Big|e^{-i\theta}\Big|\, |f(t)|\, dt = \int_a^b \Big|f(t)\Big|\, dt \\ \therefore \quad & \Big|\int_a^b f(t)\, dt\Big| \le \int_a^b \Big|f(t)\Big|\, dt. \end{aligned}$$

Therefore, the result of the theorem follows. □

Theorem 5.5. [The absolute value of a complex integral]: *If a function $f(z)$ is piecewise continuous on a contour C of finite length L and if $M \in \mathbb{R}^+$ be such that $|f(z)| \le M\ \forall z \in C$. Then,*

$$\Big|\int_C f(z)\, dz\Big| \le M \cdot L. \tag{5.10}$$

This is the upper bound for moduli of contour integrals, sometimes referred to as ML-inequality.

Proof: Let $z = \gamma(t); t \in [a, b]$ be the parametrization of the contour C. Then, using Eq. (5.5),

$$\begin{aligned} \Big|\int_C f(z)\, dz\Big| &= \Big|\int_a^b f(\gamma(t))\gamma'(t)\, dt\Big| \\ &\le \int_a^b \Big|f(\gamma(t))\gamma'(t)\, dt\Big| = \int_a^b \Big|f(\gamma(t))\Big|\Big|\gamma'(t)\Big|\, dt \\ &\le M \int_a^b |\gamma'(t)|\, dt = ML \end{aligned}$$

where, L is the finite length of the contour C as in Eq.(5.3). □

Example 5.3.6. *Find the upper bound for the absolute value of $\oint_C \frac{e^z}{z+1}\, dz$, where C is the circle $|z| = 4$.*

Solution. First, the length L (circumference) of the circle of radius 4 is 8π. Now, for all points z on the circle, $|z + 1| \ge |z| - 1 = 4 - 1 = 3$. Thus

$$\Big|\frac{e^z}{z+1}\Big| \le \frac{|e^z|}{|z| - 1} = \frac{|e^z|}{3}.$$

In addition, $|e^z| = |e^x(\cos y + i \sin y)| = e^x$. For points on the circle $|z| = 4$, the maximum that $x = \text{Re}(z)$ can be is 4 and so

$$\Big|\frac{e^z}{z+1}\Big| \le \frac{|e^z|}{3} \le \frac{e^4}{3}.$$

Using the ML inequality, Eq. (5.10), we have

$$\Big|\oint_C \frac{e^z}{z+1}\, dz\Big| \le \frac{8\pi\, e^4}{3}.$$

Example 5.3.7. *Let Γ be the arc of the circle $|z| = 2$ from $z = 2$ to $z = 2i$ that lies in the first quadrant. Without actually evaluating the integral, show that* $\left|\int_\Gamma \frac{dz}{z^2+1}\right| \le \frac{\pi}{3}$.

Solution. Let $f(z) = 1/(z^2+1)$. Since Γ is the circular arc of radius 2 (Fig. 5.14) lying in the first quadrant, the length L of Γ is given by $L = \frac{1}{4}(2\pi \times 2) = \pi$. Also, on Γ,

$$|z^2+1| = |z^2 - (-1)| \ge |z|^2 - |-1| = |z|^2 - 1 = 3 \Rightarrow |z^2+1| \ge 3,$$

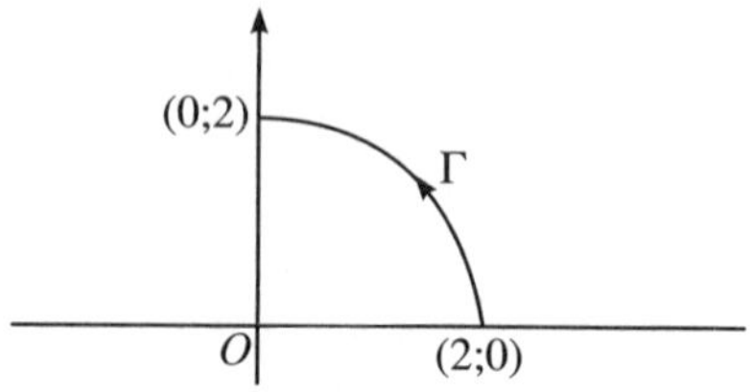

Figure 5.14

so that $M = 1/3$. Using, Eq. (5.10), we get

$$\left|\int_\Gamma \frac{dz}{z^2+1}\right| \le \frac{1}{3}\cdot\pi = \frac{\pi}{3}.$$

Example 5.3.8. *Find the upper bound for the absolute value of* $\int_C \frac{(z^2+3)e^{iz}\,\text{Log}\,z}{z^2-2}\,dz$, *where* $C = \{z : z = 2e^{i\theta}, 0 \le \theta \le \pi/3\}$.

Solution. Let $f(z) = \{(z^2+3)e^{iz}\,\text{Log}\,z\}/(z^2-2)$. The contour (Fig. 5.15) is the part of the circular arc $C = \{z : z = 2e^{i\theta}, 0 \le \theta \le \pi/3\}$. Then

$$\begin{aligned}
|f(z)| &= \left|\frac{(z^2+3)e^{iz}\,\text{Log}\,z}{z^2-2}\right| \le \frac{(|z|^2+3)e^{-y}|e^{ix}||\ln r + i\,\text{Arg}\,z|}{||z|^2-2|} \\
&\le \frac{(|z|^2+3)e^{-y}|\ln r + i\,\text{Arg}\,z|}{||z|^2-2|} \\
&\le \frac{(2^2+3)\cdot 1\cdot \ln(2+\pi/3)}{|2^2-2|} = \frac{7(\ln 2+\pi/3)}{2}, \quad \text{as } e^{-y} \le 1 \text{ for } 0 \le y \le \sqrt{3} \\
&\le \frac{7}{6}(3\ln 2 + \pi) = M.
\end{aligned}$$

Also, $L = \int_0^{\pi/3} 2\,d\theta = 2\pi/3$. Hence using Eq. (5.10) we get

$$\left|\int_C f(z)dz\right| \le M\,L = \frac{7\pi}{9}(3\ln 2 + \pi).$$

Example 5.3.9. *Find the upper bound for the absolute value of the integral* $\int_C (e^{2z} - z^2)\,dz$, *where* C *is the contour given in Fig. 5.16.*

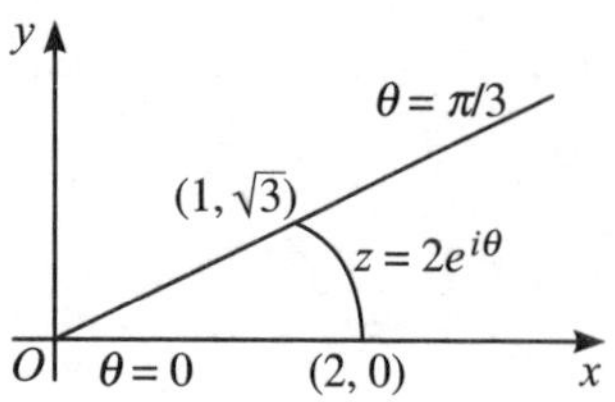

Figure 5.15

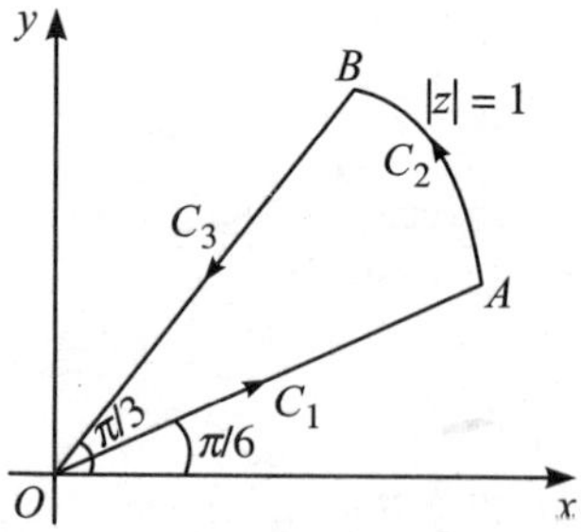

Figure 5.16

Solution. Let $f(z) = e^{2z} - z^2$. Using $z = x + iy$, we get

$$\begin{aligned}|f(z)| &= \left|e^{2z} - z^2\right| = \left|e^{2(x+iy)} - (x+iy)^2\right| \\ &\leq \left|e^{2(x+iy)}\right| + |x + iy|^2 \leq e^{2x} + 1.\end{aligned}$$

The contour C consists of three paths: C_1, C_2 and C_3.

(i) The equation of the line C_1 joining the points $O(0,0)$ and $A(\frac{\sqrt{3}}{2}, \frac{1}{2})$ is $x = \sqrt{3}\, y (0 \leq x \leq \sqrt{3}/2)$. The length of $OA = L_1 = 1$. Now $M_1 = \max|f(z)| = e^{\sqrt{3}} + 1$. Therefore

$$\left|\int_{C_1} f(z)\, dz\right| \leq M_1\, L_1 = (e^{\sqrt{3}} + 1) \cdot 1 = e^{\sqrt{3}} + 1.$$

(ii) Along the contour C_2, we have $x = \cos\theta$, $y = \sin\theta$ $(\pi/6 \leq \theta \leq \pi/3)$. So

$$\begin{aligned}L_2 &= \text{length of arc } AB \\ &= \int_{\pi/6}^{\pi/3} \sqrt{\left(\frac{dx}{d\theta}\right)^2 + \left(\frac{dy}{d\theta}\right)^2}\, d\theta = \int_{\pi/6}^{\pi/3} d\theta = \frac{\pi}{3} - \frac{\pi}{6} = \frac{\pi}{6}.\end{aligned}$$

Now $M_2 = \max|f(z)| = e^{2\cos\pi/6} + 1 = e^{\sqrt{3}} + 1$. Therefore

$$\left|\int_{C_2} f(z)\, dz\right| \leq M_2\, L_2 = (e^{\sqrt{3}} + 1) \cdot \frac{\pi}{6} = \frac{\pi}{6}(e^{\sqrt{3}} + 1).$$

(iii) The equation of the line C_3 joining the points $B(\frac{1}{2}, \frac{\sqrt{3}}{2})$ and $O(0,0)$ is $y = \sqrt{3}\, x (0 \leq x \leq 1/2)$. The length of $OB = L_3 = 1$. Now $M_1 = \max|f(z)| = e + 1$. Therefore,

$$\left|\int_{C_3} f(z)\, dz\right| \leq M_3\, L_3 = (e + 1) \cdot 1 = e + 1.$$

Using the property of contour integral, we have:

$$\int_C f(z)\, dz = \int_{C_1} f(z)\, dz + \int_{C_2} f(z)\, dz + \int_{C_3} f(z)\, dz$$

$$\begin{aligned}\left|\int_C f(z)\, dz\right| &\leq \left|\int_{C_1} f(z)\, dz\right| + \left|\int_{C_2} f(z)\, dz\right| + \left|\int_{C_3} f(z)\, dz\right| \\ &\leq e^{\sqrt{3}} + 1 + \frac{\pi}{6}(e^{\sqrt{3}} + 1) + e + 1 = (e + 1) + (1 + \frac{\pi}{6})(e^{\sqrt{3}} + 1).\end{aligned}$$

5.4 Winding Number

We wish to give a general global criterion when the integral of an analytic function along a closed path is 0. In practice, we meet two types of properties of paths: (i) properties of homotopy, and (ii) properties having to do with integration, relating the number of times a curve 'winds' around a point when we evaluate the integral

$$\int_C \frac{1}{z - z_0}\, dz$$

along the circle C with centred at z_0.

Definition 5.4.1. Let Γ be a closed contour in $\mathbb{C}$ which does not pass through a point z_0. The *index* of z_0 with respect to Γ, denoted by $\mathrm{Ind}(\Gamma; z_0)$ or $\mathrm{Ind}_\Gamma(z_0)$, or the *winding number* of Γ around z_0, denoted by $n(\Gamma; z_0)$ or $W_N(\Gamma, z_0)$, is defined by the integral

$$W_N(\Gamma; z_0) = \frac{1}{2\pi i}\int_\Gamma \frac{dz}{z - z_0}; \quad z_0 \notin \Gamma. \tag{5.11}$$

If γ is a curve defined on an interval $[a, b]$, then the integral in Eq. (5.11) can be written in the form

$$\int_\gamma \frac{dz}{z - z_0} = \int_a^b \frac{\gamma'(t)}{\gamma(t) - z_0} dt. \tag{5.12}$$

Theorem 5.6. *For a closed curve Γ in $\mathbb{C}$ and a point z_0 not on Γ, $W_N(\Gamma; z_0)$ is an integer.*

Proof: Let $\phi(t), t \in [a, b]$ be a parametric equation of the closed curve Γ. Then $\phi(a) = \phi(b)$ and

$$\frac{1}{2\pi i}\int_\Gamma \frac{dz}{z - z_0} = \frac{1}{2\pi i}\int_a^b \frac{\phi'(t)}{\phi(t) - z_0}\, dt.$$

Consider the function $F : [a, b] \to \mathbb{C}$ defined by

$$F(s) = \int_a^s \frac{\phi'(t)}{\phi(t) - z_0}\, dt \ \forall s \in [a, b].$$

Then $F(a) = 0$ and F is continuous on $[a, b]$ and $F'(s) = \frac{\phi'(s)}{\phi(s)-z_0} \forall s \in [a, b]$. Consider another function $\psi : [a, b] \to \mathbb{C}$, defined by

$$\psi(s) = \Big[\phi(s) - z_0\Big]e^{-F(s)}; \ \forall s \in [a, b]$$

Then $$\psi'(s) = \phi'(s)\, e^{-F(s)} - \Big[\phi(s) - z_0\Big]e^{-F(s)} \cdot F'(s)$$

$$\Rightarrow \quad \psi'(s) = \phi'(s)\, e^{-F(s)} - \Big[\phi(s) - z_0\Big]e^{-F(s)} \cdot \frac{\phi'(s)}{\phi(s) - z_0} = 0$$

at any point s where $\phi'(s)$ is continuous. Hence there is a constant C such that

$$\Big[\phi(s) - z_0\Big]e^{-F(s)} = C \Rightarrow \phi(s) - z_0 = Ce^{F(s)}.$$

Since Γ is a closed path, $\phi(b) = \phi(a)$, and

$$Ce^{F(b)} = \phi(b) - z_0 = \phi(a) - z_0 = Ce^{F(a)}.$$

Since $\phi(a) - z_0 \neq 0$, we conclude that $C \neq 0$, so that

$$e^{-F(a)} = e^{-F(b)} \Rightarrow e^{F(b)} = e^{F(a)}$$

$$\Rightarrow \quad F(b) = F(a) + 2\pi i\, k = 2\pi i k \text{ for some } k \in \mathbb{Z}$$

as $F(a) = 0$, thereby proving the theorem. □

Geometrically, $W_N(\Gamma; z_0)$ is an integer representing the total number of rounds of Γ around the point z_0, and the sign depends on the orientation of Γ. Counterclockwise(clockwise) winding is assigned as positive(negative) winding number. If Γ does not enclose the point z_0, then $W_N(\Gamma; z_0) = 0$. Some closed contours with their winding numbers are given in Fig. 5.17.

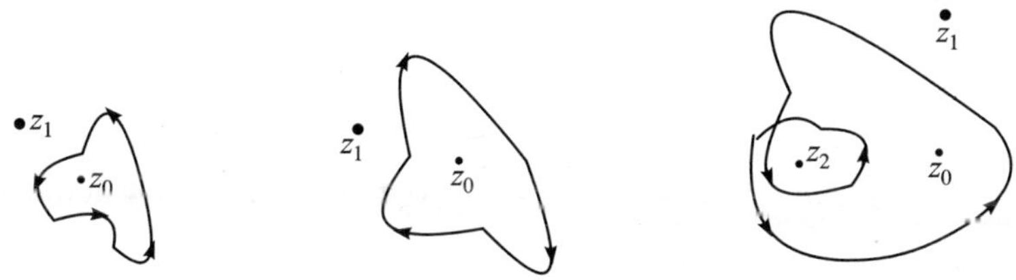

$W_N(\Gamma, z_0) = 1$ $\quad$ $W_N(\Gamma; z_0) = -1$ $\quad$ $W_N(\Gamma; z_0) = 1, W_N(\Gamma; z_1) = 0$

$W_N(\Gamma, z_1) = 0$ $\quad$ $W_N(\Gamma; z_1) = 0$ $\quad$ $W_N(\Gamma; z_2) = 2$

Figure 5.17 *Description of winding number*

Theorem 5.7. *Let Γ be a closed contour in $\mathbb{C}$. Then the mapping ξ defined by*

$$\xi \mapsto W_N(\Gamma; \xi) = \frac{1}{2\pi i}\int_\Gamma \frac{dz}{z-\xi}$$

is a continuous function of ξ at any point $\xi \notin \Gamma$.

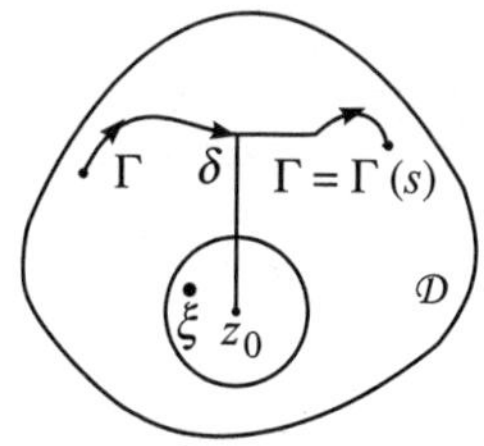

Figure 5.18 *The open set $\mathcal{D}$*

Proof: Let $\mathcal{D}$ be an open set containing no point of Γ, and $z_0 \in \mathcal{D}$ (Fig. 5.18). The function $s \mapsto |z_0 - s|;\, s \in \Gamma$ is continuous and not 0. Hence it has a minimum, the minimum distance between z_0 and the path, say

$$\delta = \inf\Big\{d(z_0, s) : s \in \Gamma\Big\} > 0.$$

Then for $z \in \Gamma$, $|z - z_0| \geq \delta$. Now for all h with $|h| < \delta/2$, we have

$$\Big|z - z_0 - h\Big| \geq |z - z_0| - |h| \geq \delta - \delta/2 = \delta/2.$$

Using the definition of winding number,

$$\begin{aligned}\Big|W_N(\Gamma; z_0 + h) - W_N(\Gamma; z_0)\Big| &= \left|\frac{1}{2\pi i}\int_\Gamma \left[\frac{1}{z - z_0 - h} - \frac{1}{z - z_0}\right] dz\right| \\ &\leq \frac{|h|}{2\pi}\int_\Gamma \frac{1}{|(z - z_0 - h)(z - z_0)|}\,|dz| \\ &\leq \frac{|h|}{2\pi}\,\frac{2}{\delta\cdot\delta}\int_\Gamma |dz| = \frac{|h|}{\pi\delta^2}L \to 0 \text{ as } h \to 0,\end{aligned}$$

where L denotes the length of Γ. As z_0 is arbitrary, the continuity of $W_N(\Gamma; \xi)$ is proved. □

Property 5.4.1. Let Γ be a close curve in $\mathbb{C}$ and $z_0 \in \mathbb{C}$ not in Γ. Then $W_N(-\Gamma; z_0) = -W_N(\Gamma; z_0)$.

Proof: Here Γ is a close curve in $\mathbb{C}$ and $z_0 \in \mathbb{C}$ is not in Γ. Let $\phi(t), t \in [a, b]$ be the parametric equation of Γ and $\psi(t) = \phi(a + b - t); t \in [a, b]$ be the parametric equation of $-\Gamma$. Then for $z_0 \notin \Gamma$,

$$\begin{aligned}W_N(-\Gamma; z_0) &= \frac{1}{2\pi i}\int_{-\Gamma}\frac{dz}{z - z_0} = \frac{1}{2\pi i}\int_a^b \frac{\psi'(t)}{\psi(t) - z_0}\,dt \\ &= -\frac{1}{2\pi i}\int_a^b \frac{\phi'(a + b - t)}{\phi(a + b - t) - z_0}(-dt);\ \ a + b - t = s \\ &= -\frac{1}{2\pi i}\int_a^b \frac{\phi'(s)}{\phi(s) - z_0}\,ds = -\int_\Gamma \frac{dz}{z - z_0} = -W_N(\Gamma; z_0),\end{aligned}$$

thereby proving the theorem. □

Property 5.4.2. Let Γ be a close curve in $\mathbb{C}$ and $z_0 \in \mathbb{C}$ be not in Γ. Then $W_N(\Gamma; z_0) = W_N(\Gamma; z_1)$ if z_0 and z_1 belong to a same region determined by Γ.

Proof: Here Γ is the closed curve in $\mathbb{C}$ and $z_0 \in \mathbb{C} - \{\Gamma\}$. First, we show that if the line segment $L_{z_0z_1}$ joining z_0 and z_1 is completely contained in the same region determined by Γ. We know that

$$\frac{d}{dz}\left\{\log\left(\frac{z-z_0}{z-z_1}\right)\right\} = \frac{1}{z-z_0} - \frac{1}{z-z_1},$$

provided the expression $(z - z_0)/(z - z_1)$ belongs to the domain of logarithm, where it is analytic. Now $\log z$ is analytic on $\mathcal{D} = \mathbb{C} - \{(x, 0); x \leq 0\}$. As

$$\begin{aligned}\frac{z-z_0}{z-z_1} \text{ is real and } \leq 0 \;\Rightarrow\; & \frac{z-z_0}{z-z_1} = -k, \text{ for some } k > 0\\ \Rightarrow\; & z = \frac{z_0+kz_1}{1+k} \Rightarrow z \text{ lies on } L_{z_0z_1}.\end{aligned}$$

Hence $\log\{(z - z_0)/(z - z_1)\}$ is differentiable on the complement of $L_{z_0z_1}$. Since $L_{z_0z_1}$ does not intersect Γ, we obtain that Γ belongs to the complement of $L_{z_0z_1}$. Thus, we get

$$\begin{aligned}& \int_\Gamma \left[\frac{1}{z-z_0} - \frac{1}{z-z_1}\right]dz = \int_\Gamma \frac{d}{dz}\left\{\log\left(\frac{z-z_0}{z-z_1}\right)\right\}dz = 0\\ \Rightarrow\quad & \frac{1}{2\pi i}\int_\Gamma \frac{dz}{z-z_0} = \frac{1}{2\pi i}\int_\Gamma \frac{dz}{z-z_1}\\ \Rightarrow\quad & W_N(\Gamma; z_0) = W_N(\Gamma; z_1),\end{aligned}$$

thereby proving the theorem. □

Property 5.4.3. Let γ be a closed rectifiable curve in $\mathbb{C}$ and $z_0 \in \mathbb{C}$ be not on γ. Then $W_N(\gamma; z_0)$ is constant for a belonging to a component of $G = \mathbb{C} - \{\gamma\}$. Also, $W_N(\gamma, z_0) = 0$ if z_0 belongs to the bounded region determined by γ.

Proof: Define a mapping $f : G \to \mathbb{C}$ by $f(z_0) = W_N(\gamma; z_0)$. Now, for each component $\mathcal{D}$ of G, $f(\mathcal{D})$ is connected. But since $f(G)$ is contained in the set of integers $\mathbb{Z}$, it follows that $f(\mathcal{D})$ reduces to a single point. Thus f is constant on $\mathcal{D}$.

Fix $z_0 \in G$ and let $r = d(z_0, \{\gamma\})$. If $|z_0 - z_1| < \delta < r/2$, then

$$\begin{aligned}\left|f(z_0) - f(z_1)\right| &= \frac{1}{2\pi}\left|\int_\gamma \left[\frac{1}{z-z_0} - \frac{1}{z-z_1}\right]dz\right|\\ &= \frac{1}{2\pi}\left|\int_\gamma \frac{z_0-z_1}{(z-z_0)(z-z_1)}\right|\\ &= \frac{|z_0-z_1|}{2\pi}\int_\gamma \frac{|dz|}{|z-z_0|\,|z-z_1|}.\end{aligned}$$

But for $|z_0 - z_1| < \frac{1}{2}r$ and z on $\{\gamma\}$, we have that

$$|z - z_0| \geq r > \frac{1}{2}r \text{ and } |z - z_1| > \frac{1}{2}r.$$

It follows that

$$f(z_0) - f(z_1)| < \frac{2|z_0 - z_1|}{r^2\pi}\delta.$$

So, corresponding to an $\varepsilon > 0$, by choosing $\delta = \min\{\frac{1}{2}r, \frac{\pi r^2 \varepsilon}{2\delta|z_0 - z_1|}\} > 0$, we see that

$$\left|f(z_0) - f(z_1)\right| < \varepsilon \text{ whenever } |z_0 - z_1| < \delta,$$

which shows that f must be continuous.

Let U be the unbounded component of G. Therefore, we choose $M > 0$ such that $\Gamma \subseteq S(0, M)$.

Let z_0 be the given point belonging to the unbounded region determined by Γ. Then for any $w_0 \notin S(0, M)$, z_0 and z_1 belong to the same region determined by Γ. Since $1/(z - z_1)$ is analytic function in $S(0, M)$, we have

$$W_N(\Gamma; z_0) = W_N(\Gamma; z_1) = \frac{1}{2\pi i}\int_\Gamma \frac{dz}{z - z_0} = 0.$$

Since $W_N(\Gamma; z_0)$ is constant on U, it must be zero. □

RESULT **40.** (i) Let $\mathcal{D}$ be a region in $\mathbb{C}$. Then $\mathcal{D}$ is simply connected if and only if $W_N(\Gamma; z_0) = 0$, i.e., $\Gamma \sim 0$ for every $z_0 \notin \mathcal{D}$ and for every Γ in $\mathcal{D}$.

(ii) The concept of the winding number is useful to characterize what is meant by the inside (interior) and the outside (exterior) of a closed curve Γ, respectively, as

$\text{Int}(\Gamma) = \{z \notin \Gamma : W_N(\Gamma; z) \neq 0\}$ and

$\text{Ext}(\Gamma) = \{z \notin \Gamma : W_N(\Gamma; z) = 0\}$.

(iii) A closed curve $\Gamma : [a, b] \to \mathbb{C}$ is said to be *positively (or negatively) oriented* if $W_N(\Gamma; z) > 0$ (or < 0) for every z inside Γ.

(iv) If Γ_1 and Γ_2 are closed paths in $\mathcal{D}$ and are *homotopic*, then they are *homologous*.

(v) If Γ_1 and Γ_2 are closed paths in $\mathcal{D}$ and are close together, then they are *homologous*.

Definition 5.4.2. A cycle Γ is a finite union of closed curves $\cup_{j=1}^m \Gamma_j$. We define

$$\int_\Gamma f(z)\, dz = \sum_{j=1}^m \int_{\Gamma_j} f(z)\, dz,$$

so that we get $W_N(\Gamma; z_0) = \sum\limits_{j=1}^m W_N(\Gamma_j; z_0)$.

Example 5.4.1. *If Γ be a circle $|\xi - z_0| = \varepsilon$, evaluate $W_N(\Gamma; z)$.*

Solution. The parametric equation of Γ is $\xi = z_0 + \varepsilon e^{i\theta}$, where $\theta \in [0, 2\pi]$. Thus

$$\frac{1}{2\pi i}\int_\Gamma \frac{d\zeta}{\zeta - z_0} = \frac{1}{2\pi i}\int_0^{2\pi} \frac{i\varepsilon e^{i\theta}}{\varepsilon e^{i\theta}}\, d\theta = 1.$$

If $|z - z_0| < \varepsilon$ and z_0 and z are lying in the same region determined by Γ, then

$$W_N(\Gamma; z) = W_N(\Gamma; z_0) = 1.$$

If $|z - z_0| > \varepsilon$, we get $W_N(\Gamma; z) = 0$. Thus

$$W_N(\Gamma; z) = \begin{cases} 1 & \text{if } |z - z_0| < \varepsilon \\ 0 & \text{if } |z - z_0| > \varepsilon \end{cases}$$

Example 5.4.2. *For each integer j, let $\gamma_j(t) = j + \frac{1}{4}e^{2\pi i t}$; $t \in [0, 1]$. For a positive integer N, set $\gamma(t) = (N + \frac{1}{2})e^{2\pi i t}$; $t \in [0, 1]$ and $\Gamma_1 = \sum_{j=-N}^N \gamma_j$; $\Gamma = \gamma - \Gamma_1$. Find $Ind(\gamma; z)$, $Ind(\Gamma_1; z)$ and $Ind(\Gamma; z)$ for each value of z which is an integer.*

Solution. The paths γ and γ_j are circles traversed once in the counterclockwise direction. For such a path we know the index is 1 inside the circle and 0 outside the circle. Thus, if k is an integer,

$$\operatorname{Ind}(\gamma; k) = \begin{cases} 1 & \text{if } |k| \le N \\ 0 & \text{if } |k| > N \end{cases} \quad \text{and } \operatorname{Ind}(\gamma_j; k) = \begin{cases} 1 & \text{if } k = j \\ 0 & \text{if } k \ne j \end{cases}$$

Thus,

$$\operatorname{Ind}(\Gamma_1; k) = \sum_{j=-k}^{k} \operatorname{Ind}(\gamma_j; k) = \begin{cases} 1 & \text{if } |k| \le N \\ 0 & \text{if } |k| > N \end{cases}$$

Therefore, $\operatorname{Ind}(\Gamma; k) = \operatorname{Ind}(\gamma; k) - \operatorname{Ind}(\Gamma_1; k) = 0$ for any integer k.

Definition 5.4.3. (i) Let Γ_1 and Γ_2 be two closed curves in an open subset $\mathcal{D}$ of $\mathbb{C}$. Then they are called *homologous* in $\mathcal{D}$ if

$$W_N(\Gamma_1; z_0) = W_N(\Gamma_2; z_0)$$

for every point z_0 in the complement of $\mathcal{D}$.

(ii) Let Γ be a closed curve in an open subset $\mathcal{D}$ of $\mathbb{C}$. We say that Γ is *homologous to* 0 in $\mathcal{D}$ if

$$W_N(\Gamma; z_0) = \frac{1}{2\pi i} \int_\Gamma \frac{dz}{z - z_0} = 0$$

for every z_0 in the complement of $\mathcal{D}$. In this case we write $\Gamma \sim 0$ in $\mathcal{D}$.

For a path γ, intuitively, the statement that γ is homologous to 0 in $\mathcal{D}$ means that γ does not go around any points of the complement of $\mathcal{D}$, that is, it does not go around any 'holes' in $\mathcal{D}$. For cycles, the meaning is more complicated. For example, the cycle Γ of Example 5.4.2 is homologous to 0 in the complement of the integers in $\mathbb{C}$.

Example 5.4.3. *Let* $\gamma_1(t) = e^{2\pi i t}$ *and* $\gamma_2(t) = 2e^{2\pi i t}$ *for* $t \in [0, 1]$. *Show that* $\Gamma = \gamma_2 - \gamma_1$ *is homologous to* 0 *in* $U_1 = \{z \in \mathbb{C} : 1/2 < |z| < 3\}$ *but is not homologous to* 0 *in the set* $U_2 = U_1/\{3/2\}$.

Solution. The curve γ_1 has index 1 about the points inside the unit disc $\mathcal{S}(0; 1)$ and index 0 about points exterior to the closed unit disc $\mathcal{S}[0; 1]$, while γ_2 has index 1 about points in $\mathcal{S}(0; 2)$ and 0 about points exterior to the closure of this disc. The complement of U_1 is

$$\mathcal{S}\left[0; \frac{1}{2}\right] \bigcup \left(\mathbb{C}/\mathcal{S}(0; 3)\right).$$

Since $\operatorname{Ind}_{\gamma_1} = \operatorname{Ind}_{\gamma_2} = 1$ on $\mathcal{S}[0; \frac{1}{2}]$ and $\operatorname{Ind}_{\gamma_1} = \operatorname{Ind}_{\gamma_2} = 0$ on $\mathbb{C}/\mathcal{S}(0; 3)$, we conclude that $\Gamma = \gamma_2 - \gamma_1$ is homologous to 0 in U_1.

However, at the point $z = 3/2$, γ_1 has index 0 and γ_2 has index 1. Since $3/2$ is in complement of $U_2 = \mathbb{C}/\{3/2\}$, Γ is not homologous to 0 in the set $U_2 = U_1/\{3/2\}$.

5.5 Cauchy's Theorem

In this section we shall prove that the integral of an analytic function over a simple closed contour is zero, which is due to French mathematician Louis-Augustin Cauchy. This result is one of the most fundamental theorems in complex analysis.

Theorem 5.8. *Let $f = u + iv$ be an analytic function in a simply-connected region $\mathcal{D}$ such that $f'(z)$ is continuous in $\mathcal{D}$. Let γ be a simple closed contour orienting positively such that γ^* and the inside of γ are continuous in Ω. Then $\int_\gamma f(z)dz = 0$.*

Proof: Since $z = x + iy$, we have

$$f(z)dz = (u + iv)(dx + idy) = (u\,dx - v\,dy) + i(v\,dx + u\,dy).$$

Also, let $\mathcal{D} =$ Int γ, the interior of γ. The integral of f over γ can be written as

$$\int_\gamma f(z)\,dz = \int_\gamma (u\,dx - v\,dy) + i\int_\gamma (v\,dx + u\,dy).$$

Since f' is continuous in $\mathcal{D}$, the partial derivatives of u and v are also continuous in $\mathcal{D}$. By applying Green's theorem, we obtain

$$\begin{aligned}\int_\gamma f(z)\,dz &= \int_\gamma (u\,dx - v\,dy) + i\int_\gamma (v\,dx + u\,dy) \\ &= \int\int_{\mathcal{D}} \Big(- v_x - u_y \Big) dx\,dy + i \int\int_{\mathcal{D}} \Big(u_x - v_y\Big) dx\,dy, \\ &= 0\end{aligned}$$

which is the well-known *Cauchy's theorem.* □

In 1883 the Franch mathematician Edouard Goursat proved that the assumption of continuity of f' is not necessary to reach the conclusion of Cauchy theorem. The resulting modified version of Cauchy's theorem is known as *Cauchy–Goursat theorem.*

Theorem 5.9. [Cauchy–Goursat theorem]: *If f is analytic in a simply-connected domain $\mathcal{D}$ and γ is any simple, closed, rectifiable contour in $\mathcal{D}$, then $\oint_\gamma f(z)\,dz = 0$.*

5.5.1 Cauchy's Theorem for Triangles

Theorem 5.10. *Let Δ be any triangular region in the z-plane contained in a simply-connected domain $\mathcal{D}$. Let $\partial\Delta$ denote a simple closed contour which is the boundary of Δ with orientation such that Δ is inside of $\partial\Delta$. If f is an analytic function on an open set containing Δ, then $\int_{\partial\Delta} f(z)\,dz = 0$.*

Proof: Here $\mathcal{D}$ is simply connected, so the interior of $\triangle$ also belong to $\mathcal{D}$ (Fig. 5.19). To achieve the desired result, join the midpoints D, E and F of the sides AB, AC and BC respectively to form four triangular regions indicated briefly by $\triangle_I, \triangle_{II}, \triangle_{III}$ and $\triangle_{IV}$ whose respective boundaries are $\partial\triangle_I, \partial\triangle_{II}, \partial\triangle_{III}$ and $\partial\triangle_{IV}$.

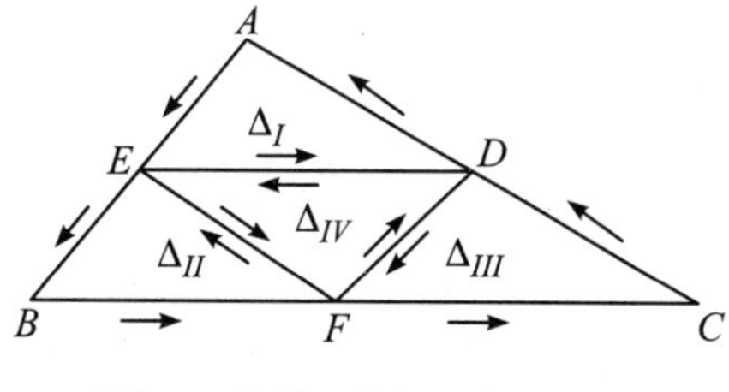

Figure 5.19 *Triangular region*

If $f(z)$ is analytic inside and on triangle ABC, we have,

$$\begin{aligned}\oint_{ABCA} f(z)dz &= \int_{DAE} f(z)dz + \int_{EBF} f(z)dz + \int_{FCD} f(z)dz \\ &= \Big[\{\int_{DAE} + \int_{ED}\} + \{\int_{EBF} + \int_{FE}\} + \{\int_{FCD} + \int_{DF}\} + \{\int_{DE} + \int_{EF} + \int_{FD}\}\Big] f(z)dz\end{aligned}$$

$$= \int_{DAED} f(z)dz + \int_{EBFE} f(z)dz + \int_{FCDF} f(z)dz + \int_{DEFD} f(z)dz$$
$$= \oint_{\triangle_I} f(z)dz + \oint_{\triangle_{II}} f(z)dz + \oint_{\triangle_{III}} f(z)dz + \oint_{\triangle_{IV}} f(z)dz$$

$$\text{or}\left|\oint_{\partial\Delta} f(z)dz\right| \le \left|\oint_{\partial\Delta_I} f(z)dz\right| + \left|\oint_{\partial\Delta_{II}} f(z)dz\right| + \left|\oint_{\partial\Delta_{III}} f(z)dz\right| + \left|\oint_{\partial\Delta_{IV}} f(z)dz\right|. \tag{5.13}$$

Let Δ_1 be the triangle corresponding to that term on the right hand side of Eq. (5.13) having the largest value. (If there are two or more such terms, then Δ_1 is any of the associated triangle.) Then,

$$\left|\oint_{\partial\Delta} f(z)dz\right| \le 4\left|\oint_{\partial\Delta_1} f(z)dz\right|. \tag{5.14}$$

By joining the midpoints of the sides of triangle Δ_1, we obtain similarly a triangle Δ_2 such that

$$\left|\oint_{\partial\Delta_1} f(z)dz\right| \le 4\left|\oint_{\partial\Delta_2} f(z)dz\right|$$
$$\Rightarrow \left|\oint_{\partial\Delta} f(z)dz\right| \le 4^2\left|\oint_{\partial\Delta_2} f(z)dz\right| \text{ by Eq. (5.14).} \tag{5.15}$$

Proceeding in this way, after n steps we obtain a triangle $\triangle_n$ such that

$$\left|\oint_{\partial\Delta} f(z)dz\right| \le 4^n\left|\oint_{\partial\Delta_n} f(z)dz\right|. \tag{5.16}$$

Now, $\triangle, \triangle_1, \triangle_2, \cdots$ is a sequence of nested triangles each of which is contained in the preceding triangle (i.e., a sequence of nested triangles). There exists a point z_0 which lies in every triangle of the sequence. Since z_0 lies inside or on the boundary of $\partial\Delta$, it follows that $f(z)$ is analytic at z_0. Then,

$$f(z) = f(z_0) + (z - z_0)f'(z_0) + (z - z_0)\eta, \tag{5.17}$$

where for any $\varepsilon > 0$, we can find δ such that $|\eta| < \varepsilon$ whenever $|z - z_0| < \delta$. Since $\partial\Delta_n$ is a closed contour, using the antiderivative theorem 5.3, we get

$$\oint_{\partial\Delta_n} f(z_0)dz = f(z_0)\oint_{\partial\Delta_n} dz = 0$$
$$\oint_{\partial\Delta_n} (z - z_0)f'(z_0)dz = f'(z_0)\oint_{\partial\Delta_n} zdz - z_0 f'(z_0)\oint_{\partial\Delta_n} dz = 0.$$

Thus, by integration of both sides of (5.17), we obtain

$$\oint_{\partial\Delta_n} f(z)dz = \oint_{\partial\Delta_n} f(z_0)dz + \oint_{\partial\Delta_n} (z - z_0)f'(z_0)dz + \oint_{\partial\Delta_n} (z - z_0)\eta dz$$
$$= \oint_{\partial\Delta_n} (z - z_0)\eta dz.$$

Now, let P be the perimeter of Δ. Then the perimeter of Δ_n is $P_n = \frac{1}{2^n}P$. If z is any point in Δ_n, then we must have, $|z - z_0| < \frac{P}{2^n} < \delta$. Hence

$$\Big| \oint_{\partial \Delta_n} f(z)dz \Big| = \Big| \oint_{\partial \Delta_n} (z - z_0)\eta dz \Big| \leq \varepsilon \frac{P}{2^n} \cdot \frac{P}{2^n} = \frac{\varepsilon P^2}{4^n}$$

$$\Rightarrow \quad \Big| \oint_{\partial \Delta} f(z)dz \Big| \leq \varepsilon P^2 \text{ from Eq. (5.16).}$$

Since ε can be made arbitrarily small, it follows that $\oint_{\partial \Delta} f(z)dz = 0$. □

5.5.2 Cauchy's Theorem for Rectangles

Theorem 5.11. *Let $\mathcal{R}$ be a rectangular region $\mathcal{R} = \{x + iy : a \leq x \leq b, c \leq y \leq d\}$ in the a z-plane. Let $\partial\mathcal{R}$ denote a simple closed contour which is the boundary of $\mathcal{R}$ with orientation such that $\mathcal{R}$ is inside of $\partial\mathcal{R}$. If f is an analytic function on an open set containing $\mathcal{R}$, then $\int_{\partial\mathcal{R}} f(z)\, dz = 0$.*

Proof: We subdivide $\mathcal{R}$ into four congruent rectangles $\mathcal{R}^{(1)}$, $\mathcal{R}^{(2)}$, $\mathcal{R}^{(3)}$ and $\mathcal{R}^{(4)}$ by joining the midpoints of opposite sides by the line segments as shown in the Fig. 5.20.

The respective boundaries of the rectangular regions are $\partial\mathcal{R}^{(1)}$, $\partial\mathcal{R}^{(2)}$, $\partial\mathcal{R}^{(3)}$ and $\partial\mathcal{R}^{(4)}$, so that

$$\partial\mathcal{R} = \partial\mathcal{R}^{(1)} + \partial\mathcal{R}^{(2)} + \partial\mathcal{R}^{(3)} + \partial\mathcal{R}^{(4)}$$

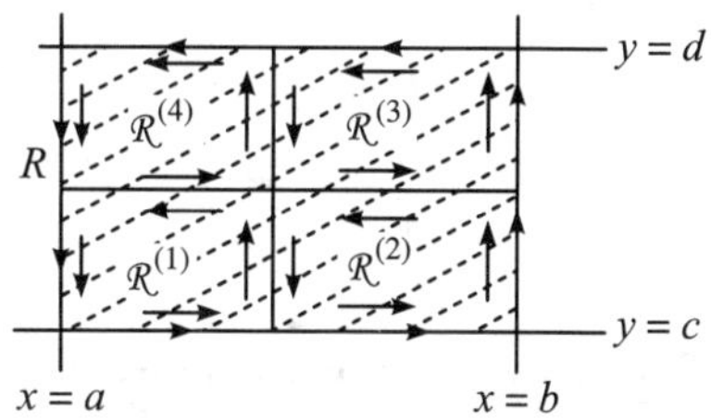

Figure 5.20 *Rectangular region*

as the dividing segments cancel in pairs. So,

$$\int_{\partial\mathcal{R}} f(z)\, dz = \Big[\int_{\partial\mathcal{R}^{(1)}} + \int_{\partial\mathcal{R}^{(2)}} + \int_{\partial\mathcal{R}^{(3)}} + \int_{\partial\mathcal{R}^{(4)}} \Big] f(z)\, dz. \tag{5.18}$$

There is some $k (1 \leq k \leq 4)$ such that

$$\Big| \int_{\partial\mathcal{R}^{(k)}} f(z)\, dz \Big| \geq \frac{1}{4} \Big| \int_{\partial\mathcal{R}} f(z)\, dz \Big|. \tag{5.19}$$

Call this $\partial\mathcal{R}^{(k)}$ as $\mathcal{R}_1$. (In case there is more than one k for which Eq.(5.19) holds, then choose one of them arbitrarily as $\mathcal{R}_1$.) Now, $\mathcal{R}, \mathcal{R}_1, \mathcal{R}_2, \cdots$ is a sequence of nested rectangles $\mathcal{R} \supset \mathcal{R}_1 \supset \mathcal{R}_2 \supset \cdots$ and

$$\begin{aligned} \Big| \int_{\partial\mathcal{R}_n} f(z)\, dz \Big| &\geq \frac{1}{4} \Big| \int_{\partial\mathcal{R}_{n-1}} f(z)\, dz \Big| \\ &\geq \frac{1}{4^2} \Big| \int_{\partial\mathcal{R}_{n-2}} f(z)\, dz \Big| \geq \cdots \geq \frac{1}{4^n} \Big| \int_{\partial\mathcal{R}} f(z)\, dz \Big|. \end{aligned} \tag{5.20}$$

There exists a common point z_0 which lies in every closed rectangle of the sequence. Since z_0 lies inside or on the boundary of $\partial\mathcal{R}$, it follows that, $f(z)$ is analytic at z_0. Then

$$f(z) = f(z_0) + (z - z_0)f'(z_0) + (z - z_0)\eta, \tag{5.21}$$

where for any $\varepsilon > 0$, we can find δ such that $|\eta| < \varepsilon$ whenever $|z - z_0| < \delta$. Since $\partial\mathcal{R}_n$ is a closed contour, using the antiderivative theorem 5.3, we get

$$\oint_{\partial\mathcal{R}_n} f(z_0)dz = f(z_0)\oint_{\partial\mathcal{R}_n} dz = 0$$
$$\oint_{\partial\mathcal{R}_n} (z-z_0)f'(z_0)dz = f'(z_0)\oint_{\partial\mathcal{R}_n} zdz - z_0 f'(z_0)\oint_{\partial\mathcal{R}_n} dz = 0.$$

Thus, by integration of both sides of (5.21), we obtain

$$\begin{aligned}\oint_{\partial\mathcal{R}_n} f(z)dz &= \oint_{\partial\mathcal{R}_n} f(z_0)dz + \oint_{\partial\mathcal{R}_n} (z-z_0)f'(z_0)dz + \oint_{\partial\mathcal{R}_n} (z-z_0)\eta dz \\ &= \oint_{\partial\mathcal{R}_n} (z-z_0)\eta dz.\end{aligned}$$

Now, let P be the perimeter of $\mathcal{R}$. Then the perimeter of $\mathcal{R}_n$ is $P_n = \frac{1}{2^n}P$. If z is any point in $\mathcal{R}_n$, then we must have, $|z-z_0| < \frac{P}{2^n} < \delta$. Hence

$$\Big|\oint_{\partial\mathcal{R}_n} f(z)dz\Big| = \Big|\oint_{\partial\mathcal{R}_n} (z-z_0)\eta dz\Big| \le \varepsilon\frac{P}{2^n}\cdot\frac{P}{2^n} = \frac{\varepsilon P^2}{4^n}$$
$$\Rightarrow \quad \Big|\oint_{\partial\mathcal{R}} f(z)dz\Big| \le \varepsilon P^2 \text{ from Eq. (5.20).}$$

Since ε can be made arbitrarily small, it follows that $\oint_{\partial\mathcal{R}} f(z)dz = 0$. □

5.5.3 Cauchy's Theorem for Discs

Theorem 5.12. *If $f(z)$ is analytic in an open disc $\mathcal{S}(z_0;\delta) = \{z \in \mathbb{C} : |z-z_0| < \delta\}$, then $\int_\gamma f(z)\,dz = 0$, for any closed contour γ in $\mathcal{S}(z_0;\delta)$.*

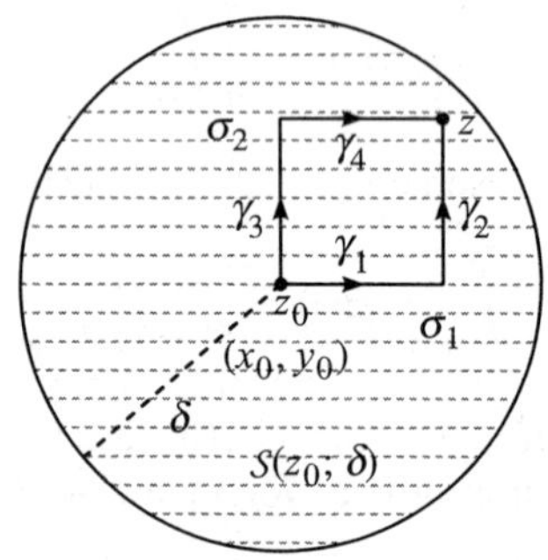

Figure 5.21 *Cauchy's theorem for discs*

Proof: Within the open disc $\mathcal{S}(z_0;\delta)$, let us take a rectangular region which consists of four lines, $\gamma_1, \gamma_2, \gamma_3$ and γ_4, as depicted in Fig. 5.21.

Let σ_1 be a curve within the disc, consisting of γ_1 joining from $z_0 = (x_0, y_0)$ and (x, y_0) and γ_2 joining (x, y_0) to $z = (x, y)$ as directed. So, $\sigma_1 = \gamma_1 + \gamma_2$. Similarly, σ_2 consists of two directed lines, γ_3 and γ_4, as shown in the Fig. 5.21. So, $\sigma_2 = \gamma_3 + \gamma_4$. Now, $\sigma_1 - \sigma_2$ is the boundary of a rectangular region. So by Cauchy's theorem for a rectangle, we get

$$\int_{\sigma_1-\sigma_2} f(z)\,dz = 0 \Rightarrow \int_{\sigma_1} f(z)\,dz = \int_{\sigma_2} f(z)\,dz. \tag{5.22}$$

We define a function $F(z)$ by $F(z) = \int_{\sigma_1} f(\zeta)d\zeta$ for $z \in \mathcal{S}(z_0;\delta)$. Therefore,

$$\begin{aligned}F(z) &= \int_{\gamma_1+\gamma_2} f(z)dz = \int_{\gamma_1} f(z)\,dz + \int_{\gamma_2} f(z)\,dz \\ &= \int_{t=x_0}^{x} f(t+iy_0)dt + \int_{t=y_0}^{y} f(x+it)\,i\,dt. \\ \therefore \frac{\partial F}{\partial y} &= 0 + f(z)\,i = if(z).\end{aligned}$$

Also, from Eq. (5.22), it follows that $F(z) = \int_{\sigma_2} f(\zeta)d\zeta$. Thus

$$\begin{aligned} F(z) &= \int_{\gamma_3+\gamma_4} f(z)dz = \int_{\gamma_3} f(z)\,dz + \int_{\gamma_4} f(z)\,dz \\ &= \int_{t=y_0}^{y} f(x_0+it)\cdot i\,dt + \int_{t=x_0}^{x} f(t+iy)\,dt. \\ \therefore \frac{\partial F}{\partial x} &= f(z)+0 = f(z). \end{aligned}$$

Thus, $\frac{\partial F}{\partial y} = i\frac{\partial F}{\partial x}$ and hence F satisfies the C-R equations and the partial derivatives of F are continuous. Therefore, F is analytic and $F'(z) = f(z)$.

Thus, f has an antiderivative in the open disc $\mathcal{S}(z_0;\delta)$ and by the antiderivative theorem $\int_\gamma f(z)\,dz = 0$ for any closed contour γ in $\mathcal{S}(z_0;\delta)$. □

5.5.4 Cauchy's Theorem for a Closed Polygon

Let γ be any closed polygon with n sides contained in $\mathcal{D}$ (Fig. 5.22). We subdivide the region of γ into $(n-2)$ triangles $\triangle_1, \triangle_2, \cdots, \triangle_{n-2}$ by drawing diagonals from a fixed vertex, say A_1, to the other vertices. Then by Cauchy's theorem for a triangle (as in the previous case) and using the fact that the integrals along A_1A_3 and A_3A_1, A_1A_4 and $A_4A_1, \cdots, A_1A_{n-1}$, and $A_{n-1}A_1$ cancel, we see that

$$\oint_\gamma f(z)dz = \int_{\partial\triangle_1} f(z)dz + \int_{\partial\triangle_2} f(z)dz + \cdots + \int_{\partial\triangle_{n-2}} f(z)dz = 0,$$

where we suppose that $f(z)$ is analytic inside and on the polygon.

5.5.5 Cauchy's Theorem for a Simple Closed Curve

Let γ be any simple closed curve which is contained in a region $\mathcal{D}$ in which $f(z)$ is analytic (Fig. 5.23). Choose n points of subdivision $z_1, z_2, \cdots, z_n$ on the curve γ, where for convenience of notation we consider $z_0 = z_n$. Construct polygon P by joining these points. Let us define a sum

$$S_n = \sum_{k=1}^{n} f(z_k)\Delta z_k \;:\; \Delta z_k = z_k - z_{k-1}.$$

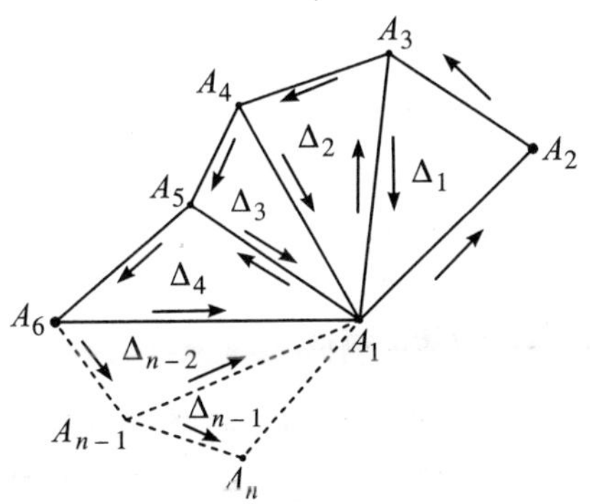

Figure 5.22 *Closed polygon*

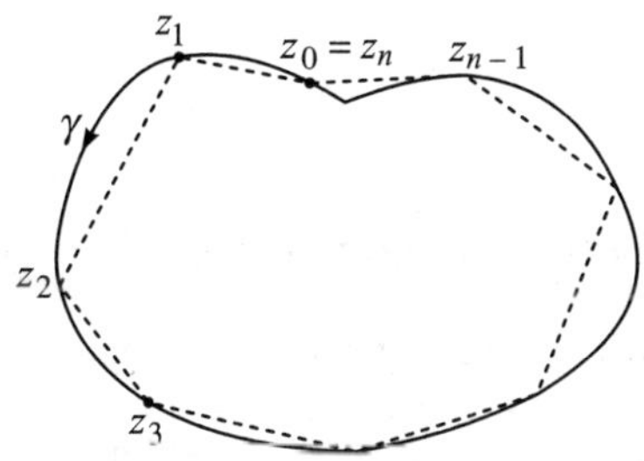

Figure 5.23 *Simple closed curve*

Since $\lim_{n\to\infty} S_n = \oint_\gamma f(z)dz$, so corresponding to an $\varepsilon > 0, \exists\ m \in \mathbb{N}$, such that

$$\left|\oint_\gamma f(z)dz - S_n\right| < \frac{\varepsilon}{2} \text{ whenever } n > m. \tag{5.23}$$

Consider now the integral along polygon P. By the previous theorem for closed polygons, we have

$$\begin{aligned}
\oint_\gamma f(z)dz &= 0 = \int_{z_0}^{z_1} f(z)dz + \int_{z_1}^{z_2} f(z)dz + \cdots + \int_{z_{n-1}}^{z_n} f(z)dz \\
&= \int_{z_0}^{z_1} \Big[f(z) - f(z_1) + f(z_1)\Big]dz + \cdots + \int_{z_{n-1}}^{z_n} \Big[f(z) - f(z_n) + f(z_n)\Big]dz \\
&= \int_{z_0}^{z_1} \Big[f(z) - f(z_1)\Big]dz + \cdots + \int_{z_{n-1}}^{z_n} \Big[f(z) - f(z_n)\Big]dz + S_n \\
\text{or, } S_n &= \int_{z_0}^{z_1} \Big[f(z_1) - f(z)\Big]dz + \cdots + \int_{z_{n-1}}^{z_n} \Big[f(z_n) - f(z)\Big]dz. \qquad (5.24)
\end{aligned}$$

Let us now choose an m so large that on the lines joining z_0 and z_1, z_1 and $z_2, \cdots, z_{n-1}$ and z_n,

$$\Big|f(z_1) - f(z)\Big| < \frac{\varepsilon}{2L}, \cdots, \Big|f(z_n) - f(z)\Big| < \frac{\varepsilon}{2L}, \tag{5.25}$$

where L is the length of Γ. From Eq. (5.24) and (5.25) we get

$$\begin{aligned}
|S_n| &\leq \left|\int_{z_0}^{z_1} \{f(z_1) - f(z)\}dz\right| + \cdots + \left|\int_{z_{n-1}}^{z_n} \{f(z_n) - f(z)\}dz\right| \\
&\leq \int_{z_0}^{z_1} |f(z_1) - f(z)||dz| + \cdots + \int_{z_{n-1}}^{z_n} |f(z_n) - f(z)||dz| \\
&\leq \frac{\varepsilon}{2L}\Big[|z_1 - z_0| + \cdots + |z_n - z_{n-1}|\Big] = \frac{\varepsilon}{2}.
\end{aligned}$$

Therefore,

$$\begin{aligned}
\left|\oint_\gamma f(z)dz\right| &= \left|\oint_\gamma f(z)dz - S_n + S_n\right| \\
&\leq \left|\oint_\gamma f(z)dz - S_n\right| + \Big|S_n\Big| < \frac{\varepsilon}{2} + \frac{\varepsilon}{2} = \varepsilon.
\end{aligned}$$

Since, ε is arbitrary, it follows that $\oint_\gamma f(z)dz = 0$ as required. □

5.5.6 Cauchy's Theorem for Multiply-connected Regions

Let us consider a multiply-connected region $\mathcal{D}$ bounded by simple closed curves γ_1 and γ_2 as indicated in Fig. 5.24. Construct cross-cut AH. Then the region bounded by $ABDEFGAHJIHA$ is simply connected, so that by the previous theorem, we get,

$$\int_{ABDEFGAHJIHA} f(z)\,dz = 0$$

or, $$\int_{ABDEFGA} f(z)dz + \int_{AH} f(z)dz + \int_{HJIH} f(z)dz + \int_{HA} f(z)dz = 0$$

or, $$\int_{ABDEFGA} f(z)dz + \int_{HJIH} f(z)dz = 0;$$

$$\left[\text{as} \int_{AH} f(z)dz = -\int_{HA} f(z)dz\right]$$

or, $$\oint_{\gamma} f(z)dz = 0,$$

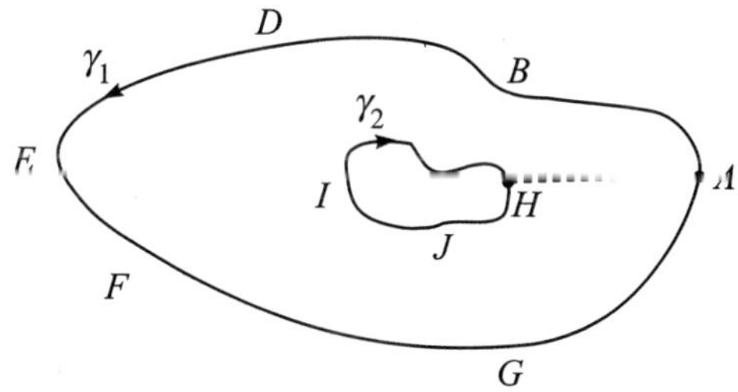

Figure 5.24 *Multiply-connected region*

where γ is the complete boundary of $\mathcal{D}$ (consisting of $ABDEFGA$ and $HJIH$) traversed in the sense that an observer walking on the boundary always has the region $\mathcal{D}$ on his left-hand side. □

RESULT **41.** If f is analytic on a region $\mathcal{D}$ and γ is a cycle which is homologous to zero in $\mathcal{D}$, then $\int_{\gamma} f(z)\, dz = 0$.

Example 5.5.1. *Verify Cauchy's theorem for $f(z) = 3z^2 + iz - 4$, where Γ is a square with vertices $1 \pm i$, $-1 \pm i$.*

Solution. Let Γ be the square $ABCD$ as shown in the Fig. 5.25. Then, since $f(z)$ is analytic in Γ, $\int_{\Gamma} f(z)dz = 0$ by Cauchy's theorem. Again,

$$\int_{\Gamma} f(z)\, dz = \int_{AB} f(z)dz + \int_{BC} f(z)dz + \int_{CD} f(z)dz + \int_{DA} f(z)dz.$$

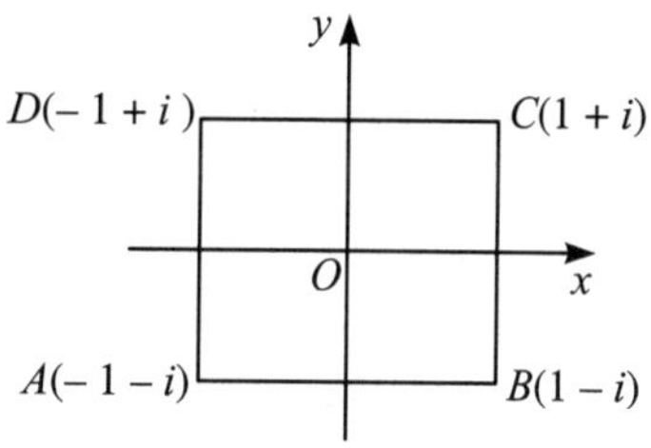

Figure 5.25

Now,

$$\int_{AB} f(z)dz = \int_{1+i}^{-1+i} (3z^2 + iz - 4)dz = 14$$

$$\int_{BC} f(z)dz = 4 + 2i \ ; \ \int_{CD} f(z)dz = -(14) \ ; \ \int_{DA} f(z)dz = -4 - 2i$$

$$\therefore \quad \int_{AB+BC+CD+DA} f(z)\, dz = 14 + 4 + 2i - 14 - 4 - 2i = 0.$$

Thus, $\int_{\Gamma} f(z)dz = 0$ and therefore, the Cauchy's theorem is verified.

Example 5.5.2. *By evaluating $\int_{\Gamma} e^z\, dz$ around the circle $|z| = 1$, show that*

$$\int_0^{2\pi} e^{\cos\theta} \cos(\theta + \sin\theta)\, d\theta - \int_0^{2\pi} e^{\cos\theta} \sin(\theta + \sin\theta)\, d\theta = 0.$$

Solution. On the circle $|z| = 1$, $z = e^{i\theta}$, $dz = ie^{i\theta}d\theta$, $0 \le \theta \le 2\pi$. Again, e^z is an analytic function everywhere within the unit circle and also on the circle. Therefore, by Cauchy's theorem,

$$\begin{aligned} 0 &= \int_\Gamma e^z\, dz = \int_0^{2\pi} e^{\cos\theta + i\sin\theta} \cdot e^{i\theta} \cdot i\, d\theta \\ &= i\int_0^{2\pi} e^{\cos\theta}\, e^{i(\theta+\sin\theta)}\, d\theta \\ &= \int_0^{2\pi} e^{\cos\theta}\Big[-\sin(\theta+\sin\theta) + i\cos(\theta+\sin\theta)\Big] d\theta \\ &= -\int_0^{2\pi} e^{\cos\theta} \cdot \sin(\theta+\sin\theta)\, d\theta + \int_0^{2\pi} e^{\cos\theta} \cdot \cos(\theta+\sin\theta)\, d\theta. \end{aligned}$$

$$\therefore \quad \int_0^{2\pi} e^{\cos\theta} \cos(\theta+\sin\theta)\, d\theta = 0 = \int_0^{2\pi} e^{\cos\theta} \sin(\theta+\sin\theta)\, d\theta.$$

5.5.7 Deformation Theorem

Let γ_1 be a simple closed contour that can be continuously deformed into another simple closed contour γ_2 without passing through a point where a given function f is not analytic. Here we shall show that the integral of f along γ_1 can be replaced by the integral of the same function along a more amenable path γ_2.

Theorem 5.13. *Let γ_1 and γ_2 be two positively-oriented piecewise smooth simple closed curves and let $\gamma_2 \subset$ Int(γ_1). Let $A = \gamma_1^* \cup ($ Int$(\gamma_1) \cap O(\gamma_2)) \cup \gamma_2^*$. Let f be a complex function on an open set Ω containing A. Then*

$$\oint_{\gamma_1} f(z)\, dz = \oint_{\gamma_2} f(z)\, dz \tag{5.26}$$

Proof: Assume that γ_1 and γ_2 are simple closed curves (Fig. 5.26). Here A is a compact subset of $\mathbb{C}$. So by the ε-neighbourhood theorem, there exists $\varepsilon > 0$ such that

$$A \subset \bigcup_{x\in A} S(x;\varepsilon) \subset \Omega.$$

Consider a tiling of the complex plane by squares (whose sides are parallel to the real and imaginary axis) of side $s < \varepsilon/\sqrt{2}$. Consider the intersection of this tiling with A. The intersection of a square tile with A can be:

(i) the whole of square tile such that the square tile is contained in the interior of A. (Call such a tile a *normal tile*.)

(ii) a compact portion of a square tile with area less than area of the square. (Call such a portion of a tile a *special tile*.)

Let us denote the boundary of A by ∂A (where $\partial A = \gamma_1^* \cup \gamma_2^*$). A special tile is possible when a square tile intersects ∂A. Let us orient the boundaries of these tiles, both normal and special, in the positive sense.

For a normal square tile S contained in the interior of A, $S \subset \Omega$. So, by Cauchy's theorem, for a rectangle,

$$\int_{\partial S} f(z)\, dz = 0.$$

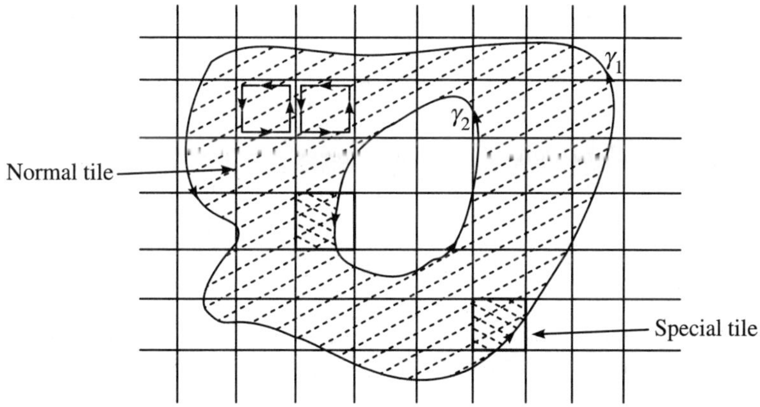

Figure 5.26 $A = \gamma_1^* \cup \Big(Int(\gamma_1) \cap O(\gamma_2) \Big) \cup \gamma_2^*$

For a special tile T contained in A, ∂T is a finite collection of positively-oriented simple closed curves, and if $x \in \partial T \bigcap \partial A$, then $T \subset S(x; \varepsilon)$. Since $s < \varepsilon/\sqrt{2}$, so for any two points $y_1, y_2 \in$ a square tile:

$$d(y_1, y_2) < \sqrt{2}\, s < \varepsilon,$$

so that by Cauchy's theorem for a disc, $\int_{\partial T} f(z)\, dz = 0$. Now, consider

$$\sum \int_{\partial T} f(z)\, dz = 0, \tag{5.27}$$

Over where the summation is taken over all special and normal tiles T. On the left-hand side of Eq. (5.27), we have the following observations:

(i) Every side of a normal square tile appears twice with opposite orientation.

(ii) In the special tiles everywhere, a portion of the side of a square tile appears twice with opposite orientation.

(iii) In special tiles, a portion of γ_1 appears once with the same orientation as γ_1. Also, every point of γ_1^* appears on the boundary of some special tile.

(iv) The portion of γ_2 appears once with opposite orientation of γ_2 and all points of γ_2^* appear in some special tile.

So the intersection along the sides of squares cancels each other. Thus from Eq. (5.27), we get

$$\sum \int_{\partial T} f(z)\, dz = \oint_{\gamma_1} f(z)\, dz + \oint_{-\gamma_2} f(z)\, dz = 0$$

$$\Rightarrow \quad \oint_{\gamma_1} f(z)\, dz = \oint_{\gamma_2} f(z)\, dz.$$

This completes the proof. This result is sometimes called the *principle of deformation of contours.* □

Example 5.5.3. *Evaluate* $I = \int_\gamma \frac{dz}{z}$, *where* $\gamma(t) = i + 29e^{it}$; $0 \le t \le 2\pi$.

Solution. Here $1/z$ is analytic in a domain consisting of the complex plane excluding $z = 0$. Thus

$$I = \int_0^{2\pi} \frac{29e^{it}}{i + 29e^{it}}\, idt = 29i \int_0^{2\pi} \frac{\cos t + i \sin t}{29\cos t + i(29\sin t + 1)}\, dt,$$

which is very difficult to evaluate. Let us consider a unit circle $\gamma_2(t) = e^{it}$. Then

$$\int_{\gamma_2} \frac{1}{z}\, dz = \int_0^{2\pi} \frac{1}{e^{it}}\, ie^{it}\, dt = 2\pi i.$$

Then, by the consequence of (5.26), it follows that

$$\int_{\gamma} \frac{dz}{z} = \int_{\gamma_1} \frac{dz}{z} = 2\pi i,$$

which is the required value.

Example 5.5.4. *Compute the integrals (a) $\int_{\Gamma_1} \frac{1}{z^2-1}\, dz$ and (b) $\int_{\Gamma_2} \frac{e^z}{z^2-9}\, dz$, where the closed contour Γ_1 and Γ_2 are depicted in the Fig. 5.27(a) and Fig. 5.27(b) respectively.*

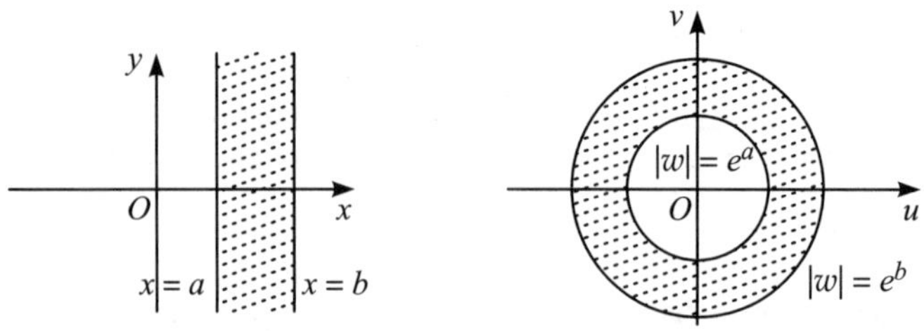

Figure 5.27 *Contour deformation*

Solution. (a) The integrand $\frac{1}{z^2-1}$ fails to be analytic at $z = \pm 1$ Now, we see that the point $z = 1$ lies outside Γ_1. We see that without passing through these points, Γ_1 can be continuously deformed into a small positively-oriented circle γ_r around $z = -1$. Using partial fraction expansion, we find

$$\begin{aligned}\int_{\Gamma} \frac{dz}{z^2-1} &= \int_{\gamma_r} \frac{dz}{z^2-1} = \int_{\gamma_r} \left[\frac{1}{2(z-1)} - \frac{1}{2(z+1)}\right] dz \\ &= 0 - \frac{1}{2} \cdot 2\pi i = -\pi i.\end{aligned}$$

(b) The integrand $e^z/(z^2-9)$ is analytic everywhere except at $z = \pm 3$, where the denominator vanishes. From Fig. 5.27(b), we immediately see that the contour can be shrunk to a point in the domain of analyticity and thus the value of the integral is zero.

Example 5.5.5. *Compute the integral $\int_{\Gamma} \frac{5z-2}{z^2-z}\, dz$, where the closed contour Γ is as depicted in the Fig. 5.28(a).*

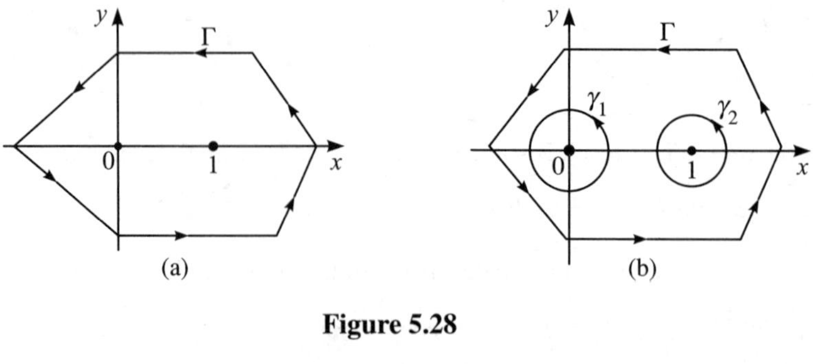

Figure 5.28

Solution. (a) The integrand $\frac{5z-2}{z^2-z}$ fails to be analytic at the points $z = 0$ and $z = 1$. Both these points lie inside Γ. Let γ_1 and γ_2 be two small positively-oriented circles enclosing these points as depicted in the Fig. 5.28(b). Using partial fraction expansion, we find

$$\begin{aligned}\int_{\Gamma} \frac{5z-2}{z^2-z}\,dz &= \int_{\gamma_1} \frac{5z-2}{z^2-z}\,dz + \int_{\gamma_2} \frac{5z-2}{z^2-z}\,dz \\ &= \int_{\gamma_1} \left[\frac{2}{z} + \frac{3}{z-1}\right]dz + \int_{\gamma_2} \left[\frac{2}{z} + \frac{3}{z-1}\right]dz \\ &= 2 \cdot 2\pi i + 0 + 0 + 3 \cdot 2\pi i = 10\pi i.\end{aligned}$$

5.5.8 Homotopy Version of Cauchy's Theorem

In this section we give a more thorough version of Cauchy's theorem. We now link Cauchy's theorem to curve homotopy.

RESULT **42.** [Cauchy's theorem in homotopy version]: Let $f : G \to \mathbb{C}$ be analytic on an open set $G \subset \mathbb{C}$ and let $\gamma_0, \gamma_1 : [a, b] \to G$ be two closed rectifiable curves in G which are homotopic in G. Then

$$\int_{\gamma_0} f(z)\ dz = \int_{\gamma_1} f(z)\ dz.$$

RESULT **43.** [Cauchy's theorem in winding number version]: If $f : G \to \mathbb{C}$ is analytic on an open set $G \subset \mathbb{C}$ and if Γ is a chain in G with the property that $\text{Ind}(\Gamma; w) = 0$, $\forall w \in \mathbb{C}/G$, then

$$\frac{1}{2\pi i} \oint_{\Gamma} \frac{f(z)}{z - z_0}\ dz = 0, \ \ \forall z_0 \in G/\Gamma. \tag{5.28}$$

Theorem 5.14. [Morera's theorem]: *Let $\mathcal{D}$ be a domain and let $f : \mathcal{D} \to \mathbb{C}$ be continuous function. Let $\oint_{\gamma} f(z)\ dz = 0$, where γ is any rectifiable closed Jordan curve in $\mathcal{D}$. Then f is analytic in $\mathcal{D}$.*

Proof: Let z_0 be a fixed point and z any variable point in the domain $\mathcal{D}$. Then, by the antiderivative theorem 5.3, the value of the integral $\int_{z_0}^{z} f(z)dz$ is independent of the curve joining z_0, z. Let us write $F(z) = \int_{z_0}^{z} f(\zeta)d\zeta$. Let $z + h$ be any point in $\mathcal{D}$ near the point z. Then,

$$F(z+h) - F(z) = \int_{z_0}^{z+h} f(\zeta)d\zeta - \int_{z_0}^{z} f(\zeta)d\zeta = \int_{z}^{z+h} f(\zeta)d\zeta. \tag{5.29}$$

Now, as the integral (5.29) is path-independent and may be taken along the straight line segment joining z to $z + h$, hence

$$\begin{aligned}\frac{F(z+h) - F(z)}{h} - f(z) &= \frac{1}{h}\int_{z}^{z+h} f(\zeta)d\zeta - f(z) \\ &= \frac{1}{h}\Big[\int_{z}^{z+h} f(\zeta)d\zeta - f(z)\int_{z}^{z+h} d\zeta\Big] \\ &= \frac{1}{h}\int_{z}^{z+h} \Big[f(\zeta) - f(z)\Big]d\zeta.\end{aligned} \tag{5.30}$$

Now, given that $f(\zeta)$ is continuous in $\mathcal{D}$, it is continuous at z and thus by definition, corresponding to an $\varepsilon > 0$, $\exists$ a $\delta > 0$ such that

$$\Big|f(\zeta) - f(z)\Big| < \varepsilon \text{ whenever } |\zeta - z| < \delta. \tag{5.31}$$

We now choose h such that $|h| < \delta$. Then the inequality (5.31) is satisfied for every point ζ on the line segment joining z to $z+h$. Hence from (5.30) and (5.31) we get,

$$\begin{aligned}\left|\frac{F(z+h)-F(z)}{h} - f(z)\right| &\leq \frac{1}{|h|}\int_z^{z+h} [f(\zeta)-f(z)]|d\zeta| \\ &< \frac{1}{|h|}\int_z^{z+h} \varepsilon|d\zeta| = \varepsilon \text{ whenever } |\zeta - z| < \delta.\end{aligned}$$

Since ε is arbitrary, we get,

$$\lim_{h\to 0}\frac{F(z+h)-F(z)}{h} = f(z).$$

Hence $F'(z)$ exists and $F'(z) = f(z)$. Since the inequality (5.31) holds for every $z \in \mathcal{D}$, so $F(z)$ possesses derivative $f(z)$ at every point $z \in \mathcal{D}$ and consequently, $F(z)$ is analytic in $\mathcal{D}$. Since the derivative of an analytic function is also analytic, it follows that the function $f(z)$ is also analytic in $\mathcal{D}$. □

Note: (i) The Cauchy theorem gives a necessary condition that the function f is analytic in a simply-connected domain $\mathcal{D}$.
(ii) Morera's theorem 5.14 gives the sufficient condition that a continuous function f in a domain $\mathcal{D}$ is analytic in that domain. In this sense, this theorem is the *converse of Cauchy's theorem.*
(iii) Consider $f(z) = \int_0^1 e^{-z^2\zeta}\, d\zeta = \frac{1-e^{-z^2}}{z^2}; z \neq 0$. Let γ be any simple closed contour in the complex plane $\mathbb{C}$, then

$$\int_\gamma f(z)\, dz = \int_\gamma \Big[\int_0^1 e^{-z^2\zeta}\, d\zeta\Big] dz = \int_0^1 \Big[\int_\gamma e^{-z^2\zeta}\, dz\Big]\, d\zeta = 0.$$

Hence, in the view of Morera's theorem 5.14, f is analytic.

5.6 Cauchy's Integral Formula

Theorem 5.15. *Let Γ be a simple, closed, positively-oriented contour within a simply-connected domain $\mathcal{D}$. Let $f : \mathcal{D} \to \mathbb{C}$ be an analytic function. If z_0 be any point inside Γ, then*

$$f(z_0) = \frac{1}{2\pi i}\oint_\Gamma \frac{f(z)}{z - z_0}\, dz. \tag{5.32}$$

Proof: Since $f(z)$ is analytic in $\mathcal{D}$, it is analytic at z_0 and therefore, it is derivable and continuous there. About the point $z = z_0$, let us describe a positively-oriented small circle γ of radius r, defined by $\gamma : z - z_0 = re^{i\theta}$, where $\theta \in [0, 2\pi]$ (Fig.5.29). Consider

$$\psi(z) = \frac{f(z)}{z - z_0} \text{ for } z \in \mathcal{D} \text{ and } z \neq z_0$$

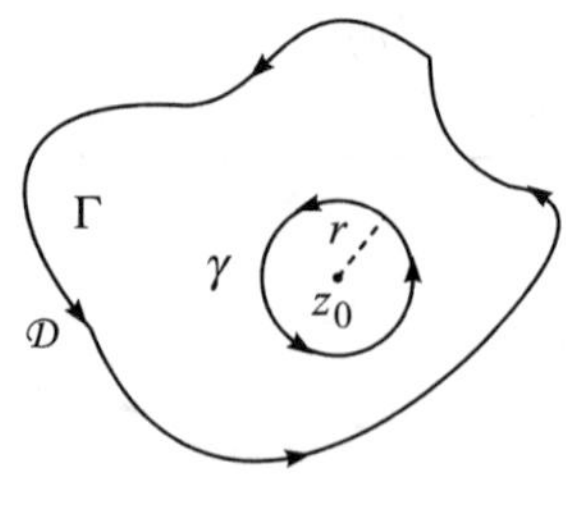

Figure 5.29 *Cauchy's integral formula*

which is analytic in the doubly-connected domain bounded externally by Γ and internally by γ. Hence, by the principle of deformation of contours, we get

$$\oint_\Gamma \psi(z)\ dz = \oint_\gamma \psi(z)\ dz$$

$$\therefore \oint_\Gamma \psi(z)dz - \oint_\Gamma \frac{f(z)}{z-z_0}dz = \int_0^{2\pi} \frac{f(z_0+re^{i\theta})}{re^{i\theta}} ire^{i\theta}d\theta$$

$$= i\int_0^{2\pi} f(z_0+re^{i\theta})d\theta. \qquad (5.33)$$

It is evident that the integral on the right-hand side of Eq.(5.33) is independent of r and hence the value of the integral does not change if we allow $r \to 0$. Taking the limit $r \to 0$ of both sides, and making use of the continuity of $f(z)$, we have,

$$\oint_\Gamma \psi(z)dz = \lim_{r\to 0} i\int_0^{2\pi} f(z_0+re^{i\theta})d\theta$$

$$= i\int_0^{2\pi} \lim_{r\to 0} f(z_0+re^{i\theta})d\theta = 2\pi i f(z_0)$$

$$\text{or, } f(z_0) = \frac{1}{2\pi i}\oint_\Gamma \psi(z)\ dz = \frac{1}{2\pi i}\oint_\Gamma \frac{f(z)}{z-z_0}dz.$$

This is known as *Cauchy's integral formula.* □

The formula (5.32) tells us that if a function f is to be analytic within and on a simple closed contour Γ, the values of f interior to Γ are completely determined by the values of f on Γ.

Example 5.6.1. *Compute the integrals (a)* $\int_{\Gamma_1} \frac{\cos z+\sin z}{z^2-9}dz$ *and (b)* $\int_{\Gamma_2} \frac{e^z}{z(z-2)}dz$, *where the contours* Γ_1 *and* Γ_2 *are given in Fig. 5.30.*

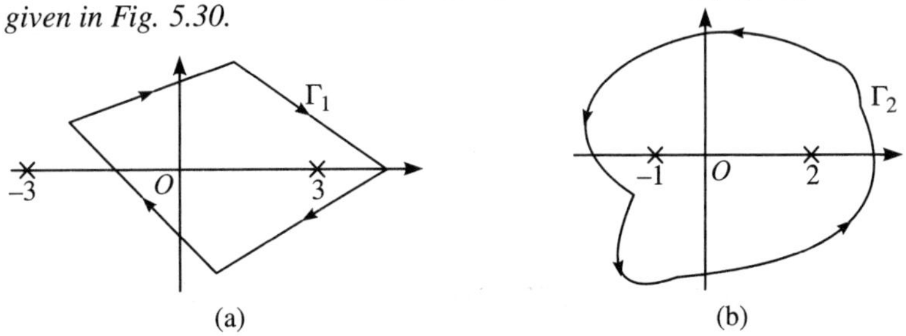

Figure 5.30 *Contours of Example 5.6.1*

Solution. (a) The integrand fails to be analytic at points $z = \pm 3$. However, only $z = 3$ lies inside Γ_1 (Fig. 5.30(a)). Using Cauchy's integral formula (5.32), we get

$$\int_{\Gamma_1} \frac{\cos z+\sin z}{z^2-9}dz = 2\pi i \times \frac{1}{2\pi i}\int_{\Gamma_1} \frac{(\cos z+\sin z)/(z+3)}{z-3}dz$$

$$= 2\pi i\frac{\cos 3+\sin 3}{3+3} = \frac{\pi i}{3}\Big[\cos 3+\sin 3\Big].$$

(b) The integrand fails to be analytic at points $z = 0$ and $z = 2$ and both of them lie inside Γ_2 (Fig. 5.30(b)). Using, Cauchy's integral formula (5.32), we get

$$\int_{\Gamma_2} \frac{e^z}{z(z-2)}dz = 2\pi i\Big[\frac{1}{2\pi i}\int_{\Gamma_2} \frac{e^z/2}{z-2}dz - \frac{1}{2\pi i}\int_{\Gamma_2} \frac{e^z/2}{z}dz$$

$$= 2\pi i\Big[\frac{e^2}{2} - \frac{e^0}{2}\Big] = \pi i(e^2-1).$$

RESULT **44.** [Cauchy's formula in winding number version] :

(i) If $f : G \to \mathbb{C}$ is an analytic function on an open set $G \subset \mathbb{C}$. If Γ is a closed, rectifiable curve in G such that $\text{Ind}(\Gamma;\zeta) = 0, \forall \zeta \in \mathbb{C}/G$, then

$$\frac{1}{2\pi i}\oint_\Gamma \frac{f(z)}{z-z_0}\,dz = W_N(\Gamma;z_0)\,f(z_0) = \text{Ind}(\Gamma;z_0)\,f(z_0), \tag{5.34}$$

where $\forall z_0 \in G/\Gamma$.

(ii) If $\Gamma_1, \Gamma_2, \cdots, \Gamma_m$ are closed, rectifiable curves in G such that $\text{Ind}(\Gamma_1;\zeta) + \cdots + \text{Ind}(\Gamma_m;\zeta) = 0$; $\zeta \in \mathbb{C}/\Gamma$, then for $\forall z_0 \in G/\Gamma$

$$f(z_0)\sum_{k=1}^{m}\text{Ind}(\Gamma_k;\zeta_0) = \sum_{k=1}^{m}\frac{1}{2\pi i}\int_{\Gamma_k}\frac{f(z)}{z-z_0}dz.$$

This is a more general version of Cauchy's integral formula that involves more than one curve, which is needed, for example in dealing with an annulus where one needs a formula involving two curves.

Example 5.6.2. *Using Cauchy's integral formula, evaluate the integral*

$$\textit{(i)}\int_\Gamma \frac{e^{2z}}{(z-1)(z-2)}\,dz, \text{ where } \Gamma : |z| = 3;\ \textit{(ii)}\int_\Gamma \frac{\cosh(\pi z)}{z(z^2+1)}\,dz; \Gamma : |z| = 2.$$

Solution. (i) The function $f(z) = e^{2z}$ is analytic within the circle $\Gamma : |z| = 3$ and the singular points $z_0 = 1$ and $z_0 = 2$ lie inside Γ. Therefore,

$$\begin{aligned}
\int_\Gamma \frac{e^{2z}}{(z-1)(z-2)}\,dz &= \int_\Gamma e^{2z}\left[\frac{1}{z-2} - \frac{1}{z-1}\right]dz \\
&= 2\pi i \times \left[\frac{1}{2\pi i}\int_\Gamma \frac{e^{2z}}{z-2}\,dz - \frac{1}{2\pi i}\int_\Gamma \frac{e^{2z}}{z-1}\,dz\right] \\
&= 2\pi i[f(2) - f(1)], \text{ using Eq. (5.32)} \\
&= 2\pi i e^4 - 2\pi i e^2 = 2\pi i(e^4 - e^2).
\end{aligned}$$

(ii) The function $f(z) = \cosh(\pi z) = \cos(i\pi z)$ is analytic within the circle $\Gamma : |z| = 2$ and the singular points $z_0 = 0$, $z_0 = i$ and $z_0 = -i$ lie inside Γ. Therefore,

$$\begin{aligned}
\int_\Gamma \frac{\cosh(\pi z)}{z(z^2+1)}\,dz &= \int_\Gamma \left[\frac{1}{z} + \frac{-1/2}{z-i} + \frac{-1/2}{z+i}\right]\cosh(\pi z)\,dz \\
&= 2\pi i \times \left[\frac{1}{2\pi i}\int_\Gamma \frac{\cos(i\pi z)}{z} - \frac{1}{2}\frac{1}{2\pi i}\int_\Gamma \frac{\cos(i\pi z)}{z-i} - \frac{1}{2}\frac{1}{2\pi i}\int_\Gamma \frac{\cos(i\pi z)}{z+i}\right] \\
&= 2\pi i\Big[f(0) - \frac{1}{2}f(i) - \frac{1}{2}f(-i)\Big], \text{ using Eq. 5.32} \\
&= 2\pi i\Big[\cos 0 - \frac{1}{2}\cos(i^2\pi) - \frac{1}{2}\cos^2(-i^2\pi)\Big] \\
&= 2\pi i\Big[1 + \frac{1}{2} + \frac{1}{2}\Big] = 4\pi i.
\end{aligned}$$

Theorem 5.16. [Cauchy's integral formula for derivatives]: *Let $f : \mathcal{D} \to \mathbb{C}$ be analytic within and on a closed contour Γ of a simply-connected domain $\mathcal{D}$ and z_0 be any interior point in $\mathcal{D}$. Then $f^{(n)}(z)$ is also analytic and*

$$f^{(n)}(z_0) = \frac{n!}{2\pi i}\oint_\Gamma \frac{f(z)}{(z-z_0)^{n+1}}dz \ : \ n = 1, 2, 3, \cdots \tag{5.35}$$

where $f^{(n)}(z_0)$ is the nth derivative of $f(z)$ at $z = z_0$.

Proof: Let $z_0 + h$ be the neighbouring point of z_0 within Γ. Then by Cauchy's integral formula,

$$f(z_0+h) = \frac{1}{2\pi i}\oint_\Gamma \frac{f(z)}{z-z_0-h}dz; \quad f(z_0) = \frac{1}{2\pi i}\oint_\Gamma \frac{f(z)}{z-z_0}dz$$

and so

$$\begin{aligned}
\frac{f(z_0+h)-f(z_0)}{h} &= \frac{1}{2\pi i}\oint_\Gamma \frac{1}{h}\left[\frac{1}{z-z_0-h} - \frac{1}{z-z_0}\right] f(z)dz \\
&= \frac{1}{2\pi i}\oint_\Gamma \frac{f(z)}{(z-z_0-h)(z-z_0)}dz \\
&= \frac{1}{2\pi i}\oint_\Gamma \frac{f(z)dz}{(z-z_0)^2} + \frac{h}{2\pi i}\oint_\Gamma \frac{f(z)dz}{(z-z_0-h)(z-z_0)^2}.
\end{aligned} \tag{5.36}$$

Let us consider a small circle $\gamma : |z - z_0| = r$, lying entirely in Γ. Then by principle of deformation of contours, we have

$$\frac{h}{2\pi i}\oint_\Gamma \frac{f(z)dz}{(z-z_0-h)(z-z_0)^2} = \frac{h}{2\pi i}\oint_\gamma \frac{f(z)dz}{(z-z_0-h)(z-z_0)^2}.$$

We now choose h in such a way that the point $z_0 + h$ lies within γ and that $|h| < r/2$. Since $\gamma : |z - z_0| = r$, for any point z on γ, we have

$$|z-z_0-h| \geq |z-z_0| - |h| \geq r - \frac{r}{2} = \frac{r}{2}.$$

Since $f(z)$ is analytic in $\mathcal{D}$, it is bounded in $\mathcal{D}$, so that we can find a positive constant M such that $|f(z)| \leq M$. Using these facts, we get, from Eq. (5.36),

$$\begin{aligned}
\left|\frac{f(z_0+h)-f(z_0)}{h} - \frac{1}{2\pi i}\oint_\gamma \frac{f(z)dz}{(z-z_0)^2}\right| &= \left|\frac{h}{2\pi i}\oint_\gamma \frac{f(z)dz}{(z-z_0)^2(z-z_0-h)}\right| \\
&\leq \frac{|h|}{2\pi}\oint_\gamma \frac{|f(z)|}{|z-z_0|^2|z-z_0-h|}|dz| \\
&\leq \frac{|h|}{2\pi}\frac{M}{r^2\cdot(r/2)}\cdot 2\pi r = \frac{2|h|M}{r^2}
\end{aligned}$$

whenever $|h| < r/2$. Therefore, by definition,

$$\lim_{h\to 0}\frac{f(z_0+h)-f(z_0)}{h} = \frac{1}{2\pi i}\oint_\Gamma \frac{f(z)dz}{(z-z_0)^2}$$

$$\Rightarrow \quad f'(z_0) = \frac{1}{2\pi i}\oint_\Gamma \frac{f(z)dz}{(z-z_0)^2}.$$

Since z_0 is arbitrary, this implies that $f'(z)$ is analytic for all $z \in \mathcal{D}/\Gamma$. Therefore, the formula (5.35) is true for $n = 1$. To prove the general result we must proceed by induction on n. Suppose the result is true for $n = m - 1$, $m \geq 2$. Then

$$\begin{aligned} f^{(m-1)}(z_0+h) - f^{(m-1)}(z_0) &= \frac{(m-1)!}{2\pi i}\oint_\Gamma \Big[\frac{1}{(z-z_0-h)^m} - \frac{1}{(z-z_0)^m}\Big] f(z)dz \\ &= \frac{(m-1)!}{2\pi i}\oint_\Gamma \Big[m\int_{[z_0,z_0+h]} \frac{du}{(z-u)^{m+1}}\Big] f(z)dz \\ &\because K \int\limits_{[z_0,z_0+h]} \frac{du}{(z-u)^{m+1}} = \frac{1}{(z-z_0-h)^m} - \frac{1}{(z-z_0)^m} \\ &= \frac{m!}{2\pi i}\oint_\Gamma \Big[\int_{[z_0,z_0+h]} \frac{du}{(z-u)^{m+1}}\Big] f(z)dz. \end{aligned}$$

Therefore,

$$\begin{aligned} &\frac{f^{(m-1)}(z_0+h) - f^{(m-1)}(z_0)}{h} - \frac{m!}{2\pi i}\oint_\Gamma \frac{f(z)dz}{(z-z_0)^{m+1}} \\ &= \frac{m!}{2\pi i}\oint_\Gamma \Big[\frac{1}{h}\int_{[z_0,z_0+h]} \frac{du}{(z-u)^{m+1}} - \frac{1}{(z-z_0)^{m+1}}\Big] f(z)dz \\ &= \frac{m!}{2\pi i}\oint_\Gamma \frac{1}{h}\Big[\int_{[z_0,z_0+h]} \Big\{\frac{1}{(z-u)^{m+1}} - \frac{1}{(z-z_0)^{m+1}}\Big\} du\Big] f(z)dz \\ &= \frac{m!}{2\pi i\cdot h}\oint_\Gamma \Big[\int_{[z_0,z_0+h]} (m+1)\Big\{\int_{[z_0,u]} \frac{dv}{(z-v)^{m+2}}\Big\} du\Big] f(z)dz \\ &= \frac{(m+1)!}{2\pi i\cdot h}\oint_\Gamma \Big[\int_{[z_0,z_0+h]} \Big\{\frac{dv}{(z-v)^{m+2}}\Big\} du\Big] f(z)dz. \end{aligned}$$

Also, we can write

$$\begin{aligned} &\Big|\frac{f^{(m-1)}(z_0+h) - f^{(m-1)}(z_0)}{h} - \frac{m!}{2\pi i}\oint_\Gamma \frac{f(z)dz}{(z-z_0)^{m+1}}\Big| \\ &= \Big|\frac{(m+1)!}{2\pi i\cdot h}\oint_\Gamma \Big[\int_{[z_0,z_0+h]} \Big\{\frac{dv}{(z-v)^{m+2}}\Big\} du\Big] f(z)dz\Big|, \end{aligned} \tag{5.37}$$

where $\gamma : |z - z_0| = r$ lies entirely within Γ. Again, $v \in [z_0, u] \subset [z_0, z_0 + h]$ and so for all z on γ, we have

$$|z - v| \geq |z - z_0| - |v - z_0| \geq |z - z_0| - |h| = r - r/2 = r/2.$$

Hence from Eq. (5.37), we have

$$\begin{aligned} &\Big|\frac{f^{(m-1)}(z_0+h) - f^{(m-1)}(z_0)}{h} - \frac{m!}{2\pi i}\oint_\Gamma \frac{f(z)dz}{(z-z_0)^{m+1}}\Big| \\ &= \Big|\frac{(m+1)!}{2\pi|h|}\frac{M|h|\cdot|h|}{(r/2)^{m+2}}\cdot 2\pi r = \frac{2^{m+2}(m+1)!M|h|}{r^{m+1}} \end{aligned}$$

whenever $|h| < r/2$. Hence by definition

$$\lim_{h\to 0}\frac{f^{(m-1)}(z_0+h) - f^{(m-1)}(z_0)}{h} = \frac{m!}{2\pi i}\oint_\Gamma \frac{f(z)dz}{(z-z_0)^{m+1}}$$

$$\Rightarrow \quad f^{(m)}(z_0) = \frac{m!}{2\pi i}\oint_\Gamma \frac{f(z)dz}{(z-z_0)^{m+1}}.$$

Thus the result is true for $n = m$ if it is true for $n = m - 1$. Therefore, by mathematical induction, the result is true for all values of $n \in \mathbb{N}$. Hence $f^{(n)}(z)$ exists at $z = z_0$, which is arbitrary. Thus $f^{(n)}(z)$ exist for all $z \in \mathcal{D}/\Gamma$. This completes the proof. □

Example 5.6.3. *Evaluate* $\int_\Gamma \frac{z^3}{(2z+i)^3}dz$, *where* Γ *is the unit circle.*

Solution. Let $f(z) = z^3$. Then $f'(z) = 3z^2$ and $f''(z) = 6z$. Also, the point $z = -i/2$ lies inside Γ. Hence using Eq. (5.35) we get

$$\begin{aligned}\int_\Gamma \frac{z^3}{(2z+i)^3}dz &= \frac{1}{8}\int_\Gamma \frac{z^3}{(z+i/2)^3}dz = \frac{1}{8}\frac{2\pi i}{2!}\frac{2!}{2\pi i}\int_\gamma \frac{f(z)dz}{(z+i/2)^3}\\ &= \frac{1}{8}\frac{2\pi i}{2!}f''(-i/2) = \frac{2\pi i}{16}(-3i) = \frac{3\pi}{8}.\end{aligned}$$

Example 5.6.4. *Evaluate* $\int_\Gamma \frac{e^z + z\sinh z}{(z-\pi i)^2}dz$, *where* Γ *is the circle* $|z| = 4$.

Solution. Let $f(z) = e^z + z\sinh z$. Then $f'(z) = e^z + z\cosh z + \sinh z$. Also, the point $z = \pi i$ lies inside γ. Hence using Eq. (5.35), we get

$$\begin{aligned}\int_\Gamma \frac{e^z + z\sinh z}{(z-\pi i)^2}dz &= 2\pi i \times \frac{1!}{2\pi i}\int_\Gamma \frac{f(z)}{(z-\pi i)^2}dz\\ &= 2\pi i f'(\pi i) = 2\pi i\Big[e^{\pi i} + \pi i\cosh(\pi i) + \sinh(\pi i)\Big]\\ &= 2\pi i(-1-\pi i) = -2\pi i(1+\pi i).\end{aligned}$$

Example 5.6.5. *If* $f(\xi) = \displaystyle\int_\Gamma \frac{4z^2+z+5}{z-\xi}\,dz$, *where* Γ *is the ellipse* $\frac{x^2}{4} + \frac{y^2}{9} = 1$, *find* $f(i)$, $f'(-1)$ *and* $f''(-i)$.

Solution. Let $g(z) = 4z^2 + z + 5$. Since $g(z)$ is analytic within Γ and the singular points $\xi = i, -1, -i$ all lie inside Γ, by Cauchy's integral formula (5.32), we get

$$\begin{aligned}& f(\xi) = \frac{1}{2\pi i}\int_\Gamma \frac{f(z)}{z-\xi}\,dz\\ \therefore\quad & \int_\Gamma \frac{4z^2+z+5}{z-\xi}\,dz = 2\pi i(4\xi^2+\xi+5)\\ \Rightarrow\quad & f(\xi) = 2\pi i(4\xi^2+\xi+5)\\ \Rightarrow\quad & f'(\xi) = 2\pi i(8\xi+1) \text{ and } f''(\xi) = 16\pi i.\end{aligned}$$

Thus, $f(i) = 2\pi(i-1)$, $f'(-1) = -14\pi i$ and $f''(-i) = 16\pi i$.

Theorem 5.17. [Cauchy's estimate] *Let* f *be analytic within a positively-oriented circle* $\Gamma = \{z \in \mathbb{C} : |z - z_0| = r\}$. *Then the derivatives of* f *at* z_0 *satisfy*

$$\left|f^{(n)}(z_0)\right| \le \frac{M \cdot n!}{r^n};\quad n = 0, 1, 2, \cdots \tag{5.38}$$

where M *is a constant such that* $M = \sup_{z\in\Gamma}|f(z)|$.

Proof: Since Γ is closed and bounded, it is compact. Hence there exists some $M \in \mathbb{R}^+$ such that $|f(z)| \le M$ on Γ. By using Cauchy's integral formula (5.35),

$$|f^{(n)}(z_0)| \le \frac{n!}{2\pi}\oint \frac{|f(z)|}{|z-z_0|^{n+1}}|dz|.$$

Now, parametrizing the equation of the circle Γ as $z - z_0 = re^{it} : 0 \le t < 2\pi$,

$$\begin{aligned}|f^{(n)}(z_0)| &\le \frac{n!}{2\pi}\oint_\Gamma \frac{|f(z)|}{|z-z_0|^{n+1}}|dz| \\ &\le \frac{n!}{2\pi}\int_0^{2\pi}\frac{M\cdot r}{r^{n+1}}dt = \frac{M\cdot n!}{r^n}.\end{aligned}$$

This proves *Cauchy's inequality* that gives the maximum upper bound of $|f^{(n)}(z_0)|$. □

Example 5.6.6. *Let the function $f(z)$ be analytic within and on $\Gamma : |z-2| = 3$, taken in the positive sense. If $\max|f(z)| = 2$ on Γ, find the upper bound of $|f^{(4)}(2)|$.*

Solution. In the Cauchy's inequality (5.38), let us take $n = 4$, $z_0 = 2$, $M = 2$ and $r = 3$. We get

$$|f^{(4)}(2)| \le \frac{4!\,2}{3^4} = \frac{16}{27}.$$

Example 5.6.7. *Let $f(z)$ be analytic on the unit disc and suppose that $|f(z)| \le \frac{1}{1-|z|}$ for each $z \in \mathcal{D}$. Show that $|f^{(n)}(0)| \le (n+1)!\cdot e$*

Solution. Let γ be a positively-oriented circle $|z| = 1 - \frac{1}{k}$. Then on γ,

$$\left|f(z)\right| \le \frac{1}{1-\left(1-\frac{1}{k}\right)} = k.$$

Using Cauchy's integral formula (5.35) for derivatives, we get

$$f^{(n)}(0) = \frac{n!}{2\pi i}\oint_\gamma \frac{f(z)}{z^{n+1}}dz$$

$$\therefore \quad \left|f^{(n)}(0)\right| \le \frac{n!}{2\pi}\oint_\gamma \frac{|f(z)|}{|z|^{n+1}}|dz| = \frac{n!}{2\pi}\oint_\gamma \frac{|f(z)|}{\left(1-\frac{1}{k}\right)^{n+1}}|dz|$$

$$\le \frac{n!}{2\pi}\int_0^{2\pi}\frac{k}{\left(1-\frac{1}{k}\right)^{n+1}}\left(1-\frac{1}{k}\right)d\theta = \frac{n!}{2\pi}\int_0^{2\pi}\left(\frac{k}{k-1}\right)^n k\,d\theta.$$

Let us take $k = n+1$. Then

$$\begin{aligned}\left|f^{(n)}(0)\right| &\le \frac{(n+1)!}{2\pi}\int_0^{2\pi}\left(\frac{n+1}{n}\right)^n d\theta \\ &\le \frac{(n+1)!}{2\pi}\left(1+\frac{1}{n}\right)^n\cdot 2\pi \le (n+1)!\cdot e\end{aligned}$$

Theorem 5.18. [Liouville's theorem]: *A bounded entire function in the complex plane $\mathbb{C}$ is constant throughout $\mathbb{C}$.*

Proof: Suppose f is analytic and bounded by some number M over the whole complex plane $\mathbb{C}$, i.e., $\exists M \in \mathbb{R}^+$ such that $|f(z)| \le M\ \forall z \in \mathbb{C}$. Let z_1 and z_2 be any two points in the z-plane. Take the contour $\Gamma = \{z \in \mathbb{C} : |z - z_1| = r\}$ to be a large circle (Fig. 5.31) such that z_2 is interior to Γ. Then, by Cauchy's integral formula (5.35),

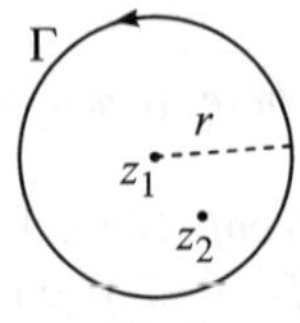

Figure 5.31

$$f(z_2) - f(z_1) = \frac{1}{2\pi i}\oint_\Gamma \frac{f(z)dz}{z-z_2} - \frac{1}{2\pi i}\oint_\Gamma \frac{f(z)dz}{z-z_1}$$

$$= \frac{z_1 - z_2}{2\pi i} \oint_{\Gamma} \frac{f(z)dz}{(z-z_1)(z-z_2)}. \tag{5.39}$$

Also, from the equation of Γ, we have,

$$\begin{aligned} |z - z_2| &= |z - z_1 + z_1 - z_2| \geq |z - z_1| - |z_1 - z_2| \\ &\geq r - |z_1 - z_2| \geq r - \tfrac{r}{2} = \tfrac{r}{2} \end{aligned}$$

if we choose r so large that $|z_1 - z_2| < r/2$. From Eq. (5.39) we get,

$$\begin{aligned} \left|f(z_2) - f(z_1)\right| &= \frac{|z_1 - z_2|}{2\pi} \left| \oint_{\Gamma} \frac{f(z)dz}{(z-z_1)(z-z_2)} \right| \\ &\leq \frac{|z_1 - z_2|}{2\pi} \oint_{\Gamma} \frac{|f(z)|}{|z-z_1| \cdot |z-z_2|} dz \\ &\leq \frac{|z_1 - z_2|}{2\pi} \oint_{\Gamma} \frac{M}{r \cdot \frac{r}{2}} dz = \frac{M|z_2 - z_1|}{\pi r^2} \cdot 2\pi r \\ &\leq \frac{2|z_2 - z_1|M}{r}. \end{aligned}$$

Letting $r \to \infty$, we see that the right-hand side tends to zero and consequently

$$f(z_1) - f(z_2) = 0, \qquad \Rightarrow f(z_1) = f(z_2). \tag{5.40}$$

Since, Eq. (5.40) holds for all values of z_1 and z_2, therefore $f(z)$ is constant. Therefore the only bounded entire functions are the constant functions. □

(i) Let us consider a bounded entire function f in the whole complex plane $\mathbb{C}$ so that f is analytic in $\mathbb{C}$. By taking $n = 1$ in Cauchy's inequality (5.38), we get

$$|f'(z_0)| \leq \frac{M}{r}, \text{ where } |f(z)| \leq M$$

r being arbitrarily large. Let $r \to \infty$. We have

$$|f'(z_0)| = 0 \Rightarrow f'(z_0) = 0, \quad \text{for any } z_0.$$

Hence, f' vanishes everywhere, i.e., f must be a constant. □

(ii) Every non-constant entire function must be unbounded. There is no non-constant meromorphic function which is analytic on the Riemann sphere.

(iii) Let the function $f(z)$ be entire and $u(x, y) = \text{Re}[f(z)]$ be bounded as $|z| \to \infty$. Let $|u(x, y)| \leq u_0 \; \forall (x, y) \in \mathbb{C}$. Now the function $g(z) = e^{f(z)}$ is entire and

$$|g(z)| = e^{u(x,y)} \leq e^{u_0},$$

i.e., $|g(z)|$ is bounded as $|z| \to \infty$. Thus, by Liouville's theorem, $g(z)$ is constant, i.e. $e^{f(z)}$ is constant $\Rightarrow f(z)$ is constant.

(iv) Let $f = u + iv$ be entire and $u^2 \leq v^2 + 2017$ on $\mathbb{C}$. Now, $g(z) = e^{f^2(z)}$ is entire and

$$\begin{aligned} |g(z)| &= \left|e^{f^2(z)}\right| = \left|e^{(u+iv)^2}\right| = \left|e^{(u^2-v^2)+2uvi}\right| \\ &= e^{u^2-v^2}|\cos(2uv) + i\sin(2uv)| = e^{u^2-v^2} \text{ on } \mathbb{C}. \end{aligned}$$

By Liouville's theorem, g and hence f is constant.

(v) Let f be entire such that f does not assume values in a disc, say Γ, for some $c \in \mathbb{C}$ and $M > 0$. Then

$$|f(z) - c| > M \ \ \forall z \in \mathbb{C}.$$

Hence by Liouville's theorem f is constant.

(vi) Let us consider a non-constant polynomial with complex coefficients as

$$P(z) = a_0 z^n + a_1 z^{n-1} + a_2 z^{n-2} + \cdots + a_n,$$

where $n \geq 1$ and $a_0 \neq 0$. Suppose $P(z)$ has no zero in $\mathbb{C}$, i.e., $P(z) \neq 0 \ \forall z \in \mathbb{C}$. Then the function $f(z) = 1/P(z)$ is defined for all z and analytic on $\mathbb{C}$. On the other hand, writing

$$P(z) = a_0 z^n \left[1 + \frac{b_1}{z} + \cdots + \frac{b_n}{z^n}\right]$$

with appropriate constants $b_1, b_2, \cdots, b_n$, we see that $\lim\limits_{z \to \infty} |P(z)| \longrightarrow \infty$ and hence $\lim\limits_{z \to \infty} |f(z)| \to 0$. For sufficiently $R > 0$,

$$|f(z)| \leq 1 \text{ whenever } z \in \mathcal{S}^c(0, R) : |z| > R.$$

Also, $|f(z)|$ has a maximum on disc $\mathcal{S}(0, R)$ since the disc is compact. Thus, $f(z)$ is bounded entire and therefore, by Liouville's theorem, it is constant. Thus $P(z)$ is constant function and hence it is a polynomial of degree zero, which is a contradiction. Hence $P(z)$ must have at least one zero in $\mathbb{C}$.

This is known as the *fundamental theorem of algebra.*

RESULT **45.** [Generalised version of Liouville's theorem]: An entire function f, which satisfies the inequality

$$|f(z)| \leq M|z|^k \text{ for some } k \geq 0, M > 0,$$

and for all sufficiently large $|z|$, reduces to a polynomial of degree n, where n is the largest integer such that $n \leq k$.

Example 5.6.8. *Let $f \in \mathbb{C} - \{0\}$ be a holomorphic function such that $|f(z)| < a|z|^{1/2} + b|z|^{-1/2}$ for some $a, b > 0$. Show that f is constant.*

Solution. From the given inequality, we see that

$$|zf(z)| < a|z|^{3/2} + b|z|^{1/2} \to 0 \text{ as } z \to 0,$$

so that the meromorphic function $zf(z) \in \mathbb{C} - \{0\}$ has a removable singularity at $z = 0$. Define

$$g(z) = \begin{cases} zf(z) & \text{for } z \neq 0 \\ 0 & \text{for } z = 0. \end{cases}$$

Then $g(z)$ is entire and satisfies the inequality

$$|g(z)| < a|z|^{3/2} + b|z|^{1/2} \leq M|z|^{3/2}$$

for sufficiently large $|z|$. Hence, by Result 45, $g(z)$ is a polynomial of degree 1, say $g(z) = c_1 z + c_2$. But $g(0) = 0$ gives $c_2 = 0$. Therefore, $g(z) = c_1 z$ and consequently $f(z) = c_1$, so that f is constant.

Theorem 5.19. [Gauss's mean-value theorem]: *If $f(z)$ is analytic function within and on a positive-oriented circle $\Gamma = \{z \in \mathbb{C} : |z - z_0| = r\}$, then $f(z_0)$, the average mean of the values of $f(z)$ around Γ, is given by*

$$f(z_0) = \frac{1}{2\pi}\int_0^{2\pi} f(z_0 + re^{i\theta})\, d\theta. \tag{5.41}$$

Proof: Here Γ is a closed contour with winding number 1. By Cauchy's integral formula (5.35), we have

$$\begin{aligned} f(z_0) &= \frac{1}{2\pi i}\int_\Gamma \frac{f(z)}{z - z_0}\, dz = \frac{1}{2\pi i}\int_\Gamma \frac{f(z_0 + re^{i\theta})}{re^{i\theta}}\, rie^{i\theta}\, d\theta \\ &= \frac{1}{2\pi}\int_0^{2\pi} f(z_0 + re^{i\theta})\, d\theta. \end{aligned}$$

This theorem asserts that the value of an analytic function at the centre of a circle is equal to the arithmetic mean of its value on the circle, known as *Gauss's mean-value for analytic functions.* □

DEDUCTION **11.** In particular, if we write the equation of a circle in the form where L is the length of the circumference of the circle, then $s = r\theta$ and s varies from 0 to L.

$$\therefore f(z_0) = \frac{1}{2\pi\, r}\int_0^{L} f(z_0 + re^{i\theta})\, ds = \frac{1}{L}\int_0^{L} f(z)\, ds.$$

Thus the value of the function at the centre is equal to the mean of the value of the function on the circumference.

Example 5.6.9. *Evaluate* $\frac{1}{2\pi}\int_0^{2\pi} \sin^2\left(\frac{\pi}{6} + 2e^{i\theta}\right) d\theta.$

Solution. Here the circle Γ has the centre at $z_0 = \pi/6$ and radius $r = 2$ (comparing with Eq. (5.41)). Thus using Cauchy's integral formula,

$$\begin{aligned} \frac{1}{2\pi}\int_0^{2\pi} \sin^2\left(\frac{\pi}{6} + 2e^{i\theta}\right) d\theta &= \frac{1}{2\pi i}\int_\Gamma \frac{\sin^2\left(\frac{\pi}{6} + 2e^{i\theta}\right)}{2e^{i\theta}} 2ie^{i\theta}\, d\theta \\ &= \frac{1}{2\pi i}\int_\Gamma \frac{f(z)}{z - \pi/6}\, dz, \text{ where } f(z) = \sin^2 z \\ &= f(\pi/6) = \sin^2 \pi/6 = 1/4. \end{aligned}$$

5.7 Modulus Theorems

We often get interested about the concept of maximum and minimum while dealing with functions of real variables. But a similar argument cannot be applied to the functions of complex variables since $\mathbb{C}$ is not a ordered field.

Let $\mathcal{D}$ be a bounded domain in $\mathbb{C}$ with $\partial\mathcal{D}$ being the boundary. Then $\bar{\mathcal{D}} = \mathcal{D} \cup \partial\mathcal{D}$ denotes the closed and bounded domain in the complex plane $\mathbb{C}$.

Theorem 5.20. [Optimum values of moduli of analytic function]: *Let $\mathcal{D} \subset \mathbb{C}$ be a domain and $f : \mathcal{D} \to \mathbb{C}$ is an analytic function such that there is a point $z_0 \in \mathcal{D}$ with*

$$|f(z)| \leq |f(z_0)|;\ \forall z \in \mathcal{D}.$$

Then $f(z)$ has a constant value $f(z_0)$ throughout $\mathcal{D}$.

Proof: Since $z_0 \in \mathcal{D}$ and $\mathcal{D}$ is open, there exists $\varepsilon > 0$ such that $\mathcal{S}(z_0;\varepsilon) \subset \mathcal{D}$. Draw a small circle $\Gamma = \{z : |z - z_0| = \varepsilon\}$ such that $\mathcal{S}(z_0;\varepsilon) \subset \mathcal{D}$. Then f is analytic inside and on the circle Γ. Let the parametric equation of Γ be $z = z_0 + \varepsilon e^{it}; 0 \le t \le 2\pi$. Using Cauchy's integral formula, we get

$$\begin{aligned} f(z_0) &= \frac{1}{2\pi i}\int_\Gamma \frac{f(z)}{z - z_0}dz = \frac{1}{2\pi}\int_0^{2\pi} f(z_0 + \varepsilon e^{it})dt \\ \therefore |f(z_0)| &\le \frac{1}{2\pi}\int_0^{2\pi} \left|f(z_0 + \varepsilon e^{it})\right|dt \\ &\le |f(z_0)|;\ \text{as} \left|f(z_0 + \varepsilon e^{it})\right| \le |f(z_0)|\ \forall t \in [0, 2\pi]. \end{aligned} \tag{5.42}$$

From these inequalities, it is now evident that

$$|f(z_0)| = \frac{1}{2\pi}\int_0^{2\pi} \left|f(z_0 + \varepsilon e^{it})\right|dt$$

$$\text{or,} \quad \int_0^{2\pi} \left[|f(z_0)| - \left|f(z_0 + \varepsilon e^{it})\right|\right]dt = 0.$$

The integrand in the last integral is a continuous function of f, in variable t, and it is also positive by virtue of Eq.(5.42) so that the integral can vanish only if the integrand is identically zero. Thus

$$\left|f(z_0 + \varepsilon e^{it})\right| = |f(z_0)|; \quad 0 \le t \le 2\pi.$$

This shows that $|f(z)| = |f(z_0)|$ for all points z on the circle Γ. It means that the analytic function $f(z)$ in Γ has a constant modulus and therefore, it is constant throughout Γ. Hence by the uniqueness theorem, f is constant on the whole of $\mathcal{D}$. □

For example, let f be analytic, and satisfy the conditions $|f(z)| \le 2$ on $\Gamma : |z| = 1$ and $f(0) = \sqrt{3} + i$. Since $f(0) = \sqrt{3} + i$, so, $|f(0)| = 2$. Thus $|f(z)|$ attains its maximum value 2 at $z = 0$, which is the interior point of $\mathcal{D}$. Therefore, by the above theorem $f(z)$ is constant.

Theorem 5.21. [Maximum modulus theorem/principle]: *Let $f : \mathcal{D} \to \mathbb{C}$ be non-constant analytic function in a region $\mathcal{D}$ and continuous on $\bar{\mathcal{D}}$. Then the maximum value of $|f(z)|$ occurs on $\partial\mathcal{D}$ and never in the interior of $\mathcal{D}$.*

Proof: Since f is continuous on a closed and bounded set, $|f(z)|$ must reach its maximum value M(say) at some point on or within $\Gamma = \partial\mathcal{D}$ (Fig. 5.32). If possible, let $|f(z)|$ have maximum value M at a point z_0 lying within $\Gamma = \partial\mathcal{D}$ and not on Γ. Then

$$|f(z_0)| = M, \text{ where } z_0 \text{ lies within } \Gamma.$$

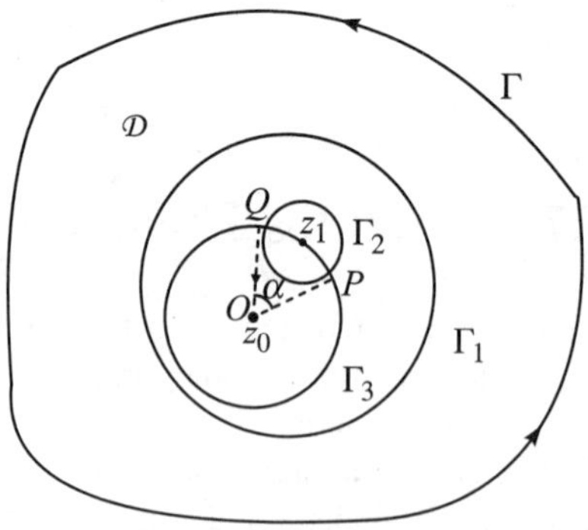

Figure 5.32

Since $\mathcal{D}$ is connected, we can cover it by a sequence of discs. Draw a small circle Γ_1 with its centre at z_0 and lying within Γ. Thus the existence of a maximum value implies that all the values in the domain are the same, and therefore $f(z)$ is constant. If $f(z)$ is not constant inside Γ_1, there must exist a point, say z_1, inside Γ_1 such that $|f(z_1)| < M$. Let

$$|f(z_1)| = M - \varepsilon, \text{ where } \varepsilon > 0.$$

Due to continuity of $|f(z)|$ at z_1, corresponding to an $\varepsilon > 0, \exists\, \delta > 0$ such that

$$\Big||f(z)| - |f(z_1)|\Big| < \frac{1}{2}\varepsilon \text{ whenever } |z - z_1| < \delta$$

or, $$|f(z)| - |f(z_1)| \leq \Big||f(z)| - |f(z_1)|\Big| < \frac{1}{2}\varepsilon \text{ whenever } |z - z_1| < \delta$$

$\therefore$ $$|f(z)| < |f(z_1)| + \frac{1}{2}\varepsilon = M - \varepsilon + \frac{1}{2}\varepsilon = M - \frac{1}{2}\varepsilon \text{ whenever } |z - z_1| < \delta$$

for all points z inside $\Gamma_2 : |z - z_1| = \delta$. Now, construct a circle $\Gamma_3 : |z - z_0| = |z_1 - z_0| = r$, (say). The arc $\widehat{PQ}$ of the circle Γ_3 lies within Γ_2 so that on this arc, $|f(z)| < M - \frac{1}{2}\varepsilon$. On the remaining arc on Γ_2, we have $|f(z)| \leq M$. Let α be measured counterclockwise from OP and let $\angle POQ = \alpha$. By the Cauchy's integral formula, we have

$$\begin{aligned}
f(z_0) &= \frac{1}{2\pi i}\int_{C_1} \frac{f(z)}{z - z_0}dz = \frac{1}{2\pi}\int_0^{2\pi} f(z_0 + re^{i\theta})d\theta;\ z - z_0 = re^{i\theta} \\
&= \frac{1}{2\pi}\int_0^{\alpha} f(z_0 + re^{i\theta})d\theta + \frac{1}{2\pi}\int_{\alpha}^{2\pi} f(z_0 + re^{i\theta})d\theta \\
\text{or, } |f(z_0)| &\leq \frac{1}{2\pi}\int_0^{\alpha} |f(z_0 + re^{i\theta})|d\theta + \frac{1}{2\pi}\int_{\alpha}^{2\pi} |f(z_0 + re^{i\theta})|d\theta \\
&\leq \frac{1}{2\pi}\int_0^{\alpha} (M - \frac{\varepsilon}{2})d\theta + \frac{1}{2\pi}\int_{\alpha}^{2\pi} Md\theta \\
&\leq \frac{\alpha}{2\pi}(M - \frac{\varepsilon}{2}) + \frac{M}{2\pi}(2\pi - \alpha) = M - \frac{\alpha\varepsilon}{4\pi} \\
\text{or, } |f(z_0)| &= M < M - \frac{\alpha\varepsilon}{4\pi}
\end{aligned}$$

which is absurd. This contradiction shows that $|f(z)|$ cannot attain its maximum value at any interior point Γ and so must attains its maximum on Γ. □

(i) Suppose f is analytic in a bounded domain $\mathcal{D}$ and continuous on $\bar{\mathcal{D}}$. Then, each of $\text{Re}f(z)$, $-\text{Re}f(z)$, $\text{Im}f(z)$ and $-\text{Im}f(z)$ attains its maximum at some point on the boundary $\partial\mathcal{D}$ of $\mathcal{D}$.

(ii) Maximum modulus theorem, usual version : The absolute value of a non-constant analytic function on a open connected set $U \subset \mathbb{C}$ cannot have a local maximum point in U.

Example 5.7.1. *Find the maximum of $|f(z)|$ in $\mathcal{S}[0,1] = \{z : |z| \leq 1\}$ for the functions $f(z)$: (a) $z^2 + 2z - 3$; (b) $\cos 3z$; and (c) $(2z + 1)/(2z - 1)$.*

Solution. (a) Let $f(z) = z^2 + 2z - 3$. Being polynomial, it is analytic in the disc $\mathcal{S}(0,1)$ and continuous on the closed disc $\mathcal{S}[0,1]$. By maximum modulus theorem, the maximum value of $|f(z)|$ is attained on the boundary $\partial\mathcal{S}(0,1) = \{z : |z| = 1\}$. Parameterizing the circle by $z(t) = e^{it}, 0 \leq t \leq 2\pi$, we get

$$\begin{aligned}
\Big|f(z)\Big|^2 &= \Big|e^{2it} + 2e^{it} - 3\Big|^2 = \Big(e^{2it} + 2e^{it} - 3\Big)\overline{\Big(e^{2it} + 2e^{it} - 3\Big)} \\
&= \Big(e^{2it} + 2e^{it} - 3\Big) \cdot \Big(e^{-2it} + 2e^{-it} - 3\Big) \\
&= 14 - 6\cos 2t - 8\cos t.
\end{aligned}$$

The function $g(t) = 14 - 6\cos 2t - 8\cos t$ attains its maximum when $\cos t = -2/3$ and $M = \max\limits_{|z|\leq 1} |z^2 - 3z + 2| = 8/\sqrt{3}$.

(b) Let $z = x + iy$. Then $f(z) = \cos 3z = \cos 3x\ \cosh 3y - i \sin 3x\ \sinh 3y$. Therefore

$$\begin{aligned} |f(z)|^2 &= \cos^2 3x\ \cosh^2 3y + \sin^2 3x\ \sinh^2 3y \\ &= \cos^2 3x\ \cosh^2 3y + (1 - \cos^2 3x)\ \sinh^2 3y \\ &= \cos^2 3x + \sinh^2 3y. \\ \therefore\ |f(z)| &= \sqrt{\cos^2 3x + \sinh^2 3y}. \end{aligned}$$

Now, f is analytic in the disc $\mathcal{S}(0,1)$ and continuous on the closed disc $\mathcal{S}[0,1]$. By maximum modulus theorem, the maximum value of $|f(z)|$ is attained on the boundary $\partial\mathcal{S}(0,1) = \{z : |z| = 1\}$. As $|\cos 3x| \leq 1$, the maximum value of $\cos 3x$ is 1 occurs at $x = 0$. Thus

$$g(y) = \sqrt{1 + \sinh^2 3y}; \qquad -1 \leq y \leq 1.$$

Since $\sinh 3y$ is an increasing function of y, the maximum value of $\sinh^2 3y$ occurs at $y = \pm 1$ and $g(y)$ has the maximum $\sqrt{1 + \sinh^2 3}$ at the points $z = \pm i$. Thus

$$\max_{|z|\leq 1} \left|f(z)\right| = \sqrt{1 + \sinh^2 3y} = \cosh 3.$$

(c) Let $z = x + iy$. Then

$$\begin{aligned} f(z) &= \frac{2z+1}{2z-12} = \frac{2(x+iy)+1}{2(x+iy)-1} = \frac{(2x+1)+2yi}{(2x-1)+2yi} \\ \Rightarrow |f(z)|^2 &= \frac{(2x+1)^2 + 4y^2}{(2x-1)^2 + 4y^2} = \frac{5+4x}{5-4x} = g(x), \text{ say; as } x^2 + y^2 = 1. \end{aligned}$$

Now $g'(x) = 40/(5-4x)^2$. $g(x)$ is an increasing function of x, and so the maximum value of $g(x)$ is 9 and it occurs at $x = 1$, i.e., $z = 1$.

$$\therefore M = \max_{|z|\leq 1} |f(z)| = \sqrt{9} = 3.$$

RESULT **46.** Let us consider an entire function $f(z) = e^{-iz^2}$ in the positive quadrant $\mathcal{H} = \{z = x + iy : x \geq 0, y \geq 0\}$. Now

$$|f(z)| = \left|e^{-iz^2}\right| = \left|e^{-i(x^2-y^2)+2xy}\right| = e^{2xy}.$$

On the boundary of $\mathcal{H}$, $|f(z)| = 1$; however $|f(z)| \to \infty$ as $y = x \to \infty$. Thus the maximum-modulus principle is not applicable to unbounded domains.

Theorem 5.22. [Minimum-modulus theorem]: *Let $\mathcal{D}$ be a bounded domain. Let function $f(z)$ be non-constant analytic within and on a closed contour Γ lying in $\mathcal{D}$. If $f(z) \neq 0$ inside Γ, then $|f(z)|$ must assume its minimum value on Γ.*

Proof: Since $f(z)$ is analytic in Γ and $f(z) \neq 0, \forall z \in \Gamma$, then the function $\frac{1}{f(z)}$ is analytic in Γ. By maximum modulus principle, $\frac{1}{|f(z)|}$ does not attain its maximum value in Γ unless $f(z)$ is constant, i.e., $|f(z)|$ does not attain its minimum value in Γ unless $f(z)$ is constant. □

RESULT **47.** Suppose that f is analytic in a domain $\mathcal{D}$ with boundary Γ, $f(z) \neq 0$ on $\mathcal{D}$ and $z_0 \in \mathcal{D}$ be such that $|f(z)| \geq |f(z_0)|$ holds for all $z \in \mathcal{D}$. Then by minimum modulus theorem, $f(z)$ is constant.

Example 5.7.2. *Verify that the maximum and minimum modulus theorems hold for the function $f(z) = e^z$, where Γ is the domain $|z| \leq 1$.*

Solution. We have $f(z) = e^z \Rightarrow |f(z)| = e^x$, where x takes the values in $[-1, 1]$. Now, e^x is maximum when $x = 1$ and minimum when $x = -1$; therefore, the maximum value is e and the minimum value is e^{-1}. The points $x = 1$ and $x = -1$ correspond to the points $z = 1$ and $z = -1$, respectively, which lies on the boundary of Γ. Hence, the maximum and minimum modulus theorems hold for $f(z) = e^z$.

5.8 Series Expansions

In this section we will be concerned with a new kind of series expansions of a complex function f, which is analytic at a point z_0 and in an annular region $a < |z - z_0| < b$. We prove that

(i) if f is analytic at z_0, then it can be expanded as a power series called *Taylor's series* consisting of non-negative powers of $z - z_0$, and the expansion is valid in some neighbourhood of z_0.

(ii) if f is analytic in an annular region $a < |z - z_0| < b$, Taylor's theorem cannot be applied; so in such case f has another series expansion consisting of positive and negative powers of $z - z_0$, called *Laurent series* expansion.

Finally, the concept of *singular points,* their classifications for f and the behaviour of f in the neighbourhood of a singularity have been discussed using series expansions.

5.8.1 Taylor's Series

Theorem 5.23. [Taylor's theorem]: *Let $f(z)$ be analytic inside a circle Γ with centre at z_0 and radius r. Then for every point z within Γ,*

$$\begin{aligned} f(z) &= f(z_0) + (z - z_0)f'(z_0) + \frac{(z-z_0)^2}{2!}f''(z_0) + \cdots + \frac{(z-z_0)^n}{n!}f^{(n)}(z_0) + \cdots \\ &= f(z_0) + \sum_{n=1}^{\infty} \frac{(z-z_0)^n}{n!} f^{(n)}(z_0). \end{aligned} \tag{5.43}$$

Proof: Let z be any point inside the circle Γ with center z_0 and radius r (Fig. 5.33). Let $|z - z_0| = \rho_0$ and γ be the circle with centre z_0 and radius ρ such that $\rho_0 < \rho < r$, so that the point z lies inside γ. Hence, by Cauchy's integral formula, we get,

$$f(z) = \frac{1}{2\pi i} \oint_\gamma \frac{f(w)}{w - z} dw. \tag{5.44}$$

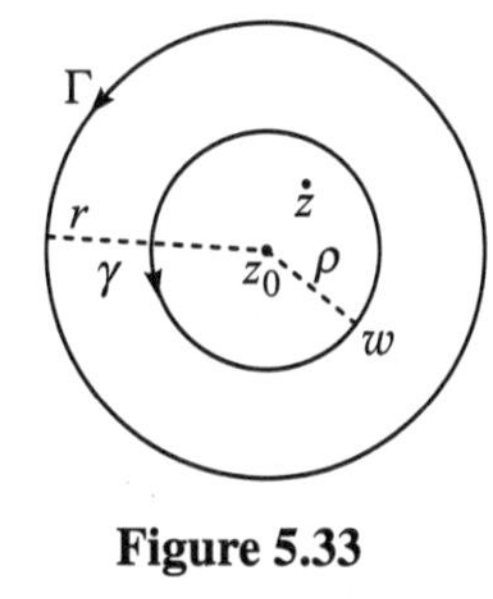

Figure 5.33

To obtain the required result, consider the identity,

$$\begin{aligned} \frac{1}{w-z} &= \frac{1}{(w-z_0) - (z-z_0)} = \frac{1}{w-z_0}\left[1 - \frac{z-z_0}{w-z_0}\right]^{-1} \\ &= \frac{1}{w-z_0}\Bigg[1 + \frac{z-z_0}{w-z_0} + \left(\frac{z-z_0}{w-z_0}\right)^2 + \cdots + \left(\frac{z-z_0}{w-z_0}\right)^{n-1} \\ &\qquad + \left(\frac{z-z_0}{w-z_0}\right)^n + \left(\frac{z-z_0}{w-z_0}\right)^{n+1} + \cdots\Bigg] \end{aligned}$$

$$\begin{aligned}
&= \frac{1}{w-z_0}\Big[1+\frac{z-z_0}{w-z_0}+\Big(\frac{z-z_0}{w-z_0}\Big)^2+\cdots+\Big(\frac{z-z_0}{w-z_0}\Big)^{n-1} \\
&\qquad +\Big(\frac{z-z_0}{w-z_0}\Big)^n\Big\{1+\Big(\frac{z-z_0}{w-z_0}\Big)+\Big(\frac{z-z_0}{w-z_0}\Big)^2+\cdots\Big\}\Big] \\
&= \frac{1}{w-z_0}\Big[1+\frac{z-z_0}{w-z_0}+\Big(\frac{z-z_0}{w-z_0}\Big)^2+\cdots \\
&\qquad +\Big(\frac{z-z_0}{w-z_0}\Big)^{n-1}+\Big(\frac{z-z_0}{w-z_0}\Big)^n\frac{1}{1-\{(z-z_0)(w-z_0)\}}\Big].
\end{aligned}$$

Therefore,

$$\frac{1}{w-z} = \frac{1}{w-z_0}+\frac{z-z_0}{(w-z_0)^2}+\cdots+\frac{(z-z_0)^{n-1}}{(w-z_0)^n}+\Big(\frac{z-z_0}{w-z_0}\Big)^n\frac{1}{w-z}. \tag{5.45}$$

Multiplying both sides of (5.45) by $f(w)$ and using (5.44), we get,

$$\begin{aligned}
f(z) &= \frac{1}{2\pi i}\oint_\gamma\Big[\frac{1}{w-z_0}+\frac{z-z_0}{(w-z_0)^2}+\cdots+\Big(\frac{z-z_0}{w-z_0}\Big)^n\frac{1}{w-z}\Big]f(w)dw \\
&= \frac{1}{2\pi i}\oint_\gamma\frac{f(w)dw}{w-z_0}+\frac{z-z_0}{2\pi i}\oint_\gamma\frac{f(w)dw}{(w-z_0)^2} \\
&\qquad +\cdots+\frac{(z-z_0)^{n-1}}{2\pi i}\oint_\gamma\frac{f(w)dw}{(w-z_0)^n}+R_n,
\end{aligned} \tag{5.46}$$

where, the remainder term R_n is given by

$$R_n = \frac{1}{2\pi i}\oint_\gamma\Big(\frac{z-z_0}{w-z_0}\Big)^n\frac{f(w)}{w-z}dw.$$

Using the Cauchy's integral formula (5.32) and derivative of Cauchy's integral formula (5.35), equation (5.46) becomes,

$$f(z) = f(z_0)+(z-z_0)f'(z_0)+\frac{(z-z_0)^2}{2!}f''(z_0)+\cdots+\frac{(z-z_0)^{n-1}}{(n-1)!}f^{(n-1)}(z_0)+R_n.$$

Now, we are to show that $R_n \to 0$ as $n \to \infty$. To prove this, we observe that, $|z-z_0| = \rho_0,\ |w-z_0| = \rho$, and so

$$|w-z| = |(w-z_0)-(z-z_0)| \geq |w-z_0|-|z-z_0| = \rho-\rho_0$$

$$\text{and } \left|\frac{z-z_0}{w-z_0}\right| = \frac{\rho_0}{\rho} = k<1\ ;\ \text{where } k = \text{ constant.}$$

Also, M denotes the maximum value of $f(w)$ on γ, i.e., $|f(w)|<M$, where M is a constant. Then,

$$\begin{aligned}
|R_n| &= \left|\frac{(z-z_0)^n}{2\pi i}\oint_\gamma\frac{f(w)}{(w-z_0)^n(w-z)}dw\right| \\
&\leq \frac{|z-z_0|^n}{2\pi}\oint_\gamma\frac{|f(w)|}{|w-z_0|^n\cdot|w-z|}|dw|
\end{aligned}$$

$$\leq \frac{M\rho_0^n}{2\pi\rho^n}\oint_\gamma \frac{|dw|}{\rho-\rho_0} = \frac{M}{2\pi(\rho-\rho_0)}\left(\frac{\rho_0}{\rho}\right)^n \times 2\pi\rho$$

$$\leq \frac{M\rho}{\rho-\rho_0}k^n \to 0 \text{ as } n\to\infty \text{ ; since } k=\frac{\rho_0}{\rho}<1.$$

Therefore, $f(z)$ is represented by the infinite series

$$f(z) = f(z_0) + \sum_{n=1}^{\infty}\frac{(z-z_0)^n}{n!}f^{(n)}(z_0),$$

which is the well-known *Taylor's series.* In particular, when, $z_0 = 0$, equation (5.43) reduces to

$$f(z) = f(0) + \sum_{n=1}^{\infty}\frac{z^n}{n!}f^{(n)}(0), \tag{5.47}$$

which is the well-known *Maclaurin's series.* □

Example 5.8.1. *Expand* $\sin z$ *into a Taylor's series about the point* $z = \pi/4$ *and determine the region of convergence.*

Solution. Let $f(z) = \sin z$. Then $f(\pi/4) = 1/\sqrt{2}$. Also $f'(z) = \cos z$, $f'(\pi/4) = 1/\sqrt{2}$; $f''(z) = -\sin z$, $f''(\pi/4) = -1/\sqrt{2}$; $f'''(z) = -\cos z$, $f'''(\pi/4) = -1/\sqrt{2}, \cdots$. Therefore, using Eq. (5.43), the Taylor's series for $\sin z$ about $z = \pi/4$ is

$$\begin{aligned} f(z) &= f(\pi/4 + z - \pi/4) = \sin(\pi/4 + z - \pi/4) \\ &= f(\pi/4) + \frac{(z-\pi/4)}{1!}f'(\pi/4) + \frac{(z-\pi/4)^2}{2!}f''(\pi/4) + \cdots \\ &= \frac{1}{\sqrt{2}} - \frac{(z-\pi/2)}{1!}\left(\frac{1}{\sqrt{2}}\right) - \frac{(z-\pi/2)^2}{2!}\left(\frac{1}{\sqrt{2}}\right) + \cdots \\ &= \sum_{n=0}^{\infty}\sin\left(\frac{\pi}{4}+\frac{n\pi}{2}\right)\frac{1}{n!}\left(z-\frac{\pi}{4}\right)^n. \end{aligned}$$

The expansion is valid in the entire complex plane.

Example 5.8.2. *Find the Taylor's series to represent* $\frac{z^2-1}{(z+2)(z+3)}$ *in* $|z| < 2$.

Solution. Using partial fraction method, we get

$$\begin{aligned} f(z) &= \frac{z^2-1}{(z+2)(z+3)} = 1 + \frac{3}{z+2} - 8\frac{8}{z+3} \\ &= 1 + \frac{3}{2}\left(1+\frac{z}{2}\right)^{-1} - \frac{8}{3}\left(1+\frac{z}{3}\right)^{-1} \\ &= 1 + \frac{3}{2}\left[1 - \frac{z}{2} + \frac{z^2}{2^2} - \frac{z^3}{2^3} + \cdots\right] - \frac{8}{3}\left[1 - \frac{z}{3} + \frac{z^2}{3^2} - \frac{z^3}{3^3} + \cdots\right] \\ &= \left(1+\frac{3}{2}-\frac{8}{3}\right) + \left(-\frac{3}{2^2}+\frac{8}{3^2}\right)z + \left(\frac{3}{2\cdot 2^2} - \frac{8}{3\cdot 3^2}\right)z^2 + \cdots \\ &= -\frac{1}{6} + \sum_{n=1}^{\infty}(-1)^{n+1}\left(\frac{8}{3^{n+1}} - \frac{3}{2^{n+1}}\right)z^n. \end{aligned}$$

The expansion is valid in $|z| < 2$.

Theorem 5.24. [Laurent's theorem]: *Let a function $f(z)$ be analytic in a ring shaped $\mathcal{D}$ bounded by two concentric circles Γ_1 and Γ_2 with centre z_0 and radii r_1 and r_2 respectively $(r_1 > r_2)$. Then for all z in $\mathcal{D}$,*

$$f(z) = \sum_{n=0}^{\infty} a_n(z-z_0)^n + \sum_{n=1}^{\infty} b_n(z-z_0)^{-n} \tag{5.48}$$

where,

$$\left.\begin{aligned} a_n &= \frac{1}{2\pi i}\oint_{\Gamma_1} \frac{f(w)}{(w-z_0)^{n+1}}dw; \qquad n = 0,1,2,\cdots \\ b_n &= \frac{1}{2\pi i}\oint_{\Gamma_2} (w-z_0)^{n-1}f(w)dw; \quad n = 1,2,3,\cdots \end{aligned}\right\}. \tag{5.49}$$

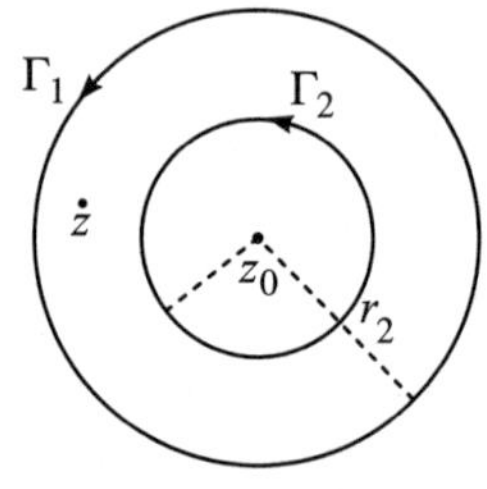

Figure 5.34 *The Laurent's series*

Proof: Let z be any point of the ring shaped region $\mathcal{D}$ bounded by two concentric circles Γ_1 and Γ_2 with center z_0 and radii r_1 and r_2 respectively $(r_1 > r_2)$ (Fig. 5.34). Then, by Cauchy's integral formula for doubly-connected region, we have,

$$f(z) = \frac{1}{2\pi i}\oint_{\Gamma_1} \frac{f(w)}{w-z}dw - \frac{1}{2\pi i}\oint_{\Gamma_2} \frac{f(w)}{w-z}dw. \tag{5.50}$$

Consider the first integral of (5.50) for any point w on Γ_1. Using the identity (5.45), we get,

$$\begin{aligned} \frac{1}{2\pi i}\oint_{\Gamma_1} \frac{f(w)dw}{w-z} &= \frac{1}{2\pi i}\oint_{\Gamma_1} \frac{f(w)dw}{w-z_0} + \frac{z-z_0}{2\pi i}\oint_{\Gamma_1} \frac{f(w)dw}{(w-z_0)^2} \\ &\quad + \cdots + \frac{(z-z_0)^{n-1}}{2\pi i}\oint_{\Gamma_1} \frac{f(w)dw}{(w-z_0)^n} + R_n \\ &= a_0 + a_1(z-z_0) + \cdots + a_{n-1}(z-z_0)^{n-1} + R_n \end{aligned} \tag{5.51}$$

where a_k's and the remainder R_n are given by,

$$\left.\begin{aligned} a_k &= \frac{1}{2\pi i}\oint_{\Gamma} \frac{f(w)}{(w-z_0)^{k+1}}dw; \qquad k = 0,1,2,\cdots,n-1 \\ R_n &= \frac{1}{2\pi i}\oint_{\Gamma_1} \left(\frac{z-z_0}{w-z_0}\right)^n \frac{f(w)dw}{w-z} \end{aligned}\right\}. \tag{5.52}$$

Now we are to show that $R_n \to 0$ as $n \to \infty$. Let $|z - z_0| = \rho$ so that $r_2 < \rho < r_1$. Also, $|w - z| = r_1$; it follows that

$$|w-z| = |(w-z_0)-(z-z_0)| \geq |w-z_0| - |z-z_0| = r_1 - \rho.$$

Also, M_1 denotes the maximum value of $f(z)$ on Γ_1, i.e., $|f(w)| < M_1$ on Γ_1, where M_1 is a constant.Then,

$$\begin{aligned} |R_n| &= \left|\frac{(z-z_0)^n}{2\pi i}\oint_{\Gamma_1} \frac{f(w)}{(w-z_0)^n(w-z)}dw\right| \\ &\leq \frac{|z-z_0|^n}{2\pi}\oint_{\Gamma_1} \frac{|f(w)|}{|w-z_0|^n \cdot |w-z|}|dw| \\ &\leq \frac{M_1 r_1}{r_1-\rho}\left(\frac{\rho}{r_1}\right)^n \to 0 \text{ as } n \to \infty \text{ ; since } \frac{\rho}{r_1} < 1. \end{aligned}$$

Consequently, it follows from (5.51) that,

$$\begin{aligned}\frac{1}{2\pi i}\oint_{\Gamma_1}\frac{f(w)}{w-z}dw &= a_0+a_1(z-z_0)+a_2(z-z_0)^2+\cdots\\ &= \sum_{n=0}^{\infty}a_n(z-z_0)^n. \qquad (5.53)\end{aligned}$$

We now consider the second integral of (5.50). For any point w on Γ_2, we consider the identity

$$\begin{aligned}-\frac{1}{w-z} &= \frac{1}{(z-z_0)-(w-z_0)}=\frac{1}{z-z_0}\left[1-\left(\frac{w-z_0}{z-z_0}\right)\right]^{-1}\\ &= \frac{1}{z-z_0}\left[1+\frac{w-z_0}{z-z_0}+\left(\frac{w-z_0}{z-z_0}\right)^2\right.\\ &\quad \left.+\cdots+\left(\frac{w-z_0}{z-z_0}\right)^{n-1}+\left(\frac{w-z_0}{z-z_0}\right)^n\cdot\frac{1}{1-\frac{w-z_0}{z-z_0}}\right]\\ &= \frac{1}{z-z_0}+\frac{w-z_0}{(z-z_0)^2}+\frac{(w-z_0)^2}{(z-z_0)^3}\\ &\quad +\cdots+\frac{(w-z_0)^{n-1}}{(z-z_0)^n}+\left(\frac{w-z_0}{z-z_0}\right)^n\cdot\frac{1}{z-w}. \qquad (5.54)\end{aligned}$$

Multiplying both sides of (5.54) by $f(w)$ as using the Cauchy's integral formula (5.44), we get,

$$\begin{aligned}-\frac{1}{2\pi i}\oint_{\Gamma_2}\frac{f(w)}{w-z}dw &= \frac{1}{2\pi i}\oint_{\Gamma_2}\frac{f(w)}{z-z_0}dw+\frac{1}{2\pi i}\oint_{\Gamma_2}\frac{w-z_0}{(z-z_0)^2}f(w)dw\\ &\quad +\cdots+\frac{1}{2\pi i}\oint_{\Gamma_2}\frac{(w-z_0)^{n-1}}{(z-z_0)^n}f(w)dw+S_n\\ &= \frac{b_1}{z-z_0}+\frac{b_2}{(z-z_0)^2}+\cdots+\frac{b_n}{(z-z_0)^n}+S_n \qquad (5.55)\end{aligned}$$

where b_k's and the remainder term S_n are given by

$$\begin{aligned} b_k &= \frac{1}{2\pi i}\oint_{\Gamma_2}(w-z_0)^{k-1}f(w)dw; \quad k=1,2,\cdots,n\\ S_n &= \frac{1}{2\pi i}\oint_{\Gamma_2}\left(\frac{w-z_0}{z-z_0}\right)^n\frac{f(w)}{z-w}dw. \end{aligned} \qquad (5.56)$$

Now, we are to show that $S_n\to 0$ as $n\to\infty$. We observed that $|z-z_0|=r$, $|w-z_0|=r_2$ and $r_2<r$, so that

$$|z-w|=|(z-z_0)-(w-z_0)|\geq|z-z_0|-|w-z_0|=r-r_2.$$

Also, M_2 denotes the maximum value of $f(w)$ on Γ_2, i.e., $|f(w)|<M_2$ on Γ_2, where M_2 is a constant.Then

$$\begin{aligned}|S_n| &= \frac{1}{2\pi}\left|\oint_{\Gamma_2}\left(\frac{w-z_0}{z-z_0}\right)^n\frac{f(w)}{z-w}dw\right|\\ &\leq \frac{1}{2\pi}\oint_{\Gamma_2}\left|\frac{w-z_0}{z-z_0}\right|^n\frac{|f(w)|}{|z-w|}|dw|\\ &\leq \frac{1}{2\pi}\left(\frac{r_2}{r}\right)^n\frac{M_2}{r-r_2}2\pi r_2\leq\frac{M_2r_2}{r-r_2}\left(\frac{r_2}{r}\right)^n\to 0\end{aligned}$$

as $n \to \infty$ and $\frac{r_2}{r} < 1$. Therefore, $\lim\limits_{n\to\infty} S_n = 0$. Hence from (5.55), we have,

$$-\frac{1}{2\pi i}\oint_{\Gamma_2} \frac{f(w)}{w-z}dw = \frac{b_1}{z-z_0} + \frac{b_2}{(z-z_0)^2} + \cdots + \frac{b_n}{(z-z_0)^n} + \cdots$$

$$= \sum_{n=1}^{\infty} b_n(z-z_0)^{-n}; b_n = \frac{1}{2\pi i}\oint\limits_{\Gamma_2}(w-z_0)^{n-1}f(w)dw. \tag{5.57}$$

Therefore, from (5.50), (5.53) and (5.57), we have,

$$f(z) = \frac{1}{2\pi i}\oint_{\Gamma_1}\frac{f(w)}{w-z}dw - \frac{1}{2\pi i}\oint_{\Gamma_2}\frac{f(w)}{w-z}dw$$

$$= \sum_{n=0}^{\infty} a_n(z-z_0)^n + \sum_{n=1}^{\infty} b_n(z-z_0)^{-n},$$

where a_n and b_n are given by (5.49). Therefore, the proof is complete. □

DEDUCTION **12.** The right-hand side of Eq. (5.48) consists of two parts: The first part $\sum_{n=0}^{\infty} a_n(z-z_0)^n$ of (5.48) is called the *analytic part* while the remaining part $\sum_{n=1}^{\infty} b_n(z-z_0)^{-n}$, is called the *principal part*.

DEDUCTION **13.** Take $b_n = a_{-n}$. Hence if Γ is any circle of radius r with centre z_0 such that $r_2 < r < r_1$. Then, since the integrand is analytic in the annulus $r_2 < |w-z_0| < r_1$. Then,

$$a_n = \frac{1}{2\pi i}\oint_{\Gamma}\frac{f(w)}{(w-z_0)^{n+1}}dw \quad \text{and} \quad a_{-n} = \frac{1}{2\pi i}\oint_{\Gamma}(w-z_0)^{n-1}f(w)dw.$$

Therefore, the Laurent's series (5.48) can be written in the compact form as

$$f(z) = \sum_{n=-\infty}^{\infty} a_n(z-z_0)^n \text{ ; where } a_n = \frac{1}{2\pi i}\oint_{\Gamma}\frac{f(w)}{(w-z_0)^{n+1}}dw \tag{5.58}$$

DEDUCTION **14.** The Laurent's series (5.58) is unique. Suppose we had obtained in any manner or as the definition of $f(z)$ the formula

$$f(z) = \sum_{n=-\infty}^{\infty} c_n(z-z_0)^n \text{ ; } r_2 < |z-z_0| < r_1.$$

Let Γ be the circle $|z-z_0| = r$, where $r_2 < r < r_1$. Then the coefficient a_n in the Laurent's series expansion (5.58) is given by,

$$a_n = \frac{1}{2\pi i}\oint_{\Gamma}\frac{f(w)dw}{(w-z_0)^{n+1}} = \frac{1}{2\pi i}\oint_{\Gamma}\frac{1}{(w-z_0)^{n+1}}\sum_{m=-\infty}^{\infty} c_m(w-z_0)^m dw$$

$$= \frac{1}{2\pi i}\sum_{m=\ \infty}^{\infty} c_m \oint_{\Gamma}\frac{(w-z_0)^m}{(w-z_0)^{n+1}}dw$$

$$= \frac{1}{2\pi i}\sum_{m=-\infty}^{\infty} c_m \int_0^{2\pi}\frac{\rho^m e^{im\theta}}{\rho^{n+1}e^{i(n+1)\theta}}i\rho e^{i\theta}d\theta \text{ ; } w-z_0 = \rho e^{i\theta}$$

$$= \frac{1}{2\pi} \sum_{m=-\infty}^{\infty} c_m \int_0^{2\pi} \rho^{m-n} e^{i(m-n)\theta} d\theta$$

$$= \begin{cases} \dfrac{1}{2\pi} c_n \displaystyle\int_0^{2\pi} d\theta - c_n ; & \text{when } m = n \\ \dfrac{1}{2\pi} \displaystyle\sum_{m=-\infty}^{\infty} c_m \rho^{m-n} \int_0^{2\pi} e^{i(m-n)\theta} d\theta ; & \text{when } m \neq n. \end{cases}$$

This shows that the given series is identical with the Laurent's series expansion (5.48) for $f(z)$.

DEDUCTION **15.** We can associate the Laurent's series for the annular domain $r_2 < 1 < r_1$ with the Fourier series expansion. Let f be analytic in a domain $1 - \delta < |z| < 1 + \delta, \delta > 0$ and this domain contains the unit disc $z = e^{i\theta}; 0 \leq \theta \leq 2\pi$. Now by Eq. (5.58), when $z_0 = 0$, we have

$$f(z) = \sum_{n=-\infty}^{\infty} a_n z^n \text{ ; where } a_n = \frac{1}{2\pi} \int_0^{2\pi} f(e^{i\theta})\, e^{-in\theta}\, d\theta.$$

Taking $f(e^{it}) = F(t)$ and $z = e^{it}$ on the unit circle

$$F(t) = \sum_{n=-\infty}^{\infty} a_n e^{int}; \quad a_n = \frac{1}{2\pi} \int_0^{2\pi} F(t)\, e^{-int}\, dt,$$

which is called *Fourier expansion* of $F(t)$ in the complex form.

Example 5.8.3. *Find the Laurent's series expansion of $f(z) = z^2 e^{1/z}$ about $z = 0$.*

Solution. Here $f(z) = z^2 e^{1/z}$ is analytic at all points $z \neq 0$, but not analytic at the isolated singular point $z = 0$ and hence can be expressed in the Maclaurin series. Now

$$\begin{aligned} f(z) &= z^2\left[1 + \frac{1}{z} + \frac{1}{2!z^2} + \frac{1}{3!z^3} + \cdots\right] \\ &= z^2 + z + \frac{1}{2} + \frac{1}{3!z} + \frac{1}{4!z^2} + \cdots. \end{aligned}$$

The analytic part of the series converges for $|z| < \infty$. The principal part is valid for $|z| > 0$. Thus the series converges for z except at $z = 0$. This representation is the required Laurent's series expansion valid for $0 < |z| < \infty$.

Example 5.8.4. *Find the Laurent's series expansion of $f(z) = \sin z / z^4$ about $z = 0$.*

Solution. Here $f(z) = \sin z / z^4$ is analytic at all points $z \neq 0$, but not analytic at the isolated singular point $z = 0$ and hence can be expressed in the Maclaurin series. Now

$$\begin{aligned} f(z) &= \frac{\sin z}{z^4} = \frac{1}{z^4}\left[z - \frac{z^3}{3!} + \frac{z^5}{5!} - \frac{z^7}{7!} + \cdots\right] \\ &= \frac{1}{z^3} - \frac{1}{3!z} + \frac{z}{5!} - \frac{z^3}{7!} + \cdots. \end{aligned}$$

The analytic part of the series converges for $|z| < \infty$. The principal part is valid for $|z| > 0$. Thus the series converges for z except at $z = 0$. This representation is the required Laurent's series expansion valid for $0 < |z| < \infty$.

Example 5.8.5. *Expand $f(z) = 1/(z-3)$ in a Laurent's series valid for $0 < |z| < 3$ and $|z| > 3$.*

Solution. Here $f(z)$ is analytic at every point in the open disc $0 < |z| < 3$ and hence can be expressed in a Maclaurin series. Thus

$$\begin{aligned}\frac{1}{z-3} &= -\frac{1}{3}\frac{1}{1-(z/3)} = -\frac{1}{3}\left[1-\frac{z}{3}\right]^{-1}; \quad \frac{|z|}{3}<1\\ &= -\frac{1}{3}\left[1+\frac{z}{3}+\left(\frac{z}{3}\right)^2+\left(\frac{z}{3}\right)^3+\cdots\right]\\ &= -\frac{1}{3}-\frac{1}{4}z-\frac{1}{27}z^2-\frac{1}{81}z^3-\cdots.\end{aligned}$$

Since there is no principal part, it is actually a Maclaurin series. The domain $|z| > 3$ consists of all points exterior to the circle $|z| = 3$. Rewriting $f(z)$, we get

$$\begin{aligned}f(z) &= \frac{1}{z-3} = \frac{1}{z}\cdot\frac{1}{1-(3/z)} = \frac{1}{z}\left[1-\frac{3}{z}\right]^{-1}\\ &= \frac{1}{z}\left[1+\left(\frac{3}{z}\right)+\left(\frac{3}{z}\right)^2+\left(\frac{3}{z}\right)^3+\cdots\right]\\ &= \frac{1}{z}\left[1+\frac{3}{z}+\frac{9}{z^2}+\frac{27}{z^3}+\cdots\right] = \frac{1}{z}+\frac{3}{z^2}+\frac{9}{z^3}+\cdots.\end{aligned}$$

Example 5.8.6. *Expand* $f(z) = \frac{1}{(z+1)(z+3)}$ *in a Laurent's series valid for the region (i)* $|z| < 1$;*(ii)* $1 < |z| < 3$; *and (iii)* $|z| > 3$ *(iv)* $0 < |z+1| < 2$.

Solution. Resolving $f(z)$ its partial fractions, we get,

$$f(z) = \frac{1}{(z+1)(z+3)} = \frac{1}{2(z+1)} - \frac{1}{2(z+3)}.$$

(i) If $|z| < 1$, we have, $|\frac{z}{3}| < \frac{1}{3} < 1$. Therefore, the Laurent's series expansion is valid for $|z| < 1$. Hence

$$\begin{aligned}f(z) &= \frac{1}{2}(1+z)^{-1}-\frac{1}{6}\left(1+\frac{z}{3}\right)^{-1}\\ &= \frac{1}{2}[1-z+z^2-z^3+\cdots]-\frac{1}{6}\left[1-\frac{z}{3}+\left(\frac{3}{z}\right)^2-\left(\frac{3}{z}\right)^3+\cdots\right]\\ &= \left(\frac{1}{2}-\frac{1}{6}\right)-\left(\frac{1}{2}-\frac{1}{18}\right)z+\left(\frac{1}{2}-\frac{1}{54}\right)z^2-\left(\frac{1}{2}-\frac{1}{162}\right)z^3+\cdots\\ &= \frac{1}{3}-\frac{4}{9}z+\frac{13}{27}z^2-\frac{40}{81}z^3+\cdots\end{aligned}$$

Since there is no principal part, so, this is a Taylor's series.

(ii) When $|z| > 1$, we have, $\frac{1}{|z|} < 1$, therefore,

$$\begin{aligned}\frac{1}{2(z+1)} &= \frac{1}{2z\left(1+\frac{1}{z}\right)} - \frac{1}{2z}\left(1+\frac{1}{z}\right)^{-1}\\ &= \frac{1}{2z}\left[1-\frac{1}{z}+\frac{1}{z^2}-\frac{1}{z^3}+\cdots\right] = \frac{1}{2z}-\frac{1}{2z^2}+\frac{1}{2z^3}-\frac{1}{2z^4}+\cdots.\end{aligned}$$

For $|z| < 3$, we have,

$$\frac{1}{2(z+3)} = \frac{1}{6}\left(1+\frac{z}{3}\right)^{-1} = \frac{1}{6}\left[1-\frac{z}{3}+\left(\frac{z}{3}\right)^2-\left(\frac{z}{3}\right)^3+\cdots\right].$$

Then the required Laurent's series expansion valid for both $|z| > 1$ and $|z| < 3$, i.e., $1 < |z| < 3$, is given by

$$\begin{aligned} f(z) &= \frac{1}{2(z+1)} - \frac{1}{2(z+3)} = \frac{1}{2z}\left(1+\frac{1}{z}\right)^{-1} - \frac{1}{6}\left(1+\frac{z}{3}\right)^{-1} \\ &= \frac{1}{2z} - \frac{1}{2z^2} + \frac{1}{2z^3} - \cdots - \frac{1}{6}\left[1-\frac{z}{3}+\left(\frac{z}{3}\right)^2-\left(\frac{z}{3}\right)^3+\cdots\right] \\ &= \cdots + \frac{1}{2z^3} - \frac{1}{2z^2} + \frac{1}{2z} - \frac{1}{6} + \frac{z}{18} - \frac{z^2}{54} + \frac{z^3}{162} - \cdots. \end{aligned}$$

which contains both analytic and principal parts.

(iii) Let us consider $|z| > 3$. Then $\frac{1}{|z|} < \frac{1}{3} < 1$ and $\frac{3}{|z|} < 1$. The required Laurent's series expansion becomes;

$$\begin{aligned} f(z) &= \frac{1}{2(z+1)} - \frac{1}{2(z+3)} = \frac{1}{2z}\left(1+\frac{1}{z}\right)^{-1} - \frac{1}{2z}\left(1+\frac{3}{z}\right)^{-1} \\ &= \frac{1}{2z} - \frac{1}{2z^2} + \frac{1}{2z^3} - \frac{1}{2z^4} + \cdots - \frac{1}{2z}\left[1-\frac{3}{z}+\left(\frac{3}{z}\right)^2-\left(\frac{3}{z}\right)^3+\cdots\right] \\ &= \left(\frac{3}{2}-\frac{1}{2}\right)\frac{1}{z^2} + \left(\frac{1}{2}-\frac{9}{2}\right)\frac{1}{z^3} + \left(-\frac{1}{2}+\frac{27}{2}\right)\frac{1}{z^4} + \cdots \\ &= \frac{1}{z^2} - \frac{4}{z^3} + \frac{13}{z^4} - \cdots \end{aligned}$$

which contains only the principal parts ($b_1 = 0$) of the Laurent's series expansion of $f(z)$.

(iv) Let $z+1 = u$. Then $0 < |z+1| < 2$ implies $0 < |u| < 2$, $u \neq 0$. Therefore, $\left|\frac{u}{2}\right| < 1$ and so,

$$\begin{aligned} f(z) &= f(z) = \frac{1}{(z+1)(z+3)} = \frac{1}{u(u+2)} = \frac{1}{2u}\left(1+\frac{u}{2}\right)^{-1} \\ &= \frac{1}{2u}\left[1-\frac{u}{2}+\left(1+\frac{u}{2}\right)^2-\left(1+\frac{u}{2}\right)^3+\cdots\right] \\ &= \frac{1}{2u} - \frac{1}{4} + \frac{u}{8} - \frac{u^2}{16} + \cdots \\ &= \frac{1}{2(z+1)} - \frac{1}{4} + \frac{1}{8}(z+1) - \frac{1}{16}(z+1)^2 + \cdots \end{aligned}$$

which contain only one principal part of the Laurent's series expansion of $f(z)$ about $z = -1$.

Example 5.8.7. *Let $f \in \mathbb{C} - \{0\}$ be a holomorphic function such that $|f(z)| < a|z|^2 + b|z|^{-2}$ for $z \in \mathbb{C} - \{0\}$ and for some $a, b > 0$. If f is an odd function of z, what form must the Laurent series of f have?*

Solution. From the given inequality, we see that

$$|z^3 f(z)| < a|z|^5 + b|z| \to 0 \text{ as } z \to 0,$$

so that the meromorphic function $z^3 f(z) \in \mathbb{C} - \{0\}$ has a removable singularity at $z = 0$. Define

$$g(z) = \begin{cases} z^3 f(z) & \text{for } z \neq 0 \\ 0 & \text{for } z = 0. \end{cases}$$

Then $g(z)$ is entire and satisfies the inequality

$$|g(z)| < a|z|^5 + b|z| \leq M|z|^5$$

for sufficiently large $|z|$. Hence, by Result 45, $g(z)$ is a polynomial of degree at most 5,with $g(0) = 0$. If f is odd, then f must be of the form $f(z) = c_1 z + c_2/z$.

5.9 Singularities of a Function

In Chapter 3 we learnt about zeros and different types of singularities of analytic functions in brief. The concept of singularities of a complex-valued function f very much depends upon the domain of f. Here, Laurent's series expansion of the function $f(z)$ can be used to classify the isolated singular points.

Let $f(z)$ fails to be analytic in the neighbourhood of its isolated singular point z_0 but is analytic in the annulus $0 < |z - z_0| < R$. In this region, we can expand the function $f(z)$ as a Laurent's series. Here, we shall discuss the kind or nature of singular points of the complex function $f(z)$ in the light of the principal part of the series expansion of $f(z)$ about the singular point.

5.9.1 Zeros of Analytic Functions

Definition 5.9.1. Let $f : \mathcal{D} \to \mathbb{C}$ be analytic in an open connected set $\mathcal{D} \subset \mathcal{N}(z_0, \delta)$ and $z_0 \in \mathcal{D}$. If $f(z_0) = 0$ and there exists $r \in \mathbb{N}$ such that $f'(z_0) = f''(z_0) = \cdots = f^{(r-1)}(z_0) = 0$ and $f^{(r)}(z_0) \neq 0$, then the analytic function f is said to have a *zero of order* r at z_0.

Theorem 5.25. *Let $f : \mathcal{D} \to \mathbb{C}$ be analytic in an open connected set $\mathcal{D}$. Then $f(z)$ has a zero z_0 of order r if and only if there is an analytic function $g : \mathcal{D} \to \mathbb{C}$ such that*

$$f(z) = (z - z_0)^r g(z); \text{ where } g(z_0) \neq 0.$$

Proof [Necessary condition] : Let $f : \mathcal{D} \to \mathbb{C}$ be analytic in an open connected set $\mathcal{D}$ and $z_0 \in \mathcal{D}$. Suppose f has a zero of order r at z_0.Then, $f'(z_0) = f''(z_0) = \cdots = f^{(r-1)}(z_0) = 0$. The Taylor's series of $f(z)$ about $z = z_0$ is therefore given by

$$\begin{aligned} f(z) &= \frac{(z - z_0)^r}{r!} f^{(r)}(z_0) + \frac{(z - z_0)^{r+1}}{(r+1)!} f^{(r+1)}(z_0) + \cdots \\ &= (z - z_0)^r \left[\frac{f^{(r)}(z_0)}{r!} + \frac{f^{(r+1)}(z_0)}{(r+1)!}(z - z_0) + \cdots \right] \\ &= (z - z_0)^r g(z), \quad z \in \mathcal{D} \end{aligned}$$

where, $g(z) = \frac{f^{(r)}(z_0)}{r!} + \frac{f^{(r+1)}(z_0)}{(r+1)!}(z - z_0) + \cdots$, $z \in \mathcal{D}$. Being a polynomial, $g(z)$ is analytic function in $\mathcal{D}$ with the property $g(z_0) = \frac{f^{(r)}(z_0)}{r!} \neq 0$.

Sufficient condition: Let $f(z) = (z-z_0)^r g(z)$, where $g(z)$ is analytic at z_0 and $g(z_0) \neq 0$. As $g(z)$ is analytic at z_0, therefore its Taylor series representation is given by

$$g(z) = g(z_0) + \frac{g'(z_0)}{1!}(z-z_0) + \frac{g''(z_0)}{2!}(z-z_0)^2 + \cdots$$

in $\mathcal{D}$ containing z_0. Therefore,

$$\begin{aligned} f(z) &= (z-z_0)^r g(z) = (z-z_0)^r \Big[g(z_0) + \frac{g'(z_0)}{1!}(z-z_0) + \frac{g''(z_0)}{2!}(z-z_0)^2 + \cdots\Big] \\ &= g(z_0)(z-z_0)^r + \frac{g'(z_0)}{1!}(z-z_0)^{r+1} + \frac{g''(z_0)}{2!}(z-z_0)^{r+2} + \cdots, \end{aligned} \tag{5.59}$$

when $z \in \mathcal{D}$. We see that $\frac{f^{(r)}(z_0)}{r!} + \frac{f^{(r+1)}(z_0)}{(r+1)!}(z-z_0) + \cdots$ is Taylor series representation for $f(z)$ and hence $f'(z_0) = f''(z_0) = \cdots = f^{(r-1)}(z_0) = 0$ and $f^{(r)}(z_0) = r!g(z_0) \neq 0$. Hence, f has a zero of order r at z_0. □

For example, consider:

(i) $f(z) = \sin z$. We know that

$$\sin z = z - \frac{z^3}{3!} + \frac{z^5}{5!} - \cdots = z\Big[1 - \frac{z^2}{3!} + \frac{z^4}{5!} - \cdots\Big] = z\, g(z),$$

where $g(z) = 1 - \frac{z^2}{3!} + \frac{z^4}{5!} - \cdots$ is analytic and $g(0) = 1 \neq 0$. Therefore, $z = 0$ is a simple zero of $\sin z$. Similarly, for $f(z) = z^2 \sin z$, $z \sin z^2$, $z = 0$ is a zero of order 3.

(ii) $f(z) = z(e^z - 1)$. Then

$$\begin{aligned} f(z) &= z\Big[\Big(1 + \frac{z}{1!} + \frac{z^2}{2!} + \cdots\Big) - 1\Big] \\ &= z^2\Big[1 + \frac{z}{1!} + \frac{z^2}{2!} + \cdots\Big] = z^2 e^z = (z-0)^2 g(z), \end{aligned}$$

where, $g(z) = e^z$, with $g(0) = 1 \neq 0$. This means that $f(z)$ has a zero of order 2 at $z = 0$.

Theorem 5.26. *Let $\mathcal{D}$ be a domain and $f : \mathcal{D} \to \mathbb{C}$ be an analytic function. Then the set $Z_f = \{z \in \mathcal{D} : f(z) = 0\}$ of all zeros of f is a set of isolated points.*

Proof: Let for a $\delta > 0$ there be an open disc $\mathcal{S}(z_0; \delta) \subseteq \mathcal{D}$. Let $z_0 \in Z_f$ be a zero of order r for f. Then

$$f(z) = (z-z_0)^r\, g(z); \qquad z \in \mathcal{S}(z_0; \delta)$$

where $g(z)$ is analytic at z_0 and $g(z_0) \neq 0$. Suppose $z_1 \neq z_0$ is another zero for f in $\mathcal{S}(z_0; \delta)$. Then $f(z_1) = 0$ and $|z_1 - z_0| < \delta$. Therefore,

$$\Big(z_1 - z_0\Big)^r g(z_1) = 0 \Rightarrow g(z_1) = 0; \text{ as } z_1 \neq z_0.$$

Now, since g is analytic at z_0, it is continuous there. Thus, corresponding to an $\varepsilon = \frac{|g(z_0)|}{2}$, we can find a $\delta > 0$ such that

$$\begin{aligned} |z_1 - z_0| < \delta \Rightarrow \quad & \Big|g(z_1) - g(z_0)\Big| < \frac{|g(z_0)|}{2} \\ \Rightarrow \quad & \Big|0 - g(z_0)\Big| < \frac{|g(z_0)|}{2}, \end{aligned}$$

which is a contradiction. Thus $\mathcal{S}(z_0, \delta)$ contains no other zero of f and hence the set Z_f of all zeros of f is isolated. □

Theorem 5.27. [Identity or uniqueness theorem]: *Let $\mathcal{D}$ be a domain and $f : \mathcal{D} \to \mathbb{C}$ be an analytic function. If the set $Z_f = \{z \in \mathcal{D} : f(z) = 0\}$ of all zeros of f has a limit point z^* in $\mathcal{D}$, then f is constant, i.e., f is identically zero.*

Proof: Let $\langle z_n \rangle$ be a sequence of zeros of f in $\mathcal{D}$ such that $z_n \to z^*$. Then $f(z_n) = 0$ for all n. Since $\mathcal{D}$ is a domain (open, connected), so $z^* \in \mathcal{D}$. Since f is analytic in $\mathcal{D}$, it is continuous and hence

$$f(z^*) = f\Big(\lim_{n\to\infty} z_n\Big) = \lim_{n\to\infty} f(z_n) = 0.$$

Since the set $Z_f = \{z \in \mathcal{D} : f(z) = 0\}$ of all zeros of f is a set of isolated points,

(i) either $f(z) \equiv 0$ in a neighbourhood of z^*, or

(ii) $f(z) \neq 0$ in some punctured neighbourhood $\mathcal{S}(z^*; \delta)/\{z^*\} \subset \mathcal{D}$ of the limit point z^*.

But (ii) contradicts $f(z_n) = 0$, as for sufficiently large n, $z_n \in \mathcal{S}(z^*; \delta)/\{z^*\}$. Therefore, we must have $f(z) \equiv 0$ in any disc $\mathcal{S}(z^*; \delta) \subseteq \mathcal{D}$. Now, we are to show that $f(z) \equiv 0$ in the whole of $\mathcal{D}$. For this, let

$$\mathcal{A} = \Big\{\alpha \in \mathcal{D} : \alpha \text{ is a limit point of } Z_f\Big\} \text{ and } \mathcal{B} = \Big\{\alpha \in \mathcal{D} : \alpha \notin \mathcal{A}\Big\}.$$

Then $\mathcal{D} = \mathcal{A} \cup \mathcal{B}$ and $\mathcal{A} \cap \mathcal{B} = \phi$. Since $z^* \in Z_f$, $\mathcal{A}$ is nonempty. Let $\zeta \in \mathcal{A}$. Then ζ is a limit point of Z_f, $f(z) \equiv 0$ in a neighbourhood of ζ. Thus each point of $\mathcal{A}$ is an interior point of $\mathcal{A}$. Indeed, if $\zeta^* \in \mathcal{S}(\zeta; \delta)$ for some $\delta > 0$, then $|\zeta' - \zeta| < \delta$. Let $\varepsilon = \delta - |\zeta' - \zeta| > 0$, and

$$z \in \mathcal{S}(\zeta'; \varepsilon) \qquad \text{i.e.,} \qquad |z - \zeta'| < \varepsilon.$$

Then using triangle inequality, we get

$$\begin{aligned} & \Big|z - \zeta\Big| \leq \Big|z - \zeta'\Big| + \Big|\zeta' - \zeta\Big| < \varepsilon + \Big|\zeta' - \zeta\Big| = \delta \\ \Rightarrow \quad & \mathcal{S}\Big(\zeta'; \varepsilon\Big) \subset \mathcal{S}\Big(\zeta; \delta\Big). \end{aligned}$$

Thus $\mathcal{A}$ open and is nonempty by hypothesis.

Let $z' \in \mathcal{B}$. Then z' is not a limit point of Z_f. By the neighbourhood property of continuity, $\exists\, \delta > 0$ such that

$$f(z) \neq 0 \text{ in } z \in \mathcal{S}(z'; \delta) \subset \mathcal{D}.$$

Since the domain $\mathcal{D}$ is nonempty and connected, it cannot be written as the union of two nonempty disjoint open sets. Thus we must have either $\mathcal{A} = \phi$ or $\mathcal{B} = \phi$. But $z^* \in \mathcal{A}$, so $\mathcal{A} \neq \phi$ and thus $\mathcal{B} = \phi$.

Therefore $\mathcal{A} = \mathcal{D}$, and every $\alpha \in \mathcal{D}$ is a limit point of the zeros of f so that $f \equiv 0$ in $\mathcal{D}$. □

In particular, this observation implies the following assertions:

DEDUCTION **16.** Let $\mathcal{D} \subset \mathbb{C}$ be a domain. Let f_1 and f_2 be two analytic functions in $\mathcal{D}$. If Z_f, the set of zeros of $f_1 - f_2$ in $\mathcal{D}$, has a limit point z^* in $\mathcal{D}$, then considering a function $\psi(z) = f_1(z) - f_2(z)$ in $\mathcal{D}$, by uniqueness theorem we have $f_1(z) = f_2(z)$ in $\mathcal{D}$.

DEDUCTION **17** [Weierstrass's theorem]: Let $\mathcal{R}$ be a simply-connected region in $\mathbb{C}$ and $\mathcal{G} \subseteq \mathcal{R}$ be a discrete subset of $\mathcal{R}$. Then there exists a non-constant map $f : \mathcal{R} \to \mathcal{R}$ such that $Z_f = \mathcal{G}$.

In particular, this observation implies the following assertions:

(i) Consider $\{\frac{1}{n} : n \in \mathbb{N}\} \cup \{0\}$: it is not a zero set of analytic functions.

(ii) Consider $\{1 + \frac{1}{n} : n \in \mathbb{N}\}$: it is not also a zero set of analytic functions.

(iii) Consider $\{n + \frac{1}{n} : n \in \mathbb{N}\}$: Since it is a discrete set, it is a zero set of analytic functions.

(iv) Consider $\mathbb{R}$, which is everywhere dense:, it is not a zero set of analytic functions.

DEDUCTION **18.** Let $\mathcal{D} \subset \mathbb{C}$ be a domain. Let f_1 and f_2 be two analytic functions in $\mathcal{D}$ such that $f_1(z) \cdot f_2(z) = 0$ for each $z \in \mathcal{D}$. Let $f_1(a) \neq 0$ for some $a \in \mathcal{D}$. Due to continuity of f_1, there exists a disc $\mathcal{S}(a;\delta) \subseteq \mathcal{D}$ such that

$$f_1(z) \neq 0 \text{ in } \mathcal{S}(a;\delta) \Rightarrow f_2(z) = 0 \text{ for all } z \in \mathcal{S}(a;\delta).$$

By the uniqueness theorem, $f_2(z) = 0$ in $\mathcal{D}$. Thus, if $f_1(z) \cdot f_2(z) = 0$ for each $z \in \mathcal{D}$, then either $f_1(z) = 0$ for $z \in \mathcal{D}$ or $f_2(z) = 0$ for each $z \in \mathcal{D}$.

RESULT **48.** The above theorem (Theorem 5.27) is not true if domain $\mathcal{D}$ is not connected. For example, let $\mathcal{D} = \{z : |z| < 1\} \cup \{z : |z| > 3\}$. Clearly, $\mathcal{D}$ is not a connected subset of $\mathbb{C}$ so that $\mathcal{D}$ is not a region. Let $f : \mathcal{D} \to \mathbb{C}$ be defined by

$$f(z) = \begin{cases} 0 & \text{if } |z| < 1 \\ 1 & \text{if } |z| > 3 \end{cases}$$

Clearly, f is analytic and $Z_f = \{z \in \mathbb{C} : |z| < 1\}$ has a limit point of $\mathcal{D}$, but f is not identically zero in $\mathcal{D}$.

Example 5.9.1. *Suppose f is an entire function such that $f'(0) = 0$ and $f''(1 + \frac{1}{n}) = 7 - \frac{3}{n}$ for each $n = 1, 2, 3, \cdots$. Find all f that satisfy these properties.*

Solution. The given relation can be written as

$$f''\left(1 + \frac{1}{n}\right) - 10 + 3\left(1 + \frac{1}{n}\right) = 0; \quad n = 1, 2, 3, \cdots .$$

Let us define $g(z) = f''(z) - 10 + 3z$, so that $g(z)$ is holomorphic and $g(1 + \frac{1}{n}) = 0$. Hence, by uniqueness theorem, $f''(z) - 10 + 3z = 0$.

$$\therefore f'(z) = 10z - \frac{3}{2}z^2 + c_1.$$

Given $f'(0) = 0$. So $c_1 = 0$. Therefore,

$$f(z) = 5z^2 - \frac{1}{2}z^3 + c; \quad c = \text{ a constant.}$$

5.9.2 Isolated Singular Points

An isolated singular point $z = z_0$ of a complex function f is given a classification depending on whether the principal part of its Laurent expansion contains a zero, a finite number, or an infinite number of terms. Isolated singular points are further classified into three types: (i) removable singular point, (ii) pole, (iii) essential singular point.

5.9.3 Removable Singularity

Consider the domain $\mathcal{D} = \mathcal{S}(z_0, R) = \{z \in \mathbb{C} : d(z, z_0) < R\}$.

Definition 5.9.2. Let f be an analytic function on the annulus $\mathcal{D}^* = \mathcal{D} - \{z_0\}$ for some $z_0 \in \mathcal{D}$. Let all the coefficients b_n of the principal part of the Laurent's series expansion (5.48) of f be identically zero. Then, the Laurent's series expansion of $f(z)$ about z_0 is given by

$$f(z) = \sum_{n=0}^{\infty} a_n (z - z_0)^n; \quad z \in \mathcal{D}^*. \tag{5.60}$$

Then the singular point z_0 of f is called *isolated removable singularity* of $f(z)$.

If we define $f(z_0) = a_0$, the expansion (5.60) becomes valid throughout the disc $|z - z_0| < R$. As a power series represents an analytic function inside the circle of convergence, therefore f is analytic at z_0 when it is assigned the value a_0 there, and the singularity z_0 is hence removed. For example:

(i) for the function $f(z) = \frac{\sin z}{z}$; $z \neq 0$ has a removable singularity at $z = 0$ because in the Laurent's series expansion

$$\frac{\sin z}{z} = \frac{1}{z}\left\{z - \frac{z^3}{3!} + \frac{z^5}{5!} - \frac{z^7}{7!} + \cdots\right\} = 1 - \frac{z^2}{3!} + \frac{z^4}{5!} - \frac{z^6}{7!} + \cdots$$

there is no principle part (i.e., $b_n = 0$; $n = 1, 2, \cdots$). So, at $z = 0$, $f(z)$ has the removable singularity. When $f(0) = 1$ is assigned, f becomes entire.

(ii) for the function $f(z) = \frac{1-\cos z}{z^2}$; $z \neq 0$, has a removable singularity at $z = 0$ because in the Laurent's series expansion

$$\frac{1 - \cos z}{z^2} = \frac{1}{z^2}\left[1 - \left\{1 - \frac{z^2}{2!} + \frac{z^4}{4!} - \frac{z^6}{6!} + \cdots\right\}\right] = \frac{1}{2!} - \frac{z^2}{4!} + \frac{z^4}{6!} - \cdots$$

there is no principle part (i.e., $b_n = 0$; $n = 1, 2, \cdots$). So, at $z = 0$, $f(z)$ has the removable singularity. When $f(0) = 1/2$ is assigned, f becomes entire.

Theorem 5.28. *Let f be an analytic function on a punctured domain $\mathcal{D}^* = \mathcal{D} - \{z_0\}$ for some $z_0 \in \mathcal{D}$. Then, z_0 is a removable singularity of f if and only if there exist an analytic function g such that $f(z) = g(z)$ for every $z \in \mathcal{D}^*$.*

Proof: Assume that there exists an analytic function g on $\mathcal{D}$ such that $f(z) = g(z)$ for every $z \in \mathcal{D}^*$. Then, g is continuous at z_0. Therefore

$$\lim_{z \to z_0} f(z) = \lim_{z \to z_0} g(z) = g(z_0)$$

which exists in $\mathbb{C}$.

Conversely, assume that z_0 is a removable singularity. We choose $R > 0$ such that $\mathcal{S}(z_0; R) \subset \mathcal{D}$. Then for each $z \in \mathcal{S}(z_0; R)^* = \mathcal{S}(z_0; R) - \{z_0\}$, we define

$$g_z(\zeta) = \frac{f(\zeta) - f(z)}{\zeta - z} \quad \text{for every } \zeta \in \mathcal{S}(z_0; R)^*.$$

Then, g_z is analytic on $\mathcal{S}(z_0;R)^*$

$$\lim_{\zeta\to z} g_z(\zeta) = \lim_{\zeta\to z}\frac{f(\zeta)-f(z)}{\zeta-z} = f'(z)$$

and $$\lim_{\zeta\to z_0} g_z(\zeta) = \lim_{z\to z_0}\frac{f(\zeta)-f(z)}{\zeta-z} = \frac{\lim_{\zeta\to z_0} f(\zeta)-f(z)}{z_0-z}.$$

Therefore, g_z is bounded on some neighbourhood of z_0 and z. Using Cauchy's theorem for open disc with exceptional points, if $0 < R_1 < R$ and $0 < |z-z_0| < R_1$, then we get

$$\begin{aligned}
&\int_{|\zeta-z_0|=R_1} g_z(\zeta)d\zeta = 0 \Rightarrow \frac{1}{2\pi i}\int_{|\zeta-z_0|=R_1}\frac{f(\zeta)-f(z)}{\zeta-z}d\zeta = 0\\
\Rightarrow\quad & \frac{1}{2\pi i}\int_{|\zeta-z_0|=R_1}\frac{f(\zeta)}{\zeta-z}d\zeta = \frac{f(z)}{2\pi i}\int_{|\zeta-z_0|=R_1}\frac{d\zeta}{\zeta-z}\\
\Rightarrow\quad & f_1(z) = f(z).
\end{aligned}$$

Since f is continuous on $|z-z_0| = R_1$, then f_1 is analytic on $\mathcal{S}(z_0;R)$ and $f_1(z) = f(z); \forall z \in \mathcal{S}(z_0;R)$. Hence, if $g : \mathcal{D} \to \mathbb{C}$ is defined by

$$g(z) = \begin{cases} f(z) & \text{if } z \neq z_0\\ f_1(z_0) & \text{if } z = z_0,\end{cases}$$

then g is analytic on $\mathcal{D}$ and $g(z) = f(z)$ for every $z \in \mathcal{D}^*$. □

Theorem 5.29. [Riemann's theorem]*: Let z_0 be an isolated singularity of $f(z)$ and if $f(z)$ is bounded and analytic in some deleted neighbourhood $\mathcal{N}'(z_0;\delta) = \{z \in \mathbb{C} : 0 < |z-z_0| < \delta\}$, then either $f(z)$ is analytic at z_0, or else z_0 is a removable singular point of f.*

Proof: Let $f(z)$ be bounded on $\mathcal{N}'(z_0;\delta)$. Then $|f(z)| \leq M < \infty$, where M is the maximum value of $f(z)$ on a positively-oriented circle $\Gamma : |z-z_0| = R$, where $R < \delta$, i.e., Γ lies entirely within $\mathcal{N}'(z_0;\delta)$. The Laurent's theorem gives

$$\begin{aligned}
|b_n| &= \frac{1}{2\pi}\left|\oint_\Gamma (z-z_0)^{n-1}f(z)dz\right|\\
&\leq \frac{1}{2\pi}\oint_\Gamma |z-z_0|^{n-1}|f(z)|\,|dz|\\
&\leq \frac{1}{2\pi}R^{n-1}M\cdot 2\pi R = MR^n. \qquad (5.61)
\end{aligned}$$

Since it is true for every R such that $0 < R < \delta$. Taking limit $R \to 0$, we get $b_n = 0 (n > 0)$, so that the Laurent's series for $f(z)$ has no principal part. Consequently, the singularity at z_0 is removable. □

5.9.4 Poles

Consider the domain $\mathcal{D} = \mathcal{S}(z_0, R) = \{z \in \mathbb{C} : d(z, z_0) < R\}$.

Definition 5.9.3. Let f be an analytic function on the annulus $\mathcal{D}^* = \mathcal{D} - \{z_0\}$ for some $z_0 \in \mathcal{D}$. Let the coefficients b_n of the principal parts of the Laurent's series expansion (5.48) of f have only a finite number of non-zero terms. Then, the Laurent's series expansion of $f(z)$ about z_0 is given by

$$\sum_{n=1}^{m}\frac{b_n}{(z-z_0)^n} = \frac{b_1}{z-z_0} + \frac{b_2}{(z-z_0)^2} + \cdots + \frac{b_m}{(z-z_0)^m}, \qquad (5.62)$$

where $b_m \neq 0$. Then the singularity $z = z_0$ is called a *pole* of order m of the function $f(z)$.

Thus, if f has a pole of order $m \in \mathbb{N}$ at z_0, then m is the smallest integer such that $(z - z_0)^m f(z)$ is holomrphic at z_0 and has no zero at z_0, For example:

(i) $f(z) = \sin z/z^2$ has the Laurent series expansion about the singularity $z = 0$ given by

$$\frac{\sin z}{z^2} = \frac{1}{z} - \frac{z}{3!} + \frac{z^3}{5!} - \cdots,$$

for $0 < |z| < \infty$. From this series, we see that $b_1 \neq 0$ and so $z = 0$ is a simple pole of the function $f(z)$.

(ii) $f(z) = \sinh z/z^4$ has the Laurent series expansion about the singularity $z = 0$ given by

$$f(z) = \frac{\sinh z}{z^4} = \frac{1}{z^3} + \frac{1}{3!z} + \frac{z}{5!} + \frac{z^3}{7!} + \cdots, 0 < |z| < \infty.$$

Therefore, $z = 0$ is a pole of $f(z)$ of order 3.

(iii) $f(z) = 1/(z-1)^2(z-3)$ has the Laurent series expansion valid for $0 < |z-1| < 2$ given by

$$f(z) = -\frac{1}{2(z-1)^2} - \frac{1}{4(z-1)} - \frac{1}{8} - \frac{z-1}{16} - \cdots.$$

Since $b_2 = -1/2 \neq 0$, we conclude that $z = 1$ is a pole of order 2.

Theorem 5.30. *Let f be an analytic function on the annulus $\mathcal{D}^* = \mathcal{D} - \{z_0\}$, where z_0 is a pole of order m of f. Then f can be expressed in the form $f(z) = (z - z_0)^{-m}\, g(z)$, where $g(z)$ is analytic at z_0 with $g(z_0) \neq 0$.*

Proof: Since $f(z)$ has a pole of order m at $z = z_0$, the function $f(z)$ can be written as

$$\begin{aligned} f(z) &= \sum_{n=0}^{\infty} a_n(z-z_0)^n + \frac{b_1}{z-z_0} + \frac{b_2}{(z-z_0)^2} + \cdots + \frac{b_m}{(z-z_0)^m}; b_m \neq 0 \\ &= \frac{1}{(z-z_0)^m}\left[\sum_{n=0}^{\infty} a_n(z-z_0)^{n+m} + \sum_{n=1}^{m} b_n(z-z_0)^{m-n}\right] \\ &= \frac{1}{(z-z_0)^m}\,\phi(z) = (z-z_0)^{-m}\,\phi(z), \text{ say,} \end{aligned} \tag{5.63}$$

where $\phi(z)$ is the polynomial in $(z - z_0)$ given by

$$\phi(z) = \sum_{n=0}^{\infty} a_n(z-z_0)^{n+m} + \sum_{n=1}^{m} b_n(z-z_0)^{m-n}. \tag{5.64}$$

Being polynomial, $\phi(z)$ is analytic in $\mathcal{D}$ and $\lim_{z \to z_0} \phi(z) = b_m \neq 0$. □

Thus the poles of a meromorphic function $f(z)$ are all isolated singular points.

Theorem 5.31. *If $f(z)$ has a pole of order m at z_0, then $1/f(z)$ has a zero of order m at z_0.*

Proof: Since $f(z)$ has a pole of order m at $z = z_0$, $1/f(z)$ can be expressed as

$$\frac{1}{f(z)} = (z - z_0)^m \cdot \frac{1}{\phi(z)} = (z - z_0)^m \, \psi(z), \tag{5.65}$$

where $\psi(z_0) = 1/\phi(z_0)$ is non-zero finite, as $\phi(z_0)$ is so. Hence, $1/f(z)$ has a zero at z_0 of order m.□

Note: Let f be meromorphic and z_0 a pole of f of order m. Then $f(z) = (z - z_0)^{-m} g(z)$, where g is analytic at z_0, $g(z_0) \neq 0$. This gives $f'(z) = (z - z_0)^{-m-1}\phi(z)$, where $\phi(z) = (z - z_0)g'(z) - mg(z)$ is analytic at z_0 and $\phi(z_0) = -mg(z_0) \neq 0$. Thus f and f' have the same poles and the order of the poles of f' increases by one.

Theorem 5.32. *If a function $f(z)$ is of the form $f(z) = p(z)/q(z)$, where $p(z)$ and $q(z)$ are analytic at z_0, $p(z_0) \neq 0$ and $q(z)$ has a zero of order m at z_0, then $f(z)$ has a pole of order m at z_0.*

Proof: Since $q(z)$ has a zero of order m at z_0, so by Theorem 5.27, there exists a deleted neighbourhood of z_0 in which $q(z) \neq 0$. This implies that z_0 is an isolated singular point of the quotient of $p(z)/q(z)$. Also, as $q(z)$ has a zero of order m at z_0, we have

$$q(z) = (z - z_0)^m \, \phi(z), \quad \phi(z_0) \neq 0;$$

also, $\phi(z)$ is analytic at z_0. Thus

$$f(z) = \frac{p(z)}{q(z)} = \frac{p(z)}{(z - z_0)^m \, \phi(z)} = \frac{\phi_1(z)}{(z - z_0)^m}, \text{ where } \phi_1(z) = \frac{p(z)}{\phi(z)}.$$

Since, $\phi_1(z)$ is analytic at z_0 and $\phi_1(z_0) \neq 0$, so z_0 is a pole of order m of $f(z)$. □

This theorem gives a very simple method for finding the order of any pole.

Theorem 5.33. *If $f(z)$ has a pole at $z = z_0$, then $\lim_{z \to z_0} f(z) = \infty$.*

Proof: Let $f(z)$ be analytic inside and on a circle Γ of radius R except at the isolated singularity $z = z_0$ taken to be the center of Γ. Let $z = z_0$ be the pole of order m for the function $f(z)$. Then the Laurent's expansion becomes

$$f(z) = \sum_{n=0}^{\infty} a_n (z - z_0)^n + \sum_{n=1}^{m} b_n (z - z_0)^{-n}, \tag{5.66}$$

where the coefficients a_k are given by (5.49). In particular,

$$b_n = \frac{1}{2\pi i} \oint_\Gamma (z - z_0)^{n-1} f(z) dz \quad ; n = 1, 2, \cdots, m$$

where Γ is a circle of radius R in which $f(z)$ is analytic. Now, if $f(z)$ is bounded, i.e., $|f(z)| < M$, where M ia a constant, then as in Eq. (5.61), we get

$$|b_n| \leq \frac{1}{2\pi} R^{n-1} M \cdot 2\pi R = MR^n.$$

Since R can be made arbitrarily small, we have $b_n = 0$; $n \geq 1$ and the Laurent's series reduces to a Taylor's series about $z = z_0$ so that z_0 is not a singularity, which is a contradiction. This shows that $f(z)$ cannot be bounded in the neighborhood of an isolated singularity. Therefore, if $f(z)$ has a pole at $z = z_0$, then $\lim_{z \to z_0} f(z) = \infty$. □

5.9.5 Essential Singularity

Definition 5.9.4. If the principal part of the Laurent's series expansion (5.48) of the function contains an infinite number of terms, i.e., $b_m \neq 0$ are infinite in number, then the isolated singular point z_0 is called an essential singular point of $f(z)$.

In this case, $\lim_{z \to z_0} f(z)$ does not exist. For example,

(i) the function $\sin \frac{1}{z-1}$ has essential singularity at $z = 1$ because

$$\sin \frac{1}{z-1} = \frac{1}{z-1} - \frac{1}{3!} \frac{1}{(z-1)^3} + \frac{1}{5!} \frac{1}{(z-1)^5} - \cdots$$

has infinite number of the terms in negative powers of $(z-1)$.

(ii) $f(z) = e^{3/z}$ has the Laurent series expansion valid for $0 < |z| < \infty$ given by

$$e^{3/z} = 1 + \frac{3}{z} + \frac{3^2}{2!z^2} + \frac{3^3}{3!z^3} + \cdots.$$

It contains an infinite number of non-zero terms. Also, $z = 0$ is neither a pole nor a removable singularity; hence it is an essential singularity of f.

The following theorem describes the behaviour of a holomorphic function in the neighbourhood of an essential singularity.

Theorem 5.34. [Casorati–Weierstrass theorem]: *Let $\mathcal{D} \subset \mathbb{C}$ be a domain containing the number z_0, and a function f that is holomorphic on $\mathcal{D}/\{z_0\}$, but has an essential singularity at z_0. If U is any deleted (punctured) neighbourhood of z_0 contained in $\mathcal{D}$, then $f(U)$ is dense in $\mathbb{C}$, i.e., for any $\varepsilon > 0$, $\delta > 0$ and a complex number α, there exists a complex number $z \in \mathcal{D}$ with*

$$|z - z_0| < \delta \text{ and } |f(z) - \alpha| < \varepsilon.$$

Proof: Assume that z_0 is an essential singularity of f. Let $U \subset \mathcal{D}$ be a punctrued neighbourhood of z_0 and $\alpha \in \mathbb{C}$. We have to show that α is a limit point of $f(U)$.

Suppose it is not, then $\exists$ an $\varepsilon > 0$ such that

$$\Big|f(z) - \beta\Big| \geq \varepsilon \text{ for all } z \in U.$$

Consider a new function $g : U \longrightarrow \mathbb{C}$, by

$$g(z) = \frac{1}{f(z) - \beta}; \quad z \in U$$

which is holomorphic on U with zeros at the pole of f. Also,

$$|g(z) = \frac{1}{\Big|f(z) - \beta\Big|} \leq \frac{1}{\varepsilon} \text{ for every } z \in U,$$

i.e., bounded by ε^{-1}. It can therefore be analytically continued (or holomorphically extended) to all of U by Riemann's analytic continuation theorem, Theorem 5.29 of removable singularity of g. Now

$$g(z) - \frac{1}{f(z) - \beta} \Rightarrow f(z) = \frac{1}{g(z)} + \beta; \quad z \in U.$$

Consider the two possible cases for $\lim_{z \to z_0} g(z)$:

(i) If $\lim_{z \to z_0} g(z) = 0$, then f has a pole at z_0.

(ii) If $\lim_{z \to z_0} g(z) \neq 0$, then z_0 is a removable singularity of f.

These contradict our assumption that z_0 is an essential singularity, which means that α must be a limit point of $f(U)$. The argument holds for all $\alpha \in |C$; so $f(U)$ is dense in $\mathbb{C}$ for any punctured neighbourhood U of z_0.

To prove the converse, assume that $f(U)$ is dense in $\mathbb{C}$ for any punctured neighbourhood U of z_0. If z_0 is a removable singularity, then f is bounded near z_0, and if z_0 is a pole, $f(z) \to \infty$ as $z \to z_0$. Either of these possibilities contradicts the assumption that the image of any punctured neighbourhood of z_0 under f is dense in $\mathbb{C}$. So z_0 must be an essential singularity of f. □

Note: The expansion $\sum_{n=0}^{\infty} \frac{z^n}{3^n} + \sum_{n=1}^{\infty} \frac{1}{z^n}$, $1 < |z| < 3$ has infinite number of negative powers of z but $z = 0$ is not an essential singularity. This is because the domain of convergence is not a deleted neighbourhood of $z = 0$. This point must be kept in mind while deciding for essential singular point. In fact, the mentioned expansion is the Laurent series expansion of the function $2z/\{(1-z)(z-3)\}$ in the annular domain $1 < |z| < 3$, and thus the function has the simple poles at $z = 1$ and $z = 3$.

Theorem 5.35. [Picard's great theorem]: *In the neighbourhood of an essential singular point, a function takes every finite value, with at most a single exception, infinite number of times.*

Let z_0 be an essential singular point of f which is analytic in the punctured disc $\mathcal{D}^* = \{z \in \mathbb{C} : 0 < |z - z_0| < R\} \subset \mathbb{C}$. Then $\mathbb{C}/f(\mathcal{D}^*)$ is a singleton set.

For example, using the fact $e^z = -1$, when $z = (2n+1)\pi i, n \in \mathbb{N}$, we see that $e^{1/z}$ assumes the value -1 infinite number of times in each neighbourhood of origin. Precisely,

$$e^{1/z} = -1 \Rightarrow z = -\frac{i}{(2n+1)\pi}, \quad n \in \mathbb{N},$$

and if n is large enough, such points lie in any given neighbourhood of the origin. Zero is evidently the exceptional value in the Picard's theorem.

RESULT **49.** This is a substantial strengthening of the *Casorati–Weierstrass theorem*, which only guarantees that the range of f is dense in the complex plane. A result of the *Picard's great theorem* is that any entire, non-polynomial function attains all possible complex values infinitely often, with at most one exception. The 'single exception' is needed in both theorems, as demonstrated here:

(i) e^z is an entire non-constant function that is never 0.

(ii) $e^{1/z}$ has an essential singularity at 0, but still never attains 0 as a value.

5.9.6 Singularities at Infinity

Let $z = \frac{1}{\xi}$ in $f(z)$. We then obtain the function $f(z) = f(\frac{1}{\xi}) = g(\xi)$, say. Then, the nature of the singularity at $z = \infty$ (the point of infinity) is defined to be the same as that of $g(\xi)$ at $\xi = 0$.

Theorem 5.36. *If a single-valued function $f(z)$ has no singularities other than poles either in the finite part of the plane or at infinity, then $f(z)$ is a rational function.*

Proof: Suppose that the poles of $f(z)$ in the finite part of the z-plane are $z_1, z_2, \cdots z_r$, of orders $\alpha_1, \alpha_2, \cdots, \alpha_r$ respectively. Then $f(z)$ can be expressed as

$$f(z) = \frac{\phi(z)}{(z-z_1)^{\alpha_1}(z-z_2)^{\alpha_2}\cdots(z-z_r)^{\alpha_r}}$$

$$\text{i.e.,} \qquad \phi(z) = (z-z_1)^{\alpha_1}(z-z_2)^{\alpha_2}\cdots(z-z_r)^{\alpha_r} \cdot f(z),$$

where $\phi(z)$ is analytic for all finite values of z. Therefore, by Taylor's theorem

$$\phi(z) = \sum_{n=0}^{\infty} a_n z^n = a_0 + a_1 z + a_2 z^2 + \cdots .$$

This $\phi(z)$ may have singularity at infinity. Putting $z = 1/\xi$, we have

$$\phi\left(\frac{1}{\xi}\right) = \sum_{n=0}^{\infty} a_n \left(\frac{1}{\xi}\right)^n = a_0 + \frac{a_1}{\xi} + \frac{a_2}{\xi^2} + \cdots .$$

As the singularity of $\phi(z)$ at $z = \infty$ is a pole, therefore the singularity of $\phi(1/\xi)$ at $\xi = 0$, is also a pole. This means that the series for $\phi(1/\xi)$ as well as $\phi(z)$ must terminate, and as such $\phi(z)$ is a polynomial. Therefore,

$$f(z) = \frac{\phi(z)}{(z - z_1)^{\alpha_1}(z - z_2)^{\alpha_2} \cdots (z - z_r)^{\alpha_r}} = \text{Quotient of two polynomials.}$$

This implies that $f(z)$ is a rational function.

Conversely, let $f(z)$ be a rational function. Then it must be expressed as $f(z) = \psi(z)/\phi(z)$, where $\psi(z)$ and $\phi(z)$ are two polynomials having no common factors.

Singularities of $f(z)$ are given by $\phi(z) = 0$, i.e., at the zeros of $\phi(z)$. But zeros of $\phi(z)$ are poles of $1/\phi(z)$. Hence $1/\psi(z)$ has pole at the zero of $f(z)$. Consequently, $f(z)$ has poles at the zeros of $\phi(z)$.

That is, a rational function has no other singularities except the poles. □

Theorem 5.37. *If $f(z)$ is an entire function which has no singularity at infinity. Then $f(z)$ is constant.*

Proof:Let $f(z)$ be entire function. Then it has a power series expansion of the form

$$f(z) = \sum_{k=0}^{\infty} a_k z^k = a_0 + a_1 z + a_2 z^2 + \cdots ; 0 < |z| < \infty \tag{5.67}$$

which converges absolutely for all $z \in \mathbb{C}$. Since $f(z)$ has no singularity at $z = \infty$, the function $g(z) = f(1/z)$ is analytic at $z = 0$. If it is not so, then $g(z) = f(1/z)$ has an essential singularity at $z = 0$, which is a contradiction. Thus $f(1/z)$ is analytic at $z = 0$. Thus, $f(1/z)$ is analytic at $z = 0$ only if $a_n = 0, n \geq 1$. Thus, $f(1/z) = a_0$ or $f(z) = a_0$. □

DEDUCTION **19.** Let an entire function $f : \mathbb{C} \to \mathbb{C}$ have $z = \infty$ as a removable singularity. Then $g(z) = f(1/z)$ has a removable singularity at $z = 0$ and therefore, $g(z)$ is bounded in the deleted neighbourhood of 0. Evidently, $f(z)$ is bounded on $|z| > R$ for some $R > 0$. Thus the Taylor's series expansion for $f(z)$ contains no powers of z, i.e., $f(z) = a_0$. Hence by Liouville's theorem, $f(z)$ reduces to a constant.

Theorem 5.38. *If the only singularity of an analytic function $f(z)$ in the $\mathbb{C}_\infty$ is a pole of order m, then $f(z)$ is a polynomial of degree m.*

Proof: As the only singularity of the function $f(z)$ is a pole of order m at $z = \infty$, therefore, $f(1/z)$ has a pole of order m at $z = 0$. Consider the power series (5.67). Replacing z by $1/z$, we get

$$f\left(\frac{1}{z}\right) = \sum_{n=0}^{\infty} a_n \left(\frac{1}{z}\right)^n = a_0 + \frac{a_1}{z} + \frac{a_2}{z^2} + \cdots , 0 < |z| \leq \infty.$$

Since $f(1/z)$ has a pole of order m at $z = 0$, the terms $z^{-(m+1)}, z^{-(m+2)}, \cdots$ should vanish, i.e., $a_n = 0, \forall n \geq m$. Therefore

$$f\left(\frac{1}{z}\right) = \sum_{n=0}^{m} a_n \left(\frac{1}{z}\right)^n = a_0 + \frac{a_1}{z} + \frac{a_2}{z^2} + \cdots + \frac{a_m}{z^m}, \quad 0 < |z| \leq \infty$$

$$f(z) = a_0 + a_1 z + a_2 z^2 + \cdots + a_m z^m$$

which shows that $f(z)$ is a polynomial of degree m. □

Example 5.9.2. *Prove that an entire function $f(z)$ has a pole of order n at infinity if and only if $f(z)$ is a polynomial of degree n.*

Solution. Suppose $f(z)$ has a pole of order n at infinity. Then $g(z) = f(1/z)$ has a pole of order n at the origin. It follows that $z^n g(z)$ is bounded near 0, i.e., $z^{-n} f(z)$ is bounded near ∞.

That is, $f(z)$ is entire such that $|f(z)| \leq M|z|^n$ for $|z| > R$. Thus $f(z)$ admits a Taylor's series expansion around 0:

$$f(z) = \sum_{k=0}^{\infty} a_k\, z^k; \quad a_k = \frac{1}{2\pi i} \int_{|z|=R} \frac{f(z)}{z^{k+1}}\, dz, \quad R > 0$$

$$\Rightarrow \quad |a_k| \leq \frac{1}{2\pi} \int_{|z|=R} \frac{|f(z)|}{|z|^{k+1}}\, |dz| = \frac{MR^n}{R^{k+1}} 2\pi R = MR^{n-k}.$$

Since R can be made arbitrarily large, we must have $a_k = 0; k > n$. Thus f becomes a polynomial of degree greater than n.

The converse of this part is trivial.

Example 5.9.3. *If $f(z)$ is a non-constant entire function, prove that $e^{f(z)}$ has an essential singularity at $z = \infty$.*

Solution. If $f(z)$ is a non-constant entire function, then $g(z) = e^{f(z)}$ is also entire.

(i) If $g(z)$ has a removable singularity at ∞, then $F(z) = g(1/z)$ has a removable singularity at $z = 0$ and therefore $F(z)$ is bounded in a deleted neighbourhood of 0. Evidently, $g(z)$ is bounded on $|z| > R$ for some $R > 0$. Thus the Taylor's series expansion

$$F(z) = g(1/z) = \sum_{k \geq 0} a_k z^{-k}; \text{ for } |z| > 0,$$

contains no powers of z, i.e., $g(z) = a_0$. Thus by Liouville's theorem $g(z)$ reduce to a constant and consequently $f(z)$ is a constant.

(ii) If $g(z)$ has a pole at ∞, then $1/g(z)$, being a bounded entire function, is a constant.

In either case, $f(z)$ is a constant, which is a contradiction. Thus, ∞ is neither a removable singularity nor a pole of g. Therefore, $g(z) = e^{f(z)}$ has an essential singularity at $z = \infty$ if $f(z)$ is a non-constant entire function.

Theorem 5.39. [Limit point of zeros of a function]: *The limit point of a sequence of zeros of a function is an isolated essential singular point.*

Proof: Let $\langle z_n \rangle$ be a sequence of zeros of $f(z)$ having at least one limit point, say, z_0. This limit point z_0 may or may not belong to $\langle z_n \rangle$.

If $z_0 \in \langle z_n \rangle$, it must be a zeros of $f(z)$. Also, z_0 being the limit point, its any small neighbourhood must contain an infinite number of members of $\{z_n\}$, which goes against the fact that the zeros of a function are isolated. Hence, z_0 cannot be a zero of $f(z)$ and hence it does not belong to $\{z_n\}$ unless $f(z)$ is identically zero.

Thus, if $f(z)$ is not zero identically, z_0 is not a zero of $f(z)$ but is surrounded by many zeros of $f(z)$. This means that z_0 is a singularity of $f(z)$. This singularity z_0 is not a pole as $f(z)$ does not tend to infinity in the neighbourhood of z_0. In fact, $f(z)$ tends to zero there. So, z_0 is an essential singularity of $f(z)$. Also, z_0 is isolated, as in its neighbourhood $f(z)$ is analytic.

Hence z_0 is an isolated essential singularity of $f(z)$. □

For example,

(i) Let $f(z) = (z^2 + 4)e^{-z}$. Everywhere in the complex z-plane, $f(z)$ is defined. Therefore, $f(z)$ has no singularity in the finite part of the complex plane. Zeros of $f(z)$ are given by

$$(z^2 + 4)e^{-z} = 0 \Rightarrow z = \pm 2i, z = 2n\pi i; n \in \mathbb{Z}.$$

Obviously, $z = \infty$ is the limit point of these zeros. Hence $z = \infty$ is an isolated essential singularity.

(ii) Let $f(z) = \cos z - \sin z$. Everywhere in the complex z-plane, $f(z)$ is defined. Therefore, $f(z)$ has no singularity in the finite part of the complex plane. Zeros of $f(z)$ are given by

$$\cos z - \sin z = 0 \Rightarrow \tan z = 1 \Rightarrow z = n\pi + \frac{\pi}{4}; \quad n = 0, \pm 1, \pm 2, \cdots .$$

Obviously, $z = \infty$ is the limit point of these zeros. Hence $z = \infty$ is an isolated essential singularity of $f(z)$.

(iii) Let $f(z) = \sin(\frac{1}{1-z})$. Here, the zeros of $f(z)$ are given by

$$\sin\left(\frac{1}{1-z}\right) = 0 \Rightarrow \frac{1}{1-z} = n\pi \Rightarrow z = 1 - \frac{1}{n\pi}; n = 0, \pm 1, \pm 2, \cdots .$$

The limit point of all these zeros is $z = 1$, which is clearly a singularity of $f(z)$. This singularity is an isolated essential singularity of $f(z)$.

(iv) Let $f(z) = \frac{z-2}{z^2} \sin \frac{1}{z-1}$. The singularities of $f(z)$ are at points, where $z^2 = 0$, i.e., $z = 0$. This singularity $z = 0$ of $f(z)$ is a pole of order 2, since in this case $f(z) = \frac{1}{z^2}\phi(z)$, where $\phi(z) = (z-2)\sin\frac{1}{z-1}$ is being regular at $z = 0$.

The zeros of $f(z)$ are at points where

$$(z-2)\sin\frac{1}{z-1} = 0 \Rightarrow z = 2 \text{ and } z = 1 + \frac{1}{n\pi}; n = 0, \pm 1, \pm 2, \cdots ,$$

which are the zeros of $f(z)$. The limit point of the zeros of $f(z)$ is $z = 1$. Therefore, $z = 1$ is an isolated essential singularity of $f(z)$.

(v) Let $f(z) = \sec(1/z) = 1/\cos(1/z)$. Here $f(0)$ does not exist. So, $z = 0$ is a singular point of $f(z)$. The singular points, other than $z = 0$, are given by the equation

$$\cos\left(\frac{1}{z}\right) = 0 \Rightarrow \frac{1}{z} = 2n\pi + \frac{\pi}{2} \Rightarrow z = \frac{1}{2n\pi + \frac{\pi}{2}}; \quad n = 0, \pm 1, \pm 2, \cdots .$$

These points are poles of $\sec(1/z)$. Thus the set of poles of the function is $\{z_n\} = \left\{\dfrac{1}{2n+\frac{\pi}{2}}\right\}$, where n is an integer. $\{z_n\}$ has the limit point $z_0 = z = 0$. Clearly, $z = 0$ is a singularity of $f(z)$. It is non-isolated and is not a pole but an essential singularity of $f(z)$.

RESULT **50.** Let $f(z)$ be an analytic function in a simply-connected domain $\mathcal{D}$ and $\langle z_n\rangle$ be a sequence of zeros having z_0 as its limit point, where $z_0 \in \mathcal{D}$. Then $f(z)$ vanishes identically in $\mathcal{D}$.

Theorem 5.40. [Limit point of poles of a function]: *The limit point of a sequence of poles of an analytic function $f(z)$ is a non-isolated essential singularity of $f(z)$.*

Proof: Let $\langle z_n\rangle$ be a sequence of poles of $f(z)$ having at least one limit point, say, z_0. This limit point z_0 may or may not belong to $\langle z_n\rangle$.

(i) If $z_0 \in \langle z_n\rangle$, it is a pole of $f(z)$. By virtue of z_0 being a limit point, there must exist a cluster of poles, of $f(z)$ around it. This goes against the definition of poles as poles are isolated singular points. Therefore, $z_0 \notin \langle z_n\rangle$.

(ii) Also, z_0 cannot be zero of $f(z)$ because $f(z)$ is not analytic within the neighbourhood of z_0.

Therefore, z_0 is an essential singularity of $f(z)$. This singularity is not isolated because its neighbourhood contains an infinite number of poles of the function.

Therefore, the limit point of the poles of a function is a non-isolated essential singularity. □

For the validity of the theorem, consider the following examples:

(i) Let $f(z) = \frac{\cot \pi z}{(z-1)^2}$. The pols of $f(z)$ are at points where

$$(z-1)\sin \pi z = 0 \Rightarrow z = 1 \text{ and } z = n; \quad n = 0, \pm 1, \pm 2, \cdots .$$

This means that $f(z)$ has poles at $z = 0$, $z = 1$(a triple pole), and $z = -1, \pm 2, \pm 3, \cdots$ in which $z = \infty$ is the limit point. Hence, $z = \infty$ is a non-isolated essential singularity for $f_1(z)$.

(ii) Let $f_2(z) = \tan \frac{1}{z} = \frac{\sin 1/z}{\cos 1/z}$. Poles of $f_2(z)$ are given by

$$\cos\frac{1}{z} = 0 \Rightarrow \frac{1}{z} = 2n\pi + \frac{\pi}{2} \Rightarrow z = \frac{1}{(2n+1/2)\pi}; \quad n = 0, \pm 1, \pm 2, \cdots .$$

Obviously $z = 0$ is the limit point of these poles of $f_2(z)$ and so it is a non-isolated essential singularity.

(iii) Let $f_3(z) = \cot z = \frac{\cos z}{\sin z}$. Poles of $f_3(z)$ are given by

$$\sin z = 0 \Rightarrow z = n\pi; \quad n = 0, \pm 1, \pm 2, \cdots .$$

Obviously $z = \infty$ is the limit point of the sequence of these poles. Hence $z = \infty$ is a non-isolated essential singularity.

(iv) Let $f_4(z) = z\text{cosec } z = \frac{z}{\sin z}$. Poles of $f_4(z)$ are given by $\sin z = 0$. As in previous, $z = \infty$ is the limit point of the sequence of these poles. Hence $z = \infty$ is a non-isolated essential singularity.

(v) Let $f_5(z) = \text{cosec }(\frac{1}{z}) = 1/\sin(1/z)$. Here $f_5(0)$ does not exist. So, $z = 0$ is a singular point of $f_5(z)$. The singular points, other than $z = 0$, are given by

$$\sin\left(\frac{1}{z}\right) = 0 \Rightarrow \frac{1}{z} = n\pi \Rightarrow z = \frac{1}{n\pi}; \quad n = 0, \pm 1, \pm 2, \cdots .$$

which are poles of $f_5(z)$. Thus the sequence of poles of $f_5(z)$ is $\{z_n\} = \{\frac{1}{n\pi}\}$, where $n \in \mathbb{Z}$ which has the limit point $z = 0$. Hence $z = 0$ is a non-isolated essential singularity.

(vi) Poles of $f_6(z) = \frac{\cot \pi z}{(z-z_0)^2}$ are given by $(\sec \pi z)(z - z_0)^2 = 0$, which gives

$$\Rightarrow \pi z = n\pi \text{ and } z = z_0 \Rightarrow z = n \text{ and } z = z_0 \ ; \ n \in \mathbb{Z}.$$

Here $z = z_0$ is a double pole. Obviously, $z = \infty$ is the limit point of these poles $z = n : n \in \mathbb{Z}$. Hence $z = \infty$ is a non-isolated essential singularity.

5.10 Schwarz's Lemma and Its Consequences

Schwarz's lemma implies that every conformal equivalence between Δ and itself is implemented by a Möbius transformation. Indeed, suppose f maps Δ conformally onto itself. Fix a Möbius transformation T which stands $f(0)$ to 0 and maps Δ into itself. This lemma tells us that a α so that $(T \circ f)(z) = \alpha z$, for all $z \in \Delta$. In particular, $T \circ f$ is the restriction of a Möbius transformation to Δ. Thus the same is true for f, as desired.

Theorem 5.41. [Schwarz's Lemma]: *Let $\Delta = \mathcal{S}(0,1)$ and $\bar{\Delta} = \mathcal{S}[0,1] = \mathcal{D} \cup \partial\mathcal{D}$. Let $f : \Delta \to \bar{\Delta}$ be an analytic function, having a zero of order n at the origin. Also, $|f(z)| \leq 1, \forall z \in \Delta$. Then*

$$|f(z)| \leq |z|^n, \quad \forall z \in \Delta \tag{5.68}$$

$$\left|f^{(n)}(0)\right| \leq n!. \tag{5.69}$$

Further, equality holds in Eq. (5.68) for some non-zero $z \in \Delta$ or in Eq. (5.69) if and only if $f(z)$ is of the form $f(z) = kz^n$, where k is a constant of unit modulus.

Proof: Let $f : \Delta \to \bar{\Delta}$ be analytic on Δ and have an nth order zero at the origin. By using the fact $f(0) = f'(0) = f''(0) = \cdots = f^{(n-1)}(0) = 0$ and $f^{(n)}(0) \neq 0$ in Taylor expansion, we have

$$f(z) = \frac{z^n}{n!} f^{(n)}(0) + \frac{z^{n+1}}{(n+1)!} f^{(n+1)}(0) + \cdots ; \text{ for } z \in \Delta.$$

For $z = 0$, the result is obvious. Let $z \neq 0$. Then from the above expansion we have

$$\frac{f(z)}{z^n} = \frac{1}{n!} f^{(n)}(0) + \frac{z}{(n+1)!} f^{(n+1)}(0) + \cdots ; \text{ for } z \in \Delta^*,$$

where $\Delta^* = \Delta - \{0\}$ is the punctured unit disc. The function $f(z)/z^n$ has a removable singularity at the origin, and so, let us define

$$g(z) = \begin{cases} z^{-n} f(z) & \text{for } z \in \Delta^* \\ \dfrac{1}{n!} f^{(n)}(0) & \text{for } z = 0. \end{cases}$$

Then, $g(z)$ is analytic and continuous on Δ. We claim that $|g(z)| \leq 1$ for all $z \in \Delta$. Let $\Delta_R = \mathcal{S}(0, R)$, where $0 < R < 1$, so that Δ_R is a bounded domain. Now,

(i) g is analytic on the bounded domain Δ_R.

(ii) g is continuous on the compact set $\bar{\Delta}_R$.

Thus, the maximum modulus principle is applicable and

$$|g(z)| \leq \max_{z\in\mathcal{D}_R^*} |g(z)| = \max_{|z|=R} \left|\frac{f(z)}{z^n}\right| \leq \frac{1}{R^n} \text{ for all} z \in \Delta^*$$

as $|f(z)| < 1$, for every $z \in \Delta$. The above inequality is true for arbitrary $R < 1$. Thus by letting $R \to 1_-$, we have

$$|g(z)| = \frac{|f(z)|}{|z|^n} \leq 1 \Rightarrow |f(z)| \leq |z|^n, \quad z \in \Delta,$$

which yields Eq. (5.68). Also, as $g(0) = \frac{f^{(n)}(0)}{n!}$ and $|g(0)| \leq 1$, we get the inequality (5.69).

If $|f(z_0)| = |z_0|^n$ for some $z_0 \neq 0$, then $g(z_0) = 1$ and by the strict maximum modulus principle, $g(z)$ is constant, say, $g(z) = k \Rightarrow f(z) = kz^n$ for some constant with modulus 1, say $g(z) = e^{i\alpha} \Rightarrow f(z) = e^{i\alpha}z^n = kz^n$, where, $|k| = 1$. Similarly, we can prove $f^{(n)}(0) = n!$. □

Corollary 5.10.1. Any holomorphic automorphism of $\Delta = \{z \in \mathbb{C} : |z| < 1\}$ from origin is a rotation.

Proof: Let $f : \Delta \to \Delta$ be holomorphic isomorphism with $f(0) = 0$. All the conditions of Schrawz's lemma is satisfied. Applying this lemma (Lemma 5.41),

$$\left|f(z)\right| \leq |z| \text{ for every } z \text{ with } |z| < 1. \tag{5.70}$$

Since f is holomorphic isomorphism, so is f^{-1} and $f^{-1}(\infty) = f(0)$. Since Schwarz's lemma can be applied to f^{-1},

$$\begin{aligned} &\left|f^{-1}(w)\right| \leq |w| \text{ for every } w \text{ with } |w| < 1 \\ \Rightarrow \quad & w = f(z) \Rightarrow |z| \leq \left|f(z)\right|. \end{aligned} \tag{5.71}$$

From Eqs. (5.70) and (5.71), it follows that $\left|f(z)\right| = |z|$ for every z with $|z| < 1$, and by Schwarz's lemma, f is a rotation. From this, we conclude that, every member of the holomorphic automorphisms of $\triangle$ which fixes the origin is given by the mapping $f(z) = e^{i\theta}\, z$ for some $\theta \in \mathbb{R}$. □

Theorem 5.42. *Let $\Delta = \mathcal{S}(0,1)$. For a fixed $a \in \Delta$ with $|a| < 1$, define a transformation $\phi_a : \Delta \to \Delta$ as*

$$\phi_a(z) = \frac{a - z}{1 - \bar{a}z} \text{ for } z \in \Delta. \tag{5.72}$$

Then ϕ_a is an analytic bijection of Δ onto itself; the inverse of ϕ_a is ϕ_{-a}. Furthermore, ϕ_a maps $\partial\Delta$ onto $\partial\Delta$, with $\phi_a(a) = 0$, $\phi_a'(0) = -(1 - |a|^2)$ and $\phi_a'(a) = -(1 - |a|^2)^{-1}$.

Proof: Clearly, ϕ_a is a Möbius transformation. It is analytic on $\mathbb{C} - \{\bar{a}^{-1}\}$ if $a \neq 0$ and $\phi_0(z) = -z$. In particular, for each $a \in \Delta$, $\phi_a(z)$ is analytic in an open disc containing the closure $\bar{\Delta} = \Delta \cup \partial\Delta = \mathcal{S}[0,1]$ with $\phi_a(a) = 0$. Now,

$$1 - \left|\phi_a(z)\right|^2 = \frac{|1 - \bar{a}z|^2 - |a - z|^2}{|1 - \bar{a}z|^2} = \frac{(1 - |a|^2)(1 - |z|^2)}{|1 - \bar{a}z|^2}$$

so that $|\phi_a(z)| = 1$ if $|z| = 1$ and $|\phi_a(z)| < 1$ for $|z| < 1$. On $\partial\Delta$, $z = e^{i\theta}; \theta \in \mathbb{R}$. Then

$$\left|\phi_a(e^{i\theta})\right| = \left|\frac{a - e^{i\theta}}{1 - \bar{a}e^{i\theta}}\right| = |e^{-i\theta}|\left|\frac{a - e^{i\theta}}{e^{-i\theta} - \bar{a}}\right| = \left|\frac{a - e^{i\theta}}{\overline{a - e^{i\theta}}}\right| = 1.$$

This shows that $\phi_a(\partial\Delta) = \partial\Delta$ and this is true as every Möbius transformation maps circles to circles. Thus by the maximum modulus principle, $|\phi_a(z)| < 1$ on Δ.

Since ϕ_a has a unique singularity $\bar{a}^{-1} \notin \bar{\Delta}$; ϕ_a is an analytic function on Δ; and hence $\phi_a(\Delta)$ is a connected subset of $\{z \in \mathbb{C} : |z| \neq 1\}$. Therefore,

$$\text{either } \phi_a(\Delta) = \Delta \text{ or } \phi_a(\Delta) = \{z \in \mathbb{C} : |z| > 1\}.$$

Also for $|z| < 1$

$$\phi_a\Big(\phi_a(z)\Big) = \frac{a - \phi_a(z)}{1 - \bar{a}\phi_a(z)} = \frac{a - \frac{a-z}{1+\bar{a}z}}{1 - \bar{a}\frac{a-z}{1+\bar{a}z}} = \frac{(1 - |a|^2)z}{1 - |a|^2} = z.$$

Hence ϕ_a maps $\Delta \to \Delta$ in a one-one and onto fashion. This shows that ϕ_a is invertible and ϕ_a itself is the inverse for ϕ_a. Since $\phi_a(a) = 0$ and $a \in \Delta$, we get $\phi_a(\Delta) = \Delta$. As every linear factional transform is a one-to-one map, $\phi_a : \Delta \to \Delta$ is a bijection, satisfying $\phi_a(a) = 0$, and so ϕ_a^{-1} given by

$$\phi_a^{-1}(w) = \frac{a - w}{1 - \bar{a}\, w}$$

is also an analytic map on Δ satisfying $\phi_a^{-1}(0) = a$. Since $\phi_a(z)$ is analytic, so

$$\phi_a'(z) = \frac{(1 - \bar{a}z)\cdot(-1) - (a - z)\cdot(-\bar{a})}{(1 - \bar{a}z)^2} = -\frac{1 - |a|^2}{(1 - \bar{a}z)^2}.$$

It follows that $\phi_a'(0) = -(1 - |a|^2)$ and $\phi_a'(a) = -(1 - |a|^2)^{-1}$. □

Corollary 5.10.2. Let f be a non-constant analytic function in the neighbourhood of closed unit disc $\bar{\Delta} = \mathcal{S}[0, 1]$ and $z_0 \in \partial\Delta$ be such that $|f(z_0)| = \max_{z\in\bar{\Delta}} |f(z)|$. Then $f'(z_0) \neq 0$.

Proof: Let $f'(z_0) \neq 0$. Using maximum modulus principle, $f(\Delta) \subset \Delta$. Let us consider a function g, defined by $g(z) = (\phi_a \circ f)(z)$, where $\phi_a(z)$ is given by Eq. (5.72). Let $f(0) = a \neq 0$. Then $g : \Delta \to \Delta$ is homomorphic isomorphism with $g(0) = 0$. Now

$$\begin{aligned}\Big|g'(z_0)\Big| &= \Big|\phi_a'(f(z_0))\cdot f'(z_0)\Big| = \Big|\phi_a'(e^{i\theta}z_0)\Big|\cdot\Big|z_0\Big| \\ &= \frac{1 - |a|^2}{\Big|1 - \bar{a}e^{i\theta}z_0\Big|^2}\cdot|z_0| \geq 1.\end{aligned}$$

A direct calculation shows that

$$g'(z) = \phi_a'(f(z))\, f'(z) = -\frac{1 - |a|^2}{[1 - \bar{a}f(z)]^2}\, f'(z)$$

$$\Rightarrow \quad f'(z_0) = -\frac{[1 - \bar{a}f(z_0)]^2}{1 - |a|^2} g'(z_0) = -\frac{[1 - \bar{a}z_0]^2}{1 - |a|^2} g'(z_0).$$

Since $|g'(z_0)| \geq 1$, the above result gives

$$\Big|f'(z_0)\Big| \geq \frac{|1 - \bar{a}z_0|^2}{1 - |a|^2} g'(z_0) \geq \frac{[1 - |a||z_0|]^2}{1 - |a|^2}\cdot 1$$

Since $z_0 \in \partial\Delta$, we have $|z_0| = 1$, so,

$$\Big|f'(z_0)\Big| \geq \frac{[1 - |a|^2]}{1 - |a|^2} = \frac{1 - |a|}{1 + |a|} > 0.$$

This completes the proof. In this corollary, the behaviour of an analytic function at the point z_0 where it assumes its maximum modulus will be reflected. □

The Schwarz's lemma leads to the hyperbolic metric and hyperbolic geometry on the unit disc.

Definition 5.10.1. Now, for $\zeta_0, \zeta_1 \in \Delta$, we consider the hyperbolic metric $\rho_{hy}(\zeta_0, \zeta_1)$. Let us define $H : \Delta \longrightarrow \Delta$ as

$$H(\zeta) = e^{i\beta}\Big(\frac{\zeta - \zeta_0}{1 - \bar{\zeta}_0\zeta}\Big); \quad \text{where, } \beta = -\arg\Big(\frac{\zeta_1 - \zeta_0}{1 - \bar{\zeta}_0\zeta_1}\Big).$$

Then the *hyperbolic metric* $\rho_{hy}(\zeta_0, \zeta_1)$ is defined by

$$\begin{aligned}\rho_{hy}(\zeta_0, \zeta_1) &= \rho_{hy}\Big(H(\zeta_0), H(\zeta_1)\Big) = \rho_{hy}\Big(0, H(\zeta_1)\Big) = \frac{1}{2}\ln\Big(\frac{1 + H(\zeta_1)}{1 - H(\zeta_1)}\Big) \\ &= \text{Radial distance from origin to a point on the real axis.}\end{aligned}$$

5.10.1 Some Consequences of Schwarz's Lemma

Theorem 5.43. [A maximal problem]: *Let f be analytic and satisfies $|f(z)| \leq 1$ on the unit disc $\Delta = \mathcal{S}(0,1)$. If $f(a) = b$, for some $a, b \in \Delta$, then the maximum possible value of $|f'(a)|$ is given by*

$$|f'(a)| \leq \frac{1 - |b|^2}{1 - |a|^2} = \frac{1 - |f(a)|^2}{1 - a^2}. \tag{5.73}$$

Proof: If, $|f(z)| = 1$, for some point $z \in \Delta$, then f is a constant and Eq. (5.73) is trivial.

Let us assume that $|f(z)| < 1$ for all $z \in \Delta$. For fixed $a \in \Delta$, we define $g : \Delta \to \Delta$ by $g = \phi_b \circ f \circ \phi_a^{-1}$, where $\phi_a(z)$ is given by Eq. (5.72). Then g satisfies

$$g(0) = \Big(\phi_b \circ f\Big)\phi_a^{-1}(0) = \Big(\phi_b \circ f\Big)(a) = \phi_b(f(a)) = \phi_b(b) = 0.$$

Thus g satisfies all the conditions of Schwarz's lemma and hence we obtain $|g'(0)| \leq 1$. To obtain an explicit form $g'(0)$, apply chain rule of differentiation. We get

$$\begin{aligned}g'(0) &= \Big(\phi_a \circ f\Big)'(\phi_a(0)) \cdot \Big[\phi_a^{-1}\Big]'(0) = \Big(\phi_a \circ f\Big)'(a) \cdot \Big(1 - |a|^2\Big) \\ &= \phi_a'(f(a)) \cdot f'(a) \cdot \Big(1 - |a|^2\Big) \\ &= \phi_a'(b) \cdot f'(a)\Big(1 - |a|^2\Big) = \frac{1 - |a|^2}{1 - |b|^2}\, f'(a).\end{aligned}$$

But $|g'(0)| \leq 1$, so the inequality (5.73) is verified. Moreover, equality will occur exactly when $|g'(0)| = 1$. In that case, by virtue of Schwarz's Lemma, there is a constant $k \in \mathbb{C}$ with $|k| = 1$ and $g(z) = kz = I_k(z)$. Hence

$$f(z) = \phi_b^{-1} \circ I_k(z) = \phi_b^{-1}\Big(k\phi_a(z)\Big) \text{ for } z \in \Delta \tag{5.74}$$

This proves the theorem and is known as *Schwarz–Pick lemma* . □

In particular, this observation implies the following results:

(i) There is no analytic function $f : \Delta \to \bar{\Delta}$ such that $f(1/2) = 3/4$ and $f'(1/2) = 3/4$. Indeed, let $a = 1/2$ and $b = 3/4$ in Eq. (5.73). According to Eq. (5.73), with $f(a) = b$, it follows that

$$|f'(1/2)| \leq \frac{1 - \frac{9}{16}}{1 - \frac{1}{4}} = \frac{7}{12} \Rightarrow \frac{3}{4} \leq \frac{7}{12}$$

as given $f'(1/2) = 3/4$, which is not possible.

(ii) There exists an analytic function $f : \triangle \to \triangle$ such that $f(1/2) = 0$ and $|f'(1/2)| \leq 4/3$.

DEDUCTION **20.** Let $f : \Delta \to \bar{\Delta}$ be analytic such that $f(0) = a_1$ for some point $a_1 \in \Delta$. Using Eq. (5.73), we observe that

$$|f'(0)| \leq \frac{1 - |f(0)|^2}{1 - 0^2} = 1 - |a_1|^2.$$

Let there be a complex number a_2 satisfying $|a_2| \leq 1 - |a_1|^2$. Thus, there exists an f such that $f'(0) = a_2$ for some constant $a_2 \in \mathbb{C}$.

For example:

(i) The function $\phi_a(z)$, as in Eq. (5.72), satisfies $\phi_a(0) = a$ and $\phi_a'(0) = -(1 - |a|^2)$ and so, f is defined by $f(z) = k\phi_a(z)$, with $|k| \leq 1$.

(ii) Let $a_1 = -1/2$ and $f(0) = a_1$. Then there exists an analytic function $f : \triangle \to \triangle$ such that $f(0) = a_1$ and $f'(0) = 3/4$. Thus the function is

$$f(z) = \phi_{a_1}(z) = \frac{z - \frac{1}{2}}{1 - \frac{1}{2}z} = \frac{2z - 1}{2 - z}.$$

(iii) There is no analytic function $f : \triangle \to \triangle$ such that $f(0) = \pm 1/2$ and $f'(0) = 4/5$.

RESULT **51.** Let us define a metric

$$\rho(z, a) = |\phi_a(z)| = \left| \frac{z - a}{1 - \bar{a}z} \right| \text{ for } a, z \in \Delta.$$

If we apply Schwarz lemma on $g = \phi_a \circ f \circ \phi_a^{-1}$, it follows that $|g'(0)| \leq 1$ and $|g(z)| \leq |z|$ on Δ.

$$\therefore \left| \left(\phi_b \circ f \circ \phi_a^{-1} \right)(z) \right| \leq \left| z \right|, \quad z \in \Delta$$

Since $\phi_a(\phi_a(z)) = z$, for each $z \in \Delta$, put z in $\phi_a(z)$. It follows that

$$\left| \left(\phi_b \circ f \right)(z) \right| \leq \left| \phi_a(z) \right|, \quad z \in \Delta \tag{5.75}$$

In particular, if we take $z = a_1$, then the inequality (5.75) is equivalent to

$$\left| \phi_b(b_1) \right| \leq \left| \phi_a(a_1) \right| \Rightarrow \rho(b, b_1) \leq \rho(a, a_1)$$

$$\Rightarrow \quad \rho\Big(f(a), f(a_1) \Big) \leq \rho(a, a_1). \tag{5.76}$$

The inequality (5.76) for the analytic functions $f : \triangle \to \triangle$ reveals the contraction property of f with respect to the metric ρ. Furthermore, assume that $a_1 \neq a_2$ are elements of $\partial\triangle$ and that $f(a_1) = b_1$ and $f(a_2) = b_2$. Then

$$\left| \frac{b_2 - b_1}{1 - \bar{b}_1 b_2} \right| \leq \left| \frac{a_2 - a_1}{1 - \bar{a}_1 a_2} \right|, \tag{5.77}$$

By Schrawz lemma, equality (5.77) holds if and only if $g(z) = kz$ with $|k| = 1$, i.e.,

$$f(z) = \phi_b^{-1} \circ I_k(z) \circ \phi_a(z); \quad z \in \Delta.$$

As a consequence, there is also a pair of uniqueness statements:

$$\text{either } f'(a) = \frac{1-|b|^2}{1-|a|^2} \text{ or } \left|\frac{b_2 - b_1}{1-\bar{b}_1 b_2}\right| = \left|\frac{a_2 - a_1}{1-\bar{a}_1 a_2}\right|. \tag{5.78}$$

For $a_1 \neq a_2$, then f is a conformal self-map of the unit disc Δ. □

DEDUCTION **21.** Let $f : \Delta \to \bar{\Delta}$ be analytic such that $f(a_1) = 0$ and $f(a_1 - 1) = a_2$ for some $a_1 \in (0,1)$ and $a_2 \in \Delta$. Using Eq. (5.77), we have

$$|a_2| \leq \frac{1}{1 - a_1(a_1 - 1)} = \frac{1}{1 + a_1 - a_1^2}. \tag{5.79}$$

From this result we can make the following assertions:

(i) Let $a_1 = \frac{1}{4}$ and $a_2 = \frac{17}{20}$. Then $1 + a_1 - a_1^2 = \frac{19}{16}$. Thus there exists no function f, analytic in Δ such that $f(1/4) = 0$ and $f(-3/4) = 17/20$, as $\frac{17}{20} \not< \frac{16}{19}$.

(ii) Let $a_1 = \frac{1}{3}$ and $a_2 = \frac{5}{6}$. Then $1 + a_1 - a_1^2 = \frac{11}{9}$. As $\frac{5}{6} \not< \frac{9}{11}$, so there exists no function f that is analytic in Δ such that $|f(z)| \leq 1$, $f(1/3) = 0$ and $f(-2/3) = 5/6$.

Theorem 5.44. *Let $f : \Delta \to \Delta$ be a conformal, analytic bijection and $f(a) = 0$, where $a \in \Delta$. Then there is a constant k, with $|k| = 1$ such that $f(z) = k\phi_a(z)$, where $\phi_a(z)$ is defined in Eq. (5.72)*

Proof: Since f is conformal and bijective, we can have inverse of f as $g : \Delta \to \Delta$ defined by

$$g\Big(f(z)\Big) = z \text{ for } |z| < 1,$$

which is an analytic function and $g(0) = a$. Applying chain rule of differentiation, we get

$$g'(f(a)) \cdot f'(a) = 1 \Rightarrow g'(0) \cdot f'(a) = 1.$$

Applying to inequality (5.73), f and g have to satisfy

$$|f'(a)| \leq \Big(1 - |a|^2\Big)^{-1} \text{ and } |g'(0)| \leq 1 - |a|^2.$$

But since $g'(0)f'(a) = 1$, it follows that $|f'(a)| = (1 - |a|^2)^{-1}$. Hence applying the result (5.74), we find that $f = k\phi_a$ for some $k \in \mathbb{C}$ with $|k| = 1$. □

This is an important consequence of Schwarz's lemma, which may be taken as the converse form of Theorem 5.43.

Theorem 5.45. *Let $f : \bar{\Delta} \to \bar{\Delta}$ be analytic such that $|f(z)| \leq 1$ on $\partial\Delta = \{z \in \mathbb{C} : |z| = 1\}$; $f(a_1) = 0$ and $f(-a_1) = a_2$ for some $a_1 \in (0,1)$ and $a_2 \in (0,1]$. Then the maximum value of $|a_2|$ is $2a_1/(1 + a_1^2)$.*

Proof: Using relation (5.76), we get

$$\begin{aligned} & \rho\Big(f(a_1), f(-a_1)\Big) \leq \rho(a_1, -a_1) \\ \Rightarrow \quad & \left|\frac{f(a_1) - f(-a_1)}{1 - \bar{f(a_1)} f(-a_1)}\right| \leq \left|\frac{a_1 - (-a_1)}{1 - \bar{a}_1(-a_1)}\right| \\ \Rightarrow \quad & \left|\frac{0 - a_2}{1 - 0 \cdot a_2}\right| \leq \left|\frac{2a_1}{1 + |a_1|^2}\right| \Rightarrow |a_2| \leq \frac{2a_1}{1 + a_1^2}. \end{aligned}$$

Thus the maximum value of $|a_2|$ is $2a_1/(1 + a_1^2)$, and hence the theorem. □

From that result, we can make the followings observations:

(i) Let $a_1 = 1/3$ and $a_2 = 13/20$. Then $\frac{2a_1}{1+a_1^2} = \frac{3}{5}$. Since $\frac{13}{20} \not< \frac{3}{5}$, so there exists no analytic function $f : \bar{\Delta} \to \bar{\Delta}$ such that $|f(z)| \le 1$ on $\partial\Delta$, $f(1/3) = 0$ and $f(-1/2) = 13/20$.

(ii) Let $a_1 = 1/2$. Then there exists an analytic function $f : \bar{\Delta} \to \bar{\Delta}$ such that $f(1/2) = 0$ and $f(-1/2) = 4/5$.

Theorem 5.46. *Let f be analytic in an unit disc Δ with $f(a) = 0$ and $|f(z)| \le |e^{iz}|$ for all $z \in \partial\Delta$. The the maximum value of $|f(w)|$ is $\left|e^{iw}\ \phi_a(w)\right|$ for all w satisfying $|w| \le 1$.*

Proof: Let $F(z) = e^{-iz} f(z)$. Then $|F(z)| \le 1$ for $z \in \partial\Delta$. Let us define a function g as $g = F \circ \phi_a$, where $\phi_a(z)$ is defined in Eq. (5.72). Then $g(a) = 0$ and by the maximum modulus principle, $|g(z)| \le 1$ for $|z| \le 1$. Thus by Schwarz's lemma,

$$|g(z)| \le |z| \quad \text{i.e.,} \quad \left|e^{-i\phi_a(z)}\ f\Big(\phi_a(z)\Big)\right| \le |z|$$

$$\text{or,} \quad \left|e^{-iw}\ f(w)\right| \le \left|\phi_a^{-1}(w)\right| = \left|\frac{a-w}{1-\bar{a}w}\right|$$

$$\text{or,} \quad |f(w)| \le \left|e^{iw}\ \frac{a-w}{1-\bar{a}w}\right| = \left|e^{iw}\ \phi_a(w)\right| \tag{5.80}$$

for all w satisfying $|w| < 1$. The equality sign holds if $f(w) = e^{i\beta}\ e^{iw}\ \phi_a(w)$ for some β. □

DEDUCTION **22.** Let $f : \Delta \to \bar{\Delta}$ be analytic such that $f(a_1) = 0$ for some point $a_1 \in \Delta$. Using Eq. (5.80), we have

$$\Big|f(z)\Big| \le \Big|\phi_{a_1}(z)\Big|, \quad \phi_{a_1}(z) = \frac{a_1 - z}{1 - \bar{a}_1 z} \quad \text{for } z \in \Delta.$$

In particular, for some $a_2 \in \Delta$, we have $|f(a_2)| \le |\phi_{a_1}(a_2)|$, which is the estimate for $|f(a_2)|$ for $a_2 \in \Delta$. The maximum is achieved when $f(z) = \phi_{a_1}(z)$.

For example, let $a_1 = 1/3$ and $a_2 = 1/4$. If $f(1/3) = 0$ then the estimate for $f(a_2)$ is given by

$$f(a_2) \le \left|\frac{a_1 - a_2}{1 - \bar{a}_1 \cdot a_2}\right| = \left|\frac{\frac{1}{3} - \frac{1}{4}}{1 - \frac{1}{3}\cdot\frac{1}{4}}\right| = \frac{1}{11}$$

and the maximum is attained for $\phi_{1/3}(z)$.

Exercises

Multiple-choice Questions

1. For the positively oriented unit circle $\oint_{|z|=1} \frac{2\text{Re}(z)}{z+2} dz$

   (a) 0 (b) πi (c) $2\pi i$ (d) $4\pi i$

2. The value of $\int_0^{2\pi} \exp(e^{i\theta} - i\theta) d\theta$ equals to

   (a) $2\pi i$ (b) 2π (c) π (d) πi

3. The value of the integral $\int_{-\pi}^{\pi} \frac{dt}{4+R^2-4R\cos(\theta-t)}$, $-\pi < \theta \le \pi, 0 < R < 2$ is

   (a) 0 (b) $2\pi R^2(\cos\theta - \sin\theta)$ (c) $\frac{4-R^2}{2\pi}$ (d) $\frac{2\pi}{R^2-4}$

4. If the function $f(z) = \frac{(1-\cos z)^k}{z^7}$ has a simple pole at the origin, then the value of k is:

   (a) 1 (b) 2 (c) 3 (d) 4

5. The value of $\int_\Gamma \bar{z}\, dz$, where Γ is parametrized by $z(t) = 3t + it^2$, $-1 \leq t \leq 4$, is
   (a) $159 + 65i$ (b) $195 + 65i$ (c) $195 + 56i$ (d) $159 + 56i$

6. Let $\Gamma = \{z \in \mathbb{C} : |z| = 2\}$ be oriented in the counter clockwise direction. Let $I = \frac{1}{2\pi i} \oint_\Gamma z^7 \cos(z^{-2}) dz$. Then, the value of I is equal to
   (a) 0.01 (b) 0.02 (c) 0.03 (d) 0.04

7. Suppose function f is continuous in a domain $\mathcal{D}$. Then among the following statements, namely: P. f has primitive in $\mathcal{D}$; Q. the integral of $f(z)$ along any path lying in $\mathcal{D}$ between any two fixed points in $\mathcal{D}$ is independent of path; R. the integral of $f(z)$ along every closed contour in $\mathcal{D}$ is zero:
   (a) P implies R but not Q. (b) Q implies R but not implied by P.
   (c) P implies Q and R but not implied by either of Q or R.
   (d) All the statements are equivalent.

8. If $f(z) = (z^2 + 5z + 6)/(z - 2)$ and the path of integration is a circle with origin at the centre and radius R, the Cauchy's theorem is applicable whenever R equals
   (a) 1 (b) 2 (c) 3 (d) 1 and 2

9. The value of the integral $\int_\Gamma (4x^3 dx + 3y^2 z^2 dy + 2y^3 z\, dz)$, where Γ is any path joining $A(-1, 1, 0)$ to $B(1, 2, 1)$, is
   (a) 1 (b) 0 (c) -8 (d) 8

10. The value of $\int_\Gamma \frac{4z^2+z+5}{z-4} dz$, where $\Gamma : 9x^2 + 4y^2 = 36$, is
   (a) -1 (b) 1 (c) 2 (d) 0

11. The value of the integral $\int_\Gamma \frac{z^2+1}{(z+1)(z+2)} dz$, where $\Gamma : |z| = 3/2$, is
   (a) πi (b) 0 (c) $2\pi i$ (d) $4\pi i$

12. Let Γ be the contour $|z| = 2$, oriented in the anti clockwise direction. The value of the integral $\int_\Gamma z e^{3/z} dz$ is
   (a) $3\pi i$ (b) $5\pi i$ (c) $7\pi i$ (d) $9\pi i$

13. Let $\Omega = \{z \in \mathbb{C} : \text{Im}(z) > 0\}$ and Γ be a smooth curve lying in Ω with initial point $-1 + 2i$ and final point $1 + 2i$. The value of $\int_\Gamma \frac{1+2z}{1+z}\, dz$ is
   (a) $4 - \ln\frac{2}{2} + \frac{\pi}{4}$ (b) $-4 - \ln\frac{2}{2} + \frac{\pi i}{4}$ (c) $4 + \ln\frac{2}{2} + \frac{\pi}{4}$ (d) $-4 + \ln\frac{2}{2} + \frac{\pi i}{4}$

14. If $a \in \mathbb{C}$ with $|a| < 1$, then the value of $\frac{1-|a|^2}{\pi} \int_\Gamma \frac{|dz|}{|z+a|^2}$, where Γ is the simple closed curve $|z| = 1$ taken with the positive orientation, is
   (a) 0 (b) 1 (c) 2 (d) 3.

15. The value of the integral $\int_\Gamma \frac{z+1}{z^3-2z^2} dz$, where Γ, is the circle $|z| = 1$ is equal to
   (a) $2\pi i$ (b) $-2\pi i/3$ (c) zero (d) $-3\pi i/2$

16. The maximum of $|\sin z|$ on the square $\{x + iy : 0 \leq x \leq 2\pi, 0 \leq y \leq 2\pi\}$ is
   (a) 1 (b) $\sinh 2\pi$ (c) $\cosh 2\pi$ (d) ∞

17. The coefficient of $(z - \pi)^2$ in the Taylor series expansion of

$$f(z) = \begin{cases} \sin z/(z - \pi) & \text{if } z \neq \pi \\ -1 & \text{if } z = \pi \end{cases}$$

   around π is
   (a) $1/2$ (b) $-1/2$ (c) $1/6$ (d) $-1/6$

18. In the Laurent's series expansion of $f(z) = \frac{1}{z-1} - \frac{1}{z-2}$ valid in the region $|z| > 2$, the coefficient of $1/z^2$ is

    (a) -1 (b) 0 (c)' 1 (d) 2.

19. Let $f(z) = \frac{1}{z^2-3z+2}$. Then, the coefficient of z^{-3} in the Laurent series expansion of $f(z)$ for $|z| > 2$ is

    (a) 0 (b) 1 (c) 3 (d) 5

20. The poles of the function $f(z) = \frac{\cos z}{z^2+1}$ are

    (a) Integral multiple of 2π (b) 0 only (c) i and $-i$ (d) ± 1

21. The function $f(z) = (\sin z/z^2)$ has

    (a) pole of order 2 at origin with residue 1 (b) pole of order 1 at origin with residue 1
    (c) pole of order 1 at origin with residue 2 (d) an essential singularity at origin with residue 1.

22. The function $f(z) = \sin z/z^2$ has

    (a) no singularities (b) an essential singularity at $z = 0$
    (c) a removable singularity at $z = 0$ (d) a pole at order 1 at $z = 0$.

23. The function $f(z) = e^{z/\sin z}$ has

    (a) no singularities (b) an essential singularity at $z = k\pi$, $k \in \mathbb{Z} - \{0\}$
    (c) a removable singularity at $z = 0$ (d) has an uncountable number of zeros.

24. The function $f(z) = \frac{e^z-1}{z(z-2)}$ has

    (a) no singularities (b) an essential singularity at $z = 0$
    (c) a removable singularity at $z = 0$ (d) a pole at order 1 at $z = 0$.

25. The function $f(z) = \frac{e^z-1}{z(z-2)}$ has

    (a) no singularities (b) an essential singularity at $z = \infty$.
    (c) a removable singularity at $z = \infty$ (d) a pole at order 1 at $z = 0$.

26. The function $f(z) = \frac{e^z-1}{z(z-2)}$ has

    (a) no singularities (b) an essential singularity at $z = 2$
    (c) a removable singularity at $z = 2$ (d) a pole at order 1 at $z = 2$.

27. The function $f(z) = \sin(1 - \frac{1}{z})$ has

    (a) a countable number of zeros (b) an uncountable number of zeros
    (c) a pole at $z = 0$ (d) an essential singularity at $z = 0$.

28. The function $f(z) = e^z - 2i$

    (a) has no zeros (b) has only one zero
    (c) has a countable number of zeros (d) has an uncountable number of zeros

29. The function $f(z) = \tan z$

    (a) has poles at $z = (2n+1)\pi/2$, $n \in \mathbb{Z}$ (b) has no zeros in $\mathbb{C}$
    (c) is an entire function (d) has removable singularity at $z = 0$

30. At $z = 0$, the function $f(z) = \frac{e^z+1}{e^z-1}$ has

    (a) a removable singularity (b) a pole
    (c) an essential singularity (d) the residue of $f(z)$ at $z = 0$ is 2.

31. Let f be an entire function such that $\lim_{|z|\to\infty} |f(z)| = \infty$, then

(a) $f(1/z)$ has an essential singularity at 0 (b) f cannot be a polynomial
(c) f has finitely many zeros (d) $f(1/z)$ has a pole at 0.

32. For the function $f(z) = \sin\frac{1}{z}$, $z = 0$ is a

(a) removable singularity (b) simple pole
(c) essential singularity (d) branch point

33. The nature of the singularities of $f(z) = \frac{e^{2z}}{(z-1)^4}$ is

(a) A removable singularity (b) A pole
(c) An essential singularity (d) Branch point

34. The nature of the singularities of $f(z) = (z-3)\sin\frac{1}{z+2}$, is

(a) A removable singularity (b) A pole
(c) An essential singularity (d) Branch point

35. The nature of the singularities of $f(z) = \frac{z-\sin z}{z^2}$, is

(a) A removable singularity (b) A pole
(c) An essential singularity (d) Branch point

36. An example of a function with a non-isolated essential singularity at $z = 2$ is

(a) $\tan\frac{1}{z-2}$ $\sin\frac{1}{z-2}$ (c) e^{z-2} (d) $\tan\frac{z-2}{z}$

37. At $z = 0$, the function $f(z) = \exp\left(\frac{z}{1-\cos z}\right)$ has

(a) A removable singularity (b) A pole (c) An essential singularity
(d) The Laurent's expansion at $f(z)$ around $z = 0$ has infinitely many positive and negative powers of z.

38. Consider the function $f(z) = z^2(1-\cos z)$, $z \in \mathbb{C}$. Which of the following are correct?
(a) The function f has zeros at orders 2 at 0.
(b) The function f has zeros of order 1 at $2k\pi$, $k = \pm 1, \pm 2, \cdots$
(c) The function f has zeros of order 4 at 0.
(d) The function f has zeros of order 2 at $2k\pi$, $k = \pm 1, \pm 2, \cdots$

39. Let $f(z) = \frac{z-1}{\exp(2\pi i/z)-1}$. Then

(a) f has an isolated singularity at $z = 0$ (b) f has infinitely many poles
(c) f has a removable singularity at $z = 1$ (d) Each pole of f is of order 1

40. For the function $f(z) = \frac{1-e^{-z}}{z}$, the point $z = 0$ is

(a) An essential singularity (b) A pole of order zero
(c) A pole of order one (d) A removable singularity

41. Let $f(z) = u(x,y) + iv(x,y)$ be an entire function having Taylors series expansion $\sum_{n=0}^{\infty} a_n z^n$. If $f(x) = u(x,0)$ and $f(iy) = iv(0,y)$, then

(a) $a_{2n} = 0, \forall n$ (b) $a_0 = a_1 = a_2 = a_3 = 0, a_4 \neq 0$
(c) $a_{2n+1} = 0, \forall n$ (d) $a_0 \neq 0$ but $a_2 = 0$

42. $\int_{|z+1|=2} \frac{z^2}{4-z^2}\, dz =$

(a) 0 (b) $-2\pi i$ (c) $2\pi i$ (d) 1

43. The maximum modulus of e^{z^2} on the set $S = \{z \in \mathbb{C} : 0 \leq \text{Re}(z) \leq 1, 0 \leq \text{Im}(z) \leq 1\}$ is

(a) $2/e$ (b) e (c) $e+1$ (d) e^2

44. Let $f(z)$ be defined on the domain $E : |z - 2i| < 3$ and on its boundary ∂E. Then, which one of the following statements is always true?

(a) If $f(z)$ is analytic on E and $f(z) \neq 0$ for any z in E, then $|f|$ attains its maximum on ∂E.
(b) If $f(z)$ is analytic on $E \cup \partial E$, then $|f|$ attains on ∂E.
(c) If $f(z)$ is analytic on E and continuous on $E \cup \partial E$, then $|f|$ attains its maximum and minimum on ∂E.
(d) If $f(z)$ is analytic on $E \cup \partial E$ and $f(z) \neq 0$ for any z in $E \cup \partial E$, then $|f|$ attains its minimum on ∂E.

45. Let f and g be meromorphic functions on $\mathbb{C}$. If f has a zero of order k at $z = a$ and g has a pole of order m at $z = 0$. Then $g\{f(z)\}$ has

(a) A zero of order km at $z = a$ (b) A pole of order km at $z = a$
(c) A zero of order $|k - m|$ at $z = a$ (d) A pole of order $|k - m|$ at $z = a$

46. For $z \in \mathbb{C}$, define $f(z) = \frac{e^z}{e^z - 1}$. Then

(a) f is entire (b) f has infinitely many poles on the imaginary axis
(c) only singularities of f are poles (d) each pole of f is simple.

47. Let $f(z) = a_0 + a_1 z + \cdots + a_n z^n$ and $g(z) = b_1 z + b_2 z^2 + \cdots + b_n z^n$ be complex polynomials. If a_0, b_1 are non-zero complex numbers, then residue of $f(z)/g(z)$ at $z = 0$ is equals to

(a) $\dfrac{a_0}{b_1}$ (b) $\dfrac{b_1}{a_0}$ (c) $\dfrac{a_1}{b_1}$ (d) $\dfrac{a_0}{a_1}$

48. Let $\mathcal{D} = \{z \in \mathbb{C} : |z| < 1\}$ be the unit disc. Let $f : \mathcal{D} \to \mathbb{C}$ be an analytic function satisfying $f(1/n) = \frac{2n}{3n+1}$ for $n \geq 1$. Then

(a) $f(0) = 2/3$ (b) f has a simple pole at $z = 3$
(c) $f(3) = 1/3$ (d) No such f exists.

49. Let f be an entire function such that $|f(z)| <\leq A|z|^p$ for every $z \in \mathbb{C}$ and for some $p(\geq 0)$, A being a positive constant. Then

(a) f is a polynomial of degree not exceeding p (b) f is transcendental
(c) f is a polynomial of degree greater than p (d) f is a polynomial of degree p
[**Hint:** Generalised Liouville's theorem]

50. Let f be an entire function. Which of the following statements are correct?
(a) f is constant if the range of f is contained in a straight line.
(b) f is constant if f has uncountably many zeros.
(c) f is constant if f is bounded on $\{z \in \mathbb{C} : \text{Re}(z) \leq 0\}$
(d) f is constant if the real part of f is bounded.

51. Schwarz's lemma states that:

(a) If $\triangle = \{z \in \mathbb{C} : |z| < 1\}$ be the open unit disc, then $|f'(0)| \leq 1$ and $|f'(z)| \leq |z|$ for all $z \in \triangle$.

(b) Suppose f is analytic on $\triangle$ with $|f(z)| \leq 1$ for $z \in \triangle$ and $f(0) = 0$. Then $|f'(0)| \leq 1$ and $|f(z)| \leq |z|$ for all $z \in \triangle$.

(c) f is analytic on $\triangle$ (d) None of these

52. If $f(z)$ is analytic for $|z| < 1$ such that $|f(z)| \le 1$, $f(0) = 0$. Then

(a) $|f(z)| > |z|$ (b) $|f(z)| < |z|$ (c) $|f(z)| = |z|$ (d) $f(z) = z$

53. Let $\triangle = \{z \in \mathbb{C} : |z| < 1\}$ be the open unit disc in $\mathbb{C}$ and $f : \mathbb{C} \to \triangle$ be analytic with $f(10) = 1/2$. Then $f(10 + i)$ is

(a) $\dfrac{1+i}{2}$ (b) $\dfrac{1-i}{2}$ (c) $\dfrac{1}{2}$ (d) $\dfrac{i}{2}$

54. Let $\triangle = \{z \in \mathbb{C} : |z| < 1\}$ be the open unit disc. Then there exists a holomorphic function $f : \triangle \to \bar{\triangle}$ with $f(0) = 0$ with the property

(a) $f'(0) = 1/2$ (b) $|f(1/3)| = 1/4$ (c) $f(1/3) = 1/2$ (d) $|f'(0)| = \sec\frac{\pi}{6}$

55. Let $\bar{\triangle} = \{z \in \mathbb{C} : |z| \le 1\}$ be the closed unit disc. Let $f : \bar{\triangle} \to \bar{\triangle}$ be a holomorphic function satisfying $f(0) = 1/2$ and $f(1/2) = 0$. Which of the following is/are not correct?

(a) $|f'(0)| \le 3/4$ (b) $f(z) = z$ for $z \in \bar{\triangle}$

(c) $|f'(0)| \le 3/4$ and $|f'(1/2)| \le 4/3$ (d) $|f'(1/2) \le 4/3$

56. Let f be analytic function defined on the unit open disc $\triangle = \{z \in \mathbb{C} : |z| < 1\}$ in $\mathbb{C}$. Then f is constant if

(a) $f\left(\dfrac{1}{n}\right) = 0$ for all $n \ge 1$ (b) $f(z) = 0$ for all $|z| = \frac{1}{2}$

(c) $f\left(\dfrac{1}{n^2}\right) = 0$ for all $n \ge 1$ (d) $f(z) = 0$ for all $z \in (-1, 1)$

57. Let $\triangle = \{z \in \mathbb{C} : |z| < 1\}$ be the open unit disc. Let $f : \triangle \to \mathbb{C}$ be an analytic function satisfying $f(\frac{1}{n}) = \frac{2n}{3n+1}$ for $n \in \mathbb{N}$. Then

(a) $f(0) = 2/3$ (b) f has a pole at $z = -3$ (c) $f(3) = 1/3$ (d) No such f exists

58. Let $\triangle = \{z \in \mathbb{C} : |z| < 1\}$ be the open unit disc and U be an open subset of $\mathbb{C}$ containing $\triangle$. If $f : U \to \mathbb{C}$ is the open map given by $f(z) = e^{i\theta}\frac{z-a}{1-\bar{a}z}$ for some $a \in \triangle$, then

(a) $|f(e^{i\theta})| = 1$ for $0 \le \theta \le 2\pi$ (b) f maps $\triangle$ onto itself

(c) f is analytic automorphism (d) f is one-to-one

59. The transformation $w = e^{i\theta}\frac{z-a}{\bar{a}z-1}$, where a is a constant, maps $|z| < 1$ onto

(a) $|w| < 1$ if $|a| < 1$ (b) $|w| > 1$ if $|a| > 1$

(c) $|w| = 1$ if $|a| = 1$ (b) $|w| = 3$ if $a = 0$

60. Let $f(z)$ be analytic on the open unit disc $\triangle = \{z \in \mathbb{C} : |z| < 1\}$ such that $f(1) = 1$. If $f(z) = f(z^2)$ for all $z \in \triangle$, then which one of the following statements is not correct?

(a) $f(z) = [f(z)]^2$ for all $z \in \triangle$ (b) $f\left(\dfrac{z}{2}\right) = \dfrac{1}{2}f(z)$ for all $z \in \triangle$

(c) $f(z^3) = [f(z)]^3$ for all $z \in \triangle$ (d) $f'(1) = 0$

Answers

1(a); 2(b); 3(d); 4(c); 5(b); 6(d); 7(d); 8(a); 9(d); 10(d); 11(d); 12(d); 13(a); 14(c); 15(d); 16(c); 17(c); 18(a); 19(c); 20(c); 21(b); 22(b); 23(b); 24(c); 25(b); 26(d); 27(d); 28(a); 29(a); 30(d); 31(d); 32(c); 33(b); 34(c); 35(a); 36(a); 37(d); 38(c,d); 39(d); 40(c); 41(a); 42(c); 43(b); 44(d); 45(b); 46(b,c,d); 47(a); 48(a,b,c); 49(a); 50(a,b,c); 51(b); 52(c); 53(a); 54(a,b); 55(b); 56(a,b,c,d); 57(a,b,c); 58(a,c); 59(a); 60(a,d)

Short-answer Type Questions

1. For each of the following domains, determine whether it is simply-connected(SC), multiply-connected(MC), or neither (NC, not connected):

(*i*) $\{z \in \mathbb{C} : 0 < |2z-1| < 3\}$ (*ii*) $\{z \in \mathbb{C} : \text{Arg } z \neq 0\}$

(*iii*) $\{z \in \mathbb{C} : |z-i| > 1, |z-3| > 3 \text{ and } |z| < 10\}$ (*iv*) $\{z : \mathcal{I}m\, z \neq 0\}$

(*v*) $\{z : |z| < 1 \text{ and } \mathcal{R}e\, z < 0\}$ (*vi*) $\mathbb{C} - \{x+iy : 0 \leq x \leq 3\}$

2. For each of the following smooth curves give an admissible parametrization that is consistent with the indicated direction:

(a) The line segment from $z = 1+i$ to $z = -2-3i$
(b) The circle $|z-2i| = 4$ traversed once in the clockwise direction, starting from the point $z = 4+2i$
(c) The arc of the circle $|z| = R$ lying in the second quadrant, from $z = iR$ to $z = -R$
(d) The segment of the parabola $y = x^2$ from $(1,1)$ to the point $(3,9)$

3. Show that if $\gamma_1 : [a,b] \longrightarrow \mathbb{C}$ is a path, then there is an equivalence path $\gamma_2 : [0,1] \longrightarrow \mathbb{C}$.
4. Show that the ellipse $\frac{x^2}{a^2} + \frac{y^2}{b^2} = 1$ is a smooth curve by producing an admissible parametrization.
5. Show that the range of the function $z(t) = t^3 + it^6$; $-1 \leq t \leq 1$, is a smooth curve even though the given parametrization is not admissible.
6. Let $\gamma(t) = e^{(-1+i)t^{-1}}$ for $0 < t \leq 1$ and $\gamma(0) = 0$. Show that γ is a rectifiable path and find $|\gamma(t)|$. Give a rough sketch of the trace of γ.
7. Show that $\gamma : [0,1] \longrightarrow \mathbb{C}$, defined by $\gamma(t) = t + it\sin\frac{1}{t}$ for $t \neq 0$ and $\gamma(0) = 0$, is a path but is not rectifiable. Sketch this path.
8. Let $G = \mathbb{C}/\{0\}$ and show that every closed curve in G is homotopic to a closed curve whose trace is contained in $\{z : |z| = 1\}$.
9. Let $I(r) = \int_\gamma \frac{e^{iz}}{z} dz$, where $\gamma : [0,\pi] \longrightarrow \mathbb{C}$ is defined by $\gamma(t) = re^{it}$. Show that $\lim\limits_{r \longrightarrow \infty} I(r) = 0$.
10. Define $\gamma : [0,2\pi] \longrightarrow \mathbb{C}$ by $\gamma(t) = e^{int}$ where n is some integer. Show that $\int_\gamma z^{-1} dz = 2\pi ni$.
11. Evaluate the following integrals ($n \in \mathbb{N}$):

(a) $\displaystyle\int_\gamma \frac{e^{iz}}{z^2} dz;\ \gamma(t) = e^{it}, 0 \leq t \leq 2\pi$ (b) $\displaystyle\int_\gamma \frac{dz}{z-a} dz;\ \gamma(t) = a + re^{it}, 0 \leq t \leq 2\pi$

(c) $\displaystyle\int_\gamma \frac{\sin z}{z^3} dz;\ \gamma(t) = e^{it}, 0 \leq t \leq 2\pi$ (d) $\displaystyle\int_\gamma \frac{\log z}{z^n} dz;\ n \geq 0\ \gamma(t) = 1 + \tfrac{1}{2}e^{it}, 0 \leq t \leq 2\pi$

(e) $\displaystyle\int_\gamma \frac{\sin z}{z} dz;\ \gamma(t) = e^{it}, 0 \leq t \leq 2\pi$ (f) $\displaystyle\int_\gamma \frac{z^{1/m}}{(z-1)^m} dz;\ \gamma(t) = 1 + \tfrac{1}{2}e^{it}, 0 \leq t \leq 2\pi$

(g) $\displaystyle\int_\gamma \frac{dz}{z^2+1};\ \gamma(t) = 2e^{it}, 0 \leq t \leq 2\pi$ (h) $\displaystyle\int_\gamma \frac{e^z - e^{-z}}{z^n} dz;\ \gamma(t) = e^{it}, 0 \leq t \leq 2\pi$

(i) $\displaystyle\int_\gamma \frac{dz}{(z-\frac{1}{2})^n};\ \gamma(t) = e^{it}, 0 \leq t \leq 2\pi$ (j) $\displaystyle\int_\gamma \left(\frac{z}{z-1}\right)^n dz;\ \gamma(t) = 1 + e^{it}, 0 \leq t \leq 2\pi$

(k) $\displaystyle\int_\gamma z^n dz;\ \gamma(t) = e^{it}, 0 \leq t \leq 2\pi$ (l) $\displaystyle\int_\gamma \frac{dz}{z^2+1};\ \gamma(t) = 2|\cos 2t|e^{it}, 0 \leq t \leq 2\pi$

12. Find all possible values of $\int_\gamma \frac{dz}{1+z^2}$, where γ is any closed rectifiable curve in $\mathbb{C}$ not passing through $\pm i$.
13. Evaluate $\int_\gamma \frac{z^2+1}{z(z^2+4)}dz$, where $\gamma(t) = re^{it}$; $0 \le t \le 2\pi$, for all possible values of r, $0 < r < 23$ and $2 < r < \infty$.
14. Let $f(z) = [(z-\frac{1}{2}-i)\cdot(z-1-\frac{3}{2}i)(z-1-\frac{i}{2})(z-\frac{3}{2}-i)]^{-1}$ and γ be the polygon $[0, 2, 2+2i, 2i, 0]$. Find $\int_\gamma f$.
15. Prove that a function which has complex derivative identically 0 on a convex open set U is constant on U.
16. Calculate $\text{Ind}_\gamma(z_0)$ if $\gamma(t) = z_0 + e^{int}$; $t \in [0, 2\pi]$ and n is any integer.
17. Prove that if γ is a closed path whose complement has just two components and if γ has at least one simple point, then $\text{Ind}_\gamma(z) = \pm 1$ on the bounded component.
18. Suppose γ_1 and γ_2 are closed paths and z is not either of the paths. Show that $\text{Ind}_{\gamma_1+\gamma_2}(z) = \text{Ind}_{\gamma_1}(z) + \text{Ind}_{\gamma_2}(z)$ and $\text{Ind}_{-\gamma_1} = -\text{Ind}_{\gamma_1}(z)$.
19. Evaluate $\int_\gamma (z^2-1)^{-1}dz$, where
    (a) $\gamma(t) = 1 + e^{it}$ for $0 \le t \le 2\pi$ (b) $\gamma(t) = 2e^{it}$ for $-\pi \le t \le \pi$.
20. Evaluate the given integral along the indicated contour:
    (a) $\displaystyle\int_C (z+3)\,dz$, where C is $x = 2t, y = 4t - 1; 1 \le t \le 3$
    (b) $\displaystyle\int_C z^2\,dz$, where C is $z(t) = 3t + 2it;\ -2 \le t \le 2$
    (c) $\displaystyle\int_C \frac{z+1}{z}\,dz$, where C is the right half of the circle $|z| = 1$ from $z = -i$ to $z = i$
    (d) $\displaystyle\oint_C \text{Re}(z)\,dz$, where C is the circle $|z| = 1$
    (e) $\displaystyle\int_C (x^2 + iy^3)\,dz$, where C is the line from 1 to i
21. Let $f_1(z)$ and $f_2(z)$ be analytic inside and on a simple loop Γ. Prove that if $f_1(z) = f_2(z)$ for all $z \in \Gamma$, then $f_1(z) = f_2(z)$ for all z inside Γ.
22. Show that if f_1 and f_2 are primitives for $f : G \longrightarrow \mathbb{C}$ and G is connected, then there is a constant c such that $f_1(z) = f_2(z) + c$ for each $z \in G$.
23. Let f and g be analytic in G and let γ be a rectifiable curve from a to b in G. Then prove $\int_\gamma fg' = f(a)g(b) - f(a)g(a) - \int_\gamma f'g$.
24. Compute the integral
    (a) $\int_\Gamma \sin z\,dz$, where Γ is the contour as shown in the Fig. 5.9.
    (b) $\int_\Gamma \frac{3z-2}{z^2-z}\,dz$, where the closed contour Γ is depicted in the Fig. 5.28(a).
    (c) $\int_\Gamma \frac{\cosh z}{z(z+1)^2}dz$, where the closed contour Γ is depicted in the Fig. 5.30(b).
    (d) $\int_\Gamma \frac{\cos z}{z^2(z-3)}dz$ along the closed contour Γ as depicted in the Fig. 5.35(a)
    (e) $\int_\Gamma \frac{3z+1}{z(z-2)^2}dz$ along the closed contour Γ as depicted in the Fig. 5.35(b)
25. Evaluate $\int_\gamma \frac{z+2}{z}\,dz$ where, the equation of γ is

    (i) $z = 2e^{i\theta}; 0 \le \theta \le \pi$ (ii) $z = 2e^{i\theta}, \pi \le \theta \le 2\pi$ (iii) $z = 2e^{i\theta}; -\pi \le \theta \le \pi$

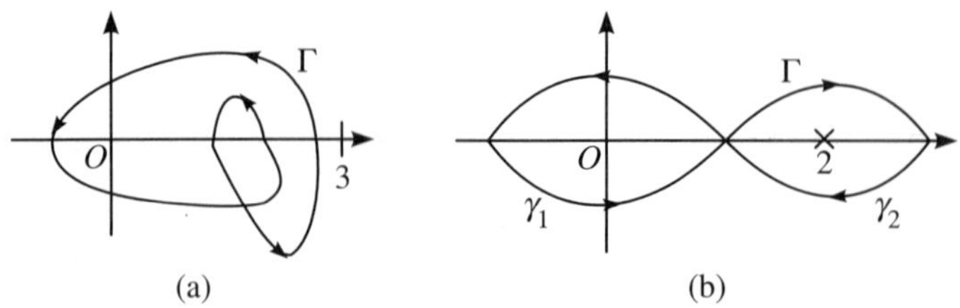

Figure 5.35 *Contour of the exercise*

26. Evaluate the following integrals using Cauchy's integral formula:

(i) $\displaystyle\int_{|z|=3} \frac{z\,e^z}{z-1}\,dz.$ (ii) $\displaystyle\int_{|z-(2+i)|=2} \frac{z}{z-(1+i)}\,dz$

(iii) $\displaystyle\int_{|z-2|=3} \frac{e^{z+1}}{(z+6)^4}\,dz$ (iv) $\displaystyle\int_{|z|=2} \frac{e^z}{(z+1)(z-3)^2}\,dz$

(v) $\displaystyle\int_{|z+2i|=2} \frac{dz}{z^2+1}$ (vi) $\displaystyle\int_{|z-1|=2} \frac{(z^2+5z+6)\,dz}{(z^2+1)(z-1+i)}$

(vii) $\displaystyle\int_{|z|=2} \frac{\sin z dz}{z+i}$ (viii) $\displaystyle\int_{|z-1|=2} \frac{z+1}{z^2-9z+20}\,dz$

(ix) $\displaystyle\int_{|z|=1} \frac{e^z}{(z-2)^3}\,dz$ (x) $\displaystyle\int_{|z|=3} \frac{z^3+1}{z^2+3z-10}\,dz$

(xi) $\displaystyle\int_{|z|=5} \frac{(3z+1)dz}{(z-1)(z-2)^2}$ (xii) $\displaystyle\int_{|z|=5} \frac{2z+3}{z^2-2z-3}\,dz$

(xiii) $\displaystyle\int_{|z-3|=1} \frac{z^2+1}{z-1}\,dz$ (xiv) $\displaystyle\int_{|z-1|=4} \frac{(z^2-3)\,dz}{(z^2+3z+2)(z+6)}$

27. If $f(z)$ is analytic

(a) and $|f(z)| \le \frac{1}{1-|z|}$ in $|z|<1$, show that $f'(0)| \le 4$.

(b) on $S(0,1)$ and suppose that $|f(z)| \le 1$ for $|z|<1$, show that $|f'(0)| \le 1$.

28. Without actually evaluating the value of the integral, show that

(a) $\displaystyle\left|\int_C \frac{1}{z^4}\,dz\right| \le 4\sqrt{2}$, where C is the line segment from $z=i$ to $z=1$.

(b) $\displaystyle\left|\int_C \frac{z}{z^2+1}\,dz\right| \le \frac{1}{2}$, where C is the straight line segment from $z=2$ to $z=2+i$.

(c) $\displaystyle\left|\int_C \frac{1}{z^2+1}\,dz\right| \le \frac{\pi}{3}$, where C is the arc of the circle $|z|=2$ from $z=2$ to $z=2i$.

(d) $\displaystyle\left|\int_C \frac{z+4}{z^3-1}\,dz\right| \le \frac{6\pi}{7}$, where C is the arc of the circle $|z|=2$ from $z=2$ to $z=2i$.

(e) $\displaystyle\left|\int_C \frac{\cos z}{z}\,dz\right| \le 2\pi e$, where C is the path that traces the unit circle once.

29. Find the upper bound for the absolute value of the integral

(a) $\displaystyle\int_C \sin(z^2)\, dz$, where C is the circle $|z| = 2$.

(b) $\displaystyle\int_C \frac{(z-1)e^{2z} \operatorname{Log} z}{z^2 - 7}\, dz$, where $C = \left\{z : z = 2e^{i\theta}, \pi/3 \le \theta \le \pi/2\right\}$.

(c) $\displaystyle\int_C e^{z^2}\, dz$, where C denotes the broken lines from $z = 0$ to $z = 1$ and then from $z = 1$ to $z = 1 + i$.

(d) $\displaystyle\int_C \frac{\operatorname{Log} z}{z^2} dz$, where $C : |z| = R$.

30. Prove that the following:

(i) $\int_{|z|=1} \frac{\sin^2 z}{(z-\pi i/6)^3} dz = \pi i$ (ii) $\int_{|z|=1} \frac{\sin 2z}{(z-\pi i/4)^4} dz = -\frac{8i\pi}{3} \cosh(\pi/2)$

(iii) $\int_{|z|=2} \frac{e^{2z}}{(z+1)^4} dz = -\frac{8i\pi}{3} e^{-2}$ (iv) $\int_{|z|=2} \frac{e^{iz}}{z^3} dz = -\pi i$

(v) $\int_{|z|=3} \frac{e^z\, dz}{(z+2)(z+1)^2} = 2\pi i e^{-2}$ (vi) $\int_{|z|=4} \frac{e^z}{(z^2+\pi^2)^2} dz = i/\pi$.

31. If $f(z) = \int_\Gamma \frac{\zeta^2-\zeta+2}{\zeta - z}\, d\zeta$, where Γ is the ellipse $\frac{x^2}{4} + \frac{y^2}{9} = 1$, traversed once in the positive direction, find $f(1)$, $f'(i)$ and $f''(-i)$.

32. Let Γ be a simple-closed positively-oriented contour that passes through the point $2 + 3i$. Set $f(z) = \frac{1}{2\pi i} \int_\Gamma \frac{\cos\zeta}{\zeta - z} dz$. Find the following limits:

(a) $\lim\limits_{z\to 2+3i} f(z)$, where z approaches $2 + 3i$ from inside Γ

(b) $\lim\limits_{z\to 2+3i} f(z)$, where z approaches $2 + 3i$ from outside Γ

33. Find the maximum value of $|f(z)|$, where

(a) $f(z) = \cos z$ on the square $\{x + iy : 0 \le x \le 2\pi, 0 \le y \le 2\pi\}$

(b) $f(z) = e^{z^3}$ on the closed circle $S[0, 1]$

(c) $f_1(z) = ze^z + z^2$ and $f_2(z) = iz^2 - 2z$ on the set $\{z : |z| \le 1 \text{ and } \operatorname{Im} \ge 0\}$

34. Find the Laurent series expansions of the given functions in the specified regions:

(a) $f(z) = \frac{1}{z^2(1-z)}$ in $0 < |z| < 1$ (b) $f(z) = \dfrac{1}{z^2 - 5z + 4}$ in $|z| > 4$

(c) $f(z) = \dfrac{1}{z^2 - 5z + 6}$ in $0 < |z| < 2$ (d) $f(z) = \dfrac{(1-z)^3}{z-2}$ in $|z - 1| > 1$

35. Let C be the semicircular path $z = Re^{i\theta} (0 \le \theta \le \pi)$, where $R > 1$ and $z^{1/2}$ be the branch $z^{1/2} = e^{\log z/2} = \sqrt{r}\, e^{i\theta/2}$ $(r > 0, -\pi/2 < \theta < 3\pi/2)$ of the square root function. Without actually finding the value of the integral, show that $\lim\limits_{R\to\infty} \int_{C_R} \frac{z^{1/2}}{z^2+1}\, dz = 0$.

36. Let γ be a simple closed contour described in the positive sense, and write $f(z) = \int_\gamma \frac{\zeta^3+7\zeta}{(\zeta-z)^3} d\zeta$. Show that $f(z) = 6\pi i$ when z is inside γ and that $f(z) = 0$ when z is outside γ.

37. Show that Cauchy's inequality becomes an equality iff $f(z)$ is a constant multiple of a power of z.

38. Let $f(z)$ be analytic within and on a circle $\gamma : |z| = 1$. Using the Cauchy's inequality, find the bound on $f^n(0)$ where $f(z) = e^{3z}$.

39. Expand $\sinh z$ in Taylor's series at $z_0 = \pi i$, and show that $\lim\limits_{z\to\pi i} \dfrac{\sinh z}{z - \pi i} = -1$.

40. Find the Maclaurin series for the error function erf $(x) = \frac{2}{\sqrt{\pi}} \int_0^z e^{-t^2} dt$.

41. Find the Maclaurin series expansion of the following functions:
(a) $z^2 e^{3z}$ (b) $\sinh z$ (c) $\cos z$ (d) $z \cosh(z^2)$ (e) $z(z^2 + 9)^{-1}$

42. Find the Laurent expansion for $f(z)$ in $\mathbb{C}/\{0\}$, where $f(z) =$

(a) $z^3 e^{1/z}$ (b) $z^{-3} e^z$ (c) $\dfrac{\log(1+z)}{z^4}$ (d) $\sin \dfrac{1}{z}$ (e) $e^z + e^{1/z}$ (f) $e^{z+\frac{1}{z}}$

43. Obtain the Laurent series expansion for each of the following functions in the specified domains:

(a) $f(z) = \dfrac{1}{z - z^2}$ in the annulus $\mathcal{D} = \{z \in \mathbb{C} : 0 < |z| < 1\}$
(b) $f(z) = \dfrac{z}{z^2 + 1}$ in the annulus $\mathcal{D} = \{z \in \mathbb{C} : 0 < |z - i| < 2\}$
(c) $f(z) = \dfrac{z + 1}{z(z - 4)^3}$ in the annulus $\mathcal{D} = \{z \in \mathbb{C} : 0 < |z - 4| < 4\}$
(d) $f(z) = \dfrac{1}{e^z - 1}$ in the annulus $\mathcal{D} = \{z \in \mathbb{C} : 0 < |z| < 2\pi\}$
(e) $f(z) = \operatorname{cosec} z$ in the annulus $\mathcal{D} = \{z \in \mathbb{C} : 0 < |z| < \pi\}$
(f) $f(z) = \dfrac{\sin 2z}{z^3}$ in the domain $\mathcal{D} = \{z \in \mathbb{C} : |z| > 0\}$
(g) $f(z) = z^2 \cos \dfrac{1}{3z}$ in the domain $\mathcal{D} = \{z \in \mathbb{C} : |z| > 0\}$
(h) $f(z) = \dfrac{e^{1/z}}{z^2 - 1}$ in the domain $\mathcal{D} = \{z \in \mathbb{C} : |z| > 1\}$
(i) $f(z) = \dfrac{\sinh z}{z^2}$ in the domain $\mathcal{D} = \{z \in \mathbb{C} : 0 < |z| < \infty\}$
(j) $f(z) = z^2 \sin \dfrac{1}{z^2}$ in the domain $\mathcal{D} = \{z \in \mathbb{C} : 0 < |z| < \infty\}$
(k) $f(z) = \dfrac{1}{4z - z^2}$ in the domain $\mathcal{D} = \{z \in \mathbb{C} : 0 < |z| < 4\}$
(l) $f(z) = \dfrac{e^z}{(z + 1)^2}$ in the domain $\mathcal{D} = \{z \in \mathbb{C} : 0 < |z + 1| < \infty\}$

44. Show that the function $f(z)$, whose Laurent's series is $\sum_{k=-\infty}^{-1} z^k + \sum_{k=1}^{\infty} \frac{z^k}{3^k}$, does not have an essential singularity at $z = 0$, although it contains infinitely many negative powers of z.

45. Does the principal branch $\sqrt{z}$ have a Laurent series expansion in the domain $\mathbb{C}/\{0\}$?

46. Find all functions f analytic in $\mathcal{S}(0, R) = \{z \in \mathbb{C} : |z| < R\}$ that satisfy $f(0) = i$ and $|f(z)| \leq 1$ for all $z \in \mathcal{S}(0, R)$.

47. If $P(z) = a_0 + a_1 z + \cdots + a_n z^n$ is a polynomial and $\max |P(z)| = M$ for $|z| = 1$, show that each coefficient a_k is bounded by M.

48. For each of the following, construct a function f, analytic in the plane except for isolated singularities, that satisfies the given conditions:

(a) f has a zero of order 2 at $z = i$ and a pole of order 5 at $z = 2 - 3i$.
(b) f has a simple zero at $z = 0$ and an essential singularity at $z = 1$.
(c) f has a removable singularity at $z = 0$, a pole of order 6 at $z = 1$, and an essential singularity at $z = i$.

(d) f has a pole of order 2 at $z = 1 + i$ and essential singularities at $z = 0$ and $z = 1$.

49. Characterise those rational functions which have a
(a) Removable singularity at infinity (b) Pole of order m at infinity.

50. For each of the following, determine whether the statement made is always true or sometimes false.
    (a) If f and g have a pole at z_0, then $f + g$ has a pole at z_0.
    (b) If f has an essential singularity at z_0 and g has a pole at z_0, then $f + g$ has an essential singularity at z_0.
    (c) If $f(z)$ has a pole of order m at $z = 0$, then $f(z^2)$ has a pole of order $2m$ at $z = 0$.
    (d) If f has a pole at z_0 and g has an essential singularity at z_0, then the product $f \cdot g$ has a pole at z_0.
    (e) If f has a zero of order m at z_0 and g has a pole of order $n(n \leq m)$, at z_0, then the product $f \cdot g$ has a removable singularity at z_0.

51. Does there exist a function $f(z)$ having an essential singularity at z_0 that is bounded along some line segment emanating from z_0?

52. If $f : G \to \mathbb{C}$ is analytic except for poles, show that the poles of f cannot have a limit point in G.

53. If $f(z)$ is analytic in $\mathcal{D} : 0 < |z| \leq 1$, and $z^k \cdot f(z)$ is unbounded in $\mathcal{D}$ for every integer k, then what kind of singularity does $f(z)$ have at $z = 0$?

54. Let G be a region and suppose $f : G \to \mathbb{C}$ is analytic such that $f(G)$ is a subset of a circle. Show that f is constant.

55. Let G be a region and suppose $f : G \to \mathbb{C}$ is analytic and $\alpha \in G$ such that $|f(\alpha)| \leq |f(z)|$ for all $z \in G$. Show that either $f(\alpha) = 0$ or f is a constant.

56. Show that if f and g are analytic functions on a region G such that fg is analytic, then either f is constant or $g \equiv 0$.

57. Show that if f and g are analytic functions on a region G such that $f(z)g(z) = 0$ for all $z \in G$, then either $f \equiv 0$ or $g \equiv 0$.

58. Prove that if f is an entire function
    (a) and suppose that Re $f(z)$ or Im $f(z)$ has upper bounds for all z, then f must be a constant function. [**Hint:** Apply Liouville's theorem on e^f.]
    (b) its real part is a non-positive is constant.
    (c) which satisfies $|f(z)| \geq 1$ on the entire plane, then f is constant.
    (d) with $f(z) = f(z + 1) = f(z + i)$ for all $z \in \mathbb{C}$, then f is constant.
    (e) such that $|f'(z)| \leq |z|$, then $f(z) = a + bz^2$ with some constants $a, b \in \mathbb{C}$ such that $|b| \leq 1$.
    (f) and Re $f(z)$ or Im $f(z)$ have no zeros, then f is constant.
    (g) and $f(z)$ has no zeros and $f(z) \to c \neq 0$ as $z \to \infty$, then f is constant.
    (h) and $|f(z)| \leq K|z|$ for all $z \in \mathbb{C}$, where K is a positive constant, then $f(z) = Cz$ for some constant C.
    (i) and $|f(z)| \leq K|e^z|$ for all $z \in \mathbb{C}$, where K is a positive constant, then $f(z) = Ce^z$, for some constant C.
    (j) and that $|f(z)| \leq |z|^2$ for all sufficiently large values of $|z|$, say $|z| > R$, then f must be a polynomial of degree at most 2.
    (k) and suppose $f^{(5)}$ is bounded in the whole complex plane, then f must be a polynomial of degree at most 5.

(l) and $f(z) \to \infty$ as $z \to \infty$, then f must have at least one zero.

59. Show that the image of a non-constant entire function is dense in $\mathbb{C}$.
60. Find the entire function $f(z)$ such that

    (a)$|f(z)| \leq |z|^2, f(1) = 2i$ and $f(2) = i$ (b)$|f(z)| = 1$ on $|z| = 1$
61. Find where the function $f(z)$ attains its maximum modulus on the closed unit disc $\mathcal{S}[0, 1]$, where

    (a) $f(z) = z^2 - z$ (b) $f(z) = z^2 - 1$ (c) $f(z) = e^z$
62. Find where the function $(z-1)^2$ attains maximum modulus on the triangle with vertices at $0, 1+i$ and $1-i$.
63. Prove that if f is analytic on the disc $\mathcal{S}(z_0; R)$ and $|f(z)| \leq M$ on $\mathcal{S}(z_0; R)$, then $|f'(z_0)| \leq M/R$.
64. Suppose $p(z) = a_0 + a_1 z + a_2 z^2 + a_3 z^3$ is a polynomial of degree 3. If $|p(z)| \leq 1$ on the unit circle $\partial\mathcal{S}(0; 1)$, then show that $|a_3| \leq 1$.
65. Analyze each singularity z_0 of f. Is it removable, a pole, or essential?

    (a) $\dfrac{1}{z - z^3}$ (b) $\sin \dfrac{1}{z}$ (c) $\dfrac{e^z - 1 - z}{z^2}$ (d) $\dfrac{1}{e^z - 1} - \dfrac{1}{z}$ (e) $\dfrac{\log z}{(1-z)^2}$,
    where log is the principal branch of log function.
66. Investigate the behaviour of the following functions at infinity:

    (a) $\dfrac{1}{z^2(z^3+1)^2}$ (b) $\dfrac{1-e^z}{1+e^z}$ (c) ze^z (d) $e^{\cot z}$ (e) e^z (f) $\cosh z$

    (g) $\dfrac{z-1}{z+1}$ (h) $\dfrac{z}{z^3+i}$ (i) $\dfrac{z^3+i}{z}$ (j) $e^{\sinh z}$ (k) $\dfrac{\sin z}{z^2}$ (l) $\dfrac{1}{\sin z}$ (m) $e^{\tan \frac{1}{z}}$

Answers

1. (i) MC; (ii) SC; (iii) MC; (iv) NC; (v) SC; (vi) NC

20. (a) $-28 + 84i$; (b) $-48 + \frac{736}{3}i$; (c) $(2+\pi)i$; (d) πi; (e) $-\frac{7}{12} + \frac{1}{12}i$

25. (i) $-4 + 2\pi i$; (ii) $4 + 2\pi i$; (iii) $4\pi i$

26. (i) $2\pi ie$; (ii) $2\pi(-1+i)$; (iii) 0; (iv) $\pi ie^{-1}/8$; (v) $-\pi$; (vi) $(6-8i)\pi$; (vii) $2\pi\sinh(1)$; (viii) 0; (ix) 0; (x) $18\pi i/7$; (xi) 0; (xii) $4\pi i$; (xiii) 0; (xiv) $-3\pi i/40$

29. (a) $2\pi(1+e^4)$; (b) $\pi^2 e/36$; (c) $2e$; (d) $2\sqrt{2}\pi\frac{\log R}{R}, R > e^\pi$

33. (a) $\cosh 2\pi$; (b) e; (c) $e + 1, 3$

34. (a) $\sum\limits_{n=-2}^{\infty} 1/z^n$; (b) $\frac{1}{3}\sum\limits_{n=0}^{\infty} \frac{4^n}{z^{n+1}}$; (c) $\sum\limits_{n=1}^{\infty}\left(\frac{1}{2^n} - \frac{1}{3^n}\right)z^{n-1}$; (d) $-(z-1)^3\sum\limits_{n=1}^{\infty}\frac{1}{(z-1)^n}$

38. $e^3\, n!$ 60. (a) $-iz + 3i$; (b) $f(z) = e^{i\alpha}z, \alpha \in \mathbb{R}$

Long-answer Type Questions

1. Parametrize the contour consisting of the perimeter of the square with vertices $-1-i, 1-i, 1+i$ and $-1+i$, traversed once in that order. What is the length of that contour?
2. Parametrize the contours in the following figures. Also, give a parametrization for the opposite contour.

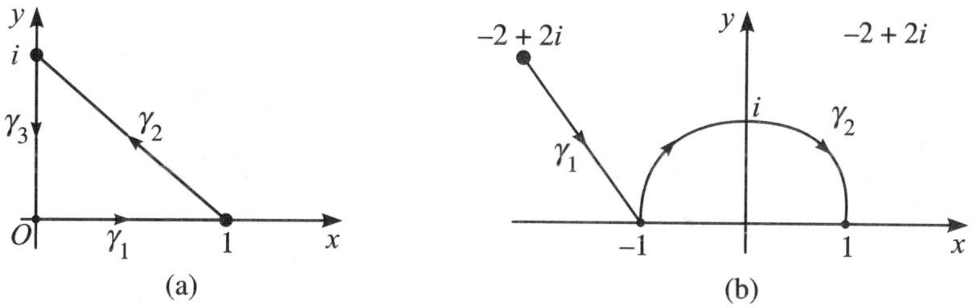

Figure 5.36

3. Let γ be a directed smooth curve. Show that if $z = z(t)$; $a \leq t \leq b$, is an admissible parametrization of γ consistent with the ordering of γ, then the same is true of

$$z_1(t) = z\left(\frac{b-a}{d-c}t + \frac{ad-bc}{d-c}\right);\ c \leq t \leq d.$$

4. Calculate $\mathrm{Ind}_\gamma(1+i)$ if γ is the path which traces the line from 0 to 2, then proceeds counter-clockwise around the circle $|z| = 2$ from 2 to $2i$ and then traces the line from $2i$ to 0. What is the answer if this path is traversed in the opposite direction ?

5. Determine the value of $\mathrm{Ind}_\gamma(z)$ in each component of $\mathbb{C}/\gamma(I)$ if γ is a curve of Fig. 5.37.

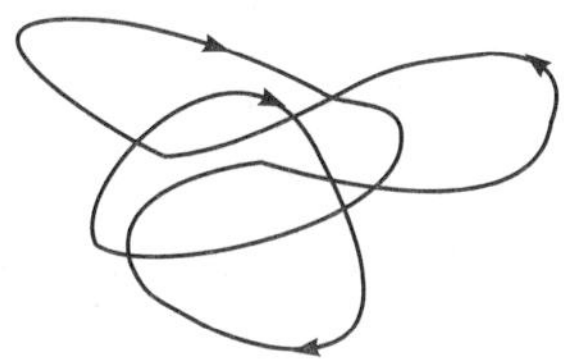

Figure 5.37

6. Let γ_1 and γ_2 be two polygons $[1, i]$ and $1, 1+i, i]$. Express γ_1 and γ_2 as paths and calculate $\int_{\gamma_1} f$ and $\int_{\gamma_2} f$, where $f(z) = |z|^2$.

7. Show that if $\triangle$ is a triangle in the plane of diameter d (length of its longest side), and if f is a continuous function on $\triangle$ with $|f|$ bounded by M on $\triangle$, then $\left|\int_{\partial\triangle} f(z)dz\right| \leq 3Md$.

8. Show that if f is a bounded continuous function on $\mathbb{C}$, and γ_R is the path $\gamma_R(z) = Re^{it}$ for $t \in [0, 2\pi]$, then $\lim\limits_{R\to\infty} \int_{\gamma_R} \frac{f(z)}{(z-w)^2}dz = 0$ for each $w \in \mathbb{C}$.

9. If $\sqrt{z}$ is defined by $\sqrt{z}e^{(\log z)/2}$ for the branch of the log function defined by the condition $-\frac{\pi}{2} \leq \arg z \leq \frac{3\pi}{2}$, find an antiderivative for $\sqrt{z}$ and then find $\int_\gamma \sqrt{z}dz$, where γ is any path from -1 to 1 which lies in the upper half- plane.

10. Prove that if f is analytic in an open set containing a rectangle R, then the path integral of f around the boundary of this rectangle is 0.

11. Calculate $\int_\gamma z^n dz$, where γ be a path that traces the straight line from 1 to $1+i$, followed by the straight line from $1+i$ to i and then the straight line from i to 0.

12. Let $\gamma(t) = te^{it}$ for $0 \leq t \leq 2\pi$ and $\gamma(t) = 4\pi - t$ for $2\pi \leq t \leq 4\pi$. Evaluate $\int_\gamma \frac{dz}{z^2+\pi^2}$.

13. Evaluate $\int_\gamma \frac{e^z - e^{-z}}{z^4}dz$, where γ is one of the curves as depicted in the Fig. 5.38.

(a) (b) (c)

Figure 5.38 *Curves of Exercise 13*

14. Let $\triangle$ be the triangle with vertices $1-i$, i and $-1-i$ and S is the square with vertices $1-i, 1+i$, $-1+i$ and $-1-i$. If f is any function which is analytic on $\mathbb{C}/\{0\}$, prove that $\int_{\partial\triangle} f(z)\,dz = \int_{\partial S} f(z)\,dz$, where $\partial\triangle$ and ∂S are traversed in the counterclockwise direction.

15. Suppose f is analytic in an open set containing the triangle with vertices $a, b, c \in \mathbb{C}$. Show that

$$\int_a^c f(z)dz - \int_a^b f(z)dz = \int_b^c f(z)dz$$

16. Let γ be a closed polygon $[1-i, 1+i, -1+i, -1-i, 1+i]$. Find $\int_\gamma z^{-1}dz$.

17. Let f be a function which is analytic on the open unit disc $S(0,1)$ and γ_1 and γ_2 two closed paths in $S(0,1)$ each of which has index 1 about 0. If $\Gamma = \gamma_1 - \gamma_2$, compute $\int_\Gamma \frac{f(z)}{z}dz$.

18. Give an example of a closed rectifiable curve γ in $\mathbb{C}$ such that for any integer k, there is a point $\alpha \notin \{\gamma\}$ with $n(\gamma;\alpha) = k$.

19. Fix $\alpha = re^{i\theta} \neq 0$ and let γ be a rectifiable path in $\mathbb{C}/\{0\}$ from 1 to α. Show that there is an integer k such that $\int_\gamma z^{-1}dz = \log r + i\theta + 2\pi i k$.

20. Let $P(z)$ be a polynomial of degree n and let $R > 0$ be sufficiently large so that $P(z)$ never vanishes in $\{z \in \mathbb{C} : |z| > R\}$. If $\gamma(t) = Re^{it}, 0 \le t \le 2\pi$, show that $\int_\gamma \frac{P'(z)}{P(z)}dz = 2\pi i n$.

21. If n is an integer, and $\gamma(t) = e^{2n\pi i t}$ and $\gamma_1(t) = 2e^{2\pi i t}$ for $t \in [0,1]$. Show that the cycle $\gamma - n\gamma_1$ is homologous to 0 in $A = \{z \in \mathbb{C} : 0 < |z| < 3\}$.

22. Find a cycle Γ which is homologous to 0 in the annulus $\{z : 1 < |z| < 4\}$, but has index 1 about every point in the annulus $\{z : 2 < |z| < 3\}$.

23. For $t \in [0,1]$, let $\gamma(t) = 1 + e^{2\pi i(t+1/2)}$ and $\gamma_2(t) = - + e^{2\pi i t}$. Find a single closed path γ with parameter interval $[0,1]$, such that γ is equivalent to $\gamma_1 - \gamma_2$.

24. For positive numbers $r < R$, set $\gamma_1(t) = re^{2\pi i t}$ and $\gamma_2(t) = Re^{2\pi i t}$ for $t \in [0,1]$ and set $\Gamma = \gamma_2 - \gamma_1$. Prove that if f is analytic in an open set containing the set $\{z : r \le |z| \le R\}$, then

$$f(z) = \frac{1}{2\pi i}\int_\Gamma \frac{f(\zeta)}{\zeta - z}d\zeta.$$

25. Prove that if an entire function f is not constant, then the range $f(\mathbb{C})$ is dense in $\mathbb{C}$, meaning that the closure of $f(\mathbb{C})$ is $\mathbb{C}$. [**Hint :** If $f(\mathbb{C})$ is not dense, then there is a point z_0 and a disc $S(z_0;r)$ such that $f(\mathbb{C}) \cap S(z_0;r) = \phi$.]

26. Evaluate the following integrals using Cauchy's integral formula:

(a) $\int_{|z|=4} \frac{e^z}{(z^2+1)^3}\, dz$ (b) $\int_{|z-1|=4} \frac{z+i}{(z-1)^2(z-2)(z-6)}\, dz$

(c) $\int_{|z-(1-i)|=3} \frac{e^z\,(3z^2+4)}{(z-1)^3}\, dz$ (d) $\int_{|z-1|=2} \frac{z^2+5z+6}{(z^2+1)(z-1+i)}\, dz$

(e) $\int_{|z|=4} \frac{e^z}{z(z-\pi/2)(z-\pi)}\, dz$

(f) $\int_{\Gamma} \frac{z+i}{(z+4)(z-1-i)}\, dz$, where$\Gamma$is the boundary of $[0,2] \times [0,3]$

(g) $\int_{|z+\frac{1}{2}|=1} \frac{z^2\, e^z}{(z+1)(z-2)}\, dz$

27. Prove Cauchy's integral formulae for all positive integral values on n. [**Hints :** Use principle of mathematical induction.]

28. Let $f(z) = \frac{1}{(1-z)^2}$ and let $0 < R < 1$. Verify that $\max\limits_{|z|=R} |f(z)| = \frac{1}{(1-R)^2}$, and also show that $f^{(n)}(0) = (n+1)!$, so that by the Cauchy estimates $(n+1)! \leq \frac{n!}{R^n(1-R)^2}$.

29. Suppose f is analytic in $|z| < 1$ and $|f(z)| < \frac{1}{1-|z|}$. Prove that

$$\left|f^{(n)}(0)\right| \leq \frac{n!}{R^n(1-R)}; \quad 0 < R < 1$$

and show that the upper bound is the smallest when $R = n/(n+1)$.

30. Let f be analytic and bounded by M in $|z| \leq r$. Prove that $|f^{(n)}(z)| \leq \frac{n!M}{(R-|z|)^n}$; $|z| < R$.

31. If f is analytic in the annulus $1 \leq |z| \leq 2$ and $|f(z)| \leq 3$ on $|z| = 1$ and $|f(z)| \leq 12$ on $|z| = 2$, prove that $|f(z)| \leq 3|z|^2$ for $1 \leq |z| \leq 2$.

32. Let f and g be analytic in the bounded domain $\mathcal{D}$ and continuous up to and including its boundary $\partial\mathcal{D}$. Suppose that g never vanishes. Prove that the inequality $|f(z)| \leq |g(z)|$ holds for all $z \in \partial\mathcal{D}$.

33. Let R be a rectangular region $\{(x,y) : |x| \leq 4, |y| \leq 3\}$ and f be analytic in R. If f satisfies $|f(z)| \leq 1$ on the boundary of R, then show that $|f'(0)| \leq \frac{14}{9\pi}$.

34. Use Cauchy's integral formula to prove that if f is a function which is analytic in an open set containing the closed unit disc $\mathcal{S}[0,1]$, and $\partial\mathcal{S} = \{z : |z| = 1\}$ is a unit circle, then $|f(0)| \leq M$, where M is the maximum value of $|f|$ on $\partial\mathcal{S}$.

35. Use Morera's theorem to show that if f is continuous on an open set U and analytic on U/E, where E is either a point or a line-segment, then f is actually analytic on all of U.

36. Use Morera's theorem to prove that if U is an open subset of $\mathbb{C}$, $I = [a,b]$ an interval on the real line and $g(z,t)$ a continuous function on $U \times I$, which is analytic in z for each $t \in I$, then the function $f(z) = \int_a^b g(z,t)dt$ is analytic in U.

37. Let $f(z)$ and $g(z)$ be entire functions. If $g(z)$ is never 0 and $|f(z)| \leq |g(z)|, \forall z$, show that there is a constant c such that $f(z) = cg(z)$.

38. With the help of Gauss mean value theorem, prove that

$$(i)\quad \frac{1}{2\pi}\int_0^{2\pi} \sin^2\left(\frac{\pi}{6} + 2e^{i\theta}\right) d\theta = \frac{1}{4}; \quad (ii)\quad \int_0^{\pi} \ln\sin\theta\; d\theta = -\pi\ln 2.$$

39. Suppose f is analytic in an open set U. Also, suppose $z \in U$ and the distance from z to the complement of U is d. If $|f(\zeta)| \leq M$ for all $\zeta \in U$, find estimates, similar to Cauchy's estimates, on the size of $\left|f^{(n)}(z)\right|$ in terms of M and d.

40. Suppose f is analytic on a disc $\mathcal{S}(z_0; R)$ and unbounded (there is no M such that $|f(z)| \leq M$ on $\mathcal{S}(z_0; R)$). Then prove that the radius of convergence of the power series expansion of f about z_0 is r.

41. An entire function f is a polynomial of degree at most n only if there are positive constants A and B such that $|f(z)| \leq A + B|z|^n$ for all $z \in \mathbb{C}$.

42. Show that if f is an entire function and $\lim_{z\to\infty} f(z) = \infty$, then f must have a zero somewhere in $\mathbb{C}$.

43. Let $f(z)$ be analytic on the open disc $\mathcal{S}(0, R)$ and $|f(z)| \leq M < \infty$. Show that
(a) $\left|f(z) - f(0)\right| \leq \dfrac{2Mr}{R}$ for $|z| = r < R$ (b) $|f'(z_0)| \leq \dfrac{R}{M}\dfrac{M^2 - |f(z_0)|^2}{R^2 - |z_0|^2}$; $z_0 \in \mathcal{S}(0, R)$

44. Show that for sufficiently large R, the polynomial $P(z) = a_0 z^n + a_1 z^{n-1} + \cdots + a_n$, where the degree $n \geq 1$ and $a_0 \neq 0$, satisfies the inequality $|P(z)| < 2\,|a_n|\,|z|^n$ whenever $|z| \geq R$.

45. Find the maximum and minimum of $|f|$, where

(a) $f(z) = \begin{cases} \sin z/z & z \neq 0 \\ 1 & z = 0 \end{cases}$ is defined on the rectangle with the vertices 0, π, i, $\pi + i$, and determine where these values occur.

(b) The function $f(z) = z + 1$ is a closed triangle with vertices at $z = 0$, $z = 2$ and $z = i$.

46. Define $f(z)$ as in Example 45a.

(a) Using the Maclaurin expansion for $\sin z$, show that for all z, $f(z) = 1 - \frac{z^2}{3!} + \frac{z^4}{5!} - \cdots$ is an entire function.
(b) Explain why $f(z)$ is analytic at the origin.
(c) Show that $\lim_{z\to 0} f(z) = 1$.
(d) Find $f^{(3)}(0)$ and $f^{(4)}(0)$.

47. Let f be continuous on the real interval $[-1, 2]$, and define $g(z) = \int_{-1}^{2} f(t)\sin(zt)dt$.

(a) Prove that g is entire and find its power series expansion around the origin.
(b) Prove that $g'(z) = \int_{-1}^{2} tf(t)\cos(zt)dt$ for all z.

48. Suppose $f(z)$ and $g(z)$ are continuous in $\mathcal{S}[0, r]$ and analytic in $\mathcal{S}(0, r)$ and that $f(z) \neq 0$ and $g(z) \neq 0$ for all $z \in \mathcal{S}[0, r]$. If $|f(z)| = |g(z)|$ for $|z| = r$, show that there exists a constant k such that $|c| = 1$ and $f(z) = kg(z)$ for all $z \in \mathcal{S}[0, r]$.

49. Let $f(z) = \sum\limits_{k=0}^{\infty} \frac{k^3}{3^k} z^k$. Compute the following:

(a) $f^{(6)}(0)$ (b) $\displaystyle\oint_{|z|=1} \frac{f(z)}{z^4}dz$ (c) $\displaystyle\oint_{|z|=1} e^z f(z)dz$ (d) $\displaystyle\oint_{|z|=1} \frac{f(z)\sin z}{z^2}dz$

50. **[Parseval's formula]** Let $f(z) = \sum\limits_{k=0}^{\infty} a_k(z - z_0)^k$ be Taylor's expansion of the analytic function $f(z)$ on $\mathcal{S}(z_0, R)$. Show that, for any $0 < r < R$,

$$\sum_{k=0}^{\infty} |a_k|^2 r^{2k} = \frac{1}{2\pi}\int_0^{2\pi} \left|f(z_0 + re^{it})\right|^2 dt.$$

51. Let $f(z)$ be analytic in the annulus $r < |z - z_0| < R$ and bounded by M there. Prove that the coefficients a_k and b_k of the Laurent expansion of $f(z)$ in the annulus satisfy

$$|a_k| \leq \frac{M}{R^k}, \qquad |b_k| \leq mr^k; k = 0, 1, 2, \cdots$$

52. Expand $f(z) = \frac{1}{z(z^2-3z+2)}$ for the regions (i) $0 < |z| < 1$; (ii) $1 < |z| < 2$; (iii) $|z| > 2$

53. Expand $f(z) = \frac{z+3}{(z^2-z-2)z}$ for the regions:

    (i) $0 < |z| < 1$; (ii) $1 < |z| < 2$; (iii) $|z| > 2$

54. Expand $f(z) = \frac{z^2-1}{(z+2)(z+3)}$ for the regions:

    (i) $|z| < 2$; (ii) $2 < |z| < 3$; (iii) $|z| > 3$

55. Obtain a general formula for the Laurent expansion of $f_n(z) = \frac{1}{(z-\alpha)^n}; n \in \mathbb{N}$, that is valid for $|z| > |\alpha|$.

56. Prove that if $f(z)$ has a Laurent series expansion of the form $\sum_{k=1}^{\infty} a_k z^k$ in $0 < |z| < \rho$, then $\lim_{z \to 0} f(z)$ exists.

57. Find the Laurent series expansions of the given functions in the specified regions:

    (a) $f(z) = 1/(1+z)$ in (i) $0 < |z - i| < \sqrt{2}$; (ii) $|z - i| > \sqrt{2}$
    (b) $f(z) = \frac{1}{z^2-5z+4}$ in (i) $0 < |z| < 1$; (ii) $1 < |z| < 4$; (iii) $|z| > 4$
    (c) $f(z) = \frac{z+1}{z^2-z-6}$ in (i) $0 < |z| < 2$; (ii) $2 < |z| < 3$; (iii) $|z| > 3$
    (d) $f(z) = \frac{z}{z^2-8z+15}$ in (i) $0 < |z - 2| < 1$; (ii) $1 < |z - 2| < 3$; (iii) $|z - 2| > 3$
    (e) $f(z) = \frac{z+5}{z^2-6z+8}$ in $1 < |z + 1| < 3$.

58. Suppose f, analytic in the annulus $A = \{z \in \mathbb{C} : R < |z|\}$, satisfies the inequality $|f(z)| \leq |z|^k$ in this set. Prove that the Laurent expansion of f in A has no terms with positive exponent greater than k.

59. Define the function $f(z)$ by $f(z) = \frac{1}{\sin z} - \frac{1}{z} + \frac{2z}{z^2-\pi^2}$.

    (a) Show that $f(z)$ is analytic in the disc $|z| < 2\pi$, except for removable singularities at $z = 0, \pm\pi$.
    (b) Find the first four terms of the Taylor series about $z = 0$ for $f(z)$. What is the radius of convergence of this series?
    (c) Use the result of part (59b) to obtain the first few coefficients in Laurent's series expansion for cosec $z = 1/\sin z$, valid in the annulus $\pi < |z| < 2\pi$.

60. Prove that an entire function has a removable singularity at infinity if and only if it is constant.

61. Let $f(z)$ be a function having z_0 as an isolated singular point. Then prove that the following are equivalent:

    (a) z_0 is a pole of order m for $f(z)$.
    (b) $f(z)$ can be written in the form $f(z) = (z - z_0)^{-m}\phi(z)$, where $\phi(z)$ has a removable singularity at $z = z_0$ and $\lim_{z \to z_0} \phi(z) \neq 0$.
    (c) z_0 is a zero of order m for $1/f(z)$.

62. Let a non-constant function f be analytic in the bounded domain $\mathcal{D}$ and continuous up to and including its boundary $\partial\mathcal{D}$. Prove that if $|f(z)|$ is constant on $\partial\mathcal{D}$, then f must have at least one zero in $\mathcal{D}$.

63. Show that $f(z) = \tan z$ is analytic in $\mathbb{C}$ except for simple poles at $z = \pi/2 + n\pi$ for each integer n. Determine the singular part of f at each of these poles.

64. Prove that if f is an analytic function on an open connected set U and if K is a compact subset of U, then there can be at most finitely many zeros of f in K.

65. Suppose U is an open connected set and $z_0 \in U$. Prove that if f is a non-constant analytic function on $U/\{z_0\}$ and f takes on a certain value c at least once in every deleted neighbourhood of z_0, then f has an essential singularity at z_0.

66. If f has a pole at z_0, show that $\text{Re}f$ and $\text{Im}f$ take on arbitrarily large positive as well as negative values in any punctured neighbourhood of z_0.

67. Prove that if $f(z)$ has a pole of order m at z_0, then $g(z) = f'(z)/f(z)$ has a simple pole at z_0. What is the coefficient of $(z - z_0)^{-1}$ in the Laurent expansion for $g(z)$?

68. Show that if f is a non-constant analytic function on an open connected set U and if f has no zeros on U, then there are no points of U where $|f(z)|$ has a local minimum.

69. Show that if f is a non-constant, continuous function on $\mathcal{S}[0,1]$, which is analytic on $\mathcal{S}(0,1)$ and $|f(z)| = 1$ for all z on the unit circle, then f has a zero somewhere in $\mathcal{S}(0,1)$.

70. Prove that if f is a non-constant entire function and $c > 0$ is a constant such that the closed set $K = \{z \in \mathbb{C} : |f(z)| \leq c\}$ is bounded, then the open set $U = \{z \in \mathbb{C} : |f(z)| < c\}$ contains at least one zero of f.

71. Determine the regions in which the functions $f(z) = (\sin \frac{1}{z})^{-1}$ and $g(z) = \int_0^1 (\zeta - z)^{-1} d\zeta$ are analytic. Do they have any isolated singularities? Do they have any singularities that are not isolated?

72. Let f be analytic in a region $G = \{z \in \mathbb{C} : 0 < |z - a| < R\}$. Show that if $\int\int_G \left|f(x+iy)\right|^2 dxdy < \infty$, then f has a removable singularity at $z = a$.

73. Suppose f is an analytic function on the unit disc $\mathcal{S}(0,1)$. Prove that if $|f(z)| \leq 1$ on $\mathcal{S}(0,1)$ and f has a zero of order 2 at 0, then $|f(z)| \leq |z|^2$ for all $z \in \mathcal{S}(0,1)$.

74. Suppose that $f(z)$ is analytic in an open set containing the closed unit disc $|z| \leq 1$ and that $|f(z)| \leq 1$ for all $|z| = 1$. If $f(1/2) = 0$, show that $|f(3/4)| \leq 2/5$.

75. Let $g(t)$ be a continuous complex-valued function of a real variable on $[0, \pi]$. Show that the function $f(z) = \int_0^\pi g(t)\sin(tz)\, dt$ is entire and find its power series around the origin.

Answers

26. (a) $[2\sin 1 - 3\cos 1]\pi/4$; (b) $\pi[1 - 6i]/50$; (c) $25\pi ie$; (d) $(6 - 8i)\pi$;
(e) $4i[1 - 2e^{\pi/2} + e^{\pi}]/\pi$; (f) $(-9 + 7i)\pi/13$; (g) $-2\pi ie^{-1}/3$

45. (a) $\sinh 1, 0$ at i, π; (b) $3, 1$ at $2, 0$

52. (i) $\frac{1}{2z} + \frac{3}{4} + \frac{7}{8}z + \frac{15}{16}z^2 + \frac{31}{32}z^3 + \cdots$; (ii) $-\frac{1}{z^3} - \frac{1}{z^2} - \frac{1}{2z} - \frac{1}{4}[1 + \frac{z}{2} + \left(\frac{z}{2}\right)^2 + \cdots]$
(iii) $(2-1)\frac{1}{z^3} + (2^2-1)\frac{1}{z^4} + (2^3-1)\frac{1}{z^5} + \cdots$

53. (i) $-\frac{3}{2z}\sum_{n=0}^{\infty}\left[\frac{2}{3}(-1)^n - \frac{5}{12}\left(\frac{1}{2}\right)^n\right]z^n$; (ii) $-\frac{3}{2z} + \frac{2}{3z}\sum_{n=0}^{\infty}\frac{1}{z^n} - \frac{5}{12}\sum_{n=0}^{\infty}\left(\frac{z}{2}\right)^n$;
(iii) $-\frac{3}{2z} + \frac{2}{3z}\sum_{n=0}^{\infty}(-1)^n\frac{1}{z^n} - \frac{5}{62}\sum_{n=0}^{\infty}\left(\frac{2}{z}\right)^n$.

54. (i) $1 + \sum\limits_{n=0}^{\infty} (-1)^n \left\{\frac{3}{2^{n+1}} - \frac{8}{3^{n+1}} z^n\right\} z^n$; (ii) $1 + \frac{3}{2} + \sum\limits_{n=0}^{\infty} (-1)^n \frac{2^n}{z^n} - \frac{8}{3} \sum\limits_{n=0}^{\infty} (-1)^n \left(\frac{z}{3}\right)^n$;

(iii) $1 + \frac{3}{z} + \sum\limits_{n=0}^{\infty} (-1)^n \left(\frac{2}{z}\right)^n - \frac{8}{z} \sum\limits_{n=0}^{\infty} (-1)^n \left(\frac{3}{z}\right)^n$.

57a. (i) $\sum\limits_{n=0}^{\infty} (-1)^n \left(\frac{z-i}{1+i}\right)^n$; (ii) $\sum\limits_{n=1}^{\infty} (-1)^{n+1} \left(\frac{1+i}{z-i}\right)^n$

57b. (i) $\frac{1}{3} \sum\limits_{n=0}^{\infty} \left(1 - \frac{1}{4^{n+1}}\right) z^n$; (ii) $-\frac{1}{3}\left[\sum\limits_{n=0}^{\infty} \frac{1}{4^{n+1}} z^n + \sum\limits_{n=1}^{\infty} z^{-n}\right]$; (iii) $\frac{1}{3} \sum\limits_{n=0}^{\infty} \frac{4^n}{z^{n+1}}$

57c. (i) $-\frac{1}{10} + \frac{1}{10} \sum\limits_{n=0}^{\infty} (-1)^{n+1} \left(\frac{z}{2}\right)^{n+1}$; (ii) $-\frac{4}{15} + \frac{4}{15} \sum\limits_{n=1}^{\infty} 3^{-n} z^n + \frac{1}{10} \sum\limits_{n=1}^{\infty} (-1)^n 2^n z^{-n}$;

(iii) $\sum\limits_{n=1}^{\infty} \left[\frac{4}{5} 3^{n-1} + (-1)^n \frac{1}{10} 2^n\right] z^{-n}$

57d. (i) $\frac{2}{3} + \sum\limits_{n=1}^{\infty} \left(\frac{3}{2} - \frac{5}{6} 3^{-n}\right)(z-2)^k$; (ii) $-\frac{5}{6} - \frac{5}{6} \sum\limits_{n=1}^{\infty} 3^{-n}(z-2)^n + \frac{1}{2} \sum\limits_{n=1}^{\infty} (z-2)^{-n}$;

(iii) $\frac{5}{6} \sum\limits_{n=1}^{\infty} 3^{-n}(z-2)^n + \frac{1}{2} \sum\limits_{n=1}^{\infty} (z-2)^{-n}$

57e. $-1 - \frac{7}{10} \sum\limits_{n=1}^{\infty} 5^{-n}(z+1)^n + \frac{3}{2} \sum\limits_{n=1}^{\infty} (-1)^n (z+1)^{-n}$

6 Linear Fractional Transformations

In the geometric interpretation of a function $f : \mathbb{C} \to \mathbb{C}$, we saw how the nature of such a function can be displayed graphically, to some extent, by the manner in which it maps certain curves and regions from one complex plane to another.

In this chapter we shall consider a more global aspect of analytic functions, describing geometrically what their effect is on various curves and regions. Specifically, we develop the theory of linear fractional transformation and explain the concept of conformal mapping with the help of some frequently used elementary functions.

6.1 Transformation

Let $w = f(z)$ be a complex valued function of complex variable z. A function $f : \mathbb{C} \to \mathbb{C}$, defined by $f(z) = u(x, y) + iv(x, y)$, when interpreted geometrically, is called a *transformation* or *mapping*. The transformation equations

$$u = u(x, y), \qquad v = v(x, y) \tag{6.1}$$

establishes a correspondence between points in w-plane and z-plane. The corresponding points of the two planes are called *images* of each other. Hence the nature of a complex function can be described by the manner in which it maps region and curves from complex plane to another. Let $w = f(z)$ map a closed region D_z of z-plane into a closed region D_w of w-plane. If $u(x, y)$ and $v(x, y)$ are continuously differential functions, then

$$\frac{\partial(u, v)}{\partial(x, y)} = \begin{vmatrix} \dfrac{\partial u}{\partial x} & \dfrac{\partial u}{\partial y} \\ \dfrac{\partial v}{\partial x} & \dfrac{\partial v}{\partial y} \end{vmatrix} = \frac{\partial u}{\partial x} \cdot \frac{\partial v}{\partial y} - \frac{\partial u}{\partial y} \cdot \frac{\partial v}{\partial x} \tag{6.2}$$

is called *Jacobian of transformation* of Eq. (6.1). If, in particular, $f(z)$ is analytic, then by using Eq. (3.27) we have $\frac{\partial(u,v)}{\partial(x,y)} = \left| \frac{\partial u}{\partial x} + i \, \frac{\partial v}{\partial x} \right|^2 = |f'(z)|^2$.

Example 6.1.1. Find the Jacobian of the transformation $w = 2z^2 - iz + 3 - i$.

Solution. Let $z = x + iy$ and $w = u + iv$. Then the transformation can be written as

$$\begin{aligned} w &= u + iv = 2(x + iy)^2 - i(x + iy) + 3 - i \\ \Rightarrow u &= 2x^2 - 2y^2 + y + 3 \text{ and } v = 4xy - y - 1. \end{aligned}$$

The Jacobian of transformation as in Eq.(6.2)is given by

$$\begin{aligned}\frac{\partial(u,v)}{\partial(x,y)} &= \begin{vmatrix} 4x & -4y+1 \\ 4y-1 & 4x \end{vmatrix} = 16x^2 + (4y-1)^2 \\ &= 16(x^2+y^2) - 8y + 1 = |4z - i|^2\end{aligned}$$

Example 6.1.2. *Determine the curve in the w-plane which is the image of the straight line $x + 2y = 1$ under the map $w = (z+1)^2$.*

Solution. Let $z = x + iy$ and $w = u + iv$. Then

$$\begin{aligned}& w = u + iv = \{(x+1) + iy\}^2 = (x+1)^2 - y^2 + 2iy(x+1) \\ \Rightarrow\ & u(x,y) = (x+1)^2 - y^2 \text{ and } v(x,y) = 2y(x+1).\end{aligned}$$

Given the equation of a straight line $x + 2y = 1$ in z-plane as in the Fig. 6.1. From the Fig. 6.1, we have $C_1 : x + 2y = 1$ corresponds to

$$\begin{aligned}& \Gamma_1 : u = (x+1)^2 - \frac{1}{4}(1-x)^2 \text{ and} \\ & v = 1 - x^2;\ 0 \le x \le 1 \\ \Rightarrow\ & \Gamma_1 : (4u + 3v - 6)^2 = 100(1 - v).\end{aligned}$$

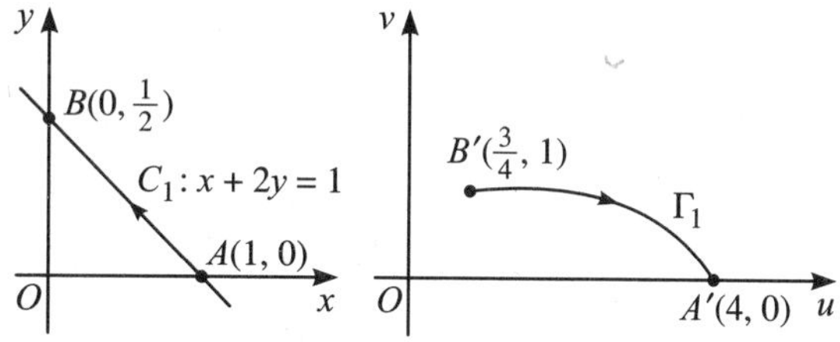

Figure 6.1 *The transformation $w = (z+1)^2$*

The points A, B in z-plane map into A', B' in w-plane.

6.2 Conformal Mapping

In this section we introduce the basic notion of conformal mappings, which is a direct application of the complex derivative. This special type of mapping has important applications in solving boundary value problems involving Laplace equation with the aid of harmonic functions.

Definition 6.2.1. [Isogonal mapping]: Let $w = f(z)$ be single-valued and analytic function of $z \in \mathcal{D}$ and $z_0 \in \mathcal{D}$. If the mapping accomplished by this transformation preserves the magnitude and angles between two intersecting curves meeting at $P(z_0)$, even after transformation but not necessarily the sense, the mapping is called *isogonal* or *indirectly conformal.*

For example, consider the mapping $w = \overline{z}$. Geometrically, it represents reflection about the real axis and $\frac{dw}{dz}$ does not exist. Thus it preserves the angle in magnitude but reverse in direction, so it is an isogonal mapping.

Definition 6.2.2. [Conformal mapping]: If the mapping accomplished by transformation $w = f(z)$ preserves both the magnitude and the sense of rotation of the angle between two intersecting curves at $z_0 \in \mathcal{D}$ even after transformation, the mapping is said to be *conformal.*

A mapping which is conformal at every point of region $\mathcal{D}$ is called a *conformal mapping* in $\mathcal{D}$. For example,

(i) the exponential map $f(z) = e^z$ is conformal at all points $z \in \mathbb{C}$.

(ii) the map $f(z) = z^2$ is conformal everywhere except at the origin.

Conformal mappings have its applications in Dirichlet and Neumann problems, fluid flow, electrostatics, etc.

Theorem 6.1. [Necessary condition for conformality]: *Let $f(z)$ be an analytic function of z defined in a region $\mathcal{D}$ of z-plane and $z_0 \in \mathcal{D}$. If $f'(z_0) \neq 0$ for any $z_0 \in \mathcal{D}$, then $w = f(z)$ determines a conformal transformation.*

Proof: Let C be a smooth regular curve (Fig. 6.2) parameterized by $z = z(t); \alpha \le t \le \beta$ lying in the region $\mathcal{D}$. Then $z'(t)$ is continuous function of t and $z'(t) \neq 0$ for any $t \in (\alpha, \beta)$. Let $w = f(z)$ be a mapping for all points on C. Now, $w(t) = f(z(t))$, $\alpha \le t \le \beta$ is the parametric representation of image C' of the arc C (Fig. 6.2). Suppose C passes through the point $z_0 = z(t_0) \in \mathcal{D}, \alpha < t_0 < \beta$ and w_0 is the image of z_0, i.e., $w_0 = f(z(t_0)) = f(z_0)$. Now $f(z)$ is analytic at z_0, given by

$$f'(z_0) = \lim_{\Delta z \to 0} \frac{\Delta w}{\Delta z}.$$

$$\therefore \quad \left| \lim_{\Delta z \to 0} \frac{\Delta w}{\Delta z} \right| = \lim_{\Delta z \to 0} \left| \frac{\Delta w}{\Delta z} \right| = |f'(z_0)| \tag{6.3}$$

and $f'(z_0) \neq 0$. Then $f'(z_0) = \psi_0$(say), $0 \le \psi_0 \le 2\pi$, and has an unique value. According to the rule of differentiation of the composite functions, we have

$$w'(t_0) = f'(z(t_0))\, z'(t_0) = f'(z_0) z'(t_0)$$

$$\therefore \quad \arg w'(t_0) = \arg f'(z(t_0)) + \arg z'(t_0). \tag{6.4}$$

Since $f'(z_0) \neq 0$, for a regular curve $z'(t_0) \neq 0$, so that $w'(t_0) \neq 0$. This relation (6.4) is useful in connecting the direction of C and its image C' at z_0 and w_0 respectively.

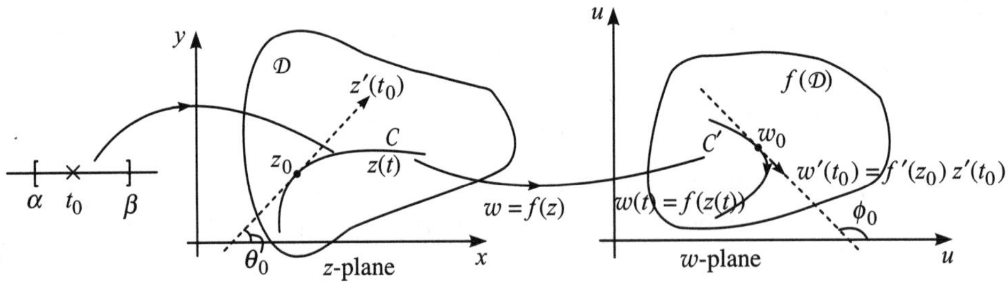

Figure 6.2 *w-plane*

Let $\theta_0 = \arg z'(t_0)$. Then the inclination of the tangent to the arc C at z_0 is θ_0. Let $\arg w'(t_0) = \phi_0$. Then from Eq. (6.4), we get

$$\arg w'(t_0) = \phi_0 = \psi_0 + \theta_0. \tag{6.5}$$

ϕ_0 represents the angle made by the tangent to the curve C' at w_0 with the positive direction of the u-axis in the w-plane. Hence it follows from Eq. (6.5) that the tangent to the regular curve C at z_0 is rotated through the angle ψ_0 by the transformation.

Now, let C_1 and C_2 be two regular curves (Fig.6.3) passing through $z_0 \in \mathcal{D}$ and lying in $\mathcal{D}$ and C_1', C_2' be their image curves under the mapping $w = f(z)$. Let θ_1 and θ_2 be the angles made by the tangents to the curves C_1 and C_2 respectively at z_0 with the positive direction of the x-axis in the z-plane. Let ϕ_1, ϕ_2 be the angles made by the tangents to the curves C_1' and C_2' respectively at w_0 in the w-plane. Therefore,

$$\phi_1 = \psi_0 + \theta_1 \text{ and } \phi_2 = \psi_0 + \theta_2 \Rightarrow \phi_2 - \phi_1 = \theta_2 - \theta_1.$$

The angle $\phi_2 - \phi_1$ from C_1' to C_2' is the same in magnitude and sense as angle $\theta_2 - \theta_1$ from C_1 to C_2. Hence the function f preserves angle between the curves C_1 and C_2 at z_0 both in magnitude and sense. Hence f is a conformal at z_0. □

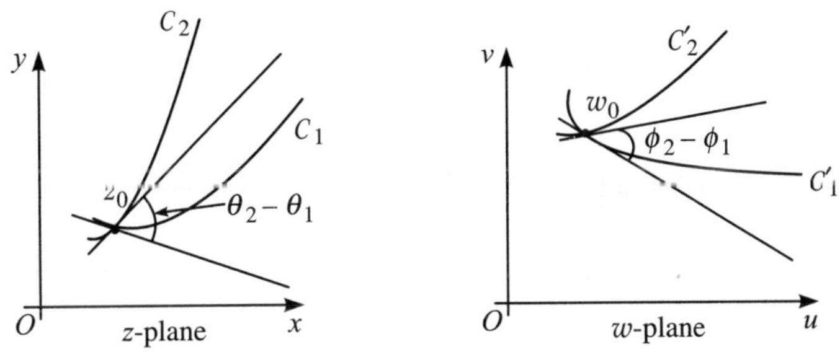

Figure 6.3 *w-plane*

If $w = f(z) = e^z$, then f is analytic throughout $\mathbb{C}$ and $f'(z) = e^z \neq 0$, at any point $z \in \mathbb{C}$. Thus it is a conformal mapping. Now, we have the following observations:

(i) Small distances in the z-plane in the neighbourhood of z_0 are magnified or reduced in the w-plane by an amount given approximately by $|f'(z_0)|$. This scale factor is called the *linear magnification factor*.

(ii) For conformal mappings, small figures in the neighbourhood of a point z_0 in the z-plane map into similar small figures in the w-plane and are magnified or reduced by a amount given approximately by $|f'(z_0)|^2$. This scale factor is called the *area magnification factor.*

(iii) Using Eq. (3.27), the condition for conformality at a point z_0, i.e., $f'(z_0) \neq 0$ can be written as $|f'(z_0)| = \frac{\partial(u,v)}{\partial(x,y)} \neq 0$.

(iv) Let $f(z)$ be an non-constant analytic function defined in $\mathcal{D}$ and let $z_0 \in \mathcal{D}$. If $f'(z_0) = 0$, then z_0 is called a *critical point of* f(z).

RESULT **52.** Let us take a point on the Re z-axis. Then $\theta_0 = 0$, so that, from Eq. (6.5) we have $\phi_0 = \psi_0 =$ constant. This shows that the direction of the tangent vector to C' keeps it fixed. This means that C' is a segment of a straight line. This phenomena can be used to derive the Schrawz–Christoffel transformations.

Example 6.2.1. *Find the coefficient of magnification and angle of rotation under the transformation $w = (1 - i)z$ at any point z.*

Solution. Here

$$\arg[f'(z)] = \arg(1 - i) = \tan^{-1}(-1) = -\frac{\pi}{4}; \quad \text{and } |f'(z)| = \sqrt{1+1} = \sqrt{2}.$$

Thus the coefficient of magnification is $\sqrt{2}$ and the angle of rotation is $-\pi/4$ at any point z.

Example 6.2.2. *Find the points where the mapping $w = z + 1/z$ is conformal. Also, find the critical points if any.*

Solution. Let $z = x + iy \in \mathbb{C}$. Then the transformation becomes

$$\begin{aligned} w &= z + \frac{1}{z} = x + iy + \frac{1}{x+iy} = x + iy + \frac{x - iy}{x^2 + y^2} \\ &= \frac{x(x^2 + y^2 + 1)}{x^2 + y^2} + i\,\frac{y(x^2 + y^2 - 1)}{x^2 + y^2} = u + iv, \end{aligned}$$

which is the relation between (x, y) in z-plane and (u, v) in w-plane. Here $f(z)$ is not analytic at $z_0 = 0$, so that $f(z)$ is not conformal at 0. Now,

$$f'(z) = 0 \Rightarrow 1 - 1/z^2 = 0 \Rightarrow z = \pm 1$$

so that the critical points of $f(z)$ are ± 1.

RESULT **53.** If a function is analytic at a point z_0 but $f'(z_0) = 0$, then the mapping $w = f(z)$ need not be conformal at z_0. For example, let $w = f(z) = z^2$. Since $f'(z) = 2z$ at point $z_0 = 0$, $f'(z)$ vanishes. Take two rays (Fig. 6.4) that starts at the point $z_0 = 0$, say $\arg z = \theta = \alpha$ and $\theta = \beta$.

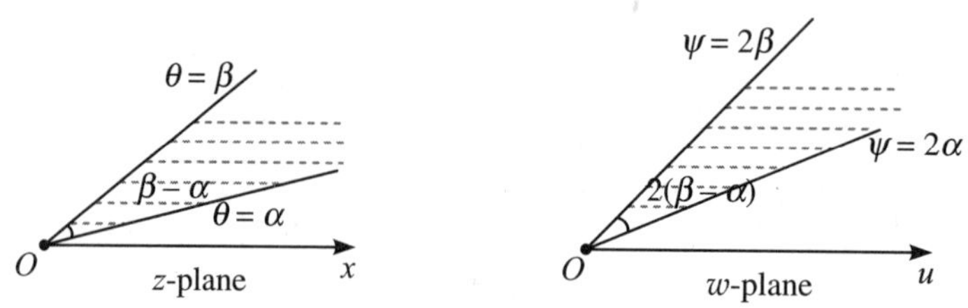

Figure 6.4 *The transformation $w = z^2$*

The function $w = z^2$ maps these two rays into two rays: $\arg w = \psi = 2\alpha$ and $\psi = 2\beta$ respectively. The initial rays form an angle $\beta - \alpha$, while their images form an angle $2(\beta - \alpha)$. We see that the angles are doubled, i.e., the mapping $w = z^2$ at a point $z_0 = 0$ is not conformal.

6.3 Elementary Transformations

In this section we shall present some elementary transformations, such as translation, rotation, stretching and inversion. We shall see further examples of how various curves and regions are mapped by elementary analytic functions.

6.3.1 Translation

The general form of simplest linear transformation

$$w = f(z) = z + \alpha \ \forall z \in \mathbb{C}_\infty \tag{6.6}$$

for some fixed $\alpha \in \mathbb{C}$ is known as *translation* or *parallel displacement* with the displacement vector α. If $\alpha = 0$. Then it is the *identity transformation*. Let $z = x + iy, w = u + iv$ and $\alpha = \alpha_1 + i\alpha_2$. Then

$$\begin{aligned} & w = u + iv = (x + \alpha_1) + i(y + \alpha_2) \\ \Rightarrow \quad & u = x + \alpha_1, v = y + \alpha_2. \end{aligned} \tag{6.7}$$

Thus the image of the point (x, y) in the z-plane is the point $(x + \alpha_1, y + \alpha_2)$ in the w-plane. ∞ is the only fixed point of the transformation when $\alpha \neq 0$.

Example 6.3.1. *What is the region of the w-plane into which the rectangular region in z-plane bounded by $x = 0, y = 0, x = 1$ and $y = 2$ is mapped under the transformation $w = z + (2 - i)$?*

Solution. Let $z = x + iy \in \mathbb{C}$. The given transformation can be written in the form

$$w = u + iv = x + iy + (2 - i) = (x + 2) + i(y - 1)$$
$$\Rightarrow \quad u(x, y) = x + 2 \text{ and } v(x, y) = y - 1.$$

The rectangular region in the z-plane bounded by the lines $x = 0, y = 0, x = 1$ and $y = 2$ is given in Fig. 6.5. Now, from the figure,

$$y = 0 \text{ corresponds to } v + 1 = 0; \quad x = 1 \text{ corresponds to } u - 2 = 1$$
$$y = 2 \text{ corresponds to } v + 1 = 2; \quad x = 0 \text{ corresponds to } u - 2 = 0.$$

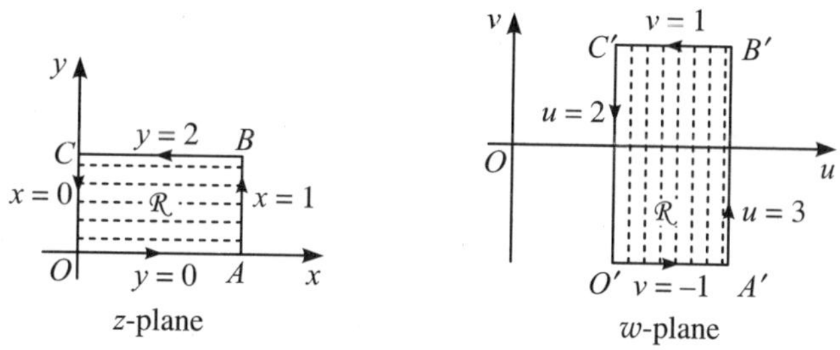

Figure 6.5 $w = z + (2 - i)$.

Under this transformation, the points O, A, B, C on z-plane map onto the points O', A', B', C' on w-plane (Fig. 6.5). Since the sense of rotation as well as magnitude of the angle is preserved, so it is a conformal mapping.

6.3.2 Rotation–Dilation

Case I : Let us consider the general form of transformation

$$w = f(z) = \alpha z; \quad z(\neq 0) \in \mathbb{C}_\infty, \tag{6.8}$$

where α is a non-zero complex constant. We write z and α in exponential form as

$$z = r_1 e^{i\theta}; |z| = r_1 \text{ and } \alpha = r_2 e^{i\psi}; |\alpha| = r_2.$$

Then the transformation becomes

$$w = \alpha z = r_1 r_2 e^{i(\theta+\psi)} = |z||\alpha| e^{i(\theta+\psi)} \tag{6.9}$$
$$\Rightarrow \quad |w| = |\alpha||z| \text{ and } \arg(w) = \arg(\alpha) + \arg(z). \tag{6.10}$$

By the transformation (6.9), a point z with polar coordinates (r, θ) in the z-plane is mapped to the point $(r, \theta + \psi)$ in the w-plane, and hence it represents a rotation through an angle $\psi = \arg(\alpha)$ about the origin and at the same time dilated in the ratio $|\alpha| : 1$. Such a transformation is called a *homothetic transformation*.

Case II : Let us consider the general form of transformation

$$w = f(z) = \alpha z + \beta, \tag{6.11}$$

where $\alpha, \beta(\alpha, \beta \neq 0)$ are complex constants. Such type of transformation is called *affine transformation*. If $w_1 = \alpha z = f_1(z)$ and $w_2 = z + \beta = f_2(z)$, then

$$w = f(z) = f_2(\alpha z) = f_2(f_1(z))$$

Hence, w consists of a rotation of the finite z-plane through the angle $\arg(\alpha)$ (Fig. 6.6) and a magnification by a factor $|\alpha|$, followed by a translation through the vector β.

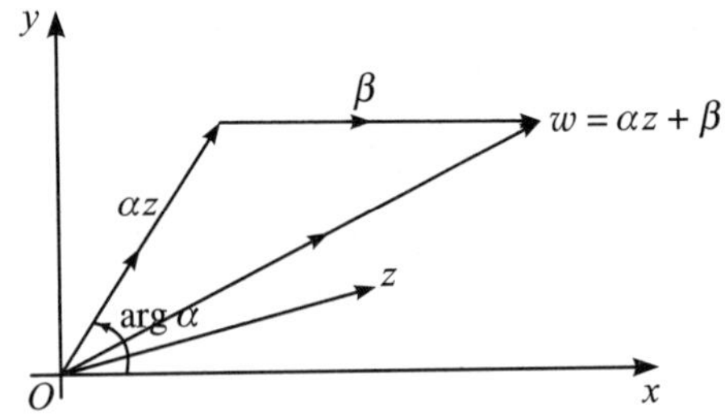

Figure 6.6 $w = \alpha z + \beta$.

Example 6.3.2. *Determine the region in the w-plane corresponding to the triangle region bounded by the lines $x = 0, y = 0$ and $x + y = 1$ in the w-plane under the transformation $w = e^{i\pi/4}z$.*

Solution. The given transformation can be written in the form

$$\begin{aligned}
& w = f(z) = u + iv = e^{i\pi/4}(x + iy) = (1/\sqrt{2} + i/\sqrt{2})(x + iy) \\
\Rightarrow \quad & u(x, y) = (x - y)/\sqrt{2} \text{ and } v(x, y) = (x + y)/\sqrt{2} \\
\Rightarrow \quad & x = (u + v)/\sqrt{2} \text{ and } y = -(u - v)/\sqrt{2}.
\end{aligned}$$

The triangular region OAB in the z-plane bounded by $x = 0, y = 0$ and $x + y = 1$, is given in Fig. 6.7. Now, using the transformation, the lines $y = 0$, $x + y = 1$ and $x = 0$ is transformed as $y = 0 \longrightarrow u - v = 0$; $x + y = 1 \longrightarrow v = 1/\sqrt{2}$; $x = 0 \longrightarrow u + v = 0$, as shown in the Fig. 6.7. Under this given transformation, the region OAB in z-plane is transformed into a triangular region $O'A'B'$ in w-plane. Since $|\alpha| = |e^{i\pi/4}| = 1$ and $\arg(\alpha) = 45°$, each side gets rotated through an angle 45° without magnification.

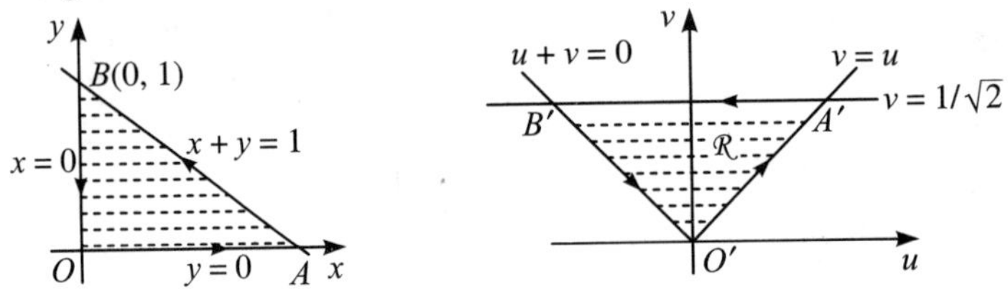

Figure 6.7 $w = e^{i\pi/4}\, z$.

6.3.3 Magnification or Contraction

Consider the transformation

$$w = f(z) = \lambda\, z; \text{ where } \lambda(\neq 0) \in \mathbb{R}. \tag{6.12}$$

Let $z = (r, \theta)$ be the coordinates of a point in z-plane. Then

$$w = f(z) = \lambda\, r\, e^{i\theta}. \tag{6.13}$$

By this transformation, figures in the z-plane are stretched or contracted in the direction of z if $\lambda > 1$ or $0 < \lambda < 1$.

Example 6.3.3. *Let the rectangular region in the z-plane be bounded by the lines $x = 0, y = 0; x = 2$, and $y = 1$. Determine the region in w-plane under the transformation $w = \sqrt{2}e^{i\pi/4}z + (1 - 2i)$.*

Solution. The given transformation can be written in the form

$$\begin{aligned}
& w = u + iv = \sqrt{2} \cdot \left(\frac{1}{\sqrt{2}} + i\frac{1}{\sqrt{2}}\right)(x + iy) + (1 - 2i) \\
\Rightarrow \quad & u = x - y + 1 \text{ and } v = x + y - 2 \\
\Rightarrow \quad & x = (u + v + 1)/2 \text{ and } y = (v - u + 3)/2.
\end{aligned}$$

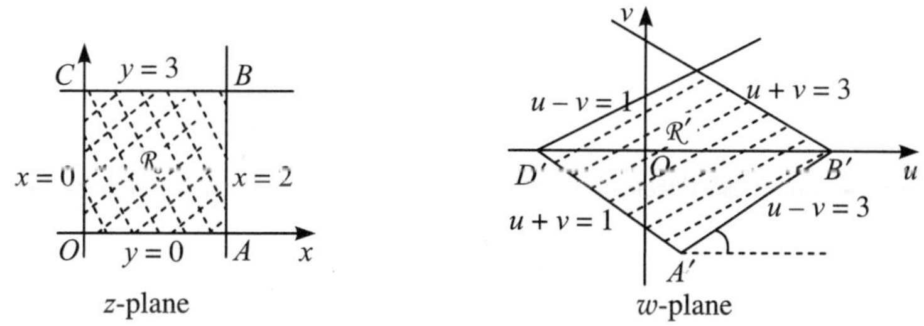

Figure 6.8 $w = \sqrt{2}\, e^{i\pi/4} z + (1 - 2i)$

From the Fig. 6.8, $y = 0 \longrightarrow v - u + 3 = 0$; $x = 2 \longrightarrow u + v - 3 = 0$; $y = 1 \longrightarrow v - u + 1 = 0$; $x = 0 \longrightarrow u + v + 1 = 0$. The mapping accomplishes a rotation of $\mathcal{R}$ through angle $\pi/4$ or $45°$, a stretching of lengths (of magnitude $\sqrt{2}$) and a translation.

6.3.4 Inversion

Consider the reciprocal transformation

$$w = f(z) = 1/z, \tag{6.14}$$

which establishes a one to one mapping between the point in $\mathbb{C} - \{0\}$ of the z-plane and the w-plane. If $z = re^{i\theta}$, where $r > 0$, then

$$w = 1/z = e^{-i\theta}/r. \tag{6.15}$$

Let us consider $f_1(z) = \frac{1}{r} e^{i\theta} = \frac{z}{|z|^2}$ and $f_2(z) = re^{-i\theta} = \bar{z}$. Then

$$(f_1 f_2)(z) = f_1(f_2(z)) = f_1(re^{-i\theta}) = \frac{1}{r}\, e^{-i\theta} = \frac{1}{z}. \tag{6.16}$$

Thus the transformation (6.14) can be expressed as a product of two transformations, f_1 and f_2. The first transformation f_1 is an *inversion* with respect to the unit circle $|z| = 1$, i.e. the image of a non-zero point z is the point Z with the properties

$$|Z| = 1/|z| \text{ and } \arg Z = \arg z.$$

The second transformation f_2 is the *reflection* about $\operatorname{Im} z = 0$. Thus the transformation (6.14) is the result of an inversion with respect to the unit circle followed by the reflection about the real axis. Points on the unit circle are reflected by Eq. (6.15) in real axis into points on the unit circle.

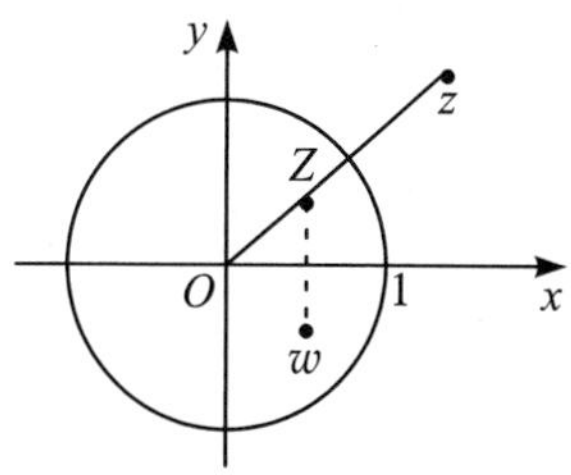

Figure 6.9 $w = 1/z = 1/|z|^2\, \bar{z} = e^{-i\theta}/r$

RESULT **54.** Here we can define f as in Eq. (6.14) at the points 0 and ∞, so as to be continuous on the extended complex plane $\mathbb{C}_\infty$. Now,

$$\lim_{z\to 0} f(z) = \infty \quad \text{since} \quad \lim_{z\to 0} \frac{1}{f(z)} = \lim_{z\to 0} z = 0, \tag{6.17}$$

$$\lim_{z\to 0} f(z) = 0 \quad \text{since} \quad \lim_{z\to 0} f\left(\frac{1}{z}\right) = \lim_{z\to 0} z = 0. \tag{6.18}$$

In order to make f continuous on the extended plane, then, we write

$$f(0) = \infty,\ f(\infty) = 0 \text{ and } f(z) = 1/z;\ z \in \mathbb{C}_\infty/\{0, \infty\}. \tag{6.19}$$

More precisely, the first of limits (6.17) and (6.18) tells us that the limit

$$\lim_{z\to z_0} f(z) = f(z_0),$$

which is clearly true when $z_0 \neq 0$ and when $z_0 \neq \infty$, is also true for those two values of z_0.

RESULT **55.** Let $z = x + iy \neq 0 \in \mathbb{C}$. When a point $w = u + iv$ in the image of the non-zero point z in the finite plane under the transformation (6.14), then

$$w = u + iv = \frac{\bar{z}}{z\bar{z}} = \frac{1}{|z|^2}\bar{z} = \frac{x}{x^2+y^2} - i\,\frac{y}{x^2+y^2}.$$

Therefore, in terms of Cartesian coordinates the transformation (6.14) can be expressed in the form

$$u = \frac{x}{x^2+y^2}, \quad v = -\frac{y}{x^2+y^2} \tag{6.20}$$

$$\Rightarrow \quad x = \frac{u}{u^2+v^2}, \quad y = -\frac{v}{u^2+v^2}. \tag{6.21}$$

RESULT **56.** Let Γ be an arbitrary circle or an extended line given by

$$\Gamma : a(x^2+y^2) + bx + cy + d = 0, \tag{6.22}$$

where a, b, c and d are real, with $a \neq 0$ for a circle and $a = 0$ for a line. Let $z(\neq 0) \in \mathbb{C}$ be a point on Γ. When $a \neq 0$, Eq. (6.22) can be written as

$$\left(x + \frac{b}{2a}\right)^2 + \left(y + \frac{c}{2a}\right)^2 = \left(\frac{\sqrt{b^2+c^2-4ad}}{2a}\right)^2,$$

so that $b^2 + c^2 > 4ad$. Now using Eq.(6.20), we get from Eq.(6.22),

$$a + b\cdot\frac{x}{x^2+y^2} + c\cdot\frac{y}{x^2+y^2} + \frac{d}{x^2+y^2} = 0$$

$$\text{or,} \quad \Gamma_1 : d(u^2+v^2) + bu - cv + a = 0, \tag{6.23}$$

which determines a circle (if $d \neq 0$) and an extended line (if $d = 0$). Again using Eq. (6.20), we see that if $(u, v) \neq (0, 0)$ satisfies Eq. (6.23), then $z = 1/w$ satisfies Eq. (6.22). Thus

(i) a circle $(a \neq 0)$ not passing through the origin $(d \neq 0)$ in the z-plane is transformed into a circle not passing through the origin in the w-plane;

(ii) a circle $(a \neq 0)$ not passing through the origin $(d = 0)$ in the z-plane is transformed into a line does not passing through the origin in the w-plane;

(iii) a line $(a = 0)$ not passing through the origin $(d \neq 0)$ in the z-plane is transformed into a circle not passing through the origin in the w-plane;

(iv) a line $(a = 0)$ through the origin $(d = 0)$ in the z-plane is transformed into a line through the origin in the w-plane.

If we regard the straight lines as special cases of circles, i.e., circles passing through infinity, then we may conclude on basis of the above workout that the reciprocal transformation indeed maps circles onto circles.

Example 6.3.4. *Show that by means of the inversion the circle $C : |z-3| = 5$ is mapped into the circle $C' : |w + 3/16| = 5/16$. Into what region is the interior on C mapped?*

Solution. Using $w = u + iv$, we get

$$\begin{aligned}
&\left|\frac{1}{w} - 3\right| = 5 \Rightarrow \left|\frac{1}{u+iv} - 3\right| = 5 \\
\Rightarrow\quad & |(1 - 3u) - 3iv| = 5|u + iv| \\
\Rightarrow\quad & (1 - 3u)^2 + 9v^2 = 25(u^2 + v^2) \\
\Rightarrow\quad & u^2 + v^2 + \frac{6}{16}u - \frac{1}{16} = 0.
\end{aligned}$$

This is the equation of the circle with centre $(-3/16, 0)$ and radius $= 5/16$ units. So the image circle in the w-plane is given by the equation $|w + 3/16| = 5/16$ (Fig. 6.10). Also, every interior point on C corresponds to an exterior point of C'.

Example 6.3.5. *Find the image of the strip* $2 < x < 3$ *under the map* $w = 1/z$.

Solution. The transformation $w = 1/z$ can be written in the Cartesian coordinates as in Eq. (6.21). Now, we find that

$$x > 2 \Rightarrow \frac{u}{u^2 + v^2} > 2 \Rightarrow \left(u - \frac{1}{4}\right)^2 + v^2 < \frac{1}{16}.$$

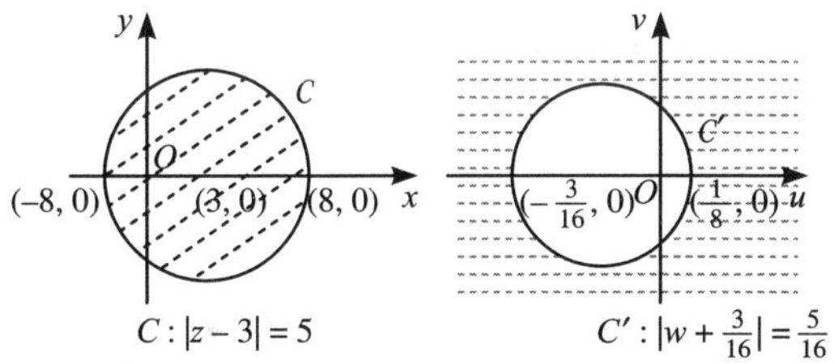

Figure 6.10 *Example 6.3.4*

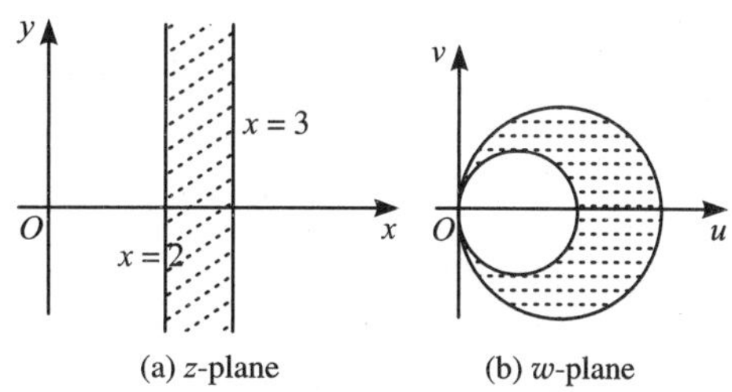

Figure 6.11 *Example 6.3.5*

Thus the region $x > 2$ is mapped into the interior of the circle with centre $(1/4, 0)$ and radius $1/4$ (Fig. 6.11). Also,

$$x < 3 \Rightarrow \frac{u}{u^2 + v^2} < 3 \Rightarrow \left(u - \frac{1}{6}\right)^2 + v^2 > \frac{1}{36}.$$

Thus the region $x < 3$ is mapped onto the exterior of the circle with centre $(1/6, 0)$ and radius $1/6$ (Fig. 6.11). The strip $2 < x < 3$ is mapped onto the region bounded by the circles $u^2 + v^2 = u/2$ and $u^2 + v^2 = u/3$ in the w-plane.

6.4 Linear Fractional Transformations

Definition 6.4.1. Let a, b, c, d be complex numbers which form a matrix $\begin{pmatrix} a & b \\ c & d \end{pmatrix}$. A nonconstant rational function of first order of the form

$$w = f(z) = \frac{az + b}{cz + d};\quad ad - bc \neq 0 \tag{6.24}$$

is called a *linear fractional* or *Möbius transformations* after A. F. Möbius who introduced it.

If $\alpha \in \mathbb{C} - \{0\}$, then Eq. (6.24) can be written as

$$f(z) = \frac{(\alpha a)z + (\alpha b)}{(\alpha c)z + (\alpha d)}; \quad \alpha(ad - bc) \neq 0,$$

which shows that the coefficients a, b, c, d are not unique.

The condition $ad - bc \neq 0$ states that $az + b$ and $cz + d$ do not have a common zero. The constant $ad - bc$ is called the *determinant of the transformation.* The transformation (6.24) is said to be *normalized* if $ad - bc = 1$. Equation (6.24) can be written in the form

$$\begin{aligned} & czw - az + dw - b = 0 \\ \Rightarrow \quad & Azw + Bz + Cw + D = 0; \ AD - BC \neq 0 \end{aligned} \tag{6.25}$$

and, conversely, any equation of the type (6.25) can be put in the form (6.24).

Example 6.4.1. *If* $w_1 = \frac{2z+1}{4z+3}$ *and* $w_2 = \frac{(3+2i)z+i}{-13iz+(3-2i)}$, *which is the linear fractional transformation and why?*

Solution. For w_1, the determinant of a transformation is

$$\begin{vmatrix} a & b \\ c & d \end{vmatrix} = \begin{vmatrix} 2 & 1 \\ 4 & 4 \end{vmatrix} = 2 \neq 0.$$

So, w_1 is a linear fractional transformation. For w_2, the determinant of transformation is

$$\begin{vmatrix} a & b \\ c & d \end{vmatrix} = \begin{vmatrix} 3+2i & i \\ -13i & 3-2i \end{vmatrix} = 0.$$

So this transformation w_2 is not linear fractional.

Example 6.4.2. *Find the linear fractional transformation which transforms the points* $-i, 0, i$ *in the* z*-plane respectively into the points* $-1, i, 1$ *of the* w*-plane. Into what curve is the imaginary axis* $x = 0$ *transformed?*

Solution. Since $-i, 0, i$ in the z-plane corresponds to the points $-1, i, 1$ of the w-plane, we have, using Eq. (6.24),

$$\begin{aligned} -1 &= \frac{-ia + b}{-ic + d} \Rightarrow -i(a - c) + b + d = 0; i = b/d \ \Rightarrow b = di \\ 1 &= \frac{ia + b}{ic + d} \quad \Rightarrow i(a - c) = d - b. \end{aligned}$$

The solutions are $a = -id, b = id$ and $c = d$. Therefore

$$w = \frac{-idz + id}{c(z + 1)} = \frac{id(1 - z)}{ia(1 + z)} = i\,\frac{1 - z}{1 + z},$$

which is the required linear fractional transformation. The inverse transformation is

$$(1 + z)w = i(1 - z) \Rightarrow z = \frac{1 + iw}{1 - iw}.$$

Let $z = x + iy$ and $w = u + iv$. Then

$$\begin{aligned} x + iy &= \frac{1 + i(u + iv)}{1 - i(u + iv)} = \frac{1 - v^2 - u^2}{(1 + v)^2 + u^2} + i\,\frac{u + v}{(1 + v)^2 + u^2} \\ \Rightarrow \quad x &= \frac{1 - v^2 - u^2}{(1 + v)^2 + u^2} \text{ and } y = \frac{u + v}{(1 + v)^2 + u^2}. \end{aligned}$$

Therefore, $x = 0$ corresponds to $u^2 + v^2 = 1$, which is the equation of the circle in w-plane.

DEDUCTION **23.** The condition $ad - bc \neq 0$ in the transformation (6.24) holds if and only if w is not constant. Let us suppose first w is constant, i.e.,

$$\begin{aligned} & f(z) = \frac{az+b}{cz+d} = k \Rightarrow az+b = kcz + kd \\ \Rightarrow \quad & a = kc \text{ and } b = kd \Rightarrow ad - bc = kc\,d - kd\,c = 0, \end{aligned}$$

which is a contradiction. Conversely, let $ad - bc = 0$. Then $ad = bc$. Not both c and d are zero. If $c \neq 0$, then $b = ad/c$ and hence

$$f(z) = \left(az + \frac{ad}{c}\right) \cdot \frac{1}{cz+d} = \frac{a}{c},$$

which is a constant function and is therefore not linear. Similarly, if $d \neq 0$, then $w = b/d$, which is a constant. When c, d are non-zero, then

$$\begin{aligned} & \frac{a}{c} = \frac{b}{d} = k(\text{ say }),\ \text{constant} \\ \Rightarrow \quad & f(z) = \frac{az+b}{cz+d} = \frac{kcz+kd}{cz+d} = k = \frac{a}{c};\ \forall z \in \mathbb{C} - \{-d/c\}, \end{aligned}$$

which is a trivial transformation. Thus (in either case) w is constant, a contradiction. Therefore, it follows that *a constant function is not a linear fractional transformation.*

DEDUCTION **24.** Let $c = 0$. Then $ad \neq 0$. Thus the Eq. (6.24) becomes

$$f(z) = \frac{a}{d}\,z + \frac{b}{d} = \alpha\,z + \beta;\ \text{where } \alpha = \frac{a}{d}, \beta = \frac{b}{d}, \tag{6.26}$$

which is a non-constant linear transformation, sometimes called *entire linear transformation*, defined on $\mathbb{C}$. As $f'(z) = \alpha \neq 0$, f is a conformal mapping of the whole z-plane.

DEDUCTION **25.** The inversion $w = 1/z$ is evidently a special case of the transformation (6.24) when $c \neq 0$. Also, let $c \neq 0$ in Eq.(6.24). Then

$$\begin{aligned} f(z) &= \frac{a}{c} \cdot \frac{z + \frac{b}{a}}{z + \frac{d}{c}} = \frac{a}{c} \cdot \frac{(z + \frac{d}{c}) + (z + \frac{b}{a}) - (z + \frac{d}{c})}{z + \frac{d}{c}} \\ &= \frac{a}{c}\left[1 + \frac{\frac{b}{a} - \frac{d}{c}}{z + \frac{d}{c}}\right] = \frac{a}{c} + \frac{bc - ad}{c} \cdot \frac{1}{cz+d};\ ad - bc \neq 0. \end{aligned} \tag{6.27}$$

Therefore, once again, the condition $ad - bc \neq 0$ ensures that we do not have a constant function. Now,

$$\begin{aligned} f'(z) &= \frac{a(cz+d) - (az+b)c}{(cz+d)^2} = \frac{ad - bc}{(cz+d)^2} \\ &= \frac{ad-bc}{c^2} \cdot \frac{1}{(z-p)^2}; p = -d/c;\ ad - bc \neq 0, \end{aligned} \tag{6.28}$$

which exists for every $z \in \mathbb{C} - \{p\}$. The magnification factor at z is

$$|f'(z)| = \frac{|ad-bc|}{c^2} \cdot \frac{1}{|z-p|^2}.$$

The points $z = -d/c, z = \infty$ are called *critical points* , where the conformal property does not hold good. Now, we see that $f'(z) \neq 0$ for all $z \in \mathbb{C} - \{-d/c\}$.

DEDUCTION **26.** Every linear fractional transformation may be written in any one of the forms:

$$\begin{aligned} f(z) &= \alpha\, z + \beta, \text{ where, } \alpha \neq 0 \\ f(z) &= \alpha + \frac{\beta}{z+p}, \text{ where, } \beta \neq 0. \end{aligned}$$

Equation (6.27) reveals that when $c \neq 0$, a linear fractional transformation is a composition of mappings

$$Z = cz + d, \; W = \frac{1}{Z}, \; w = \frac{a}{c} + \frac{bc - ad}{c} W; \; ad - bc \neq 0.$$

DEDUCTION **27.** Solving Eq.(6.24) for z, we find that

$$f^{-1}(w) = z = \frac{dw - b}{-cw + a} \text{ and } \begin{vmatrix} d & -b \\ -c & a \end{vmatrix} = da - bc \neq 0. \tag{6.29}$$

The Eq. (6.29) is also linear fractional. Thus there is an inverse function associated with the matrix $\begin{pmatrix} d & -b \\ -c & a \end{pmatrix}$. Also,

$$f\left(f^{-1}(w)\right) = w \text{ and } f^{-1}(f(z)) = z,$$

and hence f is bijective and the inverse of f is f^{-1}. Observe that the inverse function f^{-1} is not defined at $z = a/c$. Thus f gives an isomorphism of $\mathbb{C} - \{-d/c\}$ with $\mathbb{C} - \{a/c\}$.

Example 6.4.3. *Find the image of the lines $x = 2$ and $y = 1$ in the z-plane under the transformation $w = (4z + 1)/(z - 2 - i)$.*

Solution. For the given transformation, the inverse transformation is given by

$$w = \frac{4z+1}{z-2-i} \Rightarrow z = \frac{(2+i)w+1}{w-4}.$$

Let $z = x + iy$ and $w = u + iv$. We have

$$\begin{aligned} x + iy &= \frac{(2+i)(u+iv)+1}{u+iv-4} = \frac{\{(2u - v + 1) + i(u + 2v)\}\{(u-4) - iv\}}{(u+4)^2 + v^2} \\ &= \frac{2(u^2+v^2) - 7u + 4v - 4}{(u-4)^2 + v^2} + i\frac{u^2 + v^2 - 4u - 9v}{(u-4)^2 + v^2} \\ \therefore x &= \frac{2(u^2+v^2) - 7u + 4v - 4}{(u-4)^2 + v^2}, \quad y = \frac{u^2 + v^2 - 4u - 9v}{(u-4)^2 + v^2} \end{aligned}$$

The image of the line $x = 2$ is obtained as

$$2(u^2 + v^2) - 7u + 4v - 4 = 2\{(u-4)^2 + v^2\} \Rightarrow 9u + 4v - 36 = 0.$$

The image of the line $y = 1$ is obtained as

$$u^2 + v^2 - 4u - 9v = (u-4)^2 + v^2 \Rightarrow 4u - 9v - 16 = 0.$$

DEDUCTION **28.** [Conformal mapping of the extended plane]: We can, however, enlarge the domain of definition of transformation (6.24) in order to define a linear fractional transformation f on $\mathbb{C}_\infty$ such that the point $w = a/c$ is the image of $z = \infty$ when $c \neq 0$, as

$$\lim_{z\to\infty} f(z) = \lim_{z\to\infty} \frac{az+b}{cz+d} = \lim_{z\to\infty} \frac{a+\frac{b}{z}}{c+\frac{d}{z}} = \frac{a}{c}.$$

So, there is no z in $\mathbb{C}$ such that $f(z) = a/c$, as $ad - bc \neq 0$. We may record the explicit rule for the map f as

$$f(z) = \begin{cases} \dfrac{az+b}{cz+d} & \text{if } z \in \mathbb{C} \\ a/c & \text{if } z = \infty. \end{cases}$$

The points $z = -d/c$ and $w = a/c, c \neq 0$, are known as *exceptional points*. These exceptional points are mapped into the point $w = \infty$ and $z = \infty$ respectively. Therefore,

$$f(z) = \begin{cases} \infty & \text{if } z = \infty \text{ and } c = 0 \\ \infty & \text{if } z = -d/c \text{ and } c \neq 0 \\ a/c & \text{if } z = \infty \text{ and } c \neq 0 \\ \dfrac{az+b}{cz+d} & \text{elsewhere.} \end{cases} \tag{6.30}$$

By the definition, f is continuous on the $\mathbb{C}_\infty$. The transformation (6.24) is one-to-one and an onto map from the extended z-plane to the extended w-plane. So, if the complex plane is closed by the addition of the point ∞, then a linear fractional transformation establishes a one-to-one correspondence between the z-plane closed with the point ∞ and the w-plane closed with ∞, i.e.,

$$f : z\{a+ib, \infty\} \to w\{c+id, \infty\},$$

then $f(z)$ is one-to-one. This implies that inverse transformation $f^{-1} : \mathbb{C}_\infty \to \mathbb{C}_\infty$ exists and is defined as

$$f^{-1}(w) = z \text{ if and only if } f(z) = w,$$

and is given in Eq. (6.29). Evidently, f^{-1} is itself a linear fractional transformation, where

$$f^{-1}(w) = \begin{cases} \infty & \text{if } w = \infty \text{ and } c = 0 \\ \infty & \text{if } w = a/c \text{ and } c \neq 0 \\ -d/c & \text{if } w = \infty \text{ and } c \neq 0 \\ \dfrac{-dw+b}{cw-a} & \text{elsewhere.} \end{cases} \tag{6.31}$$

If f and g are two linear fractional transformations, so is the composition $g(f(z))$, and in particular, $f^{-1}[f(z)] = z$ for each z in the extended plane.

Example 6.4.4. *Find the linear fractional transformation which transforms the points $1, 0, -1$ in the z-plane respectively into the points $i, \infty, 1$ of the w-plane. Find also the critical points, if any.*

Solution. Since $w_2 = \infty$ corresponds to $z = 0$ in Eq. (6.24), we have $d = 0$. So Eq. (6.24) becomes

$$w = f(z) = \frac{az+b}{cz}; \quad bc \neq 0.$$

Because, 1 is mapped into i and -1 onto 1, we have

$$ic = a + b \text{ and } -c = a + b \Rightarrow b = \frac{c}{2}(i-1) \text{ and } a = \frac{c}{2}(i+1).$$

$$\Rightarrow \quad w = f(z) = \frac{\frac{c}{2}(i+1)z + \frac{c}{2}(i-1)}{cz} = \frac{(1+i)z + (i-1)}{2z}; \quad \text{as } c \neq 0,$$

which is the required linear fractional transformation. Now

$$f'(z) = 0 \Rightarrow z = \infty,$$

which is the critical point of the transformation.

DEDUCTION **29.** [The Steiner circles]: Let us choose two arbitrary points $z = z_1, z_2 (\neq \infty, -d/c)$ in the z-plane and denote their images under the transformation (6.24) by $w = w_1, w_2$ respectively. Then,

$$\frac{w - w_1}{w - w_2} = \frac{cz_2 + d}{cz_1 + d}\,\frac{z - z_1}{z - z_2} = \lambda\,\frac{z - z_1}{z - z_2}, \tag{6.32}$$

where λ is a non-zero, finite, complex constant. Therefore,

$$\frac{|w - w_1|}{|w - w_2|} = |\lambda|\,\frac{|z - z_1|}{|z - z_2|} \tag{6.33}$$

$$\arg\frac{w - w_1}{w - w_2} = \arg\lambda + \arg\frac{z - z_1}{z - z_2} (\bmod 2\pi). \tag{6.34}$$

Let ρ be a positive number. Then $|z - z_1|/|z - z_2| = \rho$, the locus of all points z for which the ratio of distances from the 'limiting points' z_1, z_2 is constant, is the equation of a circle known as the *circle of Apollonius*. As z moves around the circle, the point w, according to Eq. (6.33), must move on the circle with limiting points w_1 and w_2 and radius $\rho|\lambda|$. The expression

$$\theta = \arg\frac{z - z_1}{z - z_2} = \arg(z - z_1) - \arg(z - z_2)$$

gives the angle between the vectors $z - z_2$ and $z - z_1$ (Fig. 6.12). When z moves from z_1 to z_2 along a circular arc, α, θ remains constant and equals the angle θ_0 between the tangent to this arc at z_1 and the vector $z_1 - z_2$. According, to (6.34),

$$\psi = \arg\frac{w - w_1}{w - w_2} = \arg\lambda + \theta_0 = \text{ constant.}$$

Thus, the image point w also moves along a circular arc β, which makes the angle $\theta_0 + \arg\lambda$ with the vector $w_1 - w_2$. If z now moves along the complementary arc of the arc $\alpha, \theta = \theta_0 + \pi$ and, consequently, $\psi = \theta_0 + \arg\lambda + \pi$. Hence, w also moves along the complementary arc of β. The circle through the points z_1 and z_2 therefore corresponds to a circle through the points w_1 and w_2.

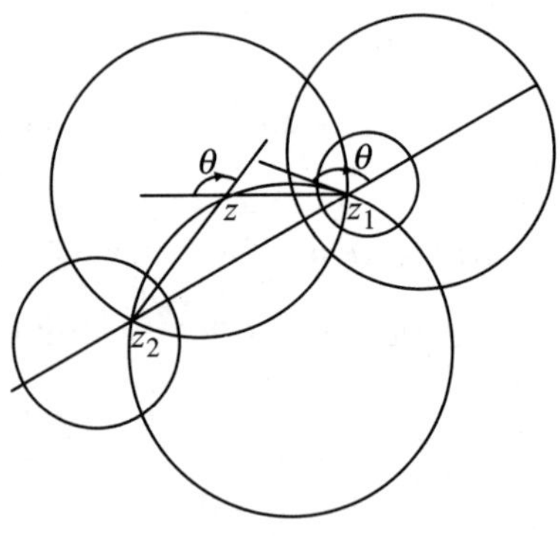

Figure 6.12 *The Steiner circles*

For constructing 'Steiner circles' which belong to the limiting points z_1 and z_2, we shall distinguish the following two kinds of circles:

(i) The first kind consists of the Apollonius circles (circles about the limiting points) which correspond to different values of ρ.

(ii) The second kind consists of the circles of the circles through the limiting points which correspond to different values of θ (circles through the limiting points).

Under the transformation (6.34), circles of the first kind are mapped onto circles of the first kind, and circles of the second kind are mapped onto circles of the second kind. From elementary geometry it is known that through every point in the plane there passes exactly one circle of the first kind and one circle of the second kind in the Steiner system belonging to any two arbitrary limiting points. The two kinds of circles intersect orthogonally.

6.4.1 Properties of Linear Fractional Transformation

Below are some important properties of linear fractional transformation.

Property 6.4.1. The conformal maps, i.e., translation, rotation–dilation and inversion are some of the particular cases of the linear fractional transformation.

Proof: We consider the linear fractional transformation (6.24).

(i) If $c = 0$, $b \neq 0$ and $a = d \neq 0$, we get the translation $w = z + b/a$.

(ii) If $b = 0 = c$ and $a, d \neq 0$, we get the rotation-dilation $w = az/d$.

(iii) If $b = c \neq 0$ and $a = d = 0$, we retrieve the complex inversion $w = 1/z$.

(iv) If $b = c = 0$ and $a = d \neq 0$, we have the identity map $w = z$.

Therefore the elementary transformations are the particular cases of the linear fractional transformation.□

Property 6.4.2. Product of two linear fractional transformations is also a linear fractional transformation.

Proof: Let us consider two linear fractional transformations, f_1 and f_2, given by

$$\zeta = f_1(z) = \frac{a_1 z + b_1}{c_1 z + d_1}; \quad \begin{vmatrix} a_1 & b_1 \\ c_1 & d_1 \end{vmatrix} \neq 0.$$

$$w = f_2(\zeta) = \frac{a_2 \zeta + b_2}{c_2 \zeta + d_2}; \quad \begin{vmatrix} a_2 & b_2 \\ c_2 & d_2 \end{vmatrix} \neq 0.$$

Now the transformation $f_2 f_1$, called the *resultant* or the *product of two transformations* f_1 and f_2, in that order, is given by

$$\begin{aligned} (f_2 f_1)(z) &= f_2(f_1(z)) = f_2\Big(\frac{a_1 z + b_1}{c_1 z + d_1}\Big); \quad \text{by Eq. (6.35)} \\ &= \Big[a_2\Big(\frac{a_1 z + b_1}{c_1 z + d_1}\Big) + b_2\Big] \Big/ \Big[c_2\Big(\frac{a_1 z + b_1}{c_1 z + d_1}\Big) + d_2\Big] \\ &= \frac{(a_1 a_2 + b_2 c_1) z + (a_2 b_1 + b_2 d_1)}{(c_2 a_1 + d_2 c_1) z + (c_2 b_1 + d_2 d_1)} = \frac{\alpha z + \beta}{\gamma z + \delta} \ (\text{ say}), \end{aligned} \tag{6.35}$$

where, $\alpha = a_1 a_2 + b_2 c_1$, $\beta = a_2 b_1 + b_2 d_1$, $\gamma = c_2 a_1 + d_2 c_1$ and $\delta = c_2 b_1 + d_2 d_1$. The determinant of the transformation (6.35) is

$$\begin{aligned} \alpha\delta - \beta\gamma &= (a_1 a_2 + b_2 c_1)(c_2 b_1 + d_2 d_1) - (a_2 b_1 + b_2 d_1)(c_2 a_1 + d_2 c_1) \\ &= a_1 a_2 d_1 d_2 + b_1 b_2 c_1 c_2 - a_2 b_1 d_2 c_1 - b_2 d_1 c_2 a_1 \\ &= (a_1 d_1 - b_1 c_1)(a_2 d_2 - b_2 c_2). \end{aligned}$$

$$\Rightarrow \begin{vmatrix} \alpha & \beta \\ \gamma & \delta \end{vmatrix} = \begin{vmatrix} a_1 & b_1 \\ c_1 & d_1 \end{vmatrix} \cdot \begin{vmatrix} a_2 & b_2 \\ c_2 & d_2 \end{vmatrix} \neq 0.$$

Thus Eq. (6.35) represents a linear fractional transformation by the relation $w = f_2(f_1(z))$. □

Property 6.4.3. If f is a linear fractional transformation, then f is the composition of the elementary transformations with simple geometric properties.

Proof (*Case I*): Let $c \neq 0$. Using Eq. (6.27), the linear fractional transformation (6.24) can be written as

$$w = \frac{a}{c} + \frac{bc - ad}{c^2} \cdot \frac{1}{z + \frac{d}{c}} = \lambda + \frac{\mu}{z + \gamma} \quad \text{(say)}, \tag{6.36}$$

which is a non-constant function, where $\lambda = a/c, \mu = (bc - ad)/c^2$, and $\gamma = d/c$. Let us define the following auxiliary basic transformations:

$$f_1(z) = z + \frac{d}{c} = z + \gamma; \quad \text{translation} \tag{6.37}$$

$$f_2(z) = \frac{1}{z}; \quad \text{inversion} \tag{6.38}$$

$$f_3(z) = \mu\, z; \quad \text{rotation–dilation} \tag{6.39}$$

$$f_4(z) = z + \lambda; \quad \text{translation} \tag{6.40}$$

The resultant of the above transformations give

$$\begin{aligned}\Big(f_4 f_3 f_2 f_1\Big)(z) &= \Big(f_4 f_3 f_2\Big)(f_1(z)) = \Big(f_4 f_3 f_2\Big)(z + \gamma)\\ &= \Big(f_4 f_3\Big)(f_2(z + \gamma)) = \Big(f_4 f_3\Big)\Big(\frac{1}{z + \gamma}\Big)\\ &= f_4\Big(f_3\Big(\frac{1}{z + \gamma}\Big)\Big) = f_4\Big(\frac{\mu}{z + \gamma}\Big) = \frac{\mu}{z + \gamma} + \lambda = f(z).\end{aligned} \tag{6.41}$$

This shows that Eq. (6.24) is the resultant of a translation given by Eq. (6.37), inversion in the real axis and unit circle given by Eq. (6.38), a rotation and magnification given by Eq. (6.39), and the translation given by Eq. (6.40).

Case II: Let $c = 0$. Then the linear fractional transformation (6.24) reduces to an equation of the type (6.26), provided $d \neq 0$. Taking, $f_1(z) = \alpha z$ and $f_2(z) = z + \beta$, where $\alpha = a/d$ and $\beta = b/d$ are constants, we get

$$(f_2 f_1)(z) = f_2(f_1(z)) = f_2(\alpha z) = \alpha\, z + \beta = f(z),$$

which is a composition of rotation–dilation followed by translation. □

Example 6.4.5. *Express $w = (3z + 2i)/(iz + 6)$ as a product of elementary transformations.*

Solution. Here $a = 3, b = 2i, c = i$ and $d = 6$. As described by Eq. (6.36), we have

$$\lambda = \frac{a}{c} = \frac{3}{i} = -3i, \mu = \frac{bc - ad}{c^2} = \frac{-2 - 18}{-1} = 20, \gamma = \frac{d}{c} = \frac{6}{i} = -6i.$$

Let us define the following auxiliary basic transformations:

$$f_1(z) = z + \gamma = z - 6i;\ f_2(z) = 1/z;\ f_3(z) = \mu\, z = 20z;\ f_4(z) = z + \lambda = z - 3i.$$

Thus, w is the composition of elementary transformations.

Property 6.4.4. A set of all linear fractional transformation from $\mathbb{C}_\infty$ to itself form a non-abelian group under usual composition of mappings.

Proof: Let M denote the collection of all linear fractional transformations of the form (6.24). The set M forms a group if the following four axioms are satisfied:

(i) *Closure axiom:* If $f_1, f_2 \in M$ as in Eqs. (6.35) and (6.35), then $f_2 f_1 \in M$. Thus closure axiom holds with respect to product composition of mappings.

(ii) *Associative axiom:* If $f_1, f_2, f_3 \in M$, the associative law $f_1(f_2 f_3) = (f_1 f_2) f_3$ follows from the fact that this law holds for arbitrary mappings.

(iii) *Existence of identity:* Let us define a mapping I as

$$w = f(z) = z; \quad b = c = 0 \text{ and } a = d \neq 0, \tag{6.42}$$

so that the transformation determinant is

$$\begin{vmatrix} a & b \\ c & d \end{vmatrix} = \begin{vmatrix} a & 0 \\ 0 & d \end{vmatrix} = a^2 \neq 0.$$

So the mapping I, defined in Eq.(6.42), is a linear fractional transformation, i.e., $I \in M$. Also,

$$(fI)(z) = f(z) = (If)(z) \ \forall f \in M.$$

Thus the mapping I defined in Eq. (6.42) serves as an identity element in M.

(iv) *Existence of inverse :* The inverse of a linear fractional transformation (6.24), defined by Eq. (6.29) is a linear fractional transformation as $ad - bc \neq 0$, i.e., $f^{-1} \in M$. Also,

$$\begin{aligned} (f^{-1}f)(z) &= f^{-1}(f(z)) = f^{-1}\Big(\frac{az+b}{cz+d}\Big) \\ &= \Big[d\Big(\frac{az+b}{cz+d}\Big) - b\Big] \Big/ \Big[-c\Big(\frac{az+b}{cz+d}\Big) + a\Big] \\ &= \frac{(ad-bc)}{(ad-bc)} z = z = I(z) \quad \text{as } ad - bc \neq 0. \end{aligned}$$

Similarly, $(ff^{-1})(w) = w$, so that $ff^{-1} = f^{-1}f = I \ \forall f \in M$. Thus $f^{-1} \in M$ defined in Eq. (6.29) is the inverse of f, and so the law of inverse holds.

Therefore, the set of all linear fractional transformations forms a group. Since in general, $(f_1 f_2)(z) \neq (f_2 f_1)(z)$, this group is non-abelian. □

For example, consider $f_1(z) = z/(z+1)$ and $f_2(z) = (z+2)/(z+3)$ as two linear fractional transformations. Then

$$f_1\Big(f_2(z)\Big) = f_1\Big(\frac{z+2}{z+3}\Big) = \frac{z+2}{2z+5}; \quad f_2\Big(f_1(z)\Big) = f_2\Big(\frac{z}{z+1}\Big) = \frac{3z+2}{4z+3}.$$

Example 6.4.6. *Find the linear fractional transformation that maps $\mathcal{H} = \{z : \text{Im}(z) > 0\}$ onto the open disc $\mathcal{D} = \{w : |w| < 1\}$ and the boundary real axis $\text{Im}(z) = 0$ onto the boundary $\partial\mathcal{D} = \{w \in \mathbb{C} : |w| = 1\}$ of the disc.*

Solution. Let us consider the transformation (6.24), which can be written as

$$w = f(z) = \frac{az+b}{cz+d} = \frac{a}{c} \cdot \frac{z+\frac{b}{a}}{z+\frac{d}{c}}; \quad \begin{vmatrix} a & b \\ c & d \end{vmatrix} \neq 0. \tag{6.43}$$

We first note that $c \neq 0$; otherwise the points at infinity will correspond. This transforms $\text{Im}(z) = 0$ onto the unit circle $|w| = 1$ (Fig. 6.13). So points w and $\bar{w}^{-1}$, inverse with respect to $|w| = 1$, transform respectively into the points z and $\bar{z}$. Since $\text{Im}(z) > 0$ is mapped onto $|w| < 1$, $\exists$ a point $\alpha \in \text{Im}(z)$ that corresponds to $w = 0$, i.e.,

$$\frac{a\alpha + b}{c\alpha + d} = 0 \Leftrightarrow b = -a\alpha.$$

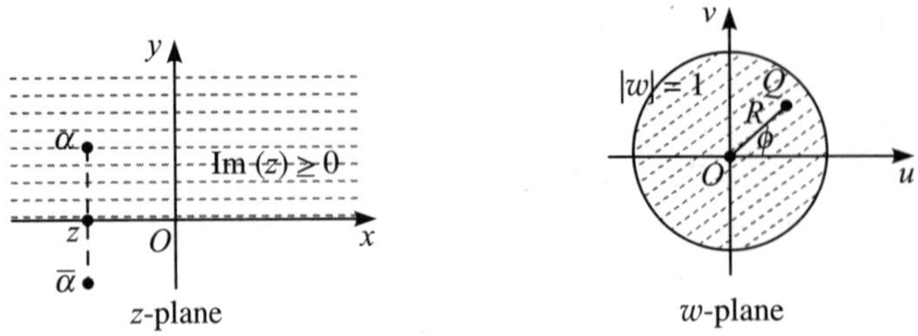

Figure 6.13 *Example 6.4.6*

Since, with respect to the line $\mathrm{Im}(z) = 0$, the inverse point of α is $\bar{\alpha}$, the image of $\bar{\alpha}$ under the mapping is the inverse point of $w = 0$ with respect to $|w| = 1$.

$$\therefore \quad c\bar{\alpha} + d = 0 \Rightarrow d = -c\bar{\alpha}.$$

Thus the linear fractional transformation Eq. (6.43) becomes

$$w = \frac{a}{c} \cdot \frac{z-\alpha}{z-\bar{\alpha}} = \lambda_1 \cdot \frac{z-\alpha}{z-\bar{\alpha}}; \text{ where } \lambda_1 = \frac{a}{c}.$$

Again, as real line $\mathrm{Im}(z) = 0$ is mapped onto the circle $|w| = 1$, for points z on the real axis, $|z-\alpha| = |z-\bar{\alpha}|$, we have $|\lambda_1| = 1$, say $\lambda_1 = e^{i\lambda}$, where $\lambda = \arg(\lambda_1)$ is real. Consequently,

$$w = \frac{z-\alpha}{z-\bar{\alpha}}\, e^{i\lambda}; \text{ where } \lambda \in \mathbb{R} \text{ and } \mathrm{Im}(z) > 0. \tag{6.44}$$

Further, using Eq. (6.44), we get

$$w\bar{w} - 1 = \frac{z-\alpha}{z-\bar{\alpha}}\, e^{i\lambda} \cdot \frac{\bar{z}-\bar{\alpha}}{\bar{z}-\alpha}\, e^{-i\lambda} - 1 = \frac{1}{|z-\bar{\alpha}|^2}\Big[(z-\bar{z})(\alpha-\bar{\alpha})\Big]$$

$$\text{or,} \quad |w|^2 - 1 = \frac{2i\,\mathrm{Im}(z)\cdot 2i\mathrm{Im}(\alpha)}{|z-\bar{\alpha}|^2} = -4\frac{\mathrm{Im}(z)\,\mathrm{Im}(\alpha)}{|z-\bar{\alpha}|^2}.$$

Since $w = 0$ corresponds to α, so $\mathrm{Im}(z) > 0$. Hence $|w|^2 - 1 < 0$ for $\mathrm{Im}(z) > 0$. Hence $|w| < 1$ corresponds to $\mathrm{Im}(z) > 0$. Therefore, the Eq. (6.44) is the required linear fractional transformation.

Example 6.4.7. *Find the linear fractional transformation that maps* $\mathcal{H}_1 = \{z : \mathrm{Re}(z) > 0\}$ *onto the open disc* $\mathcal{D} = \{w \in \mathbb{C} : |w| < 1\}$ *and the boundary* $\mathrm{Re}(z) = 0$ *onto the boundary* $\partial\mathcal{D} = \{w \in \mathbb{C} : |w| = 1\}$.

Solution. Let the required transformation be of the form (6.43). Here $c \neq 0$; otherwise the points at infinity will correspond. This transforms $\mathrm{Re}(z) = 0$ into $|w| = 1$ (Fig. 6.14). Hence the points w and $\bar{w}^{-1}$, inverse with respect to $|w| = 1$ in w-plane, transform respectively into the points $z, -\bar{z}$, inverse with respect to $\mathrm{Re}(z) = 0$. Particularly, the points 0 and ∞ which are inverse points with respect to $|w| = 1$ correspond to the points α and $\bar{\alpha}$ respectively.

$$\therefore -\frac{b}{a} = \alpha, \quad -\frac{d}{c} = -\bar{\alpha}, \text{ so that } w = \frac{a}{c}\cdot\frac{z-\alpha}{z+\bar{\alpha}}.$$

To find a/c, we make use of the fact that any point on $\mathrm{Re}(z) = 0$ must correspond to a point on $|w| = 1$. In particular, the point $z = 0$ must correspond to a point on $|w| = 1$ so that

$$\left|\frac{a}{c}\right|\left|\frac{0-\alpha}{0+\bar{\alpha}}\right| = 1 \Rightarrow \left|\frac{a}{c}\right| = 1, \text{ as } |\alpha| = |\bar{\alpha}|$$

$$\text{or,} \quad \frac{a}{c} = e^{i\lambda}; \quad \lambda \in \mathbb{R}.$$

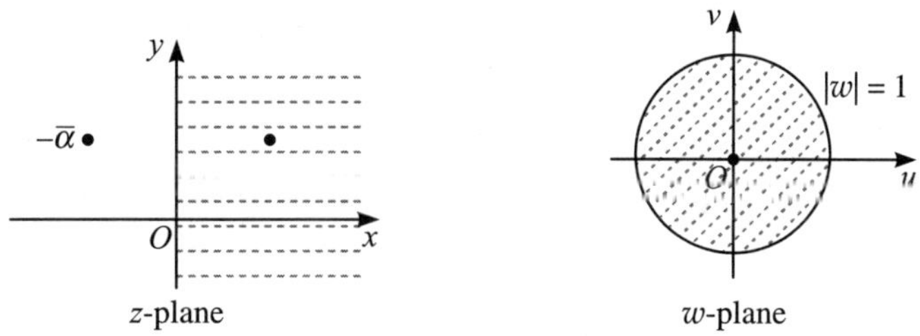

Figure 6.14 *Example 6.4.7*

Thus the transformation which maps $\text{Re}(z) > 0$ into $|w| = 1$ is given by

$$w = \frac{z - \alpha}{z + \bar{\alpha}}\, e^{i\lambda}; \text{ where } \lambda \in \mathbb{R}.$$

Further, under the transformation, $z = \alpha$ corresponds to $w = 0$, which is the interior point (the centre) of the circle $|w| = 1$; the point $z = \alpha$ must be a point of the right half plane. Therefore $\text{Re}(\alpha) > 0$, and

$$w = \frac{z - \alpha}{z + \bar{\alpha}}\, e^{i\lambda}; \text{ where } \lambda \in \mathbb{R} \text{ and } \text{Re}(z) > 0, \tag{6.45}$$

which is a transformation that maps $\text{Im}(z) = 0$ into $|w| = 1$. Further,

$$w\bar{w} - 1 = \frac{z - \alpha}{z + \bar{\alpha}}\, e^{i\lambda} \cdot \frac{\bar{z} - \bar{\alpha}}{\bar{z} + \alpha}\, e^{-i\lambda} - 1 = -\frac{1}{|z + \bar{\alpha}|^2}\Big[(z + \bar{z})(\alpha + \bar{\alpha})\Big]$$

$$\text{or,} \qquad |w|^2 - 1 = \frac{2\,\text{Re}(z) \cdot 2\text{Re}(\alpha)}{|z + \bar{\alpha}|^2} = -4\frac{\text{Re}(z)\,\text{Re}(\alpha)}{|z + \bar{\alpha}|^2} < 0, \text{ as } \text{Re}(\alpha) > 0,$$

provided $\text{Re}(z) > 0$. Hence $|w| < 1$ corresponds to $\text{Re}(z) > 0$. Therefore Eq. (6.45) is the required linear fractional transformation.

Example 6.4.8. *Find the linear fractional transformation that transforms the interior in a circle of the z-plane into interior of a circle of the w-plane.*

Solution. Refer to the Fig. 6.15. The points $w = 0$ and $w = \infty$, which are inverse points with respect to the circle $|w| = 1$, correspond to $z = -b/a$, $z = -d/c$ respectively. Hence $-b/a$ and $-d/c$ must be the inverse points with respect to the circle $|z| = 1$ so that

$$-\frac{b}{a} = \alpha \text{ and } -\frac{d}{c} = \frac{1}{\bar{\alpha}}.$$

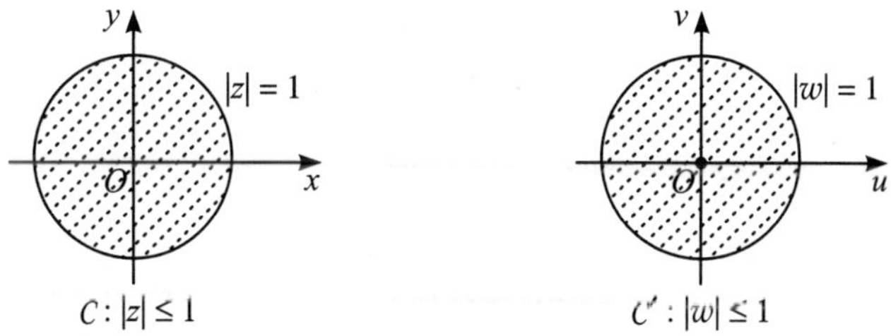

Figure 6.15 *Example 6.4.8*

Therefore, the required linear fractional transformation can be written as

$$w = \frac{a}{c} \cdot \frac{z-\alpha}{z-(1/\bar{\alpha})} = \frac{a\,\bar{\alpha}}{c} \cdot \frac{z-\alpha}{z\bar{\alpha}-1}.$$

Also, the points of $|z| = 1$ correspond to $|w| = 1$. Hence

$$\left|\frac{a\,\bar{\alpha}}{c}\right| \cdot \left|\frac{1-\alpha}{\bar{\alpha}-1}\right| = |w| = 1$$

or, $$\left|\frac{a\,\bar{\alpha}}{c}\right| = 1 \text{ as } |1-\alpha| = |\bar{\alpha}-1|.$$

Let $a\bar{\alpha}/c = e^{i\lambda}$, where $\lambda \in \mathbb{R}$. Hence the transformation may be written as

$$w = \frac{z-\alpha}{z\bar{\alpha}-1}\, e^{i\lambda};\ \ \lambda \in \mathbb{R}. \tag{6.46}$$

Since $|z| = 1$, i.e., $z\bar{z} = 1$, we have

$$\begin{aligned} |w| &= \left|\frac{z-\alpha}{z\bar{\alpha}-1}\right| \left|e^{i\lambda}\right| = \left|\frac{z-\alpha}{z\bar{\alpha}-z\bar{z}}\right| \\ &= \frac{1}{|z|}\left|\frac{z-\alpha}{\bar{z}-\bar{\alpha}}\right|, \text{ as } |z| = 1 \text{ and } |z-\alpha| = |\bar{z}-\bar{\alpha}|. \end{aligned}$$

Further, under the transformation we have

$$w\bar{w} - 1 = \frac{z-\alpha}{\bar{\alpha}z-1}\, e^{i\lambda} \cdot \frac{\bar{z}-\bar{\alpha}}{\alpha\bar{z}-1}\, e^{-i\lambda} - 1 = -\frac{z\bar{z} - z\bar{\alpha} - \bar{z}\alpha + \alpha\bar{\alpha}}{(z\bar{\alpha}| - 1)(\alpha\bar{z} - 1}$$

or, $$|w|^2 - 1 = \frac{(z\bar{z}-1)(1-\alpha\bar{\alpha}}{|z\bar{\alpha}|-1|^2} = \frac{(|z|^2-1)(1-|\alpha|^2)}{|z\bar{\alpha}|-1|^2}.$$

The transformation will map $|z| < 1$ on $|w| < 1$ or on $|w| > 1$ conformally according as $|\alpha| < 1$ or $|\alpha| > 1$. Thus, it is shown that the transformation maps $|z| \le 1$ onto $|w| \le 1$, provided $|\alpha| < 1$. Hence (6.46) is the required linear fractional transformation.

6.4.2 Fixed Points

Definition 6.4.2. Let $w = f(z)$ be a mapping. If a point is carried by the mapping into itself, then it is called a *fixed point*. Obviously, every such point is a solution of the equation

$$w = f(z) = z. \tag{6.47}$$

The set of all fixed points of f is denoted by $\mathcal{F}$. For example $f(z) = z^2$ has exactly three fixed points, $0, 1$ and ∞; $f(z) = z - 2$ has no fixed point in $\mathbb{C}$, and $f(z) = \bar{z}$ has all points on $\mathbb{R}$ as its fixed points.

Now, we consider the linear fractional transformation (6.24), or clarify it according to the behaviour of its fixed point (or points).

Case I : Let $c \neq 0$. In this case, $f(\infty) = a/c \neq \infty$. So infinity is not a fixed point. Also, $f(p) = \infty \neq p$, so that p is not a fixed point as well. Let us take $z \neq p, z \neq \infty$. Then using Eq.(6.47) in Eq. (6.24) gives

$$\frac{az+b}{cz+d} = z \ \Rightarrow\ cz^2 - (a-d)z - b = 0;\ \ ad - bc \neq 0 \tag{6.48}$$

$$\Rightarrow\ z = \frac{1}{2c}\Big[(a-d) \pm \sqrt{(a-d)^2 + 4bc}\Big]. \tag{6.49}$$

Consider the following two important subcases:

(i) Let $\Delta = (a-d)^2 + 4bc \neq 0$. Then Eq. (6.49) ensures that a linear fractional transformation can have at most two fixed points, say ζ_1 and ζ_2.

(ii) Let $\Delta = (a-d)^2 + 4bc = 0$. There is only one fixed point $z = (a-d)/2c$.

From deduction 29, it follows that the system of Steiner circles with limiting points $z_1 = \zeta_1$ and $z_2 = \zeta_2$ remains fixed under the transformation (6.24), with circles of the first kind going over into circles of the first kind, and circles of the second kind going over into circles of the second kind (Fig. 6.16).

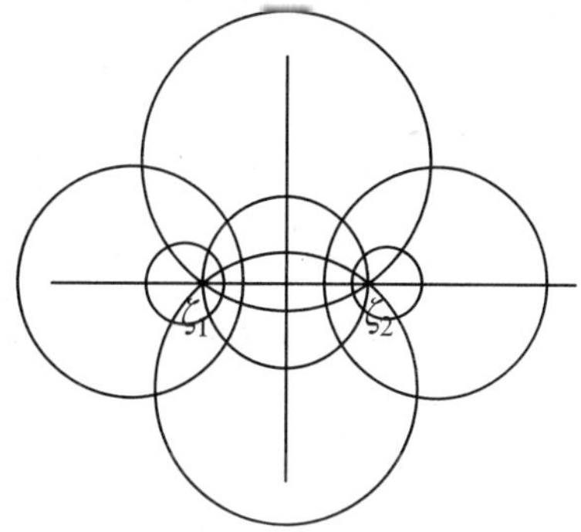

Figure 6.16 *Fixed points*

Case II : Let $c = 0$, $\Delta \neq 0$. Then Eq. (6.24) becomes a non-constant linear transformation of the type (6.26). Obviously, $f(\infty) = \infty$. Thus the fixed points are

$$z_1 = \zeta_1 = \frac{\beta}{1-\alpha} = \frac{b}{d-a} \text{ and } \zeta_2 = \infty$$

and the mapping of the Steiner net is the same as in the case $c \neq 0$. The circles of the first kind are concentric circles about ζ_1, while the circles of the second kind are straight lines through ζ_1. The transformation (6.24) assumes the form

$$w - \zeta_1 = \lambda(z - \zeta_1); \quad \lambda = a/d.$$

This is a *similarity* (or *homothetic*) *transformation* with the fixed point ζ_1 as centre.

If $a - d = 0$ and $\beta \neq 0$, obviously, there is no finite fixed point.

RESULT **57.** Any linear fractional transformation other than the identity transformation has two fixed points. However, the transformation $w = f(z) = \bar{z}$ has infinitely many fixed points, namely all real numbers. Hence $f(z) = \bar{z}$ is not a linear fractional transformation.

Example 6.4.9. *Find all the linear fractional transformation whose fixed points are i and $-i$.*

Solution. Since i and $-i$ are the fixed points of Eq. (6.24), so

$$\begin{aligned} & i = \frac{ai+b}{ci+d} \text{ and } -i = \frac{-ai+b}{-ci+d} \\ \Rightarrow \quad & (a-d)i + (b+c) = 0 \text{ and } i(d-a) + (b+c) = 0 \\ \Rightarrow \quad & a - d = 0 \text{ and } b + c = 0 \Rightarrow a = d, b = -c. \end{aligned}$$

Thus the required linear fractional transformation is

$$w = \frac{az+b}{cz+d} = \frac{az-c}{cz+a}; \qquad a^2 + c^2 \neq 0.$$

Corollary 6.4.1. A linear fractional transformation can have at most two fixed points unless $f(z) = z$ for all z.

Proof *(Case I)*: If ∞ is a fixed point of f, then $f(\infty) = \infty \Rightarrow c = 0$, i.e., $ad \neq 0$. Hence, the fixed points other than ∞ are the roots of the equation

$$z = \frac{az+b}{d} \Rightarrow (a-d)z + b = 0.$$

However, this linear relation in z can be satisfied by more than one value of z if it were an identity, i.e., if $a = d$ and $b = 0$, implying that $f(z) = z \; \forall z \in \mathbb{C}_\infty$.

Case II : If ∞ be not a fixed point of z, all the fixed points of the linear fractional transformation (6.24) are the roots of the quadratic equation

$$cz^2 - (a-d)z - b = 0.$$

If this quadratic equation is satisfied by more than two values of z, then it would be an identity and hence $c = 0, b = 0$ and $a = d \neq 0$. This reduces the transformation to the identity map $f(z) = z$. □

6.4.3 Normal/Canonical Form

One Fixed Point

Let the linear fractional transformation (6.24) have only finite fixed point, say α. Then equation (6.48) has only one root, so that

$$cz^2 - (a-d)z - b = c(z-\alpha)^2; \quad \alpha = (a-d)/2c.$$

Equating coefficients of like powers in z on both sides of the above identity, we have

$$d - a = -2\alpha c \text{ and } -b = c\alpha^2 \Rightarrow a = d + 2\alpha c \text{ and } b = -c\alpha^2.$$

Hence, Eq. (6.24) can be written as

$$\begin{aligned}
& w = \frac{az+b}{cz+d} = \frac{(d+2\alpha c)z - c\alpha^2}{cz+d} \\
\text{or,} \quad & cwz + wd = zd + 2\alpha cz - c\alpha^2 \\
\text{or,} \quad & c(w-\alpha)(z-\alpha) + cw\alpha + c\alpha z - c\alpha^2 + wd = zd + 2\alpha cz - c\alpha^2 \\
\text{or,} \quad & c(w-\alpha)(z-\alpha) + (d+c\alpha)(w-\alpha) + \alpha d + c\alpha z = zd + 2\alpha cz - c\alpha^2 \\
\text{or,} \quad & c(w-\alpha)(z-\alpha) + (d+c\alpha)(w-\alpha) = zd + \alpha cz - d\alpha - c\alpha^2 \\
\text{or,} \quad & c(w-\alpha)(z-\alpha) + (d+c\alpha)(w-\alpha) = (d+c\alpha)(z-\alpha) \\
\text{or,} \quad & \frac{1}{w-\alpha} = \frac{1}{z-\alpha} + \lambda; \text{ where } \lambda = \frac{c}{d+c\alpha} = \frac{2c}{a+d} \in \mathbb{C},
\end{aligned} \tag{6.50}$$

which is the normal form of the linear fractional transformation. If we set $z' = \frac{1}{z-\alpha}$ and $w' = \frac{1}{w-\alpha}$, then Eq. (6.24) reduces to the form

$$w' = z' + \lambda. \tag{6.51}$$

This is a parallel displacement of the z'-plane. Under this mapping all straight lines L_1' in the direction w' are left invariant. The lines L_2', which cut them orthogonally, go over into one another, being displaced by an amount equal to the magnitude of λ.

Definition 6.4.3. A linear fractional transformation with only one finite fixed point is called *parabolic*. Under parabolic transformation, the circles L_1 appear as invariant 'streamlines'; the orthogonal family of circles L_2 remains unchanged as a whole, but individual circles are transformed into other circles of the family.

Example 6.4.10. *Find the fixed points of* $w = (3iz+1)/(z+i)$. *By reducing its canonical form, determine the nature of the transformation.*

Solution. The fixed points of the transformation is given by

$$(3iz+1)/(z+i) = z \Rightarrow z^2 - 2iz - 1 = 0 \Rightarrow z = i, i.$$

Since the finite fixed point i is unique, so the transformation is parabolic. Now,

$$\begin{aligned} w - i &= \frac{3iz+1}{z+i} - i = \frac{2iz+2}{z+i} \\ \Rightarrow \quad \frac{1}{w-i} &= \frac{z+i}{2i(z-i)} = \frac{1}{2i}\left[1 + \frac{2i}{z-i}\right] = \frac{1}{z-i} - \frac{i}{2}, \end{aligned}$$

which is of the form (6.50) and is the required canonical form.

Two Fixed Points

If α and β are two given fixed points of the linear fractional transformations (6.24). Then these will be the roots of the equation $cz^2 - (a-d)z - b = 0$.

$$\begin{aligned} c\alpha^2 - (a-d)\alpha - b = 0 &\Leftrightarrow c\alpha^2 - a\alpha = b - d\alpha \\ c\beta^2 - (a-d)\beta - b = 0 &\Leftrightarrow c\beta^2 - a\beta = b - d\beta \end{aligned}$$

Now, using Eq. (6.24), we get

$$\begin{aligned} w - \alpha &= \frac{az+b}{cz+d} - \alpha = \frac{(a-\alpha c)z + (b - d\alpha)}{cz+d} \\ &= \frac{(a-\alpha c)z + (c\alpha^2 - a\alpha)}{cz+d} = \frac{(a-\alpha c)(z-\alpha)}{cz+d}. \end{aligned}$$

Similarly, we get $w - \beta = (a - \beta c)(z-\beta)/(cz+d)$. Thus

$$\frac{w-\alpha}{w-\beta} = \frac{(a-\alpha c)(z-\alpha)}{(a-\beta c)(z-\beta)} = \frac{a-\alpha c}{a-\beta c}\,\frac{z-\alpha}{z-\beta} = \lambda\,\frac{z-\alpha}{z-\beta}.$$

$$\therefore \quad \frac{w-\alpha}{w-\beta} = \lambda\,\frac{z-\alpha}{z-\beta}; \text{ where } \lambda = \frac{a-\alpha c}{a-\beta c}(\neq 0) \in \mathbb{C}, \tag{6.52}$$

which is the normal form of linear fractional transformation.

Definition 6.4.4. Consider a non-constant linear fractional transformation with two finite fixed points α and β.

(i) If λ is real, the transformation is called *hyperbolic*. A comparison of the arguments of the left and right sides of Eq. (6.52) shows that in the Steiner system belonging to the limiting points α and β, circles of the first kind go over into one another, while circles of the second kind remain fixed. The transformation therefore represents a 'flow' along the circle of the second kind.

(ii) If $|\lambda| = 1$, the transformation is called *elliptic*. If $|\lambda| = 1$, then it follows from Eq. (6.52) that every circle of the first kind in the Steiner system with limiting points α and β is left invariant. Under the mapping, then, the points z 'flow' along these invariant circles in such a way that circles of the second kind pass over into one another.

(iii) The general mapping (6.52) can always be thought of as the result of composing an elliptic transformation with a hyperbolic transformation. Let $\arg\lambda = \varphi$. We first effect the elliptic transformation in which λ is replaced by $\lambda_1 = \cos\varphi + i\sin\varphi$ $(|\lambda_1| = 1)$, and then the hyperbolic transformation with the λ-value $\lambda_2 = |\lambda|$. The resulting general linear transformation is called *loxodromic* if λ can be expressed in the form $\lambda = k\,e^{i\theta}$; $k(\neq 1) \in \mathbb{R}^+$, and $\theta(\neq 0) \in \mathbb{R}$.

Example 6.4.11. *Find the fixed points and normal form of* $w = \{(2+i)z - 2\}/(z+i)$. *Hence, determine the nature of the transformation.*

Solution. For fixed point(s) of the given transformation, we have

$$f(z) = z \Rightarrow \frac{(2+i)z-2}{z+i} = z \Rightarrow z = 1 \pm i.$$

$\therefore \alpha = 1+i$ and $\beta = 1-i$ are two distinct fixed points of the transformation. Now,

$$\begin{aligned} w-(1+i) &= \frac{(2+i)z-2}{z+i} - (1+i) = \frac{z-(1+i)}{z+i} \\ w-(1-i) &= \frac{(1+2i)z-(3+i)}{z+i} = (1+2i)\left[\frac{z-(3+i)/(1+2i)}{z+i}\right] \\ &= (1+2i)\frac{z-(1-i)}{z+i}. \end{aligned}$$

Dividing, we get

$$\frac{w-(1-i)}{w-(1+i)} = (1+2i)\,\frac{z-(1-i)}{z-(1+i)},$$

which is the required canonical form as in Eq. (6.52). Here, $\lambda = 1+2i = k\,e^{i\theta}$, so that $k = \sqrt{5}(>0, \neq 1)$ and $\theta = \tan^{-1} 2 \neq 0$. Hence, the given transformation is loxodromic.

Example 6.4.12. *What is the form of the linear fractional transformation which has one finite fixed point* α *and the other fixed point* ∞.

Solution. If this linear fractional transformation (6.24) has one finite fixed point α and other fixed point infinity, then $c = 0$ and $a \neq d$, so that it takes the form

$$w = \frac{a}{d}\,z + \frac{b}{d} \Rightarrow \alpha = \frac{a}{d}\,\alpha + \frac{b}{d}$$

or, $$w - \alpha = \frac{a}{d}(z-\alpha) = \lambda(z-\alpha), \text{ where } \lambda = \frac{a}{d}.$$

If ∞ is the only fixed point of the transformation, then $c = 0$ and $a = d$, so that in this case the transformation takes the form $w = z + \frac{b}{d}$, which is a translation.

Corollary 6.4.2. If three distinct points z_1, z_2, z_3 have the same image under the two linear fractional transformations $f_1(z)$ and $f_2(z)$, then they will be identical in $\mathbb{C}_\infty$.

Proof: Let f_1 be a linear fractional transformation and z_1, z_2, z_3 distinct points in $\mathbb{C}_\infty$ with $w_1 = f_1(z_1)$, $w_2 = f_1(z_2)$ and $w_3 = f_1(z_3)$. Suppose $f_2(z)$ is another map with this property. Due to group property, $f_2^{-1} f_1$ is also a linear fractional transformation and has z_1, z_2, z_3 as fixed points. Therefore $f_2^{-1} f_1$ is an identity map, i.e.,

$$(f_2^{-1} f_1)(z) = z \Rightarrow f_1(z) = f_2(z)\ \forall z \in \mathbb{C}_\infty.$$

Thus, it follows that a linear fractional transformation is uniquely determined by its action at three points in $\mathbb{C}_\infty$. □

6.5 Cross Ratio

Definition 6.5.1. The *cross ratio* of four distinct points z_1, z_2, z_3, z_4 taken in order in $\mathbb{C}_\infty$ is defined by the ratio

$$\left(\frac{z_1 - z_2}{z_2 - z_3}\right) \Big/ \left(\frac{z_4 - z_1}{z_3 - z_4}\right) = \frac{(z_1 - z_2)(z_3 - z_4)}{(z_2 - z_3)(z_4 - z_1)} \tag{6.53}$$

and is denoted by $\langle z_1, z_2, z_3, z_4\rangle$.

For example, let $1 - i, i, 1 + i, 3i$ be the four distinct points taken in order, Then their cross ratio is given by

$$\langle 1 - i, i, 1 + i, 3i\rangle = \frac{(1 - i - i)(1 + i - 3i)}{\{i - (1 + i)\}\{3i - (1 - i)\}} = \frac{4i + 3}{4i - 1}.$$

By permuting the four points z_1, z_2, z_3, z_4, we can obtain $4! = 24$ cross ratios. But they are equivalent to only six different ratios

$$\langle z_1, z_2, z_3, z_4\rangle = \langle z_2, z_1, z_4, z_3\rangle = \langle z_3, z_4, z_1, z_2\rangle = \langle z_4, z_3, z_2, z_1\rangle.$$

RESULT **58.** Let z_2, z_3, z_4 be three distinct finite points in $\mathbb{C}_\infty$. Define $f : \mathbb{C}_\infty \to \mathbb{C}_\infty$ by

$$f(z) = \frac{(z - z_2)(z_3 - z_4)}{(z_2 - z_3)(z_4 - z)} \text{ if } z_2, z_3, z_4 \in \mathbb{C}. \tag{6.54}$$

Then $f(z_2) = 0, f(z_3) = 1$, and $f(z_4) = \infty$ and f is the only transformation having this property. Also, the image of a point $z_1 \in \mathbb{C}_\infty$ under Eq. (6.54) is

$$f(z_1) = \frac{(z_1 - z_2)(z_3 - z_4)}{(z_2 - z_3)(z_4 - z_1)} = \langle z_1, z_2, z_3, z_4\rangle.$$

Thus the cross ratio $\langle z_1, z_2, z_3, z_4\rangle$ is defined as the image of $z_1 \in \mathbb{C}_\infty$ under the unique linear fractional transformation which transforms z_2, z_3, z_4 into $0, 1, \infty$ respectively.

Theorem 6.2. *Cross ratio is invariant under linear fractional transformation.*

Solution. Let w_1, w_2, w_3, w_4 in w-plane be the images of the four distinct points z_1, z_2, z_3, z_4 respectively in the z-plane under a linear fractional transformation (6.24). Then,

$$w_k = f(z_k) = \frac{az_k + b}{cz_k + d}; \quad k = 1, 2, 3, 4.$$

$$\therefore \quad w_1 - w_2 = \frac{az_1 + b}{cz_1 + d} - \frac{az_2 + b}{cz_2 + d} = \frac{(ad - bc)(z_1 - z_2)}{(cz_1 + d)(cz_2 + d)};$$

$$w_3 - w_4 = \frac{az_3 + b}{cz_3 + d} - \frac{az_4 + b}{cz_4 + d} = \frac{(ad - bc)(z_3 - z_4)}{(cz_3 + d)(cz_4 + d)}.$$

Multiplying the above two, we get

$$(w_1 - w_2)(w_3 - w_4) = \frac{(ad - bc)^2(z_1 - z_2)(z_3 - z_4)}{(cz_1 + d)(cz_2 + d)(cz_3 + d)(cz_4 + d)}.$$

In a similar manner we get

$$(w_2 - w_3)(w_4 - w_1) = \frac{(ad - bc)^2(z_2 - z_3)(z_4 - z_1)}{(cz_1 + d)(cz_2 + d)(cz_3 + d)(cz_4 + d)}.$$

Hence, dividing the last two equations, we get

$$\frac{(w_1 - w_2)(w_3 - w_4)}{(w_2 - w_3)(w_4 - w_1)} = \frac{(z_1 - z_2)(z_3 - z_4)}{(z_2 - z_3)(z_4 - z_1)}$$

$$\therefore \quad \langle w_1, w_2, w_3, w_4 \rangle = \langle z_1, z_2, z_3, z_4 \rangle. \tag{6.55}$$

Therefore, any linear fractional transformation preserves cross ratio and hence may be regarded as an intrinsic quantity associated with the quadruplet $\langle z_1, z_2, z_3, z_4 \rangle$.

Example 6.5.1. *The cross ratio of four distinct points z_1, z_2, z_3, z_4 is real if and only if the four points are concyclic or collinear.*

Solution. Let the four distinct points $A(z_1)$, $B(z_2)$, $C(z_3)$ and $P(z_4)$ lie on a circle. In Fig. 6.17 (a), $P(z_4)$ lies on the semicircular are $\widehat{ACB}$ and in Fig. 6.17 (b), $P(z_4)$ does not lie on the semicircular arc $\widehat{ACB}$. Then $\angle ACB = \angle APB$ in Fig. 6.17 (a) and $\angle ACB + \angle APB = \pi$ in Fig. 6.17 (b). Also, from Fig. 6.17 (a)

$$\angle ACB = \arg\frac{z_3 - z_2}{z_3 - z_1} \text{ and } \angle APB = \arg\frac{z_4 - z_2}{z_4 - z_1}$$

$$\Rightarrow \quad \arg\frac{z_3 - z_2}{z_3 - z_1} = \arg\frac{z_4 - z_2}{z_4 - z_1} \Rightarrow \arg\frac{z_3 - z_2}{z_3 - z_1} - \arg\frac{z_4 - z_2}{z_4 - z_1} = 0$$

$$\Rightarrow \quad \arg\left(\frac{z_3 - z_2}{z_3 - z_1}\right)\Big/\left(\frac{z_4 - z_2}{z_4 - z_1}\right) = 0 \Rightarrow \arg\frac{(z_3 - z_2)(z_4 - z_1)}{(z_3 - z_1)(z_4 - z_2)} = 0$$

$$\Rightarrow \quad \arg\langle z_2, z_3, z_1, z_4 \rangle = 0 \Rightarrow \langle z_2, z_3, z_1, z_4 \rangle \text{ is real.}$$

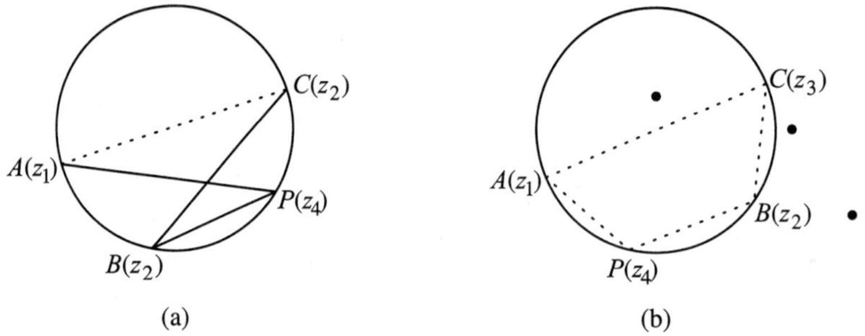

Figure 6.17

From Fig. 6.17 (b), we get

$$\arg\frac{z_3 - z_2}{z_3 - z_1} + \arg\frac{z_4 - z_1}{z_4 - z_2} = \pi \Rightarrow \arg\frac{(z_3 - z_2)(z_4 - z_1)}{(z_3 - z_1)(z_4 - z_2)} = \pi$$

$$\Rightarrow \quad \arg\left\langle z_2, z_3, z_1, z_4 \right\rangle = \pi \Rightarrow \left\langle z_2, z_3, z_1, z_4 \right\rangle \text{ is real.}$$

Example 6.5.2. *Determine a linear fractional transformation which transforms three distinct points in the z-plane into three specified distinct points.*

Solution. Let z_1, z_2, z_3 and w_1, w_2, w_3 be three distinct finite points in the z and w-planes respectively. If one of the points, say z_4, is taken as variable z and w as its image, then

$$\frac{(w - w_1)(w_2 - w_3)}{(w - w_3)(w_1 - w_2)} = \frac{(z - z_1)(z_2 - z_3)}{(z - z_3)(z_1 - z_2)}$$

$$\text{or,} \quad \langle w, w_1, w_2, w_3 \rangle = \langle z, z_1, z_2, z_3 \rangle = f(z), \text{ say,} \tag{6.56}$$

which is itself a Möbius transformation as Eq. (6.56) can be written in the form

$$f(z) = \frac{z_2 - z_3}{z_2 - z_1} \cdot \frac{z - z_1}{z - z_3}.$$

The solution of Eq. (6.56) in terms of w is rather cumbersome, but it gives the required linear fractional transformation. In fact, we write Eq. (6.56) in the form

$$(w - w_1)(w_2 - w_3)(z - z_3)(z_1 - z_2) = (z - z_1)(z_2 - z_3)(w - w_3)(w_1 - w_2). \tag{6.57}$$

If $z = z_1$, the right-hand side of Eq. (6.57) vanishes and consequently $w = w_1$. Again, if $z = z_2$, the right-hand side of Eq. (6.57) vanishes and consequently $w = w_2$. Finally, when $z = z_2$, then the Eq. (6.57) becomes

$$\begin{aligned} & (w - w_1)(w_2 - w_3) = (w - w_3)(w_1 - w_2) \\ \Rightarrow \quad & w(w_1 - w_3) = w_2(w_1 - w_3) \Rightarrow w = w_2. \end{aligned}$$

Thus the transformation (6.56) is obviously unique. Now, the determinant of the transformation is given by

$$\begin{vmatrix} a & b \\ c & d \end{vmatrix} = \begin{vmatrix} z_2 - z_3 & z_1(z_3 - z_2) \\ z_2 - z_1 & z_3(z_1 - z_2) \end{vmatrix} = (z_1 - z_2)(z_2 - z_3)(z_3 - z_1) \neq 0,$$

as z_i's are distinct. Thus Eq. (6.56) is the required unique linear fractional transformation.

Example 6.5.3. *Find the linear fractional transformation which maps the points $(2, i, -2)$ in z-plane into the points $(1, i, -1)$ in w-plane.*

Solution. Let the image of any point z under the required transformation be w. From Eq. (6.56) we get

$$\frac{(w-1)(i+1)}{(w+1)(1-i)} = \frac{(z-2)(i+2)}{(z+2)(2-i)}$$

$$\text{or,} \quad \frac{(w-1)(i+1)^2}{(w+1)(1-i)(1+i)} = \frac{(z-2)(i+2)^2}{(z+2)(2-i)(2+i)}$$

$$\text{or,} \quad \frac{(w-1)\cdot 2i}{2(w+1)} = \frac{(z-2)(3+4i)}{5(z+2)} = \frac{3z - 6 + 4iz - 8i}{5z + 10}$$

$$\text{or,} \quad \frac{(w-1)+(w+1)}{(w-1)-(w+1)} = \frac{(3z - 6 + 4iz - 8i) + (5iz + 10i)}{(3z - 6 + 4iz - 8i) - (5iz + 10i)}$$

$$\text{or,} \quad w = \frac{(3z + 2i) + 3i(3z + 2i)}{iz(1 + 3i) + 6(1 + 3i)} = \frac{3z + 2i}{iz + 6},$$

which is the required linear fractional transformation.

Example 6.5.4. *Find the linear fractional transformation which transforms the unit circle $|z| = 1$ onto real axis such that the points $z_1 = 1, z_2 = i$ and $z_3 = -i$ are mapped onto the points $w_1 = 0, w_2 = 1$ and $w_3 = \infty$ respectively. Into which regions the interior and exterior of the circle is mapped ?*

Solution. The linear fractional transformation (6.56) can be written as

$$\frac{(w - w_1)(w_2/w_3 - 1)}{(w/w_3 - 1)(w_1 - w_2)} = \frac{(z - z_1)(z_2 - z_3)}{(z - z_3)(z_1 - z_2)}.$$

$$\Rightarrow \quad \frac{(w - 0)(1/\infty - 1)}{(w/\infty - 1)(0 - 1)} = \frac{(z - 1)(i + i)}{(z + i)(1 - i)}$$

$$\Rightarrow \quad w = -\frac{2i}{1 - i}\frac{z - 1}{z + i} = (1 - i)\frac{z - 1}{z + i}.$$

The inverse transformation is $z = -i\{w-(1+i)\}/\{w-(1-i)\}$. Now,

$$\begin{aligned}
|z| = 1 \;\Leftrightarrow\; & \left|w-(1+i)\right|^2 = \left|w-(1-i)\right|^2; \quad w = u+iv \\
\Leftrightarrow\; & |(u-1)+i(v-1)|^2 = |(u-1)+i(v+1)|^2 \\
\Leftrightarrow\; & (u-1)^2+(v-1)^2 = (u-1)^2+(v+1)^2 \\
\Leftrightarrow\; & -4v = 0 \Rightarrow v = 0.
\end{aligned}$$

Thus the unit circle $|z| = 1$ is transformed into the real axis of w-plane. Also,

$$\begin{aligned}
|z| < 1 \;\Leftrightarrow\; & (u-1)^2+(v-1)^2 < (u-1)^2+(v+1)^2 \\
\Leftrightarrow\; & -4v < 0 \Rightarrow v > 0.
\end{aligned}$$

Thus, the interior of the unit circle $|z| = 1$ is transformed into the upper half of w-plane. Similarly, the exterior of the unit circle $|z| = 1$ is transformed into the lower half of w-plane.

Example 6.5.5. *Show that every linear fractional transformation with two distinct finite fixed points, α and β, can be put in the form (6.52), where $\lambda \in \mathbb{C}$, a parameter.*

Solution. Consider a non-constant linear fractional transformation with α, β as fixed points and suppose it transforms a point γ into a point δ. We know that if z_1, z_2, z_3, z be four points in the z-plane and w_1, w_2, w_3, w their respective images in w-plane, and if we identify z_1 and z_3 as fixed points, i.e., $z_1 = \alpha = w_1$ and $z_3 = \beta = w_3$, then since cross ratio remains invariant, i.e., $\langle w, \alpha, \delta, \beta\rangle = \langle z, \alpha, \gamma, \beta\rangle$, therefore,

$$\frac{(w-\alpha)(\delta-\beta)}{(\beta-w)(\alpha-\delta)} = \frac{(z-\alpha)(\gamma-\beta)}{(\beta-z)(\alpha-\gamma)}$$

$$\text{or,}\qquad \frac{w-\alpha}{w-\beta} = \frac{(\alpha-\delta)(\beta-\gamma)}{(\beta-\delta)(\alpha-\gamma)}\cdot\frac{z-\alpha}{z-\beta} = \lambda\,\frac{z-\alpha}{z-\beta},$$

which is the required form, where $\lambda = (\alpha-\delta)(\beta-\gamma)/(\beta-\delta)(\alpha-\gamma)$ is a complex parameter as it changes with the choice of z_2.

Definition 6.5.2. Consider a non-constant linear fractional transformation with two finite distinct fixed points α and β and suppose it transforms a point γ into a point δ. Define

$$\lambda = \frac{(\alpha-\delta)(\beta-\gamma)}{(\beta-\delta)(\alpha-\gamma)}, \text{ where } \lambda(\neq 0, 1) \in \mathbb{C}. \tag{6.58}$$

The type of the transformation (elliptic, hyperbolic or loxodromic), depending on λ, can be determined from definition 6.4.4.

Example 6.5.6. *Determine the linear fractional transformation which maps $0, 1, \infty$ into $i, -1, -i$ respectively. Find also the inverse transformation. Find the fixed points of the transformation and hence determine its nature.*

Solution. Here $z_1 = 0, z_2 = 1, z_3 = \infty$ and $w_1 = i, w_2 = -1, w_3 = -i$. The linear fractional transformation (6.56) can be written in the form

$$\frac{(w-i)(-1+i)}{(-i-w)(i+1)} = \frac{(z-z_1)\left(\frac{z_2}{z_3}-1\right)}{\left(1-\frac{z}{z_3}\right)(z_1-z_2)} = \frac{(z-0)(0-1)}{(1-0)(0-1)}$$

$$\begin{aligned}
\Rightarrow\qquad z &= \frac{(w-i)(-1+i)}{-(i+w)(1+i)} = \frac{(w-i)(-1+i)(i-i)}{-(w+i)(1+1)} \\
&= \frac{-(w-i)(1-2i-1)}{-2(w+i)} = -\frac{i(w-i)}{w+i} = -\frac{iw+1}{w+i}.
\end{aligned}$$

Its inverse transformation is $w = \{-i(z-i)\}/(z+i)$. For this given transformation,

$$w = z \Rightarrow z^2 + 2zi + 1 = 0 \Rightarrow z = -i, -i.$$

Since the transformation has unique fixed point, so it is parabolic.

Example 6.5.7. *Determine the linear fractional transformation which maps* $-1, 0, 1$ *into* $-1, -i, 1$ *respectively.*

(i) Show that under this transformation the upper half of the z*-plane maps onto the interior of the unit circle of the* w*-plane.*

(ii) Find the invariant points of the transformation and hence find the nature.

Solution. Let the image of any point z under the required transformation be w. Then Eq. (6.56) can be written in the form

$$\frac{(w+1)(-i-1)}{(1-w)(-1+i)} = \frac{(z+1)(0-1)}{(1-z)(-1-0)}$$

$$\Rightarrow \quad \frac{(1+w)(1+i)}{(1-w)(1-i)} = \frac{1+z}{1-z} \Rightarrow w = i\,\frac{z-i}{z+i},$$

which is the required transformation. Now, let, $z = x+iy$ and $w = u+iv$. Then

$$u + iv = i\,\frac{x+iy-i}{x+iy+i} = i\left[\frac{x^2+y^2-1}{x^2+(y+1)^2} - i\frac{-2x}{x^2+(y+1)^2}\right]$$

$$\therefore u = \frac{-2x}{x^2+(y+1)^2}, \qquad v = \frac{x^2+y^2-1}{x^2+(y+1)^2}.$$

(i) The equation of the upper half of the w-plane is given by $\operatorname{Im} w = v > 0$.

$$\therefore \frac{x^2+y^2-1}{x^2+(y+1)^2} > 0 \Leftrightarrow x^2+y^2-1 < 0 \Leftrightarrow |z| < 1.$$

Hence, the upper half of the z-plane maps onto the interior of the unit circle of the w-plane.
(ii) To find fixed points, $w = z$. So

$$i\left(\frac{z-i}{z+i}\right) = z \Rightarrow z = \pm 1.$$

Thus $\alpha = 1$ and $\beta = -1$ are two fixed point of the transformation. Let $\gamma = i$, which corresponds to point $\delta = 0$. Now,

$$\lambda = \frac{(\alpha-\delta)(\beta-\gamma)}{(\beta-\delta)(\alpha-\gamma)} = \frac{(1-0)(-1-i)}{(-1-0)(1-i)} = \frac{1+i}{1-i} = i.$$

Since $|\lambda| = 1$, the given transformation is elliptic.

6.6 Mappings by Elementary Functions

In the previous section we have studied the properties of linear fractional transformations and proved that circles and straight lines are mapped into circles and straight lines. In this section we shall discuss some of most useful transformations with graphical representations which can be effected by elementary functions (algebraic and transcendental) and which are of great importance in hydrodynamics, heat conduction, electrostatics, etc. For this, we will need two complex planes representing respectively the domain and the image of the function.

6.6.1 The Mapping $w = z^2$

Let $w = f(z) = z^2$, which is an algebraic function. This mapping is not one-one but is onto. Now $f'(z) = 2z \neq 0$, when $z \neq 0$. Hence f is conformal at all points in $\mathbb{C} - \{0\}$.

Let $z = x + iy$ and $w = u + iv$. Then the transformation can be written as

$$u + iv = (x + iy)^2 = (x^2 - y^2) + 2xyi \Rightarrow u = x^2 - y^2 \text{ and } v = 2xy. \tag{6.59}$$

Let $z = re^{i\theta}$ and $w = \rho e^{i\phi}$. Then in polar coordinates the transformation becomes

$$\rho e^{i\phi} = r^2 e^{2i\theta} \Rightarrow \rho = r^2 \text{ and } \phi = 2\theta. \tag{6.60}$$

We now proceed to discuss how curves and region are transformed under the transformation $w = z^2$ using Eqs. (6.59) and (6.60).

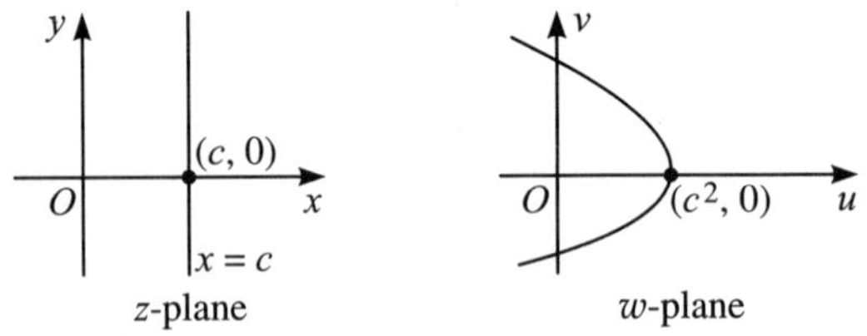

Figure 6.18 $v^2 = -4c^2(u - c^2)$

(i) Let $x = \text{constant} = c(> 0)$. Then the image of this line is given by

$$u = c^2 - y^2 \text{ and } v = 2cy \Rightarrow v^2 = -4c^2(u - c^2), \tag{6.61}$$

which is the equation of the parabola (Fig. 6.18) in w-plane with vertex at $(c^2, 0)$, focus at the origin, negative u axis as the axis of the parabola, and which is symmetrical about the real axis. Observe that the line $x = -c(c > 0)$ is transformed into the same parabola.

It is relevant to point out that as $c \to 0$ the parabolas, as in Eq. (6.61), gets closer and closer to negative real axis. In particular, the x-axis$(y = 0)$ is mapped into $u = c^2$ and $v = 0$, which is positive u-axis in the w-plane.

(ii) Similarly, the image of the line $y = d(\neq 0)$ is given by

$$u = x^2 - d^2 \text{ and } v = 2xd \Rightarrow v^2 = 4d^2(u + d^2), \tag{6.62}$$

which is again a parabola. Thus the y-axis $(x = 0)$ is mapped into $u = -d^2$ and $v = 0$, which is the negative u axis in the w-plane. These parabolas as in Eqs. (6.61) and (6.62) intersect at $(c^2 - d^2, \pm 2|cd|)$.

(iii) Now, let us consider an infinite strip $\{(x, y) : a < x < b; a, b \in \mathbb{R}^+\}$. This strip region between $x = a$ and $x = b$, if $0 < a < b$, is mapped conformally onto a parabolic strip region on the w-plane included between the parabolas

$$v^2 = -4a^2(u - a^2) \text{ and } v^2 = -4b^2(u - b^2) \tag{6.63}$$

as depicted in Fig. 6.19. If we make $b \to \infty$, we see that the half plane $\operatorname{Re} z > a > 0$ is mapped conformally on the exterior of the parabola $v^2 = -4a^2(u - a^2)$ not containing the origin.

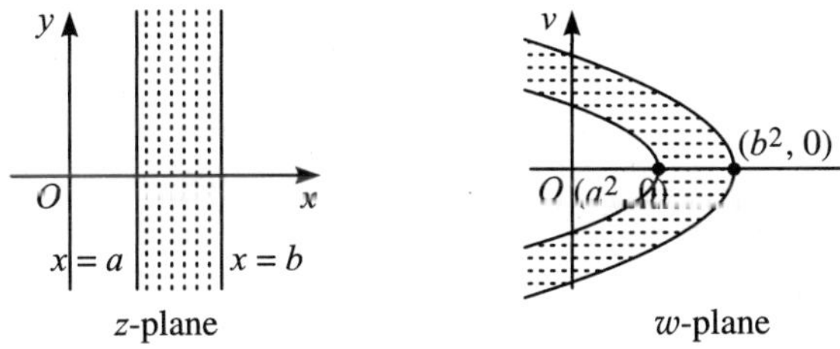

Figure 6.19 *Infinite strip* $\{(x, y) : a < x < b; a, b \in \mathbb{R}^+\}$

(iv) The image of the strip $0 \leq \operatorname{Re} z \leq a$ is the whole interior of the parabola $v^2 = -4a^2(u - a^2)$ together with a cut along the negative part of w-axis.

Putting $x = 0$, we get $v = 0, y^2 = -u$. This means that the image of the line $x = 0$ exists if $u \leq 0$. Thus the image of Re $z = 0$ is the negative part of the Re ew-axis.

(v) Let us now consider the rectangular region $\{(x, y) : a \leqslant x \leqslant b \text{ and } c \leqslant y \leqslant d\}$. The rectangular strip is mapped into the region bounded by the four parabolas

$$v^2 = -4a^2(u - a^2); \quad v^2 = -4b^2(u - b^2); \quad v^2 = 4c^2(u + c^2); \quad v^2 = 4d^2(u + d^2)$$

as shown in Fig. 6.20. There are two regions bounded by these four parabolas. The region to which the given rectangle is mapped can be determined by locating the image of any one of the vertices of the rectangle.

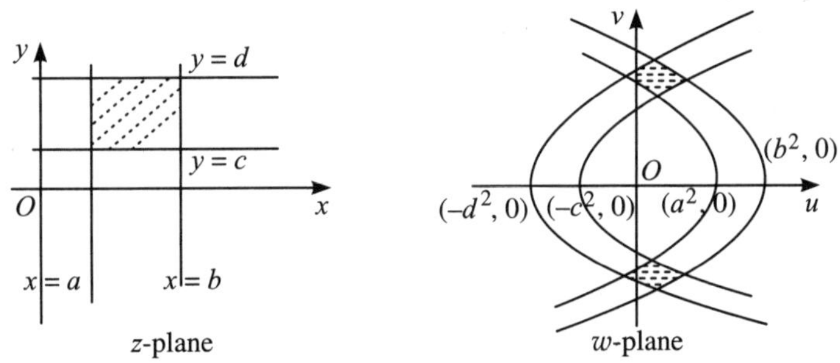

Figure 6.20 *Rectangular region*

(vi) Now, let z be any point in the first quadrant. Then $0 \leq \arg z \leq \pi/2$. By Eq. (6.60), $0 \leq \arg w = 2 \arg z \leq \pi$ so w lies in the upper half of the w-plane. Thus the first quadrant in the z-plane is mapped into the upper half plane in the w-plane. Also, the upper half of the z-plane is mapped into the entire w-plane.

(vii) Now we consider the sector $S_1(\alpha, \beta) = \{re^{i\theta} : \alpha \leq \theta \leq \beta\}$ as shown in Fig. 6.4. This region is mapped into the region $\{\rho e^{i\phi} : 2\alpha \leq \phi \leq 2\beta\}$. The restriction of f to $S_1(\alpha, \beta)$ will be one-one exactly when $\beta - \alpha < \pi$. Thus the circle of radius R about the origin is mapped onto the circle of radius R^2 in a two-to-one fashion.

(viii) Lastly, we may consider the curves $u = c_1$ and $v = c_2$. They are orthogonal and are called level curves of u and v. Using Eq. (6.59), the image curves of the level curves are given by

$$x^2 - y^2 = c_1 \text{ and } 2xy = c_2. \tag{6.64}$$

Example 6.6.1. *By the transformation $w = z^2$, show that the circle $|z - a| = c; a \neq c$ in the z-plane corresponds to the limacon in the w-plane.*

Solution. The equation of the circle can be put in the polar form as $z = ce^{i\theta} + a$.

$$\therefore \quad w = z^2 \Rightarrow w = (ce^{i\theta} + a)^2$$
$$\Rightarrow \quad w - a^2 + c^2 = c^2 + 2ace^{i\theta} + c^2 e^{i2\theta}$$
$$\Rightarrow \quad w - (a^2 - c^2) = ce^{i\theta}(ce^{-i\theta} + ce^{i\theta} + 2a) = 2ce^{i\theta}(a + c\cos\theta).$$

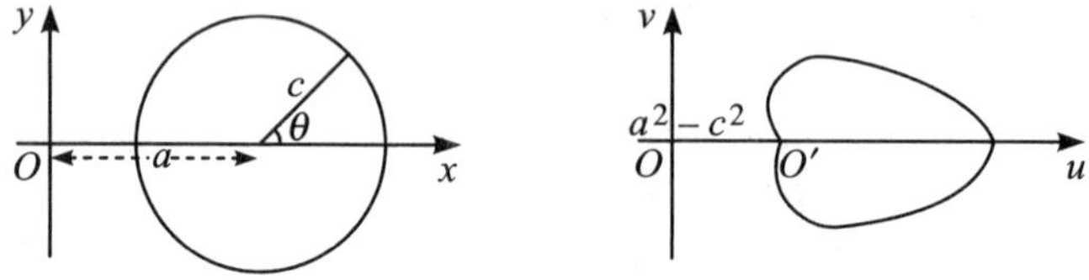

Figure 6.21 *The transformation $w = z^2$*

By writing $w-(a^2-c^2) = \rho e^{i\phi}$, i.e. by taking the pole at $w = a^2-c^2$, we get $\rho e^{i\phi} = 2ce^{i\theta}(a+c\cos\theta)$. Therefore, equating modulus and argument on both sides, we get

$$\rho = 2c(a + c\cos\theta) \text{ and } \phi = \theta \Rightarrow \rho = 2c(a + c\cos\phi),$$

which is the limacon in the w-plane (Fig. 6.21). In particular, when $a = c$, the limacon becomes a cardioid, $\rho = 2a^2(1 + \cos\phi)$.

6.6.2 The Transformation $w = \frac{1}{2}\left(z + \frac{1}{z}\right)$

Let $w = f(z) = \frac{1}{2}\left(z + \frac{1}{z}\right), z \neq 0$, and $z = x + iy$. Then

$$w = \frac{x}{2}\left(1 + \frac{1}{x^2 + y^2}\right) + i\,\frac{y}{2}\left(1 - \frac{1}{x^2 + y^2}\right), \tag{6.65}$$

known as *Joukowski's transformation*, which occurs when solving a variety of applied problems, particularly in aerodynamics. Now,

$$f'(z) = 0 \Rightarrow \frac{1}{2}\left(1 - \frac{1}{z^2}\right) = 0 \Rightarrow z = \pm 1.$$

Therefore the mapping is conformal at all points in $\mathbb{C}/\{0, -1, 1\}$. Let $w = u + iv$ and $z = re^{i\theta}$. Then

$$\begin{aligned} w &= u + iv = \frac{1}{2}\Big[r(\cos\theta + i\sin\theta) + \frac{1}{r}(\cos\theta - i\sin\theta)\Big] \\ &= \frac{1}{2}\left(r + \frac{1}{r}\right)\cos\theta + i\frac{1}{2}\left(r - \frac{1}{r}\right)\sin\theta \end{aligned}$$
$$\Rightarrow \quad u = \frac{1}{2}\left(r + \frac{1}{r}\right)\cos\theta \text{ and } v = \frac{1}{2}\left(r - \frac{1}{r}\right)\sin\theta. \tag{6.66}$$

We now discuss this transformation in some detail.

(a) *When r is constant :* Eliminating θ, we get the equation of ellipse as

$$\frac{u^2}{(r + 1/r)^2} + \frac{v^2}{(r - 1/r)^2} = \frac{1}{4}, \tag{6.67}$$

which is the image of transformation in the w-plane.

(i) Consider the unit circle $|z| = 1$, in z-plane, i.e., $r = 1$. Eq.(6.66) becomes $u = \cos\theta$ and $v = 0$ (Fig. 6.22). The image of the unit circle in z-plane is the line segment $-1 \leq w \leq 1$ of the real axis in the w-plane. Since $0 \leq \theta \leq 2\pi$, we see that $|u| \leq 1$; u varies from 1 to -1 as $0 \leq \theta \leq \pi$. Further, as $\pi \leq \theta \leq 2\pi$, u varies from -1 to 1, again on the real w-axis.

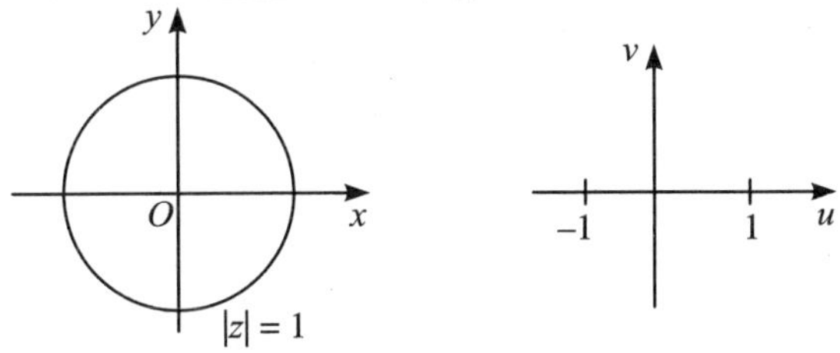

Figure 6.22 $u = \cos\theta, v = 0$

(ii) Next we consider the circle $|z| = \lambda(\neq 1)$ in the z-plane. The equation of the image of the circle $|z| = \lambda$ in w-plane is given by

$$u = \frac{1}{2}\left(\lambda + \frac{1}{\lambda}\right)\cos\theta \text{ and } v = \frac{1}{2}\left(\lambda - \frac{1}{\lambda}\right)\sin\theta : \theta \in [0, 2\pi]$$

$$\Rightarrow \quad \frac{u^2}{\frac{1}{4}(\lambda + \frac{1}{\lambda})^2} + \frac{v^2}{\frac{1}{4}(\lambda - \frac{1}{\lambda})^2} = 1. \tag{6.68}$$

Equation (6.68) represents an ellipse with semi major axis $\frac{1}{4}(\lambda + \frac{1}{\lambda})$ along u-axis; semi minor axis $\frac{1}{4}(\lambda - \frac{1}{\lambda})$ along v-axis; eccentricity $\frac{2}{\lambda + 1/\lambda}$ and foci $(\pm 1, 0)$. We observe that the image of the circle $|z| = 1/\lambda$ $(\lambda \neq 1)$ is also the same as the ellipse given in Eq. (6.68).

(iii) The region bounded by $\lambda < |z| < \mu$ where $\lambda, \mu > 1$ is mapped conformally into the region bounded by the two conformal ellipses

$$\frac{u^2}{\frac{1}{4}(\lambda + \frac{1}{\lambda})^2} + \frac{v^2}{\frac{1}{4}(\lambda - \frac{1}{\lambda})^2} = 1 \text{ and } \frac{u^2}{\frac{1}{4}(\mu + \frac{1}{\mu})^2} + \frac{v^2}{\frac{1}{4}(\mu - \frac{1}{\mu})^2} = 1 \tag{6.69}$$

as shown in Fig. 6.23. Thus, the region of two circles is mapped conformally into the region by

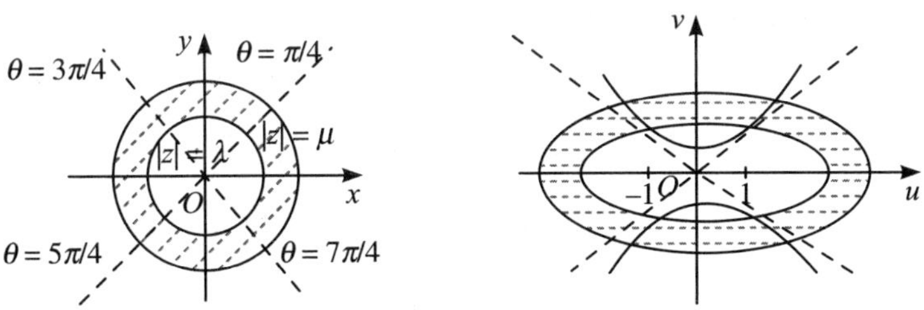

Figure 6.23 $\lambda < |z| < \mu$

two confocal ellipses. Also, the region between $|z| = 1/\lambda$ and $|z| = 1/\mu$ transform onto the same annular region between the above ellipses.

(iv) Taking $\mu \to \infty$, the region exterior to circle $|z| = \lambda$ and the region interior to circle $|z| = 1/\lambda$ are both separately mapped onto the entire region in the w-plane exterior to the ellipse represented by Eq. (6.68).

(b) *When θ is constant :* The image of the transformation in Eq.(6.66) is given by

$$\frac{u^2}{\cos^2\theta} - \frac{v^2}{\sin^2\theta} = 1 \tag{6.70}$$

This represents a hyperbola in the w-plane.

(i) Consider the ray $\theta = \alpha$ where $0 < \alpha < \pi/2$ in the z-pane. The equation of the image of that ray in the w-plane is given by

$$u = \frac{1}{2}\left(r + \frac{1}{r}\right)\cos\alpha \text{ and } v = \frac{1}{2}\left(r - \frac{1}{r}\right)\sin\alpha$$

$$\Rightarrow \quad \frac{u^2}{\cos^2\alpha} - \frac{v^2}{\sin^2\alpha} = 1; 0 < r < \infty. \tag{6.71}$$

This represents a hyperbola with foci at $(\pm 1, 0)$. Since $u > 0$ when $0 < \alpha < \pi/2$, the image is actually the right-hand branch of the hyperbola (6.71). As r varies from 0 to 1 on $\theta = \alpha$, v varies from $-\infty$ to 0 and u varies from ∞ to $\cos\alpha$. So the lower part of the hyperbola lying to the right side of the Im(w)-axis is described.

Also, on $\theta = \alpha$, as r varies from 1 to ∞, v varies from 0 to ∞ while u varies from $\cos\alpha$ to ∞, and thus the upper part of the hyperbola lying to the right side of the Im (w)-axis is described. This branch of the hyperbola is also a complete image of $\theta = -\alpha$.

Similarly, the complete image of the ray $\theta = \pi - \alpha$ and $\theta = \pi + \alpha$ is the second brunch of the same hyperbola (6.70) lying to the left side of Im(w)-axis.

(ii) Let us consider the wedge $\alpha < \theta < \beta$, where $0 < \alpha, \beta < \pi/2$ as seen in Fig. 6.24 in the z-plane. The image of this wedge is the region bounded by the two branches lying to the right of the imaginary axis of the hyperbola

$$\frac{u^2}{\cos^2\alpha} - \frac{v^2}{\sin^2\alpha} = 1 \text{ and } \frac{u^2}{\cos^2\beta} - \frac{v^2}{\sin^2\beta} = 1$$

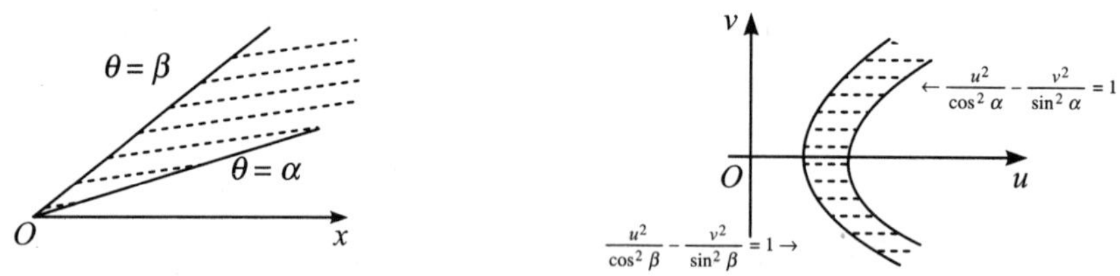

Figure 6.24 *Wedge $\alpha < \theta < \beta$*

as shown in Fig. 6.24. As $\alpha < \beta$, $\cos\alpha > \cos\beta$ and $\sin\alpha < \sin\beta$. So when v increases from α to β, the length of the transverse axes decrease and that of the conjugate axes increase in the wedges $\theta = \alpha$ and $\theta = \beta$. Therefore, when θ increases, i.e., revolves about the origin of the z-plane in the anticlockwise direction, the hyperbolic images go on falling outside of the preceding ones but remain confocal with their focii being at $w = \pm 1$.

(iii) The image in the w-plane of the ray $\theta = \pi/2$ is given by the equations $u - 0$ and $v = \frac{1}{2}(r - \frac{1}{r})$. This represents the complete imaginary axis in the w-plane as r varies from 0 to ∞, v varies from $-\infty$ to ∞, i.e., the entire Im w-axis. Similarly, we get the image for the ray $\theta = -\pi/2$. Hence the y-axis is mapped onto the v-axis.

(iv) The image in the w-plane of the ray $\theta = 0$ is given by the equations $u = \frac{1}{2}(r + \frac{1}{r})$ and $v = 0$, where $0 < r < \infty$. We observe that $u > 1$ for all values of r and hence the image is a part of the u-axis $u \geq 1$. In this case, the image line segment is described twice when r varies from 0 to ∞.

Therefore, the image in w-plane of the wedge $0 < \arg z < \beta$ corresponds to the domain interior to the hyperbola $\frac{u^2}{\cos^2\beta} - \frac{v^2}{\sin^2\beta} = 1$ with a slit along the real w-axis from 1 to ∞.

(v) The image of the line $\theta = \pi$ is given by $u = -\frac{1}{2}(r + \frac{1}{r})$ and $v = 0$. The line $\theta = \pi$ in the z-plane is mapped onto the part $-\infty < u < -1$ of the real u-axis in w-plane.

(vi) Hence the region $\alpha < \theta < \pi/2$ is mapped conformally onto the domain bounded by the imaginary axis and the branch of the hyperbola on the right as shown in Fig. 6.25. Since the ray $\theta = \alpha$ cuts the circle $r = \lambda$ orthogonally in the z-plane, their images which are two families of ellipses and hyperbolas also from an orthogonal system in the w-plane.

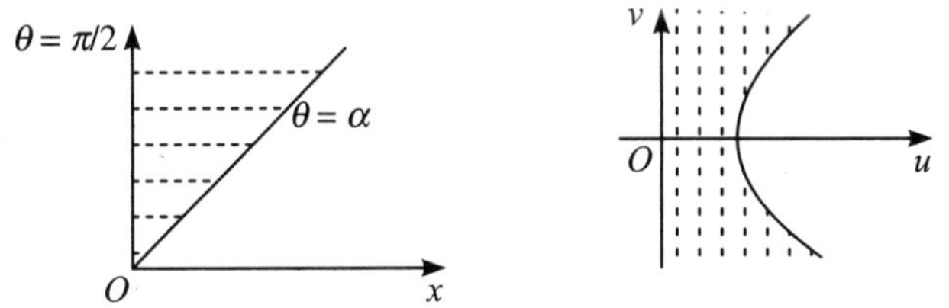

Figure 6.25 $\alpha < \theta < \pi/2$

(vii) Let us discuss the image of the wedge $\alpha < \arg z < \pi - \alpha$: Under the transformation, the image of $\theta = \alpha$ is the branch of the hyperbola $\frac{u^2}{\cos^2\alpha} - \frac{v^2}{\sin^2\alpha} = 1$ that is towards the right of the imaginary w-axis (Fig. 6.26). As θ increases to $\pi/2$,

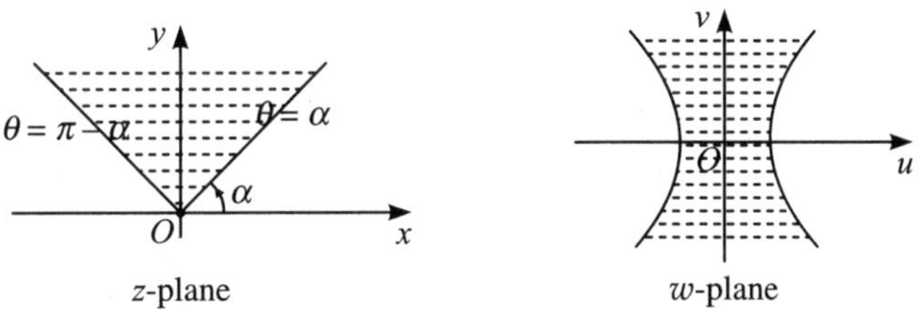

Figure 6.26 $\alpha < \theta < \pi - \alpha$

its image is the entire imaginary w-axis and when θ further increases to $\theta = \pi - \alpha$, the image of it is the other branch of the same hyperbola to the left side of the imaginary w-axis. Therefore, under this transformation, the wedge $\alpha < \arg z < \pi - \alpha$ in the w-plane corresponds to the region lying between the two branches of the hyperbola.

6.6.3 The Transformation $w = e^z$

Consider the exponential transformation $w = f(z) = e^z$. This mapping is conformal at all points, since $f'(z) = e^z$ is never zero for any point in the z-plane.

Let $z = x + iy$ and $w = \rho e^{i\phi}$. Then the transformation becomes

$$w = e^z = e^x \cdot e^{iy}; \text{ for each } z \in \mathbb{C}. \tag{6.72}$$

$$\therefore \quad \rho = |w| = e^x \text{ and } \phi = \arg w = y. \tag{6.73}$$

From these equations we obtain the following results.

(i) The line $x = a$ is mapped into the circle $\rho = |w| = e^a$(Fig. 6.27), i.e., $u^2 + v^2 = e^{2a}$. Thus f maps the line $x = a$ onto a circle of radius e^a with centre at the origin. The infinite strip bounded by $x = a$ and $x = b$ in z-plane is transformed into the region in the w-plane between the concentric circles.

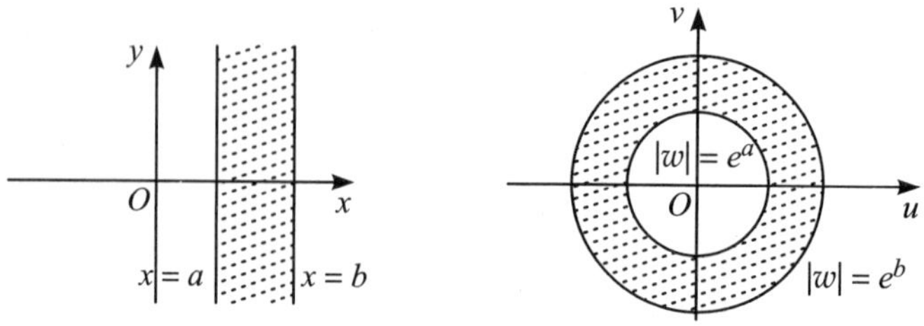

Figure 6.27 *The transformation* $y = e^z$

(ii) We know that $e^x (= \rho)$ increases from 0 to ∞ monotonically as x takes values from $-\infty$ to ∞. Thus the line $y = b$ is mapped onto the infinite ray $\{\rho e^{id} : 0 < \rho < \infty\}$ (Fig. 6.28). In particular the x-axis($y = 0$) is mapped onto the ray $\phi = 0$ with $u > 0$, while the line $y = \pm\pi$ is mapped onto the ray $\phi = \pi$ with $u < 0$.

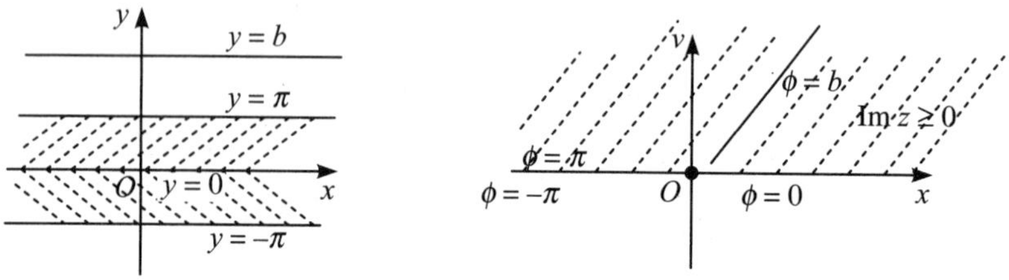

Figure 6.28 *Infinite ray* $\{\rho e^{id} : 0 < \rho < \infty\}$

(iii) Consider the region $c < \text{Im}\, z < d$; $(d - c) < \pi$ (Fig. 6.29). The infinite strip bounded by the lines $y = c$ and $y = d$ is transformed into the wedge shaped region in the w-plane bounded by the radial line $\phi = c$ and $\phi = d$. In particular, the infinite strip bounded by the lines $y = 0$ and $y = \pi$ in the z-plane is mapped to the the real w-plane. (Fig. 6.28).

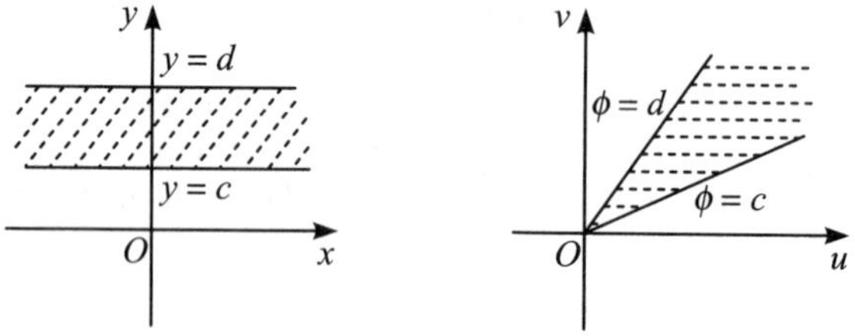

Figure 6.29 $c < \text{Im}\, z < d$; $(d - c) < \pi$

(iv) As x varies from $-\infty$ to 0, ρ varies from 0 to 1. It follows that the infinite strip on the negative side of the x-axis, i.e., the strip $0 \le y \le \pi$, $-\infty \le x \le 0$, is transformed into the unit semi-circle $\rho = 1, 0 \le \phi \le \pi$ in the upper half plane. Similarly, the strip $-\infty < x < \infty$ and $-\pi < y < 0$ is mapped onto the lower half plane $\phi > 0$ and $-\pi < \phi < 0$.

We further note that the whole strip $0 \le y \le 2\pi$ is mapped onto the entire w-plane with a slit along the positive real axis from 0 to ∞. The line $y = 0$ corresponds to the upper edge of the slit and the line $y = 2\pi$ corresponds to the lower edge.

(v) We know e^z is one-to-one on any horizontal strip of width $< 2\pi$. Thus f maps the strip $G = \{z \in \mathbb{C} : -\pi < y \leq \pi\}$ onto $\Omega = f(G) = \mathbb{C} - \{z : z \leq 0\}$. Now, e^z is a periodic function with period $2k\pi i$ for each z in $\mathbb{C}$. Thus for each integer k, the strip

$$\Big\{z : (2k-1)\pi < y \leq (2k+1)\pi\Big\}$$

is mapped onto $\mathbb{C} - \{0\}$ in the same way as the strip in G. For this reason, the strip in G is called the *fundamental strip*.

(vi) f maps vertical segments $\{z = a + iy; -\pi < y < \pi\}$ onto the part of the circle $\{e^a \cdot e^{i\theta}; -\pi < y < \pi\}$, and a horizontal line $y = d, -\pi < y < \pi$, goes onto the ray making an angle d with the positive real axis.

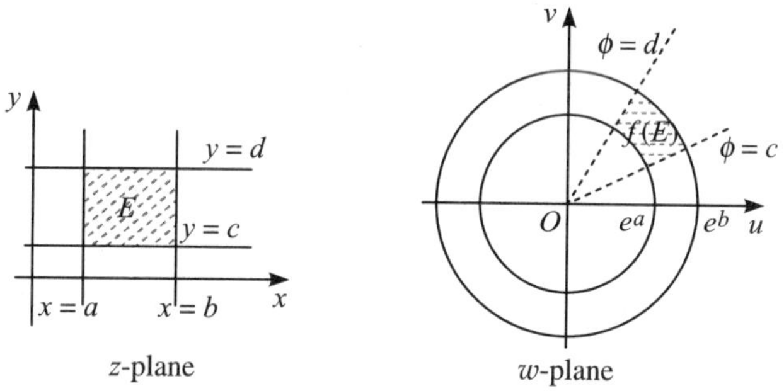

Figure 6.30 $\{w : e^a \leq |w| \leq e^b \text{ and } c \leq \phi \leq d\}$

(vii) Let E be a rectangular region in the z-plane as $a \leq x \leq b, c \leq y \leq d$ (Fig. 6.30). This region in the z-plane is mapped into the image region in w-plane bounded by the circles and rays given by

$$f(E) = \Big\{w : e^a \leq |w| \leq e^b \text{ and } c \leq \phi \leq d\Big\}.$$

(viii) Let $w = \rho e^{i\phi}$, $z = re^{i\theta}$. The transformation becomes

$$\rho\, e^{i\phi} = e^{re^{i\theta}} = e^{r\cos\theta}\, e^{ir\sin\theta} \Rightarrow \rho = e^{r\cos\theta}; \quad \phi = r\sin\theta.$$

Eliminating r, we get $\rho = e^{\phi\cot\theta}$. Hence the straightlines in the z-plane defined by $\theta = \alpha$ for varying values of α corresponds to logarithmic spirals. Again, eliminating θ, we obtain

$$(\log\rho)^2 + \phi^2 = r^2.$$

For constant values of r, this equation describes the curves of the w-plane which corresponds to arcs of circles with centres at the origin of the w-plane.

6.6.4 The Transformation $w = \log z$

For the transformation $w = \log z$, $\frac{dw}{dz} = \frac{1}{z}$ so that the derivative is infinite at $z = 0$. Thus, this transformation is conformal for every point in the z-plane except at origin.

Let $z = x + iy$ and $w = u + iv$. Then the transformation becomes

$$u + iv = \log(x + iy) \Rightarrow x + iy = e^{u+iv} = e^u\Big[\cos v + i\,\sin v\Big]$$

$$\Rightarrow \quad x = e^u\,\cos v \text{ and } y = e^u\,\sin v. \tag{6.74}$$

Again, if we write $z = r\, e^{i\theta}$ and $w = u + iv$, then the transformation becomes

$$u + iv = \ln r + i\,\theta \Rightarrow u = \ln r \text{ and } v = \theta. \tag{6.75}$$

(i) The straight lines $x = c_1$ and $y = c_2$ in the z-plane are mapped onto the curves

$$c_1 = e^u \cos v \text{ and } c_2 = e^u \sin v$$

in the w-plane.

(ii) Using Eq. (6.75), the lines $\theta = \alpha$ for different values of α are mapped onto the lines $v = \alpha$. In fact, the sectorial region bounded by rays $\theta = \alpha_1$ and $\theta = \alpha_2$ is mapped onto the infinite strip parallel to the real axis between the lines $v = \alpha_1$ and $v = \alpha_2$.

(iii) In particular. the whole z-plane cut along the positive real axis from 0 to ∞ is mapped onto the infinite strip $0 \le v \le 2\pi$. Also, every parallel strip

$$n \cdot 2\pi \le v \le (n+1) \cdot 2\pi, \quad n \in \mathbb{Z}$$

is the image of the same cut plane.

(iv) Also, the image of every circle defined by $r = r_1$ in the z-plane is the straightline $u = \log r_1$, which is parallel to y-axis.

6.6.5 The Transformation $w = \sin z$

Let $w = f(z) = \sin z$, a trigonometrical transformation. Since $\sin(z + 2k\pi) = \sin z$, it is clear that $\sin z$ is transcendental. Now,

$$\frac{dw}{dz} = \cos z = 0 \Rightarrow z = n\pi + \pi/2; \quad n \in \mathbb{Z}.$$

Thus the mapping $w = \sin z$ is conformal at all points in the complex plane $\mathbb{C}$ except at $z = n\pi + \pi/2$, where $n \in \mathbb{Z}$. Let $z = x + iy$ and $w = u + iv$. Then

$$w = \sin z = \sin(x + iy) = \sin x \cosh y + i \cos x \sinh y,$$

$$\Rightarrow \quad u = \sin x \cosh y \text{ and } v = \cos x \sinh y. \tag{6.76}$$

(i) Under the transformation (6.76), the image of the vertical line $x = c$, where $0 < c < \pi/2$, is given by

$$\frac{u^2}{\sin^2 c} - \frac{v^2}{\cos^2 c} = 1$$

and is the right-hand branch of the hyperbola whose focii are at points $w = \pm 1$ (Fig. 6.31). Any point on $x = c$ is (c, y), where $-\infty < y < \infty$. Then from the Eq. (6.76) it is obvious that as (c, y) moves upward along the entire length of the line, its image moves upward along the entire branch of the hyperbola, as depicted in Fig. 6.31. This mapping is one-to-one because when $y > 0, v > 0$ and when $y < 0, v < 0$.

(ii) Further, if $-\frac{\pi}{2} < c < 0$, line $x = c$ is mapped onto the left-hand branch of the same hyperbola described in the upward direction as shown in the Fig. 6.31.

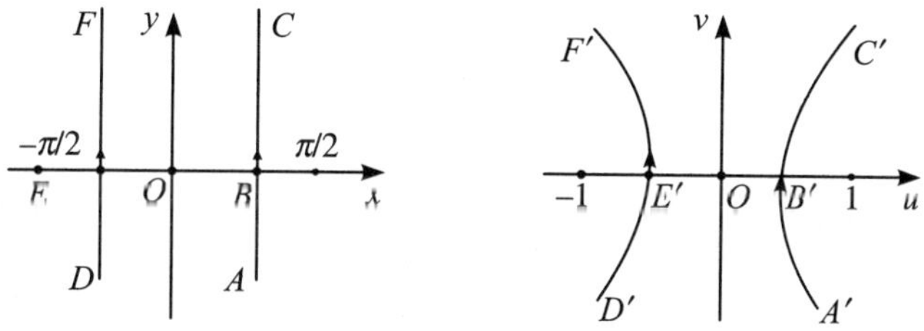

Figure 6.31 *Region under transformation* $w = \sin z$

(iii) From Eq. (6.76) we see that if $x = 0$, then $u = 0$, so that under this transformation the y axis of z-plane is mapped onto the v-axis of the w-plane. The image of any point $(0, y)$ on the y-axis is $(0, \sinh y)$. Also, if $y > 0$, $\sinh y > 0$. Therefore, when y-axis is described upward along its entire length, its image is v-axis and is also described upward along its entire length.

(iv) Under the transformation (6.76), the image of the horizontal line $y = c$, $c > 0$ is considered here. The image is given by the parametric equation $u = \sin x \cosh c$ and $v = \cos x \sinh c$. This image, therefore, represents the curve

$$\frac{u^2}{\cosh^2 c} + \frac{v^2}{\sinh^2 c} = 1,$$

which is an ellipse with focii at points where $w = \pm 1$. The image of the points (x, c) on $y = c$ moving from left to right makes a circuit around the ellipse in the clockwise direction.

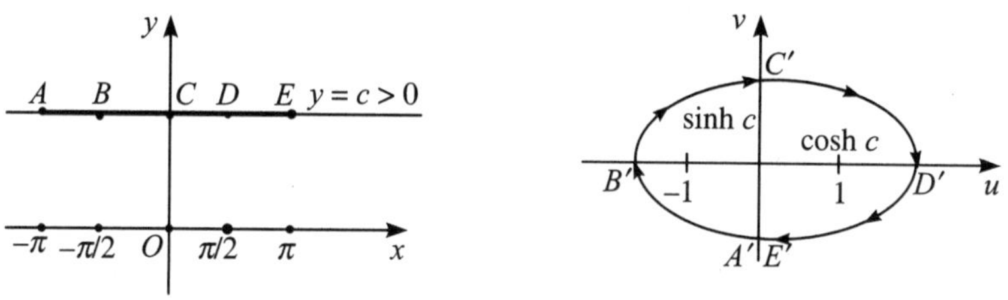

Figure 6.32 *Region under transformation* $w = \sin z$

(v) Let us take the line segment S_c given by $S_c = \{|x| \leq \pi/2 \text{ and } y = c; c > 0\}$ (Fig. 6.32). Now, $\sinh c > 0$ for $c > 1$ and $\cos x \geq 0$ for $|x| \leq \pi/2$, so $v \leq 0$. Thus the segment S_c in z-plane is mapped onto the semi ellipse E_c in w-plane, given by

$$\frac{u^2}{\cosh^2 c} + \frac{v^2}{\sinh^2 c} = 1, v \leq 0. \tag{6.77}$$

Now as c varies over the positive reals, S_c sweeps out the strip $\{(x, y) : |x| \geq \pi/2 \text{ and } 0 < y < \infty\}$ and E_c sweeps out the half plane $\mathcal{H} - E_0$, where $\mathcal{H} = \{w \in \mathbb{C} : \text{Im}w \geq 0\}$ and $E_0 = \{(u, v) : |u| \leq 1; v = 0\}$. In particular, the image of $S = \{|x| \leq \pi; y = c\}$ represents an ellipse as depicted in the Fig. 6.32.

The image of $y = 0$, is given by the parametric equations $u = \sin x$, $v = 0$. From Eq.(6.76), the segment $S_0 = \{|x| \leq \pi/2; y = 0\}$ is mapped onto the segment E_0.

(vi) Let us consider the rectangular region $-\pi \leq x \leq \pi$ and $0 < c_1 < y < c_2$ (Fig.6.33) in the z-plane. The region is mapped onto the region in w-plane bounded by two confocal ellipses. The boundary of the rectangle is mapped into the closed curve consisting of the boundaries of the two ellipses and the part of the u-axis traversed twice.

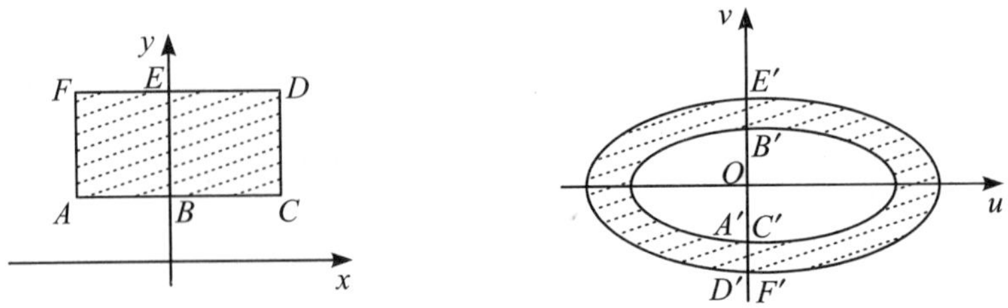

Figure 6.33 $|x| \le \pi, c_1 \le y \le c_2$

(vii) The rectangular region $-\pi/2 \le x \le \pi/2$ and $0 \le y \le c$ is mapped onto a semi elliptic region.

6.6.6 The Transformation $w = \cosh z$

Let $w = f(z) = \cosh z$, which is a transcendental function. Now,

$$\frac{dw}{dz} = \sinh z = 0 \Rightarrow z = n\pi i : n \in \mathbb{Z}.$$

Therefore, the mapping is conformal at all points except $z = n\pi i$, where $n \in \mathbb{Z}$. Now

$$\begin{aligned} w &= \cosh z = \cos i(x+iy) = \cosh x \cos y + i \sinh x \sin y. \\ \therefore u &= \cosh x \cos y \text{ and } v = \sinh x \sin y. \end{aligned} \tag{6.78}$$

(i) When $y = 0$, we have $u = \cosh x$ and $v = 0$. Further, the image of $(0,0) = (\cosh 0, 0) = (1,0)$ when $x \ge 0$, $1 \ge \cosh x < \infty$. Thus the positive real axis is mapped onto the interval $(1, \infty)$.

(ii) When $y = c$, we have

$$u = \cosh x \cos c \text{ and } v = \sinh x \sin c.$$

$$\Rightarrow \quad \frac{u^2}{\cos^2 c} - \frac{v^2}{\sin^2 c} = 1 \quad \text{as } \cosh^2 x - \sinh^2 x = 1. \tag{6.79}$$

Hence, the straightline $y = c$, where $c \ne n\pi/2$, is mapped onto the hyperbola (6.79).

(iii) Similarly, the straightline $x = c$, where $c \ne 0$, is mapped onto the ellipse

$$\frac{u^2}{\cosh^2 c} + \frac{v^2}{\sinh^2 c} = 1.$$

Now, $w = \cosh z = \cos iz = \sin(iz + \pi/2)$. Hence the image under transformation $w = \cosh z$, can be obtained from the corresponding results for the transformation $w = \sin z$.

Also, $w = \sinh z = -i \sin(iz)$, and hence the images of the regions under $w = \sinh z$ can also be obtained from the corresponding results for the transformation $w = \sin z$.

6.6.7 The Transformation of $w = \tan z$

Let $z = x + iy$ and $w = u + iv$. Then the transformation of $w = \tan z$ becomes

$$\begin{aligned} u + iv &= \tan(x+iy) = \frac{\sin(x+iy)}{\cos(x+iy)} \\ &= \frac{2\sin(x+iy)\,\cos(x-iy)}{\cos(x+iy)\,\cos(x-iy)} = \frac{\sin 2x + i\,\sinh 2y}{\cos 2x + \cosh 2y}. \\ \Rightarrow u &= \frac{\sin 2x}{\cos 2x + \cosh 2y} \text{ and } v = \frac{\sinh 2y}{\cos 2x + \cosh 2y}. \end{aligned} \tag{6.80}$$

Now, the points on the segment $x = x_1$ are mapped into the points on the curve

$$u = \frac{\sin 2x_1}{\cos 2x_1 + \cosh 2y} \text{ and } v = \frac{\sinh 2y}{\cos 2x_1 + \cosh 2y}, \tag{6.81}$$

where y is a parameter. The equation (6.81) can be written as

$$u \cosh 2y = \sin 2x_1 - u \cos 2x_1 \tag{6.82}$$

and

$$\begin{aligned} \sinh 2y &= v \cos 2x_1 + v \cosh 2y \\ &= v \cos 2x_1 + v\left(\frac{\sin 2x_1}{u} - \cos 2x_1\right) = \frac{v \sin 2x_1}{u}. \end{aligned} \tag{6.83}$$

Squaring and subtracting the equations (6.82) and (6.83), we get

$$\begin{aligned} & u^2(\cosh^2 2y - \sinh^2 2y) = \sin^2 2x_1 + u^2\cos^2 2x_1 - 2u \sin 2x_1 \cos 2x_1 - v^2 \sin^2 2x_1 \\ \Rightarrow\quad & u^2 - u^2\cos^2 2x_1 + v^2\sin^2 2x_1 + 2u\sin 2x_1 \cos 2x_1 - \sin^2 2x_1 = 0 \\ \Rightarrow\quad & u^2\sin^2 2x_1 + v^2 \sin^2 2x_1 + +2u \sin 2x_1 \cos 2x_1 - \sin^2 2x_1 = 0 \\ \Rightarrow\quad & u^2 + v^2 + 2u \cot 2x_1 - 1 = 0 \\ \Rightarrow\quad & (u + \cot 2x_1)^2 + v^2 = 1 + \cot^2 2x_1 = \text{cosec}\,^2 2x_1, \end{aligned}$$

which represents a family of circles for varying values of x_1 with centre $(-\cot 2x_1, 0)$ and radius cosec $2x_1$. All these circles pass through the points $(0, \pm 1)$, i.e., through $w = \pm i$. The image of the infinite strip $x_1 < x < x_2$ $(0 < x_2 - x_1 \leq \pi)$ parallel to the x-axis in the z-plane is a circular line with angles $2(x_2 - x_1)$ radians and vertices at $w = \pm i$ (Fig. 6.34). Similarly, the image of family of straight lines $y = y_1$ for varying values of y_1 is the family of circles $u^2 + v^2 - 2v \coth 2y_1 + 1 = 0$.

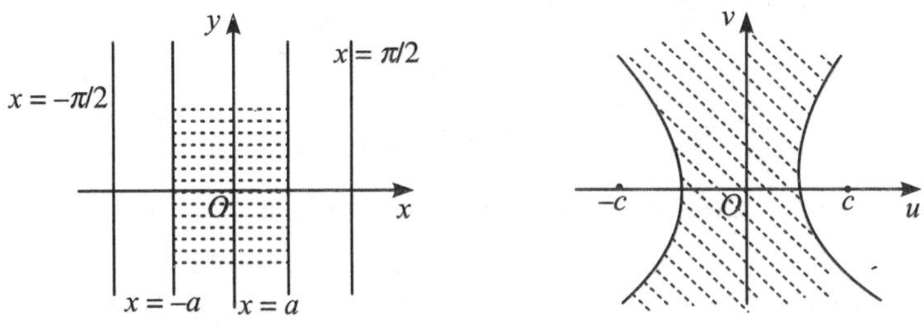

Figure 6.34 $(u + \cot 2x_1)^2 + v^2 = 1 + \cot^2 2x_1 = \text{cosec}\,^2 2x_1$

6.7 The Schawarz–Cristoffel Transformations

The Riemann mapping theorem guarantees the existence of an analytic bijection between a simply-connected domain to the open unit ball, but did not give any procedure for constructing such conformal mapping. Here, we shall provide an explicit formula for the derivative of a conformal mapping that maps the upper half plane $\mathcal{H} = \{z \in \mathbb{C} : \text{Im } z > 0\}$ of the z-plane onto a given bounded or unbounded polygonal region (boundary contains a finite number of line segments). Such mappings are often applied in physical problems such as in heat conduction, fluid mechanics, and electrostatics potential theory.

6.7.1 Transformation of the Real Axis onto a Polygon

Here, we want to construct a mapping that maps real axis onto a polygon, using the fact as derived in the Result 52. Let us consider a transformation

$$w = g(z) = (z - x_1)^{\alpha/\pi};\ \ 0 < \alpha < 2\pi, \tag{6.84}$$

where x_1 is a point on the real axis of the z-plane. Consider the translation $T(z) = z - x_1$ and real power function $G(z) = z^{\alpha/\pi}$. Then

$$(GT)(z) = G(T(z)) = G(z - x_1) = (z - x_1)^{\alpha/\pi} = w.$$

This shows that the mapping (6.84) is a composition of two transformations: a ray emanating from x_1 (Fig. 6.35) and making an angle θ with the real axis is mapped onto a ray emanating from the origin and making an angle $\alpha\theta/\pi$ with the real axis. The image of the half-plane $\mathcal{H}$ is the point $w = 0$, together with the wedge $0 < \arg(w) < \alpha$. Since

$$g'(z) = \frac{\alpha}{\pi}\,(z - x_1)^{(\alpha/\pi)-1} \neq 0,$$

it follows that mapping (6.84) is conformal at any point $z \in \mathcal{H}$. In what follows, we shall use the form of the derivative of the function $g(z)$ to describe a conformal mapping of the upper half-plane $\mathcal{H}$ onto an arbitrary polygonal region.

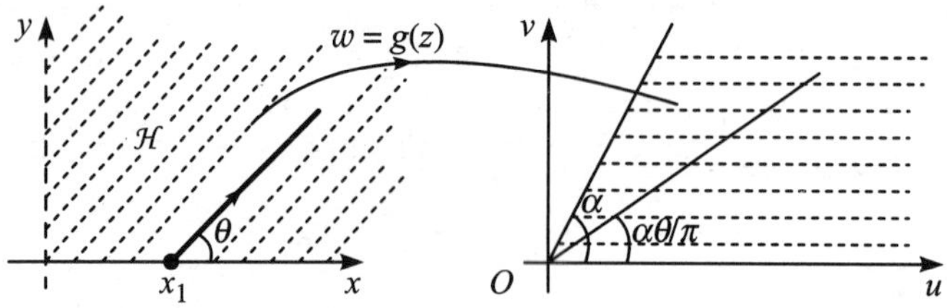

Figure 6.35 *Transformation* $w = g(z)$

For this, we shall first analyse the mapping $h(z)$, which is analytic in the domain $\mathcal{H}$ and whose derivative is

$$h'(z) = A(z - x_1)^{(\alpha_1/\pi)-1}\,(z - x_2)^{(\alpha_2/\pi)-1};\ \ 0 < \alpha_1, \alpha_2 < 2\pi \tag{6.85}$$

where A is a complex number, and $x_1 < x_2$ are real. Under the mapping $w = h(z)$, we shall find the images of the intervals $(-\infty, x_1)$, (x_1, x_2) and (x_2, ∞). On the real axis $z = x$, from Eq. (6.85), it follows that

$$\arg[h'(z)] = \arg A + \left(\frac{\alpha_1}{\pi} - 1\right)\arg(x - x_1) + \left(\frac{\alpha_2}{\pi} - 1\right)\arg(x - x_2).$$

Thus, if $z = x$ and $-\infty < x < x_1$, then $\arg(x - x_1) = \arg(x - x_2) = \pi$. Therefore,

$$\begin{aligned}\arg[h'(z)] &= \arg A + \left(\frac{\alpha_1}{\pi} - 1\right)\pi + \left(\frac{\alpha_2}{\pi} - 1\right)\pi \\ &= \arg A + \alpha_1 + \alpha_2 - 2\pi,\end{aligned}$$

which is a constant for all x. Hence, the interval $(-\infty, x_1)$ is mapped into a line segment L_1 defined by $w = h(z)$ (Fig. 6.36). Similarly, if $x_1 < x < x_2$, then $\arg(x - x_1) = 0$ and $\arg(x - x_2) = \pi$ and so,

$$\arg[h'(z)] = \arg A + \alpha_2 - 2\pi;$$

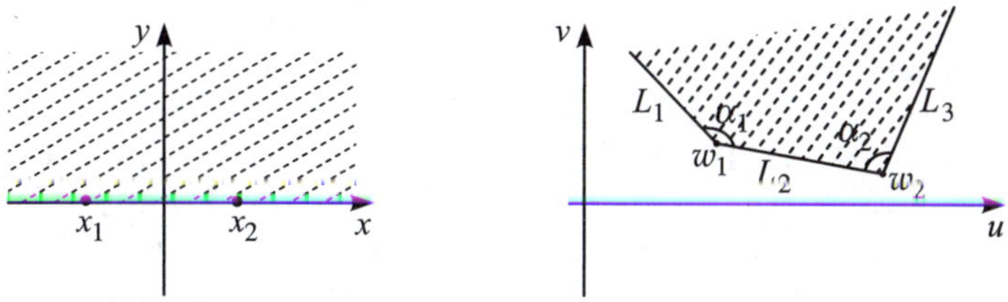

Figure 6.36 *Transformation of the real axis onto a polygon*

i.e., the interval (x_1, x_2) is mapped into a line segment L_2 defined by $w = h(z)$, which changes the exterior angle by $\pi - \alpha_1$ from L_1. Now,

$$\arg[h'(z)] = \arg A; \text{ if } -\infty < x < x_2;$$

i.e., the interval (x_2, ∞) is mapped into a line segment L_3 defined by $w = h(z)$, which changes the exterior angle by $\pi - \alpha_2$ from L_2 (Fig. 6.36).

Since $h(z)$ is analytic, the image of the half-plane $\operatorname{Im} z \geq 0$ is an unbounded polygonal region (Fig. 6.36). In the following result, we generalise the discussion above for the derivative $f'(z)$ of a function $f(z)$ that maps the half-plane $\operatorname{Im} z \geq 0$ onto a polygonal region with any number of sides.

6.7.2 Transformation of Schawarz–Cristoffel

In this section, we construct a one-to-one analytic conformal mapping, known as *Schawarz–Cristoffel transformation*, which maps the upper half plane $\mathcal{H}$ of the z-plane onto a given simple closed polygon and its interior in the w-plane. For each integer $j = 1$ to $n + 1$, we consider the points x_j on the real axis in the z-plane as

$$x_1 < x_2 < \cdots < x_n; \quad x_{n+1} = \infty,$$

whose images $w_j = f(x_j) : j = 1, 2, \cdots, n$ and $w_{n+1} = f(\infty)$ are to be the vertices $w_1, w_2, \cdots, w_n$, w_{n+1} of a polygon in the w-plane, with corresponding interior angles $\alpha_1, \alpha_2, \cdots, \alpha_n$ respectively (Fig. 6.37). The function f should be such that $\arg f'(z)$ jumps from one constant value to another at the points $z = x_j$ as the points z trace out of the x-axis.

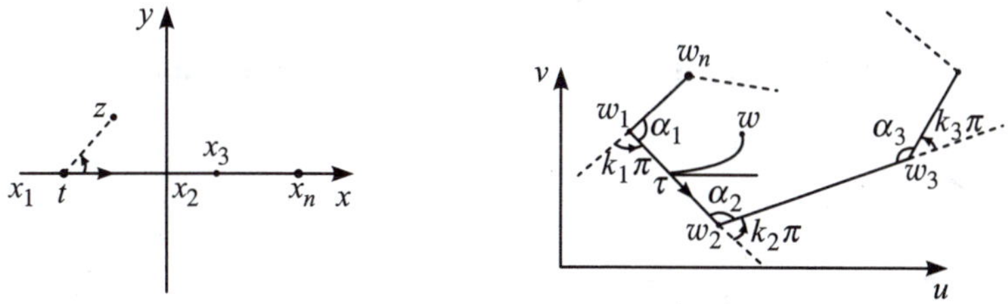

Figure 6.37 *The Schawarz–Cristoffel transformation*

Let $f(z)$ be analytic in the domain $\mathcal{H}$ and have the derivative

$$f'(z) = A(z - x_1)^{-k_1}(z - x_2)^{-k_2} \cdots (z - x_n)^{-k_n}, \tag{6.86}$$

where A is a complex constant and each k_j is a real constant. Then the argument of $f'(z)$ changes in the prescribed manner as z describes the real axis. Now

$$\arg f'(z) = \arg A - k_1 \arg(z - x_1) - k_2 \arg(z - x_2) - \cdots - k_n \arg(z - x_n). \tag{6.87}$$

When $z = x$ and $x < x_1$, $\arg(z - x_1) = \arg(z - x_2) = \cdots = \arg(z - x_n) = \pi$. When $x_1 < x < x_2$, $\arg(z - x_1) = 0$ and each of the arguments is π. According to Eq.(6.87), then, $\arg f'(z)$ increases abruptly to $k_1\pi$ as z moves to the right through the point $z = x_1$. It jumps the value by amount $k_2\pi$ as z passes through the point x_2, etc. Those angles, $k_j\pi$, are the exterior angles of the polygon described by the point w.

The exterior angles can be limited to angles between $-\pi$ and π, in which case $-1 < k_j < 1$. We assume that the sides of the polygon never cross one another and that the polygon is given a positive, or anticlockwise orientation. The sum of the exterior angles of a closed polygon is then 2π, and the exterior angle at the vertex w_{n+1} can be written as

$$k_{n+1}\pi = 2\pi - (k_1 + k_2 + \cdots + k_n)\pi.$$

Thus the real numbers k_j must necessarily satisfy the conditions

$$\sum_{j=1}^{n+1} k_j = 2 \text{ and } -1 < k_j < 1 \;\; (j = 1, 2, \cdots, n+1). \tag{6.88}$$

Note that w_{n+1} is not a vertex and the polygon has n vertices. In integral form, Eq. (6.86) can be written as

$$\begin{aligned} w &= A\int_{z_0}^{z} (s - x_1)^{-k_1}(s - x_2)^{-k_2}\cdots(s - x_n)^{-k_n}\, ds + B \\ &= A\int_{z_0}^{z} \Big(s - x_1\Big)^{\frac{\alpha_1}{\pi}-1}\Big(s - x_2\Big)^{\frac{\alpha_2}{\pi}-1}\cdots\Big(s - x_n\Big)^{\frac{\alpha_n}{\pi}-1} ds + B, \end{aligned} \tag{6.89}$$

which is the *Schwarz–Christoffel transformation*. Here, in the integral (6.89), z_0 can be chosen arbitrarily. Clearly, from Eqs. (6.86) and (6.89), it follows that $(x_1, 0), (x_2, 0), \cdots, (x_n, 0)$ are the images in the z-plane of the vertices $w_1, w_2, \cdots, w_n$ of the polygon in the w-plane. If $z = \infty$ is mapped into one vertex, say w_j, then the term containing $(z - x_j)$ is absent in Eq. (6.86). If the polygon is closed, then the last vertex w_n is the same as the initial vertex w_1, and hence instead of n interior angles we have only $n - 1$ interior angles. The size and orientation of the polygon are determined by the complex constants A and B. This is usually fixed by choosing some of the points x_j on the x-axis.

6.7.3 Schwarz–Christoffel Transformation for Triangles

Let the polygon be a triangle Q with vertices w_1, w_2 and w_3 (Fig. 6.38). By Riemann mapping theorem, there exists a function f that maps $\mathcal{H}$ into the triangle Q. The transformation (6.89) can be written as

$$w = A\int_{z_0}^{z} (s - x_1)^{-k_1}(s - x_2)^{-k_2}(s - x_3)^{-k_3}\, ds + B, \tag{6.90}$$

where $k_1 + k_2 + k_3 = 2$; A, B are complex constants; z_0 is arbitrary. Then πk_j $(j = 1, 2, 3)$ denotes the exterior angle at w_j. In terms of the interior angles α_j,

$$\pi k_j + \alpha_j = \pi; \;\; j = 1, 2, 3.$$

Here we have taken all three points, x_1, x_2, x_3 $(x_1 < x_2 < x_3)$, as finite points on the real axis of z-plane. Arbitrary values can be assigned to each of them. The complex constants, A and B, can be determined so that upper half is mapped onto the given triangular region.

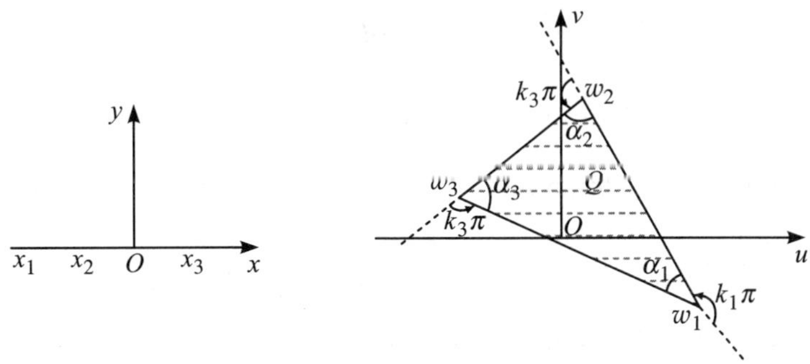

Figure 6.38 *Real axis⟶ triangle*

Let us consider the triangle with vertices $w_1 = f(x_1)$, $w_2 = f(x_2)$ and $w_3 = f(\infty)$. Then the transformation (6.90) becomes

$$w = A\int_{z_0}^{z} (s - x_1)^{-k_1}(s - x_2)^{-k_2}\, ds, \tag{6.91}$$

where arbitrary real values can be assigned to x_1 and x_2 ($x_1 < x_2$), z_0 being arbitrary. The integrals in Eqs. (6.90) and (6.91) do not represent elementary functions unless the triangle is degenerate with one or two vertices at infinity.

For an equilateral triangle, $k_1 = k_2 = 2/3$. Eq. (6.91) becomes

$$w = f(z) = \int_{a}^{z} (s + a)^{-2/3}(s - a)^{-2/3}\, ds$$

if we take $z_0 = a$, $x_1 = -a$ and $x_2 = a$, where $a > 0$. Then

$$w_2 = f(a) = \int_{a}^{a} (s + a)^{-2/3}(s - a)^{-2/3}\, ds = 0.$$

To obtain $w_1 = f(-a)$, we take our contour to be the segment of the x-axis from $x = a$ to $x = -a$ (Fig. 6.39). Since Arg$(x - x_j)$ is π or 0 according as x is to the left or the right of x_3 on the real axis, it follows that for $-a < x < a$,

$$(x + a)^{-2/3} = |x + a|^{-2/3}e^{0i} = (a + x)^{-2/3} : \text{ as } x + a > 0,$$
$$(x - a)^{-2/3} = |x - a|^{-2/3}e^{-2/3\,\pi i} = (a - x)^{-2/3}e^{-2/3\,\pi i} \text{ as } x - a < 0.$$

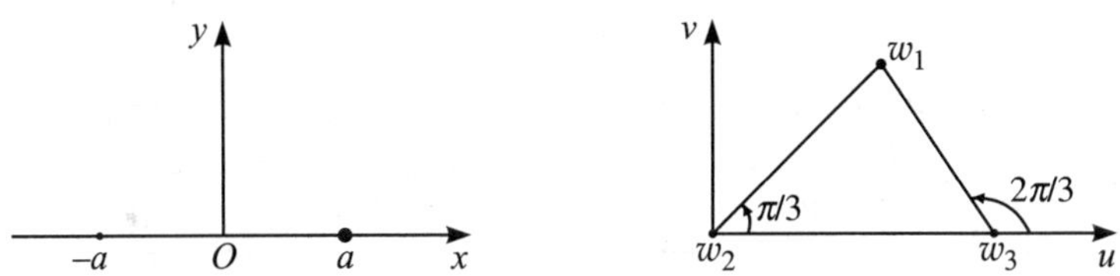

Figure 6.39 *Real axis⟶ equilateral triangle*

The vertex w_1 is, therefore,

$$w_1 = f(-a) = \int_{a}^{-a} (x + a)^{-2/3}(x - a)^{-2/3}dx$$

$$
\begin{aligned}
&= -\int_{-a}^{a}(a+x)^{-2/3}\Big[(a-x)^{-2/3}e^{-2/3\,\pi i}\Big]dx;\ -1=e^{\pi i}\\
&= e^{i\pi}\int_{-a}^{a}(a^2-x^2)^{-2/3}e^{-2/3\,\pi i}dx = 2e^{\pi i/3}\int_0^a\frac{dx}{(a^2-x^2)^{2/3}}.
\end{aligned}\tag{6.92}
$$

Since $k_1+k_2=\frac{2}{3}+\frac{2}{3}=\frac{4}{3}\neq 2$, if follows that $w_3=f(\infty)$ is a vertex of our triangle. Therefore,

$$
w_3=\int_a^{\infty}(x+a)^{-2/3}(x-a)^{-2/3}dx=\int_a^{\infty}\frac{dx}{(x^2-a^2)^{2/3}},\tag{6.93}
$$

which implies that w_3 is on the positive u-axis since $(x^2-a^2)^{-2/3}>0$. Also,

$$
\begin{aligned}
w_3 &= \int_a^{-\infty}(x+a)^{-2/3}(x-a)^{-2/3}dx\\
&= w_1+\int_{-a}^{-\infty}(x+a)^{-2/3}(x-a)^{-2/3}dx \text{ by Eq. (6.92)}\\
&= w_1+\int_{-a}^{-\infty}\Big[|x+a|^{-2/3}e^{-2/3\,\pi i}\Big]\Big[|x-a|^{-2/3}e^{-2/3\,\pi i}\Big]dx\\
&= w_1+e^{-4/3\,\pi i}\int_{-a}^{-\infty}\frac{dx}{(x^2-a^2)^{2/3}}\ ;\ x=-r\\
&= w_1+e^{-\pi i/3}\int_a^{\infty}\frac{dr}{(r^2-a^2)^{2/3}}=w_1+e^{-\pi i/3}w_3 \quad \text{by Eq.(6.93)}\\
\Rightarrow w_3 &= \frac{w_1}{1-e^{-\pi i/3}}=\frac{e^{\pi i/3}}{1-e^{-\pi i/3}}\int_{-a}^{a}\frac{dx}{(a^2-x^2)^{2/3}} \quad \text{by Eq.(6.92)}\\
&= \int_{-a}^{a}\frac{dx}{(a^2-x^2)^{2/3}}\ ;\ \text{as } \frac{e^{\pi i/3}}{1-e^{-\pi i/3}}=1.
\end{aligned}
$$

Hence f maps the extended real axis onto the equilateral triangle with vertices

$$
w_2=0, w_3=\int_{-a}^{a}\frac{dx}{(a^2-x^2)^{2/3}},\ \text{and } w_1=e^{\pi i/3}w_3.\tag{6.94}
$$

Furthermore, f maps the upper half plane onto the interior of this triangle. In particular, when $a=1$, we get from Eq.(6.94)

$$
w_2=0, w_3=\beta(1/2,1/3) \text{ and } w_1=e^{i\pi/3}\beta(1/2,1/3).
$$

Example 6.7.1. *Find the transformation which maps the upper half of the z-plane onto the shaded portion of the w-plane as shown in the Figure 6.40.*

Solution. The region in the w-plane is unbounded. Let the image under transformation of the points $Q(x_1=-1)$, $S(x_2=0)$ and $T(x_3=1)$ on the z-plane be $Q'(w_1=-b)$, $S'(w_2=ai)$ and $T'(w_3=b)$, respectively. Then the interior angles of the polygon at Q', T' and S' are $\alpha_1=\pi-\alpha$, $\alpha_2=\pi-\alpha$ and $\alpha_3=2\pi-(\pi-2\alpha)=\pi+2\alpha$, respectively. The Schwarz–Christoffel transformation (6.89) is

$$
\begin{aligned}
w &= f(z)=A_1\int_0^z(s+1)^{\frac{\pi-\alpha}{\pi}-1}(s-0)^{\frac{\pi+2\alpha}{\pi}-1}(s-1)^{\frac{\pi-\alpha}{\pi}-1}\,ds+B\\
&= A_1\Big(-1\Big)^{-\alpha/\pi}\int_0^z s^{2\alpha/\pi}\Big(1-s^2\Big)^{-\alpha/\pi}ds+B\\
&= A\int_0^z s^{2\alpha/\pi}\,(1-s^2)^{-\alpha/\pi}\,ds+B.
\end{aligned}
$$

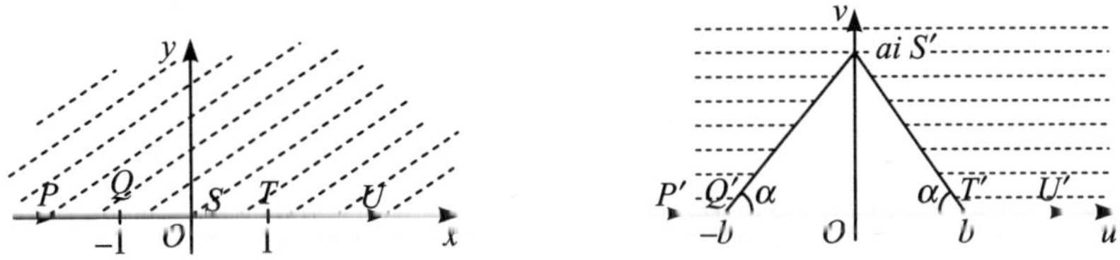

Figure 6.40

When, $z=0$, $w=ai$. So $B=ai$. Also, when $z=1$, $w=b$. Hence

$$\begin{aligned} b &= A\int_0^1 s^{2\alpha/\pi}\left(1-s^2\right)^{-\alpha/\pi} ds + ai;\, s^2=t \\ &= \frac{A}{2}\int_0^1 t^{\left(\frac{\alpha}{\pi}+\frac{1}{2}\right)-1}\,(1-t)^{\left(-\frac{\alpha}{\pi}+1\right)-1}\,dt+ai \\ &= \frac{A}{2}\beta\left(\frac{\alpha}{\pi}+\frac{1}{2},1-\frac{\alpha}{\pi}\right)+ai \end{aligned}$$

Thus the required transformation is

$$w=\frac{2(b-ai)}{\beta\left(\frac{\alpha}{\pi}+\frac{1}{2},1-\frac{\alpha}{\pi}\right)}\int_0^z s^{2\alpha/\pi}\left(1-s^2\right)^{-\alpha/\pi}ds+ai.$$

6.7.4 Schwarz–Christoffel Transformation for Rectangles

To map the extended real axis onto a polygon such as a rectangle, we choose $k_j=1/2$. If we choose $\pm a$ and $\pm b$ as the points x_j whose images are vertices (Fig. 6.41) and write

$$g(z)=(z+b)^{-1/2}\,(z+a)^{-1/2}\,(z-b)^{-1/2}\,(z-a)^{-1/2};\ \ 0<a<b,$$

where $0\le \arg(z-x_j)\le\pi$, the Schwarz–Christoffel transformation becomes

$$w=f(z)=\int_0^z g(s)\ ds. \tag{6.95}$$

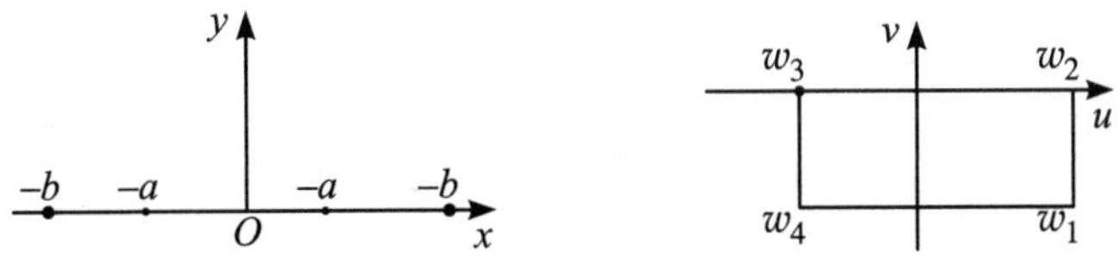

Figure 6.41 *Real axis⟶ rectangle*

Let w_3 be the image of x_3. Then

$$\begin{aligned} w_3 &= f(x_3)=f(a)=\int_0^a g(x)\ dx \\ &= \int_0^a (x+b)^{-1/2}\,(x+a)^{-1/2}\Big[\big|x-a\big|^{-1/2}e^{-i\pi/2}\Big]\Big[\big|x-b\big|^{-1/2}e^{-i\pi/2}\Big]dx \\ &= -\int_0^a\frac{dx}{\sqrt{\left(a^2-x^2\right)\left(b^2-x^2\right)}}<0. \end{aligned} \tag{6.96}$$

The vertex w_2, which is the image of x_2, is given by

$$\begin{aligned} w_2 &= f(x_2) = f(-a) = \int_0^{-a} g(x)\ dx \\ &= \int_0^{-a} (x+b)^{-1/2}(x+a)^{-1/2}\left[\left|x-a\right|^{-1/2} e^{-i\pi/2}\right]\left[\left|x-b\right|^{-1/2} e^{-i\pi/2}\right]dx \\ &= -\int_0^{-a} \frac{dx}{\sqrt{(a^2-x^2)(b^2-x^2)}} \\ &= \int_0^{a} \frac{dx}{\sqrt{(a^2-x^2)(b^2-x^2)}} = -w_3 > 0. \end{aligned} \tag{6.97}$$

Similarly, the vertex w_4 as the image of x_4 is

$$\begin{aligned} w_4 &= f(x_4) = f(b) = \int_0^{a} g(x)\ dx + \int_a^{b} g(x)\ dx \\ &= w_3 + \int_a^{b} (x+b)^{-1/2}(x+a)^{-1/2}\left|x-a\right|^{-1/2}\left|x-b\right|^{-1/2} e^{-i\pi/2}dx \\ &= w_3 - \int_0^{-a} \frac{dx}{\sqrt{(a^2-x^2)(b^2-x^2)}} = w_3 - \int_a^{b} \frac{idx}{\sqrt{(x^2-a^2)(b^2-x^2)}} \\ &= w_3 - i\,p;\quad \text{where } p = \int_a^{b} \frac{dx}{\sqrt{(x^2-a^2)(b^2-x^2)}}. \end{aligned} \tag{6.98}$$

Finally, the vertex w_1 as the image of x_1 is

$$\begin{aligned} w_1 &= f(x_1) = f(-b) = \int_0^{-b} g(x)\ dx = \int_0^{-a} g(x)dx + \int_{-a}^{-b} g(x)dx \\ &= w_2 + \int_{-a}^{-b} (x+b)^{-1/2}\left[\left|x+a\right|^{-1/2} e^{-i\pi/2}\right]\left[\left|x-a\right|^{-1/2} e^{-i\pi/2}\right] \\ &\qquad \left[\left|x-b\right|^{-1/2} e^{-i\pi/2}\right]dx = w_2 - i\,p. \end{aligned} \tag{6.99}$$

Thus the transformation (6.95) maps the extended real axis onto the rectangle with vertices w_1, w_2, w_3 and w_4 as in Eqs. (6.99), (6.97), (6.96) and (6.98) respectively.

Example 6.7.2. [Degenerate polygon]: *Find the Schwarz–Christoffel transformation which maps* $\{z : Im(z) \geq 0\}$ *onto* $\{w : Im(w) \geq 0$ *and* $|Re(w)| \leq b\}$, *where* $b > 0$.

Solution. Let $w_1 = -b, w_2 = b$ and $w_3 = b + iy$, where $y > 0$ be the vertices of a right triangle (Fig. 6.42). The infinite half strip is the limiting region of the interior of this triangle with vertices w_1, w_2 and w_3, as the imaginary part $w_3 \to \infty$. The exterior angle at w_1 has the measure $k_1\pi$, which approaches to $\pi/2$ as $w_3 \to \infty$, so $k_1 \to 1/2$. The limiting values of the exterior angles are $k_1\pi = k_2\pi = \pi/2$ and $k_3\pi = \pi$.

We choose the points $x_1 = -1, x_2 = 1$ and $x_3 = \infty$ as the points whose images are the vertices. Let the transformation be

$$w = f(z) = A\int_0^{z} (t+1)^{-\frac{1}{2}}(t-1)^{-\frac{1}{2}}dt + B, \tag{6.100}$$

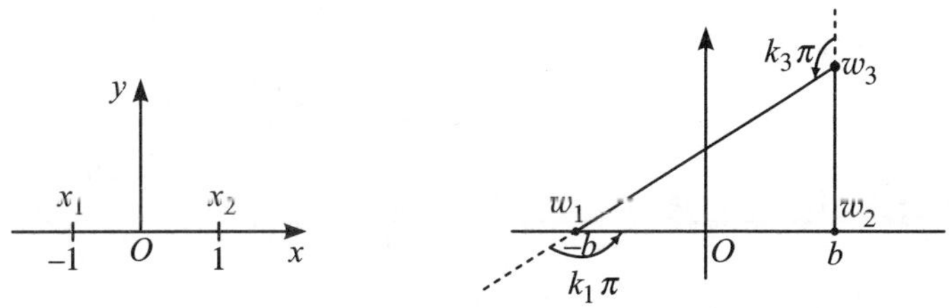

Figure 6.42

where A and B are to be determined so that $f(-1) = -b$ and $f(1) = b$. Thus

$$\begin{aligned} b = w_2 = f(1) &= A\int_0^1 |t+1|^{-\frac{1}{2}}\left[|t-1|^{-\frac{1}{2}}e^{-\frac{i\pi}{2}}\right]dt + B \\ &= -Ai\int_0^1 \frac{dt}{\sqrt{1-t^2}} + B = -\frac{A\pi}{2}i + B. \\ -b = w_1 = f(-1) &= A\int_0^{-1} |t+1|^{-\frac{1}{2}}\left[|t-1|^{-\frac{1}{2}}e^{-\frac{i\pi}{2}}\right]dt + B \\ &= -Ai\int_0^{-1} \frac{dt}{\sqrt{1-t^2}} + B = -\frac{A\pi}{2}i + B. \end{aligned}$$

Solving, we get $B = 0, A = \frac{2b}{\pi}i$. Thus the transformation (6.100) becomes

$$\begin{aligned} w = f(z) &= \frac{2bi}{\pi}\int_0^z (t+1)^{-\frac{1}{2}}(t-1)^{-\frac{1}{2}}dt \\ &= \frac{2bi}{\pi}\frac{1}{i}\int_0^z \frac{dt}{(t+1)^{-\frac{1}{2}}(t-1)^{-\frac{1}{2}}} = \frac{2b}{\pi}\sin^{-1} z. \end{aligned}$$

Therefore, the solution for z is $z = \sin\dfrac{\pi w}{2b}$.

Example 6.7.3. *Sketch the image of the extended real axis under the transformation*

$$w = f(z) = \int_0^z (t+1)^{-1/2}\, t^{-1/3}\, (t-1)^{-2/3}\, dt.$$

Solution. The given transformation can be written as

$$w = f(z) = \int_0^z (t+1)^{-1/2}\, (t-0)^{-1/3}\, (t-1)^{-2/3}\, dt,$$

where, the exterior angles are $\pi/2$, $\pi/3$, $2\pi/3$ and $2\pi - (\pi/2 + \pi/3 + 2\pi/3) = \pi/2$. Thus, for this transformation, $k_1 = 1/2, k_2 = 1/3$ and $k_3 = 2/3$. Hence from

$$-1 < k_4 = 2 - \frac{1}{2} - \frac{1}{3} - \frac{2}{3} = \frac{1}{2} < 1,$$

it follows that f maps the extended real axis onto a closed polygon P with four sides (Fig. 6.43). The vertices of the rectangle P is given by

$$w_2 = 0,\ \ w_1 = \int_0^1 \frac{dx}{x^{1/3}(1-x)^{1/2}(1+x)^{2/3}},$$

$$w_3 = -\frac{1+3i}{2}\int_0^1 (x-1)^{-1/2}x^{-1/3}(1-x)^{-2/3}dx.$$

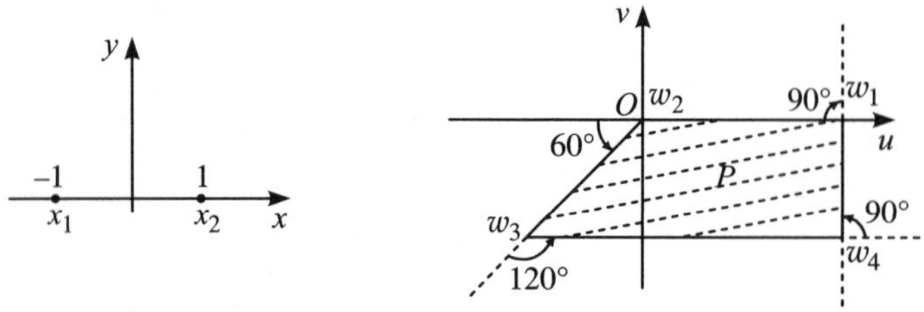

Figure 6.43 *Real axis⟶ polygon*

Exercises

Multiple-choice Questions

1. Let $f = u + iv = z^3 - 3z$. Then the value of $\begin{pmatrix} u_x & u_y \\ v_x & v_y \end{pmatrix}$ at $z = i + \sqrt{2}$ equals:
   (a) 75 (b) 63 (c) 64 (d) 72
2. The image of the region $\{z \in \mathbb{C} : \text{Re}(z) > \text{Im}(z) > 0\}$ under the mapping $z \to e^{z^2}$ is
   (a) $\{w \in \mathbb{C} : \text{Re}(w) > 0, \text{Im}(w) > 0\}$ (b) $\{w \in \mathbb{C} : \text{Re}(w) > 0, \text{Im}(w) > 0, |w| > 1\}$
   (c) $\{w \in \mathbb{C} : |w| > 1\}$ (d) $\{w \in \mathbb{C} : \text{Im}(w) > 0, |w| > 1\}$
3. The straightlines $L_1 : x = 0$, $L_2 : y = 0$ and $L_3 : x + y = 1$ are mapped by the transformation $w = z^2$ into the curves C_1, C_2 and C_3 respectively. The angle of intersection between the curves at $w = 0$ is
   (a) 0 (b) $\pi/4$ (c) $\pi/2$ (d) π
4. The image of a line segment in z-plane through $z = 0$ and $z = 1 + i$ under the transformation $w = (z - 1)(z + 1)$ is
   (a) $v^2 = 2(u + 1)$ (b) $u^2 = 2(v + 1)$ (c) $u^2 - (v + 1)^2 = 2$ (d) $u^2 + (v + 1)^2 = 2$
5. The image of the half plane $z > 0$ under the transformation $w = i(z + 1)$ is
   (a) $u > 0$ (b) $v > 0$ (c) $u > 1$ (d) $v > 1$
6. Let $f : \mathbb{C}/\{3i\} \to \mathbb{C}$ be defined by $f(z) = \frac{z-i}{iz+3}$. Which of the following statements about f is false?
   (a) f is conformal on $\mathbb{C}/\{3i\}$. (b) f maps circles in $\mathbb{C}/\{3i\}$ onto the circles in $\mathbb{C}$.
   (c) All the fixed points of f are in the region $\{z \in \mathbb{C} : \text{Im}(z) > 0\}$.
   (d) There is no straight line in $\mathbb{C}/\{3i\}$ which is mapped onto a straight line in $\mathbb{C}$ by f.
7. The image of the curve $|z - 2i| = 2$ under the transformation $w = 1/z$ is
   (a) $4u + 1 = 0$ (b) $4v + 1 = 0$ (c) $u + 4 = 0$ (d) $v + 4 = 0$
8. The image of the line $y - x + 1 = 0$ under the transformation $w = 1/z$ is
   (a) $u^2 + v^2 + u + v = 0$ (b) $u^2 + v^2 + u - v = 0$ (c) $u^2 + v^2 - u + v = 0$ (d) $u^2 + v^2 - u - v = 0$
9. The region onto which the half plane $y > 0$ is mapped under the transformation $w = (1 + i)z$ is
   (a) $u = v$ (b) $u > v$ (c) $u < v$ (d) $u + v = 0$
10. A linear transformation which carries i onto $-i$ and maps $1 + 2i$ onto itself is
    (a) $w = (1 + 2i)z + (1 - 3i)$ (b) $w = (2 + i)z + (1 - 3i)$
    (c) $w = (2 + i)z + (1 + 3i)$ (d) $w = (2 + i)z + (3 - i)$

11. The linear fractional transformation w which maps the points $0, 1, \infty$ in the z-plane into the points $-i, \infty, 1$ in the w-plane is
(a) $\frac{z-1}{z+i}$ (b) $\frac{z-i}{z+1}$ (c) $\frac{z+i}{z-1}$ (d) $\frac{z+1}{z-i}$

12. The image of the strip $0 < x < 1$ under the transformation $w = iz$ is
(a) $0 < u < v < 1$ (b) $0 < v < u < 1$ (c) $0 < u < 1$ (d) $0 < v < 1$

13. The image of the region $y > 1$ under the transformation $w = iz + 1$ is
(a) $u + v > 2$ (b) $u - v > 2$ (c) $-u + v > 2$ (d) $u + v < 2$

14. A Möbius transformation is completely determined by its effect on how many points:
(a) two (b) three (c) four (d) five

15. The linear fractional transformation $f(z) = (3z - 4)/(z - 1)$ is
(a) Parabolic (b) Elliptic (c) Hyperbolic (d) Loxodromic

16. The linear fractional transformation $f(z) = \{(2 + i)z - 2\}/(z + i)$ is
(a) Parabolic (b) Elliptic (c) Hyperbolic (d) Loxodromic

17. The linear fractional transformation having 0 and ∞ as fixed points can be written as
(a) $w = az$ (b) $w = az^{-1}$ (c) $w = a + z$ (d) $w = a + \frac{1}{z}$

18. The linear fractional transformation having 0 as the only fixed point can be written as
(a) $w = az$ (b) $w = az + b$ (c) $w = (az + b)/z$ (d) $w = z/(cz + d)$

19. The linear fractional transformation having ∞ as the only fixed point can be written as
(a) $w = az$ (b) $w = z + b$ (c) $w = a + b/z$ (d) $w = z/(cz + d)$.

20. The linear fractional transformation having α and ∞ as fixed points can be written as
(a) $w - \alpha = a(z - \alpha)$ (b) $w + \alpha = a(z + \alpha)$
(c) $w - \alpha = a/(z - \alpha)$ (d) $w + \alpha = a/(z - \alpha)$.

21. If $f_1(z) = (z + 2)/(z + 3)$ and $f_2(z) = z/(z + 1)$, then $f_2^{-1}f_1(z)$ is
(a) $\frac{2z+5}{z+2}$ (b) $\frac{z(z+2)}{(1-z)(z+3)}$ (c) $z + 2$ (d) $\frac{(z+1)(z+2)}{z(z+3)}$

22. The region into which the circular sector $0 < \theta < \pi/4, r < 1$ mapped by the transformation $w = z^2$ is
(a) $-\pi/4 < \theta < \pi/4,\ r < 1$ (b) $0 < \theta < \pi/4,\ r < 1$
(c) $-\pi/2 < \theta < \pi/2,\ r < 1$ (d) $0 < \theta < \pi/2,\ r < 1$

23. The region of the w-plane into which the first quadrant of z-plane is mapped by the transformation $w = z^2$ is
(a) Entire w-plane (b) Upper half of w-plane
(c) Lower half of w-plane (d) A quadrant of w-plane

24. Under the transformation $w = e^z$ the infinite strip $0 \leq y \leq \pi$ is mapped onto the
(a) Entire w-plane (b) Upper half $u > 0$ of w-plane
(c) Lower half $u < 0$ of w-plane (d) First quadrant of w-plane

25. Under the transformation $w = c \sin z, c \in \mathbb{R}$ the region $0 \leq x \leq \pi/2$ is mapped onto the
(a) Entire w-plane (b) Upper half of w-plane
(c) Lower half of w-plane (d) First quadrant of w-plane

26. Under the transformation $w = \sin z$, the region $0 \leq x \leq \pi/2, y \geq 0$ is mapped onto the
(a) Entire w-plane (b) Upper half of w-plane
(c) Lower half of w-plane (d) First quadrant of w-plane

27. The function $f(z) = z^2$ maps the first quadrant onto
(a) Itself (b) Upper half plane (c) Third quadrant (d) Right half plane.

28. The transformation $w = \sin^2 z$ maps the strip $0 \le x \le \pi/2, y \ge 0$ onto
(a) Entire w-plane (b) Upper half of w-plane
(c) Lower half of w-plane (d) First quadrant of w-plane.

29. For the value of α for the function $g(z) = \sinh(z - \pi/2\alpha), \alpha \neq 0$ is not conformal at a point
(a) $\frac{\pi}{6}(1+3i)$ (b) $\frac{\pi}{6}(3+i)$ (c) $\frac{2\pi}{3}$ (d) $\frac{i\pi}{2}$

30. Under the transformation $w = \sqrt{\frac{1-iz}{z-i}}$, the region $\mathcal{D} = \{z \in \mathbb{C} : |z| < 1\}$ is transformed to
(a) $\{z \in \mathbb{C} : 0 < \arg(z) < \pi\}$ (b) $\{z \in \mathbb{C} : -\pi < \arg(z) < 0\}$
(c) $\{z \in \mathbb{C} : 0 < \arg(z) < \pi/2 \text{ or } 0 < \arg(z) < 3\pi/2\}$
(d) $\{z \in \mathbb{C} : \pi/2 \arg(z) < \pi \text{ or } 3\pi/2 < \arg(z) < 2\pi\}$

31. If f be a linear fractional transformation that maps -1 to 1, i to ∞ and $-i$ to 0, then $f(1)$ equals to
(a) -2 (b) -1 (c) i (d) $-i$

32. Let f be a linear fractional transformation that maps $-1, 0$ and 1 to $-i$ and 1 to i, respectively. Then, $f(1-i)$ equals
(a) $-1+2i$ (b) $-2+i$ (c) $2i$ (d) $-1+i$

33. The image of $\{z \in \mathbb{C} : \text{Re } z > 1/2\}$ under the map $w = z^{-1}$ is:
(a) $\{w \in \mathbb{C} : \text{Re } w > 1/2\}$ (b) $\{w \in \mathbb{C} : |w-1| > 1\}$
(c) $\{w \in \mathbb{C} : |w-1| < 1/2\}$ (d) $\{w \in \mathbb{C} : |w-1| < 1\}$

34. Let G_1 and G_2 be the images of the disc $\{z \in \mathbb{C} : |z+1| < 1\}$ under the transformation $w = \frac{(1-i)z+2}{(1+i)z+2}$ and $w = \frac{(1+i)z+2}{(1-i)z+2}$ respectively. Then
(a) $G_1 = \{w \in \mathbb{C} : \text{Im}(w) < 0\}$ and $G_2 = \{w \in \mathbb{C} : \text{Im}(w) > 0\}$
(b) $G_1 = \{w \in \mathbb{C} : \text{Im}(w) > 0\}$ and $G_2 = \{w \in \mathbb{C} : \text{Im}(w) < 0\}$
(c) $G_1 = \{w \in \mathbb{C} : |w| > 2\}$ and $G_2 = \{w \in \mathbb{C} : |w| < 2\}$
(d) $G_1 = \{w \in \mathbb{C} : |w| < 2\}$ and $G_2 = \{w \in \mathbb{C} : |w| > 2\}$

35. Let $f = u + iv$ and $g = v + iu$ be non-zero analytic functions on $|z| < 1$. Then, it follows that
(a) $f' = 0$ (b) f is conformal on $|z| < 1$
(c) $f = kg$ for some real k (d) f is one-to-one

36. Let $w(z) = (az+b)/(cz+d)$ and $f(z) = (\alpha z + \beta)/(\gamma z + \delta)$ be Möbius transformation. Then the following is also a linear fractional transformation:
(a) $f(z)w(z)$ (b) $f(w(z))$ (c) $f(z) + w(z)$ (d) $f(z) + \frac{1}{w(z)}$

37. The fixed points of $f(z) = (2iz+5)/(z-2i)$ are
(a) $1 \pm i$ (b) $1 \pm 2i$ (c) $2i \pm 1$ (d) $i \pm 1$

38. The transformation $w = e^{i\theta} \frac{z-\rho}{\bar{\rho}\, z-1}$, where ρ is a constant, maps $|z| < 1$ onto
(a) $|w| < 1$, if $|\rho| < 1$ (b) $|w| > 1$, if $|\rho| > 1$
(c) $|w| = 1$, if $|\rho| = 1$ (d) $|w| = 3$, if $|\rho| = 1$

39. The cross ratio $\langle z_1, z_2, z_3, z_4 \rangle$ of four points in the complex plane is real if and only if the four points are:
(a) The corners of a square (b) On a circle
(c) On a straight line (d) On a circle or on a straight line

40. The value of the cross ration $\langle 7+i, 1, 0, \infty\rangle$ is:
(a) $6+i$ (b) $-6+i$ (c) $6-i$ (d) $-6-i$

41. The horizontal strip in the z-plane bounded by the rays $y=0, x \leq 0$ and $y=\pi, x \leq 0$ is mapped by $w-e^z$ in the w-plane into:
(a) A vertical strip (b) A horizontal strip (c) An oblique strip (d) A semi-circular disc

42. The image of $\{z : \text{Re } z < 0 : |\text{Im } z| < r\}$ under the exponential function is:
(a) interior of unit circle $-\{z : \text{Re } z = 0\}$ (b) Lower half plane $-\{z : \text{Im } z = 0\}$
(c) Interior of unit circle $-\{z : \text{Re } z \leq 0 \;\&\; \text{Im } z = 0\}\}$ (d) Upper half plane $-\{z : \text{Im } z = 0\}$

43. The function $\sin z$ maps:
(a) Horizontal lines other than x-axis to parabolas
(b) Horizontal lines other than x-axis to ellipes
(c) Vertical lines other than y-axis to parabolas
(d) Vertical lines other than y-axis to ellipes

44. The image of the line $y=\pi/2$ under the transformation $w=\cosh z$ is
(a) u-axis (b) Segment of u-axis (c) v-axis (d) Segment of v-axis

45. The image of the line $y=-\pi/2$ under the transformation $w=\cosh z$ is
(a) u-axis (b) Segment of u-axis (c) v-axis (d) Segment of v-axis

46. The transformation $w=\cosh z$ maps the semi-infinite strip $x \geq 0, 0 \leq y \leq \pi/2$ in the z-plane on
(a) $u \geq 0, v \geq 0$ (b) $u \leq 0, v \geq 0$ (c) $u \leq 0, v \leq 0$ (d) $u \geq 0, v \leq 0$

47. The transformation $w=\cosh z$ maps the segment $z=iy(0 \leq y \leq \pi/2)$ onto the segment
(a) $-1 \leq u \leq 0$ (b) $0 \leq u \leq 1$ (c) $-1 \leq v \leq 0$ (d) $0 \leq v \leq 1$.

48. Let $H=\{z \in \mathbb{C} : \text{Im } z > 0\}$ be the upper half plane and $\triangle=\{z \in \mathbb{C} : |z| < 1\}$ be the open unit disc. Suppose that f is a Möbius transformation, which maps H conformally onto D. Suppose that $f(2i)=0$. Pick the correct statement from below:
(a) f has simple pole at $z=-2i$ (b) f satisfies $f(i)\,\overline{f(-i)}=1$
(c) f has an essential singularity at $z=-2i$ (d) $|f(2+2i)|=\frac{1}{\sqrt{5}}$

Answers

1(d); 2(c); 3(c); 4(c); 5(d); 6(d); 7(b); 8(d); 9(c); 10(b); 11(d); 12(d); 13(a); 14(b); 15(a); 16(a); 17(a); 18(d); 19(b); 20(a); 21(c); 22(d); 23(b); 24(b); 25(d); 26(d); 27(b); 28(b); 29(a); 30(d); 31(b); 32(b); 33(d); 34(b); 35(a); 36(b); 37(c); 38(a); 39(b); 40(d); 41(c); 42(b); 43(c); 44(c); 45(c); 46(a); 47(b); 48(c)

Review Questions [Short-answer Type]

1. Show that the linear transformation is conformal at each point $z \in \mathbb{C}$.
2. Show that if $a \neq 0$, then $w=az^2+bz+c$ is conformal at each point $z_0 \in \mathbb{C}$ such that $z_0 \neq -b/2a$.
3. Argue that a one-to-one analytic function will map simply-connected domains to simply-connected domains.
4. Explain why the complex plane and the open unit disc cannot be conformally equivalent.
5. What happens to angles at the origin under the mapping $f(z)=z^\alpha$ for $\alpha > 1$? For $0 < \alpha < 1$?
6. Find all functions $f(z)$ analytic in the unit disc $S(0,1)$ that assume only pure imaginary values in $S(0,1)$.

7. What is the image, under the mapping $w = (z+i)/(z-i)$, of the third quadrant?
8. Let $P(z) = (z-\alpha)(z-\beta)$, and let L be any straightline through $(\alpha+\beta)/2$. Prove that P is one-to-one on each of the half-planes determined by L.
9. Explain why the complex plane and the open unit disc cannot be conformally equivalent.
10. If $f(z)$ is analytic at z_0 and $f'(z_0) \neq 0$, show that the function $F(z) = \overline{f(z)}$ preserves the magnitude but reverses the sense of the angle at z_0.
11. Find a linear transformation mapping the circle $|z| = 1$ onto the circle $|w-5| = 3$, taking the point $z = i$ to $w = 2$.
12. What is the image of the strip $0 < \operatorname{Im} z < 1$ under the mapping $w = (z-i)/z$?
13. Discuss the image of the circle $|z-2| = 1$ and its interior under the following transformations:
(a) $w = z - 2i$ (b) $w = 3iz$ (c) $w = \frac{z-2}{z-1}$ (d) $w = \frac{z-4}{z-3}$ (e) $w = z^{-1}$
14. Show that as z moves on the real axis from -1 to $+1$, the point $w = \frac{1-iz}{z-i}$ moves on part of the unit circle with centre at origin.
15. Find a conformal map of the unit disc $|z| < 1$ onto itself, taking $z = i/2$ to the origin.
16. Let $w = f(z)$ be the Möbius transformation mapping the points $0, \lambda, \infty$ to $-i, 1, i$, respectively, where λ is real. For what values of λ is the upper half-plane mapped onto $|w| < 1$?
17. Show that if $a \neq 0$, then $w = az^3 + bz^2 + cz + d$ is not conformal at some point.
18. Construct a linear transformation which
(a) Carries i onto $-i$ and maps $1+2i$ onto itself. (b) Maps $|z-1| = 1$ onto $|w-3i| = 2$.
19. Determine the angle of rotation, coefficient of magnification and the scale factor for the transformations:

(i) $w = 1/z; z = 1$ (ii) $w = 1/z; z = i$ (iii) $w = z^2; z = 2+i$
(iv) $w = z^2; z = 1+i$ (v) $w = z^3 + 8iz; z = 1-i$ (vi) $w = z^3; z = 1+i$
(vii) $w = (1-i)z; z$

20. Find the points where the functions are conformal. Also, find the critical points, if any:

(i) $w = z^n, n \in \mathbb{N}$ (ii) $w = 1/z$ (iii) $w = z + 1/z$ (iv) $w = e^z$
(v) $w = \sin z$ (vi) $w = \cosh z$ (vii) $w = \sinh z$ $(viii)$ $w = \cos z$

21. Find the image of the square region with vertices $(0,0), (2,0), (2,2), (0,2)$ under the transformation $w = (1+i)z + (2+i)$.
22. Find the image of the line $y = x + 1$ under the linear fractional transformation $w = \frac{z+i}{z-1}$. Find the center and the of the image circle.
23. Find the image of the line $x = 2$ under the linear fractional transformation $w = \frac{2z-i}{z+i}$.
24. Show that every involutory linear transformation is elliptic.
25. Show that the line $3y = x$ is mapped onto the circle under the linear fractional transformation $w = (iz+2)/(4z+i)$. Find the centre and radius of the image circle.
26. Find all the linear fractional transformations whose fixed points are -1 and 1.
27. Determine a, b, c, d such that the Möbius transformation coincides with its inverse.
28. For $f(z)$ as in Eq. (6.24), show that $f(\mathbb{R}_\infty) = \mathbb{R}_\infty$ if and only if we can choose a, b, c and d to be real numbers.

29. Find a Möbius transformation mapping, the
    (a) Lower half-plane to the disc $|w+1|<1$
    (b) Unit disc $|z|<1$ onto the right half plane and taking $z=-i$ to the origin
    (c) The interior of the unit disc $|z|<1$ to the exterior takes the form (6.46) with $|a|>1$
30. Find all the abelian subgroups of the group of Möbius transformations.
31. Prove each of the following statements
    (a) The linear fractional transformation $w=(az+b)/(cz+d)$ is conformal in $\mathbb{C}-\{-d/c\}$.
    (b) The identity mapping $w=z$ is the only linear fractional mapping with more than two fixed points.
    (c) Every linear fractional transformation has at least one fixed point in $\mathbb{C}_\infty$.
    (d) Every linear fractional transformation having ∞ as the only fixed point is a translation.
    (e) Any linear fractional transformation which has zero and ∞ as fixed points of the form $w=az$.
    (f) The linear fractional transformation which has zero and ∞ as fixed points and which maps $1+i$ onto $2+3i$.
    (g) A linear fractional transformation preserves inverse points.
32. Show that a Möbius transformation f satisfies $f(0)=\infty$ and $f(\infty)=0$ if and only if $f(z)=\alpha z^{-1}$ for some $\alpha\in\mathbb{C}$.
33. Let f be a Möbius transformation, $f\neq$ the identity. Show that a Möbius transformation g commutes with f if g and f have the same fixed points.
34. Evaluate the following cross ratios:
    (a) $\langle 7+i,1,0,\infty\rangle$; (b) $\langle 2,1-i,1,1+i\rangle$; (c) $\langle 0,1,i,-1\rangle$; (d) $\langle i-1,\infty,1+i,0\rangle$.
35. Find the fixed points of a dilation, a translation and an inversion on $\mathbb{C}_\infty$.
36. Discuss the mapping properties of $(1-z)^i$.
37. Discuss the application of the transformation $w=z^2$ to the region in the first quadrant of the z-plane bounded by the axes and the circle $|z|=a$, $|z|=b$ (where $a>b>0$).
38. Find the image/region of
    (a) The circle $|z-3|=5$ under the transformation $w=1/z$.
    (b) the half plane $y>0$ under the transformation $w=(1+i)z$.
    (c) of the circle $|z|=2$ under the transformation $w=z+3+2i$.
39. Describe geometrically the transformation: (a) $w=1/z-1$ (b) $w=i/z$.
40. Find the region into which the circular sector $0<\theta<\pi/4, r<1$ is mapped by the transformation $w=z^2$.
41. Find the image of the circle with centre origin and radius r under $w=z^2$.
42. Find the image of the straight line $y=mx+c$ under the transformation $w=e^z$.
43. Find the image of the strip $0\le y\le\pi$, $-\infty<x\le\infty$ on the negative side of the x-axis under $w=e^z$.
44. Find the image of the following lines under the transformation $w=\cosh z$: (i) $y=0$ (ii) $y=\pi/2$ (iii) $y=-\pi/2$ (iv) $y=\pi$ (v) $x=0$
45. Show that the transformation $w=\cosh z$ maps the semi-infinite strip $x\ge 0, 0\le y\le\pi/2$ in the z-plane on the first quadrant $u\ge 0, v\ge 0$ in the w-plane.

46. Show that the transformation $w = \cosh z$ maps the segment $z = iy(0 \le y \le \pi/2)$ onto the segment $0 \le u \le 1$ of the u-axis.

47. Show that the image of the region $x \le 0$ and $0 \le y \le \pi$ under the transformation $w = e^z$ is interior and the boundary of unit semicircle above the real axis (except at origin)

48. Find the image of the rectangle bounded by the lines $x = 0, y = 0, x = 1$ and $y = \pi$ under the transformation $w = e^z$.

49. Sketch the region of the w-plane into which the interior of the triangle T in the z-plane with vertices at $i, 1-i, 1+i$ is mapped under the transformation $w = z + 1/z$.

Answers

18. (a) $w = (2+i)z + (1-3i)$; (b) $w = 2z - 2(z-3i)$
19. (i) π; (ii) 0; (iii) $2\sqrt{5}$; (iv) $2\sqrt{2}$; (v) $\sqrt{68}$; (vi) $6, \pi/2$; (vii) $\sqrt{2}, -\pi/4$
20. (i) $z \ne 0, 0$; (ii) $z \ne 0$, no; (iii) $z \ne 0, 1, -1$; (iv) entire plane, no; (v) $\pi/2 + n\pi, n \in \mathbb{Z}$, all; (vi) $n\pi i, n \in \mathbb{Z}$, all; (vii) $(2n-1)/2\pi i, n \in \mathbb{Z}$, all; (viii) $n\pi, n \in \mathbb{Z}$, all
22. $(1, -1/2), 1/2$; 23. Circle; 26. $w = (az+b)/(bz+a), a^2 - b^2 \ne 0$
31. (f) $2w = (5+i)z$; 38. (a) $|w + 3/16| = 5/16$; (b) $v > u$; (c) $(u-3)^2 + (v-i)^2 = 4$
40. $0 < \theta < \frac{\pi}{2}, r < 1$ 43. Unit semicircle $\rho = 1, 0 \le \phi < \pi$
44. (i) The segment of the u-axis $u \ge 1$; (ii) The v-axis.; (iii) The v-axis; (iv) The segment of the u-axis $u \le -1$; (v) The segment of the u-axis $-1 \le u \le 1$

Review Questions [Long-answer Type]

1. Prove the *Noshiro-Warschawski theorem* : Let f be analytic in a convex domain $\mathcal{D}$. If $\text{Re} f'(z) > 0$ for all $z \in \mathcal{D}$, then f is one-to-one in $\mathcal{D}$. [**Hint:** Write $f(z_2) - f(z_1)$ as an integral of f'.]

2. Determine points where each of the following mappings is conformal:

   (a) $z + kz^2$ (b) $z + kz^3$ (c) $z^{e^{z^3}} + 1$ (d) $e^{-z} + z - 5$

3. Determine points where each of the following mappings fails to be conformal:

   (a) $z^5 + 1$ (b) $z^2 - e^{\pi z^2}$ (c) $\sinh z$ (d) $\cosh z$

4. Prove that the quadrant $|z| < 1, 0 < \arg z < \pi/2$ is mapped conformally on the domain $\mathcal{D}$ in the w-plane by the transformation $w = f(z) = 4/(z+1)^2$. Find $\mathcal{D}$ and determine the length of the boundary.

5. (a) What is the region of the w-plane onto which the rectangular region in the z-plane bounded by the lines $x = 1, y = 0, x = 3$ and $y = 1$ is mapped under the transformation $w = z + (1+2i)$?

   (b) Determine the region in the w-plane onto which the square region bounded by the lines $x = 0, y = 0, x = 1$ and $y = 1$ in the z-plane is mapped under the transformation $w = \sqrt{2}e^{i\pi/4}\, z$.

6. Let f be a Möbius transformation with fixed points z_1 and z_2. If g is a Möbius transformation, show that $g^{-1}fg$ has fixed points $g^{-1}z_1$ and $g^{-1}z_2$.

7. Show that the rational mapping $w = f(z) = P(z)/Q(z)$, where $P(z)$ and $Q(z)$ are polynomials, is conformal at any simple zero of $Q(z)$ and also at $z = \infty$ if the equation $f(z) = f(\infty)$ has no multiple roots.

8. Let $f_1(z) = \frac{z}{2z-1}$, $f_2(z) = \frac{2z}{3z-1}$ and $f_3(z) = \frac{(2+i)z-2}{z+i}$. Show that 0 and 1 are the fixed points for f_1 as well as f_2, whereas $1 \pm i$ are fixed points of f_3, and f_3 is loxodromic.

9. Prove that the Möbius transformation taking z_k to w_k $(k = 1, 2, 3)$ can be expressed in the determinant form:

$$\begin{vmatrix} 1 & z & w & zw \\ 1 & z_1 & w_1 & z_1w_1 \\ 1 & z_2 & w_2 & z_2w_2 \\ 1 & z_3 & w_3 & z_3w_3 \end{vmatrix} = 0.$$

10. Under the transformation $w = iz + i$, show that the half plane $x > 0$ maps onto the half plane $v > 1$.

11. Show that the region in the z-plane given by $x > 0$ and $0 < y < 2$ is mapped into the region in the w-plane given by $-1 < u < 1$ and $v > 0$ under the transformation $w = iz + 1$.

12. For $f(z)$ as in Eq. (6.24), find the necessary and sufficient conditions that $f(\partial\mathcal{S}(0, 1)) = \partial\mathcal{S}(0, 1)$, where $\partial\mathcal{S}(0, 1)$ is the boundary of the unit circle $\partial\mathcal{S}(0, 1) = \{z \in \mathbb{C} : |z| = 1\}$.

13. Find the linear fractional transformation which maps the points

    (a) $(1, 0, -1)$ in z-plane into the points $(i, 0, -i)$ in w-plane

    (b) $(0, i, -i)$ in z-plane into $(1, -1, 0)$ in w-plane

    (c) $(2, 1, 0) \longrightarrow (1, 0, i)$ (d) $(0, -i, -1) \longrightarrow (i, 1, 0)$

14. Find the linear fractional transformation which maps the vertices $1 + i, -i, 2 - i$ of a triangle T of the z-plane into the points $0, 1, i$ of the w-plane. Sketch the region into which the interior of the triangle T is mapped under the transformation.

15. If z_0 is the upper half of the z-plane, show that the linear fractional transformation $w = e^{i\theta_0}(\frac{z-z_0}{z-\bar{z}_0})$ maps the upper half of the z-plane into the set $\{w : |w| \leq 1\}$.

16. Express each of the following linear fractional transformations as product of elementary transformations:
    (i) $w = \frac{z-1}{z+1}$ (ii) $w = \frac{3z+2i}{iz+6}$ (iii) $w = \frac{1+i-3z}{2-i+iz}$

17. Find the fixed points and the normal form of the following linear fractional transformations. Hence, determine the nature of the transformations:

(i)	$w = z + 3$	(ii)	$w = 2z + 3$	(iii)	$w = \frac{z-1}{z+1}$	(iv)	$w = \frac{z}{2-z}$
(v)	$w = \frac{3z-4}{z-1}$	$vi)$	$w = \frac{3iz+1}{z+i}$	(vii)	$w = \frac{(2+i)z-2}{z+i}$		
$(viii)$	$w = \frac{z-3}{z+1}$	(ix)	$w = \frac{3z+2i}{iz+6}$	(x)	$w = \frac{1+i-3z}{2-i+iz}$		

    Express each of the above transformations as a product of elementary transformations.

18. Using the inverse property of the cross-ratio, find a Möbius transformation f in each of the following cases:
    (a) $\{1, i, -1\}$ onto $\{2i, -2, -2i\}$ (b) $\{\infty, i, 0\}$ onto $\{1, i, -1\}$
    (c) $\{0, 1, -1\}$ onto $\{0, 1, \infty\}$ (d) $\{0, 4, 2 - 2i\}$ onto $\{-\frac{3}{4}, \infty, \frac{11i}{4}\}$

19. Show that the linear fractional transformation which maps the unit circle $|z| = 1$ onto the unit circle $|w| = 1$ can be put into the form $w = e^{i\lambda}(\frac{az+b}{\bar{b}z+\bar{a}})$; where λ is real. Further this transformation maps the circular disc $|z| \leq 1$ onto the circular disc $|w| \leq 1$ if and only if $|a| > |b|$. Also, if the point $z = 1$ is the only invariant point show that the transformation may be written as $\frac{1}{w-1} = \frac{1}{z-1} + \frac{1}{k}$, where $k = 1 + \frac{\bar{a}}{b}$.

20. Using the cross-ratio notation, write an equation defining a Möbius transformation that maps the half-plane below the line $y = 2x - 3$ onto the interior of the circle $|w - 4| = 2$. Repeat for the exterior of this circle.

21. Does there exist a Möbius transformation f that maps the real axis onto the unit circle $|w| = 1$ and satisfies $f(i) = 2$, $f(-i) = -1/2$?

22. Let f be a Möbius transformation such that $f(1) = \infty$ and f maps the imaginary axis onto the unit circle $|w| = 1$. What is the value of $f(-1)$?

23. Prove that if z_2, z_3 and z_4 are distinct points in the extended complex plane and f is any Möbius transformation, then $\langle z_1, z_2, z_3, z_4\rangle = \langle f(z_1), f(z_2), f(z_3), f(z_4)\rangle$ for any point z_1 in the extended plane.

24. For $|z| < 1$, define $g(z)$ by $g(z) = \exp[-i \log\{i(\frac{1+z}{1-z})\}^{1/2}]$.

    (a) Show that g maps $S(0, 1) = \{z : |z| < 1\}$ conformally onto an annulus G.

    (b) Find all Möbius transformation $f(z)$ that map $S(0, 1)$ onto $S(0, 1)$ and such that $g(f(z)) = g(z)$ when $|z| < 1$.

25. Let z_1, z_2 and z_3 be three distinct points that lie on the circle (or line) C. Prove that z and z^* are symmetric with respect to C if and only if $\langle z^*, z_1, z_2, z_3\rangle = \overline{\langle z, z_1, z_2, z_3\rangle}$

26. Discuss the mapping properties of:
    (a) z^n (b) $z^{1/n}; n \geq 2$ (c) $\cos z$ (d) $\sin z$ (e) $\frac{1}{2}(z + \frac{1}{z})$

27. Find the image of the semi-infinite strip $1 < x < 2, y > 0$ under the transformation $w = 2iz$ by

    (a) finding its image under a rotation and magnification,

    (b) writing $w = u + iv = 2ix - 2y$ and showing that $u < 0, 2 < v < 4$ are the restrictions on u and v.

28. Find all Möbius transformations that map the upper half-plane onto itself.

29. Show that in two ways the image of the set $1 < x < 2, y > 0$ under $w = iz + 3i$ are $u < 0, 4 < v < 6$.

30. Under the transformation $w = 1/z$, what may be said about the image of a
    (a) Line through the origin (b) Line not through the origin
    (c) Circle through the origin (d) Circle not through the origin

31. Let c be a positive constant. Under the transformation $w = f(z) = 1/z$, show that the image of the

    (a) Extended line determined by $y = c$ is $u^2 + (v + \frac{1}{2c})^2 = \frac{1}{4c^2}$;
    (b) Extended x-axis is the extended line $v = 0$
    (c) Strip $0 < y < c$ is $v < 0$ and the disc $u^2 + (v + \frac{1}{2c})^2 > \frac{1}{4c^2}$
    (d) Half-plane $y > c$ is the disc $u^2 + (v + \frac{1}{2c})^2 < \frac{1}{4c^2}$
    (e) Half-plane $y < 0$ is the half plane $\{(u, v) : v > 0\}$
    (f) Infinite strip $0 < y < \frac{1}{2c}$ is $u^2 + (v + c)^2c^2, v < 0$

32. Sketch the region of the w-plane into which the interior of the triangle T in the z-plane with vertices at $i, 1 - i, 1 + i$ is mapped under the transformation (a) $w = z^2$ and (b) $w = iz^2 + (2 - i)z$.

33. (a) Show that the image of the quadrant $y > 0, x > c$ (where $c > 0$) under $w = 1/z$ is the half-disc $\{(u, v) : (u - \frac{1}{2c})^2 + v^2 < \frac{1}{4c^2}$ and $v < 0\}$. (b) Show that in the z-plane is mapped onto the hyperbola $u^2 - v^2 = 1$ under $w = 1/z$ is the Lemiriscate of Bernoulli, $r^2 = \cos 2\theta$.

34. Show that the image of the circular sector $0 < \theta < \pi/3, 0 < r < 2$, under the transformation $w = z^2 = r^2 e^{zi\theta}$ is

$$0 < \phi < \frac{2\pi}{3}, 0 < \rho < 4; \text{ where } w = \rho e^{i\phi}.$$

35. If $w = \tan^2 z/2$, show that the strip in the z-plane between $x = 0$ and $x = \pi/2$ is represented on the interior of the unit circle in the w-plane with a cut along the real axis from $w = -1$ to $w = 0$.

36. Show that $w = z^2$ maps the line $x = c$ (where $c \neq 0$) onto the parabola $u = c^2 - (v^2/c^2)$.

37. Let $w = f(z) = \omega^2$. Show that the set of all points z such that $f(z)$ is in the rectangular region $2 < u < 4, 2 < v < 4$ is $\{(x, y) : 2 < x^2 - y^2 < 4 \text{ and } 1 < xy < 2\}$.

38. (a) Describe geometrically the transformation $w = 1/(z - 1)$.
(b) Describe geometrically the transformation $w = i/z$. State why it transforms circles and lines into circles and lines.

39. Show that by means of the inversion $w = \frac{1}{z}$ the circle $|z - 2| = 7$ is mapped into the circle $|w + \frac{2}{45}| = \frac{7}{45}$.

40. Show that under the transformation $w = 1/z$, the image of
    (a) the circle $|z| = 1$ in z-plane is the straight line $2u = 1$ in w-plane;
    (b) the circle $|z - 2i| = 2$ is $4v + 1 = 0$;
    (c) the strip $1 < x < 2$ is the region between the circles $u^2 + v^2 - u = 0$ and $u^2 + v^2 - \frac{u}{2} = 0$.

41. Discuss the transformation of the region in the first quadrant of the z-plane bounded by the axes and the circles $|z| = a$ and $|z| = b$ (where $a > b > 0$) under the transformation $w = z^2$.

42. Determine the region of the w-plane into which the triangle formed by $x = 1, y = 1$ and $x + y = 1$ is mapped under the transformation $w = z^2$.

43. Find the region in the w-plane which correspond to the interior of the triangle in the z-plane formed by the lines $y = 0, x = 1$ and $y = x$ under the transformation $w = z^2$.

44. Find the preimage in the z-plane of the square region in the w-plane bounded by the lines $u = 1, u = 2, v = 1$ and $v = 2$ under the transformation $w = z^2$.

45. Show that the transformation $w = z^2$ transforms the circle $|z - a| = c$, where $a \neq c$, in the z-plane to the limacon in the w-plane of the form $\rho = 2c(a + c\cos\phi)$.

46. If $w = 2z + z^2$, prove that the circles $|z| = 1$ corresponds to a cardioid in the w-plane.

47. Show that the transformation $w = \sin^2 z$ maps the region $0 \leq x \leq \pi/2, y \geq 0$ onto the region $v \geq 0$ and indicate the corresponding parts of the boundaries.

48. Prove that the semi-infinite strip $x < 0, 0 \leq y \leq \pi$ is mapped onto upper semicircular disc with centre 0 and radius 1 under the transformation $w = \sin z$.

49. Describe the image of each of the following domains under the mapping $w = e^z$:
(a) The strip $0 < \text{Im } z < \pi$; (b) the slanted strip between the two lines $y = x$ and $y = x + 2\pi$; (c) the half-strip $\text{Re } z < 0, 0 < \text{Im } z < \pi$; (d) the half-strip $\text{Re } z > 0, 0 < \text{Im } z < \pi$; (e) the rectangle $1 < \text{Re } z < 2, 0 < \text{Im } z < \pi$; (f) the half-planes $\text{Re } z > 0$ and $\text{Re } z < 0$.

50. Sketch each of the following figures and its image under the map $w = e^z$:
(a) The vertical strip $0 < \text{Re } z < 1$ (b) The horizontal strip $\frac{5\pi}{3} < \text{Im } z < \frac{8\pi}{3}$
(c) The rectangle $0 < x < 1, 0 < y < \pi/4$ (d) The disc $|z| \leq \pi/2$
(e) The disc $|z| \leq \pi$ (f) The disc $|z| \leq 3\pi/2$

51. Sketch the image under the map $w = \log z$ of each of the following figures:
 (a) The right half-plane $\text{Re}\, z > 0$ (b) The half-disc $|z| < 1$, $\text{Re}\, z > 0$
 (c) The unit disc $|z| = 1$ (d) The slit annulus $\sqrt{e} < |z| < e^2$; $z \in (-e^2, -\sqrt{e})$
 (e) The horizontal line $y = e$ (f) The vertical line $x = e$
52. Describe the image of each of the following domains under the mapping $w = \cos z$:
 (a) The half-strip $0 < \text{Re}\, z < \pi$, $\text{Im}\, z < 0$ (b) The half-strip $0 < \text{Re}\, z < \pi/2$, $\text{Im}\, z > 0$
 (c) The strip $0 < \text{Re}\, z < \pi$ (d) The rectangle $0 < \text{Re}\, z < \pi$, $-1 < \text{Im}\, z < 1$
53. Show that under the transformation $w = z + 1/z$ the upper semicircular disc with centre 0 and radius 1 is mapped onto the lower half plane.
54. Show that the transformation $w = \text{Log}\, \frac{1+z}{1-z}$ maps the circle $|z| < 1$ on the horizontal strip $-\pi/2 < v < \pi/2$. Also, show that the upper semicircle is mapped onto the line $v = \pi/2$ and the lower semicircle onto the line $v = -\pi/2$.
55. Derive the most important properties of the function $\tan z$ from its mapping properties, which may be assumed as known.
56. Considering the conformal mappings to triangles:
 (a) Show that $\int_0^z s^{-k_1}(1-s)^{-k_2} ds$, with $0 < k_1, k_2 < 1, 1 < k_1 + k_2 < 2$, maps $\mathcal{H}$ to a triangle whose vertices are the images of $0, 1$, and ∞, and with angles $\alpha_1\pi, \alpha_2\pi$, and $\alpha_3\pi$, where, $\alpha_j + k_j = 1$ and $k_1 + k_2 + k_3 = 2$.
 (b) What happens when $k_1 + k_2 = 1$?
 (c) What happens when $0 < k_1 + k_2 < 1$?
57. Find the Schwarz–Christoffel transformation that maps the that maps the upper half plane $\mathcal{H} = \{z : \text{Im}(z) \geq 0\}$ in z-plane
 (a) To the inside of the triangle with vertices at $-1, 0$ and i in the w-plane.
 (b) Onto the interior of an isosceles right triangle.
 (c) Onto the polygonal region defined by the strip $u \geq 0, -1 \leq v \leq 1$ in the w-plane (Fig. 6.44).
 (d) To the set $P = \{z = x + iy : x > 0 \text{ and } y > 0\} \cup \{z = x + iy : x \leq 0 \text{ and } y > 1\}$. The boundary of P is described by the solid edges in the picture 6.45(a).

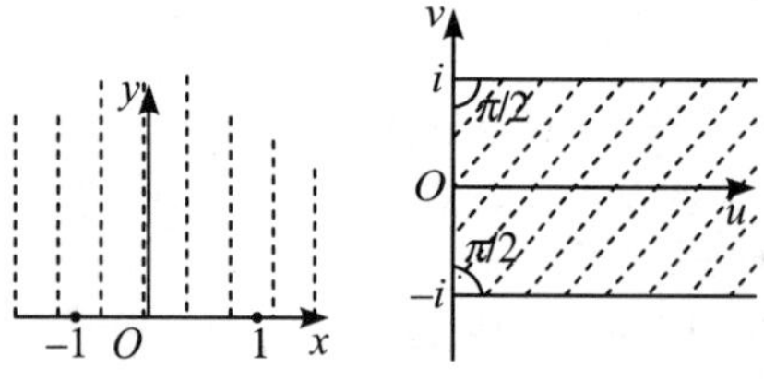

Figure 6.44

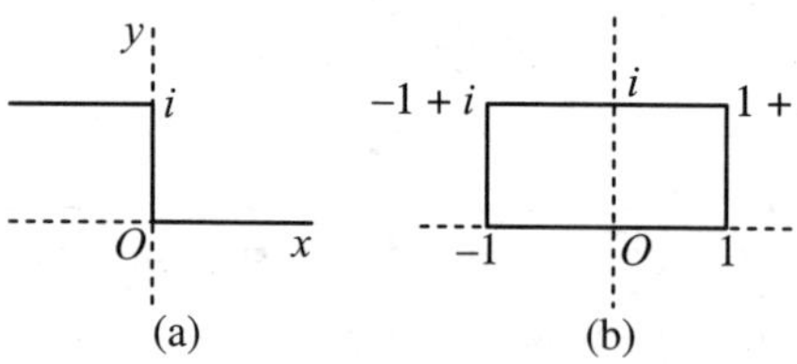

Figure 6.45

 (e) To the inside of the rectangle with vertices at ± 1 and $\pm 1 + i$. The boundary of the rectangle is described by the solid edges in the picture 6.45(b).
 (f) To the domain $P = \mathbb{C}/\{z = x \pm i : x \leq 0\}$ The boundary of the set P is described by the solid edges in the picture 6.46(b) when the point w_2 is taken to infinity along the negative real axis.
 (g) To the domain $P = \mathcal{H}/\{z = yi : y \leq 1\}$. The boundary of the set P is described by the solid edges in Fig. 6.46(a).

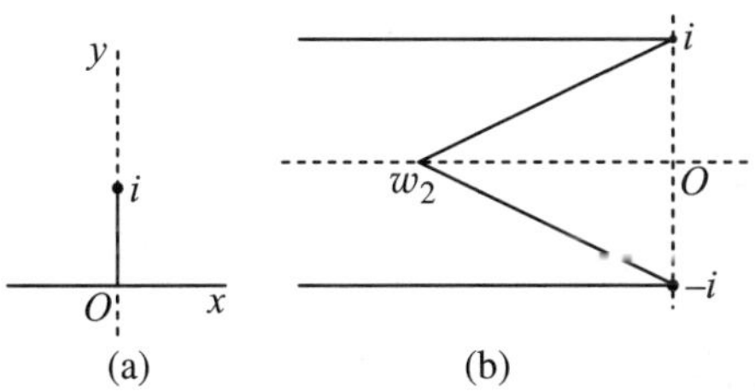

Figure 6.46

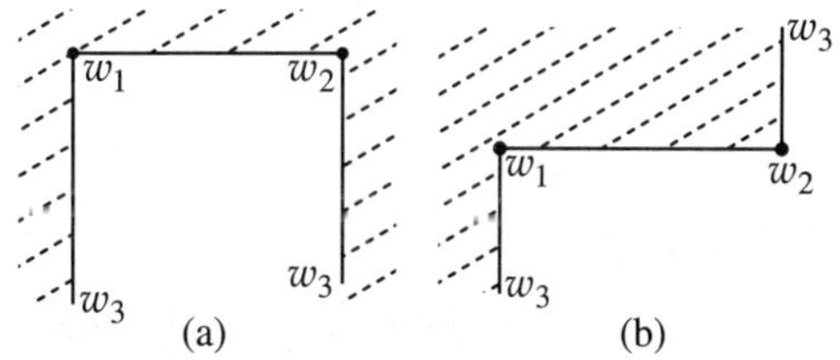

Figure 6.47

58. Find the transformations that map the upper half-plane Im ≥ 0 onto the domains in the w-plane as shown in Figure 41.5 with the following correspondences of the points:

    (a) $z(x_1 = 0, x_2 = 1, x_3 = \infty) \to w(w_1 = 0, w_2 = 1, w_3 = \infty)$ (Fig. 6.47 (a))
    (b) $z(x_1 = 0, x_2 = 1, x_3 = \infty) \longrightarrow w(w_1 = 0, w_2 = 1, w_3 = \infty)$ (Fig. 6.47 (b))
    (c) $z(x_1 = 0, x_2 = \infty, x_3 = \infty) \longrightarrow w(w_1 = 0, w_2 = 1, w_3 = \infty)$ (Fig. 6.48 (a))
    (d) $z(x_1 = 0, x_2 = \infty, x_3 = \infty) \longrightarrow w(w_1 = 0, w_2 = 1, w_3 = \infty)$ (Fig. 6.48 (b))

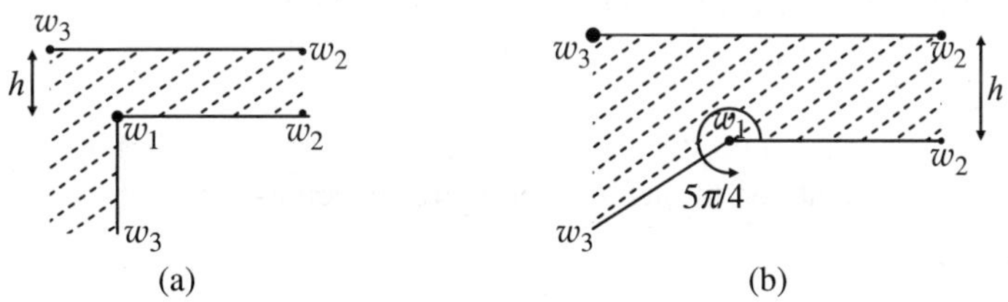

Figure 6.48

59. For each of the sets A below, find a Schwarz–Christoffel transformation that maps the upper half plane $\mathcal{H}$ onto the set A:

    (a) A is an open triangular region with vertices ± 1 and $i\sqrt{3}$.
    (b) A is the region above the polygonal path $\{z = x + i : x \leq 0\} \cup \{z = x + (1 - x)i : 0 \leq x \leq 1\} \cup \{z = x : x \geq 1\}$
    (c) $A = \{z = x + iy : y > 0 \text{ or } |x| < 1\}$.

Answers

13. (a) $w = iz$; (b) $w = (z + i)/(-3z + i)$; (c) $w = 2i(z - 1)/\{(1 + i)z - 2\}$
    (d) $w = -i(z + 1)/(z - 1)$
42. The region bounded by the parabolas $v^2 = 4(1 \pm u)$, $u^2 = 1 - 2v$.
57. (a) $w = \sqrt{2}\frac{1+i}{1-i}\int_0^z (s^2 - 1)^{-3/4} ds + \frac{1-i}{2}$.;
    (b) $f(z) = \int_0^z \frac{dt}{(t^2-1)^{3/4}}$; $-\frac{\pi}{2} < \theta_j < \frac{3\pi}{2}$. The vertices of the triangle are given by $w_1 = \frac{1+i}{\sqrt{2}}\int_0^1 \frac{dx}{(1-x^2)^{3/4}}$; $w_2 = -w_1$ and $w_3 = w_2 + b = w_1 - ib$; where $0 < b = \int_1^\infty \frac{dx}{(x^2-1)^{3/4}}$.
    (c) $w = \sin(\pi z/2i)$. (d) $f(z) = \frac{1}{\pi}[\sqrt{z^2 - 1} + \log\{z + \sqrt{z^2 - 1}\}]$
    (e) $f(z) = -\frac{2}{\pi}(\frac{z^2}{2} - \log z) + (\frac{1}{\pi} - i)$ (f) $f(z) = \sqrt{z^2 - 1}$
58. (a) $\frac{2}{\pi}[\sin^{-1}\sqrt{z} - (1 - 2z)\sqrt{z - z^2}]$; (b) $\frac{2}{\pi}[\sin^{-1}\sqrt{z} - \sqrt{z - z^2}]$;
    (c) $w = \frac{2h}{\pi}[\tanh^{-1}\sqrt{z} - \sqrt{z}]$; (d) $w = \frac{2h}{\pi}[\tanh^{-1}\sqrt[4]{z} - 2\sqrt[4]{z}]$

7 Calculus of Residues

In this chapter, we introduce the concept of residue at an isolated singular point and also at the point of infinity of a complex function $f(z)$ and methods of evaluating them. We shall use Laurent's expansion to establish Cauchy's residue theorem, which has far-reaching applications.

The Argument Principle is described in the second section. This result is then used to establish Rouché's theorem, which provides locations of the zeros and poles of meromorphic functions. We shall also prove the following five important applications of the argument principle: (a) Fundamental theorem of algebra, (b) Local mapping theorem, (c) Hurwitz's theorem, (d) Open mapping theorem, and (e) Inverse function theorem.

In the next section we shall introduce infinite products of complex numbers and functions and provide necessary and sufficient conditions for their convergence.

7.1 Residues

Before defining a residue, we consider the following lemma:

Lemma 7.1.1. Let f be analytic on and inside a positively-oriented simple closed curve γ except at a point z_0 inside γ, where f has a pole of order m. Then

$$\oint_{\gamma} f(z)\, dz = 2\pi i\, b_{m-1},$$

where, b_{m-1} is the coefficient of $(z - z_0)^{-1}$ in Eq. (5.48), in the singular part of the expression of f as powers of $(z - z_0)$ in the neighbourhood of z_0.

Definition 7.1.1. Suppose that f is analytic in $S^*(z_0, \delta) = S(z_0, \delta)/\{z_0\}$ and that f has a isolated pole at z_0. The *residue* of f at z_0 is the unique coefficient b_1 of $(z - z_0)^{-1}$ in the Laurent's series expansion (5.48) of f about z_0 and is denoted by $\text{Res}\{f; z_0\}$.

Example 7.1.1. *Determine the residue at $z = 0$ of $f(z) = z\, \cos \frac{1}{z}$.*

Solution. The Laurent's series expansion of $f(z)$ about $z = 0$ gives

$$f(z) \;=\; z\, \cos \frac{1}{z} = z - \frac{1}{2!} \cdot \frac{1}{z} + \frac{1}{4!} \cdot \frac{1}{z^3} - \cdots .$$

From this expansion we see that

$$\text{Res}\{f; 0\} = b_1 = \text{ coefficient of } z^{-1} = -1/2.$$

RESULT **59.** Let $z = z_0$ be a simple pole for $f(z)$. Then the Laurent's series expansion for $f(z)$ about $z = z_0$ is given by

$$\begin{aligned} f(z) &= \frac{b_1}{z - z_0} + a_0 + a_1(z - z_0) + a_2(z - z_0)^2 + \cdots \\ \Rightarrow \quad (z - z_0)f(z) &= b_1 + a_0(z - z_0) + a_1(z - z_0)^2 + \cdots \\ \Rightarrow \quad \lim_{z \to z_0} (z - z_0)f(z) &= b_1 = \text{Res}\{f; z_0\}. \end{aligned} \tag{7.1}$$

However, if $f(z)$ has a pole z_0 of order m, then

$$\begin{aligned} f(z) &= \frac{b_m}{(z - z_0)^m} + \frac{b_{m-1}}{(z - z_0)^{m-1}} + \cdots + \frac{b_1}{z - z_0} \\ &\quad + a_0 + a_1(z - z_0) + a_2(z - z_0)^2 + \cdots \\ \text{or,} \quad (z - z_0)^m f(z) &= b_m + b_{m-1}(z - z_0) + \cdots + b_1(z - z_0)^{m-1} \\ &\quad + a_0(z - z_0)^m + a_1(z - z_0)^{m+1} + a_2(z - z_0)^{m+2} + \cdots \\ \text{or,} \quad \frac{d^{m-1}}{dz^{m-1}}\Big[(z - z_0)^m f(z)\Big] &= (m-1)!b_1 + m(m-1)\cdots 2a_0(z - z_0) + \cdots \\ \text{or,} \quad \lim_{z \to z_0} \frac{d^{m-1}}{dz^{m-1}}\Big[(z - z_0)^m f(z)\Big] &= (m-1)!b_1 \\ \text{or,} \quad b_1 = \text{Res}\{f; z_0\} &= \frac{1}{(m-1)!} \lim_{z \to z_0} \frac{d^{m-1}}{dz^{m-1}}\Big[(z - z_0)^m f(z)\Big]. \end{aligned} \tag{7.2}$$

The relation (7.2) is the formula for calculation of residues.

Example 7.1.2. *Find the residue of* $f(z) = \frac{e^z}{z^2(z^2+9)}$ *at the poles.*

Solution. Here $z = 0$ is a double pole and $z = 3i$ and $z = -3i$ are simple poles for $f(z)$. Using, Eq.(7.2), the residue at $z = 0$ is given by

$$\begin{aligned} \text{Res}\{f; 0\} &= \frac{1}{(2-1)!} \lim_{z \to 0} \frac{d^{2-1}}{dz^{2-1}} \left[(z - 0)^2 \frac{e^z}{z^2(z^2 + 9)}\right] \\ &= \lim_{z \to 0} \frac{d}{dz}\left[\frac{e^z}{z^2 + 9}\right] = \lim_{z \to 0} \left[\frac{(z^2 + 9) - 2z}{(z^2 + 9)^2}\right] = \frac{1}{9}. \end{aligned}$$

Using Eq. (7.1), the residue at the simple poles $z = 3i$ and $z = -3i$ are given by

$$\begin{aligned} \text{Res}\{f; 3i\} &= \lim_{z \to 3i} (z - 3i) \frac{e^z}{z^2(z^2 + 9)} = \lim_{z \to 3i} \frac{e^z}{z^2(z + 3i)} = -\frac{e^{3i}}{54i} \\ \text{Res}\{f; -3i\} &= \lim_{z \to -3i} (z + 3i) \frac{e^z}{z^2(z^2 + 9)} = \lim_{z \to -3i} \frac{e^z}{z^2(z - 3i)} = -\frac{e^{-3i}}{54i}. \end{aligned}$$

RESULT **60.** Let z_0 be a simple pole for $f(z)$ and let $f(z) = \frac{g(z)}{z - z_0}$, where $g(z)$ is analytic at z_0 and $g(z_0) \neq 0$. Then, using the formula (7.1), we get

$$\text{Res}\{f; z_0\} = \lim_{z \to z_0} (z - z_0)f(z) = \lim_{z \to z_0} g(z) = g(z_0). \tag{7.3}$$

Let z_0 be a pole of order m for $f(z)$ and let $f(z) = \frac{g(z)}{(z - z_0)^m}$, where $g(z)$ is analytic at z_0 and $g(z_0) \neq 0$. Then, using the formula (7.2), we get

$$\begin{aligned}\text{Res}\{f(z); z_0\} &= \frac{1}{(m-1)!}\lim_{z\to z_0}\frac{d^{m-1}}{dz^{m-1}}\Big[(z-z_0)^m f(z)\Big]\\ &= \frac{1}{(m-1)!}\lim_{z\to z_0}\frac{d^{m-1}}{dz^{m-1}}g(z) = \frac{g^{(m-1)}(z_0)}{(m-1)!}.\end{aligned} \tag{7.4}$$

RESULT **61.** Let z_0 be a simple pole for $f(z)$. If $f(z)$ is of the form $\frac{p(z)}{q(z)}$, where $p(z)$ and $q(z)$ are analytic at z_0 and $p(z_0) \neq 0$ and $q(z_0) = 0$, then

$$\begin{aligned}\text{Res}\{f; z_0\} &= \lim_{z\to z_0}(z-z_0)f(z) = \lim_{z\to z_0}(z-z_0)\,\frac{p(z)}{q(z)}\\ &= \lim_{z\to z_0}p(z)\lim_{z\to z_0}\frac{z-z_0}{q(z)}\\ &= \lim_{z\to z_0}p(z)\lim_{z\to z_0}\frac{z-z_0}{q(z)-q(z_0)}, \quad \text{as } q(z_0)=0\\ &= p(z_0)\,\frac{1}{q'(z_0)} = \frac{p(z_0)}{q'(z_0)}.\end{aligned} \tag{7.5}$$

Eq. (7.5) cannot be applied to the function $f(z) = \tanh z/z^2$ at $z = 0$, because at $z = 0$ the numerator vanishes, i.e., $p(0) = 0$.

Example 7.1.3. *Find the residues at the poles of* $f(z) = \frac{4+z^2}{z^3+2z^2+2z}$.

Solution. Here $f(z)$ can be written in the form $\frac{p(z)}{q(z)}$, where $p(z) = z^2 + 4$ and $q(z) = z^3 + 2z^2 + 2z$, so that $q'(z) = 3z^2 + 4z + 2$. The poles of $f(z)$ are given by

$$z^3 + 2z^2 + 2z = 0 \Rightarrow z = 0, z^2 + 2z + 2 = 0 \Rightarrow z = 0, z = -1 \pm i$$

and they are simple poles. Using Eq. (7.5), the residue at the poles are given by

$$\begin{aligned}\text{Res}\{f; 0\} &= \frac{p(0)}{q'(0)} = \frac{4}{2} = 2,\\ \text{Res}\{f; -1+i\} &= \frac{p(-1+i)}{q'(-1+i)} = \frac{(i-1)^2+4}{3(i-1)^2+4(i-1)+2} = \frac{3i-1}{2},\\ \text{Res}\{f; -1-i\} &= \frac{p(-1-i)}{q'(-1-i)} = \frac{(-i-1)^2+4}{3(-i-1)^2+4(-i-1)+2} = \frac{-3i-1}{2}.\end{aligned}$$

Note: The following facts are to be noted.

(i) If z_0 is a removable singular point of $f(z)$, then there is no term in the principal part of the Laurent's series expansion of $f(z)$. Thus in this case, $b_1 = 0$ and $\oint_\Gamma f(z)\,dz = 0$.

(ii) In case z_0 is an isolated essential singular point, the residue at z_0 is the coefficient of $(z-z_0)^{-1}$ in the Laurent's series expansion. In this case, this is the only way of computing residue at z_0.

(iii) Residue is not defined for non-isolated singular points of a function.

(iv) If z_0 is an isolated singular point of $f(z)$ and $f(z)$ has even powers of $(z - z_0)$, i.e., $f(z - z_0) = f(-(z - z_0))$, then $\text{Res}\{f; z_0\} = 0$. For example, the Laurent's series expansion of $f(z) = \operatorname{cosec}^2 z$ about $z = 0$ gives

$$\begin{aligned}
f(z) &= \operatorname{cosec}^2 z = (\sin z)^{-2} = \frac{1}{z^2}\Big[1-\Big(\frac{z^2}{3!}-\frac{z^4}{5!}+\cdots\Big)\Big]^{-2} \\
&= \frac{1}{z^2}\Big[1+\Big(\frac{z^2}{3!}-\frac{z^4}{5!}+\cdots\Big)+3\Big(\frac{z^2}{3!}-\frac{z^4}{5!}+\cdots\Big)^2+\cdots\Big] \\
&= \frac{1}{z^2}\Big[1+\frac{z^2}{3}+\frac{z^4}{15}+\cdots\Big] = \frac{1}{z^2}+\frac{1}{3}+\frac{z^2}{15}+\cdots.
\end{aligned}$$

From this expansion we see that $b_1 =$ coefficient of $z^{-1} = 0$. Also, $\operatorname{cosec}^2 z$ is an even function.

(v) We know $1/z^2$ is analytic everywhere except at $z = 0$. Therefore, $f(z) = e^{1/z^2}$ is also analytic everywhere except at $z = 0$ and the isolated singular point $z = 0$ lies inside $\Gamma : |z| = 1$. Hence the Laurent's series expansion of $f(z)$ can be written as

$$f(z) = e^{1/z^2} = 1+\frac{1}{z^2}+\frac{1}{2!}\,\frac{1}{z^4}+\cdots.$$

The residue of $\oint_\Gamma e^{1/z^2}\,dz$ at $z = 0$ is the coefficient of $1/z$ in the Laurent's series expansion which is equal to 0, i.e., $\operatorname{Res}\{f;0\} = 0$. Hence

$$\oint_\Gamma e^{1/z^2}\,dz = 2\pi i \operatorname{Res}\{f(z);0\} = 0.$$

Thus the analyticity of a function within and on a simple closed contour Γ is a sufficient condition but not a necessary condition for the value of the integral around Γ to be zero.

Example 7.1.4. *If $f(z) = \frac{1}{z}\sum\limits_{k=0}^{n}(z+z_0)^{-k}$, find the residue for different values of z_0.*

Solution. For $z_0 = 0$, $\operatorname{Res}\{f(z);0\} = 1$. For $z_0 \neq 0$, we write the Laurent series expansion of $f(z)$ at the point $-z_0$:

$$\begin{aligned}
f(z) &= -\frac{1}{z_0}\Big[1-\frac{z+z_0}{z_0}\Big]^{-1}\sum_{k=0}^{n}(z+z_0)^{-k}, \quad 0<|z+z_0|<|z_0| \\
&= -\frac{1}{z_0}\Big[1+\Big(\frac{z+z_0}{z_0}\Big)+\cdots+\Big(\frac{z+z_0}{z_0}\Big)^{n-1}+\cdots\Big]\sum_{k=0}^{n}(z+z_0)^{-k}.
\end{aligned}$$

The coefficient of $1/(z+z_0)$ is $-\frac{1}{z_0}[1+\frac{1}{z_0}+\cdots+\frac{1}{z_0^{n-1}}]$ and hence

$$\operatorname{Res}\{f;-z_0\} = \begin{cases} -\frac{1-z_0^n}{z_0^n(1-z_0)}; & z_0 \neq 1 \\ -n; & z_0 = 1 \end{cases}$$

Example 7.1.5. *The function $f(z)$ has a simple pole at $z = 1$ with residue 2, a double pole at $z = 0$ with residue 2, is analytic at all other finite points of the plane and is bounded as $|z| \to \infty$. If $f(2) = 5$ and $f(-1) = 2$, find $f(z)$.*

Solution. Since $f(z)$ has a pole at $z = 1$ with residue 2, a double pole at $z = 0$ with residue 2, the principal part of $f(z)$ will be of the form $\frac{2}{z-1}+\frac{2}{z}+\frac{b}{z^2}$ and so, $f(z)$ will have a Laurent series expansion of the form

$$f(z) = \sum_{n=0}^{\infty} a_n\, z^n + \frac{2}{z-1}+\frac{2}{z}+\frac{b}{z^2}.$$

Since $f(z)$ is bounded when $|z| \to \infty$, there exists a positive number M such that $|f(z)| \leq M \ \forall z$ which implies that $f(z)$ has no singularity at $|z| = \infty$. Hence $f(1/z)$ given by

$$f(1/z) = a_0 + \frac{a_1}{z} + \frac{a_2}{z^2} + \cdots + \frac{2z}{1-z} + 2z + bz^2,$$

has no singularity at $z = 0$. This implies that there is no term in the principal part of $f(1/z)$, and so, $a_1 = a_2 = \cdots = 0$. Hence,

$$f(1/z) = a_0 + \frac{2z}{1-z} + 2z + bz^2 \Rightarrow f(z) = a_0 + \frac{2}{z-1} + \frac{2}{z} + \frac{b}{z^2}.$$

Given that $f(2) = 5$ and $f(-1) = 2$, therefore,

$$5 = a_0 + 2 + 1 + \frac{b}{4} \text{ and } 2 = a_0 - 1 - 2 + b$$

$$\Rightarrow \qquad a_0 + \frac{b}{4} = 2 \text{ and } a_0 + b = 5 \Rightarrow a_0 = 1 \text{ and } b = 4.$$

By putting, the values of a_0 and b, we get

$$f(z) = a_0 + \frac{2}{z-1} + \frac{2}{z} + \frac{b}{z^2} = 1 + \frac{2}{z-1} + \frac{2}{z} + \frac{4}{z^2}.$$

Example 7.1.6. *The only singularities of a single-valued function $f(z)$ are pole of order 1 and 2 at $z = -1$ and $z = 2$ with residues at these poles being 1 and 2 respectively. If $f(0) = 7/4$, $f(1) = 5/2$, determine the function and expand it in a Laurent series in $1 < |z| < 2$.*

Solution. Since the single-valued function $f(z)$ has pole of order 1 and 2 at $z = -1$ and $z = 2$, hence $f(z)$ has the form

$$f(z) = a_0 + \frac{1}{z+1} + \frac{2}{z-2} + \frac{b}{(z-2)^2}.$$

Given $f(0) = 7/4$, $f(1) = 5/2$ so that $\frac{7}{4} = a_0 + \frac{b}{4}$ and $\frac{5}{2} = a_0 - \frac{3}{2} + b$, giving $a_0 = 1$ and $b = 3$. By putting the values of a_0 and b, we get

$$f(z) = 1 + \frac{1}{z+1} + \frac{2}{z-2} + \frac{3}{(z-2)^2}.$$

Since $f(z)$ is analytic in the region $1 < |z| < 2$, it can be expanded as Laurent's series about any point of the region. Therefore,

$$\begin{aligned}
f(z) &= 1 + \frac{1}{z}\left(1 + \frac{1}{z}\right)^{-1} - \left(1 - \frac{z}{2}\right)^{-1} + \frac{3}{4}\left(1 - \frac{z}{2}\right)^{-2} \\
&= 1 + \frac{1}{z}\left(1 - \frac{1}{z} + \frac{1}{z^2} - \frac{1}{z^3} + \cdots\right) - \left(1 + \frac{z}{2} + \frac{z^2}{2^2} + \frac{z^3}{2^3} + \cdots\right) \\
&\quad + \frac{3}{4}\left[1 + 2\cdot\frac{z}{2} + 3\cdot\left(\frac{z}{2}\right)^2 + \cdots\right] \\
&= 1 + \frac{1}{z}\sum_{n=0}^{\infty}(-1)^n\frac{1}{z^n} - \sum_{n=0}^{\infty}\left(\frac{z}{2}\right)^n + \frac{3}{4}\sum_{n=0}^{\infty}(n+1)\left(\frac{z}{2}\right)^n \\
&= 1 + \sum_{n=0}^{\infty}\frac{1}{z^{n+1}} + \sum_{n=0}^{\infty}\left(\frac{3n-1}{3}\right)\left(\frac{z}{2}\right)^n.
\end{aligned}$$

Example 7.1.7. Find the residue(s) of the function $f(s)$, where $f(s) = \frac{\sinh x\sqrt{s}}{s \sinh a\sqrt{s}}$, where both x and a are positive and $a - x > 0$.

Solution. From the expression of $f(s)$, it appears that $s = 0$ is a branch point of $sf(s)$, but this is not true as shown below:

$$\begin{aligned} f(s) &= \frac{x\sqrt{s} + \frac{1}{3!}\left(x\sqrt{s}\right)^3 + + \frac{1}{5!}\left(x\sqrt{s}\right)^5 + \cdots}{s\left[a\sqrt{s} + \frac{1}{3!}\left(a\sqrt{s}\right)^3 + + \frac{1}{5!}\left(a\sqrt{s}\right)^5 + \cdots\right]} \\ &= \frac{x\left(1 + \frac{1}{3!}x^2 p + \frac{1}{5!}x^4 p^2 + \cdots\right)}{as\left[1 + \frac{1}{3!}a^2 p + \frac{1}{5!}a^4 p^2 + \cdots\right]}. \end{aligned}$$

This expression shows that $s = 0$ is not a branch point of $f(s)$ but a simple pole and other singularities of $f(s)$ are at the zeros of $\sinh a\sqrt{s}$ except at $s = 0$, which are all simple poles. Therefore, the singularities of $f(s)$ are given by

$$\begin{aligned} &\sinh a\sqrt{s} = 0 \Rightarrow a\sqrt{s} = \pm in\pi; n = 0, 1, 2, \cdots \\ \Rightarrow \quad &s = 0, s = -\frac{n^2\pi^2}{a^2}; n = 1, 2, \cdots \end{aligned}$$

which are simple poles. Now

$$\begin{aligned} R_0 &= \text{Res}\{F_1(s); 0\} = \lim_{s\to 0}(s - 0)\frac{\sinh x\sqrt{s}}{s \sinh a\sqrt{s}}\, e^{st} = \frac{x}{a} \\ R_n &= \text{Res}\left\{F_1(s); -\frac{n^2\pi^2}{a^2}\right\} = \lim_{s\to -\frac{n^2\pi^2}{a^2}} \left(s + \frac{n^2\pi^2}{a^2}\right)\frac{\sinh x\sqrt{s}}{s \sinh a\sqrt{s}}\, e^{st} \\ &= \lim_{s\to -\frac{n^2\pi^2}{a^2}} \frac{\sinh x\sqrt{s}}{s\cosh(a\sqrt{s}) \cdot \frac{1}{2s}a}\, e^{st} \\ &= 2\frac{\sinh\left(\pm ix\frac{n\pi}{a}\right)e^{-\frac{n^2\pi^2}{a^2}t} \cdot (\pm in\pi)}{\left(-\frac{n^2\pi^2}{a^2}\right)\cosh\left(\pm ia\frac{n\pi}{a}\right)a^2} = \frac{2}{n\pi}(-1)^n e^{-\frac{n^2\pi^2}{a^2}t}\sin\frac{n\pi x}{a}. \end{aligned}$$

7.1.1 Cauchy's Residue Theorem

Let f be an analytic function which is single-valued in a simply-connected domain $\mathcal{D}$ of the finite z-plane and which is regular there except for a finite number of isolated poles $z_1, z_2, z_3, \cdots, z_n$. Let γ be a continuous, positively oriented, closed contour lying entirely in $\mathcal{D}$ and not passing through any of the points $z_1, z_2, z_3, \cdots, z_n$. We wish to evaluate $\oint_\gamma f(z)\, dz$.

Theorem 7.1. [Cauchy's residue theorem]: *Let f be analytic within and on a positively oriented closed contour Γ, except at a finite number of poles $z_1, z_2, z_3, \cdots, z_n$ inside Γ but not on Γ. Then*

$$\oint_\Gamma f(z)dz = 2\pi i\sum_{k=1}^{n} \text{Res}\,\{f; a_k\} = 2\pi i\sum_{k=1}^{n} R_k \tag{7.6}$$

Proof: Let $z_1, z_2, z_3, \cdots, z_n$ be the finitely many poles of orders $m_1, m_2, \cdots, m_n$ respectively. f is an analytic function and on the boundary it is analytic.

Now, $\mathrm{Int}(\Gamma) \cup \Gamma$ is a compact set. If the number of isolated poles are infinite, then we consider a subset of poles to become an infinite subset and that have been a compact set, have an unique point that unique point certainly also be a pole not point of analyticity. Because of completeness, they are finite in numbers.

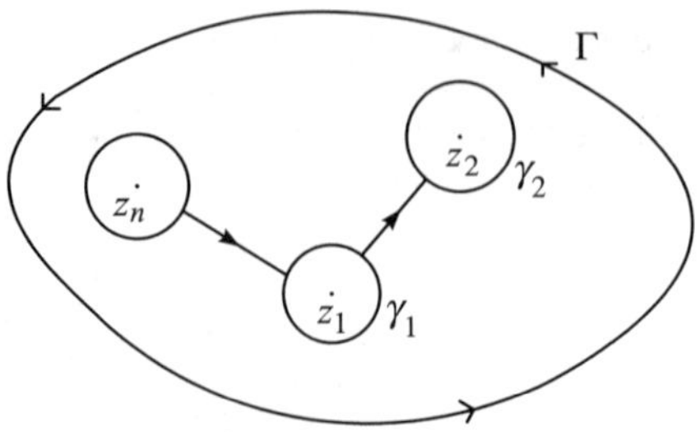

Figure 7.1 *Cauchy's residue theorem*

Let $\gamma_1, \gamma_2, \cdots, \gamma_n$ be the positively oriented small circles with respective centers $z_1, z_2, \cdots, z_n$, each of radius r (Fig. 7.1). The circles are so small that all these circles lie entirely within the interior of Γ and are mutually disjoint. Then in the multi-connected domain bounded externally by Γ and internally by $\gamma_1, \gamma_2, \cdots, \gamma_n$, the function $f(z)$ is analytic. Hence, by Cauchy's theorem in multiply-connected domain, we get

$$\int_\Gamma f(z)dz - \sum_{k=1}^{n} \int_{\gamma_k} f(z)dz = 0. \tag{7.7}$$

Since f has a pole of order m_1 at $z = z_1$, we have,

$$f(z) = \phi_1(z) + \sum_{p=1}^{m_1} R_p (z - z_1)^{-p}, \tag{7.8}$$

where $\phi_1(z)$ is regular within and on γ_1. So, $\int_{\gamma_1} \phi_1(z)dz = 0$. Therefore,

$$\begin{aligned}
\int_{\gamma_1} f(z)dz &= \int_{\gamma_1} \phi_1(z)dz + \sum_{p=1}^{m_1} \int_{\gamma_1} \frac{R_p}{(z-z_1)^p} dz \\
&= \int_{\gamma_1} \frac{R_1}{z - z_1} dz + \sum_{p=2}^{m_1} \int_{\gamma_1} \frac{R_p}{(z-z_1)^p} dz; \quad z - z_1 = re^{i\theta} \\
&= R_1 \int_0^{2\pi} \frac{rie^{i\theta}}{re^{i\theta}} d\theta + \sum_{p=2}^{m_1} \frac{iR_p}{r^p - 1} \int_0^{2\pi} e^{-i(p-1)\theta} d\theta \\
&= R_1 \times 2\pi i + \sum_{p=2}^{m_1} iR_p \times 0 = 2\pi i \times R_1.
\end{aligned}$$

Similarly, for others. Therefore, using Eq. (7.7), we get the relation (7.6). This proves the theorem. □

Example 7.1.8. *Evaluate* $\int_\Gamma \frac{z^2}{(z-2)(z+3)}\, dz$*, where* Γ *is the circle* $|z| = 4$.

Solution. Let $f(z) = z^2/(z-2)(z+3)$; $z = 2$ and $z = -3$ be simple poles for $f(z)$ and let both of them lie inside $|z| = 4$. Now

$$\begin{aligned}
\text{Res}\,\{f; 2\} &= \lim_{z \to 2} (z-2) f(z) = \lim_{z \to 2} (z-2) \Big[\frac{z^2}{(z-2)(z+3)} \Big] = \frac{4}{5} \\
\text{Res}\,\{f; -3\} &= \lim_{z \to -3} (z+3) f(z) = \lim_{z \to -3} (z+3) \Big[\frac{z^2}{(z-2)(z+3)} \Big] = -\frac{9}{5}.
\end{aligned}$$

Therefore by residue theorem (7.6), we get

$$\int_{\Gamma} \frac{z^2}{(z-2)(z+3)}\, dz = 2\pi i \Big[\frac{4}{5} + \Big(-\frac{9}{5}\Big)\Big] = -2\pi i.$$

RESULT **62.** (i) Residue theorem, homotopy version: Let $\mathcal{D} \subset \mathbb{C}$ be open and f a function analytic on $\mathcal{D}$, except perhaps for a finite number of distinct points $\alpha_1, \alpha_2, \cdots, \alpha_n \in \mathcal{D}$. (More formally, f is analytic on $\mathcal{D}/\{\alpha_1, \alpha_2, \cdots, \alpha_n\}$. Let γ be a simple closed curve in $\mathcal{D}/\{\alpha_1, \alpha_2, \cdots, \alpha_n\}$, which is null-homotopic in $\mathcal{D}$. Then

$$\oint_{\gamma} f(z)\, dz = 2\pi i \sum_{k=1}^{n} \text{Res}\{f; \alpha_k\} \text{ Ind }_{\gamma}(\alpha_k).$$

(ii) Residue theorem, simple closed curve version : Let $\mathcal{D} \subset \mathbb{C}$ be open and f a function analytic on $\mathcal{D}$, except perhaps for a finite number of distinct points, $\alpha_1, \alpha_2, \cdots, \alpha_n \in \mathcal{D}$. Suppose that γ is a simple closed curve in $\mathcal{D}/\{\alpha_1, \alpha_2, \cdots, \alpha_n\}$, oriented anticlockwise, with its interior contained in $\mathcal{D}$. Then

$$\oint_{\gamma} f(z)\, dz = 2\pi i \sum_{k:1 \le k \le n, \alpha_k \text{ inside } \gamma} \text{Res}\{f; \alpha_k\}.$$

Theorem 7.2. *Let a function $f(z)$ be analytic except for isolated singular points in $\mathbb{C}_\infty$. Then the sum of all the residues of $f(z)$ is equal to zero.*

Proof: Here $f(z)$ has only finite number of isolated singular points, otherwise, the singular points would have a limit point z_0 (possibly at infinity) and then, z_0 would be a non-isolated singular point of $f(z)$, which contradicts the hypothesis. Thus, there exists a positive number R which is large enough so that all the singularities lie inside the positively-oriented circle $\Gamma : |z| = R$ such that all finite singular points $z_1, z_2, \cdots, z_n$ lie inside Γ. So, by the Cauchy's residue theorem (7.6),

$$\int_{\Gamma} f(z)\, dz = 2\pi i \sum_{k=1}^{n} \text{Res } \{f; z_k\}, \tag{7.9}$$

But according to the point at ∞, this is the negative direction of traversing Γ.

$$\int_{\Gamma} f(z)\, dz = -2\pi i \text{ Res } \{f; \infty\}. \tag{7.10}$$

Subtracting equation (7.10) from equation (7.9), we get

$$2\pi i \Big[\sum_{k=1}^{n} \text{Res } \{f; z_k\} + \text{ Res } \{f(z); \infty\}\Big] = 0$$

$$\Rightarrow \quad \sum_{k=1}^{n} \text{Res } \{f; z_k\} + \text{ Res } \{f; \infty\} = 0. \tag{7.11}$$

Thus, theorem is verified. Since rational functions have only isolated singular points (poles), this theorem is directly applicable to rational functions. □

Example 7.1.9. *Evaluate $\int_{|z|=2} \frac{dz}{(z-3)(z^5-1)}$, where $|z| = 2$ is the positively-oriented circle.*

Solution. Let $f(z) = 1/(z-3)(z^5-1)$. We can use equation (7.11) to evaluate the integral and thus avoid the lengthier procedure otherwise.

$$\sum_{k=1}^{5} \text{Res}\,\{f(z); z_k\} + \text{Res}\,\{f(z); 3\} + \text{Res}\,\{f(z); \infty\} = 0$$

$$\Rightarrow \quad \sum_{k=1}^{5} \text{Res}\,\{f; z_k\} = -\text{Res}\,\{f(z); 3\} - \text{Res}\,\{f; \infty\}$$

$$= -\left[\frac{1}{z^5-1}\right]_{z=3} + \underset{z=0}{\text{Res}}\left[\frac{1}{z^2} f\left(\frac{1}{z}\right)\right] = -\frac{1}{242} + 0$$

$$\therefore \int_{|z|=2} \frac{1}{(z-3)(z^5-1)} dz = -\frac{2\pi i}{242} = -\frac{\pi i}{121}.$$

Example 7.1.10. *Let $f(z)$ be analytic at z_0 such that $f(z_0) \neq 0$ and let $g(z)$ have a zero of order 2 at z_0. Show that*

$$\text{Res}\left\{\frac{f}{g}; z_0\right\} = \frac{6f'(z_0)g''(z_0) - 2f(z_0)g'''(z_0)}{3[g''(z_0)]^2}.$$

Solution. Since the function $g(z)$ has a zero of order 2 at z_0, so by Theorem 5.25, we have $g(z) = (z-z_0)^2 p(z)$, where $p(z)$ is analytic at z_0 and $p(z_0) \neq 0$. Thus

$$\begin{aligned} & g'(z) = 2(z-z_0)p(z) + (z-z_0)^2 p'(z) \\ \Rightarrow \quad & g''(z) = 2p(z) + 4(z-z_0)p'(z) + (z-z_0)^2 p''(z) \\ \Rightarrow \quad & g'''(z) = 6p'(z) + 6(z-z_0)p''(z) + (z-z_0)^2 p'''(z). \end{aligned}$$

There relations give $g''(z_0) = 2p(z_0)$ and $g'''(z_0) = 6p'(z_0)$. Thus $\frac{f(z)}{g(z)} = \frac{f(z)/p(z)}{(z-z_0)^2}$ has a pole of order 2 and its residue at z_0 is

$$\begin{aligned} \text{Res}\left\{\frac{f}{g}; z_0\right\} &= \lim_{z\to z_0} \frac{d}{dz}\left[(z-z_0)^2 \frac{f(z)/p(z)}{(z-z_0)^2}\right] = \lim_{z\to z_0} \frac{d}{dz}\left[\frac{f(z)}{p(z)}\right] \\ &= \lim_{z\to z_0} \frac{f'(z)p(z) - f(z)p'(z)}{[p(z)]^2} = \frac{f'(z_0)p(z_0) - f(z_0)p'(z_0)}{[p(z_0)]^2} \\ &= \frac{6f'(z_0)g''(z_0) - 2f(z_0)g'''(z_0)}{3[g''(z_0)]^2} \end{aligned}$$

as $g''(z_0) = 2p(z_0)$ and $g'''(z_0) = 6p'(z_0)$.

Example 7.1.11. *Evaluate $I = \int_\gamma \frac{dz}{(z-5)(z^{23}-1)}$, where $\gamma : z = 3e^{i\theta}; 0 \leq \theta \leq 2\pi$.*

Solution. By Cauchy's residue theorem (7.6), we have

$$\oint_\Gamma f(z)dz = 2\pi i \sum_{k=23}^{n} R_k(f; \alpha_k),$$

where α_k are the 23 roots of unity. Obviously, this sum is not easy to compute. However, since $\text{Res}(f;\infty) = 0$ and $\text{Res}(f;5) = 1/(5^{23}-1)$, from Eq.(7.11) we get

$$I = -\Big[2\pi i\, \text{Res}\,\{f; \infty\} + \text{Res}\,\{f; 5\}\Big] = -\frac{2\pi i}{5^{23}-1}.$$

7.1.2 The Residue at Infinity

Let $f(z)$ has an isolated singularity at infinity. Let C be the circle centred at origin described in the clockwise direction with its radius so large that it encloses all the singularities at finite distance from origin. Then the residue of $f(z)$ at the point at infinity is defined by

$$\text{Res}\,\{f;\infty\}f(z) = \frac{1}{2\pi i}\oint_C f(z)\,dz = -\frac{1}{2\pi i}\oint_C f(z)\,dz \tag{7.12}$$

$$\Rightarrow \quad \oint_C f(z)\,dz = -2\pi i \;\; \text{Res}\,\{f;\infty\}.$$

Here C is a circle described in the anticlockwise sense, i.e., the positive sense, and it encloses all the singularities of $f(z)$ at finite distance from the origin. Let R be the radius of the circle C. Then the Laurent's series (5.49) of $f(z)$ that is valid in $R < |z| < \infty$ is given by

$$f(z) = \sum_{n=-\infty}^{\infty} c_n\, z^n, \text{ where } c_n = \frac{1}{2\pi i}\int_C \frac{f(z)}{z^{n+1}}dz;\;\; n \in \mathbb{Z}. \tag{7.13}$$

Replacing z by $1/\zeta$ in Eq. (7.13) and then multiplying the result by $1/\zeta^2$, we get

$$\frac{1}{\zeta^2}f\left(\frac{1}{\zeta}\right) = \sum_{n=-\infty}^{\infty} \frac{c_n}{\zeta^{n+2}} = \sum_{n=-\infty}^{\infty} \frac{c_{n-2}}{\zeta^n},\;\; 0 < |\zeta| < \frac{1}{R},$$

and c_{-1} = coefficient of $1/\zeta$ in the Laurent series expansion of $\frac{1}{\zeta^2}f(\frac{1}{\zeta}) = \underset{\zeta=0}{\text{Res}}\,[\frac{1}{\zeta^2}f(\frac{1}{\zeta})]$. But, using Eq. (7.13), we get

$$c_{-1} = \frac{1}{2\pi i}\int_C f(z)\,dz = -\;\text{Res}\,\{f;\infty\}. \tag{7.14}$$

Therefore, from Eq. (7.14), we get

$$\text{Res}\,\{f;\infty\}f(z) = -c_{-1} = -\underset{z=0}{\text{Res}}\; f(z)\left[\frac{1}{z^2}f\left(\frac{1}{z}\right)\right]. \tag{7.15}$$

The formula described in Eq. (7.15) gives an efficient method of evaluating the residue of $f(z)$ at infinity.

Example 7.1.12. *Find the residue of* $f(z) = \frac{z^3}{z^2-1}$ *at* $z = \infty$.

Solution. The function $f(z) = z^3/(z^2-1)$ has a singularity at $z \to \infty$, because $f(1/z) = 1/z(1-z^2)$ has a pole at $z = 0$. By formula in Eq. (7.15),

$$\text{Res}\,\{f;\infty\} \;=\; -\underset{z=0}{\text{Res}}\; f(z)\left[\frac{1}{z^3(1-z^2)}\right] = -\frac{1}{2!}\lim_{z\to 0}\frac{d^2}{dz^2}\left[\frac{1}{1-z^2}\right] = -1.$$

7.2 The Argument Principle

In this section we shall discuss of the important properties of meromorphic functions, called the *argument principle*, regarding the number of zeros of f. The result is then used to establish Rouché's theorem, which provides locations of zeros and poles of meromorphic functions. Using this principle, we also prove some interesting theorems like Hurwitz's theorem, open mapping theorem, inverse function theorem.

Theorem 7.3. [The argument theorem]: *Let $\mathcal{D} \subset \mathbb{C}$ be a domain with boundary $\partial\mathcal{D}$. Let $f(z)$ be analytic on $\mathcal{D} \cup \partial\mathcal{D}$ except for finite number of isolated poles in $\mathcal{D}$ (meromorphic) and let $f(z)$ have no zeros on the boundary $\partial\mathcal{D}$. If the meromorphic function $f(z)$ has finite number N_∞ of poles (a pole or zero of order m must be counted m times) and N_0 zeros inside the contour, then*

$$\frac{1}{2\pi i}\oint_{\partial\mathcal{D}} d\log f(z) = \frac{1}{2\pi i}\oint_{\partial\mathcal{D}} \frac{f'(z)}{f(z)}\,dz = N_0 - N_\infty. \tag{7.16}$$

Proof: Let $z_0, z_1, \cdots, z_{k-1}$ be the finitely many zeros of orders $p_1, p_2, \cdots, p_k$ respectively (Fig. 7.2). $f(z)$ is meromorphic function and on the boundary it is analytic. Now $\mathcal{D} \cup \partial\mathcal{D}$ is a compact set.

If the number of isolated zeros are infinite, and as $\mathcal{D} \cup \partial\mathcal{D}$ is compact, then they have a limit point and by uniqueness theorem, f is identically zero. Because of completeness, isolated zeros are finite in numbers.

Similarly, there are finite number of poles. Let $\alpha_1, \alpha_2, \cdots, \alpha_n$ be the finitely many poles of orders $m_1, m_2, \cdots, m_n$ respectively (Fig. 7.2).

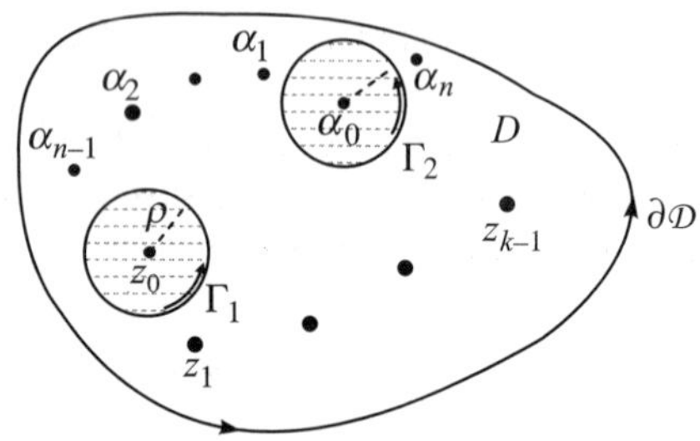

Figure 7.2 *Argument theorem*

Since $f(z)$ has a zero of order p_1, we have

$$f(z) = (z - z_0)^{p_1} g_1(z),$$

where $g_1(z)$ is analytic at z_0 and $g_1(z_0) \neq 0$. Therefore, in some punctured neighbourhood of z_0

$$\begin{aligned} d\log f(z) &= \frac{f'(z)}{f(z)}dz = \frac{p_1(z-z_0)^{p_1-1}g_1(z) + (z-z_0)^{p_1}g_1'(z)}{(z-z_0)^{p_1}g_1(z)}dz \\ &= \frac{p_1}{z-z_0}dz + \frac{g_1'(z)}{g_1(z)}dz \end{aligned}$$

where, as $g_1(z_0) \neq 0$, so $g_1'(z)/g_1(z)$ is analytic at z_0. Let us consider a closed disc Γ_1 as $\{z \in \mathcal{D} : |z - z_0| = \rho\}$, which can be parametrized as $z = z_0 + \rho e^{i\theta}$. Therefore, $d\log f(z)$ is analytic within $\{z : 0 < |z - z_0| < \rho\}$. Therefore, by Cauchy's integral theorem

$$\begin{aligned} \frac{1}{2\pi i}\oint_{\Gamma_1} d\log f(z) &= \frac{1}{2\pi i}\oint_{\Gamma_1}\frac{f'(z)}{f(z)}dz = \frac{p_1}{2\pi i}\oint_{\Gamma_1}\frac{dz}{z-z_0} + \frac{1}{2\pi i}\oint_{\Gamma_1}\frac{g_1'(z)}{g_1(z)}dz \\ &= \frac{p_1}{2\pi i}\int_0^{2\pi}\frac{\rho i e^{i\theta}}{\rho e^{i\theta}}d\theta + 0 = p_1. \end{aligned}$$

Thus if we surround a pole and compute the integral of the logarithmic derivative, it is exactly of the order of the zeros.

By similar argument, since $f(z)$ has a pole of order m_1 at $z = \alpha_0$, we have, $f(z) = (z - \alpha_0)^{-m_1} g_2(z)$, where $g_2(z)$ is analytic at α_0 and is different from 0 inside and on Γ_2.

$$\therefore \frac{1}{2\pi i}\oint_{\Gamma_2} d\log f(z) = \frac{1}{2\pi i}\oint_{\Gamma_2}\frac{g_2'(z)}{g_2(z)}dz - \frac{m_1}{2\pi i}\oint_{\Gamma_2}\frac{dz}{z-\alpha_0} = -m_1.$$

Now, if we surrounded each of the zeros and poles by small discs, then by Cauchy's residue theorem

$$\frac{1}{2\pi i}\oint_{\partial\mathcal{D}} d\log f(z) = \frac{1}{2\pi i}\oint_{\partial\mathcal{D}}\frac{f'(z)}{f(z)}dz = \sum_{r=1}^{k} p_r - \sum_{s=1}^{n} m_s = N_0 - N_\infty.$$

Hence the theorem is established. □

DEDUCTION **30.** [Nomenclature of Argument theorem]: For a particular value of z, we have

$$\log f(z) = \ln|f(z)| + i\Big[\text{Arg }(f(z)) + 2n\pi\Big].$$

The problem for $f(z)$ is that uniformly in the domain $\mathcal{D}$, $\log f(z)$ is not an analytic function. For that, what will do is split the domain for $\log f(z)$ so that it has branch. Now

$$d\log f(z) = d\ln|f(z)| + id\Big[\text{Arg }(f(z))\Big]$$

where d denotes the difference as the change z along an arc. Here we choose a particular branch of logarithm. On the boundary $\partial\mathcal{D}$, parameterized by $\gamma(t) : t \in [a,b] \subset \mathbb{R}$, we have

$$\begin{aligned}\oint_{\partial\mathcal{D}} d\log f(z) &= \oint_{\Gamma} d\log f(\gamma(t)) \\ &= \oint_{\Gamma} d\ln|f(\gamma(t))| + i\oint_{\Gamma} d\Big[\text{Arg }(f(\gamma(t)))\Big]\end{aligned}$$

where $\arg f(z)$ varies by less than 2π along Γ. Then there is a branch of $\log f(z)$ that is analytic on Γ. Now, as $|f(z)|$ is a continuous real-valued function on Γ, so $\oint_\Gamma d\ln|f(\gamma(t))| = 0$. Also

$$\oint_{\Gamma} d\Big[\text{Arg }(f(\gamma(t)))\Big] = \text{The difference in the argument of } f(z) \text{ as } z \text{ goes around once along the boundary which is not parameterized by } \Gamma \text{ from the starting point to the ending point.} = \text{Net excursion in } \arg f(z) \text{ as we go around } \Gamma.$$

$\oint_{\partial\mathcal{D}} d\log f(z)$ is evaluated over a simple contour, which is the change in the argument of the function. In all the calculations, we have taken a particular branch of the logarithm. If we take a different branch of logarithm, then the two branches of logarithm differ by a constant multiple, $2\pi i$. Therefore, the difference $\oint_\Gamma d[\text{Arg }(f(\gamma(t)))]$ is not dependent on which branch of logarithm we take. If we take Arg on one branch and Arg$'$ on another branch, then Arg $-$ Arg$'$ $= 2n\pi$; if we take the difference, then $2n\pi$ cancels out. Therefore, the quantity $\oint_\Gamma d[\text{Arg }(f(\gamma(t)))]$ is the change in the argument which also changes. So, if we calculate $\oint_{\partial\mathcal{D}} d\log f(z)$ over the boundary that actually changes in the argument is zero or depends on $f(z)$, (also zeros and poles of $f(z)$ in $\mathcal{D}$). By Cauchy's theorem, the argument function is

$$\oint_{\partial\mathcal{D}} d\log f(z) = 2\pi i\Big[N_0 - N_\infty\Big],$$

which is the argument principle.

Theorem 7.4. [The argument principle]: *Let a function $f(z)$ be analytic within and on a closed contour Γ, having N zeros inside Γ but no zero on Γ. Then*

$$N = \frac{1}{2\pi}\Delta_\Gamma \arg f(z), \tag{7.17}$$

where $\Delta_\Gamma \arg f(z)$ denotes the variation in the value of $\arg f(z)$ *as z moves round the closed contour Γ (a zero of order n must be counted n times).*

Proof: Since $f(z)$ has N zeros but no poles inside Γ, we get

$$\frac{1}{2\pi i}\int_C \frac{f'(z)}{f(z)}\,dz = N - P = N \text{ as } P = 0.$$

$$\therefore \quad N = \frac{1}{2\pi i}\Big[\log f(z)\Big]_\Gamma = \frac{1}{2\pi i}\Big[\log|f(z)| + i\arg f(z)\Big]_\Gamma.$$

Equating real parts from either sides, we get

$$N = \frac{1}{2\pi}\Big[\arg f(z)\Big]_{\Gamma} = \frac{1}{2\pi}\Delta_{\Gamma} \arg f(z), \tag{7.18}$$

where, $\Delta_\Gamma \arg f(z)$ represents change in $\arg f(z)$ as z traverses the positive closed contour C once. Clearly, the right-hand side expression $\frac{1}{2\pi}\Delta_\Gamma \arg f(z)$ is a positive integer as N is also so. This integer is sometimes called the *winding number* or *index* with respect to the origin $w = f(z) = 0$. This result is called *argument principle*. □

Example 7.2.1. *Evaluate* $\int_\Gamma \frac{f'(z)}{f(z)}\,dz$, *where* $f(z) = z^5 - 3iz^2 + i - 1$ *and* Γ *encloses the zeros of* $f(z)$.

Solution. Since $f(z) = z^5 - 3iz^2 + i - 1$, $f(z)$ has 5 zeros and no poles. So $N = 5$ and $P = 0$. Using Eq. (7.16), we have

$$\frac{1}{2\pi i}\int_\Gamma \frac{f'(z)}{f(z)}\,dz = 5 - 0 = 5 \Rightarrow \int_\Gamma \frac{f'(z)}{f(z)}\,dz = 10\pi i.$$

There is yet another way to express the variation in the argument of f along C; it involves the *image curve* $f(C)$. This is simply the image (in the w-plane) of the curve C under the mapping $w = f(z)$; that is, if C is parametrized by $z = z(t)$, $a \le t \le b$, then $f(C)$ is the curve parametrized by

$$w = f(z(t)) \quad (a \le t \le b). \tag{7.19}$$

Obviously the image curve is closed, but unlike C it need not be simple or positively oriented.

Now if we sketch the image curve as in Fig. 7.3, it is easy to follow the net change in the argument of $f(z)$. Every time $f(z(t))$ encircles the origin $w = 0$ in the positive (counterclockwise) direction, $\arg f$ increase by 2π, while it decreases by 2π for a negative circuit. Hence, since $f(C)$ is closed, Δ_C $\arg f(z)$ equals 2π multiplied by the net number of times $f(C)$ winds around $w = 0$ *in the positive sense,* that is, counterclockwise minus clockwise.

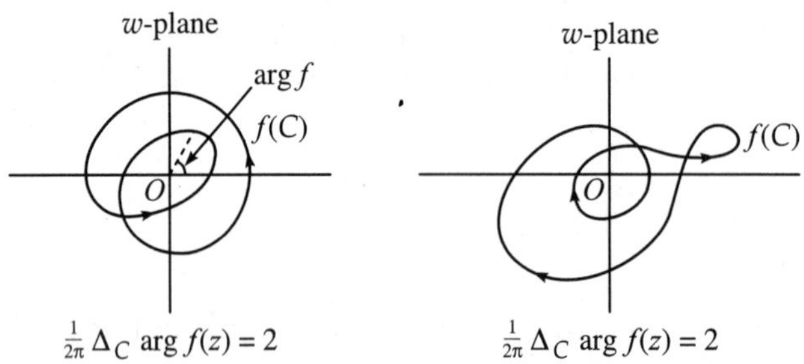

Figure 7.3 *Visualization of* $\Delta_C \arg f(z)$

As a specific illustration, consider

$$f(z) = z^3 \quad \text{and} \quad C : z = e^{it} \quad (0 \le t \le 2\pi).$$

Notice that the image of this circle under $w = f(z)$ winds around the origin three time in the positive sense (see Fig. 7.4). Hence the net change in the argument of f is 6π, and $N_0(f) - N_p(f) = 6\pi/2\pi = 3$.

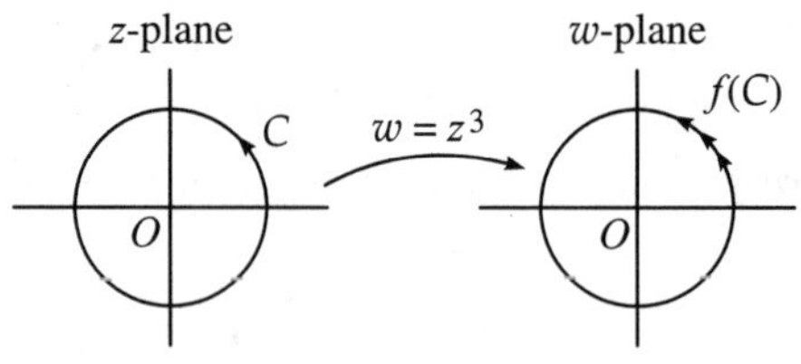

Figure 7.4 $\Delta_C \arg(z^3)$

7.2.1 Rouché's Theorem

Theorem 7.5. *Let f and g be analytic functions on $\mathcal{D} \cup \partial\mathcal{D}$. If $|g(z)| < |f(z)|$ on $\partial\mathcal{D}$ and both f and g have no zeros on $\partial\mathcal{D}$, then the number of zeros of $f(z)$ and $f(z) + g(z)$, counting their multiplicities inside $\partial\mathcal{D}$ are the same.*

Proof: We claim that $g(z) + f(z)$ has no zeros on the boundary $\partial\mathcal{D}$. On $\partial\mathcal{D}$, $|f(z)| > |g(z)| > 0$. So $f(z)$ has no zero on $\partial\mathcal{D}$. Now

$$\frac{g(z)+f(z)}{f(z)} = 1 + \frac{g(z)}{f(z)} = w(z), \text{ say, on } \partial\mathcal{D}$$

$$\therefore \quad \arg\left(\frac{g(z)+f(z)}{f(z)}\right) = \arg\left(1+\frac{g(z)}{f(z)}\right) = \arg\Big(w(z)\Big)$$

$$\Rightarrow \quad \arg\Big(g(z)+f(z)\Big) - \arg\Big(f(z)\Big) = \arg\Big(w(z)\Big)$$

$$\Rightarrow \quad \Delta_{\partial\mathcal{D}} \arg\Big(g(z)+f(z)\Big) - \Delta_{\partial\mathcal{D}} \arg\Big(f(z)\Big) = \Delta_{\partial\mathcal{D}} \arg\Big(w(z)\Big),$$

where $w(z) = 1 + \frac{g(z)}{f(z)}$. Therefore

$$\Big|w(z) - 1\Big| = \left|\frac{g(z)}{f(z)}\right| < 1 \text{ on } \partial\mathcal{D}.$$

This means that $w(z)$ is always an internal point of the unit circle $\Gamma : |w(z) - 1| < 1$, with centre at 1. This circle Γ lies totally in the right side of the imaginary w-axis. Thus the line joining any point of the circle to the origin $w = 0$ makes an angle ψ $(-\pi/2 \le \psi \le \pi/2)$ with the real w-axis. Thus, if $w = Re^{i\psi}$, then $\arg(w(z)) = \psi$, and it returns to its original value when z describes $\mathcal{D}$. Therefore

$$\Delta_{\partial\mathcal{D}} \arg\Big(w(z)\Big) = \Delta_{\partial\mathcal{D}} \arg\left(1+\frac{g(z)}{f(z)}\right) = 0$$

$$\Rightarrow \quad \Delta_{\partial\mathcal{D}} \arg\Big(g(z)+f(z)\Big) - \Delta_{\partial\mathcal{D}} \arg\Big(f(z)\Big) = 0.$$

Thus, by argument principle, the number of zeros of $f(z)$ and $f(z) + g(z)$ inside $\mathcal{D}$ are same. □

Corollary 7.2.1. Another alternative proof of Rouché's theorem.

Since $|g(z)| < |f(z)|$ on $\partial\mathcal{D}$, then $|g(z)/f(z)| < 1$ on $\partial\mathcal{D}$, i.e., $|f(z)| \ne 0$; otherwise $|g(z)/f(z)|$ will be identity and not less than 1. Also, $|f(z)| - |g(z)|$ is continuous on the compact set $\mathcal{D}$. It follows that

$$|f(z)+g(z)| > |f(z)| - |g(z)| \ne 0$$

$$\Rightarrow f(z) + g(z) \ne 0 \quad ; \text{ as } |g(z)| < |f(z)|.$$

Hence, neither $f(z)$ nor $f(z) + g(z)$ has a zero on $\partial\mathcal{D}$. Let $F(z) = \frac{g(z)}{f(z)}$. Then $|F(z)| < 1$ on $\partial\mathcal{D}$. It immediately follows that $g(z)$ and $f(z)$ are not zero on Γ. Now,

$$g(z) = F(z)f(z) \Rightarrow g'(z) = F'(z)f(z) + F(z)f'(z).$$

Let N_1 and N_2 be the finite number of zeros of $f + g$ and g respectively inside $\mathcal{D}$, and by hypothesis these functions have no poles inside $\mathcal{D}$. Hence from argument theorem (7.16), we get,

$$N_1 = \frac{1}{2\pi i}\oint_{\partial\mathcal{D}} \frac{f'+g'}{f+g}dz \text{ and } N_2 = \frac{1}{2\pi i}\oint_{\partial\mathcal{D}} \frac{f'(z)}{f(z)}dz.$$

Therefore

$$\begin{aligned}
N_1 - N_2 &= \frac{1}{2\pi i}\oint_{\partial\mathcal{D}} \frac{f'+g'}{f+g}dz - \frac{1}{2\pi i}\oint_{\partial\mathcal{D}} \frac{f'(z)}{f(z)}dz \\
&= \frac{1}{2\pi i}\oint_{\partial\mathcal{D}} \frac{f'+f'F+fF'}{f+fF}dz - \frac{1}{2\pi i}\oint_{\partial\mathcal{D}} \frac{f'(z)}{f(z)}dz \\
&= \frac{1}{2\pi i}\oint_{\partial\mathcal{D}} \left[\frac{f'}{f} + \frac{F'}{1+F}\right]dz - \frac{1}{2\pi i}\oint_{\partial\mathcal{D}} \frac{f'(z)}{f(z)}dz \\
&= \frac{1}{2\pi i}\oint_{\partial\mathcal{D}} \frac{F'}{1+F}dz. \qquad (7.20)
\end{aligned}$$

Now, $1+F \neq 0$ on $\partial\mathcal{D}$ (because $|F(z)| < 1$ by hypothesis); also $F'(z)$ is analytic on $\partial\mathcal{D}$ because F is so on $\mathcal{D}$ (derivative of an analytic function is analytic). Consequently, $F'(z)/\{1+F(z)\}$ is analytic on $\partial\mathcal{D}$. Hence by Cauchy's theorem

$$\frac{1}{2\pi i}\oint_{\partial\mathcal{D}} \frac{F'(z)}{1+F(z)}dz = 0 \Rightarrow N_1 - N_2 = 0 \text{ from (7.20)}.$$

Thus $N_1 = N_2$ as required. □

Theorem 7.6. [Fundamental theorem of algebra]: *Every polynomial* $p(z) = a_0 + a_1 z + \cdots + a_n z^n$ *with* $a_n \neq 0$ *and* $a_k \in \mathbb{C}$ *of degree* n *has exactly* n *zeros in* $\mathbb{C}$ *counted with multipliers.*

Proof: Let $f(z) = a_n z^n$ and $g(z) = a_0 + a_1 z + a_2 z^2 + \cdots + a_{n-1}z^{n-1}$. Let R be a large number. For large R, the leading term $a_n z^n$ is very large and the other terms $g(z)$ are small. Let $\Gamma : |z| = R$ be a circle, where $R(> 1)$ is large. Then on Γ, we have

$$\begin{aligned}
\left|\frac{g(z)}{f(z)}\right| &= \frac{|a_0 + a_1 z + a_2 z^2 + \cdots + a_{n-1}z^{n-1}|}{a_n z^n} \\
&\leq \frac{|a_0| + |a_1|R + |a_2|R^2 + \cdots + |a_{n-1}|R^{n-1}}{|a_n|R^n} \\
&\leq \frac{|a_0|R^{n-1} + |a_1|R^{n-1} + |a_2|R^{n-1} + \cdots + |a_{n-1}|R^{n-1}}{|a_n|R^n} \\
&\leq \frac{|a_0| + |a_1| + |a_2| + \cdots + |a_{n-1}|}{|a_n|R}.
\end{aligned}$$

Since R is arbitrary, therefore, choose R such that

$$R > \frac{|a_0| + |a_1| + |a_2| + \cdots + |a_{n-1}|}{|a_n|}$$

$$\therefore \quad \left|\frac{g(z)}{f(z)}\right| < 1 \Rightarrow |g(z)| < |f(z)| \text{ on } \Gamma.$$

Hence by Rouché's theorem $f(z) + g(z)$ has the same number of zeros as $f(z) = a_n z^n$. But, since $f(z) = a_n z^n$ has n zeros, all located at $z = 0$, so $f(z) + g(z)$ also has n zeros. □

Note: Liouville's theorem only assure the existence of at least one zero of polynomial, while Rouché's theorem assures the existence of n zeros, counting multipliers.

Example 7.2.2. *Using Rouché's theorem determine the number of zeros of the polynomial* $P(z) = z^{10} - 6z^7 + 3z^3 + 1$ *in* $|z| < 1$.

Solution. Let $f(z) = -6x^7$ and $g(z) = z^{10} + 3z^3 + 1$. Then f and g both are analytic within and on the circle $|z| = 1$. Now, on $|z| = 1$, we have

$$\left|\frac{g(z)}{f(z)}\right| = \left|\frac{z^{10} + 3z^3 + 1}{-6z^7}\right| \leq \frac{|z|^{10} + 3|z|^3 + 1}{6|z|^7} = \frac{5}{6} < 1 \text{ on } |z| = 1.$$

Applying Rouché's theorem, we get that $f(z) + g(z) = P(z)$ has the same number of zeros inside $|z| = 1$ as $f(z) = -6z^7$. But as $f(z)$ has seven zeros inside $|z| = 1$, so $P(z)$ has seven zeros inside $|z| = 1$.

Example 7.2.3. *Use Rouché's theorem to show that the equation* $z^5 + 15z + 1 = 0$ *has one root in the disc* $|z| < 3/2$ *and four roots in the annulus* $3/2 < |z| < 2$.

Solution. Let $f(z) = z^5$ and $g(z) = 15z + 1$. Then both $f(z)$ and $g(z)$ are analytic within and on the circle $|z| = 2$. On $|z| = 2$, we have

$$|g(z)| = |15z + 1| \leq 15|z| + 1 = 15 \times 2 + 1 = 31 \text{ and } |f(z)| = |z^5| = 2^5 = 32.$$

Thus, $|g(z)| < |f(z)|$ on $|z| = 2$. From, Rouché's theorem it follows that the function $f(z) + g(z) = z^5 + 15z + 1$ and $f(z) = z^5$ have the same number of zeros in $|z| = 2$. Since, $f(z)$ has a zero of order 5 at $z = 0$, all the five roots of $f(z) + g(z)$ must lie in the disc $|z| < 2$.

Now, consider $|z| = 3/2$. Let $f(z) = 15z$, and $g(z) = z^5 + 1$. Then both $f(z)$ and $g(z)$ are analytic within and on the circle $|z| = 3/2$. On $|z| = 3/2$, we have

$$|g(z)| = |z^5 + 1| \leq |z|^5 + 1 = \frac{243}{32} + 1 = \frac{275}{32} \text{ and } |f(z)| = |15z| = 15 \times \frac{3}{2} = \frac{45}{2}.$$

Thus, $|g(z)| < |f(z)|$ on $|z| = 3/2$. From Rouché's theorem it follows that the function $f(z) + g(z) = z^5 + 15z + 1$ and $f(z) = 15z$ have the same number of zeros in $|g(z)| = 3/2$. Since, $f(z)$ has exactly one zero at $z = 0$, therefore, four of the zeros of $f(z) + g(z)$ must lie in the annulus $\frac{3}{2} < |z| < 2$.

Note: [The Stronger version of Rouché's theorem]: Let $f(z)$ and $g(z)$ be analytic on $\mathcal{D} \cup \partial\mathcal{D}$ and let $\Big|f(z) + g(z)\Big| < |f(z)| + |g(z)|$ on $\partial\mathcal{D}$. Then f and g have the same number of zeros in $\mathcal{D}$.

7.2.2 Local Mapping Theorem

Theorem 7.7. *If* $f(z)$ *is analytic and univalent in a domain* $\mathcal{D}$, *then* $f'(z) \neq 0$ *in* $\mathcal{D}$.

Proof: We cannot assume that f is a constant function because, if f is a constant function and z_0 is a zero, then f is constant and by identity theorem (Result 26), f is identically zero. Let $f'(z_0) = 0$ for some $z_0 \in \mathcal{D}$. As f is univalent, so f is a non-constant analytic function. Thus the Tailor series about z_0 is:

$$f(z) - f(z_0) = \sum_{n=2}^{\infty} a_n (z - z_0)^n \text{ as } a_1 = f'(z_0) = 0.$$

Hence, $g(z) = f(z) - f(z_0)$ has a zero of order $k(\geq 2)$ at z_0. Since zeros of the non-vanishing analytic function $g(z)$ are isolated, there exists an open disc $\mathcal{S}(f(z_0), \rho) = \{|f(z) - f(z_0)| < \rho\}$, such that

$$f(z) - f(z_0) \neq 0 \text{ in } 0 < |z - z_0| < \varepsilon.$$

Now, $|g|$ is a continuous function on the boundary $\partial\mathcal{S}(f(z_0), \rho) = \{|f(z) - f(z_0)| = \rho\}$, which is compact. Since a continuous real-valued function on a compact set is uniformly continuous and attains its maximum and minimum value, so does $|g|$. Since g is non-zero, the minimum value cannot be zero. Hence,

$$\mu = \inf\left\{|f(z) - f(z_0)| : z \in \partial\mathcal{S}(f(z_0), \rho)\right\} > 0.$$

So, for any complex number w such that $0 < |w| < \mu$, we have $|w| < |f(z) - f(z_0)|$.

$$\Rightarrow \left|\left[f(z) - f(z_0) - w\right] - \left[f(z) - f(z_0)\right]\right| < \left|f(z) - f(z_0)\right|.$$

By Rouché's theorem, $f(z) - f(z_0) - w$ and $f(z) - f(z_0)$ have the same number of zeros within $\mathcal{S}(f(z_0), \rho)$ as $f(z) - f(z_0)$ has a zero of order $k(\leq 2)$ inside this circle. Thus, the equation $f(z) = f(z_0) + w$ is satisfied at two or more points which is a contradiction to the fact that f is univalent in the domain $\mathcal{D}$. Hence, $f'(z) \neq 0$ for all points in $\mathcal{D}$. □

Note: The converse of this theorem is not true. For example, $f(z) = e^z$ is analytic everywhere and $f'(z) = e^z \neq 0$ for all complex numbers. We know that e^z is a periodic function with period $2\pi i$. Thus, it is not univalent in $\mathbb{C}$.

7.2.3 Inverse Function Theorem

At a point z_0 where an analytic function has a non-vanishing derivative (Fig. 7.5), if there is a neighbourhood of z_0 restricted to which f is one-to-one (or injective), then the inverse function f^{-1} that is so defined on the image of this neighbourhood is always analytic.

Theorem 7.8. *Let f be an analytic function in a neighbourhood of z_0 satisfying $f(z_0) = w_0$ with $f'(z_0) \neq 0$. Also, $\exists\ \rho > 0$ such that $f(z) - w_0 \neq 0$ in $0 < |z - z_0| \leq \rho$. Let $\delta > 0$ be such that $|f(z) - w_0| \geq \delta$ for z with $|z - z_0| = \varepsilon$. Then*

(i) f is one-to-one on $\mathcal{S}(z_0; \rho) = \{z : |z - z_0| < \rho\}$ if and only if $f'(z_0) \neq 0$ for each z in $\mathcal{S}(z_0; \rho)$

(ii) For each w with $\mathcal{S}(w_0; \delta) = \{w : |w - w_0| < \delta\}$, $\exists$ a unique z with $|z - z_0| < \rho$ such that $f(z) = w$.

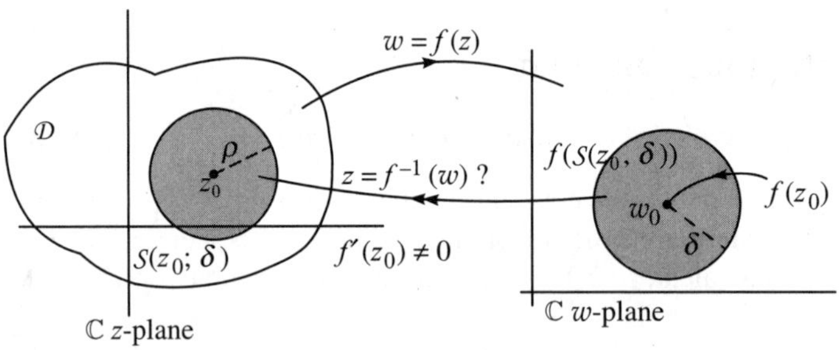

Figure 7.5 *Inverse function theorem*

In fact, if we write, $z = f^{-1}(w)$, then

$$f^{-1}(w) = \frac{1}{2\pi i}\int_{|\zeta - z_0| = \rho} \frac{\zeta f'(\zeta)}{f(\zeta) - w}\, d\zeta; \text{ for } |w - w_0| < \delta. \tag{7.21}$$

Further f^{-1} is analytic function of w for $\mathcal{S}(w_0, \delta)$.

Proof: (i) Given $f'(z_0) \neq 0$, f is a non-constant analytic function round z_0. As $w_0 = f(z_0)$, so z_0 is a zero of order 1 of $f(z) - w_0$, and so

$$\begin{aligned}\mathcal{N}(w_0) &= \text{number of zeros of } f(z) - w_0 \text{ in } |z - z_0| < \rho \\ &= \frac{1}{2\pi i}\int_{|z-z_0|=\rho} d\log\Big(f(z) - w_0\Big) \\ &= \frac{1}{2\pi i}\int_{|z-z_0|=\rho} \frac{f'(z)}{f(z) - w_0}\,dz = 1.\end{aligned}$$

Let $\mathcal{N}(w)$ = the number of times $f(z)$ takes the value w. Then

$$\mathcal{N}(w) = \frac{1}{2\pi i}\int_{|z-z_0|=\rho} d\log\Big(f(z) - w\Big) = \frac{1}{2\pi i}\int_{|z-z_0|=\rho} \frac{f'(z)}{f(z) - w}\,dz$$

for w with $|w - w_0| < \delta$ and $f(z) - w$ not vanishing on the boundary $\partial\mathcal{S}(z_0;\rho)$. Now, $\mathcal{N}(w)$ is an analytic, continuous function of w whose values are integers. Therefore, using lemma 3.3.1, it is a constant, i.e.,

$$\mathcal{N}(w) = \text{ constant } = \mathcal{N}(w_0) = 1.$$

As the image of $\mathcal{S}(w_0;\delta)$ is a connected set, the image of $\mathcal{N}(w)$ is a connected subset of integers which shows that (a) $f(\mathcal{S}(z_0;\rho)) \supset \mathcal{S}(w_0;\delta)$ and (b) f is one-to-one on $\mathcal{S}(z_0;\rho)$.

(ii) The function $g(\zeta) = \frac{\zeta f'(\zeta)}{f(\zeta)-w}$ for $|\zeta - z_0| \le \rho$ is analytic in the closed disc except for a simple pole at $\zeta = z$. Then by Cauchy's residue theorem

$$\begin{aligned}\frac{1}{2\pi i}\int_{|\zeta-z_0|=\rho} \frac{\zeta f'(\zeta)}{f(\zeta) - w}d\zeta &= \text{Res}\Big\{\frac{\zeta f'(\zeta)}{f(\zeta) - w}; z\Big\} \\ &= \lim_{\zeta\to z}\Big[(\zeta - z)\frac{\zeta f'(\zeta)}{f(\zeta) - w}\Big] = \lim_{\zeta\to z}\Bigg[\frac{\zeta f'(\zeta)}{\frac{f(\zeta)-w}{(\zeta - z)}}\Bigg] \\ &= \frac{zf'(z)}{f'(z)} = z \text{ as } f'(z) \neq 0.\end{aligned}$$

Thus, for a fixed w in $\mathcal{S}(w_0;\delta)$ set $z = f^{-1}(w)$ so that $f(z) = w$.

It remains to prove that $f^{-1}(w) = F(w)$, say, is an analytic function of w. Now

$$\begin{aligned}\frac{F(w + \Delta w) - F(w)}{\Delta w} &= \frac{1}{2\pi i}\frac{1}{\Delta w}\oint_{|\zeta-z_0|=\rho}\Big[\frac{1}{f(\zeta) - w - \Delta w} - \frac{1}{f(\zeta) - w}\Big]\zeta f'(\zeta)d\zeta \\ &= \frac{1}{2\pi i}\oint_{|\zeta-z_0|=\rho} \frac{\zeta\, f'(\zeta)}{\Big[f(\zeta) - w - \Delta w\Big]\Big[f(\zeta) - w\Big]}d\zeta. \qquad (7.22)\end{aligned}$$

Now, let $|f(\zeta) - w_0| \ge \delta_w > 0$ for $|\zeta - z_0| = \rho$ and for sufficiently small $|\Delta w|$:

$$\Big|f(\zeta) - w - \Delta w\Big| \ge \Big|f(\zeta) - w\Big| - |\Delta w| \ge \frac{\delta_w}{2} > 0$$

Since, $\zeta f'(\zeta)$ is analytic, so it is bounded. Thus we can find a positive constant M such that $|\zeta f'(\zeta)| \le M$ on $|\zeta - z_0| = \rho$. Using these facts, we get, from Eq. (7.22),

$$\Big|\frac{F(w + \Delta w) - F(w)}{\Delta w} - \frac{1}{2\pi i}\oint_{|\zeta-z_0|=\rho} \frac{\zeta f'(\zeta)d\zeta}{(f(\zeta) - w)^2}\Big|$$

$$= \left| \frac{1}{2\pi i} \oint_{|\zeta - z_0| = \rho} \left[\frac{1}{[f(\zeta) - w - \Delta w][f(\zeta) - w]} - \frac{1}{[f(\zeta) - w]^2} \right] \zeta f'(\zeta) d\zeta \right|$$

$$\leq \frac{1}{2\pi} \oint_{|\zeta - z_0| = \rho} \frac{|\Delta w| \cdot |\zeta f'(\zeta)|}{[f(\zeta) - w - \Delta w][f(\zeta) - w]^2} |d\zeta|$$

$$\leq \frac{1}{2\pi} \cdot \frac{2}{\delta_m} \cdot \frac{1}{\delta_m^2} |\Delta w| M(2\pi\rho) \to 0 \text{ as } \Delta w \to 0.$$

This shows that $F(w)$ is an analytic function. □

RESULT **63.** Theorem 7.8 says that for an analytic function $f(z)$, one-to-one is only a local property. For example, $f(z) = e^z$ is one-to-one only locally, although its derivative does not vanish anywhere. Similarly, the function z^2 is locally one-to-one at every point other than the origin.

7.2.4 Hurwitz's Theorem

Let $\mathcal{D}$ be a domain in $\mathbb{C}$ (Fig. 7.6). Suppose a sequence $\langle f_n(z) \rangle$, where $f_k : \mathcal{D} \to \mathbb{C}$, are analytic for all $k \geq 1$ and $\lim_{k \to \infty} f_k(z) = f(z)$ normally (converging uniformly not on the whole plane but on every compact subset of $\mathcal{D}$) in $\mathcal{D}$. Suppose $z_0 \in \mathcal{D}$ is a zero of f. Hurwitz's theorem tells us that z_0 is obtained as a limit of zeros of f_k, for k large.

Theorem 7.9. *Let $\langle f_n(z) \rangle$ be a sequence of analytic functions on a domain $\mathcal{D}$ that converges normally in $\mathcal{D}$ to $f(z)$. Let $z_0 \in \mathcal{D}$ be a zero of $f(z)$ of order m_0. Then $\exists \rho > 0$ and $N > 0$ such that each f_k for $k \geq N$ has precisely m_0 zeros in $|z - z_0| < \rho$ and further, all the zeros converge (as $\rho \to 0$) to z_0.*

Proof: If f_k is a one-to-one map, then the limit function f is non-constant and so it is one-to-one. Since each f_k is analytic, so f is also analytic because of the uniform convergence on the compact subset due to modulus theorem.

Suppose f is not moromorphic but analytic (so no poles), then the number of zeros inside the closed curve plays an important role. So, by argument principle,

$$\frac{1}{2\pi i} \int_{|z - z_0| = \rho} d \log f_k = \begin{array}{l} \text{number of zeros of } f_k \text{ inside } |z - z_0| < \rho, \\ \text{provided } f_k \neq 0 \text{ on } |z - z_0| = \rho \end{array}$$

$$\frac{1}{2\pi i} \int_{|z - z_0| = \rho} d \log f = \begin{array}{l} \text{number of zeros of } f \text{ inside } |z - z_0| < \rho, \\ \text{provided } f \neq 0 \text{ on } |z - z_0| = \rho. \end{array}$$

We choose ρ in such a way that f_k and f does not vanish on $|z - z_0| = \rho$. Since $f_k \to f$ normally, and each f_k is analytic, so f is analytic. Since f is analytic and the zeros of an analytic functions are isolated, so z_0 is an isolated zero, i.e., we can find a disc surrounding z_0 where there is no other zero.

We cannot assume that f is a constant function because, if f is a constant function and z_0 is zero, then f is constant and by identity theorem 5.27, f is identically zero. Since f is non-constant, so zeros are isolated and we can find always ρ. Thus choose ρ such that f does not vanish for all z satisfying $0 < |z - z_0| \leq \rho$. This shows that

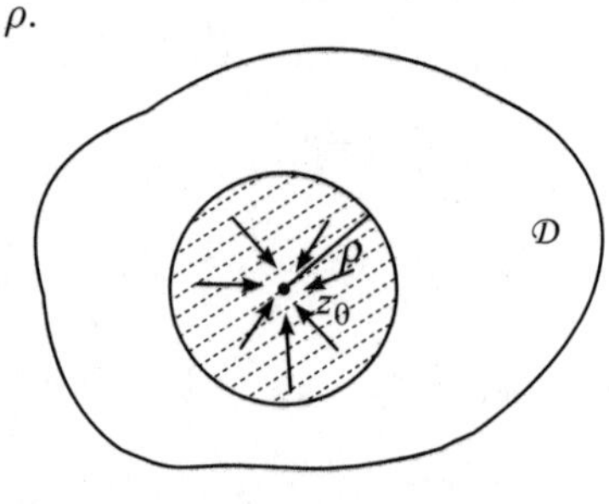

Figure 7.6 *Illustrating Hurwitz's theorem*

$$\int_{|z - z_0| = \rho} d \log f = \int_{|z - z_0| = \rho} \frac{f'}{f} dz$$

is well-defined, as the denominator f is non-zero.

Now, $|f|$ is a real, continuous function and nowhere zero on the compact set $|z - z_0| = \rho$. We know that, a real-valued continuous function on a compact set is uniformly continuous and attains its maximum and minimum value, and therefore so does $|f|$. Since f is non-zero, the minimum value cannot be zero. Thus

$$\delta = \inf\Big\{|f(z)| : |z - z_0| = \rho\Big\} > 0.$$

Thus, $|f(z)| \geq \delta$ on $|z - z_0| = \rho$ and so $\exists\, N_1 \in \mathbb{N}$ such that

$$k \geq N_1 \Rightarrow |f_k(z)| \geq \delta/2 \;\forall z \text{ such that } |z - z_0| = \rho.$$

Let M_1 and M_2 be two positive real numbers such that

$$M_1 = \sup\Big\{|f_k(z)| : |z - z_0| = \rho\Big\} \text{ and } M_2 = \sup\Big\{|f_k'(z)| : |z - z_0| = \rho\Big\}$$

Since $f_k \to f$ uniformly on compact subset $|z - z_0| \leq \rho$ of $\mathcal{D}$, $\exists\, N_2$ such that

$$k \geq N_2 \Rightarrow d\Big(f_k(z), f(z)\Big) = \Big|f_k(z) - f(z)\Big| < \frac{\varepsilon\delta^2}{4M_2}$$

for a given $\varepsilon > 0$ independent of z. Since $f_k' \to f'$ uniformly on compact subset $|z - z_0| \leq \rho$ of $\mathcal{D}$, $\exists\, N_3$ such that

$$k \geq N_3 \Rightarrow d\Big(f_k'(z), f'(z)\Big) = \Big|f_k'(z) - f'(z)\Big| < \frac{\varepsilon\delta^2}{4M_1},$$

where $\varepsilon > 0$ is independent on z. Thus, according to Morea's theorem the integrals $\frac{1}{2\pi i}\int_{|z-z_0|=\rho} d\log f_k$ and $\frac{1}{2\pi i}\int_{|z-z_0|=\rho} d\log f$ are well defined and

$$\begin{aligned}
\frac{1}{2\pi i}\int_{|z-z_0|=\rho} d\log f_k &= \frac{1}{2\pi i}\int_{|z-z_0|=\rho} \frac{f_k'(z)}{f_k(z)}dz = m_k,\\
\frac{1}{2\pi i}\int_{|z-z_0|=\rho} d\log f &= \frac{1}{2\pi i}\int_{|z-z_0|=\rho} \frac{f'(z)}{f(z)}dz = m_0.
\end{aligned}$$

It remains to show that $m_k \to m_0$. Now

$$d\log f_k - d\log f = \Big(\frac{f_k}{f_k} - \frac{f'}{f}\Big)dz$$

Let $N_0 = \max\{N_1, N_2, N_3\} \in \mathbb{N}$. Therefore, for every $z \in \mathcal{D}$ with $|z - z_0| = \rho$ and for every $k \geq N_0$, we get

$$\begin{aligned}
\Big|\frac{f_k}{f_k} - \frac{f'}{f}\Big| &= \frac{|f_k'f - f'f_k|}{|f_k|\,|f|} \leq \frac{2}{\delta^2}\Big|f_k'f - f'f_k\Big| \text{ on } |z - z_0| = \rho \text{ for } k \geq N_3\\
&\leq \frac{2}{\delta^2}\Big|f_k'f - f_k'f_k + f_k'f_k - f'f_k\Big|\\
&\leq \frac{2}{\delta^2}\Big[|f_k'||f - f_k| + |f_k||f_k' - f'|\Big]\\
&< \frac{2}{\delta^2}\Big[M_2 \cdot \frac{\varepsilon\delta^2}{4M_2} + M_1 \cdot \frac{\varepsilon\delta^2}{4M_1}\Big] = \varepsilon.
\end{aligned}$$

Therefore, $\frac{f_k'}{f_k} \to \frac{f'}{f}$ uniformly on $|z - z_0| < \rho$. Thus, for k sufficiently large

$$\frac{1}{2\pi i}\int_{|z-z_0|=\rho} \frac{f_k'}{f_k}\,dz \longrightarrow \frac{1}{2\pi i}\int_{|z-z_0|=\rho} \frac{f'}{f}$$

$\Rightarrow m_k \to m_0$ for k sufficiently large.

Now, as the sequence of integers converges to an integer, the sequence of integers must be a constant sequence. Consequently, $m_k = m_0$ and the proof is complete. □

DEDUCTION **31.** Suppose $\{f_k(z)\}$ is a sequence of univalent (one-to-one, analytic) functions on a domain $\mathcal{D}$ that converges normally on $\mathcal{D}$ to $f(z)$.

Let the limit function $f(z)$ be not constant. Suppose z_0 and ξ_0 satisfy $f(z_0) = f(\xi_0) = w_0$ for $z_0, \xi_0 \in \mathcal{D}$. Then z_0 and ξ_0 are both zeros of finite order for $f(z) - w_0$.

By the preceding Theorem 7.9, there are sequences $z_p \to z_0$ and $\xi_p \to \xi_0$ such that $f_k(z_p) - w_0 = 0$ and $f_k(\xi_p) - w_0 = 0$, with z_p and ξ_p as zeros of $f_k(z) - w_0$. Then

$$f_k(z) - w_0 \to f(z) - w_0 \text{ normally.}$$

As $f_k(z_k) = w_0 = f_k(\xi_p)$ and as f_k is univalent, this implies that

$$z_p = \xi_p \Rightarrow \lim z_p = \lim \xi_p \Rightarrow z_0 = \xi_0.$$

Thus $f(z)$ is univalent. Thus either $f(z)$ is univalent or $f(z)$ is constant.

DEDUCTION **32.**

(i) The sequence $\{e^z/n\}$ converges uniformly to zero on every compact subset of $\mathbb{C}$. Note that functions e^z/n have no zeros in the complex plane.

(ii) For each $n \geq 1$, the function $f_n(z) = z/n$ is univalent on the complex plane. The sequence $\{f_n(z)\}$ converges normally to the constant function $f(z) \equiv 0$. Thus a normal limit of univalent functions need not be univalent, and both alternatives of the theorem occur.

7.2.5 Open Mapping Theorem

Theorem 7.10. [Open mapping theorem]: *Let f be a non-constant, analytic (or holomorphic) function defined on a domain $\mathcal{D} \subset \mathbb{C}$. Then f is an open mapping, i.e., if $U \subset \mathcal{D}$ is an open subset, then $f(U) \subset \mathbb{C}$ is also an open subset. In particular, the image of f, namely $f(\mathcal{D})$ is open in $\mathbb{C}$.*

Proof: Let f be a non-constant analytic function on a domain $\mathcal{D}$. We have to show that if U is open in $\mathcal{D}$, then $f(U)$ is an open subset of $\mathbb{C}$.

Let $z_0 \in \mathcal{D}$ and $w_0 = f(z_0)$. Then z_0 is a zero of $f(z) - w_0$ (see Fig. 7.7). Thus we apply the argument principle for counting the number of zeros. Since $f(z)$ is non-constant, analytic (holomorphic), so $f(z) - w_0$ is also holomorphic and an analytic automorphism. Like f, $f(z) - w_0$ is also non-constant on $\mathcal{D}$.

Since zeros of an analytic function are isolated, $\exists \rho > 0$ such that $f(z) - w_0$ has no other zero in $0 < |z - z_0| \leq \rho$ $(\{z : |z - z_0| \leq \rho\} \subset \mathcal{D})$. We cannot assume that f is a constant function, because, if f is a constant function and z_0 is zero, then by identity theorem 5.27, f is identically zero.

If we restrict the function on the boundary circle, then $|f(z) - w_0|$ is a continuous real-valued function on the boundary of the circle (which is compact). For a continuous real-valued function on a compact connected set, the image must be an interval. Thus we have a minimum and a maximum value. Since $|f(z) - w_0|$ is continuous on the compact connected set $|z - z_0| = \rho$, $\exists \delta > 0$ such that

$$|f(z) - w_0| \geq \delta; \forall z \text{ with } |z - z_0| = \rho.$$

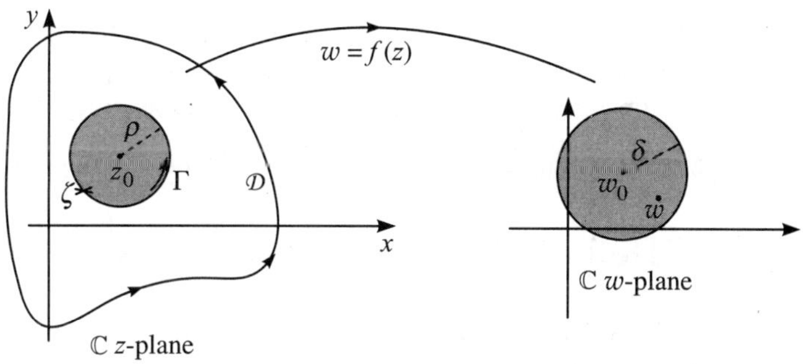

Figure 7.7 *Open mapping theorem*

δ is non-zero, positive, because the only zero z_0 is not on boundary. Let $\zeta \in \{z : |z - z_0| = \rho\}$. Then as $|f(z) - w_0| \geq \delta$, so,

$$f(\zeta) \notin \{w : |w - w_0| \leq \delta\}.$$

So for ζ with $|\zeta - z_0| = \rho$, $|f(\zeta) - w| > 0$ for any w with $|w - w_0| < \delta$. Define

$$\begin{aligned} \mathcal{N}(w) &= \frac{1}{2\pi i}\int_{|z-z_0|=\rho} d\Big(f(z) - w\Big) = \frac{1}{2\pi i}\int_{|z-z_0|=\rho} \frac{f'(z)}{f(z) - w}dz; \quad |w - w_0| < \delta \\ &= N_0 - N_\infty \text{ inside } |z - z_0| = \rho. \end{aligned}$$

Since $f(z) - w$ has no poles and is analytic, so the integral is well-defined. $\mathcal{N}(w)$ values in $\mathbb{Z} \subset \mathbb{C}$. $\mathcal{N}(w)$ is an analytic function of w for $|w - w_0| < \delta$. Suppose, we accept the above fact, then

$$\text{Image of } \mathcal{N}(w) = \mathcal{N}\left(\{|w - w_0| < \delta\}\right) \neq \phi$$

is a connected subset of a complex plane $\mathbb{C}$. This implies

$$\begin{aligned} & \mathcal{N}(w) = \text{ a constant integer (since the only connected subsets of } \mathbb{Z} \text{ are the singleton subsets)} \\ \Rightarrow\ & \mathcal{N}(w) = \mathcal{N}(w_0) = \text{ number of zeros of } f(z) \\ \Rightarrow\ & \text{Every value } w \text{ with } |w - w_0| < \delta \text{ is taken by } f(z) \text{ exactly as many times} \\ & \quad \text{as it takes the value } w_0 \text{ (which is the multiplicity of the zero } z_0 \text{ of } f(z) - w_0). \end{aligned}$$

This means that for every value w_0 of $f(z)$, $\exists$ an open disc $|w - w_0| < \delta$ contained in $\text{Im}(f) = f(\mathcal{D})$ or even in $f(\{z : |z - z_0| < \rho\})$.

$$\Rightarrow f(\mathcal{D}) \text{ or even } f\left(\{z : |z - z_0| < \rho\right) \text{ is open in } \mathbb{C}.$$

Thus a non-constant analytic function f maps open sets into open sets. □

(i) If a non-constant function is just continuous, then Theorem 7.10 does not hold. Indeed, the function $f(z) = \text{Re}\, z$ is continuous everywhere in $\mathbb{C}$, but $f(\mathbb{C}) = \mathbb{R}$ is not open in $\mathbb{C}$.

(ii) Let $\mathcal{D}$ be a non-empty domain and f be a non-constant analytic function on $\mathcal{D}$. Then by Theorem 7.10, $f(\mathcal{D})$ is open, and since f is continuous, it is clear that $f(\mathcal{D})$ is connected. Thus, $f(\mathcal{D})$ is a domain.

(iii) Suppose that $f(z)$ is non-constant. Let $z_0 \in S$ and $w_0 = f(z_0)$. The image of $\mathcal{S}(z_0, \varepsilon) \subset \mathcal{D}$ under f is a domain containing w_0, and hence contains $\mathcal{S}(w_0, \delta)$. Therefore, $\mathcal{S}(z_0, \varepsilon)$ contains a point z whose image $w = f(z)$ is farther away from the origin of the w-plane than the point w_0; i.e., a point z is such that $|f(z)| > |f(z_0)|$. Thus, the maximum modulus principle is established.

(iv) Let f be analytic in a domain $\mathcal{D}$ and let $|f|$ be constant in $\mathcal{D}$. Thus in the w-plane $f(\mathcal{D})$ is a subset of a circle. Similarly, if $\text{Re}\, f$, $\text{Im} f$, $\arg f$ are constant in $\mathcal{D}$, then in w-plane $f(\mathcal{D})$ is either a subset of $u =$ constant or $v =$ constant, or $\tan^{-1}(v/u) =$ constant, respectively. However, we note that none of these takes open sets into open sets. Thus f is a constant.

7.3 Evaluation of Definite Integrals

In this section the method of residue calculus is extensively used in the evaluation of certain type of definite integrals. Some of these integrals occur in several branches of applied mathematics, physical and engineering applications; often, they cannot be evaluated using methods of calculus.

7.3.1 Definite Integral of the Type $\int_0^{2\pi} \Phi(\cos\theta, \sin\theta) d\theta$

Here we shall apply Cauchy's residue theorem to evaluate definite integrals of the form

$$I = \int_{\alpha}^{2\pi+\alpha} \Phi(\cos\theta, \sin\theta)\, d\theta, \tag{7.23}$$

where, $\Phi(\cos\theta, \sin\theta)$ is a rational function with real coefficients of $\cos\theta$ and $\sin\theta$, and whose denominator does not vanish on $[\alpha, 2\pi+\alpha]$. We shall show that Eq. (7.23) can be identified as the parametrized form of a contour integral $\int_\Gamma f(z)dz$ of some complex function f around the positively oriented unit circle $\Gamma : \{z \in \mathbb{C} : |z| = 1\}$. Here we use the parametrization $z = e^{i\theta}, 0 \leq \theta \leq 2\pi$ which gives

$$\cos\theta = \frac{e^{i\theta} + e^{-i\theta}}{2} = \frac{1}{2}\left(z + \frac{1}{z}\right) \text{ and } \sin\theta = \frac{e^{i\theta} - e^{-i\theta}}{2i} = \frac{1}{2i}\left(z - \frac{1}{z}\right)$$

and $izd\theta = dz$. Therefore, the integral transforms to

$$I = \int_\Gamma \Phi\left\{\frac{1}{2}\left(z + \frac{1}{z}\right), \frac{1}{2i}\left(z - \frac{1}{z}\right)\right\} \frac{1}{iz} dz = \oint_\Gamma f(z)\, dz,$$

where $f(z)$ is a rational function of z, that is finite on Γ. Hence it has only removable singularities or poles. Consequently, if α_k denotes those poles of f inside Γ, then by Cauchy's residue theorem

$$I = 2\pi i \times \sum_{k=1}^{n} \text{Res}\left\{f, \alpha_k\right\}.$$

Example 7.3.1. *Evaluate* $I = \int_0^{2\pi} \frac{d\theta}{a+\cos\theta}$, *where* $a \in \mathbb{R}$.

Solution. First, observe that the denominator, $a + \cos\theta$, is never zero, and hence the integrand is finite over $[0, 2\pi]$. We transform this integral into a contour integral to obtain

$$I = \int_\Gamma \frac{dz}{iz\left[a + \frac{1}{2}(z - z^{-1})\right]} = \int_\Gamma \frac{2}{i(z^2 + 2az + 1)}\, dz = \int_\Gamma f(z)\, dz,$$

where Γ is the positively oriented unit circle $\Gamma : \{z \in \mathbb{C} : |z| = 1\}$. The poles of $f(z)$ are given by

$$z^2 + 2az + 1 = 0 \Rightarrow z_{1,2} = -a \pm \sqrt{a^2 - 1}.$$

Now $|z_2| = |-a-\sqrt{a^2-1}| > 1$. Also, as $|z_1 z_2| = 1$, it follows that $|z_1| < 1$. Hence there are no singular points on Γ and $z = z_1$ is the only simple pole inside Γ. Therefore, the residue of the integrand at z_1 is

$$\text{Res}\,\{f(z), z_1\} = \lim_{z\to z_1}\Big[(z-z_1)f(z)\Big] = \lim_{z\to z_1}\frac{1}{2z+2a} = \frac{1}{2\sqrt{a^2-1}}.$$

Thus by the residue theorem

$$I = 2\pi i \cdot \frac{2}{i}\frac{1}{2\sqrt{a^2-1}} = \frac{2\pi}{\sqrt{a^2-1}}.$$

Example 7.3.2. *Show that* $\int_0^{2\pi} \frac{d\theta}{1+a^2-2a\cos\theta} = \frac{2\pi}{1-a^2}$; $0 < a < 1$.

Solution. We transform this integral into a contour integral to obtain

$$\begin{aligned} I &= \int_0^{2\pi}\frac{d\theta}{1+a^2-2a\cos\theta} = \oint_\Gamma \frac{1}{1+a^2-a\left(z-\frac{1}{z}\right)}\frac{1}{iz}dz \\ &= -\frac{1}{ia}\oint_\Gamma \frac{dz}{(z-a)\left(z-\frac{1}{a}\right)} = \oint_\Gamma f(z)dz\ ;\ f(z) = -\frac{1}{ia}\frac{1}{(z-a)\left(z-\frac{1}{a}\right)}, \end{aligned}$$

where Γ is the positively oriented unit circle $\Gamma : \{z \in \mathbb{C} : |z| = 1\}$. The poles of $f(z)$ are given by $z = a, z = 1/a$. Since $0 < a < 1$, so $z = a$ only lies inside the positively oriented unit circle $\Gamma : \{z \in \mathbb{C} : |z| = 1\}$. Now

$$\text{Res}\,\{f(z), a\} = -\frac{1}{ia}\lim_{z\to a}\left[\frac{1}{z-\frac{1}{a}}\right] = \frac{1}{i(1-a^2)}.$$

Therefore, by Cauchy's residue theorem,

$$\oint_\Gamma f(z)dz = 2\pi i \times \frac{1}{i(1-a^2)} = \frac{2\pi}{1-a^2}.$$

Example 7.3.3. *Show that* $I = \int_0^{\pi}\frac{a\,d\theta}{a^2+\sin^2\theta} = \frac{\pi}{\sqrt{1+a^2}}$; $a > 0$.

Solution. We transform this integral into a contour integral to obtain

$$\begin{aligned} I &= \int_0^{\pi}\frac{2a\,d\theta}{2a^2+1-\cos 2\theta} = \int_0^{2\pi}\frac{a\,d\phi}{2a^2+1-\cos\phi}\ :\ 2\theta = \phi \\ &= \oint_\Gamma \frac{a}{2a^2+1-\frac{1}{2}\left(z+\frac{1}{z}\right)}\frac{1}{iz}dz\ ;\ z = e^{i\phi} \\ &= 2ai\oint_\Gamma \frac{dz}{z^2-2(2a^2+1)z+1} = \oint_\Gamma f(z)dz, \end{aligned}$$

where Γ is a positively oriented unit circle $\Gamma : \{z \in \mathbb{C} : |z| = 1\}$. The poles of $f(z)$ are given by

$$z^2 - 2(2a^2+1)z + 1 = 0 \Rightarrow z_{1,2} = (2a^2+1) \pm 2a\sqrt{a^2+1}.$$

Clearly, $|z_1| > 1$ and as $|z_1 z_2| = 1$, we have, $|z_2| < 1$. Hence $z = z_2$ is the only pole inside Γ. Now

$$\begin{aligned} \text{Res}\,\{f(z); z_2\} &= 2ai\lim_{z\to z_2}\left[\frac{z-z_2}{z^2-2(2a^2+1)z+1}\right] \\ &= 2ai\lim_{z\to z_2}\frac{z-z_2}{(z-z_1)(z-z_2)} = \frac{2ai}{z_2-z_1} = -\frac{i}{2\sqrt{a^2+1}} \end{aligned}$$

Therefore, by Cauchy's residue theorem,

$$\int_0^{\pi} \frac{a d\theta}{a^2 + \sin^2\theta} = 2\pi i \times \left(-\frac{i}{2\sqrt{a^2+1}}\right) = \frac{\pi}{\sqrt{a^2+1}}.$$

Example 7.3.4. *Evaluate* $\int_0^{2\pi} \frac{\cos n\theta}{1+2a\cos\theta+a^2}\, d\theta$, *where* $n \in \mathbb{N}$ *and* $a^2 < 1$.

Solution. Let us consider the integral $I = \int_0^{2\pi} \frac{e^{in\theta}}{1+2a\cos\theta+a^2}\, d\theta$. Let $z = e^{i\theta}, 0 \le \theta \le 2\pi$. Then $iz\, d\theta = dz$.

$$\begin{aligned}
\therefore I &= \int_\Gamma \frac{z^n}{\{1 + a(z + z^{-1}) + a^2\}} \frac{1}{iz}\, dz, \quad \text{where } \Gamma : |z| = 1 \\
&= \frac{1}{i} \int_\Gamma \frac{z^n}{\{(1 + a^2)z + az^2 + a\}}\, dz \\
&= \frac{1}{ai} \int_\Gamma \frac{z^n}{(z + a)(z + \frac{1}{a})} dz = \frac{1}{ai} \int_\Gamma f(z)\, dz;
\end{aligned}$$

where $f(z) = z^n/\{(z + a)(z + \frac{1}{a})\}$. The poles of $f(z)$ are given by

$$(z + a)(z + \frac{1}{a}) = 0 \Rightarrow z = -a, -1/a,$$

which are simple poles. Since $a^2 < 1$, only the pole $z = -a$ lies inside Γ. Now

$$\operatorname*{Res}_{z=-a} f(z) = \lim_{z\to -a} (z + a)\frac{z^n}{(z + a)(z + \frac{1}{a})} = \frac{(-1)^n a^{n+1}}{1 - a^2}.$$

Therefore, by Cauchy's residue theorem,

$$\begin{aligned}
I &= \frac{1}{ai} \int_\Gamma f(z)\, dz = \frac{1}{ai}\, 2\pi i \times \operatorname*{Res}_{z=-a} f(z) = \frac{2\pi}{a} \frac{(-1)^n a^{n+1}}{1 - a^2} \\
\Rightarrow \quad & \int_0^{2\pi} \frac{e^{in\theta}}{1 + 2a\cos\theta + a^2}\, d\theta = 2\pi \frac{(-1)^n a^n}{1 - a^2}.
\end{aligned}$$

Equating the real and imaginary parts, we get

$$\int_0^{2\pi} \frac{\cos n\theta}{1 + 2a\cos\theta + a^2}\, d\theta = \frac{2\pi(-1)^n a^n}{1 - a^2} \text{ and } \int_0^{2\pi} \frac{\sin n\theta}{1 + 2a\cos\theta + a^2}\, d\theta = 0.$$

Example 7.3.5. *Evaluate* $\int_0^{2\pi} e^{-\cos\theta} \cos(n\theta + \sin\theta)\, d\theta$, *where* n *is a positive integer.*

Solution. Let us consider the integral

$$\begin{aligned}
I &= \int_0^{2\pi} e^{-\cos\theta} \Big[\cos(n\theta + \sin\theta) - i\sin(n\theta + \sin\theta)\Big]\, d\theta \\
&= \int_0^{2\pi} e^{-\cos\theta}\, e^{-i(n\theta+\sin\theta)}\, d\theta \\
&= \int_0^{2\pi} e^{-(\cos\theta+i\sin\theta)}\, e^{-in\theta}\, d\theta = \int_0^{2\pi} e^{-e^{i\theta}}\, e^{-in\theta}\, d\theta.
\end{aligned}$$

Let $z = e^{i\theta}, 0 \le \theta \le 2\pi$. Then $iz\,d\theta = dz$, and $|z| = 1$ is the locus of z described in the positive sense when θ varies from 0 to 2π. We get

$$I = \int_\Gamma \left(e^{-z}\cdot\frac{1}{z^n}\right)\frac{dz}{iz} = \frac{1}{i}\int_\Gamma \frac{e^{-z}}{z^{n+1}}\,dz = \int_\Gamma f(z)\,dz;$$

where $f(z) = e^{-z}/z^{n+1}$ and $\Gamma : |z| = 1$. Clearly, $f(z)$ has a pole of order $(n+1)$ at $z = 0$ which lies within Γ. Thus

$$\operatorname*{Res}_{z=0} f(z) = \frac{1}{n!}\lim_{z\to 0}\frac{d^n}{dz^n}\left[z\frac{e^{-z}}{z^{n+1}}\right] = \frac{(-1)^n}{i\,n!}.$$

Therefore, by Cauchy's residue theorem,

$$I = \int_\Gamma f(z)\,dz = 2\pi i \times \operatorname*{Res}_{z=0} f(z) = 2\pi i \times \frac{(-1)^n}{i\,n!} = 2\pi\frac{(-1)^n}{n!}.$$

Equating the real and imaginary parts, we get $\int_0^{2\pi} e^{-\cos\theta}\cos(n\theta + \sin\theta)\,d\theta = 2\pi\frac{(-1)^n}{n!}$.

Definition 7.3.1. Let $f(x)$ be a function continuous on the non-negative real axis $0 \le x < \infty$. Then the improper integral of $f(x)$ over $[0, \infty)$ is defined by

$$I_1 = \int_0^\infty f(x)dx = \lim_{R\to\infty}\int_0^R f(x)\,dx.$$

If the limit exists, the integral I_1 is said to be convergent; otherwise it is divergent. Similarly, the improper integral of $f(x)$ over $(-\infty, 0]$ is defined as

$$I_2 = \int_{-\infty}^0 f(x)dx = \lim_{r\to\infty}\int_{-r}^0 f(x)\,dx.$$

If $f(x)$ is continuous for all x, the improper integral of $f(x)$ over $(-\infty, \infty)$ is defined by

$$\begin{aligned} I &= \int_{-\infty}^\infty f(x)dx = \int_{-\infty}^0 f(x)dx + \int_0^\infty f(x)dx \\ &= \lim_{r\to\infty}\int_{-r}^0 f(x)\,dx + \lim_{R\to\infty}\int_0^R f(x)\,dx = I_2 + I_1, \end{aligned}$$

provided both the limits exist.

Definition 7.3.2. The *Cauchy principal value* (p.v.) of I is defined to be the number

$$\text{p.v.}\int_{-\infty}^\infty f(x)dx = \lim_{\rho\to\infty}\int_{-\rho}^\rho f(x)dx \tag{7.24}$$

provided this limit exists.

(i) The Cauchy principal value of an integral may exist even though the integral itself is not convergent. For example, $\int_{-\rho}^\rho x\,dx = 0$, for all ρ, and hence p.v. $\int_{-\infty}^\infty x\,dx = 0$ but $\int_0^\infty x\,dx = \infty$. However, if the improper integral I exists, then clearly it must be equal to its principal value (p.v.).

(ii) If $f(x)$ is an even function, i.e., $f(-x) = f(x)$ for all $x \in \mathbb{R}$, and if the Cauchy principal value of I exists, then I exists. Moreover,

$$\frac{1}{2}\,\text{p.v.}\int_{-\infty}^\infty f(x)dx = \int_0^\infty f(x)dx.$$

7.3.2 Definite Integrals of the Type $\int_{-\infty}^{\infty} \frac{p(z)}{q(z)} dz$

Here we will evaluate improper integrals of the type

$$I = \int_{-\infty}^{\infty} f(z)\, dz; \text{ where, } f(x) = \frac{p(x)}{q(x)}, \tag{7.25}$$

$p(x)$ and that $q(x)$ are polynomials in x with real coefficients and no factors in common. We assume that where the degree of $q(x) \geq 2+$ degree of $p(x)$ and that $q(x)$ has no real zeros but has at least one zero in the upper half of the z-plane. We choose the curve Γ consisting of

(i) the interval $[-R, R]$ on the real axis and

(ii) the semicircle $C_R : |z| = R$ lying in the upper half of the plane (Fig. 7.8).

If there are no poles of $f(z)$ on the real axis, the circle C_R which is arbitrary can be taken such that there is no singularity on its circumference in the upper half of the plane, but possibly, some poles exist inside the contour Γ specified above.

In the following theorem, we use the residue theory to compute the Cauchy principal value of I for a certain class of functions $f(x)$.

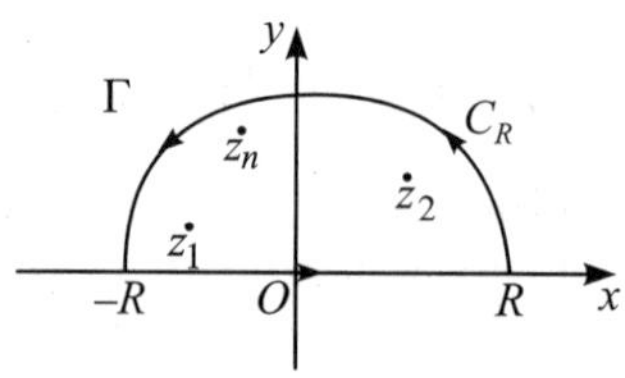

Figure 7.8 $\Gamma = [-R, R] \cup C_R$

Theorem 7.11. *Let $f(z)$ be a rational function having no real poles, and let there exist ρ such that for any $|z| > \rho$, $|f(z)| \leq M/|z|^k; k > 1$ for some $M > 0$. Then,*

$$\text{p.v.} \int_{-\infty}^{\infty} f(x)dx = 2\pi i \sum \text{Res}\, \{f(z); \alpha_j\} \tag{7.26}$$

holds, where the sum is taken over all the poles α_j of $f(z)$ that fall in the upper half-plane.

Proof: Suppose $R > \rho$ is so large that the semicircle C_R contains all the poles of $f(z)$ that are in the upper half-plane. On $\Gamma = [-R, R] \cup C_R$, by Cauchy's residue theorem, we have

$$\begin{aligned} \int_{\Gamma} f(z)\, dz &= \int_{-R}^{R} f(x)\, dx + \int_{C_R} f(z)\, dz, \\ &= 2\pi i \sum \text{Res}\, \{f(z); \alpha_j\}. \end{aligned} \tag{7.27}$$

If $|f(z)| \leq M/R^k$ on C_R, where $k > 1$ and M are constants, then using ML inequality,

$$\left| \int_{C_R} f(z)dz \right| \leq \frac{M}{R^k} \cdot \pi R = \frac{\pi M}{R^{k-1}}$$

$$\Rightarrow \quad \lim_{R\to\infty} \left| \int_{C_R} f(z)dz \right| = 0; \text{ i.e., } \lim_{R\to\infty} \int_{C_R} f(z)dz = 0.$$

Clearly, the right-hand side of Eq. (7.27) is independent of R, while for the left-hand side we find

$$\lim_{R\to\infty} \int_{-R}^{R} f(x)dx = \text{p.v.} \int_{-\infty}^{\infty} f(x)dx.$$

Hence Eq. (7.26) follows. □

Example 7.3.6. *Use the method of contour integration evaluate $\int_{-\infty}^{\infty} \frac{dx}{x^6+1}$.*

Solution. Consider $\oint_\Gamma f(z)dz$, where $f(z) = 1/(z^6+1)$ and Γ is the closed contour as in Fig. 7.8 traversed in the positive sense. The isolated singularities of $f(z)$ are given by

$$z^6 = -1 \Rightarrow z = e^{r\pi i/6};\ r = 1, 3, 5, 7, 9, 11.$$

Note that none of the simple poles lies on the real axis. Only the poles $e^{i\pi/6}, e^{3\pi i/6}, e^{5\pi i/6}$ lie within Γ. Now,

$$\begin{aligned}
\text{Res}\Big[f(z), e^{i\pi/6}\Big] &= \lim_{z\to e^{i\pi/6}}\left[(z-e^{i\pi/6})\frac{1}{z^6+1}\right] = \lim_{z\to e^{i\pi/6}}\frac{1}{6z^5} = \frac{1}{6}e^{-\frac{5\pi i}{6}} = \Re_1\\
\text{Res}\Big[f(z); e^{3\pi i/6}\Big] &= \lim_{z\to e^{3\pi i/6}}\left[(z-e^{3\pi i/6})\frac{1}{z^6+1}\right] = \lim_{z\to e^{3\pi i/6}}\frac{1}{6z^5} = \frac{1}{6}e^{-\frac{5\pi i}{2}} = \Re_2\\
\text{Res}\Big[f(z); e^{5\pi i/6}\Big] &= \lim_{z\to e^{5\pi i/6}}\left[(z-e^{5\pi i/6})\frac{1}{z^6+1}\right] = \frac{1}{6}e^{-\frac{25\pi i}{6}} = \Re_3
\end{aligned}$$

Hence by Cauchy's residue theorem, we get,

$$\begin{aligned}
\oint_\Gamma \frac{dz}{z^6+1} &= 2\pi i\Big[\Re_1+\Re_2+\Re_3\Big] = \frac{2\pi}{3}\\
\text{or,}\quad \int_{-R}^{R}\frac{dz}{z^6+1} &+ \int_{C_R}\frac{dz}{z^6+1} = \frac{2\pi}{3}. \qquad (7.28)
\end{aligned}$$

Now, along C_R let $z = Re^{i\theta}$. Then

$$\begin{aligned}
\left|\int_{C_R} f(z)dz\right| &= \left|\int_0^\pi \frac{Rie^{i\theta}}{R^6e^{6i\theta}+1}d\theta\right| \le \int_0^\pi \frac{R}{|R^6e^{6i\theta}+1|}d\theta\\
&\le \int_0^\pi \frac{R}{|R^6e^{6i\theta}|-1}d\theta;\ \text{as } |z_1+z_2| \ge |z_1|-|z_2|\\
&\le \int_0^\pi \frac{R}{R^6-1}d\theta = \frac{\pi R}{R^6-1} \to 0 \text{ as } R\to\infty.
\end{aligned}$$

Taking the limit of both sides of Eq. (7.28) as $R\to\infty$, we get,

$$\int_{-\infty}^{\infty}\frac{dz}{z^6+1} + 0 = \frac{2\pi}{3} \Rightarrow \int_0^\infty \frac{dz}{z^6+1} = \frac{2\pi}{3}.$$

Thus the required integral has the value $2\pi/3$.

Example 7.3.7. *Evaluate* $\int_{-\infty}^{\infty}\frac{x^2}{(x^2+1)(x^2+4)}dx$.

Solution. Poles of $f(z) = z^2/(z^2+1)(z^2+4)$ are given by

$$(z^2+1)(z^2+4) = 0 \Rightarrow z = \pm i, \pm 2i.$$

i and $2i$ are the only two poles which lies within the chosen contour (Fig. 7.8). The residue of $f(z)$ at these simple poles are respectively given by

$$\begin{aligned}
R_1 &= \lim_{z\to i}\left[(z-i)\frac{z^2}{(z^2+1)(z^2+4)}\right] = \frac{i}{6}\\
R_2 &= \lim_{z\to 2i}\left[(z-2i)\frac{z^2}{(z^2+1)(z^2+4)}\right] = -\frac{i}{3}.
\end{aligned}$$

Hence, by Cauchy's residue theorem, we have,

$$\oint_\Gamma f(z)dz = 2\pi i \times \text{sum of the residues within } \Gamma$$

$$\text{or,} \quad \int_{-R}^{R} f(z)dz + \int_{C_R} f(z)dz = 2\pi i \times \left[\frac{i}{6} - \frac{i}{3}\right]$$

$$\text{or,} \quad \int_{-R}^{R} \frac{z^2 dz}{(z^2+1)(z^2+4)} + \int_{C_R} \frac{z^2 dz}{(z^2+1)(z^2+4)} = \frac{\pi}{3}. \tag{7.29}$$

Now, on $C_R \ : \ |z| = R$. Let $z = Re^{i\theta}$. Then

$$\begin{aligned}\left|\int_{C_R} \frac{z^2 dz}{(z^2+1)(z^2+4)}\right| &\le \int_0^\pi \frac{|R^2 e^{2i\theta}||dz|}{(|z^2|-1)(|z^2|-4)} \\ &\le \frac{R^2 \cdot R\pi}{(R^2-1)(R^4-1)} \to 0 \text{ as } R \to \infty.\end{aligned}$$

Making $R \to \infty$ in (7.29), we get, $\int_{-\infty}^{\infty} \frac{z^2}{(z^2+1)(z^2+4)}\, dz = \pi/3$.

7.3.3 Integrals of the Form $\int_{-\infty}^{\infty} \frac{P(x)}{Q(x)} \cos mx dx$

The residue theorem will be useful in evaluating convergent improper integrals of the form

$$\int_{-\infty}^{\infty} \frac{P(x)}{Q(x)} \sin mx dx, \int_{-\infty}^{\infty} \frac{P(x)}{Q(x)} \cos mx dx; \ \ (m > 0)$$

where $P(x)$ and $Q(x)$ are polynomials satisfying

(i) degree of $Q(x)$ exceeds that of $P(x)$, i.e., deg$Q(x) \ge$ deg$P(x) + 2$

(ii) the equation $Q(x) = 0$ has no real roots and has at least one zero above the real axis.

To use the semicircular contour, we cannot use the function $\cos mz$ because it behaves too badly in the upper half plane. To obtain the integral, we substitute e^{imz} for $\cos mz$ in the contour integral, and recover the cosine integral at the end by taking real parts. Since $|e^{im(x+iy)}| = e^{-my}$, so $|e^{imz}| \le 1$, Im $z \ge 0$. So consider the integral

$$\oint_\Gamma e^{imz} f(z) \ dz; \text{ where } f(z) = P(z)/Q(z),$$

and where Γ is the contour as depicted in the Fig. 7.8. We then have by Cauchy's residue theorem,

$$\begin{aligned}\oint_\Gamma e^{imz} f(z)dz &= \int_{-R}^{R} e^{imz} f(z)dz + \oint_{C_R} e^{imz} f(z)dz \\ &= 2\pi i \times \text{ sum of the residues.}\end{aligned}$$

If deg$Q(x) \ge$ deg$P(x) + 2$, the ML-estimate shows that

$$\oint_{C_R} e^{imz} f(z)dz \to 0 \text{ as } R \to \infty.$$

These integrals occur in the theory and application of the Fourier integral. The following inequality and lemma may be found important and useful in the process of integration over the contour Γ.

RESULT **64.** [Jordan's inequality]: We know that as θ increases from 0 to $\pi/2$, $\cos\theta$ decreases steadily and consequently the ordinate of the graph of $y = \cos x$ over the range $0 \le x \le \theta$ also decreases steadily. But the mean ordinate is

$$\frac{1}{\theta}\int_0^\theta \cos x \, dx = \frac{\sin\theta}{\theta}.$$

It follows that, when $0 \le \theta \le \pi/2$,

$$\frac{2}{\pi} \le \frac{\sin\theta}{\theta} \le 1 \Rightarrow \frac{2\theta}{\pi} \le \sin\theta \le \theta. \tag{7.30}$$

Jordan's inequality is quite clear on the graph of $y = \sin\theta$ (Fig. 7.9) and the line $y = 2\theta/\pi$ between $\theta = 0$ and $\theta = \pi/2$ in the (θ, y)-plane.

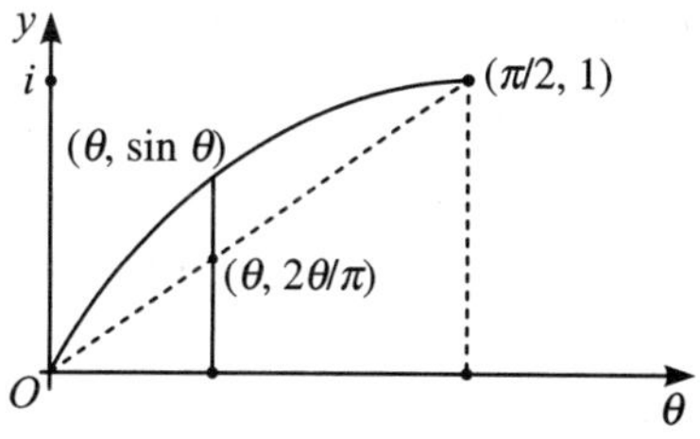

Figure 7.9 *Geometrical interpretation of Jordan's inequality*

We shall also prove Jordan's Lemma, which plays a fundamental role in the computation of integrals involving rational functions multiplied by trigonometric functions.

Lemma 7.3.1. [Jordan's lemma]: *Let a function $f(z)$ satisfy the following conditions:*

(i) $f(z)$ is analytic in upper half plane except at a finite number of poles

(ii) $f(z) \longrightarrow 0$ uniformly as $|z| \to \infty$ with $0 \le \arg z \le \pi$

(iii) m is a positive integer.

Then, the following holds:

$$\lim_{R\to\infty} \oint_{C_R} e^{imz} f(z)dz = 0; \quad m > 0, \tag{7.31}$$

where, C_R : $|z| = R$, $\text{Im}(z) > 0$ is a semi-circle.

Proof: Here we assume that $f(z)$ has no singularities on C_R for sufficiently large values of R. Since $f(z) \to 0$ as $z \to \infty$, so, corresponding to an $\varepsilon > 0$, $\exists R_0 > 0$ such that

$$|f(z)| < \varepsilon \text{ when } |z| = R \le R_0.$$

Now, on C_R, let $z = Re^{i\theta}$. Then

$$\begin{aligned}
\oint_{C_R} e^{imz} f(z)dz &= \int_0^\pi e^{imRe^{i\theta}} f(Re^{i\theta})\, Rie^{i\theta}\, d\theta \\
&= \int_0^\pi e^{imR\cos\theta} e^{-mR\sin\theta} f(Re^{i\theta})\, Rie^{i\theta}\, d\theta \\
\therefore \left|\oint_{C_R} e^{imz} f(z)dz\right| &\le \int_0^\pi \left|e^{imR\cos\theta}\right| e^{-mR\sin\theta} |f(Re^{i\theta})|\, R\, d\theta \\
&\le 2\pi R\varepsilon \int_0^{\pi/2} e^{-2mR\theta/\pi}\, d\theta; \text{ using Eq. (7.30)} \\
&\le \frac{\pi\varepsilon}{m}\left(1 - e^{-mR}\right) \to 0 \text{ as } R \to \infty.
\end{aligned}$$

Hence the result is established. □

Example 7.3.8. *Evaluate the integral $\int_0^\infty \frac{\cos mx}{a^2+x^2} dx$, $(m > 0)$.*

Solution. Let us consider $\oint_\Gamma f(z)\,dz$, where $f(z) = e^{imz}/(a^2+z^2)$ and Γ is the contour of Fig. 7.8. f has simple poles at $z = \pm ai$, but only $z = ia$ lies inside Γ. Now,

$$\text{Res}\,\{f; ia\} = \lim_{z\to ai}\left[(z-ai)\frac{e^{imz}}{a^2+z^2}\right] = \lim_{z\to ai}\frac{e^{imz}}{z+ai} = \frac{1}{2ai}e^{-am}.$$

By Cauchy's residue theorem, we have

$$\oint_\Gamma f(z)dz = \int_{-R}^{R} f(z)dz + \int_{C_R} f(z)dz = 2\pi i \times \frac{1}{2ai}e^{-am} = \frac{\pi}{a}e^{-am}. \tag{7.32}$$

Now, on C_R, let $z = Re^{i\theta}$. Then

$$\begin{aligned}
\int_{C_R} f(z)dz &= \int_0^\pi \frac{e^{imRe^{i\theta}}}{a^2+R^2e^{2i\theta}} iRe^{i\theta}\,d\theta.\\
\therefore \left|\int_{C_R} f(z)dz\right| &\le R\int_0^\pi \frac{\left|e^{imR(\cos\theta+i\sin\theta)}\right|}{|a^2+R^2e^{2i\theta}|}d\theta\\
&\le R\int_0^\pi \frac{\left|e^{-mR\sin\theta}\right|\left|e^{imR\cos\theta}\right|}{a^2-|R^2e^{2i\theta}|}d\theta;\quad |z_1+z_2|\ge|z_1|-|z_2|\\
&\le \frac{2R}{a^2-R^2}\int_0^{\pi/2} e^{-2mR\theta/\pi}d\theta \text{ as } \sin\theta \ge \frac{2\theta}{\pi} \text{ for } 0\le\theta\le\frac{\pi}{2}\\
&\le \frac{\pi}{m(a^2-R^2)}\left[1-e^{-mR}\right] \to 0 \text{ as } R\to\infty, \text{ since } m>0.
\end{aligned}$$

Thus taking the limit as $R\to\infty$, we get from (7.32),

$$\int_{-\infty}^{\infty}\frac{e^{imz}}{a^2+z^2}dz = \int_{-\infty}^{\infty}\frac{\cos mz + i\sin mz}{a^2+z^2}dz = \frac{\pi}{a}e^{-am}.$$

Equating the real parts, we get

$$\int_{-\infty}^{\infty}\frac{\cos mz}{a^2+z^2}dz = \frac{\pi}{a}e^{-am} \Rightarrow \int_0^\infty \frac{\cos mz}{a^2+z^2}dz = \frac{\pi}{2a}e^{-am}$$

for an even function.

Example 7.3.9. *Evaluate* $\int_{-\infty}^{\infty}\frac{\cos x}{(x^2+a^2)(x^2+b^2)}dx$; $a > b > 0$.

Solution. Consider the integral $\oint_\Gamma f(z)dz$; where $f(z) = e^{iz}/\{(z^2+a^2)(z^2+b^2)\}$, and Γ is the contour of Fig. 7.8. The poles of $f(z)$ are $z = \pm ai,\ \pm bi$, but $z = ai$ and $z = bi$ lie in the upper half-plane. Now,

$$\text{Res}\,\{f; ia\} = \lim_{z\to ai}\left[(z-ai)\frac{e^{iz}}{(z^2+a^2)(z^2+b^2)}\right] = \frac{e^{-a}}{2ai(b^2-a^2)} = \Re_1.$$

$$\text{Res}\,\{f; ib\} = \lim_{z\to bi}\left[(z-bi)\frac{e^{iz}}{(z^2+a^2)(z^2+b^2)}\right] = \frac{e^{-b}}{2bi(a^2-b^2)} = \Re_2.$$

By Cauchy's residue theorem we have,

$$\begin{aligned}\oint_\Gamma f(z)dz &= \int_{-R}^{R} f(z)dz + \int_{C_R} f(z)dz = 2\pi i(\Re_1+\Re_2)\\ &= 2\pi i\left[\frac{e^{-a}}{2ai(b^2-a^2)}+\frac{e^{-b}}{2bi(a^2-b^2)}\right] = \frac{\pi}{a^2-b^2}\left[\frac{e^{-b}}{b}-\frac{e^{-a}}{a}\right]. \end{aligned}\tag{7.33}$$

On the curve C_R, let $z = Re^{i\theta}$. Then

$$\begin{aligned}\int_{C_R} f(z)dz &= \int_0^\pi \frac{e^{iRe^{i\theta}}}{(R^2e^{2i\theta}+a^2)(R^2e^{2i\theta}+b^2)}Rie^{i\theta}d\theta\\ &= iR\int_0^\pi \frac{e^{-R\sin\theta}\cdot e^{iR\cos\theta}}{(R^2e^{2i\theta}+a^2)(R^2e^{2i\theta}+b^2)}e^{i\theta}d\theta.\\ \therefore \left|\int_{C_R} f(z)dz\right| &\le R\int_0^\pi \frac{e^{-R\sin\theta}}{|R^2e^{2i\theta}+a^2|\cdot|R^2e^{2i\theta}+b^2|}d\theta\\ &\le R\int_0^\pi \frac{e^{-R\sin\theta}}{(|R^2e^{2i\theta}|-a^2)\cdot(|R^2e^{2i\theta}|-b^2)}d\theta\\ &\le R\int_0^{\pi/2} \frac{2e^{-R\sin\theta}d\theta}{(R^2-a^2)(R^2-b^2)};\ \sin\theta\ge\frac{2\theta}{\pi} \text{ for } 0\le\theta\le\frac{\pi}{2}\\ &\le \frac{2R}{(R^2-a^2)(R^2-b^2)}\int_0^{\pi/2} e^{-\frac{2R\theta}{\pi}}d\theta\\ &\le \frac{2R}{(R^2-a^2)(R^2-b^2)}\frac{\pi}{2R}\left[1-e^{-R}\right]\to 0 \text{ as } R\to\infty.\end{aligned}$$

Thus taking the limit as $R\to\infty$, we get from(7.33)

$$\begin{aligned}\int_{-\infty}^{\infty}\frac{e^{iz}}{(z^2+a^2)(z^2+b^2)}dz &= \frac{\pi}{a^2-b^2}\left[\frac{e^{-b}}{b}-\frac{e^{-a}}{a}\right]\\ \text{or, }\int_{-\infty}^{\infty}\frac{\cos z+i\sin z}{(z^2+a^2)(z^2+b^2)}dz &= \frac{\pi}{a^2-b^2}\left[\frac{e^{-b}}{b}-\frac{e^{-a}}{a}\right]\\ \text{or, }\int_{-\infty}^{\infty}\frac{\cos z}{(z^2+a^2)(z^2+b^2)}dz &= \frac{\pi}{a^2-b^2}\left[\frac{e^{-b}}{b}-\frac{e^{-a}}{a}\right].\end{aligned}$$

Example 7.3.10. *Evaluate* $\int_{-\infty}^{\infty}\frac{\cos mx}{(a^2+x^2)^2}\,dx; m>0, a>0.$

Solution. Consider $\oint_\Gamma e^{imz}f(z)dz$, where $f(z)=1/(z^2+a^2)^2$ and Γ the contour of Fig. 7.8. The poles of $f(z)$ are given by $z=\pm ai$, which are double poles, but $z=ai$ lies in the upper half-plane. Now,

$$\begin{aligned}\text{Res}\,\{e^{imz}f(z);ai\} &= \lim_{z\to ai}\frac{d}{dz}\left[(z-ai)^2\frac{e^{imz}}{(z^2+a^2)^2}\right]\\ &= \lim_{z\to ai}\frac{ime^{imz}(z+ai)^2-2(z+ai)e^{imz}}{(z+ia)^4} = -\frac{i}{4a^3}(ma+1)e^{-ma}.\end{aligned}$$

By Cauchy's residue theorem, we have,

$$\begin{aligned}\oint_{\Gamma} e^{imz} f(z)\, dz &= \int_{-R}^{R} e^{imz} f(z)\, dz + \int_{C_R} e^{imz} f(z)\, dz \\ &= 2\pi i \times -\frac{i}{4a^3}(ma+1)e^{-ma} = \frac{2\pi}{2a^3}(ma+1)e^{-ma}.\end{aligned}$$

On the curve C_R let $z = Re^{i\theta}$. Then

$$\begin{aligned}\left|\int e^{imz} f(z)\, dz\right| &= \left|\int_0^{\pi} \frac{e^{imRe^{i\theta}}}{(a^2+R^2e^{2i\theta})^2} Rie^{i\theta}\, d\theta\right| \\ &= \left|\int_0^{\pi} \frac{e^{imR(\cos\theta + i\sin\theta)}}{(a^2+R^2e^{2i\theta})^2} Rie^{i\theta}\, d\theta\right| \\ &\leq \int_0^{\pi} \frac{e^{-mR\sin\theta}\, |e^{mRi\cos\theta}||Ri||e^{i\theta}|}{(|R^2e^{2i\theta} - | - a|^2)^2}\, d\theta \\ &\leq \frac{R}{(R^2-a^2)^2}\int_0^{\pi} e^{-mR\sin\theta}\, d\theta; \quad \sin\theta \geq \frac{2\theta}{\pi} \text{ for } 0 \leq \theta \leq \frac{\pi}{2} \\ &\leq \frac{2R}{(R^2-a^2)^2}\int_0^{\pi/2} e^{-mR\sin\theta}\, d\theta = \frac{\pi(1-e^{-mR})}{m(R^2-a^2)^2} \to 0 \text{ as } R \to \infty.\end{aligned}$$

Thus taking the limit as $R \to \infty$, we get from (7.34)

$$\begin{aligned}&\int_{-\infty}^{\infty} \frac{e^{iz}}{(z^2+a^2)^2} dz + 0 = \frac{2\pi}{2a^3}(ma+1)e^{-ma} \\ \Rightarrow\quad &\int_{-\infty}^{\infty} \frac{\cos mx}{(a^2+x^2)^2}\, dx + i\int_{-\infty}^{\infty} \frac{\sin mx}{(a^2+x^2)^2}\, dx = \frac{2\pi}{2a^3}(ma+1)e^{-ma} \\ \Rightarrow\quad &\int_{-\infty}^{\infty} \frac{\cos mx}{(a^2+x^2)^2}\, dx = \frac{2\pi}{2a^3}(ma+1)e^{-ma}.\end{aligned}$$

7.3.4 Poles on the Real Axis

Here we will consider such type of integrals for which the integrand $f(z)$ has poles lying on the real axis. Suppose a is a pole lying on the real axis (Fig. 7.10). In this case we indent the real axis by a semi-circle C_2 of radius ε, with centre a lying in the upper half plane, where ε is chosen to be sufficiently small (Fig. 7.10). Thus Γ is the contour consisting of

(i) the real axis from $-R$ to $-\varepsilon$

(ii) the upper half of the circle C_2, $|z-a| = \varepsilon$

(iii) the real axis from ε to R, where ε is small and R is large

(iv) The upper half circle $C_R : |z| = R$

as shown in Fig. 7.10. Such an indenting must be done for every pole of $f(z)$ lying on the real axis. The following theorems may be found important and useful in the process of integration over indented circular arcs.

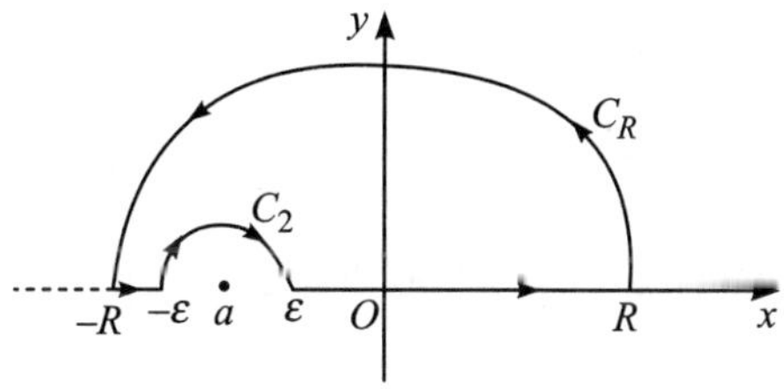

Figure 7.10 *$f(z)$ has pole a lying on the real axis*

Theorem 7.12. *Let AB be an arc $\alpha \leq \theta \leq \beta$ of the circle $|z-a| = \varepsilon$. If $\lim\limits_{z\to a}(z-a)f(z) = k$, where k is a constant, then*

$$\lim_{\varepsilon\to 0}\int_{AB} f(z)\,dz = i(\beta-\alpha)k, \tag{7.34}$$

when the integration along AB is taken in anticlockwise sense.

Proof: Since $\lim\limits_{z\to a}(z-a)f(z) = k$ and as k is a constant, we can write

$$\lim_{z\to a}\Big[(z-a)f(z)-k\Big] = 0 \Rightarrow (z-a)f(z)-k = \eta(z)$$

where, $|\eta(z)| < \xi$ and $\xi \to 0$ as $\varepsilon \to 0$. Hence

$$\begin{aligned}\int_{AB} f(z)\,dz &= \int_{AB}\frac{k+\eta(z)}{z-a}\,dz = k\int_{AB}\frac{dz}{z-a} + \int_{AB}\frac{\eta(z)\,dz}{z-a}\\ &= k\int_{\alpha}^{\beta}\frac{i\varepsilon e^{i\theta}}{\varepsilon e^{i\theta}}d\theta + \int_{AB}\frac{\eta(z)\,dz}{z-a};\ z = a+\varepsilon e^{i\theta}\\ &= i(\beta-\alpha)k + \int_{AB}\frac{\eta(z)\,dz}{z-a}.\end{aligned}$$

$$\therefore \quad \left|\int_{AB} f(z)\,dz - i(\beta-\alpha)k\right| = \left|\int_{AB}\frac{\eta(z)\,dz}{z-a}\right| < \int_{\alpha}^{\beta}\frac{\xi}{r}r\,d\theta = \xi(\beta-\alpha).$$

Thus, when $\varepsilon \to 0$, $\xi \to 0$, and therefore the result (7.34) is established. □

Theorem 7.13. *Let AB be an arc $\alpha \leq \theta \leq \beta$ of the circle $|z-a| = R$. If $\lim\limits_{z\to\infty} zf(z) = k$, where k is a constant, then*

$$\lim_{R\to\infty}\int_{AB} f(z)\,dz = i(\beta-\alpha)k \tag{7.35}$$

when the integration along AB is taken in anticlockwise sense.

Proof: Since $\lim\limits_{z\to\infty} zf(z) = k$ and as k is a constant, we can write

$$\lim_{z\to\infty}\Big[zf(z)-k\Big] = 0 \Rightarrow |zf(z)-k| < \varepsilon, \text{ for } |z| >> 1.$$

$$\therefore \quad |zf(z)-k| = |\eta(z)| < \varepsilon, \text{ if } zf(z)-k = \eta(z)$$

where, $\varepsilon \to 0$ as $|z| \to \infty$. Hence

$$\int_{AB} f(z)\,dz = \int_{AB} \frac{k+\eta(z)}{z-a}\,dz = k\int_{AB}\frac{dz}{z-a} + \int_{AB}\frac{\eta(z)\,dz}{z-a}$$

$$= k\int_{\alpha}^{\beta} \frac{i\varepsilon e^{i\theta}}{\varepsilon e^{i\theta}}d\theta + \int_{AB}\frac{\eta(z)\,dz}{z-a}; z = a+\varepsilon e^{i\theta}$$

$$= i(\beta-\alpha)k + \int_{AB}\frac{\eta(z)\,dz}{z-a}$$

$$\therefore \quad \left|\int_{AB} f(z)\,dz - i(\beta-\alpha)k\right| = \left|\int_{AB}\frac{\eta(z)\,dz}{z-a}\right| < \varepsilon(\beta-\alpha) \to 0 \text{ as } |z| \to \infty.$$

Thus, when $R \to \infty$, $\varepsilon \to 0$, and therefore the result (7.35) is established. □

It can be proved that

$$\int_{C_2} f(z)\,dz = -\pi i \text{ Res }[f(z); a].$$

By taking as $R \to \infty$ and $\varepsilon \to 0$, we obtain the value of the required definite integral.

Example 7.3.11. *Prove that* $\int_0^\infty \frac{\sin mx}{x}dx = \frac{\pi}{2}; m > 0$.

Solution. Let us consider the integral $\oint_\Gamma f(z)dz$ where $f(z) = e^{iz}/z$ and Γ is the positive *indented contour* as in Fig. 7.11. This is called *indenting at a point.* The integrand $f(z)$ has a pole at $z = 0$. Further, inside Γ, $f(z)$ has no other singularity. Since there is no singularity within Γ, so $f(z)$ is analytic within Γ. By Cauchy's residue theorem we get,

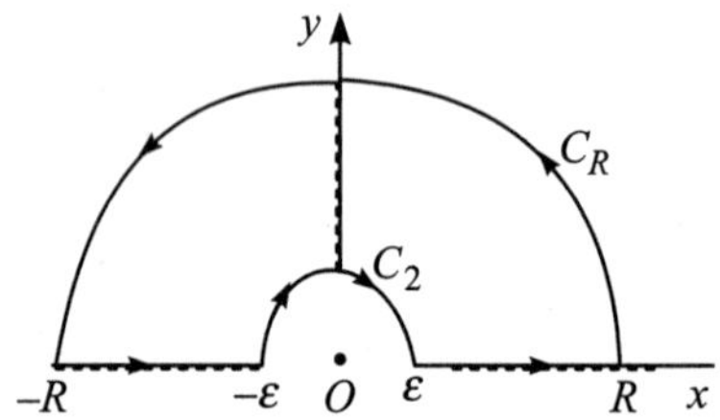

Figure 7.11 *f(z) has pole 0 lying on the real axis*

$$\oint_\Gamma f(z)dz = \int_{-R}^{-\varepsilon} f(z)dz + \int_{C_2} f(z)dz + \int_{\varepsilon}^{R} f(z)dz$$

$$+ \int_{C_R} f(z)dz = 0. \qquad (7.36)$$

Now, on $C_2 : |z| = \varepsilon$, let $z = \varepsilon e^{i\theta}$. We have,

$$\int_{C_2} f(z)dz = \int_{\pi}^{0} \frac{e^{im\varepsilon e^{i\theta}}}{\varepsilon e^{i\theta}}\varepsilon i e^{i\theta}d\theta = -i\int_0^\pi e^{im\varepsilon(\cos\theta + i\sin\theta)}d\theta$$

$$\text{or, } \lim_{\varepsilon\to 0}\int_{C_2} f(z)\,dz = -i\int_0^\pi e^0 d\theta = -\pi i.$$

On $C_R : |z| = R$, let $z = Re^{i\theta}$. Then

$$\int_{C_R} f(z)dz = \int_0^\pi \frac{e^{imRe^{i\theta}}}{Re^{i\theta}}Rie^{i\theta}d\theta = i\int_0^\pi e^{-mR\sin\theta}\cdot e^{imR\cos\theta}d\theta$$

$$\text{or, } \left|\int_{C_R} f(z)dz\right| \le \int_0^\pi e^{-mR\sin\theta}d\theta = 2\int_0^{\pi/2} e^{-mR\sin\theta}d\theta$$

$$\le 2\int_0^{\pi/2} e^{\frac{-2mR\theta}{\pi}}d\theta \;:\; \sin\theta > \frac{2\theta}{\pi} \text{ for } 0 \le \theta \le \frac{\pi}{2}$$

$$\le -\frac{\pi}{mR}\left[e^{\frac{-2mR\theta}{\pi}}\right]_0^{\pi/2} = \frac{\pi}{mR}\left[1-e^{-mR}\right] \to 0 \text{ as } R \to \infty.$$

Making $\varepsilon \to 0$ and $R \to \infty$, we get from (7.36),

$$\int_{-\infty}^{0} f(z)dz - \pi i + \int_{0}^{\infty} f(z)dz + 0 = 0$$

$$\text{or,} \quad \int_{-\infty}^{\infty} f(z)dz = \int_{-\infty}^{\infty} \frac{\cos mz + i\sin mz}{z} dz = \pi i$$

$$\text{or,} \quad \int_{-\infty}^{\infty} \frac{\sin mz}{z} dz = \pi \text{ ; equating imaginary part}$$

$$\text{or,} \quad \int_{0}^{\infty} \frac{\sin mz}{z} dz = \frac{1}{2}\int_{-\infty}^{\infty} \frac{\sin mz}{z} dz = \frac{\pi}{2}.$$

Example 7.3.12. *Prove that* $\int_0^\infty \frac{\sin x}{x(x^2+a^2)}dx = \frac{\pi}{2a^2}(1-e^{-a}); a > 0.$

Solution. Let us consider the integral $\oint_\Gamma f(z)dz$, where $f(z) = e^{iz}/z(z^2+a^2)$ and Γ is the contour as in Fig. 7.11. The poles of $f(z)$ are given by

$$z(z^2+a^2) = 0 \Rightarrow z = 0, z = \pm ai,$$

which are simple poles, of which only $z = ai$ lies within Γ. Now

$$\{f(z); ai\} = \lim_{z\to ai}\left[(z-ai)\frac{e^{iz}}{z(z^2+a^2)}\right] = \frac{e^{-a}}{-2a^2}.$$

By Cauchy's residue theorem, we get

$$\oint_\Gamma f(z)dz = \int_{-R}^{-\varepsilon} f(z)dz + \int_{C_2} f(z)dz + \int_{\varepsilon}^{R} f(z)dz + \int_{C_R} f(z)dz$$
$$= 2\pi i \times \frac{e^{-a}}{-2a^2} = -\frac{\pi i}{a^2}e^{-a}. \tag{7.37}$$

On the $C_2 : |z| = \varepsilon$, let $z = \varepsilon e^{i\theta}$. Then

$$\int_{C_2} \frac{e^{iz}}{z(z^2+a^2)}dz = \int_{\pi}^{0} \frac{e^{i\varepsilon e^{i\theta}}}{\varepsilon e^{i\theta}(\varepsilon^2 e^{2i\theta}+a^2)}\varepsilon i e^{i\theta} d\theta$$
$$= -i\int_0^\pi \frac{e^{i\varepsilon(\cos\theta + i\sin\theta)}}{\varepsilon^2 e^{2i\theta}+a^2}d\theta = -i\int_0^\pi \frac{e^{-\varepsilon\sin\theta}\cdot e^{i\varepsilon\cos\theta}}{\varepsilon^2 e^{2i\theta}+a^2}d\theta$$

$$\text{or,} \quad \lim_{\varepsilon\to 0}\int_{C_2} \frac{e^{iz}}{z(z^2+a^2)}dz = -i\int_0^\pi d\theta = -\frac{i\pi}{a^2}.$$

On the semicircle $C_R : |z| = R$, let $z = Re^{i\theta}$. Then

$$\int_{C_R} \frac{e^{iz}}{z(z^2+a^2)}dz = \int_0^\pi \frac{e^{iRe^{i\theta}}}{Re^{i\theta}(R^2e^{2i\theta}+a^2)}Rie^{i\theta}d\theta$$
$$= i\int_0^\pi \frac{e^{-R\sin\theta}}{R^2e^{2i\theta}+a^2}e^{iR\cos\theta}d\theta$$

$$\therefore \left|\int_{C_R} \frac{e^{iz}}{z(z^2+a^2)}dz\right| \leq \int_0^{\pi} \frac{e^{-R\sin\theta}}{|R^2e^{2i\theta}+a^2|}d\theta$$

$$\leq \int_0^{\pi} \frac{e^{-R\sin\theta}}{|R^2e^{2i\theta}|-a^2}d\theta \text{ ; as } |z_1+z_2| \geq |z_1|-|z_2|$$

$$\leq \frac{2}{R^2-a^2}\int_0^{\pi} e^{-\frac{2R\theta}{\pi}}d\theta \text{ ; as } \sin\theta \geq \frac{2\theta}{\pi} \text{ in } 0 \leq \theta \leq \frac{\pi}{2}$$

$$\leq \frac{\pi}{R^2-a^2}\left[1-e^{-R}\right] \to 0 \text{ as } R \to \infty.$$

Making $\varepsilon \to 0$ and $R \to \infty$, we get from (7.37),

$$\int_{-\infty}^{0} f(z)dz - \frac{i\pi}{a} + \int_0^{\infty} f(z)dz + 0 = -\frac{\pi i}{a^2}e^{-a}$$

or, $$\int_{-\infty}^{\infty} \frac{\cos z + i\sin z}{z(z^2+a^2)}dz = -\frac{\pi i}{a^2}e^{-a} + \frac{i\pi}{a}.$$

Equating the imaginary parts, we get,

$$\int_{-\infty}^{\infty} \frac{\sin z}{z(z^2+a^2)}dz = \frac{\pi}{a^2}(1-e^{-a}).$$

7.3.5 Integrals of Multi-valued Functions

For the existence of its integrals, it is necessary that the integrand must be single-valued. Accordingly, by introducing branch cut, multi-valued functions with branch point singularity can be made single-valued and analytic. Also, with these respective branch cuts at their branch point, a proper choice of contours can be made for the evaluation of their integrals.

Example 7.3.13. *Evaluate* $\int_0^{\infty} \frac{\ln z}{(z^2+1)^2}\,dz$.

Solution. We consider the integral $\int_\Gamma f(z)\,dz$, where $f(z) = \frac{\ln z}{(z^2+1)^2}$ after introducing the branch cut defined by $|z| > 0, -\pi/2 < \arg z < 3\pi/2$ to make the integrand $f(z)$ single-valued (as it has a branch point singularity at $z = 0$). Let us choose the positively oriented contour Γ (Fig. 7.12) consisting of the following: (a) The real axis from $-R$ to $-\varepsilon$; (b) the upper half of the circle C_2, $|z| = \varepsilon$; (c) the real axis from ε to R; (d) the upper half circle C_R : $|z| = R$.

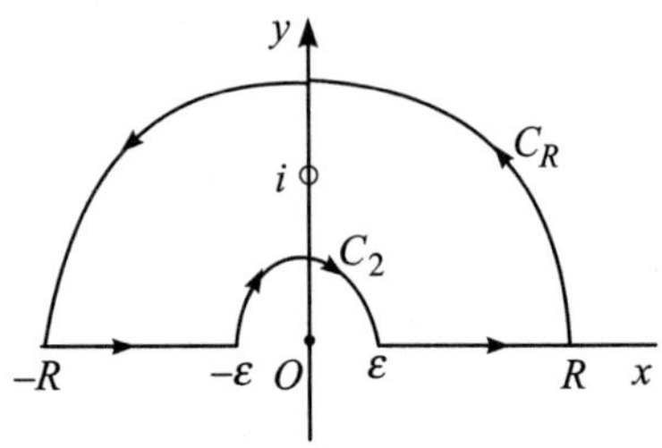

Figure 7.12 *Branch cut*

where ε is small and R is large. Thus, in the cut z-plane, f has a double pole inside Γ at $z = i$. The residue of f at $z = i$ is

$$\text{Res}\,\{f;i\} = \lim_{z\to i}\frac{d}{dz}\left[(z-i)^2\frac{\ln z}{(z^2+1)^2}\right] = \frac{\pi}{8} + \frac{i}{4}.$$

Since there is no singularity within Γ, $f(z)$ is analytic within Γ. Hence by Cauchy's residue theorem we get,

$$\oint_\Gamma f(z)dz = \int_{-R}^{-\varepsilon} f(z)dz + \int_\gamma f(z)dz + \int_\varepsilon^R f(z)dz + \int_{C_R} f(z)dz$$

$$= 2\pi i \times \left(\frac{\pi}{8} + \frac{i}{4}\right). \tag{7.38}$$

Now $\lim_{z\to 0} zf(z) = \lim_{z\to 0} \frac{z\ln z}{(z^2+1)^2} = 0 \Rightarrow \lim_{z\to 0} \int_{C_\varepsilon} f(z)\,dz = -i(\pi - 0)\cdot 0 = 0.$

Also, $\lim_{z\to\infty} zf(z) = \lim_{z\to\infty} \frac{z\ln z}{(z^2+1)^2} = \lim_{z\to\infty}\left[\frac{1}{(1+\frac{1}{z^2})^2}\cdot\frac{1}{3}\cdot\frac{\ln z^3}{z^3}\right] = 0$

$\Rightarrow \quad \lim_{R\to\infty} \int_{C_R} f(z)\,dz = -i(\pi - 0)\cdot 0 = 0.$

Making $\varepsilon \to 0$ and $R \to \infty$, we get from (7.38),

$$\int_{-\infty}^{0} \frac{\ln z}{(z^2+1)^2}\,dz + \int_0^\infty \frac{\ln z}{(z^2+1)^2}\,dz = \frac{\pi^2}{4}i - \frac{\pi}{2}$$

or, $$\int_0^\infty \frac{\ln(z\cdot e^{\pi i})}{(z^2+1)^2}\,dz + \int_0^\infty \frac{\ln z}{(z^2+1)^2}\,dz = \frac{\pi^2}{4}i - \frac{\pi}{2}$$

or, $$2\int_0^\infty \frac{\ln z}{(z^2+1)^2}\,dz + \pi i\int_0^\infty \frac{dz}{(z^2+1)^2} = \frac{\pi^2}{4}i - \frac{\pi}{2}.$$

Now, equating the real and imaginary parts of both sides of the above equation, we get

$$\int_0^\infty \frac{\ln z}{(z^2+1)^2}\,dz = -\frac{\pi}{4} \text{ and } \int_0^\infty \frac{dz}{(z^2+1)^2} = \frac{\pi}{4}.$$

Example 7.3.14. *Prove that* $\int_0^\infty \frac{z^{-a}}{z+1}\,dz = \frac{\pi}{\sin(a\pi)}$, $0 < a < 1$.

Solution. The given integral is not only improper because its upper limit of integration is infinite but also because its integrand has an infinite discontinuity at $z = 0$ and is a multivalued function for $0 < a < 1$.

We consider the integral $\int_\Gamma f(z)\,dz$ taken along the positive contour Γ (Fig. 7.13), known as *keyhole contour*, consisting of the real axis from $z = \varepsilon$ to $z = R$, the positive large circle $C_R : |z| = R$, and then R to ε of the real z-axis and finally, along the circle $C_\varepsilon : |z| = \varepsilon$ after introducing the branch cut ($|z| > 0, 0 < \arg z < 2\pi$) at the branch point $z = 0$ to make the integrand $f(z) = \frac{z^{-a}}{z+1}$ a single-valued function of z.

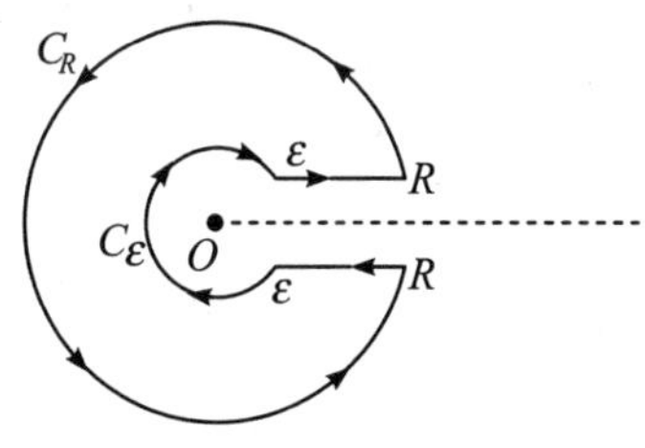

Figure 7.13 *Keyhole contour*

The singularity of the integrand $f(z)$ is at a point $z = 0$, a branch point, and at point $z = -1$, a simple pole. The branch point is avoided by the branch cut and only pole $z = -1$ is inside Γ. Therefore the residue of f at $z = -1$ is

$$\text{Res}\,\{f; -1\} = \lim_{z\to -1}\left[(z+1)\frac{z^{-a}}{z+1}\right] = e^{-i\pi a}.$$

Since there is no singularity within Γ, $f(z)$ is analytic within Γ. Hence by Cauchy's residue theorem we get,

$$\begin{aligned}\oint_\Gamma f(z)dz &= \int_\varepsilon^R f(z)dz + \int_{C_R} f(z)dz + \int_R^\varepsilon e^{-2\pi i a} f(z)dz + \int_{C_\varepsilon} f(z)dz \\ &= 2\pi i \times e^{-i\pi a}.\end{aligned} \tag{7.39}$$

Now $\lim_{z\to 0} zf(z) = \lim_{z\to 0} \dfrac{z^{1-a}}{1+z} = 0;$ as $0 < a < 1$

$\Rightarrow \quad \lim_{z\to 0} \int_{C_2} f(z)\,dz = -i(\pi - 0)\cdot 0 = 0.$

Also, $\lim_{z\to\infty} zf(z) = \lim_{z\to\infty} \dfrac{z^a}{1+z} = 0 \Rightarrow \lim_{R\to\infty} \int_{C_R} f(z)\,dz = 0.$

Making $\varepsilon \to 0$ and $R \to \infty$, we get from (7.39),

$$\int_0^\infty \frac{z^{-a}}{1+z}\,dz + \int_\infty^0 \frac{z^{-a}e^{-2\pi ia}}{1+z}\,dz = 2\pi i e^{-\pi ia}$$

$$\Rightarrow \quad \left(1 - e^{-2\pi ia}\right)\int_0^\infty \frac{z^{-a}}{1+z}\,dz = 2\pi i e^{-\pi ia}$$

$$\Rightarrow \quad \int_0^\infty \frac{z^{-a}}{1+z}\,dz = \frac{2\pi i}{e^{\pi ia} - e^{-\pi ia}} = \frac{\pi}{\sin \pi a}.$$

7.3.6 Other Types of Contours

It may not be possible to evaluate some improper integrals with the help of the contours dealt with in the above sections. In such cases, we evaluate these integrals with the help of rectangular, sector, and other type of contours.

Example 7.3.15. *Verify the Fresnel integrals* $\int_0^\infty \cos x^2\,dx = \int_0^\infty \sin x^2\,dx = \frac{1}{2}\sqrt{\frac{\pi}{2}}$.

Solution. Let Γ be as shown in Fig. 7.14, where AB is the arc of the circle C_R with centre at O and radius R. Since $f(z) = e^{iz^2}$ is analytic within and on the contour Γ (i.e., there is no singularity within the contour), hence by Cauchy's theorem

$$\oint f(z)dz = 0 \Rightarrow \left[\int_{\bar{OA}} + \int_{C_R} + \int_{BO}\right] f(z)\,dz = 0. \tag{7.40}$$

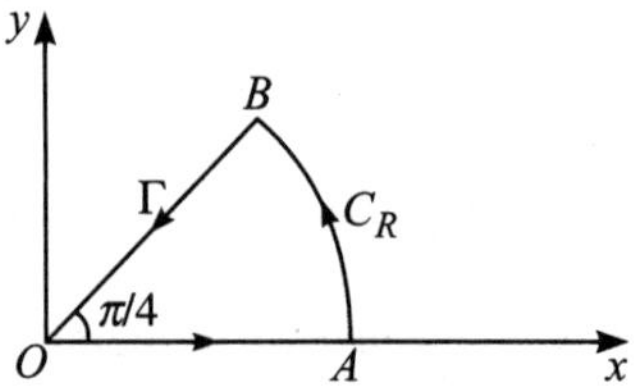

Figure 7.14 *Contour*

Now, on line OA, $z = x$(from $x = 0$ to $x = R$) so,

$$\int_{\bar{OA}} e^{iz^2}\,dz = \int_0^R e^{ix^2}\,dx = \int_0^R (\cos x^2 + i\sin x^2)dx.$$

On the arc of the circle $C_R : z = Re^{i\theta}$ (from $\theta = 0$ to $\theta = \pi/4$),

$$\begin{aligned}\int_{C_R} e^{iz^2}\,dz &= \int_0^{\pi/4} e^{iR^2e^{2i\theta}}\, iRe^{i\theta}\,d\theta \\ &= \int_0^{\pi/4} e^{iR^2\cos 2\theta - R^2\sin 2\theta} iRe^{i\theta}d\theta.\end{aligned}$$

$$\begin{aligned}\therefore \left|\int_{C_R} e^{iz^2}\,dz\right| &\le \int_0^{\pi/4} e^{-R^2\sin 2\theta} R\,d\theta = \frac{R}{2}\int_0^{\pi/2} e^{-R^2\sin\phi}\,d\phi \\ &\le \frac{R}{2}\int_0^{\pi/2} e^{-2R^2\phi/\pi} d\phi = \frac{\pi}{4R}\left(1 - e^{-R^2}\right) \to 0 \text{ as } R\to\infty,\end{aligned}$$

$$2\theta = \phi;\ \ \sin\phi \ge \frac{2\phi}{\pi}, 0 \le \phi \le \pi/2.$$

Along the line BO, $z = re^{i\pi/4}$ (from $r = R$ to $r = 0$). Then

$$\begin{aligned}\int_{BO} e^{iz^2}\, dz &= \int_R^0 e^{ir^2 e^{\pi i/2}}\, e^{i\pi/4}\, dr = -e^{i\pi/4}\int_0^R e^{-r^2}\, dr \\ &= -\sqrt{\frac{\pi}{2}}e^{i\pi/4} = -\frac{1}{2}\Big(\sqrt{\frac{\pi}{2}} + i\sqrt{\frac{\pi}{2}}\Big).\end{aligned}$$

Making $R \to \infty$, from Eq. (7.40) we get

$$\int_0^\infty (\cos x^2 + i\sin x^2)dx + 0 - \frac{1}{2}\Big(\sqrt{\frac{\pi}{2}} + i\sqrt{\frac{\pi}{2}}\Big) = 0$$

$$\Rightarrow \quad \int_0^\infty (\cos x^2 + i\sin x^2)dx = \frac{1}{2}\Big(\sqrt{\frac{\pi}{2}} + i\sqrt{\frac{\pi}{2}}\Big).$$

Equating the real and imaginary parts, we have

$$\int_0^\infty \cos x^2\, dx = \int_0^\infty \sin x^2\, dx = \frac{1}{2}\sqrt{\frac{\pi}{2}}.$$

Example 7.3.16. *Prove that* $\int_0^\infty \frac{\cosh ax}{\cosh x}\, dx = \frac{\pi}{2\cos(\pi a/2)}$, *where* $|a| < 1$.

Solution. Let $f(z) = e^{az}/\cosh z$. The poles of $f(z)$ is given by

$$\begin{aligned}& \cosh z = 0 \Rightarrow e^{iz} + e^{-iz} = 0 \Rightarrow e^{2iz} + 1 = 0 \\ \Rightarrow \quad & z = (n + \frac{1}{2})\pi i, \;\; n = 0, \pm 1, \pm 2, \cdots .\end{aligned}$$

Consider the integral $\oint_\Gamma f(z)\, dz$, where Γ is the perimeter of the rectangle $ABCD$, with corners $\mp R, \pm R + i$ (Fig. 7.15). The only pole enclosed by Γ is $\pi i/2$. The residue of $f(z)$ at $z = \pi i/2$ is

$$\begin{aligned}\text{Res}\,\{f(z); \pi i/2\} &= \lim_{z\to\pi i/2}\Big[(z - \pi i/2)\frac{e^{az}}{\cosh z}\Big] = \frac{e^{a\pi i/2}}{\sinh(\pi i/2)} \text{ using Eq. (7.5)} \\ &= \frac{e^{a\pi i/2}}{i\sin(\pi/2)} = -ie^{a\pi i/2}.\end{aligned}$$

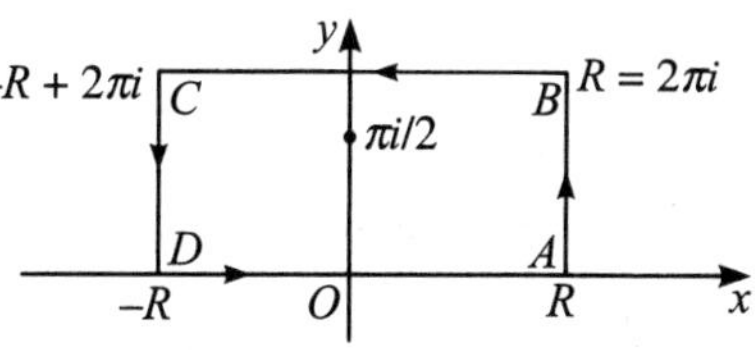

Figure 7.15 *Rectangular contour with vertices* $\mp R, \pm R + 2\pi i$.

By Cauchy's residue theorem,

$$\oint_\Gamma f(z)dz = \oint_\Gamma \frac{e^{az}}{\cosh z}\, dz = 2\pi i\Big(- ie^{a\pi i/2}\Big) = 2\pi e^{a\pi i/2}$$

$$\begin{aligned}\Rightarrow \quad & \int_{-R}^R \frac{e^{ax}}{\cosh x}\, dx + \int_0^\pi \frac{e^{a(R+iy)}}{\cosh(R + iy)}\, idy + \int_R^{-R} \frac{e^{a(x+\pi i)}}{\cosh(x + \pi i)}\, dx + \int_\pi^0 \frac{e^{a(-R+iy)}}{\cosh(-R + iy)}\, idy \\ = \quad & 2\pi e^{a\pi i/2}. \qquad (7.41)\end{aligned}$$

Consider the second integral of Eq. (7.41),

$$\int_0^{\pi} \frac{e^{a(R+iy)}}{\cosh(R+iy)}\, idy = \int_0^{\pi} \frac{2e^{a(R+iy)}}{e^{R+iy}-e^{-R-iy}}\, idy$$

$$\Rightarrow \left|\int_0^{\pi} \frac{e^{a(R+iy)}}{\cosh(R+iy)}\, idy\right| = \int_0^{\pi} \frac{2e^{aR}|\cos ay + i\sin ay|}{|e^{R+iy}+e^{-R-iy}|}\, dy$$

$$\leq \int_0^{\pi} \frac{2e^{aR}}{|e^{R+iy}|-|e^{-R-iy}|}\, dy = \int_0^{\pi} \frac{2e^{aR}}{e^{R}-e^{-R}}$$

$$\leq \int_0^{\pi} \frac{4e^{aR}}{e^{R}}\, dy = 4\pi e^{-(1-a)R} \to 0 \text{ as } R \to \infty,$$

as $|a| < 1$. In the similar manner, the fourth integral of Eq. (7.41) approaches to zero as $R \to \infty$. Therefore Eq. (7.41) becomes

$$\lim_{R\to\infty}\left[\int_{-R}^{R} \frac{e^{ax}}{\cosh x}\, dx + \int_{R}^{-R} \frac{e^{ax}\cdot e^{a\pi i}}{-\cosh x}\, dx\right] = 2\pi e^{a\pi i/2};$$

$$\text{as } \cosh(x+\pi i) = -\cosh x$$

or $$\lim_{R\to\infty}\left[1+e^{a\pi i}\right]\int_{-R}^{R} \frac{e^{ax}}{\cosh x}\, dx = 2\pi e^{a\pi i/2}$$

or $$\int_{-\infty}^{\infty} \frac{e^{ax}}{\cosh x}\, dx = \frac{2\pi e^{a\pi i/2}}{1+e^{a\pi i}} = \frac{2\pi}{e^{a\pi i/2}+e^{-a\pi i/2}}$$

or $$\int_{-\infty}^{0} \frac{e^{ax}}{\cosh x}\, dx + \int_0^{\infty} \frac{e^{ax}}{\cosh x}\, dx = \frac{\pi}{\cos(a\pi/2)}$$

or $$\int_0^{\infty} \frac{e^{-ax}}{\cosh x}\, dx + \int_0^{\infty} \frac{e^{ax}}{\cosh x}\, dx = \frac{\pi}{\cos(a\pi/2)}$$

or $$2\int_0^{\infty} \frac{\cosh ax}{\cosh x}\, dx = \frac{\pi}{\cos(\pi a/2)} \Rightarrow \int_0^{\infty} \frac{\cosh ax}{\cosh x}\, dx = \frac{\pi}{2\cos(\pi a/2)}.$$

Example 7.3.17. *Prove that* $\frac{1}{2\pi i}\int_{-\infty}^{\infty} \frac{e^{st}}{s^{\gamma+1}}\, ds = \frac{t^{\gamma}}{\Gamma(\gamma+1)}$, *where* $-1 < \gamma < 0$.

Solution. The function $F(s) = \frac{1}{s^{\gamma+1}}$ has only singularity at the point $s = 0$, which is a branch point. In the complex plane with a branch cut along negative real axis, the function $F(s)$, which is that branch point of the multiple-valued function which assumes positive values on the positive real axis, is analytic. The required function is given by

$$f(t) = \frac{1}{2\pi i}\int_{r-i\infty}^{r+i\infty} \frac{e^{st}}{s^{\gamma+1}}\, ds = \frac{1}{2\pi i}\int_{r-i\infty}^{r+i\infty} F_1(s)\, ds; \quad r > 0. \tag{7.42}$$

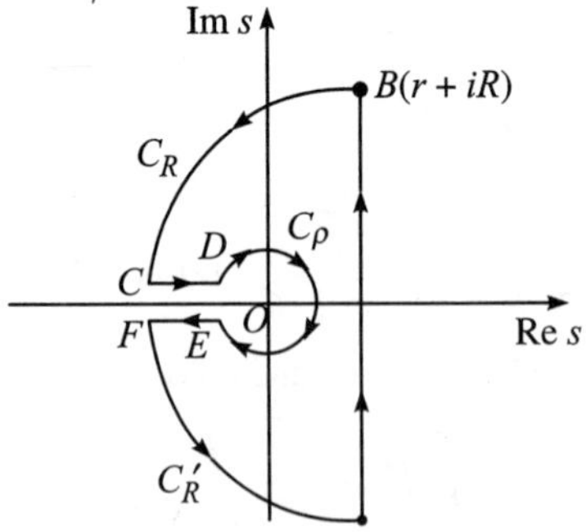

Figure 7.16 *Contour*

Consider the integral $\int_{\Gamma} F_1(s)\, ds$, where Γ is the closed contour as in the Fig. 7.16 consisting of

(i) line segment AB joining the points $r - iR$ to $r + iR$;

(ii) straight line segments CD and EF drawn just above and below the negative real axis;

(iii) circular arcs $\mathcal{C}_R : |s-r| = R, \pi/2 \le \arg(s-r) < \pi, \mathcal{C}'_R : |s-r| = R, -\pi < \arg(s-r) \le -\pi/2$ and $\mathcal{C}_\rho : |s-r| = \rho, -\pi < \arg(s) < \pi$.

As Γ is drawn in the domain in which $F_1(s)$ is analytic and single-valued, by Cauchy's integral theorem, we have,

$$\int_\Gamma F_1(s)\, ds = \int_\Gamma \frac{e^{st}}{s^{\gamma+1}}\, ds = 0$$

$$\text{or,} \quad \frac{1}{2\pi i}\left[\int_{r-iR}^{r+iR} + \int_{\mathcal{C}_R} + \int_{-R}^{-\rho} + \int_{\mathcal{C}_\rho} + \int_{-\rho}^{-R} + \int_{\mathcal{C}_R}\right] \frac{e^{st}}{s^{\gamma+1}}\, ds = 0. \tag{7.43}$$

Now, on $\mathcal{C}_R$, taking $s = r + Re^{i\theta}$, we get,

$$\left|\frac{1}{2\pi i}\int_{\mathcal{C}_R} \frac{e^{st}}{s^{\gamma+1}}\, ds\right| \le \frac{e^{rt}}{4tR^{\gamma+1}}(1-e^{-tR}) \to 0 \text{ as } R \to \infty$$

$$\Rightarrow \frac{1}{2\pi i}\int_{\mathcal{C}_R} \frac{e^{st}}{s^{\gamma+1}}\, ds \to 0 \text{ as } R \to \infty.$$

Similarly, $\frac{1}{2\pi i}\int_{\mathcal{C}'_R} \frac{e^{st}}{s^{\gamma+1}}\, ds \to 0$ as $R \to \infty$. On $\mathcal{C}_\rho$; $s = \rho e^{i\theta}$. Hence

$$\frac{1}{2\pi i}\int_{\mathcal{C}_\rho} \frac{e^{st}}{s^{\gamma+1}}\, ds = \frac{1}{2\pi}\int_{-\pi}^{\pi} \frac{e^{\rho t(\cos\theta + i\sin\theta)}}{\rho^{\gamma+1}e^{i(\gamma+1)\theta}}\rho e^{i\theta}\, d\theta$$

$$= \frac{1}{2\pi}\rho^{-\gamma}\int_{-\pi}^{\pi} \frac{e^{\rho t(\cos\theta + i\sin\theta)}}{e^{i\gamma\theta}}\, d\theta \to 0 \text{ as } \rho \to 0.$$

On CD, $s = xe^{i\pi}$ and on EF, $s = xe^{-i\pi}$. Therefore, proceeding to the limit $R \to \infty$ and $\rho \to 0$ in Eq. (7.43), we get

$$\int_{r-i\infty}^{r+i\infty} F_1(s)\, ds - \int_0^\infty \frac{e^{-xt}e^{i\pi}}{x^{\gamma+1}e^{i(\gamma+1)\pi}}dx + \int_0^\infty \frac{e^{-xt}e^{-i\pi}}{x^{\gamma+1}e^{-i(\gamma+1)\pi}}dx = 0.$$

Thus, using Eq. (7.42), we get

$$\begin{aligned}
f(t) &= \frac{1}{2\pi i}\int_{r-i\infty}^{r+i\infty} F_1(s)\, ds = \frac{1}{2\pi i}\int_{r-i\infty}^{r+i\infty} \frac{e^{st}}{s^{\gamma+1}}\, ds \\
&= \frac{1}{2\pi i}\int_0^\infty \frac{e^{-xt}}{x^{\gamma+1}}\left[\frac{e^{i\pi}}{e^{i(\gamma+1)\pi}} - \frac{e^{-i\pi}}{e^{-i(\gamma+1)\pi}}\right] dx \\
&= -\frac{2i}{2\pi i}\int_0^\infty \frac{e^{-xt}}{x^{\gamma+1}}\frac{e^{i\gamma\pi} - e^{-i\gamma\pi}}{2i}; \quad xt = u \\
&= -\frac{\sin(\gamma\pi)}{\pi}t^\gamma \int_0^\infty e^{-u}u^{-\gamma-1}\, du = -\frac{\sin(\gamma\pi)}{\pi}t^\gamma\Gamma(-\gamma) = \frac{t^\gamma}{\Gamma(\gamma+1)};
\end{aligned}$$

where we have used the relation $\Gamma(-\gamma)\Gamma(1+\gamma) = \frac{\pi}{\sin(-\gamma\pi)}$.

7.4 Estimation of Infinite Sums

Although convergence tests enable us to determine whether an infinite series converges or diverges, they do not estimate the sum of such convergent series. But the calculus of residue may have a very important application towards this direction. In this endeavour, the sum may be identified as the sum of residues of meromorphic functions.

Suppose we wish to estimate the sum $\sum_{n=-\infty}^{\infty} f(n)$. We construct a function $f(z)$ such that

(i) it has only finite number of poles at $z = a_1, a_2, \cdots, a_k$ except at $z = n$. For example, $\frac{1}{(z-1)^2} + \frac{4}{(z-3)^3} - \frac{2}{(z-2)^8}$.

(ii) $g(z)$ is another meromorphic function having infinite number of simple poles at $z = n;\ n = \pm 1, \pm 2, \cdots$ with residues at 1.

For example, $g(z) = \pi\cot(\pi z)$ is one such function. Then, clearly

$$\text{Res}\,\{f(z)\cdot g(z);\ n\} = f(n).$$

Therefore, integrating $f(z)\cdot g(z)$ around a large position circle C_N which contains all these poles, we see that

$$\begin{aligned}\int_{C_N} f(z)g(z)dz &= 2\pi i\Big[\sum_{\substack{n=-N\\ n\neq a_r}} \text{Res}\,\{f(z)\cdot g(z);\ n\} + \sum_k \text{Res}\,\{f(z)\cdot g(z);\ a_k\}\Big]\\ &= 2\pi i\sum_{n=-N}^{N} f(n) + 2\pi i\sum_k \text{Res}\,\{f(z)\cdot g(z);\ a_k\}. \qquad (7.44)\end{aligned}$$

Taking $N\to\infty$, if $\lim_{N\to\infty}\int_{C_N} f(z)g(z)dz = 0$, then (7.44) becomes

$$\sum_{n=-\infty}^{\infty} f(n) + \sum_k \text{Res}\,\{f(z)\cdot g(z);\ a_k\} = 0, \qquad (7.45)$$

from which $\sum_{n=-\infty}^{\infty} f(n)$ can be estimated.

Theorem 7.14. [Estimation of $\cot\pi z$]: *Let C_N be a square with vertices at $(N+\frac{1}{2})(\pm 1\pm i)$ as shown in Fig. 7.17. Then, there exists a positive real number A such that $|\cot\pi z| < A$ for some $A > 0$ and $|z| > N$.*

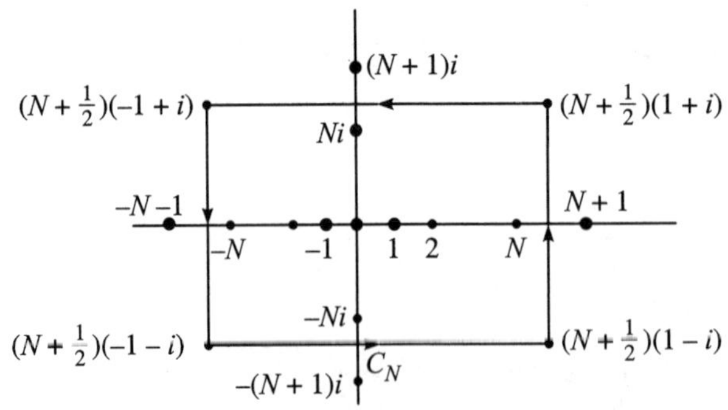

Figure 7.17 *Estimation of* $\cot\pi z$

Proof: For $z = x + iy$, we consider the following three cases: (i) $y \geq 1/2$; (ii) $y \leq -1/2$; (iii) $-1/2 \leq y \leq 1/2$.

Case-I: Since $z = x + iy$, we have, when $y \geq \frac{1}{2}$

$$
\begin{aligned}
|\cot \pi z| &= \left|\frac{e^{\pi ix-\pi y}+e^{-\pi ix+\pi y}}{e^{\pi ix-\pi y}-e^{-\pi ix+\pi y}}\right| \leq \frac{\left|e^{\pi ix-\pi y}\right|+\left|e^{-\pi ix+\pi y}\right|}{\left|e^{\pi ix-\pi y}\right|-\left|e^{-\pi ix+\pi y}\right|} \\
&\leq \frac{e^{-\pi y}+e^{\pi y}}{e^{-\pi y}-e^{\pi y}} \leq \frac{1+e^{-2\pi y}}{1-e^{-2\pi y}} \leq \frac{1+e^{-\pi}}{1-e^{-\pi}} = A_1(\text{say})
\end{aligned}
$$

Case-II: In this case, when $y \leq -\frac{1}{2}$

$$
|\cot \pi z| \leq \frac{e^{-\pi y}+e^{\pi y}}{e^{-\pi y}-e^{\pi y}} = \frac{1+e^{2\pi y}}{1-e^{2\pi y}} \leq \frac{1+e^{-\pi}}{1-e^{-\pi}} = A_1
$$

Case-III: Consider in the case, when $-\frac{1}{2} \leq y \leq \frac{1}{2}$ and $z = N + \frac{1}{2} + iy$, then we have

$$
\begin{aligned}
|\cot \pi z| &= \left|\cot \pi\Big(N+\frac{1}{2}+iy\Big)\right| = \left|\cot\Big(\frac{\pi}{2}+\pi iy\Big)\right| \\
&= |\tanh \pi y| \leq \tanh\frac{\pi}{2} = A_2(\text{say}).
\end{aligned}
$$

If we consider $z = -N - \frac{1}{2} + iy$, we have similarly

$$
|\cot \pi z| = \left|\cot \pi\Big(-N-\frac{1}{2}+iy\Big)\right| = |\tanh \pi y| \leq \tanh\frac{\pi}{2} = A_2.
$$

Therefore, if we choose $A \geq \max\{A_1, A_2\}$, is independent of N we have $|\cot \pi z| < A$ on C_N. This completes the proof. □

Note that as $A_2 < A_1$, so $|\cot \pi z| \leq \coth(\pi/2) < 2$.

Theorem 7.15. *Let the function $f(z)$ be analytic on $\mathbb{C}$ except at finitely many points $\alpha_1, \alpha_2, \cdots, \alpha_k$, none of which is a real integer. Furthermore, let there exist $p, M > 0$ such that $|z^p f(z)| \leq M$ for all $|z| > \rho$, for some $\rho > 0$. Consider the functions*

$$
g_1(z) = \pi \cot \pi z\; f(z) \text{ and } g_2(z) = \pi \text{cosec}\; \pi z\; f(z);\; z \in \mathbb{C}
$$

Then the following holds:

$$
\sum_{n=-\infty}^{\infty} f(n) = -\sum_{j=1}^{n} \text{Res}\left\{g_1; \alpha_j\right\} \text{ and } \sum_{n=-\infty}^{\infty} (-1)^n f(n) = -\sum_{j=1}^{n} \text{Res}\left\{g_2; \alpha_j\right\}. \tag{7.46}
$$

Proof: We know that the function $\sin \pi z$ vanishes at $z = n \in \mathbb{Z}$. Thus the poles of $\cot \pi z$ and cosec πz are simple and occur at $z = n \in \mathbb{Z}$. Furthermore, by hypothesis, $f(z)$ is analytic at each n. We consider the following two cases:

Case-I: Let $f(n) \neq 0$ for each n. Then both $g_1(z)$ and $g_2(z)$ have have simple poles at each $n \in \mathbb{Z}$ and singularities $\alpha_1, \alpha_2, \cdots, \alpha_k$. In this case we can choose $N > n$ so large that the rectangular path C_N of Fig. 7.17 encloses all poles of f, the integers $-n, \cdots, -1, 0, 1, \cdots, n$, and not passing through any integer.

Now, the residues of $g_1(z)$ and $g_2(z)$ at $z = n, n \in \mathbb{Z}$ are

$$
\begin{aligned}
\text{Res}\left\{g_1; n\right\} &= \lim_{z\to n}\left\{(z-n)\pi\cot \pi z f(z)\right\} = \lim_{z\to n}\pi\Big(\frac{z-n}{\sin \pi z}\Big)\cos \pi z f(z) = f(n) \\
\text{Res}\left\{g_2; n\right\} &= \lim_{z\to n}\left\{(z-n)\pi\text{cosec}\;\pi z f(z)\right\} = \lim_{z\to n}\pi\Big(\frac{z-n}{\sec \pi z}\Big)f(z) = (-1)^n f(n)
\end{aligned}
$$

using L'Hôpital's rule. We have assumed here that $f(z)$ has no poles at $z = n$, since otherwise the given series diverges. By the Cauchy's residue theorem

$$\begin{aligned}\oint_{C_N} g_1(z)\,dz &= 2\pi i\Big[\sum_{j=1}^{k} \text{Res}\,\Big\{g_1;\alpha_j\Big\} + \sum_{m=-n}^{n} \text{Res}\,\Big\{g_1;m\Big\}\Big]\\ &= 2\pi i\Big[\sum_{j=1}^{k} \text{Res}\,\Big\{g_1;\alpha_j\Big\} + \sum_{m=-n}^{n} f(m)\Big]. \qquad (7.47)\end{aligned}$$

Similarly, we find

$$\oint_{C_N} g_2(z)\,dz = 2\pi i\Big[\sum_{j=1}^{k} \text{Res}\,\Big\{g_2;\alpha_j\Big\} + \sum_{m=-n}^{n} (-1)^m f(m)\Big]. \qquad (7.48)$$

Case-II: If $f(m) = 0$ for some m, then $g_1(z)$ and $g_2(z)$ have removable singularities at m. Thus the contribution of the point m to $\oint g_1(z)dz$ and $\oint g_2(z)dz$ is zero, which is also $f(m)$. Thus, in this case also Eqs. (7.47) and (7.48) remain valid.

Let N be so large that $|\alpha_j| < N$ for all j and $\rho < N$. Then, for any z within C_N, $|z| > N > \rho$, we have

$$\Big|f(z)\Big| \le \frac{M}{|z|^p} \le \frac{M}{\rho^p} < \frac{M}{N^p}.$$

The minimum distance R_N of C_N from the origin is $(N + \frac{1}{2})\pi \to \infty$ when $N \to \infty$ and the length of C_N is $(8N\pi + 4\pi)$. On the basis of our assumption of $f(z)$, we have

$$\begin{aligned}\Big|\oint_{C_N} g_1(z)\,dz\Big| &\le \oint_{C_N} \Big|\pi\cot\pi z\Big|\,\Big|f(z)\Big|\Big|dz\Big|\\ &\le \frac{AM\pi}{N^k}(8N+4) \to 0 \text{ as } N \to \infty\\ \implies \lim_{N\to\infty}\oint_{C_N} g_1(z)\,dz &= \lim_{N\to\infty}\oint_{C_N} \pi\cot\pi z f(z)dz = 0.\end{aligned}$$

Following the same arguments, we can show that

$$\lim_{N\to\infty}\oint_{C_N} g_2(z)\,dz = \lim_{N\to\infty}\oint_{C_N} \pi\text{cosec}\,\pi z f(z)dz = 0.$$

This completes the proof of the theorem. □

Example 7.4.1. *Prove that* $\sum_{n=-\infty}^{\infty}\frac{1}{n^2+a^2} = \frac{\pi}{a}\coth\pi a$, *where* $a(>0) \in \mathbb{R}$.

Solution. Let $f(z) = \dfrac{1}{z^2+a^2}$, where a is real and $g(z) = \pi\cot\pi z$. Then f satisfies the hypothesis of Theorem 7.15. Let C_N be a simple closed contour enclosing points $z = 0, \pm1, \pm2, \cdots, \pm N$ and $z = \pm ai$. Now

$$\begin{aligned}\text{Res}\,\{f(z)g(z);ai\} &= \lim_{z\to ai}\Big[(z-ai)\frac{\pi\cot\pi z}{(z-ai)(z+ai)}\Big] = \frac{\pi\cot\pi ai}{2ai} = -\frac{\pi}{2a}\coth\pi a\\ \text{Res}\,\{f(z)g(z);-ai\} &= -\frac{\pi}{2a}\coth\pi a\\ \text{Res}\,\{f(z)g(z);0\} &= \lim_{z\to 0}\Big[(z-0)\frac{\pi\cot\pi z}{z^2+a^2}\Big] = \lim_{z\to 0}\Big[\cos\pi z\frac{\pi z}{\sin\pi z}\cdot\frac{1}{z^2+a^2}\Big] = \frac{1}{a^2}\\ \text{Res}\,\{f(z)g(z);n\} &= \lim_{z\to n}(z-n)\frac{\pi\cot\pi z}{z^2+a^2} = \lim_{z\to n}\frac{\pi(z-n)}{\sin\pi z}\cos\pi z\frac{1}{z^2+a^2} = \frac{1}{n^2+a^2}\end{aligned}$$

where $n \in \mathbb{Z} - \{0\}$. Therefore, integrating $f(z)g(z)$ around C_N, we get, by Cauchy's residue theorem,

$$\int_{C_N} f(z)g(z)dz = \int_{C_N} \frac{\pi \cot \pi z}{z^2 + a^2} dz$$

$$= 2\pi i\Big[- \frac{\pi}{2a} \coth \pi a - \frac{\pi}{2a} \coth \pi a + \frac{1}{a^2} + \sum_{n=-N, N\neq 0}^{N} \frac{1}{n^2 + a^2}\Big]$$

$$= 2\pi i\Big[- \frac{\pi}{a} \coth \pi a + \frac{1}{a^2} + \sum_{n=-N, N\neq 0}^{N} \frac{1}{n^2 + a^2}\Big].$$

Again, $\lim_{N\to\infty} \int_{C_N} f(z)g(z)dz = 0$. Hence,

$$-\frac{\pi}{a} \coth \pi a + \frac{1}{a^2} + \sum_{n=-\infty, n\neq 0}^{\infty} \frac{1}{n^2 + a^2} = 0 \Rightarrow \sum_{n=-\infty}^{\infty} \frac{1}{n^2 + a^2} = \frac{\pi}{a} \coth \pi a.$$

Example 7.4.2. *Prove that* $\sum_{n=1}^{\infty} \frac{1}{n^2} = \frac{1}{1^2} + \frac{1}{2^2} + \frac{1}{3^2} + \cdots = \frac{\pi^2}{6}$.

Solution. Let $f(z) = 1/z^2$ and $g(z) = \pi \cot \pi z$. The function f satisfies the hypothesis of the Theorem 7.15 except that it has a pole of order 2 at 0. Hence $f(z) \cdot g(z)$ has a pole of order 3 at 0. Now

$$\begin{aligned} g(z)f(z) &= \frac{\pi \cot \pi z}{z^2} = \frac{\pi \cos \pi z}{z^2 \sin \pi z} = \frac{\pi\Big(1 - \frac{\pi^2 z^2}{2!} + \frac{\pi^4 z^4}{4!} - \cdots\Big)}{\pi z^3\Big(1 - \frac{\pi^2 z^2}{3!} + \frac{\pi^4 z^4}{5!} - \cdots\Big)} \\ &= \frac{1}{z^3}\Big(1 - \frac{\pi^2 z^2}{2!} + \cdots\Big)\Big(1 + \frac{\pi^2 z^2}{3!} + \cdots\Big) \\ &= \frac{1}{z^3}\Big(1 - \frac{\pi^2 z^2}{3} + \cdots\Big). \end{aligned}$$

Therefore, $\text{Res}\,\{f(z) \cdot g(z); 0\} = -\frac{\pi^2}{3}$. Let C_N be a simple closed contour enclosing points $z = 0, \pm 1, \pm 2, \cdots, \pm N$ and $z = 0$. Then, integrate $g(z)f(z)$ around the contour as in Fig. 7.17. Thus

$$\int_{C_N} g(z)f(z)dz = \int_{C_N} \frac{\pi \cot \pi z}{z^2} dz = \sum_{n=-N}^{-1} \frac{1}{n^2} + \sum_{n=1}^{N} \frac{1}{n^2} - \frac{\pi^2}{3} = 2\sum_{n=1}^{N} \frac{1}{n^2} - \frac{\pi^2}{3}.$$

Therefore, taking $N \to \infty$, since $\int_{C_N} g(z)f(z)dz \to 0$, we get

$$2\sum_{n=1}^{\infty} \frac{1}{n^2} - \frac{\pi^2}{3} = 0 \Longrightarrow \sum_{n=1}^{\infty} \frac{1}{n^2} = \frac{\pi^2}{6}.$$

Example 7.4.3. *Prove that* $\sum_{n=-\infty}^{\infty} \frac{(-1)^n}{(n+a)^2} = \frac{\pi^2 \cos \pi a}{\sin^2 \pi a}$, *where* a *is real and different from* $0, \pm 1, \pm 2, \cdots$.

Solution. Let $f(z) = (z + a)^{-2}$ and $g(z) = \pi \text{cosec}\, \pi z$. Now, $f(z)$ has a double pole at $z = -a$. Now

$$\begin{aligned} \text{Res}\,\{f(z)g(z); -a\} &= \lim_{z\to -a} \frac{d}{dz}\Big[(z + a)^2 \frac{\pi \text{cosec}\, \pi z}{(z + a)^2}\Big] \\ &= -\pi^2 \text{cosec}\, \pi a \cos \pi a. \end{aligned}$$

The poles of $g(z) = \pi \text{cosec}\, \pi z$ are at $z = 0, \pm 1, \pm 2, \cdots$ which are simple. Thus

$$\begin{aligned} \text{Res}\,\{f(z)g(z); n\} &= \lim_{z \to n} \frac{\pi \text{cosec}\, \pi z}{(z+a)^2} = \lim_{z \to n} \frac{\pi(z-n)}{\sin \pi z} \cdot \frac{1}{(z+a)^2} \\ &= (-1)^n \frac{1}{(n+a)^2};\ n \in \mathbb{Z}. \end{aligned}$$

Let C_N be a simple closed contour enclosing points $z = 0, \pm 1, \pm 2, \cdots, \pm N$ and $z = -a$. Therefore, integrating $f(z)g(z)$ around C_N, we get, by Cauchy's residue theorem:

$$\int_{C_N} f(z)g(z)dz = 2\pi i \Big[-\pi^2 \text{cosec}\, \pi a \cos \pi a + \sum_{n=-N}^{N} \frac{(-1)^n}{(n+a)^2} \Big].$$

Again, $\lim_{N \to \infty} \int_{C_N} f(z)g(z)dz = 0$. Hence,

$$\sum_{n=-\infty}^{\infty} \frac{(-1)^n}{(n+a)^2} = \pi^2 \text{cosec}\, \pi a \cot \pi a = \frac{\pi^2 \cos \pi a}{\sin^2 \pi a}.$$

Example 7.4.4. *Prove that* $\frac{1}{1^3} - \frac{1}{3^3} + \frac{1}{5^3} - \cdots = \frac{\pi^3}{32}$.

Solution. Let $f(z) = \frac{1}{z^3}$ and $g(z) = \pi \sec \pi z$, then $f(z)g(z) = \frac{\pi \sec \pi z}{z^3}$. Again, $f(z)$ has a triple pole at $z = 0$. Further,

$$\begin{aligned} f(z)g(z) &= \frac{\pi \sec \pi z}{z^3} = \frac{\pi}{z^3 \cos \pi z} = \frac{\pi}{z^3 \left(1 - \frac{\pi^2 z^2}{2!} + \cdots\right)} \\ &= \frac{\pi}{z^3}\left(1 + \frac{\pi^2 z^2}{2} + \cdots\right) = \frac{\pi}{z^3} + \frac{\pi^3}{2z} + \cdots \end{aligned}$$

so that the residue at $z = 0$ is $\frac{\pi^3}{2}$. Also, $g(z)$ has simple poles at $z = n + \frac{1}{2};\ n = 0, \pm 1, \pm 2, \cdots$. Now,

$$\text{Res}\left\{f(z)g(z); n + \frac{1}{2}\right\} = \lim_{z \to n + \frac{1}{2}} \left[\frac{\left(z - n - \frac{1}{2}\right)\pi}{z^3 \cos \pi z} \right] = -(-1)^n \left(n + \frac{1}{2}\right)^{-3};\ n \in \mathbb{Z}.$$

Let C_N be a square with vertices $N(\pm 1 \pm i)$. Then, integrating $f(z)g(z)$ around C_N, we get, by Cauchy's residue theorem

$$\int_{C_N} f(z)g(z)dz = \int_{C_N} \frac{\pi \sec \pi z}{z^3} dz = 2\pi i \Big[\frac{\pi^3}{2} - \sum_{n=-N}^{N} (-1)^n \left(n + \frac{1}{2}\right)^{-3} \Big].$$

Again, $\lim_{N \to \infty} \int_{C_N} f(z)g(z)dz = 0$, and so

$$\begin{aligned} & -\sum_{n=-\infty}^{\infty} (-1)^n \left(n + \frac{1}{2}\right)^{-3} + \frac{\pi^3}{2} = 0 \implies \sum_{n=-\infty}^{\infty} \frac{(-1)^n}{(2n+1)^3} = \frac{\pi^3}{16} \\ & \implies \frac{1}{1^3} - \frac{1}{3^3} + \frac{1}{5^3} - \cdots = \frac{\pi^3}{32}. \end{aligned}$$

Exercises

Multiple-choice Questions

1. The function $f(z) = \frac{1}{z(z-1)^3}$ has a pole of order p and residues r, where
   (a) $p = 1, r = 1$ (b) $p = 3, r = 1/3$ (c) $p = 3, r = 1$ (d) $p = 1, r = 0$
2. The residue at $e^{1/z}$ at $z = 0$ is
   (a) 2 (b) ∞ (c) -2 (d) 1
3. Residue of $f(z) = \frac{\sin z}{z^4}$ at singularity is
   (a) 0 (b) 1 (c) $-1/6$ (d) 2
4. The residue of $\frac{\sin z}{z^4-1}$ at $z = i$ equals:
   (a) $1/4\cos 1$ (b) $1/4\cosh 1$ (c) $1/4\sinh 1$ (d) $-1/4\sinh 1$
5. The value of the integral $\oint_\Gamma \frac{dz}{z^2-1}$, $\Gamma : |z| = 4$ is equals to
   (a) πi (b) 0 (c) $-\pi i$ (d) $2\pi i$.
6. Let Γ be a simple closed curve in the complex plane. Then, the set of all possible values of $\oint_\Gamma \frac{dz}{z(1-z^2)}$ is
   (a) $\{0, \pm\pi i\}$ (b) $\{0, \pi i, 2\pi, i\}$ (c) $\{0, \pm\pi i, \pm 2\pi i\}$ (d) $\{0\}$
7. Let $I = \int_\Gamma \frac{\cot(\pi z)}{(z-i)^2} dz$, where γ is the contour $4x^2 + y^2 = 2$ (counterclockwise). Then I is equal to
   (a) 0 (b) $-2\pi i$ (c) $2\pi i(\frac{\pi}{\sinh^2 \pi} - \frac{1}{\pi})$ (d) $-\frac{2\pi^2 i}{\sinh^2 \pi}$
8. Let Γ be the curve $r = 2 + 4\cos\theta$, $0 \le \theta \le 2\pi$. If $\int_\Gamma \frac{dz}{z-1}$ and $I_2 = \int_\Gamma \frac{dz}{z-3}$, then
   (a) $I_1 = 2I_2$ (b) $I_1 = I_2$ (c) $2I_1 = I_2$ (d) $I_1 = 0, I_2 \neq 0$.
9. The number of zeros, counting multipliers, of the polynomial $z^5 + 3z^3 + z^2 + 1$ inside the circle $|z| = 2$ is
   (a) 0 (b) 2 (c) 3 (d) 5
10. The number of roots of the equation $z^5 - 12z^2 + 14 = 0$ that lie in the region $\{z \in \mathbb{C} : 2 \le |z| \le 5/2\}$ is
    (a) 2 (b) 3 (c) 4 (d) 5
11. Let $f(z)$ be an analytic function with a simple pole at $z = 1$ and a double pole at $z = 2$ with residues 1 and -2, respectively. Further, $f(0) = 0$, $f(3) = -3/4$ and f is bounded as $z \to \infty$. Then $f(z)$ must be
    (a) $z(z-3) - \frac{1}{4} + \frac{1}{z-1} - \frac{2}{(z-1)^2} + \frac{1}{(z-2)^2}$ (b) $-\frac{1}{4} + \frac{1}{z-1} - \frac{2}{z-2} + \frac{1}{(z-2)^2}$
    (c) $\frac{1}{z-1} - \frac{2}{z-2} + \frac{5}{(z-2)^2}$ (d) $\frac{15}{4} + \frac{1}{z-1} + \frac{2}{z-2} - \frac{7}{(z-2)^2}$.
12. $\int_0^{2\pi} \frac{d\theta}{13 - 5\sin\theta}$ is equal to
    (a) $-\frac{\pi}{6}$ (b) $-\frac{\pi}{12}$ (c) $\frac{\pi}{12}$ (d) $\frac{\pi}{6}$
13. The principal value of the improper integral $\int_{-\infty}^{\infty} \frac{\cos x}{1+x^2} dx$ is
    (a) $\frac{\pi}{e}$ (b) πe (c) $\pi + e$ (d) $\pi - e$

Answers

1(c); 2(d); 3(c); 4(d); 5(b); 6(b); 7(d); 8(c); 9(d); 10(a); 11(d); 12(d); 13(a)

Review Questions [Short-answer Type]

1. Suppose that $f(z)$ is an even function with an isolated singularity at 0. Show that $\text{Res}\{f(z);0\} = 0$.

2. Let f have an isolated singularity at z_0 (f analytic in a punctured neighbourhood of z_0). Show that the residue of the derivative f' at z_0 is equal to zero.

3. Is there a function f having a simple pole at z_0 with $\text{Res}\,(f;z_0) = 0$? How about a function with a pole of order 2 at z_0 and $\text{Res}\,(f;z_0) = 0$?

4. Suppose that f is analytic and has a zero of order m at the point z_0. Show that the function $g(z) = f'(z)/f(z)$ has a simple pole at z_0 with $\text{Res}\,(g;z_0) = m$.

5. If $f(z)$ has a simple pole at z_0 and $g(z)$ is analytic in an open set containing z_0, then prove that
$$\text{Res}\,\{fg, z_0\} = g(z_0)\,\text{Res}\,\{f, z_0\}.$$

6. Show that Cauchy's integral formula is a special case of the residue theorem.

7. Evaluate $\oint_{|z|=1} e^{1/z} \sin \frac{1}{z} dz$

8. Use the open mapping theorem to prove the maximum-modulus principle.

9. Prove that the sum of the residues of the function $\frac{2z^5-4z^2+5}{3z^6-8z+10}$ at all the poles is $2/3$.

10. Let $f_k \to f$ normally on a domain $\mathcal{D}$; assume that each f_k is one-to-one (univalent) on $\mathcal{D}$. If f is not constant, then prove that f is also univalent. [**Hint:** Theorem 7.9]

11. Let γ be the circle $|z+1| = 2$ traversed twice in the clockwisedirection. Evaluate $\int_\gamma \frac{dz}{(z^2+z)^2}$.

12. Find the residues at the indicated points noted against each of the following functions:

    (a) $f(z) = \frac{\cot z}{z^4}$, at $z = 0$
    (b) $f(z) = \frac{\sinh z}{z^4(1-z^2)}$, at $z = 0$
    (c) $f(z) = \frac{z-\sin z}{z}$, at $z = 0$
    (d) $f(z) = z\cos\frac{1}{z}$, at $z = 0$
    (e) $f(z) = \frac{\cot \pi z}{(z-a)^2}$ at $z = a$ and $z = n \in \mathbb{Z}$
    (f) $f(z) = \frac{z^3 e^{1/z}}{1+z^3}$ at $z = \infty$
    (g) $f(z) = \frac{z^2-z}{(z+1)^2(z^2+4)}$ at $z = -1$, $z = \pm 2i$
    (h) $f(z) = e^z \text{cosec}^2(z)$ at $z = n\pi, n \in \mathbb{Z}$
    (i) $f(z) = \frac{\cot z \coth z}{z^3}$ at $z = 0$
    (j) $f(z) = \frac{\text{cosec}\, z\ \text{cosech} z}{z^3}$ at $z = 0$
    (k) $f(z) = \frac{1}{(z^2+1)^3}$ at $z = i$.
    (l) $f(z) = \dfrac{(\log z)^3}{z^2+1}$ at $z = i$

13. Find the order of poles and value of residues of the following functions:

    (a) $f(z) = \frac{z+3}{z^2-2z}$
    (b) $f(z) = \frac{2z+1}{z^2-z-2}$
    (c) $f(z) = \frac{z+1}{z^2(z-3)}$
    (d) $f(z) = \frac{z^2}{(z-1)^2(z+2)}$
    (e) $f(z) = \frac{z^2+16}{(z-i)^2(z+3)}$
    (f) $\pi\text{cosec}\,\pi z$

14. Show that:

    (a) $\displaystyle \operatorname*{Res}_{z=\pi i} \frac{z-\sinh z}{z^2 \sinh z} = \frac{i}{\pi}$
    (b) $\displaystyle \operatorname*{Res}_{z=\pi i} \frac{e^{zt}}{\sinh z} + \operatorname*{Res}_{z=\pi i} \frac{e^{zt}}{\sinh z} = -2\cos(\pi t)$
    (c) $\operatorname*{Res}_{z=z_n} (z \sec z) = (-1)^{n+1} z_n$, where $z_n = \frac{\pi}{2} + n\pi$, $n \in \mathbb{N}$
    (d) $\operatorname*{Res}_{z=z_n} f(z)(\tanh z) = 1$, where $z_n = (\frac{\pi}{2} + n\pi)i$, $n \in \mathbb{N}$

15. Evaluate the following integrals:

(a) $\int_\gamma \frac{z^2\,dz}{(z-1)^2(z+2)}$: where $\gamma : |z| = 3$

(b) $\int_\gamma \frac{z^3\,dz}{(z-1-i)(z-2)}$: where $\gamma : |z| = 3/2$

(c) $\int_\gamma \tan z\,dz$; where $\gamma : |z| = 2$

(d) $\int_\gamma \cos\frac{1}{z^2}e^{1/z}\,dz$: where $\gamma : |z| = 1$

(e) $\int_\gamma \frac{z\,dz}{(z^2-1)^2(z^2+1)}$: where $\gamma : |z-1| = \sqrt{3}$

(f) $\oint_{|z|=3} \frac{e^{iz}}{z^2(z-2)(z+5i)}dz$

(g) $\oint_{|z|=1} \frac{dz}{z^2\sin z}$

(h) $\oint_{|z|=5}[ze^{3/z} + \frac{\cos z}{z^2(z-\pi)^3}]dz$

16. Find if possible values of $\int_\gamma e^{z^{-1}}\,dz$, where γ is any closed curve not passing through $z = 0$.

17. Find the residue at $z = \infty$ of the following functions:

(i) $\dfrac{2-z}{z^2-z}$; (ii) $\log\dfrac{z-a}{z-b}$; (iii) $\sqrt{(z-\alpha)(z-\beta)}$

18. Give an example to show that the conclusion of Rouché's theorem may be false if the strict inequality $|g(z)| < |f(z)|$ is replaced by $|g(z)| \le |f(z)|$ for z on Γ.

19. Let $P(z) = a_0 + a_1z + \cdots + a_nz^n$ where $a_n \ne 0$. Explain why for sufficiently large value of R, $\oint_{|z|=R} \frac{P'(z)}{P(z)}\,dz = 2n\pi i$.

20. Let C denotes the unit circle $|z| = 1$ described in the positive sense. Determine the value of $\triangle_C \arg f(z)$ in each of the following cases, when $f(z)$ is given by

(i) z^2; (ii) $\frac{z^3+2}{z}$; (iii) $\frac{(2z-1)^7}{z^3}$; (iv) $3z^7 + 5z - 1$

21. Consider the integral $I = \frac{1}{2\pi i}\int_C \frac{f'(z)}{f(z)}\,dz$, where C is the circle $|z-1-i| = 2$. Evaluate I when $f(z) =$

(i) $\frac{z-2}{z(z-1)}$ (ii) $\frac{z-2}{z(z-1)^2}$ (iii) $\frac{z^2-9}{z^2+1}$

22. Evaluate the integral $I = \frac{1}{2\pi i}\int_C \frac{f'(z)}{f(z)}\,dz$, where $f(z) = z^5 - 3iz^2 + i - 1$ and C encloses a zero of $f(z)$.

23. Let $f(z) = \frac{(z^2+1)}{(z^2+2z+z)^2}$. Evaluate $\frac{1}{2\pi i}\int_C \frac{f'(z)}{f(z)}dz$ where C is the circle $|z| = 4$.

Answers

12. (a) $-1/45$; (b) $7/6$; (c) 0; (d) $-1/2$; (e) $-\pi\text{cosec}^2\pi a$, $1/\pi(n-a)^2$; (f) 1; (g) $-14/25$, $(7\pm i)/25$; (h) $e^{n\pi}$; (i) $-7/45$; (j) $-1/60$; (k) $-3i/16$; (l) $-\pi^3/16$

13. (a) $z = 0, 2$(simple); $-3/2$, $5/2$; (b) $z = -1, 2$(simple); $1/3$, $5/3$; (c) $z = 0$(double), $z = 3$(simple); $-4/9$, $4/9$; (d) $z = 1$(double), $z = -2$(simple); $5/9$, $4/9$ (e) $z = i$(double), $z = -3$(simple); $-1+3i/2$, $2-3i/2$; (f) $z = n, n \in \mathbb{N}$(simple); $(-1)^n, n \in \mathbb{N}$

17. (i) 1; (ii) $a - b$ for all branches; (iii) $(\alpha-\beta)^2/8$

20. (i) 4π; (ii) -2π; (iii) 8π; (iv) 2π

21. (i) -1; (ii) -2; (iii) -1; 22. $10\pi i$; 23. -2;

Review Questions [Long-answer Type]

1. Prove that Cauchy's theorem and integral formulae can be obtained as special cases of the residue theorem.

2. Prove the residue theorem for multiply-connected regions.

3. Suppose that $f(z)$ and $g(z)$ are analytic functions with zeros of respective orders k and $k+1$ at z_0. Show that

$$\text{Res}\left\{\frac{f(z)}{g(z)}; z_0\right\} = (k+1)\,\frac{f^{(k)}(z_0)}{g^{(k+1)}(z_0)}.$$

4. Let α be a nonzero fixed complex number such that $|\alpha| = a < 1$. Show that $\int_\Gamma \frac{|dz|}{|z-\alpha|^2} = \frac{2\pi}{1-a^2}$, where Γ is the unit circle $|z| = 1$.

5. Let $f(z)$ be analytic on an open set G, and let $z_0 \in G$ and $f'(z_0) \neq 0$. Show that $\frac{2\pi i}{f'(z_0)} = \int_\gamma \frac{dz}{f(z)-f(z_0)}$, where γ is a small circle centered at z_0.

6. Show that

   (a) $\text{Res}\left\{\frac{z^{2n}}{(z+1)^n};\ -1\right\} = (-1)^{n+1}\,\frac{(2n)!}{(n-1)!(n+1)!}$

   (b) $\text{Res}\left\{\frac{z^{2n}}{(z+1)^n};\ \infty\right\} = (-1)^{n}\,\frac{(2n)!}{(n-1)!(n+1)!}$

   (c) $\text{Res}\left\{\frac{e^{1/z}}{(z+1)^n};\ \infty\right\} = \sin 1 = \text{Res}\left\{\frac{e^{1/z}}{z^2+1};\ 0\right\}$

   (d) $\text{Res}\left\{\cos\frac{1}{z}\,\frac{\sin(z-1)}{z^2+1};\ i\right\} = \frac{1}{2}\cosh 1\,\sinh(1+i)$

   (e) $\text{Res}\left\{\frac{\cos z}{z^2(z-\pi)^3};\ 0\right\} = -\frac{3}{\pi^4}$ and $\text{Res}\left\{\frac{\cos z}{z^2(z-\pi)^3};\ \pi\right\} = -\frac{6-\pi^2}{2\pi^4}$

   (f) $\text{Res}\left\{z^3\cos\frac{1}{z-2};\ 2\right\} = -\frac{143}{24}$ and $\text{Res}\left\{z^3\cos\frac{1}{z-2};\ \infty\right\} = \frac{143}{24}$

7. Find the residue of $f(z)$ at all its poles in the finite plane, where

   (a) $f(z) = \frac{z^2-z}{(z+1)^2(z^2+4)}$ at $z=-1,\ z=\pm 2i$

   (b) $f(z) = e^z\text{cosec}^{\,2}(z)$

   (c) $f(z) = \frac{z^3}{(z-1)^4(z-2)(z-3)}$,

   (d) $f(z) = \frac{1}{(z^2+a^2)^2}$,

8. Find $\text{Res}_{z=z_0} f(z)\frac{g(z)f'(z)}{f(z)}$, when $g(z)$ is analytic at z_0, while (a) $f(z)$ has a zero of order k at $z=z_0$; and (b) $f(z)$ has a pole of order k at $z=z_0$.

9. If f and g are both analytic at z_0 and $f(z_0) \neq 0$ and z_0 is a zero of first order for g. Show that

$$\text{Res}_{z=z_0} f(z)\frac{f(z)}{[g(z)]^2} = \frac{f'(z_0)g'(z_0) - f(z_0)g''(z_0)}{[g'(z_0)]^3}.$$

10. Evaluate the following integrals using residue theorem:

    (a) $\int_\Gamma \frac{\sin 3z}{(z-\pi/4)^4}\,dz,\ \Gamma = \{(x,y) : |x| \le 2, |y| \le 2\}$.

    (b) $\int_\Gamma \frac{1-2z}{z(z-1)(z-2)}\,dz,\ \Gamma : |z| = 3/2$.

    (c) $\int_\Gamma \frac{z}{(z^2+1)(z-3)^2}\,dz,\ \Gamma : |z| = 2$.

    (d) $\int_\Gamma \frac{z}{(z-1)(z-2)^2}\,dz,\ \Gamma : |z-2| = 1/2$.

    (e) $\int_\Gamma \frac{\sin\pi z^2+\cos\pi z^2}{(z-1)^2(z-2)}\,dz,\ \Gamma : |z| = 3$.

    (f) $\int_\Gamma \frac{1-\cos 2(z-3)}{(z-3)^3}\,dz,\ \Gamma : |z-3| = 1$.

    (g) $\int_\Gamma \frac{z^2-2z}{(z+1)^2(z^2+4)}\,dz,\ \Gamma : |z| = 10$.

    (h) $\int_\Gamma \frac{e^{-z}}{(z-1)^2}\,dz,\ \Gamma : |z| = 3$.

    (i) $\int_\Gamma \frac{z+1}{z^2-2z}\,dz,\ \Gamma : |z| = 3$.

    (j) $\frac{1}{2\pi i}\int_\Gamma \frac{e^{zt}}{z^2(z^2+2z+2)}\,dz,\ \Gamma : |z| = 3$.

11. If G is a region and f is analytic in G except for simple poles at $\alpha_1, \cdots, a_k$, and if g is analytic in G, then prove that

$$\frac{1}{2\pi i}\int_\gamma fg = \sum_{p=1}^{k} n(\gamma;\alpha_p)g(\alpha_p)\ \text{Res}\,\{f;\alpha_p\}$$

for any closed rectifiable curve γ not passing through $\alpha_1, \cdots, a_k$ such that $\gamma \approx 0$ in G.

12. Let f be meromorphic in the region G with zeros $z_1, z_2, \cdots, z_n$ and poles $p_1, p_2, \cdots, p_m$ counted according to multiplicity. If g is analytic in G and γ is a closed rectifiable curve in G with $\gamma \approx 0$ and not passing through z_k or p_k, then prove that

$$\frac{1}{2\pi i}\int_\gamma g\frac{f'}{f} = \sum_{r=1}^{n} g(z_r)n(\gamma;z_r) - \sum_{s=1}^{m} g(p_s)n(\gamma;p_s).$$

13. Let f be analytic in $\mathcal{S}[0,R]$ with $f(0) = 0$, $f'(0) \neq 0$ and $f(z) \neq 0$ for $0 < |z| \leq R$. Put $\rho = \min\{|f(z) : |z| = R\} > 0$. Define $g : \mathcal{S}(0;\rho) \longrightarrow \mathbb{C}$ by

$$g(z) = \frac{1}{2\pi i}\int_\gamma \frac{\zeta f'(\zeta)}{f(\zeta) - z}d\zeta,$$

where γ is the circle $|z| = R$. Show that g is analytic and discuss the properties of g.

14. In the extended complex plane, a function $f(z)$ has only singularities which are poles of orders 1 and 2 at $z = 1$ and $z = -1$, respectively and where their residues are 1 each. Also $f(0) = f(-1/2) = 0$. Find the function $f(z)$.

15. Find all functions $f(z)$ having the following properties: $f(z)$ has a double pole with residue 3 at $z = 0$; $f(z)$ has a simple pole with residue 7 at $z = 1$; and $f(z)$ is bounded in a neighbourhood of $z = \infty$. [**Hint :** By using Theorem 5.18, the function $\phi(z) = f(z) - 3/z - A/z^2 - 7/(z-1)$ is a constant.]

16. The analytic even function $f(z)$ have double poles at $z = 1, -1, \infty$ and no singularity other than these points. Find this function, given that
    (a) the residue at 1 is -1
    (b) the first non-zero coefficient in the Taylor's expansion about the origin is the coefficient of z^4, and that
    (c) $f(z)/z^2$ tends to 4 as $|z| \to \infty$. Ans: $f(z) = 4z^2 - \frac{1}{z-1} + \frac{1}{z+1} - \frac{1}{(z-1)^2} - \frac{1}{(z+1)^2}$.

17. A function $f(z)$ has a double pole $z = 0$ with residue 2 and simple pole at $z = 1$ with residue 2. It is analytic at all other points at finite distance from origin and is bounded at infinity. If $f(2) = 5$ and $f(-1) = 2$, find $f(z)$.

18. The only singularities of a single-valued function $f(z)$ are poles of order 2 and 1 at $z = 1$ and $z = 2$, with residues at these poles 1 and 3 respectively. If $f(0) = 3/2$, $f(-1) = 1$, prove that the function is $f(z) = 2 + \frac{3}{z-2} + \frac{1}{z-1} + \frac{3}{(z-1)^2}$.

19. Let f be a meromorphic function on $\mathbb{C}$ with a zero of order 1 at $z = 1$ and a pole of order 1 at $z = -1$ and no other zeros or poles. If γ is a simple closed path which does not pass through 1 or -1, what are the possible values for $\int_\gamma \frac{f'(z)}{f(z)}dz$ and what determines which value is achieved?

20. Let f be meromorphic in Γ and let there exists a positive integer n, $M > 0$ and $R > 0$ such that $|f(z) \leq M|z|^n$ for $|z| > R$. Then show that $f(z)$ is a rational function.

21. Let $f(z)$ be a non-constant function which is analytic at z_0. If $w_0 = f(z_0)$ and $f(z_0) - w_0$ have a zero of order n at z_0, then, prove that, for any given $\varepsilon > 0$, there exists a $\delta > 0$ such that $|w - w_0| < \delta\ \forall w$ and the equation $f(z) = w$ has exactly n roots in the disc $|z - z_0| < \varepsilon$.
[**Hint**: Theorem 7.7]

22. Apply Rouché's theorem to show that
    (a) to determine the number of roots of the equation $z^8 - 4z^5 + z^2 - 1 = 0$ that lie inside the circle $|z| = 1$.
    (b) the equation $z^7 - 5z^3 + 12 = 0$ lies between the circles $|z| = 1$ and $|z| = 2$.
    (c) the equation $e^z = az^n, a > e$ has n roots inside the circle $|z| = 1$.
    (d) the equation $ze^z = a$, where $a \neq 0$ is real, has infinitely many roots.

23. Carry out the details of proof of the fundamental theorem of algebra to show that the particular function $f(z) = z^4 - z^2 - 2z + 2$ has exactly four zeros. Determine the zeros of $f(z)$.

24. If f is meromorphic on G
    (a) and $\bar{f} : G \longrightarrow \mathbb{C}_\infty$ is defined by $\bar{f}(z) = \infty$ when z is a pole of f and $\bar{f}(z) = f(z)$ otherwise, show that $\bar{f}$ is continuous.
    (b) show that neither the poles nor the zeros of f have a limit point in G.

25. Prove that in the equation
    (a) $z^4 + z^3 + 1 = 0$, one of the roots lies in the first quadrant of the complex plane.
    (b) $z^5 + z^3 + 2z + 3 = 0$ has just one zero in the first quadrant of the complex plane.
    (c) $z^4 + 4(1 + i)z + 1 = 0$ has one root in each quadrant.
    (d) $z^3 + iz + 1 = 0$ has one root in each of the first, second and fourth quadrants.
    (e) $z^4 + 2z^3 + z^2 + 4z + 5 = 0$ has no real or purely imaginary roots and that it has one complex root in each quadrant.

26. Determine the number of roots (counting multiplicities) of the equation $2z^5 - 6z^2 + z + 1 = 0$ in the annulus $1 \leq |z| < 2$.

27. Prove that every non-constant analytic function on a domain is an open mapping.

28. Is a non-constant meromorphic function on a region G an open mapping of G into $\mathbb{C}$? Is it an open mapping of G into $\mathbb{C}_\infty$?

29. Let $\lambda > 1$ and show that the equation $\lambda - z - e^{-z} = 0$ has exactly one solution in the half plane $\{z \in \mathbb{C} : \text{Re}z > 0\}$. Show that this solution must be real. What happens to the solution as $\lambda \to 1$?

30. Use the open mapping property to give a quick proof of the following familiar facts: If f is analytic in a domain $\mathcal{D}$, then f is identically constant in $\mathcal{D}$ if any one of the following conditions hold:
    (a) $\text{Re}f(z)$ is constant in $\mathcal{D}$. (b) $\text{Im}f(z)$ is constant in $\mathcal{D}$. (c) $|f(z)|$ is constant in $\mathcal{D}$.

31. Prove that if $f(z)$ is analytic inside and on a simple closed contour Γ and is one-to-one on Γ, then $f(z)$ is one-to-one inside Γ.

32. Suppose that $f(z)$ is analytic on $|z| \leq 1$ and satisfies $|f(z)| < 1$ on $|z| = 1$.
    (a) Prove that the function $f(z) = z$ has exactly one root (counting multiplicity) in $|z| < 1$.
    (b) Prove that if $|z_0| \leq 1$, then the sequence $\langle z_n \rangle$ defined recursively by $z_n = f(z_{n-1})$; $n = 1, 2, \cdots$ converges to the fixed point of f.

33. Suppose $\{f_n(z)\}$ is a sequence of analytic functions on a domain $\mathcal{D}$ that converges normally on $\mathcal{D}$ to $f(z)$, and suppose that $f(z)$ has a zero of order N at z_0. Use Rouché's theorem to show that, there exists $\rho > 0$ such that for n large, $f_n(z)$ has exactly N zeros in the disc $\{|z - z_0| < \rho\}$.

34. State and prove a generalization of Rouché's theorem for meromorphic functions f and h, which concludes that $N_0(f) - N_\infty(f) = N_0(f+h) - N_\infty(f+h)$.

35. Let $f(z)$ be analytic on the closed disc $|z| \le \rho$, and suppose that $f(z) \neq w_0$ for all z on the circle $|z| = \rho$. Explain why the value of the integral

$$\frac{1}{2\pi i} \oint_{|z|=\rho} \frac{f'(z)}{f(z) - w_0}\, dz$$

equals the number of solutions of $f(z) = w_0$ inside the disc.

36. Let f be an entire function and let $a, b \in \mathbb{C}$ such that $|a| < R$ and $|b| < R$. If $\gamma(t) = Re^{it}; 0 \le t \le 2\pi$, evaluate $\int_\gamma \frac{f(z)}{(z-a)(z-b)} dz$. Use this result to give another proof of Liouville's Theorem.

37. Suppose that $\{f_n(z)\}$ is a sequence of analytic functions on a domain $\mathcal{D}$ of the complex z-plane, converging uniformly on every compact subset of $\mathcal{D}$ to a function $f(z)$. Then, prove that either

    (a) $f(z)$ is identically zero on $\mathcal{D}$, or

    (b) if $S(z_0, r)$ is any open disc in $\mathcal{D}$ such that $f(z)$ does not have zeros on its boundary, then $f_n(z)$ and $f(z)$ have the same number of zeros in $S(z_0, r)$ for all sufficiently large n.

38. Let $\{f_n(z)\}$ be a sequence of one-to-one analytic functions defined on a domain $\mathcal{D}$, and suppose $\{f_n(z)\}$ converges uniformly to $f(z)$ on every compact subset of $\mathcal{D}$. Prove that the function $f(z)$ is also one-to-one and analytic or constant on $\mathcal{D}$.

39. Let S be a family of univalent functions $f(z)$ defined on the open unit disc $\{|z| < 1\}$ that satisfy $f(0) = 0$ and $f'(0) = 1$. Show that S is closed under normal convergence, that is, if a sequence in S converge normally to $f(z)$, then $f \in S$.

40. Prove: If f is analytic and non-zero at each point of a simple closed contour Γ and is meromorphic inside Γ and if g is analytic inside and on Γ and satisfies $|g(z)| < |f(z)|$ on Γ, then f and $f + g$ have the same number of zeros inside Γ (counting multiplicity).

41. With the help of contour γ_ρ as in Fig. 7.18, show that $\int_0^\infty \frac{dx}{x^3+1} = \frac{2\pi}{3\sqrt{3}}$ by integrating $1/(1+z^3)$ around the boundary of the circular sector $\gamma_\rho = \{z = re^{it} : 0 \le t \le 2\pi/3, 0 \le r \le \rho\}$ and letting $\rho \to \infty$

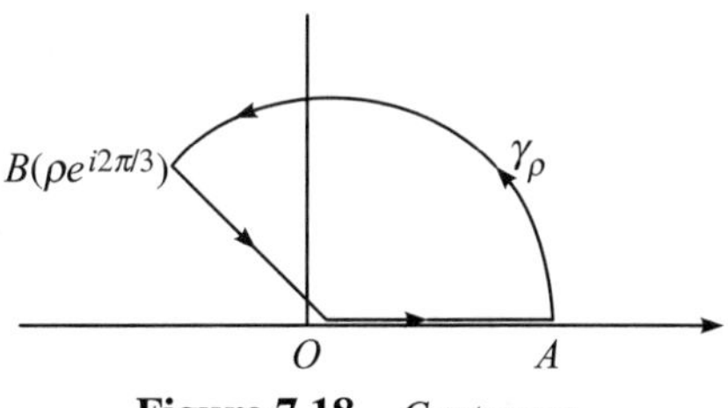

Figure 7.18 *Contour* γ_ρ

42. Apply the calculus of residues to prove the following:

(i)	$\int_0^\infty \frac{dx}{x^4+1} = \frac{\pi}{2\sqrt{2}}$	(ii)	$\int_0^\infty \frac{dx}{(x^2+1)(x^2+4)^2} = \frac{5\pi}{288}$
(iii)	$\int_0^\infty \frac{dx}{x^2+1} = \frac{\pi}{2}$	(iv)	$\int_{-\infty}^\infty \frac{dx}{(x^2+b^2)(x^2+c^2)^2} = \frac{\pi(b+c)}{2bc^3(b+c)^2}$; $b, c > 0$
(v)	$\int_0^\infty \frac{dx}{x^6+1} = \frac{\pi}{3}$	(vi)	$\int_{-\infty}^\infty \frac{x^2\,dx}{(x^2+1)(x^2+4)} = \frac{\pi}{3}$
(vii)	$\int_{-\infty}^\infty \frac{dx}{(x^2+1)^2} = \frac{\pi}{4}$	$(viii)$	$\int_{-\infty}^\infty \frac{x^2\,dx}{(x^2+9)(x^2+4)^2} = \frac{\pi}{200}$

43. Apply the calculus of residues to prove the following:

(i) $\int_0^{\pi} \frac{a\,d\theta}{a^2+\cos^2\theta} = \frac{\pi}{\sqrt{a^2+1}}$ (ii) $\int_0^{2\pi} \frac{d\theta}{a+b\cos\theta} = \frac{2\pi}{\sqrt{a^2-b^2}}; a > |b|$

(iii) $\int_0^{2\pi} \frac{\cos 3\theta}{5-4\cos\theta}\,d\theta = \frac{\pi}{12}$ (iv) $\int_0^{2\pi} \frac{\sin n\theta}{1+2a\cos\theta+a^2}\,d\theta = 0; n \in \mathbb{N}, a^2 < 1$

(v) $\int_0^{2\pi} \frac{\sin 3\theta}{5-3\cos\theta}\,d\theta = 0$ (vi) $\int_0^{2\pi} \frac{\sin^2\theta}{a+b\cos\theta}\,d\theta = \frac{2\pi}{b^2}(a - \sqrt{a^2-b^2}); a > b > 0$

44. Apply the calculus of residues to prove the following:

(i) $\int_0^{\infty} \frac{x\sin mx}{a^2+x^2}dx = \frac{\pi}{2}e^{-am}$ (ii) $\int_0^{\infty} \frac{\sin mx}{x(x^2+a^2)}\,dx = \frac{\pi}{2a^2}(1-e^{-ma}), m, a > 0$

(iii) $\int_{-\infty}^{\infty} \frac{\cos x}{a^2-x^2}\,dx = \frac{\pi \sin a}{a}; a > 0$ (iv) $\int_0^{\infty} \frac{\cos mx}{(a^2+x^2)^2}\,dx = \frac{2\pi e^{-ma}}{4a^3}(ma+1)$

(v) $\int_0^{\infty} \frac{x^4}{x^6-1}\,dx = \frac{\pi}{6}\sqrt{3}$ (vi) $\int_{-\infty}^{\infty} \frac{\sin z}{z^2+4z+5}\,dz = -\frac{\pi\sin 2}{e}$

(vii) $\int_0^{\infty} \frac{\sin \pi x}{x(1-x^2)}\,dx = \pi$ (viii) $\int_0^{\infty} \frac{\cos ax - \cos bx}{x^2}\,dx = -\frac{\pi}{2}(a-b), a, b \geq 0$

45. Let the function $f(z)$ be analytic, apart from a finite number of poles in the disc $|z| \leq 1$. Further, let $|f(z)| = 1$ for $|z| = 1$. Show that $f(z)$ is a rational function.

Answers

7. (a) $-14/25, (7 \pm i)/25$ (b) $e^{n\pi}, n \in \mathbb{Z}$ (c) $101/16, -8, 27/16$ (d) $-i/4a^3$

8. (a) $kg(z_0)$; (b) $-kg(z_0)$

10. (a) $9\pi i/\sqrt{2}$; (b) $3\pi i$; (c) $4\pi i/25$; (d) $-2\pi i$; (e) $4\pi i(\pi+1)$; (f) $4\pi i$; (g) 0; (h) $-2\pi i/e$; (i) $2\pi i$; (j) $\frac{t-1}{2} + \frac{1}{2}e^{-t}\cos t$

14. $f(z) = \frac{4}{9} + \frac{1}{z-1} + \frac{1}{z+1} - \frac{4/9}{(z+1)^2}$.

15. $f(z) = z + \frac{A}{z^2} + \frac{7}{z-1} + B$.

22. (a) 5; 23. $1, 1, -1 \pm i$; 26. 3

8 Some Relevant Theorems

In the family of analytic functions that concern geometrical orientation, conformal mapping plays an important leading role. As its consequences, we shall present here a most important result named after G.F.B. Riemann, known as the 'Riemann mapping theorem'. In this chapter we shall prove some important theorems that play an important role of advanced complex analysis.

8.1 Sequence of Functions in a Compact Set

The following elementary lemma is needed to prove certain important theorems for the sequence of complex functions in a compact set.

Lemma 8.1.1. Suppose the sequence

(i) $\langle f_n(z)\rangle$ converges uniformly to $f(z)$ on a compact set $\mathcal{D}$ and $g(z)$ is a continuous function on $\mathcal{D}$. Then $\langle g(z)f_n(z)\rangle$ converges to $g(z)f(z)$ uniformly on $\mathcal{D}$.

(ii) $\langle u_n(z)\rangle$ converges uniformly to $S(z)$ on a compact set $\mathcal{D}$ and $g(z)$ is a continuous function on $\mathcal{D}$. Then $\sum_{n=1}^{\infty} g(z)f_n(z)$ converges to $g(z)S(z)$ uniformly on $\mathcal{D}$.

Theorem 8.1. *Let $\langle f_n(z)\rangle$ be a sequence of analytic functions on a domain $\mathcal{D}$. If $f_n(z) \to f(z)$ uniformly on every compact subset of $\mathcal{D}$, then, for any $k \geq 1$, $f_n^{(k)}(z) \to f^{(k)}(z)$ for all $z \in \mathcal{D}$; i.e., the limit of the kth derivative is the kth derivative of the limit. Moreover, for each $k \geq 1$, the differentiated sequence $\left\langle f_n^{(k)}(z)\right\rangle$ converges to $f^{(k)}(z)$ uniformly on every compact subset of $\mathcal{D}$.*

Proof: Let $z_0 \in \mathcal{D}$ and Γ be a positively oriented closed contour in $\mathcal{D}$. To prove this, suppose $\varepsilon_k \to 0$ are such that $|f_k(z) - f(z)| < \varepsilon$ for $z \in \Gamma$. From the Cauchy's integral formula Eq. (5.35) for the derivative, we have

$$f_n^{(k)}(z_0) = \frac{k!}{2\pi i}\oint_\Gamma \frac{f_n(z)dz}{(z-z_0)^{k+1}} \text{ and } f^{(k)}(z_0) = \frac{k!}{2\pi i}\oint_\Gamma \frac{f(z)dz}{(z-z_0)^{k+1}}.$$

Now, since $f_n(z) \to f(z)$ uniformly on Γ and $1/(z-z_0)^{k+1}$ is continuous on Γ, from Lemma 8.1.1 it follows that

$$\frac{f_n(z)}{(z-z_0)^{k+1}} \longrightarrow \frac{f(z)}{(z-z_0)^{k+1}}$$

uniformly on Γ. Now let $\mathcal{D}_1$ be a compact subset of $\mathcal{D}$, and in $\mathcal{D}$ let Γ be a positively oriented closed contour containing $\mathcal{D}_1$ so that the shortest distance ρ between $\mathcal{D}_1$ and Γ is positive; i.e., for all $\xi \in \mathcal{D}_1$ and $z \in \mathcal{D}_1$, $|\xi - z| = \rho d > 0$. Then, from Eq. (5.35), for all $z \in \mathcal{D}_1$, it follows that

$$\begin{aligned}\left|f_n^{(k)}(z) - f^{(k)}(z)\right| &= \left|\frac{k!}{2\pi i}\oint_\Gamma \frac{f_n(\xi) - f(\xi)}{(\xi - z)^{k+1}}\,d\xi\right| \\ &\leq \frac{k!}{2\pi}\oint_\Gamma \left|\frac{f_n(\xi) - f(\xi)}{(\xi - z)^{k+1}}\,d\xi\right| \\ &\leq \frac{k!}{2\pi\rho^{k+1}}L(\Gamma)\max_{\xi\in\{\Gamma\}}\left|f_n(\xi) - f(\xi)\right| \leq \frac{k!\varepsilon_k L(\Gamma)}{2\pi\rho^{k+1}} \to 0\end{aligned}$$

By, applying Corollary 2.6.2, we obtain

$$f_n^{(k)}(z_0) = \frac{k!}{2\pi i}\oint_\Gamma \frac{f_n(z)dz}{(z - z_0)^{k+1}} \longrightarrow f^{(k)}(z_0) = \frac{k!}{2\pi i}\oint_\Gamma \frac{f(z)dz}{(z - z_0)^{k+1}}.$$

Finally, since $f_n(\xi) \to f(\xi)$ uniformly on every compact subset of $\mathcal{D}$, from the inequality above it follows that $f_n^{(k)}(z) \to f^{(k)}(z)$ uniformly on $\mathcal{D}_1$. □

RESULT **65.** Theorem 8.1 does not hold if $\mathcal{D}$ is assumed to be an arbitrary set instead of a domain. The sequence $\{\frac{\sin nx}{n}\}$ converges uniformly to zero on the real axis; however, the sequence of its derivative $\{\cos nx\}$ converges only at $x = 0$. Thus, the sequence $\{\frac{\sin nz}{n}\}$ cannot converge uniformly on any domain containing points of the real axis.

Corollary 8.1.1. Let $\langle u_n(z)\rangle$ be a sequence of analytic functions on a domain $\mathcal{D} \subseteq \mathbb{C}$. If $\sum_{n=0}^{\infty} u_n(z)$ converges to $S(z)$ uniformly on every compact subset of $\mathcal{D}$, then for any $k \geq 1$ and all $z \in \mathcal{D}$,

$$\sum_{n=0}^{\infty} u_n^{(k)}(z) = S^{(k)}(z), \tag{8.1}$$

i.e., the series may be differentiated term-by-term. Moreover, for each k, the convergence in (8.1) is uniform on every compact subset of $\mathcal{D}$.

Definition 8.1.1. A sequence $\langle f_n(z)\rangle$ of analytic functions of a domain $\mathcal{D} \subseteq \mathbb{C}$ *converges normally* to the analytic function $f(z)$ on $\mathcal{D}$ if it converges uniformly to $f(z)$ on each compact subsets in $\mathcal{D}$.

8.2 Convergence in the Space of Analytic Functions

In this section, some results are obtained by putting a metric on a set of analytic functions on a fixed domain $\mathcal{D}$, which enable us to also study of meromorphic functions. Proof of two important theorems, namely Arzelà–Ascoli theorem and the Motel's theorem, which guarantees that given a family $\mathcal{F}$ of functions in a complex domain, any sequence in $\mathcal{F}$ has a uniformly convergent subsequence.

Definition 8.2.1. [Equicontinuous]: Let $\mathcal{F}$ be a family of collection of functions, defined and continuous and complex-valued on a compact subset $E \subset \mathbb{C}$. A function $f \in \mathcal{F}$ is continuous at $z_0 \in E$ if given $\varepsilon > 0$, $\exists\,\delta = \delta(f, z_0, \varepsilon) > 0$ such that

$$|z - z_0| < \delta \text{ and } z \in E \Rightarrow \left|f(z) - f(z_0)\right| < \varepsilon.$$

If δ, independent of f and depending on z_0 and ε can be found $\forall\varepsilon > 0$, we say that the family $\mathcal{F}$ is *equicontinuous* at z_0.

Definition 8.2.2. [Uniformly bounded]: Let $\langle f_n \rangle$ be a sequence of analytic functions on a domain $\mathcal{D} \subset \mathbb{C}$ and let $U \subset \mathcal{D}$. Then, we say that $\langle f_n \rangle$ is uniformly bounded on U if $\exists$ a $M > 0$ such that

$$|f(z)| \leq M \; \forall \; z \in \mathcal{D} \text{ and } \forall \; n \in \mathbb{N}.$$

Let $\mathcal{F}$ be a family of collection of functions, defined and continuous and complex-valued on a compact subset $E \subset \mathbb{C}$. We say that $\mathcal{F}$ is *uniformly bounded on* E if

$$|f(z)| \leq M \; \forall z \in E \text{ and } \forall f \in \mathcal{F}.$$

Definition 8.2.3. [Locally bounded]: A set $\mathcal{F} \subset \mathcal{H}(\mathcal{D})$ is locally bounded if for each point $\alpha \in \mathcal{D}$, there are constants M and $\rho > 0$ such that

$$|f(z)| \leq M; |z - \alpha| < \rho \text{ for all } f \in \mathcal{F}$$

$$\text{i.e.,} \qquad \sup\Big\{|f(z)| : |z - \alpha| < \rho, f \in \mathcal{F}\Big\} < \infty.$$

That is, $\mathcal{F}$ is locally bounded if about a point $\alpha \in \mathcal{D}$, there is a disc on which $\mathcal{F}$ is uniformly bounded, which immediately extends to the requirement that $\mathcal{F}$ is uniformly bounded on a compact sets in $\mathcal{D}$.

Definition 8.2.4. [Normal convergence]: If a sequence of complex functions $\langle f_n \rangle$ converges on a compact subsets, it is called *normal convergence*. If a sequence of functions is uniformly bounded on compact subsets of the domain, it is said to be *normally bounded.*

8.2.1 Arzelà–Ascoli Theorem

In this theorem, all conditions, hypothesis and conclusions are used on the set of functions defined on the whole domain, i.e., on a compact set.

Theorem 8.2. *Let $\mathcal{F}$ be a family of collection of continuous, complex-valued function on a compact subset $E \subset \mathbb{C}$. Suppose $\mathcal{F}$ is uniformly bounded on E. Then the following are equivalent:*

(*i*) *$\mathcal{F}$ is equicontinuous at each point of E.*

(*ii*) *Every sequence $\langle f_n \rangle \subset \mathcal{F}$ has a uniformly convergent subsequence $\langle f_{n_k} \rangle$.*

Proof: Let $E_{\mathbb{Q}}$ be the set of points of E with rational coordinates as

$$E_{\mathbb{Q}} = \Big\{x + iy \in E : \; x, y \in \mathbb{Q}\Big\}.$$

Since $\mathbb{Q}$ is countable and dense subset in E, then $E_{\mathbb{Q}}$ is countable and $\overline{E}_{\mathbb{Q}} = E$. As $E_{\mathbb{Q}}$ is countable, we may enumerate the points of $E_{\mathbb{Q}}$ by $z_1, z_2, \cdots$, i.e., $E_{\mathbb{Q}} = \langle z_n \rangle_{n \geq 1}$.

Let $\langle f_n \rangle$ be a sequence of $\mathcal{F}$. Since $\mathcal{F}$ is uniformly bounded on each compact subset of E, hence it is pointwise bounded. Thus, if we consider $\langle f_n(z_1) \rangle_{n \geq 1}$, then it is a bounded sequence of complex numbers, and so $|f_n(z_1)| \leq M$. By Bolzano–Weierstrass theorem, we have a convergent subsequence $\langle f_{n_1}(z_1) \rangle_{n_1 \geq 1}$ of $\langle f_n(z_1) \rangle$.

Now, consider $\langle f_{n_1}(z_2) \rangle_{n_1 \geq 1}$. Again $|f_{n_1}(z_2)| \leq M$ and so by Bolzano–Weierstrass theorem, there exists a subsequence $\langle f_{n_2}(z_2) \rangle$ of $\langle f_{n_1}(z_2) \rangle_{n_1 \geq 1}$ which converges. Note that $\langle f_{n_2}(z_1) \rangle$ also converges.

By induction, we get for $\forall n \geq 1$, a subsequence $\langle f_{n_k} \rangle$ of $\langle f_{n_{k-1}} \rangle$ such that $\langle f_{n_k}(z_j) \rangle$ converges for $j \leq k$, and

$$\langle f_{n_1} \rangle \supseteq \langle f_{n_2} \rangle \supseteq \cdots \supseteq \langle f_{n_{k-1}} \rangle \supseteq \langle f_{n_k} \rangle \supseteq \cdots$$

We get a list of lists:

					Converge at
f_{n_1} :	$f_{i_1}(z_1) \searrow$	$f_{i_2}(z_1)$	$f_{i_3}(z_1)$	$\cdots$	z_1
f_{n_2} :	$f_{j_1}(z_2)$	$f_{j_2}(z_2) \searrow$	$f_{j_3}(z_3)$	$\cdots$	z_1, z_2
f_{n_3} :	$f_{k_1}(z_3)$	$f_{k_2}(z_3)$	$f_{k_3}(z_3) \searrow$	$\cdots$	z_1, z_2, z_3
$\vdots$	$\vdots$			$\cdots$	$\vdots$

Consider the subsequence $\langle F_l \rangle$, where $F_l = l$th member of $\langle f_{n_l} \rangle$. Then

$$\langle F_l \rangle \subset \bigcap_{k=1}^{\infty} \langle f_{n_k} \rangle$$

where $\langle f_{n_l} \rangle$ converge at the points $z_1, z_2, z_3, \cdots$. If $z \in E_{\mathbb{Q}}$, then at $z = z_l$, since $\langle f_{n_l}(z_l) \rangle$ converges, we have that the sequence $\langle F_t(z_l) \rangle_{t \geq l}$ converges.

Since $\mathcal{F}$ is equicontinuous on E, so is $\langle F_l \rangle_{l \geq 1}$. So given $z_0 \in E$ and an $\varepsilon > 0$, there exists a $\delta > 0$ with $\delta = \delta(z_0, \varepsilon)$ such that

$$\Big| f(z) - f(z_0) \Big| < \frac{\varepsilon}{3} \text{ whenever } |z - z_0| < \delta, \forall f \in \mathcal{F}. \tag{8.2}$$

The collection $\{|z - z_0| < \delta(z_0, \varepsilon)/2 : z_0 \in E\}$ is an open cover of E, which is compact, and therefore admits a finite subcover:

$$\Big\{ |z - \zeta_1| < \frac{\delta}{2}(\zeta_1, \varepsilon); |z - \zeta_2| < \frac{\delta}{2}(\zeta_2, \varepsilon); \cdots, |z - \zeta_k| < \frac{\delta}{2}(\zeta_k, \varepsilon) \Big\}.$$

Choose

$$z_{i_1} \in \Big\{ |z - \zeta_1| < \frac{\delta}{2}(\zeta_1, \varepsilon) \Big\}$$
$$z_{i_2} \in \Big\{ |z - \zeta_2| < \frac{\delta}{2}(\zeta_2, \varepsilon) \Big\}$$
$$\vdots$$
$$z_{i_k} \in \Big\{ |z - \zeta_k| < \frac{\delta}{2}(\zeta_k, \varepsilon) \Big\}$$

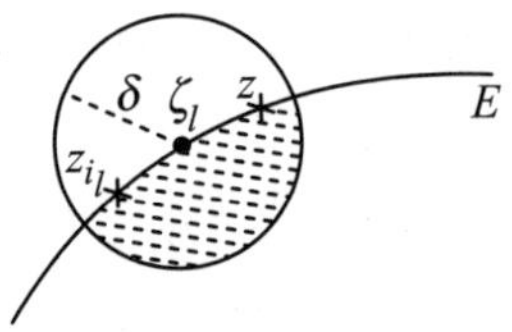

Figure 8.1 *Arzelà–Ascoli theorem*

where, $z_{i_j} \in E_{\mathbb{Q}} = \langle z_n \rangle_{n \geq 1}$. Note that

$$\begin{aligned} & |z - z_{i_l}| \leq |z - \zeta_{i_l}| + |\zeta_{i_l} - z_{i_l}| < \delta(\zeta_l, \varepsilon) \\ \Rightarrow \quad & \Big| F_m(z) - F_m(z_{i_l}) \Big| < \frac{\varepsilon}{3} \text{ and } \Big| F_n(z_{i_l}) - F_n(z) \Big| < \frac{\varepsilon}{3}. \end{aligned}$$

If $z \in E$, then $|z - \zeta_l| < \delta(\zeta_l, \varepsilon)/2$ for some l with $1 \leq l \leq k$ and z_{i_l} is also in $|z - \zeta_k| < \delta(\zeta_k, \varepsilon)/2$ (Fig. 8.1). Let m, n be two positive integers, greater than a large positive integer. Now

$$\begin{aligned} |F_m(z) - F_n(z)| &= \Big| F_m(z) - F_m(z_{i_l}) + F_m(z_{i_l}) - F_n(z_{i_l}) + F_n(z_{i_l}) - F_n(z) \Big| \\ &\leq \Big| F_m(z) - F_m(z_{i_l}) \Big| + \Big| F_m(z_{i_l}) - F_n(z_{i_l}) \Big| + \Big| F_n(z_{i_l}) - F_n(z) \Big|. \end{aligned}$$

Since $\langle F_m \rangle$ converge on $E_{\mathbb{Q}}$ and $z_{i_l} \in E_{\mathbb{Q}}$, for the given ε, $\exists \mathcal{N}(\varepsilon, z_{i_l})$ such that

$$\Big| F_m(z_{i_l}) - F_n(z_{i_l}) \Big| < \frac{\varepsilon}{3} \text{ for } m, n \geq \mathcal{N}(\varepsilon, z_{i_l})$$

Thus for $m, n \geq \max\{\mathcal{N}(\varepsilon, z_{i_l}) : 1 \leq l \leq k\}$ we have

$$|F_m(z) - F_n(z)| < \varepsilon \text{ independent of } z \in E$$

$$\Rightarrow \quad \langle F_m \rangle \text{ converges uniformly on } E$$

This proves the theorem. □

8.2.2 Montel's Theorem

In the Ascola–Arzeli theorem, whatever the conditions we put on functions on a compact set, the conclusion is also a function on a compact set. But when it is adopted in complex analysis what we do is to take analytic functions in a domain but put all the conditions on compact subsets of the domain. So, we are essentially looking for analytic functions but the continuity (and whatever conditions we put) we put on them in will be on compact subsets of the domain. This is called *normal conditions.*

Thus, as a natural version of the Ascola–Arzeli theorem for analytic functions, we can expect to have a family of analytic functions in a domain, which are normally bounded, i.e., they are bounded in a compact sets. Then equicontinuity is equivalent to normal convergence, namely normal convergence of the subsequence of a given sequence. So, if we assume a family of analytic functions in a domain, it is uniformly bounded on a compact subsets of the domain.

Thus if a family of functions is equicontinuous, then, using Montel's theorem, for any given sequence in this family, we can extract a subsequence that converges normally, i.e., converges on a compact subsets of a domain.

Lemma 8.2.1. Let $\mathcal{G}$ be a family of analytic functions on a domain $U \subset \mathbb{C}$ and $\mathcal{G}' = \{g' : g \in \mathcal{G}\}$ is uniformly bounded in the neighbourood of $z_0 \in U$. Then $\mathcal{G}$ is equicontinuous at z_0.

Proof: It suffices to prove equicontinuity of $\mathcal{G}$ on an arbitrary bounded closed disc $\Gamma = \{z \in U : 0 < |z - z_0| \leq r\}$ contained in U. Since all the derivatives $\mathcal{G}'$ are uniformly bounded in the neighbourhood of z_0, so, there exists a constant $M > 0$ such that

$$\left|g'(z)\right| \leq M \text{ in } \Gamma \; \forall g \in \mathcal{G}.$$

Now, we are to show that the family is equicontinuous.

$$\left|g(z) - g(z_0)\right| = \left|\int_{z_0}^{z} g'(\zeta)\, d\zeta\right|$$

Since g is analytic, g' is also analytic. Hence the integral on the right-hand side is path independent, and so, integrating over a straight line path z_0 to z gives

$$\left|g(z) - g(z_0)\right| \leq \int_{z_0}^{z} \left|g'(\zeta)\right| \cdot \left|d\zeta\right| \leq M\left|z - z_0\right| \quad \forall g \in \mathcal{G}.$$

So, given $\varepsilon > 0$, there exists $\delta = \varepsilon/M$ such that

$$|z - z_0| < \delta \Rightarrow \left|g(z) - g(z_0)\right| < \varepsilon \quad \forall g \in \mathcal{G}$$

$$\Rightarrow \quad g \text{ is equicontinuous at } z_0.$$

This proves the lemma. □

Lemma 8.2.1 tells us that whenever we have a uniform boundedness in the neighbourhood of a point, we have equicontinuity at that point.

Theorem 8.3. [Montel's theorem]: *Let $\mathcal{F}$ be a family of analytic functions defined on a domain $\mathcal{D} \subset \mathbb{C}$. Suppose $\mathcal{F}$ is uniformly bounded on compact subsets of $\mathcal{D}$, i.e., $\mathcal{F}$ is normally bounded on $\mathcal{D}$. Then the following are equivalent:*

(*i*) *$\mathcal{F}$ is equicontinuous (at each point of $\mathcal{D}$).*

(*ii*) *Every sequence $\langle f_n \rangle \subset \mathcal{F}$ have a subsequence $\langle f_{n_k} \rangle$ which converges normally (i.e., uniformly on compact subsets of $\mathcal{D}$).*

Proof: Take $z_0 \in \mathcal{D}$ and choose $\rho = \rho(z_0) > 0$ such that $\{z \in \mathcal{D} : |z - z_0| \le \rho\} \subseteq \mathcal{D}$. First we show that, we can find a subsequence that converges uniformly on $\{z \in \mathcal{D} : |z - z_0| \le \rho\}$. By Cauchy's integral formula (5.35) for derivatives,

$$g'(z_0) = \frac{1}{2\pi i} \int_{|z-z_0|=\rho-\varepsilon} \frac{g(\zeta)}{(\zeta - z_0)^2}\, d\zeta$$

for g analytic in $\{z \in \mathcal{D} : |z - z_0| \le \rho\}$, $\varepsilon > 0$ very small. Taking the parametric form $z = z_0 + (\rho - \varepsilon)e^{it}$; $t \in [0, 2\pi]$, we get

$$g'(z_0) = \frac{1}{2\pi i} \int_0^{2\pi} \frac{g\Big(z_0 + (\rho - \varepsilon)e^{it}\Big)}{(\rho - \varepsilon)^2 e^{2it}} (\rho - \varepsilon) i\, e^{it}\, dt$$

$$\therefore \quad |g'(z_0)| \le \frac{1}{2\pi} \int_0^{2\pi} \frac{M(\rho - \varepsilon)}{(\rho - \varepsilon)^2} dt \le \frac{M}{\rho - \varepsilon},$$

where $|g| \le M < \infty$ on $|z - z_0| = \rho - \varepsilon$. Since $\mathcal{F}$ is uniformly bounded on $|z - z_0| \le \rho$ (which is compact), there exists M such that $|f(z)| \le M$ on $|z - z_0| \le \rho$ and

$$|f'(z_0)| \le \frac{M}{\rho - \varepsilon} \quad \forall f \in \mathcal{F} \tag{8.3}$$

Thus, if the derivatives of a the family of functions are uniformly bounded, then that family is equicontinuous.

Equation (8.3) shows that the uniform boundedness of the derivatives at z_0, therefore, is also in the small neighbourhood of z_0. So by the lemma 8.2.1, $\mathcal{F}$ is equicontinuous at each point $z_0 \in \mathcal{D}$.

Now, consider $\langle f_n \rangle \subset \mathcal{F}$. For each positive integer $n \ge 1$, let

$$E_n = \Big\{z : |z| \le n\Big\} \bigcap \Big\{z \in \mathcal{D} : d(\mathcal{D}, \partial\mathcal{D}) \ge \frac{1}{n}\Big\}$$

as depicted in the Fig. 8.2. Since E_n is clearly bounded and is the intersection of two closed subsets of $\mathbb{C}$, E_n is compact. Thus E_n has the following properties:

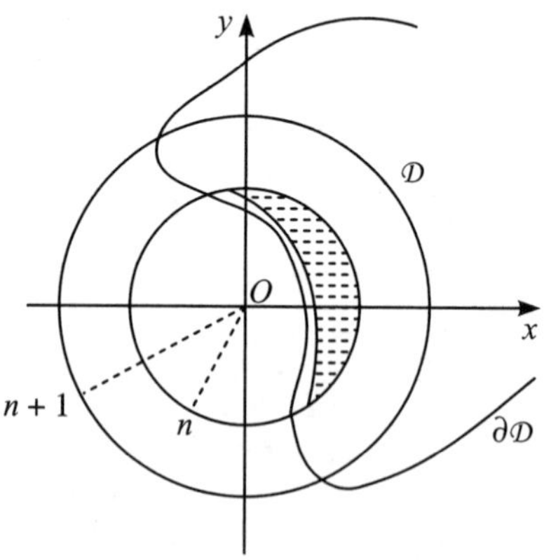

Figure 8.2 *Description of E_n*

(i) E_n is compact for all n.

(ii) Also, the set $\{z : |z| < n + 1\} \bigcap \{z : d(\mathcal{D}, \partial\mathcal{D}) > \frac{1}{n+1}\}$ is open, contains E_n and contained in E_{n+1} so, $E_n \subseteq E_{n+1}$.

(iii) $\bigcup_{n=1}^{\infty} E_n = \mathcal{D}$

(iv) Any compact subset E of $\mathcal{D}$ is contained in E_n for n sufficiently large, and $\langle E_n \rangle$ form a open cover of E.

Apply Arzelà–Ascoli theorem to sequence $\langle f_n \rangle$ on E_1 to get a subsequence $\langle f_{n_1} \rangle$ of $\langle f_n \rangle$ that converges uniformly on E_1. Now, apply to sequence $\langle f_n \rangle$ on E_2 to get a subsequence $\langle f_{n_2} \rangle \subseteq \langle f_{n_1} \rangle$ of $\langle f_n \rangle$ that converges uniformly on E_2.

By induction, we get a subsequence $\langle f_{n_k} \rangle$ which converges uniformly on $E_k; k \geq 1$, and

$$\left\langle f_{n_1} \right\rangle \supseteq \left\langle f_{n_2} \right\rangle \supseteq \cdots \supseteq \left\langle f_{n_{k-1}} \right\rangle \supseteq \left\langle f_{n_k} \right\rangle \supseteq \cdots.$$

We get a list of lists:

$$\begin{array}{lcccccc} & & & & & & \text{UC on} \\ f_{n_1}: & f_{i_1} \searrow & f_{i_2} & f_{i_3} & \cdots & E_1 \\ f_{n_2}: & f_{j_1} & f_{j_2} \searrow & f_{j_3} & \cdots & E_2 \\ f_{n_3}: & f_{k_1} & f_{k_2} & f_{k_3} \searrow & \cdots & E_3 \\ \vdots & \vdots & & & \cdots & \vdots \end{array}$$

Then, the diagonal subsequence $\langle f_{i_1}, f_{j_2}, f_{k_3}, \cdots \rangle$ converges. Define $\langle F_m \rangle$, where $F_m = m$th member of $\langle f_{n_m} \rangle$, then

$$\left\langle F_m \right\rangle \subset \bigcap_{k=1}^{\infty} \left\langle f_{n_k} \right\rangle \subset \left\langle f_k \right\rangle$$

converges uniformly on compact subsets of $\mathcal{D}$.

8.3 Riemann Mapping Theorem

In the *Riemann mapping theorem* we are to ensure the existence of unique analytic function f which defines a one-to-one mapping of a single domain $\mathcal{D}$ in the complex z-plane, with suitable boundary on the circular disc $|w| < 1$.

Lemma 8.3.1. If $\mathcal{D}_0 \subset \triangle$ is a simply-connected domain such that $\mathcal{D}_0 \neq \triangle$. If $0 \in \mathcal{D}_0$, then, there exists an analytic map $\psi : \mathcal{D}_0 \longrightarrow \triangle$ such that $\psi(0) = 0$ and $|\psi'(0)| > 1$.

Proof: Choose $\beta \in \triangle / \mathcal{D}_0$. Let us define a function h as

$$h(\zeta) = \frac{\zeta - \beta}{1 - \bar{\beta}\zeta}; \quad \text{where } h(\beta) = 0.$$

It shows that $h(\mathcal{D}_0) \neq 0$. Since the simply-connected domain $h(\mathcal{D}_0)$ does not contain 0, so we can take always an analytic branch of square root. Since h is an holomorphic automorphism, so it is homeomorphism and $h : \mathcal{D}_0 \to h(\mathcal{D}_0)$ is a topological isomorphism.

Since $\xi = h(\zeta), \exists$ analytic branch of $\ln \eta$ on $h(\mathcal{D}_0)$ and hence an analytic branch of $G(\eta)$ of $\sqrt{\eta} = e^{\frac{1}{2}\ln \eta}$. $G : h(\mathcal{D}_0) \to G(h(\mathcal{D}_0))$ is a holomorphic isomorphism.

If $\eta \mapsto \eta^2$ is a strict contraction, it implies G is a strict expansion with respect to hyperbolic i.e.

$$\rho_{\text{hy}}\Big(G(\eta_0)), (G(\eta_1)\Big) > C\rho_{\text{hy}}(\eta_0, \eta_1) \; \forall \eta_0, \eta_1 \in h(\mathcal{D}_0) \tag{8.4}$$

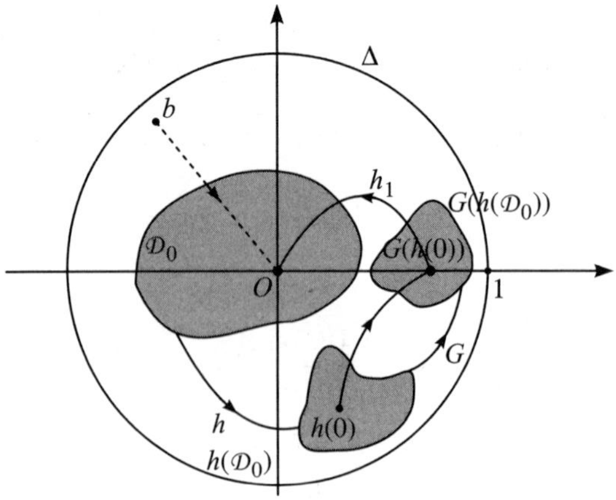

Figure 8.3 *Automorphism*

for suitable $C > 1$. In fact, $C = \frac{1+r}{2\sqrt{r}}$, where $|\eta| \leq \sqrt{r}$ contains $h(\mathcal{D}_0)$; G is an image of g. Consider $h_1(w) = \frac{w - G(h(0))}{1 - \overline{G(h(0))}w}$.

$$\mathcal{D}_0 \underset{\text{isomorphism}}{\overset{h}{\longrightarrow}} h(\mathcal{D}_0) \underset{\text{isomorphism}}{\overset{G}{\longrightarrow}} G\Big(h(\mathcal{D}_0)\Big) \overset{h_1}{\longrightarrow} h_1\Big(G(h(\mathcal{D}_0))\Big)$$

$$\zeta \mapsto h(\zeta) = \eta \mapsto G(\eta) = w \mapsto h_1(w) = \gamma \text{ (say)}.$$

$$0 \mapsto h(0) \mapsto G\Big(h(0)\Big) \mapsto 0$$

Put $\psi = h_1 \circ G \circ h$. Then $\psi(0) = 0$, $\psi(\zeta) = \gamma$. Now

$$\frac{\rho_{\text{hy}}\Big(\psi(\zeta), \psi(0)\Big)}{\rho_{\text{hy}}(\zeta, 0)} = \frac{\rho_{\text{hy}}\Big(\psi(\zeta), 0\Big)}{\rho_{\text{hy}}(\zeta, 0)} = \frac{\rho_{\text{hy}}\Big((h_1(G(h(\zeta)))), (h_1(G(h(0))))\Big)}{\rho_{\text{hy}}(\zeta, 0)}$$

$$= \frac{\rho_{\text{hy}}\Big((G(h(\zeta))), (G(h(0)))\Big)}{\rho_{\text{hy}}(\zeta, 0)}$$

as h_1 is isomorphic of $\triangle$ with respect to hyperbolic metric ρ_{hy}.

$$\geq C \frac{\rho_{\text{hy}}\Big((h(\zeta)), (h(0))\Big)}{\rho_{\text{hy}}(\zeta, 0)} = C \text{ using Eq.(8.4)}.$$

As $\zeta \to 0$, we have $\frac{\rho_{\text{hy}}(\psi(\zeta), \psi(0))}{\rho_{\text{hy}}(\zeta, 0)} \geq C > 1$. Now, using Eq. (8.6), we get

$$\rho_{\text{hy}}\Big(\psi(\zeta), \psi(0)\Big) = \Big(\psi(\zeta), 0\Big) = \frac{1}{2} \ln\left(\frac{1 + |\psi(\zeta)|}{1 - |\psi(\zeta)|}\right)$$

$$\rho_{\text{hy}}(\zeta, 0) = \frac{1}{2} \ln\left(\frac{1 + |\zeta|}{1 - |\zeta|}\right)$$

$$\therefore \quad \frac{\rho_{\text{hy}}\Big(\psi(\zeta), \psi(0)\Big)}{\rho_{\text{hy}}(\zeta, 0)} = \frac{\frac{1}{2} \ln\left(\frac{1+|\psi(\zeta)|}{1-|\psi(\zeta)|}\right)}{\frac{1}{2} \ln\left(\frac{1+|\zeta|}{1-|\zeta|}\right)} \approx \frac{|\psi(\zeta)|}{|\zeta|}, \text{ as } \zeta \text{ is small}$$

where, we use the approximation $\frac{1}{2}\ln(\frac{1+x}{1-x}) = x + O(x^2) \approx x$ as x is small. Therefore,

$$\begin{aligned} \left|\psi'(0)\right| &= \left|\lim_{\zeta\to 0}\frac{\psi(\zeta)-\psi(0)}{\zeta - 0}\right| = \left|\lim_{\zeta\to 0}\frac{\psi(\zeta)}{\zeta}\right| \\ &= \lim_{\zeta\to 0}\left|\frac{\psi(\zeta)}{\zeta}\right| = \lim_{\zeta\to 0}\frac{\rho_{\text{hy}}\Big(\psi(\zeta),\psi(0)\Big)}{\rho_{\text{hy}}(\zeta,0)} > 1. \end{aligned}$$

This proves our required statement.

Theorem 8.4. *Any simply-connected domain $\mathcal{D} \subset \mathbb{C}$ ($\mathcal{D} \neq \mathbb{C}$) is holomorphically isomorphic to $\triangle = \{w \in \mathbb{C} : |w| = 1\}$. In fact, given $z_0 \in \mathcal{D}$, a real number $\lambda \in \mathbb{R}^+$, there exists a unique holomorphic isomorphism $f : \mathcal{D} \to \triangle$ with $f(z_0) = 0$, $f'(z_0) = \lambda$.*

Proof: Here $\mathcal{D}$ is a simply-connected domain with $\mathcal{D} \neq \mathbb{C}$ and $z_0 \in \mathcal{D}$ (Fig. 8.4). The set of all holomorphic functions in an open set $\mathcal{D}$ be denoted by $\mathcal{H}(\mathcal{D})$. We divide the proof into several steps:

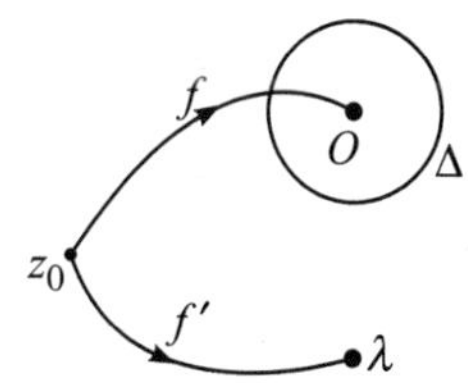

Figure 8.4

The Riemann map $f : \mathcal{D} \longrightarrow \triangle$ is unique:

Suppose $f_1, f_2 : \mathcal{D} \longrightarrow \triangle$ are holomorphic isomorphisms with $f_1(z_0) = 0 = f_2(z_0)$ and $f_1'(z_0) = \lambda = f_2'(z_0)$ for some $z_0 \in \mathcal{D}$. Then $f_1 \circ f_2^{-1} \in \mathcal{H}(\mathcal{D})$ and the analytic bijection $f_1 \circ f_2^{-1} : \triangle \longrightarrow \triangle$ is conformal (Fig. 8.5). So, by corollary (5.10.1) of Schwarz Lemma, we get

$$\left(f_1 \circ f_2^{-1}\right)(w) = e^{i\alpha}\, w \text{ for some } \alpha \in \mathbb{R}.$$

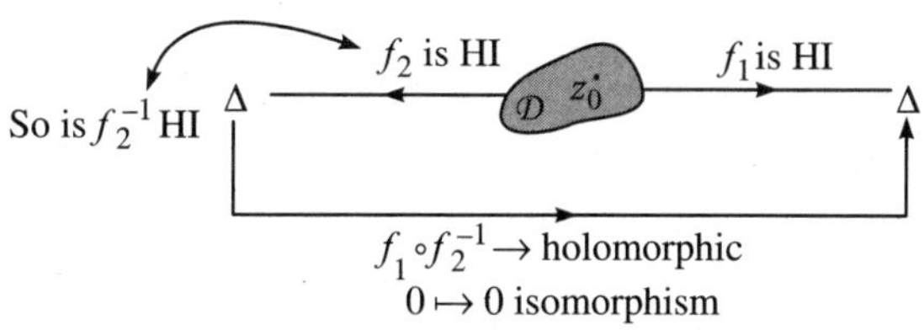

Figure 8.5 *Uniqueness of the Riemann map*

Taking differentiation, and using the chain rule of differentiation, we get

$$\begin{aligned} &\left(f_1 \circ f_2^{-1}\right)'(w) = \left(e^{i\alpha}\, w\right)' = e^{i\alpha} \\ \Rightarrow\quad & f_1'\left(f_2^{-1}(w)\right)\cdot\left(f_2^{-1}\right)'(w) = e^{i\alpha}. \end{aligned}$$

Since it is an identity, put $w = 0$:

$$\begin{aligned} & f_1'\left(f_2^{-1}(0)\right)\cdot\left(f_2^{-1}\right)'(0) = e^{i\alpha} \\ \text{or,}\quad & f_1'(z_0)\cdot\left(f_2^{-1}\right)'(0) = e^{i\alpha} \Rightarrow \lambda\cdot\left(f_2^{-1}\right)'(0) = e^{i\alpha}. \end{aligned}$$

Since $e^{i\alpha} = 1$, so, $f_1 \circ f_2^{-1}$ is an identity mapping. Therefore

$$\left(f_1 \circ f_2^{-1}\right)(w) = w \Rightarrow f_1(z) = f_2(z) \text{ for all } z \in \mathcal{D}.$$

Existence of the Riemann map $f : \mathcal{D} \longrightarrow \triangle$

Step I: Here, we are to find a holomorphic isomorphism $h : \mathcal{D} \underset{\sim}{\rightarrow} h(\mathcal{D}) \subset \triangle$. Since the domain $\mathcal{D}$ is not a whole complex plane $\mathbb{C}$, there exists a point $\alpha \in \mathbb{C}/\mathcal{D}$. So the translation (Möbius transformation) $g(z) = z - \alpha$ is non-vanishing on $\mathcal{D}$, which is simply-connected. So we have an analytic branch of $\log(z - \alpha)$ on $\mathcal{D}$.

Thus we have an analytic branch of $\sqrt{z-\alpha}$ as $e^{\frac{1}{2}\log(z-\alpha)}$ in $\mathcal{D}$. We can use this branch $\sqrt{z-\alpha}$ as the homomorphic isomorphism $h : \mathcal{D} \longrightarrow \mathbb{C}$. Let

$$\begin{aligned} h(z) &= \text{ analytic branch of } \sqrt{z-\alpha} = e^{\frac{1}{2}\log(z-\alpha)} \\ h^2(z) &= z - \alpha \Rightarrow z = h^2(z) + \alpha. \end{aligned}$$

Here $h(\mathcal{D})$ is simply connected. If $h(\mathcal{D})$ is analytic, then $-h(\mathcal{D})$ is also analytic and gives the other branch of the square root. Now

$$\int_\gamma \frac{dz}{z-\alpha} \text{ is independent of path } \gamma \text{ because of simple connectedness}$$

$$\int_{z_0}^{z} \frac{dz}{z-\alpha} = \log(z-\alpha) - \log(z_0 - \alpha)$$

$$\int_{z_0}^{z} \frac{dz}{z-\alpha} + \log(z_0-\alpha) \text{ is an analytic branch of } \log(z-\alpha) \text{ in } \mathcal{D}.$$

<u>h is one-to-one on $\mathcal{D}$</u> : Now, for $z_1, z_2 \in \mathcal{D}$, $h(z_1) = h(z_2)$ in $\mathcal{D}$ means

$$z_1 = h^2(z_1) + \alpha = h^2(z_2) + \alpha = z_2.$$

$h(z)$ is non-constant analytic and one-to-one, and so it is an isomrphism. Also, the inverse function theorem 7.8 tells us that the inverse function is also analytic. Therefore the one-to-one analytic map is an analytic isomorphism/ holomorphic isomorphism/ conformal isomorphism. Thus, due to open mapping and inverse function theorem, $h : \mathcal{D} \underset{\sim}{\rightarrow} h(\mathcal{D}) \subset \triangle$ is a holomorphic isomorphism.

Let $-h(\mathcal{D}) = \{-h(z) : z \in \mathcal{D}\}$ be the reflection of $h(\mathcal{D})$ (Fig. 8.6). Then $h(\mathcal{D}) \cap -h(\mathcal{D}) \neq \phi$, the null set. If $w_0 \in h(\mathcal{D}) \cap -h(\mathcal{D})$, then $w_0 = h(z_1) = -h(z_2)$ for $z_1, z_2 \in \mathcal{D}$.

$$\begin{aligned} & z_1 = h^2(z_1) + \alpha = h^2(z_2) + \alpha = z_2 \\ \Rightarrow \quad & h(z_1) = h(z_2) = -h(z_2) \\ \Rightarrow \quad & h(z_2) = 0 - w_0 \Rightarrow w_0 = 0. \end{aligned}$$

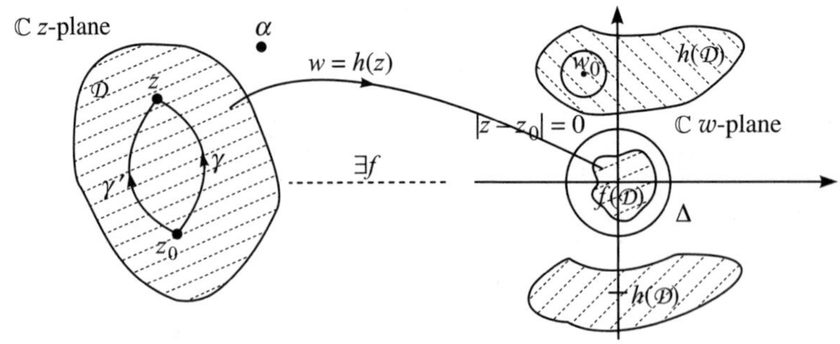

Figure 8.6

But $w_0 \neq 0$ because $h \neq 0$ (exponential function).

$$\Rightarrow h(\mathcal{D}) \cap -h(\mathcal{D}) = \phi.$$

If $|w - w_0| \leq \varepsilon$ is in $h(\mathcal{D})$, where $w_0 = h(z_0)$, then

$$d(-h(\mathcal{D}), w_0) > \varepsilon \Rightarrow \Big| - h(\mathcal{D}) - w_0\Big| > \varepsilon \, \forall z \in \mathcal{D}$$

$$\Rightarrow \quad \Big|h(\mathcal{D}) + w_0\Big| > \varepsilon \, \forall z \in \mathcal{D}.$$

Now put $f(z) = \frac{\varepsilon}{h(z)+w_0}$, which is analytic on $\mathcal{D}$ as $h(z) + w_0$ never vanishes and is analytic. Thus $f(z)$ is a one-to-one analytic function because $f(z)$ can be obtained by using a series of Möbius transformations:

$$h(z) + w_0 \to \text{ translation };\ \frac{1}{h(z) + w_0} \to \text{ inversion };\ \frac{\varepsilon}{h(z) + w_0} \to \text{ dilation.}$$

Clearly, f is analytic and one-to-one on $\mathcal{D}$ and

$$|f(z)| = \frac{\varepsilon}{|h(z) + w_0|} < 1 \text{ for all } z \in \mathcal{D}.$$

Thus f gives a holomorphic isomorphism from $\mathcal{D}$ onto $f(\mathcal{D}) \subset \triangle = \{w : |w| < 1\}$.

Step II: We now show that a simply-connected subdomain of the unit disc is also conformally or holomorphically equal to the unit disc, but not the unit disc.

Let us consider a family

$$\mathcal{F} = \Big\{f : \mathcal{D} \to \triangle : f \text{ is analytic, injective and } f(z_0) = 0\Big\}.$$

This family $\mathcal{F}$ is non-empty, as we can always find a holomorphic isomorphism of $\mathcal{D}$ onto a subdomain of $\triangle$. As $|f| < 1 \, \forall f \in \mathcal{F}$, so $\mathcal{F}$ is uniformly bounded. Therefore,

$$A = \sup\Big\{\Big|f'(z_0)\Big| : f \in \mathcal{F}\Big\} > 0,$$

which is finite due to Cauchy's estimate. Also, as f is injective, all the derivatives $|f'(z_0)|$ does not vanish, and so $A > 0$. By the definition of supremum, there is a sequence

$$\langle f_n \rangle \subset \mathcal{F}, \text{ such that } \Big|f'(z_0)\Big| \to A.$$

According to Montel's theorem (if we take any sequence in the family $\mathcal{F}$, then there will be a subsequence that converges uniformly on a compact subset, i.e., normally convergent subsequence) there exists a subsequence

$$f_{n_k} \xrightarrow{\text{normally on } \mathcal{D}} F_0,$$

where, F_0 being a limit of an analytic function, is analytic. Now

$$\Big|f'(z_0)\Big| \to A \Rightarrow \Big|F_0'(z_0)\Big| = A > 0$$

$$\Rightarrow \quad F_0 \text{ is non-constant.}$$

According to Hurwich theorem (if a sequence of injective analytic functions converges normally to a functions, then the limit function is a constant or it is injective), as the limit function F_0 is non-constant, therefore F_0 is injective. Since $|f| < 1 \, \forall f \in \mathcal{F}$,

$$\left|f_{n_k}\right| < 1 \Rightarrow |F_0| < 1.$$

Let $|F_0| = 1$. There is some point in $\mathcal{D}$ where F_0 takes the value on $\triangle$, but by open mapping theorem, we can take any boundary value. Thus, if we take $|F_0| = 1$, then we take the values outside unit disc, if we take the neighbourhood around the point on the unit circle and the neighbourhood will outside the unit circle and we take the values outside the unit circle.

Therefore, $\{|f_{n_k}|\}$ lying within the unit disc converge to a point outside the unit disc, which is a contradiction. This shows that $|F_0| < 1$.

$$0 = f_{n_k}(z_0) \to F_0(z_0) \Rightarrow F_0(z_0) = 0 \Rightarrow F_0 \in \mathcal{F}$$

i.e., F_0 is an injective analytic function on the compact family $\mathcal{F}$.

Now, for $\zeta_0, \zeta_1 \in \triangle$, we consider the hyperbolic metric $\rho_{\text{hy}}(\zeta_0, \zeta_1)$. Let us define a Möbius mapping $H : \triangle \longrightarrow \triangle$ (Fig. 8.7) as

$$H(\zeta) = e^{i\beta}\Big(\frac{\zeta - \zeta_0}{1 - \bar{\zeta}_0\zeta}\Big); \quad \text{where, } \beta = -arg\Big(\frac{\zeta_1 - \zeta_0}{1 - \bar{\zeta}_0\zeta_1}\Big). \tag{8.5}$$

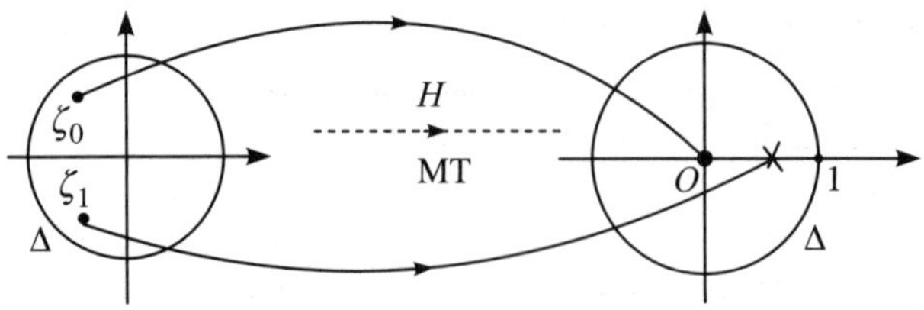

Figure 8.7 *Möbius transformation* $H : \triangle \to \triangle$

Here β is chosen to ensure that $H(\zeta_1) > 0$. The inverse of automorphisms of unit disc are isometric for the hyperbolic metric. We will use Schwarz–Pick lemma which states that if we have an analytic function that maps a unit disc, if it is not an automorphisms, then it will be a strict contraction in terms of hyperbolic metric; otherwise it will be a hyperbolic metric and an automorphism.

$$\begin{aligned} \therefore \rho_{\text{hy}}(\zeta_0, \zeta_1) &= \rho_{\text{hy}}\Big(H(\zeta_0), H(\zeta_1)\Big) = \rho_{\text{hy}}\Big(0, H(\zeta_1)\Big) = \frac{1}{2}\ln\Big(\frac{1 + H(\zeta_1)}{1 - H(\zeta_1)}\Big) \\ &= \text{Radial distance from origin to a point on the real axis.} \end{aligned} \tag{8.6}$$

Let us take a analytic map $g : \triangle \longrightarrow \triangle$ which is not an isomorphism. According to Pick's lemma, g is a strict contraction, i.e., if we take $\zeta_0, \zeta_1 \in \triangle$,

$$\rho_{\text{hy}}\Big(g(\zeta_0), g(\zeta_1)\Big) < \rho_{\text{hy}}(\zeta_0, \zeta_1).$$

If ζ_0 is fixed and $\zeta_1 \longrightarrow \zeta_0$, $\rho_{\text{hy}}(\zeta_0, \zeta_1) \to 0$ and so

$$g(\zeta_1) \longrightarrow g(\zeta_0) \Rightarrow \rho_{\text{hy}}\Big(g(\zeta_0), g(\zeta_1)\Big) \to 0.$$

$$\text{But} \qquad \frac{\rho_{\text{hy}}\Big(g(\zeta_0), g(\zeta_1)\Big)}{\rho_{\text{hy}}(\zeta_0, \zeta_1)} \le \frac{2r}{1 + r^2} < 1,$$

where $\zeta_0, \zeta_1 \in \{z \in \mathbb{C} : |z| \le r < 1\}$ and the bound is attained for $g(\zeta) = \zeta^2$.

$$\therefore \frac{\rho_{\text{hy}}(0, r^2)}{\rho_{\text{hy}}(0, r)} = \frac{\frac{1}{2}\ln\left(\frac{1+r^2}{1-r^2}\right)}{\frac{1}{2}\ln\left(\frac{1+r}{1-r}\right)}; \text{ form } \frac{0}{0} \text{ for } r \to 0$$
$$= \frac{2r}{1+r^2}, \text{ for } r \text{ very small}$$

by using L'Hôpital's rule.

In the last part, we show that $F_0(\mathcal{D}) = \triangle$. If $F_0(\mathcal{D}) \subset \triangle$, by Lemma 8.3.1, we have an injective analytic map $\psi : \mathcal{D}_0 = F_0(\mathcal{D}) \longrightarrow \triangle$ such that $\psi(0) = 0$ and $|\psi'(z_0)| > 1$. Note that $\psi \circ F_0 \in \mathcal{F}$ and $\psi \circ F_0 : \mathcal{D} \longrightarrow \triangle$, and $\psi \circ F_0$ is analytic because of composition of analytic functions, and also injective as both are injective. Also,

$$\left|\left(\psi \circ F_0\right)'(z_0)\right| = \left|\psi'(F(z_0))\right| \left|F_0'(z_0)\right| > A$$
$$\text{as } \left|\psi'(F(z_0))\right| > 1 \text{ and } \left|F_0'(z_0)\right| = A$$

which is a contradiction according to the definition of the supremum A, due to the fact that $\psi \circ F_0 \in \mathcal{F}$. This contradiction shows that $F_0(\mathcal{D}) = \triangle$. □

8.4 Harmonic Functions

In this section we shall mainly study harmonic functions and their basic properties. Poisson's integral formula, Dirichlet's problem for a disc and Harnack inequality for harmonic functions will be discussed.

8.4.1 Poisson's Integral Formula

Here we will discuss two important formulae, namely the *Poisson integral formula* for a circle and for a half-plane.

In the first, the formula gives the value of a harmonic function inside a circle in terms of its boundary values. Poisson's integral formula for the half plane exhibits the values of a harmonic function in the upper half plane in terms of the values on boundary of the half plane.

Theorem 8.5. [Poisson's integral formula for a circle] *Let $f(z)$ be an analytic function within and on the circle $\Gamma : |z| = R$ and $z_0 = re^{i\theta}$ be a point inside this region (Fig. 8.8). Then*

$$f(re^{i\theta}) = \frac{1}{2\pi}\int_0^{2\pi} \frac{R^2 - r^2}{R^2 - 2Rr\cos(\phi - \theta) + r^2} f(Re^{i\phi})\, d\phi. \tag{8.7}$$

Proof: As given the point $z_0 = re^{i\theta}$ lies inside the circle $\Gamma : |z| = R$, so, by Cauchy's integral formula (5.35), we get

$$f(z_0) = f(re^{i\theta}) = \frac{1}{2\pi i}\int_\Gamma \frac{f(z)}{z - z_0}\, dz. \tag{8.8}$$

Now, the inverse point $\bar{z}_0$ of the point z_0 with respect to Γ lies outside Γ and is given by $R^2/\bar{z}_0$. Hence the function $f(z)/\{z - (R^2/\bar{z}_0)\}$ is analytic on and within Γ. Thus by Cauchy's theorem, we have

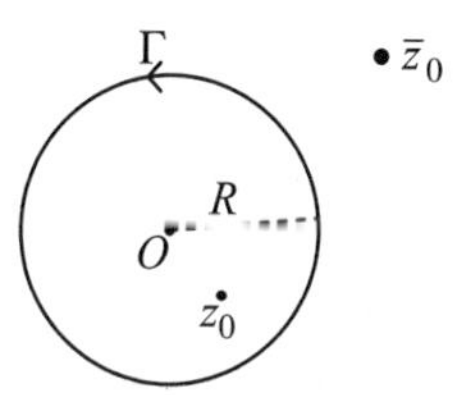

Figure 8.8 *The circle $\Gamma : |z| = R$*

$$\int_{\Gamma} \frac{f(z)}{z-(R^2/\overline{z}_0)}\, dz = 0 \Rightarrow \frac{1}{2\pi i}\int_{\Gamma} \frac{f(z)}{z-(R^2/\overline{z}_0)}\, dz = 0. \tag{8.9}$$

Subtracting Eq.(8.9) from Eq. (8.8), we obtain

$$\begin{aligned} f(z_0) &= \frac{1}{2\pi i}\int_{\Gamma} \left[\frac{f(z)}{z-z_0} - \frac{f(z)}{z-(R^2/\overline{z}_0)}\right] dz \\ &= \frac{1}{2\pi i}\int_{\Gamma} \frac{z_0-(R^2/\overline{z}_0)}{(z-z_0)(z-(R^2/\overline{z}_0))}\, f(z)\, dz. \end{aligned}$$

Let $z = Re^{i\phi}$. Substituting $z_0 = re^{i\theta} \Rightarrow \overline{z}_0 = re^{-i\theta}$, we get

$$\begin{aligned} f(z_0) &= f(re^{i\theta}) = \frac{1}{2\pi i}\int_0^{2\pi} \frac{[re^{i\theta}-(R^2/r)e^{i\theta}]f(Re^{i\phi})}{(Re^{i\phi}-re^{i\theta})(Re^{i\phi}-(R^2/r)e^{i\theta})}\, iRe^{i\phi}d\phi \\ &= \frac{1}{2\pi}\int_0^{2\pi} \frac{(r^2-R^2)e^{i(\theta+\phi)}f(Re^{i\phi})}{(Re^{i\phi}-re^{i\theta})(re^{i\phi}-Re^{i\theta})}\, d\phi \\ &= -\frac{1}{2\pi}\int_0^{2\pi} \frac{(r^2-R^2)f(Re^{i\phi})}{(Re^{i\phi}-re^{i\theta})(Re^{-i\phi}-re^{-i\theta})}\, d\phi \\ &= \frac{1}{2\pi}\int_0^{2\pi} \frac{R^2-r^2}{R^2-2Rr\cos(\phi-\theta)+r^2}\, f(Re^{i\phi})\, d\phi. \end{aligned} \tag{8.10}$$

Let $f(re^{i\theta}) = u(r,\theta)+iv(r,\theta)$ and $f(Re^{i\phi}) = u(R,\phi)+iv(R,\phi)$. Then from the above equation (8.10), we get

$$u(r,\theta)+iv(r,\theta) = \frac{1}{2\pi}\int_0^{2\pi} \frac{(R^2-r^2)[u(R,\phi)+iv(R,\phi)]}{R^2-2Rr\cos(\phi-\theta)+r^2}\, d\phi.$$

Equating the real and imaginary parts, we get

$$u(r,\theta) = \frac{1}{2\pi}\int_0^{2\pi} \frac{(R^2-r^2)u(R,\phi)}{R^2-2Rr\cos(\phi-\theta)+r^2}\, d\phi \tag{8.11}$$

$$v(r,\theta) = \frac{1}{2\pi}\int_0^{2\pi} \frac{(R^2-r^2)v(R,\phi)}{R^2-2Rr\cos(\phi-\theta)+r^2}\, d\phi. \tag{8.12}$$

□

These results are called *Poisson's integral formulae* for a circle. These formulae express the values of a harmonic function inside a circle in terms of its values on the boundary.

Equation (8.11) defines a linear transformation of $u(R,\phi)$ into $u(r,\theta)$. Excluding the factor $1/2\pi$, the real-valued function

$$\begin{aligned} \mathcal{P}(R,r,\phi-\theta) &= \frac{(R^2-r^2)u(R,\phi)}{R^2-2Rr\cos(\phi-\theta)+r^2} \\ &= \frac{R^2-r^2}{|z-z_0|^2} = \frac{z}{z-z_0} + \frac{\overline{z}_0}{\overline{z}-\overline{z}_0} \end{aligned} \tag{8.13}$$

is known as *Poisson kernel.* Now,

(i) Clearly, $\mathcal{P}(R,r,\phi-\theta)$ is an even periodic function of $\phi-\theta$ having period 2π.

(ii) Observe that when $r=0$, $\mathcal{P}(R,r,\phi-\theta) = 1$.

(iii) Since $r < R$, $\mathcal{P}(R, r, \phi - \theta)$ is a positive function.

(iv) Furthermore, we know that the real parts of $z_0/(z - z_0)$ and its complex conjugate $\bar{z}_0/(\bar{z} - \bar{z}_0)$ are the same.

$$\therefore \mathcal{P}(R, r, \phi - \theta) = \mathcal{R}e\Big(\frac{z}{z - z_0} + \frac{\bar{z}_0}{\bar{z} - \bar{z}_0}\Big) = \mathcal{R}e\Big(\frac{z + z_0}{z - z_0}\Big).$$

In terms of Poisson kernel, Eq. (8.11) can be written as

$$u(r, \theta) = \frac{1}{2\pi} \int_0^{2\pi} \mathcal{P}(R, r, \phi - \theta)\, u(R, \phi)\, d\phi; \quad r < R. \tag{8.14}$$

When $f(z) = u(r, \theta) = 1$, Eq. (8.11) becomes

$$1 = \frac{1}{2\pi} \int_0^{2\pi} \mathcal{P}(R, r, \phi - \theta)\, d\phi; \quad r < R. \tag{8.15}$$

Example 8.4.1. *Show that* $\int_0^{2\pi} \frac{e^{\cos\phi} \cos(\sin\phi)}{5 - 4\cos(\theta - \phi)}\, d\phi = \frac{2\pi}{3} e^{\cos\theta} \cos(\sin\theta)$ *using Poisson's integral formulae.*

Solution. We know $\cos(\phi - \theta) = \cos(\theta - \phi)$. Comparing the given integral with the Poisson's integral formula for the circle $|z| = R$, Eq. (8.7), we get $f(Re^{i\phi}) = e^{\cos\phi} \cos(\sin\phi)$, $R^2 + r^2 = 5$, $2Rr = 4$. Thus $R = 2$, $r = 1$ and $f(re^{i\theta}) = e^{\cos\theta} \cos(\sin\theta)$. Substituting these values in Eq.(8.7), we get

$$e^{\cos\theta} \cos(\sin\theta) = \frac{1}{2\pi} \int_0^{2\pi} \frac{(2^2 - 1^2) e^{\cos\phi} \cos(\sin\phi)}{5 - 4\cos(\theta - \phi)}\, d\phi$$

$$\Rightarrow \quad \int_0^{2\pi} \frac{e^{\cos\phi} \cos(\sin\phi)}{5 - 4\cos(\theta - \phi)}\, d\phi = \frac{2\pi}{3} e^{\cos\theta} \cos(\sin\theta).$$

Theorem 8.6. [Poisson's integral formula for a plane] *Let $f(z)$ be an analytic in the upper half $y \geq 0$ of the z-plane and let $\zeta = \xi + i\eta$ be any point in this upper half plane. Then*

$$f(\zeta) = \frac{1}{\pi} \int_{-\infty}^{\infty} \frac{\eta f(x)}{(x - \xi)^2 + \eta^2}\, dx. \tag{8.16}$$

Proof: Let Γ be the contour consisting of semicircle C_R of radius R in the upper half plane and the part of the real axis from $x = -R$ to $x = R$ (Fig. 8.9). Let Γ be the boundary of a semicircle of radius R containing ζ as an interior point. Since Γ encloses ζ but not enclose $\bar{\zeta}$, we have, by Cauchy's integral formula,

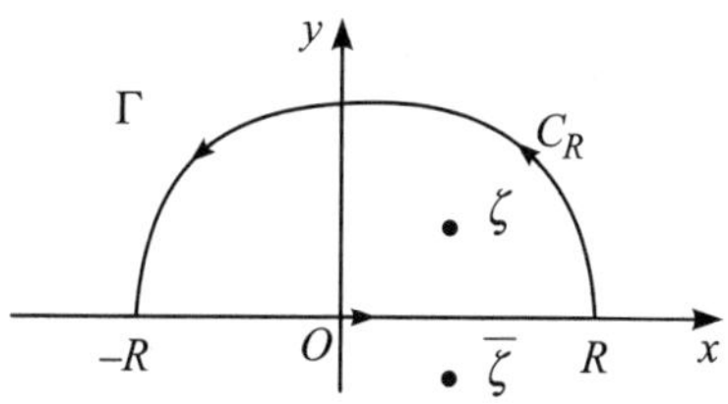

Figure 8.9

$$\begin{aligned} f(\zeta) &= \frac{1}{2\pi i} \oint_\Gamma \frac{f(z)}{z - \zeta}\, dz, \quad 0 = \frac{1}{2\pi i} \oint_\Gamma \frac{f(z)}{z - \bar{\zeta}}\, dz \\ f(\zeta) &= \frac{1}{2\pi i} \oint_\Gamma \Big[\frac{1}{z - \zeta} - \frac{1}{z - \bar{\zeta}}\Big] f(z)\, dz, \text{ on subtraction} \\ &= \frac{1}{2\pi i} \oint_\Gamma \frac{\zeta - \bar{\zeta}}{(z - \zeta)(z - \bar{\zeta})} f(z)\, dz. \end{aligned} \tag{8.17}$$

Letting $\zeta = \xi + i\eta$, so $\bar{\zeta} = \xi - i\eta$, then Eq. (8.17) can be written as

$$f(\zeta) = \frac{1}{\pi} \int_{-R}^{R} \frac{\eta f(x)}{(x - \xi)^2 + \eta^2}\, dx + \frac{1}{\pi} \int_{C_R} \frac{\eta f(z)}{(z - \zeta)(z - \bar{\zeta})}\, dz. \tag{8.18}$$

As $R \to \infty$, the second integral of Eq. (8.18) on the right-hand side approaches zero and we have

$$f(\zeta) = \frac{1}{\pi} \int_{-\infty}^{\infty} \frac{\eta f(x)}{(x-\xi)^2 + \eta^2}\, dx.$$

Let $f(\zeta) = f(\xi + i\eta) = u(\xi, \eta) + iv(\xi, \eta)$, and $f(x) = u(x, 0) + iv(x, 0)$. We obtain

$$u(\xi, \eta) = \frac{1}{\pi} \int_{-\infty}^{\infty} \frac{\eta u(x, 0)}{(x-\xi)^2 + \eta^2}\, dx, \quad v(\xi, \eta) = \frac{1}{\pi} \int_{-\infty}^{\infty} \frac{\eta v(x, 0)}{(x-\xi)^2 + \eta^2}\, dx. \quad (8.19)$$

These equations (8.19) are called *Poisson's integral formulae for the half plane.* They express the values of a harmonic function in the upper half plane in terms of the values on the x-axis (the boundary) of the half plane. □

8.4.2 Subharmonic and Superharmonic Functions

Definition 8.4.1. A real-valued continuous function $u(x, y)$ in an open set $\mathcal{D}$ of the complex plane $\mathbb{C}$ is said to be

(i) *subharmonic* if, for any $\alpha \in \mathcal{D}$ and for sufficiently small $r > 0$,

$$u(\alpha) \leq \frac{1}{2\pi} \int_0^{2\pi} u\left(\alpha + re^{i\theta}\right) d\theta.$$

(ii) *superharmonic* if, for any $\beta \in \mathcal{D}$ and for sufficiently small $r > 0$,

$$u(\beta) \geq \frac{1}{2\pi} \int_0^{2\pi} u\left(\beta + re^{i\theta}\right) d\theta.$$

From the definition it follows that every harmonic function is subharmonic as well as superharmonic.

Example 8.4.2. *If $f(z)$ is analytic on a domain $\mathcal{D}$, then $|f(z)|$ is subharmonic but not harmonic in $\mathcal{D}$ unless $f(z)$ is constant.*

Solution. Using the Cauchy's integral formula (5.32)

$$|f(\alpha)| \leq \frac{1}{2\pi} \int_0^{2\pi} \left|f\left(\alpha + re^{i\theta}\right)\right| d\theta = I(r), \text{ say,} \quad (8.20)$$

for every $\alpha \in \mathcal{D}$ and small enough $r (> 0)$. Here, equality holds only if $f(z) =$ constant. Let $0 < r_1 < r_2 < d(\alpha, \partial\mathcal{D})$, the minimum distance between α and the boundary $\partial\mathcal{D}$ of $\mathcal{D}$. Now, let $g(\theta)$ be continuous on $[0, 2\pi]$ and $F(z)$ be defined by

(i) $g(\theta) f(\alpha + r_1 e^{i\theta}) = |f(\alpha + r_1 e^{i\theta})|;\ 0 \leq \theta \leq 2\pi;$

(ii) $F(z) = \frac{1}{2\pi} \int_{)}^{2\pi} f(\alpha + ze^{i\theta}) g(\theta)\, d\theta;\ |z| \leq r_2.$

$F(z)$ is regular for $|z| \leq r_2$ and attains its maximum at the boundary of the disc, say at $z = r_2 e^{i\varphi}$. Then

$$\begin{aligned} I(r_1) &= \frac{1}{2\pi} \int_0^{2\pi} \left|f\left(\alpha + r_1 e^{i\theta}\right)\right| d\theta \\ &= \frac{1}{2\pi} \int_0^{2\pi} f\left(\alpha + r_1 e^{i\theta}\right) g(\theta)\, d\theta - F(r_1) < \left|F(r_2 e^{i\theta})\right| \\ &\leq \frac{1}{2\pi} \int_0^{2\pi} \left|f\left(\alpha + r_2 e^{i(\theta+\varphi)}\right)\right| d\theta; \quad \theta + \varphi = \psi \end{aligned}$$

$$= \frac{1}{2\pi}\int_0^{2\pi+\varphi} \left|f\left(\alpha + r_2 e^{i\psi}\right)\right| d\psi$$

$$- \frac{1}{2\pi}\left[\int_0^{2\pi} - \int_0^{\varphi} + \int_{2\pi}^{2\pi+\varphi}\right]\left|f\left(\alpha + r_2 e^{i\psi}\right)\right| d\psi$$

$$= \frac{1}{2\pi}\int_0^{2\pi} \left|f\left(\alpha + r_2 e^{i\psi}\right)\right| d\psi = I(r_2).$$

Hence equality is possible if and only if $f(z) = $ constant. Therefore $|f(z)|$ is subharmonic but not harmonic in $\mathcal{D}$ unless $f(z)$ is constant.

Exercises

Review Questions [Short-answer Type]

1. Given one conformal equivalence h from an open set U to the unit disc $\mathcal{S}(0;1)$, can you describe all the others in terms of h?
2. Is an open rectangle in a plane conformally equivalent to the unit disc $\mathcal{S}(0;1)$?
3. Why is it impossible for the unit disc $\mathcal{S}(0;1)$ to be the whole plane in the Riemann mapping theorem? [**Hint:** Appeal to Liouville's theorem.]
4. Let $u : \mathbb{C} \to \mathbb{R}$ be a harmonic function such that $u(z) \geq 0$ for all $z \in \mathbb{C}$. Prove that u is constant.
5. Is the open right half-plane, with the interval $[1, \infty)$ removed, conformally equivalent to the unit disc $\mathcal{S}(0;1)$?
6. Is the open right half-plane, with the interval $[1, 2]$ removed, conformally equivalent to the unit disc $\mathcal{S}(0;1)$?
7. Is the punctured disc $\mathcal{S}(0;1)/\{0\}$ conformally equivalent to the right half-plane $\{z : \text{Re}z > 0\}$?
8. Prove that the angle subtended at a point z of the half plane $\text{Im}z > 0$ by a given segment of the real axis is a harmonic function of z.
9. If $u(z)$ and $v(z)$ are conjugate harmonic functions, prove that the product $u(z) \cdot v(z)$ is harmonic.
10. Find a harmonic function $u(x, y)$ for which the product xu is harmonic.
11. If u is harmonic and non-negative on $\mathcal{S}(0;1)$ and $u(0) = 1$, give upper and lower bounds for $u(1/2)$.
12. Let $f : \mathbb{C} \to \mathbb{R}$ be a harmonic function such that $f(z) \geq 0$ for all $z \in \mathbb{C}$. Prove that f is constant.
13. Find all functions f harmonic in the unit disc $\mathcal{S}(0,1) = \{z \in \mathbb{C} : |z| < 1\}$ that satisfy $f(i/2) = -5$ and $f(z) \geq -5$ for all $z \in \mathcal{S}(0,1)$.
14. Does there exist an analytic function on the complex plane whose real part is given by $u(x, y) = 3x^2 + xy + y^2$?
15. Show by example that a harmonic function need not have an analytic completion in a multiply-connected domain.
16. Prove the maximum–minimum principle for harmonic functions in an arbitrary domain.
17. If $f(z) \neq 0$ is analytic in a domain $\mathcal{D}$, prove that $\log|f(z)|$ is subharmonic in $\mathcal{D}$.

Review Questions [Long-answer Type]

1. Is the set of all conformal automorphisms of the unit disc a normal family? If so, then every sequence of such maps has a convergent subsequence. Is the limit necessarily also a conformal automorphism of the unit disc?
2. Given U is a open connected set and V is a simply-connected open set which is not the whole plane. If $z_0 \in U$ and $w_0 \in V$ are fixed, let $\mathcal{F}$ be a family of all analytic functions from U to V which send z_0 to w_0. Prove that $\mathcal{F}$ is a normal family.
3. If U is a open connected subset of $\mathbb{C}$ and z_0 is a point of U, prove that the set of all analytic functions f on U with positive real part and with $f(z_0) = 1$ is a normal family.
4. Show by example that it is not true that every uniformly bounded family $\mathcal{F}$ of continuous functions on an interval on a line has the property that every sequence in $\mathcal{F}$ has a convergent subsequence.
5. Prove that the complex plane $\mathbb{C}$ is not conformally equivalent to the unit disc even though it is simply connected. Why doesn't this contradict the Riemann mapping theorem?
6. Prove that if f is analytic and satisfies $|f(z)| < 1$ on the right half-plane, and if $f(1) = 0$, then $|f(z)| \le |\frac{z-1}{z+1}|$ for all z in the right half-plane.
7. Prove that maximum principle for analytic functions is a consequence of the maximum principle for harmonic functions.
8. Prove the circumferential mean-value theorem for harmonic functions: If f is harmonic in a domain containing the disc $S[0, R] = \{z \in \mathbb{C} : |z| \le R\}$, then
$$f(0) = \frac{1}{2\pi}\int_0^{2\pi} f(Re^{i\varsigma})\, d\varsigma.$$
9. Prove the following version of the solid mean-value theorem for harmonic functions: If f is harmonic in a domain containing the closed disc $S[0, R] = \{z \in \mathbb{C} : |z| \le R\}$ with boundary $\partial S(0, R)$, then
$$f(0) = \frac{1}{\pi R^2}\int\int_{S[0,R]} f\, dx\, dy.$$
10. **Borel–Caratheodory theorem:** Let f be analytic in $|z| \le R$. For each $0 < r < R$, let $M(r) = \max\limits_{|z|=r}\{|f(z)|\}$ and $A(r) = \max\limits_{|z|=r}\{\text{Re} f(z)\}$. Then prove that
$$M(r) \le \frac{R+r}{R-r}\Big|f(0)\Big| + \frac{2r}{R-r}A(r) \tag{8.21}$$
Hence deduce that
$$\max_{|z|=r}\Big|f^{(n)}(z)\Big| \le \frac{2^{n+2}\cdot n!R}{(R-r)^{n+1}}\Big[A(R) + \Big|f(0)\Big|\Big]. \tag{8.22}$$
11. Prove Harnack's inequality: If $f(z)$ is harmonic and non-negative in a domain containing the disc $S[0, R] = \{z \in \mathbb{C} : |z| \le R\}$, then for $0 \le r < R$
$$f(0)\frac{R-r}{R+r} \le f(re^{i\theta}) \le f(0)\frac{R+r}{R-r}.$$
Hence, show that a bounded function that is harmonic in $\mathbb{C}$ is necessarily constant.
12. Prove Liouville's theorem for harmonic functions: If f is harmonic in the whole plane and bounded from above or below there, then f is constant.
13. Prove that if $\langle u_n \rangle\rangle$ is a sequence of positive harmonic functions on a open connected set U, and if $u_n(z_0) \to 0$ at one point $z_0 \in U$, then $u_n(z) \to 0$ at every point z of U.

9 Entire and Meromorphic Functions

In this chapter we shall study entire functions, their growth properties and meromorphic functions. Infinite products and their convergence will be discussed. Properties of zeros of an entire function will also be discussed.

9.1 Infinite Products

As is amply demonstrated by power series expressions, a highly useful technique in complex analysis is to express an analytic function as an infinite sum of much simpler functions. Likewise, it can be useful to express an analytic function as an infinite product of simpler functions. This is especially true in the study of the zeros of analytic functions.

9.2 Infinite Product of Complex Numbers

In this section we shall introduce *infinite products* of non-zero complex numbers and functions and provide necessary and sufficient conditions for their convergence. Since the exponential function converts sums to products, we can expect that the theory of infinite products will be closely related to the theory of infinite sums.

Definition 9.2.1. Let $\langle z_n \rangle$ be a sequence of non-zero complex numbers. The *infinite product* is an expression of the form

$$z_1 \cdot z_2 \cdot z_3 \cdots = \prod_{k=1}^{\infty} z_k \in \mathbb{C}/\{0\}. \tag{9.1}$$

If we allow any of the factors to be zero, it is evident that the infinite product would be zero regardless of the behaviour of the other terms.

Definition 9.2.2. We define the *partial product* P_n (or the nth *partial product*) as the product of n non-zero complex numbers

$$P_n = \prod_{k=1}^{n} z_k = z_1 \cdot z_2 \cdots z_n. \tag{9.2}$$

Definition 9.2.3. The infinite product as in Eq. (9.1) is said to be *convergent* if the sequence of partial products $\langle P_n \rangle$ converges to a non-zero (finite) limit P. Symbolically

$$\lim_{n\to\infty} P_n = P = \prod_{k=1}^{\infty} z_k, \text{ and } P \in \mathbb{C}/\{0\}.$$

P is called *value of the infinite product.* If P_n does not tend to a finite non-zero limit, then the product as in Eq. (9.1) is said to be *divergent.*

Note that according to the definition, $\prod_{k=1}^{\infty} \frac{1}{3}$ does not converge, as

$$\prod_{k=1}^{n} \frac{1}{3} = \frac{1}{3^n} \to 0 \text{ as } n \to \infty.$$

Therefore, the sequence $\langle z_n \rangle$ of non-zero complex numbers is said to be *multipliable* if the product as in Eq. (9.1) converges in $\mathbb{C}/\{0\}$.

RESULT **66.** The infinite product as in Eq. (9.1) converges if and only if $\sum_{k=1}^{\infty} \log z_k$ converges for a suitable branch of logarithm. In the following discussion, log will be principal branch of the log function.

RESULT **67.** Let $\langle z_n \rangle$ of non-zero complex numbers be convergent and $\lim_{n\to\infty} P_n = P \in \mathbb{C}/\{0\}$, where, P_n is as in Eq. (9.2). Then,

$$\lim_{n\to\infty} z_n = \lim_{n\to\infty} \prod_{k=1}^{n} z_k \Big/ \prod_{k=1}^{n-1} z_k = \lim_{n\to\infty} \frac{P_n}{P_{n-1}} = \frac{P}{P} = 1.$$

Thus, a necessary condition for the convergence of the infinite product as in Eq. (9.1) is $\lim_{n\to\infty} z_n = 1$. It is then better to consider the infinite product of the general term $1 + z_n$.

Definition 9.2.4. The *infinite product* of $\langle 1 + z_n \rangle$ is denoted by

$$(1 + z_1) \cdot (1 + z_2) \cdot (1 + z_3) \cdots = \prod_{k=1}^{\infty} (1 + z_k); \; z_k \neq -1. \tag{9.3}$$

Definition 9.2.5. The *nth partial product* P_n for the infinite product (9.3) is defined as

$$P_n = \prod_{k=1}^{n} (1 + z_k) = (1 + z_1) \cdot (1 + z_2) \cdots (1 + z_n); z_k \neq -1 \; \forall k \tag{9.4}$$

Definition 9.2.6. Now, if $\lim_{n\to\infty} P_n = P$, we say that the product as in Eq. (9.3) *converges to* P. Otherwise, the product *diverges.*

Also, if $z_k = -1$ for only a finite number of k while for rest $z_k \neq -1$, the infinite product as in Eq. (9.3) is said to *converge to zero.*

Thus the necessary condition for convergence becomes $\lim_{n\to\infty} z_n = 0$. Let us consider the following examples of infinite products:

(i) Consider the infinite product as in Eq. (9.3), where $z_k = 1/k \neq 0$ for any k. In this case, the partial product

$$P_n = \frac{2}{1} \cdot \frac{3}{2} \cdot \frac{4}{3} \cdots \frac{n}{n-1} \cdot \frac{n+1}{n} = n + 1 \to \infty \text{ as } n \to \infty.$$

Therefore, the infinite product is convergent.

(ii) Consider the infinite product $\prod_{k=2}^{\infty}(1-z_k)$, where $z_k = 1/k$. Here also, $1 - z_k \neq 0$ for $k \geq 2$. In this case, the partial product

$$P_n = \frac{1}{2} \cdot \frac{2}{3} \cdot \frac{3}{4} \cdots \frac{n-1}{n} \cdot \frac{n}{n+1} = \frac{1}{n+1} \to 0 \text{ as } n \to \infty.$$

Though the limit of P_n exists finitely, but it need to be a non-zero finite number for convergence of the product. Even if we redefine an infinite product $\prod_{k=1}^{\infty}(1-\frac{1}{k})$, where $1 - z_k = 0$, when $k = 1$, still the infinite product is divergent.

Theorem 9.1. *An infinite product as in Eq. (9.3) is convergent if and only if the infinite series $\sum_{k=1}^{\infty} \log(1 + z_k)$ is convergent, where we use the principal branch of the* log *function and omit, as usual, the terms with $z_k = -1$.*

Proof: Let us define P_n and S_n as follows:

$$P_n = \prod_{k=1}^{n}(1+z_k) \text{ and } S_n = \sum_{k=1}^{n} \log(1+z_k).$$

Necessary condition: We see that $\log P_n = S_n$ and $P_n = e^{S_n}$. Let the product as in Eq. (9.3) be convergent, i.e., $P_k \to P \neq 0$. By taking the principle values of logarithms, we get

$$\text{Log}P_k = \sum_{n=1}^{k} \text{Log}(1+z_n) + 2\pi i q_k = S_k + 2\pi i q_k, \tag{9.5}$$

where q_k is an integer. The principal values of logarithm of a product is not necessarily the sum of the principal values of the logarithm of its factors, so q_k is not necessarily 0.

Now, we show that for sufficiently large values of k, q_k is constant. Let $\alpha_k = \text{Im}\,\{\text{Log}\,(1+z_k)\}$ and $\beta_k = \text{Im}\,\{\text{Log}\,P_k\}$. Then equating imaginary parts on both sides of equation (9.5), we obtain

$$\beta_k = \sum_{n=1}^{k} \alpha_k + 2\pi\, q_k \Rightarrow \beta_{k+1} - \beta_k = \alpha_{k+1} + 2\pi\Big(q_{k+1} - q_k\Big)$$

$$\Longrightarrow \quad q_{k+1} - q_k = \frac{1}{2\pi}\Big[\Big(\beta_{k+1} - \beta_k\Big) - \alpha_{k+1}.\Big]. \tag{9.6}$$

As $1 + z_k = P_k/(P_{k-1})$, it follows that

$$\lim_{k\to\infty}(1+z_k) = \lim_{k\to\infty} \frac{P_k}{P_{k-1}} = 1.$$

Since we are considering the principal value, $\lim\limits_{k\to\infty} \text{Log}(1+z_k) = \text{Log}1 = 0$. This implies that $\lim_{k\to\infty} \text{Im}\{\text{Log}(1+w_k)\} = 0$ and hence $\lim\limits_{k\to\infty} \alpha_k = 0 = \lim\limits_{k\to\infty} \alpha_{k+1}$. Making $k \to \infty$ in equation (9.6), we get

$$\lim_{k\to\infty}(\beta_{k+1} - \beta_k) - 0 = \lim_{k\to\infty} 2\pi(q_{k+1} - q_k)$$

$$\therefore \quad 0 = \lim_{k\to\infty} 2\pi(q_{k+1} - q_k) \qquad \because \lim_{k\to\infty} \beta_{k+1} = \lim_{k\to\infty} \beta_k$$

$$\Rightarrow \quad \lim_{k\to\infty} q_{k+1} - \lim_{k\to\infty} q_k = q(\text{constant}),$$

where q is free from k. Since $P_k \to P$, we conclude that $S_k \to S$(say), where

$$S = \lim_{k\to\infty} S_k = \text{Log}P - 2\pi i q.$$

Thus, if the product is convergent, the series will also converge. This proves that the condition is necessary.

Sufficient condition: Let the series $S_k = \text{Log}(1+z_1) + \cdots + \text{Log}(1+z_k)$ converge to a value S so that $S_k \to S$ as $k \to \infty$. Now

$$\begin{aligned} e^{S_k} &= e^{\sum_{n=1}^{k} \text{Log}(1+z_k)} = e^{\text{Log}(1+z_1)} e^{\text{Log}(1+z_2)} \cdots e^{\text{Log}(1+z_k)} \\ &= (1+z_1)(1+z_2)\cdots(1+z_k) = P_k. \end{aligned} \tag{9.7}$$

Then exponential function being continuous, $S_k \to S$ implies that

$$\lim_{k\to\infty} P_k = \lim_{k\to\infty} e^{S_k} = e^{\lim_{k\to\infty} S_k} = e^S \neq 0.$$

This proves the sufficiency of the conditions. The theorem is thus completely established. □

Example 9.2.1. *Find the value of the infinite product* $(1-\frac{1}{2^2})(1-\frac{1}{3^2})(1-\frac{1}{4^2})\cdots$ *if it exists.*

Solution. The given infinite product can be written as $\prod_{k=2}^{\infty} z_k$, where $z_k = 1 - \frac{1}{k^2}$, $k \geq 2$. Let P_n be the partial product of n factors, then

$$\begin{aligned} P_n &= \prod_{k=2}^{\infty} \frac{k^2-1}{k^2} = \prod_{k=2}^{\infty} \frac{1+k}{k} \cdot \frac{k-1}{k} \\ &= \left(\frac{1}{2}\cdot\frac{3}{2}\right)\left(\frac{2}{3}\cdot\frac{4}{3}\right)\cdots\left(\frac{n}{n+1}\cdot\frac{n+2}{n+1}\right) = \frac{1}{2}\frac{n+2}{n+1} \to \frac{1}{2} \neq 0 \text{ as } n \to \infty. \end{aligned}$$

Thus, the given infinite product is convergent and it converges to $1/2$.

Example 9.2.2. *Prove that the infinite product* $(1+\frac{1}{2})(1-\frac{1}{3})(1+\frac{1}{4})(1-\frac{1}{5})\cdots$ *converges to* 1.

Solution. Let P_n be the partial product of n factors, then

$$P_n = \left(1+\frac{1}{2}\right)\left(1-\frac{1}{3}\right)\left(1+\frac{1}{4}\right)\left(1-\frac{1}{5}\right)\cdots\left(1+\frac{1}{n}\right)\left(1-\frac{1}{n+1}\right).$$

Case I: When n is odd. Then the factors can be grouped in pairs except the last one.

$$\begin{aligned} P_n &= \left\{\left(1+\frac{1}{2}\right)\left(1-\frac{1}{3}\right)\right\}\left\{\left(1+\frac{1}{4}\right)\left(1-\frac{1}{5}\right)\right\}\cdots\left(1-\frac{1}{n+1}\right) \\ &= \frac{3\cdot 2}{2\cdot 3}\cdot\frac{5\cdot 4}{4\cdot 5}\cdots\frac{n}{n+1} = 1\cdot 1\cdots\frac{n}{n+1} = \frac{n}{n+1} \\ \Rightarrow \lim_{n\to\infty} P_n &= \lim_{n\to\infty}\frac{n}{n+1} = 1. \end{aligned}$$

Case II: When n is even. Then the factors can be grouped in pairs.

$$\begin{aligned} P_n &= \left\{\left(1+\frac{1}{2}\right)\left(1-\frac{1}{3}\right)\right\}\left\{\left(1+\frac{1}{4}\right)\left(1-\frac{1}{5}\right)\right\}\cdots\left\{\left(1+\frac{1}{n}\right)\left(1-\frac{1}{n+1}\right)\right\} \\ &= \frac{3\cdot 2}{2\cdot 3}\cdot\frac{5\cdot 4}{4\cdot 5}\cdots\frac{(n+1)\cdot n}{(n+1)\cdot n} = 1\cdot 1\cdots 1 = 1 \\ \Rightarrow \lim_{n\to\infty} P_n &= \lim_{n\to\infty} 1 = 1. \end{aligned}$$

Thus, in either case, we get $\lim_{n\to\infty} P_n = 1$.

9.2.1 Absolute Convergence of Infinite Products

Definition 9.2.7. (i) An infinite product as in Eq. (9.1) is said to be *absolutely convergent* if $\sum_{k=1}^{\infty} |\log z_k|$ converges absolutely.

(ii) An infinite product as in Eq. (9.3) is said to be *absolutely convergent* if the infinite product $\prod_{k=1}^{\infty}(1+|z_k|)$ or the sum $\sum_{k=1}^{\infty} \left| \text{Log } (1+z_k)\right|$ is convergent.

The factors of absolutely convergent infinite product $\prod_{k=1}^{\infty}(1+z_k)$ can be rearranged in any order and it does not alter its convergence or its value.

RESULT **68.** Every absolutely convergent product is convergent.

Theorem 9.2. *A necessary and sufficient condition for the absolute convergence of* $\sum_{n=1}^{\infty} Log(1+z_n)$, $w_n \neq -1$ *is that the series* $\sum_{n=1}^{\infty} z_n$ *is absolutely convergent.*

Proof [Necessary condition]: Let $\sum_{n=1}^{\infty} \text{Log}(1+z_n)$ be absolutely convergent. Then, for $z_n \neq -1$,

$$\sum_{n=1}^{\infty} |\text{Log}(1+z_n)| \text{ is convergent } \implies \lim_{n\to\infty} |\text{Log}(1+z_n)| = 0$$

$$\implies \quad \lim_{n\to\infty}(1+z_n) = 1 \Rightarrow \lim_{n\to\infty} z_n = 0. \qquad \because \text{Log}1 = 0$$

This means that there exists a $m \in \mathbb{N}$ such that $|z_n| < 1/2 \; \forall n \geq m$. Now,

$$\text{Log}(1+z_n) = z_n - \frac{z_n^2}{2} + \frac{z_n^3}{3} - \frac{z_n^4}{4} + \cdots$$

Therefore

$$\begin{aligned}
\left|\frac{\text{Log}(1+z_n)}{z_n} - 1\right| &= \left|-\frac{z_n}{2} + \frac{z_n^2}{3} - \frac{z_n^3}{4} + \cdots\right| \\
&\leq \frac{1}{2}|z_n| + \frac{1}{3}|z_n|^2 + \frac{1}{4}|z_n|^3 + \cdots \\
&< \frac{1}{2\cdot 2} + \frac{1}{3\cdot 2^2} + \frac{1}{4\cdot 2^3} + \cdots \\
&< \frac{1}{2^2} + \frac{1}{2^3} + \frac{1}{2^4} + \cdots = \frac{1/2^2}{1-(1/2)} = \frac{1}{2}.
\end{aligned}$$

Using this result, we get

$$\begin{aligned}
&\left|\frac{\text{Log}(1+z_n)}{z_n}\right| - 1\Big| < \frac{1}{2} \implies \frac{1}{2} = 1 - \frac{1}{2} < \left|\frac{\text{Log}(1+z_n)}{z_n}\right| < 1 + \frac{1}{2} = \frac{3}{2} \\
&\implies |z_n| < 2|\text{Log}(1+z_n)| < 3|z_n| \qquad (9.8) \\
&\implies \sum_{n=1}^{\infty} |z_n| < 2\sum_{n=1}^{\infty} |\text{Log}(1+z_n)|.
\end{aligned}$$

Since $\sum_{n=1}^{\infty} |\text{Log}(1+z_n)|$ is convergent, thus by comparison test, $\sum_{n=1}^{\infty} |z_n|$ is convergent.
Sufficient condition : Let $\sum_{n=1}^{\infty} |z_n|$ be convergent. This means that there exists a $m \in \mathbb{N}$ such that $|z_n| < 1/2; \forall n \geq m$. Thus, using Eq. (9.8), we have

$$\sum_{n=1}^{\infty} |\text{Log}(1+z_n)| < \frac{3}{2}\sum_{n=1}^{\infty} |z_n|.$$

Thus by comparison test, $\sum_{n=1}^{\infty} \text{Log}(1+z_n)$ is absolutely convergent. □

Example 9.2.3. *Prove that the infinite product* $\prod_{k=1}^{\infty}(1-\frac{z^2}{k^2})$ *converges uniformly on each bounded subset of* $\mathbb{C}$.

Solution. The terms of the product vanish when $z = \pm 1, \pm 2, \cdots$. Here $w_k(z) = -\frac{z^2}{k^2}$. Now

$$\Big|w_k(z)\Big| = \left|-\frac{z^2}{k^2}\right| \le |z|^2\frac{1}{k^2} \le \frac{R^2}{k^2}$$

for all z in the disc $S(0, R)$. Since the positive termed series $\sum_{k=1}^{\infty}(R^2/k^2)$ converges, it follows that the series $\sum_{k=1}^{\infty}(|z|^2/k^2)$ converges uniformly on $S(0, R)$.

Hence, by Theorem 9.2, the given infinite product also converges uniformly and absolutely on $S(0, R)$ for each R and, hence, on each bounded subset of $\mathbb{C}$.

9.2.2 Semi-convergence

Definition 9.2.8. (i) If the series $\sum_{n=1}^{\infty} z_n$ converges and $\sum_{n=1}^{\infty}|z_n|$ diverges, then the product $\prod_{n=1}^{\infty} z_n$ is called *semi-convergent* or *conditionally convergent.*

(ii) If the series $\sum_{n=1}^{\infty}(1+w_n)$ converges and $\sum_{n=1}^{\infty}(1+|w_n|)$ diverges, then the infinite product $\prod_{n=1}^{\infty}(1+w_n)$ is called *semi-convergent.*

9.2.3 Infinite Product of Functions

So far we have considered infinite product of complex numbers. Now we shall study infinite products whose factors are functions of a complex variable. Some of the factors (finite in number) may vanish on a region considered. In that case we consider the infinite product omitting those factors. The theorems and their proof discussed in the previous section will be extended to infinite products with complex functions as their factors with some modifications.

Definition 9.2.9. Let $\mathcal{D} \subset \mathbb{C}$ be a domain in $\mathbb{C}$. The *infinite product* of $\langle 1+w_n(z)\rangle$, where each $w_k(z)$ are defined on $\mathcal{D}$, is denoted by

$$[1+w_1(z)]\cdot[1+w_2(z)]\cdot[1+w_3(z)]\cdots = \prod_{k=1}^{\infty}[1+w_k(z)]. \tag{9.9}$$

In this case, $P_n(z)$, the nth *partial product* is given by

$$P_n(z) = \prod_{k=1}^{n}\Big[1+w_k(z)\Big] = [1+w_1(z)]\cdot[1+w_2(z)]\cdots[1+w_n(z)]. \tag{9.10}$$

Definition 9.2.10. [Uniform convergence of infinite products]: An infinite product as in Eq. (9.9), where each $w_k(z)$ are defined on a domain $\mathcal{D} \subset \mathbb{C}$, is said to be *uniformly convergent* on $\mathcal{D}$ if the sequence of partial products $\langle P_n(z)\rangle = \langle \prod_{k=1}^{\infty}[1+w_k(z)]\rangle$ converges uniformly to a non-zero limit function $P(z)$ on $\mathcal{D}$.

Theorem 9.3. *The infinite product as in Eq. (9.9) is uniformly convergent on a domain* $\mathcal{D} \subset \mathbb{C}$ *if the series* $\sum_{k=1}^{\infty}|w_k(z)|$ *converges uniformly and has a bounded sum in* $\mathcal{D}$.

Proof: Since $\sum_{k=1}^{\infty}|w_k(z)|$ is uniformly convergent on a domain $\mathcal{D}$, there exists a $m \in \mathbb{N}$, independent of z such that

$$|w_{n+1}(z)| + |w_{n+2}(z)| + \cdots + |w_{n+p}(z)| < 1/2.$$

But $|w_{n+r}(z)| > 0\ \forall z$ and $\forall r > 0$.

$$\therefore \quad |w_{n+r}(z)| < \frac{1}{2};\ \forall z \in \mathcal{D} \text{ and } n \geq m.$$

For such value of n and z, we get

$$\text{Log}(1+w_n) = w_n - \frac{w_n^2}{2} + \frac{w_n^3}{3} - \cdots$$

$$\Longrightarrow \quad |\text{ Log }(1+w_n)| \leq |w_n| + \frac{|w_n|^2}{2} + \frac{|w_n|^3}{3} + \cdots$$

$$\therefore \quad |\text{ Log }[1+w_n(z)]| < |w_n| + |w_n|^2 + |w_n|^3 + \cdots = \frac{|w_n(z)|}{1-|w_n(z)|}. \tag{9.11}$$

Since we have, $|w_n(z)| < 1/2\ \forall z \in \mathcal{D}$ and for $n \geq m$,

$$1 - |w_n(z)| > 1 - \frac{1}{2} = \frac{1}{2} \Rightarrow \frac{1}{1-|w_n(z)|} < 2.$$

Thus, equation (9.11) becomes:

$$\Big|\text{ Log }[1+w_n(z)]\Big| < 2|w_n(z)|.$$

Since $\sum_{n=1}^{\infty} |w_n(z)|$ is uniformly convergent in $\mathcal{D}$, it follows that $\sum_{n=1}^{\infty} |\text{ Log }[1+w_n(z)]|$ and hence $\sum_{n=1}^{\infty} \text{Log}[1+w_n(z)]$ is uniformly convergent in $\mathcal{D}$. This implies that Log $P_n(z)$ as in Eq. (9.10) converges uniformly to a limit Log $P(z)$. Thus, $P_n(z)$ converges uniformly to $P(z)$ in $\mathcal{D}$ and hence $\prod_{n=1}^{\infty}[1+w_n(z)]$ is uniformly convergent in $\mathcal{D}$. □

Note: If the infinite product $\prod_{n=1}^{\infty}[1+w_n(z)], w_n(z) \neq -1$ converges uniformly in a domain $\mathcal{D} \subset \mathbb{C}$ and each factor of the product is analytic in $\mathcal{D}$, then the infinite product as in Eq. (9.9), is also analytic in $\mathcal{D}$.

Corollary 9.2.1. Let $\sum_{n=1}^{\infty} w_n(z)$ be a series of functions defined in a domain $\mathcal{D}$ and $\{M_n\}$ be a sequence of positive real numbers such that (i) $|w_n(z)| \leq M_n,\ \forall n$ and $z \in \mathcal{D}$ and (ii) the series $\sum_{n=1}^{\infty} M_n$ is convergent,
then, the infinite product as in Eq. (9.9), is uniformly convergent in $\mathcal{D}$.

Example 9.2.4. *Discuss the convergence of the infinite product* $\prod_{k=1}^{\infty}(1+\frac{z}{k\pi})e^{-z/k\pi}$.

Solution. By comparing the given infinite product with Eq. (9.9), we get

$$\begin{aligned} 1+w_k(z) &= \left(1+\frac{z}{k\pi}\right)\left(1-\frac{z}{k\pi}+\frac{1}{2!}\frac{z^2}{k^2\pi^2}-\cdots\right) \\ &= 1-\frac{1}{2}\frac{z^2}{k^2\pi^2}-\frac{1}{3}\frac{z^3}{k^3\pi^3}-\cdots \\ \therefore w_k(z) &= \frac{1}{2}\frac{z^2}{k^2\pi^2}-\frac{1}{3}\frac{z^3}{k^3\pi^3}-\cdots. \end{aligned}$$

Let $b_k(z) = 1/k^2$. Then

$$\lim_{k\to\infty}\left|\frac{w_k(z)}{b_k(z)}\right| = \lim_{k\to\infty}\left|\frac{1}{2}\frac{z^2}{\pi^2}-\frac{1}{3}\frac{z^3}{k\pi^3}-\cdots\right| = \frac{|z|^2}{2\pi^2},$$

which is a non-zero finite quantity for all z. According to limit form test, the series $\sum_{k=1}^{\infty}|w_k(z)|$ is convergent, i.e. the series $\sum_{k=1}^{\infty} w_k(z)$ converges absolutely as the series $\sum_{k=1}^{\infty} b_k(z)$ is convergent.

Example 9.2.5. *Discuss the convergence of the product*

$$\left(1-\frac{z}{1}\right)\left(1+\frac{z}{1}\right)\left(1-\frac{z}{2}\right)\left(1+\frac{z}{2}\right)\cdots$$

Solution. Let $\prod_{n=1}^{\infty}[1+w_n(z)] = (1-\frac{z}{1})(1+\frac{z}{1})(1-\frac{z}{2})(1+\frac{z}{2})\cdots$. We consider a bounded closed domain $\mathcal{D}$ which does not contain the points $z=\pm 1, \pm 2, \cdots$. Then

$$\sum_{n=1}^{\infty} w_n(z) = -\frac{z}{1}+\frac{z}{1}-\frac{z}{2}+\frac{z}{2}-\cdots.$$

Let $S_n(z)=\sum_{k=1}^{n} w_k(z)$. Then $S_{2n}(z)=0$ and $S_{2n+1}(z)=\frac{-z}{n+1}\to 0$ as $n\to\infty$. This implies that $\sum_{n=1}^{\infty} w_n(z)$ and consequently the given infinite product converges uniformly in $\mathcal{D}$. Now

$$\sum_{n=1}^{\infty} |w_n(z)| = \frac{|z|}{1}+\frac{|z|}{1}+\frac{|z|}{2}+\frac{|z|}{2}+\cdots = 2\left[\frac{|z|}{1}+\frac{|z|}{2}+\cdots\right].$$

Let $b_n = 1/n$.Then,

$$\lim_{n\to\infty}\left|\frac{w_n(z)}{b_n(z)}\right| = \lim_{n\to\infty}\frac{2|z|}{n}\cdot n = 2|z|,$$

which is a finite non-zero quantity for all z except $z=0$. According to limit form test, $\sum_{n=1}^{\infty}|w_n(z)|$ is divergent as $\sum_{n=1}^{\infty} b_n$, being a hyperharmonic series, is divergent. Therefore, the given product does not converge absolutely. Since $\sum_{n=1}^{\infty} w_n(z)$ is convergent and $\sum_{n=1}^{\infty}|w_n(z)|$ is divergent, the given product is semi-convergent for all z except $z=0$.

Example 9.2.6. *Discuss the convergence of the infinite product* $\prod_{n=1}^{\infty}\frac{z+z^{2n}}{1+z^{2n}}$.

Solution. By the comparing the given infinite product with Eq. (9.9), we get

$$1+w_n(z) = \frac{z+z^{2n}}{1+z^{2n}} \implies w_n(z) = \frac{z-1}{1+z^{2n}}.$$

According to geometric series, $\sum_{n=1}^{\infty} z^{-2n}$ is convergent for $|z|>1$ since

$$|w_n(z)| = \left|\frac{z-1}{1+z^{2n}}\right| < \left|\frac{1}{z^{2n}}\right|.$$

The series $\sum_{n=1}^{\infty} w_n(z)$ converges absolutely, and hence the given infinite product converges absolutely. Now, for $|z|<1$

$$\lim_{n\to\infty}[1+w_n(z)] = \lim_{n\to\infty}\frac{z(1+z^{2n-1})}{1+z^{2n}} = z$$

and $w_n(z)\not\to 0$ as $n\to\infty$. Thus, the product diverges. Hence, the given product is divergent. For $|z|=1$,

$$|w_n(z)| = \left|\frac{z-1}{1+z^{2n}}\right| \geq \frac{|z|-1}{1+|z|^{2n}} = 0.$$

In this case also, the given product is divergent. It follows that the product is absolutely convergent for $|z|>1$ and divergent for $|z|\leq 1$.

Example 9.2.7. *(i) Express $P_n(z) = \frac{n!z!}{(n+z)!}(n+1)^z$ as a product of n factors. (ii) If the number of factors is increased indefinitely, show that the product is absolutely convergent.*

Solution. (i) The expression for $P_n(z)$ can be written as

$$\begin{aligned} P_n(z) &= \frac{1\cdot 2\cdot 3\cdots n\cdot z!(n+1)^z}{z!(z+1)(z+2)\cdots(z+n)} \\ &= \frac{1}{z+1}\cdot\frac{2}{z+2}\cdot\frac{3}{z+3}\cdots\frac{n}{z+n}(n+1)^z \\ &= \Big(\frac{1}{z+1}2^z\Big)\Big(\frac{2}{z+2}\frac{3^z}{2^z}\Big)\Big(\frac{3}{z+3}\frac{4^z}{3^z}\Big)\cdots\Big(\frac{n}{z+n}\frac{(n+1)^z}{n^z}\Big) \\ &= \prod_{k=1}^{n}\frac{k}{z+k}\Big(\frac{k+1}{k}\Big)^z, \end{aligned}$$

which is the required product of n factors.
(ii) Let the product from of $P_n(z)$ be written as Eq. (9.9).Then

$$\begin{aligned} 1+w_k(z) &= \frac{k}{z+k}\Big(\frac{k+1}{k}\Big)^z = \Big(1+\frac{z}{k}\Big)^{-1}\Big(1+\frac{1}{k}\Big)^z \\ &= \Big[1-\frac{z}{k}+\frac{z^2}{k^2}-\frac{z^3}{k^3}+\cdots\Big]\Big[1+\frac{z}{k}+\frac{z(z-1)}{2!k^2}+\frac{z(z-1)(z-2)}{3!k^3}\cdots\Big] \\ &= 1+\frac{z(z-1)}{2k^2}+\frac{z(z^2-1)}{3k^3}+\cdots. \\ \therefore w_k(z) &= \frac{z(z-1)}{2k^2}+\frac{z(z^2-1)}{3k^3}+\cdots. \end{aligned}$$

Let $b_k(z) = 1/k^2$.Then

$$\lim_{k\to\infty}\left|\frac{w_k(z)}{b_k(z)}\right| = \frac{|z(z-1)|}{2},$$

which is a finite non-zero quantity for all z except $z=0,1$. Since, the series $\sum_{n=1}^{\infty} b_n(z)$ is convergent, and hence by limit form of the series $\sum_{n=1}^{\infty}|w_n(z)|$ is also convergent for all z except $z=0,1$, i.e., the series converges absolutely for all z except $z=0,1$. Consequently, the product is absolutely convergent except for $z=0,1$.

Example 9.2.8. *Prove that the function $\prod_{n=1}^{\infty}\frac{n^z+n^2+1}{n^z+n^2-1}$ represents a regular function in the domain* Re $z>2$.

Solution. Let the given product be denoted by $\prod_{n=1}^{\infty}[1+w_n(z)]$. Then, we have

$$1+w_n(z) = \frac{n^z+n^2+1}{n^z+n^2-1} \Rightarrow w_n(z) = \frac{n^z+n^2+1}{n^z+n^2-1}-1 = \frac{2}{n^z+n^2-1}$$

Now, Re $z>2 \implies x>2 \implies x=2+\delta$, where $\delta>0$.

$$\begin{aligned} |w_n(z)| = \left|\frac{2}{n^z+n^2-1}\right| &\le \frac{2}{|n^z|-|n^2|-|1|} \\ &= \frac{2}{|n^{x+iy}|-n^2-1} \\ &= \frac{2}{n^{2+\delta}-n^2-1} \quad \left[\text{Because } \begin{aligned}|n^{x+iy}| &= |n^x||n^{iy}| = |n^x||e^{iy\log n}| \\ &= |n^x|\cdot 1 = n^{2+\delta}\end{aligned}\right] \\ &= M_n \text{ (say)} \end{aligned}$$

Therefore, $|w_n(z)| \leq M_n \ \forall z$ and n such that Re $z = 2 + \delta$. By comparing with the convergent series $\sum_{n=1}^{\infty} \frac{1}{n^{2+\delta}}$, we see that the series $\sum_{n=1}^{\infty} \frac{2}{n^{2+\delta} - n^2 - 1}$ is convergent. Thus, by Weierstrass M-test, $\sum_{n=1}^{\infty} w_n(z)$ is uniformly and absolutely convergent in the domain Re $z > 2$. Consequently, $\prod_{n=1}^{\infty}[1 + w_n(z)]$ is uniformly and absolutely convergent in Re $z > 2$ and hence the given product is analytic in Re $z > 2$ as each $w_n(z)$ is analytic there.

Definition 9.2.11. [Logarithmic derivative of a product]: Let f be an analytic function on an open set U and suppose that f is not identically 0 on any component of U. Then the meromorphic function f'/f is called the *logarithmic derivative of* f on U.

RESULT **69.** Let $\langle w_k \rangle$ be a sequence of analytic functions on a domain $\mathcal{D}$. If the product $f(z) = \prod_{k=1}^{\infty} w_k(z)$ converges uniformly on compact subsets of $\mathcal{D}$ to a function f which is not identically 0, then the infinite sum $\sum_{n=1}^{\infty} \frac{w_k'(z)}{w_k(z)}$ converges uniformly to f'/f on compact subsets of $\mathcal{D}/Z_f$, where Z_f is the set of all zeros of f.

9.3 Factorization of Entire Functions

In this section we shall show that, given any sequence of points of an open set $U \subset \mathbb{C}$, with no limit point in U, there is an analytic function on U with exactly the points of the sequence as its zeros, with each zero having order equal to number of times it appears in the sequence. The analytic function will be constructed as an infinite product of certain simple functions each of which has exactly one zero.

Theorem 9.4. *Let $f(z)$ be an entire function and it never vanishes on $\mathbb{C}$. Then $f(z)$ is of the form $f(z) = e^{g(z)}$, or, more generally, $f(z) = ce^{g(z)}; c \neq 0$, constant, where $g(z)$ is also an entire function.*

Proof: Since $f(z)$ is a non-vanishing entire function on $\mathbb{C}$ and $f'(z)$ is also entire, so the function $f'(z)/f(z)$ is well defined for every $z \in \mathbb{C}$ and is also an entire function and thus the derivative of an entire function $g(z)$. Therefore, it admits a Taylors series (about the origin) expansion as

$$\frac{f'(z)}{f(z)} = \sum_{k=0}^{\infty} a_k\, z^k; \text{ for } z \in \mathbb{C}$$

converging in the whole of $\mathbb{C}$. The function possess antiderivative $g(z)$ such that

$$g'(z) = \frac{f'(z)}{f(z)} \text{ where } \sum_{k=0}^{\infty} \frac{a_k}{k+1} z^{k+1} = g(z). \tag{9.12}$$

Let ρ be the radius of convergence of the series of $g(z)$ as in Eq. (9.12). Then it is given by

$$\frac{1}{\rho} = \overline{\lim} \left|\frac{a_n}{n+1}\right|^{1/n} = \overline{\lim} \left|a_n\right|^{1/n} = 0 \Rightarrow \rho = \infty.$$

Hence $g(z)$ is an analytic function. This ensures that

$$\begin{aligned} \frac{d}{dz}\left[f(z)e^{-g(z)}\right] &= f'(z)e^{-g(z)} - f(z)e^{-g(z)}g'(z) \\ &= f(z)g(z)e^{-g(z)} - f(z)e^{-g(z)}g'(z) = 0; z \in \mathbb{C} \end{aligned}$$

which gives

$$f(z)e^{-g(z)} = c \Longrightarrow f(z) = ce^{g(z)},$$

where c is a constant which may be absorbed in $g(z)$ to have the required result. For example, e^z, e^{z^2}, e^{z+z^2}, $e^{\sin z}$, etc., are all entire functions with no zeros in $\mathbb{C}$. □

RESULT **70.** Assume that the entire function f possesses finitely many zeros, a zero of order $m > 0$ at the origin, and the non-zero ones, possibly repeated are, say $z_1, z_2, \cdots, z_n$ in $\mathbb{C}$. Let us define a function $h(z)$ as follows:

$$h(z) = \begin{cases} \dfrac{f(z)}{\prod_{k=1}^{n}(z - z_k)} & \text{for } z \neq z_k \\ \lim\limits_{z \to z_k} \dfrac{f(z)}{\prod_{k=1}^{n}(z - z_k)} & \text{for } z \neq z_k \end{cases}$$

where $k = 1, 2, \cdots, n$. Then $h(z)$ is an entire function with no zeros. Thus using the Theorem 9.4, $h(z)$ can be expressed as $e^{g(z)}$, so that

$$f(z) = \prod_{k=1}^{n}(z - z_k)\, e^{g(z)} = z^m \prod_{k=1}^{n}\left(1 - \frac{z}{z_k}\right)e^{g(z)}. \tag{9.13}$$

If the distinct zeros $z_1, z_2, \cdots, z_n$ have multiplicities $m_1, m_2, \cdots, m_n$ $(m_k \geq 0)$ respectively, then, using the Theorem 9.4, f can be expressed as

$$f(z) = \prod_{k=1}^{n}(z - z_k)^{m_k}\, e^{g(z)} = e^{g(z)} \prod_{k=1}^{n} z^{m_k}\left(1 - \frac{z}{z_k}\right)^{m_k} \tag{9.14}$$

for some entire function $g(z)$. This shows that $f(z)$ can be expressed as the product of a polynomial and a zero-free entire function. The zeros of the polynomial are exactly the same as the zeros of $f(z)$.

However, we cannot expect, in general, such a simple formula to hold in the case of infinitely many zeros. Here we have to take care of convergence problems for an infinite product.

9.3.1 Weierstrass' Primary Factor

Before starting the Weierstrass factorization theorem, we will learn about the following important result which will be helpful in proving the theorem.

Definition 9.3.1. For $p = 0, 1, 2, \cdots$, and $z \in \mathbb{C}$, we define entire functions $E(z, p)$ by

$$\left.\begin{aligned} &E(z, 0) = 1 - z \text{ and for } p > 0 \\ &E(z, p) = (1 - z)e^{z + \frac{z^2}{2} + \cdots + \frac{z^p}{p}} \end{aligned}\right\} \tag{9.15}$$

The functions $E(z, p)$ are called *Weierstrass' primary factors.*

Now the *Weierstrass' primary factors* have the following properties:

Property 9.3.1. Each primary factor $E(z, p)$ is an entire function having one simple zero at $z = 1$.

Property 9.3.2. Now, when $|z| < 1$, we have

$$\begin{aligned} \log E(z, p) &= \log(1 - z) + z + \frac{z^2}{2} + \cdots + \frac{z^p}{p} \\ &= -\frac{z^{p+1}}{p+1} - \frac{z^{p+2}}{p+2} - \cdots. \end{aligned}$$

Here, we have taken the principal branch of $\log(1-z)$. Hence if $|z| \le \frac{1}{2}$,

$$\begin{aligned}\Big|\log E(z,p)\Big| &\le \Big|z\Big|^{p+1} + \Big|z\Big|^{p+2} + \cdots = \Big|z\Big|^{p+1}\Big[1 + |z| + |z|^2 + \cdots\Big] \\ &\le \Big|z\Big|^{p+1}\Big[1 + \frac{1}{2} + \frac{1}{2^2} + \cdots\Big] = 2\Big|z\Big|^{p+1}.\end{aligned}$$

Property 9.3.3. From the expression (9.15) for $E(z,p)$,

$$\begin{aligned}1 - E(z,p) &= 1 - (1-z)e^{z+\frac{z^2}{2}+\cdots+\frac{z^p}{p}} \\ \therefore \Big[1 - E(z,p)\Big]' &= e^{z+\frac{z^2}{2}+\cdots+\frac{z^p}{p}} - E(z,p)\Big(1 + z + \cdots + z^{p-1}\Big) \\ &= z^p e^{z+\frac{z^2}{2}+\cdots+\frac{z^p}{p}} = -E'(z,p) \qquad (9.16)\end{aligned}$$

Since this has a zero of order p at $z = 0$, the function $1 - E(z,p)$ has a zero of order $p+1$ at $z = 0$.

The function (9.16) has a power series expansion about 0 with all of its coefficients non-negative real numbers, since this is true of the exponential function and the function $z + \frac{z^2}{2} + \cdots + \frac{z^p}{p}$. It follows that the function

$$g(z) = \frac{1}{z^{p+1}}\Big[1 - E(z,p)\Big]$$

also has non-negative real numbers as coefficients for its power series expansion about 0. This implies that, the maximum value achieved by $|g(z)|$ for $|z| \le 1$ is $g(1) = 1$. Therefore,

$$\Big|E(z,p) - 1\Big| < \Big|z\Big|^{p+1} \text{ for } |z| \le 1. \qquad (9.17)$$

Theorem 9.5. *Let G be a subset of $\mathbb{C}$. If $\langle z_k\rangle$ is a sequence of non-zero complex numbers and $\langle p_k\rangle$ is a sequence of integers such that*

$$\sum_{k=1}^{\infty}\Big(\frac{r}{|z_k|}\Big)^{p_k+1} = \sum_{k=1}^{\infty}\Big(\frac{r}{r_k}\Big)^{p_k+1} < \infty \textit{ for all } r > 0, \qquad (9.18)$$

then the Weierstrass product

$$f(z) = \prod_{k=1}^{\infty} E\Big(\frac{z}{z_k}; p_k\Big) \qquad (9.19)$$

converges uniformly on compact subsets of $\mathbb{C}$ to an entire function which has $\langle z_k\rangle$ as a list of zeros, counting multiplicity.

Proof: By Eq. (9.17), we have

$$\Big|E\Big(\frac{z}{z_k}; p_k\Big) - 1\Big| < \Big|\frac{z}{z_k}\Big|^{p_k+1} \text{ if } |z| \le |z_k|.$$

The condition $|z| \le |z_k|$ must be satisfied for all sufficiently large k if the series (9.18) converges. The theorem follows by applying Theorem 9.3 with $w_k(z) = E(\frac{z}{z_k}; p_k) - 1$. □

9.3.2 The Weierstrass Theorem

Let $\langle z_k \rangle$ be a sequence of non-zero complex numbers converging to ∞. The next theorem shows how to use scaled versions of the functions $E(z,p)$ to construct an entire function with the sequence as a list of its zeros, counting multiplicity. The resulting product is called a *Weierstrass product*.

Theorem 9.6. *If $\langle z_k \rangle$ is any sequence of complex numbers converging to infinity, then there is an entire function with $\langle z_k \rangle$ as a list of its zeros, counting multiplicity.*

Proof: Suppose m of the z_k are equal to 0 (m might be 0). We may as well assume these are the first m terms of the sequence. Then $\langle z_k \rangle_{k=m+1}^{\infty}$ is a sequence of non-zero complex numbers.

If $R > 0$, then, since $z_k \to \infty$, there is a $K > m$ such that $|z_k| > 2R$ for all $k \geq K$. Then the series $\sum_{k=m+1}^{\infty} |\frac{z}{z_k}|^k$ converges uniformly on $|z| \leq R$ by comparison with the geometric series with ratio $1/2$.

Thus, the hypothesis of the Theorem 9.5 are satisfied if we choose $p_k = k-1$ for each k. The resulting Weierstrass product $\sum_{k=m+1}^{\infty} E(\frac{z}{z_k}; p)$ converges uniformly on compact subsets of $\mathbb{C}$ to an entire function which has $\langle z_k \rangle_{k=m+1}^{\infty}$ as a list of its zeros, counting multiplicity. Then

$$f(z) = z^m \prod_{k=m+1}^{\infty} E\left(\frac{z}{z_k}; p\right)$$

is an entire function with $\langle z_k \rangle_{k=1}^{\infty}$ as a list of its zeros, counting multiplicity. □

Example 9.3.1. *Find an entire function which has a zero of order k at each positive integer k.*

Solution. We construct a sequence $1, 2, 2, 3, 3, 3, 4, 4, 4, 4, \cdots$ in which each positive integer n appears n times and the terms are arranged in an increasing order. Then, for this sequence $\langle z_k \rangle$ and a given positive integer p, we have

$$\sum_{k=1}^{\infty} \frac{1}{|z_k|^{p+1}} = \sum_{n=1}^{\infty} n \cdot \frac{1}{n^{p+1}} = \sum_{n=1}^{\infty} \frac{1}{n^p}.$$

If we choose $p = 2$, then the right-hand side is a hyperharmonic series with $p = 2$, and hence convergent. The Weierstrass product for the sequence $\langle z_k \rangle$ and $\langle p_k = 2 \rangle$ is

$$\prod_{k=1}^{\infty} \left[\left(1 - \frac{z}{k}\right) e^{\frac{z}{k} + \frac{z^2}{2k^2}} \right]^k.$$

By Theorem 9.5, this infinite product converges to an entire function with the required zeros.

9.3.3 Weierstrass Factorization Theorem

The Weierstrass factorization theorem is a companion to the Mittag–Leffler theorem dealt with in detail later in the chapter. The Mittag-Leffler theorem asserts that we can prescribe the poles and principal parts of a meromorphic function. The Weierstrass factorization theorem asserts that we can prescribe the zeros and poles of a meromorphic function together with their poles.

The Weierstrass theorem for the plane leads immediately to the Weierstrass factorization theorem for the entire functions.

Theorem 9.7. *Let f be an entire function which is not identically zero. Let m be the order of the zero of f at 0, and let $\langle z_k \rangle$ be a list of the non-zero zeros of f, counting multiplicity. Then there exist non-negative integers $p_1, p_2, \cdots$ and on entire function g such that*

$$f(z) = e^{g(z)}\, z^m \prod_{k=1}^{\infty} E\Big(\frac{z}{z_k}; p_k\Big). \tag{9.20}$$

The sequence $\langle p_k \rangle$ may be chosen in any way that satisfies Eq.(9.18).

Proof: The product $h(z) = z^m \prod_{k=1}^{\infty} E(\frac{z}{z_k}; p_k)$ converges uniformly on compact sets if $\langle p_k \rangle$ is chosen such that Eq.(9.18) holds ($p_k = k-1$ is one choice which always works, but there may be better choices for a given f). In view of Theorem 9.4, for an entire function $g(z)$, we have

$$\frac{f(z)}{h(z)} = e^{g(z)} \Rightarrow f(z) = e^{g(z)}\, h(z).$$

Thus, equation (9.20) follows. □

Deduction **33.** Let the entire function $f(z)$ which does not vanish at the origin have simple zeros $\{z_n : n \in \mathbb{N}\}$ arranged in the order of increasing modulus. Let $\lim\limits_{k\to\infty} |z_k| = \infty$, since $\langle z_n \rangle$ has no limiting points, in case of which the this point would be an essential singular point of $f(z)$. Let us write $f(z) = (z - z_k)h(z)$. Then, $h(z)$ is regular and non-zero in a certain neighbourhood of z_k. Therefore

$$\frac{f'(z)}{f(z)} = \frac{1}{z - z_k} + \frac{h'(z)}{h(z)},$$

and so, the only singularities of $f'(z)/f(z)$ are simple poles of residue 1 at the points $\{z_n : n \in \mathbb{N}\}$. It, therefore, follows that

$$\frac{f'(z)}{f(z)} = \sum_{k=1}^{\infty} \Big[\frac{1}{z - z_k} + \frac{1}{z_k}\Big] + \frac{f'(0)}{f(0)}$$

the series being uniformly convergent in any bounded closed region which contains none of the zeros of $f(z)$, provided that the terms are suitably bracketed. Integrating term by term, we find

$$\log f(z) = \log f(0) + \sum_{k=1}^{\infty} \Big[\log\Big(1 - \frac{z}{z_k}\Big) + \frac{z}{z_k}\Big] + z\frac{f'(0)}{f(0)}.$$

$$\therefore \quad f(z) = f(0)\, e^{zf'(0)/f(0)} \prod_{k=1}^{\infty} \Big[\Big(1 - \frac{z}{z_k}\Big) e^{z/z_k}\Big]. \tag{9.21}$$

This is a result parallel to that of *Weierstrass' factorisation theorem.*

9.3.4 Canonical Product

The representation of $f(z)$ in the form (9.20) of the Theorem 9.7 becomes all the more interesting if it is possible to choose all the m_k equal to one another.

Definition 9.3.2. Let f be an entire function with zeros $\langle z_n \rangle$. Consider the representation

$$f(z) = z^m\, e^{g(z)} \prod_{k=1}^{\infty} E\Big(\frac{z}{z_k}; p\Big) = z^m\, e^{g(z)}\, G(z)\; \forall z \in \mathbb{C}, \tag{9.22}$$

where p is the least non-negative integer for which the infinite product $G(z)$ converges. By Weierstrass' factor theorem, $G(z)$ represents an entire function having zeros precisely at the points z_n. Then $G(z)$ is called the *canonical product* associated with the sequence $\langle z_n \rangle$ and p is called *genus* of the canonical product.

If $z = 0$ is a zero of f of order m, then the canonical product is $z^m G(z)$. For example:

(i) Let $z_n = n$, then $\sum_{n=1}^{\infty} \frac{1}{n^2}$ is convergent, while $\sum_{n=1}^{\infty} \frac{1}{n}$ is divergent. Thus $p = 1$.

(ii) Let $z_n = e^n$. Then $p = 0$.

Example 9.3.2. *Prove that the Euler product for sine is* $\sin \pi z = \pi z \prod_{k=1}^{\infty}(1 - \frac{z^2}{k^2})$.

Solution. This function has a zero of order 1 at each integer and has no other zeros. As $\sin k\pi = 0$, for every $k \in \mathbb{Z}$, we have

$$\sin \pi z = z e^{g(z)} \prod_{k=-\infty, k\neq 0}^{\infty} \left(1 - \frac{z}{k}\right) e^{z/k} \; \forall z \in \mathbb{C}$$

for some entire function g. Applying logarithmic derivative on both sides, and applying the result as in Example 9.7.2 given later, we get

$$\begin{aligned} \frac{\pi \cos \pi z}{\sin \pi z} &= \frac{1}{z} + g'(z) + \sum_{k=-\infty, k\neq 0}^{\infty} \left(\frac{1}{z-k} + \frac{1}{k}\right) \\ \Rightarrow \quad & \frac{1}{z} + \sum_{k=-\infty, k\neq 0}^{\infty} \left(\frac{1}{z-k} + \frac{1}{k}\right) = \frac{1}{z} + g'(z) + \sum_{k=-\infty, k\neq 0}^{\infty} \left(\frac{1}{z-k} + \frac{1}{k}\right) \\ \Rightarrow \quad & g'(z) = 0 \Rightarrow g(z) = c \;\; \forall z \in \mathbb{C} \end{aligned}$$

for some constant $c \in \mathbb{C}$. Therefore

$$\sin \pi z = z e^c \prod_{k=-\infty, k\neq 0}^{\infty} \left(1 - \frac{z}{k}\right) e^{z/k} \; \forall z \in \mathbb{C}.$$

Using the easy fact that $\frac{\sin \pi z}{\pi z} \to 1$ as $z \to 0$, we observe that

$$1 = \lim_{z\to 0} \frac{\sin \pi z}{\pi z} = \lim_{z \to 0} \frac{1}{\pi z}\left[z e^c \prod_{k=-\infty, k\neq 0}^{\infty} \left(1 - \frac{z}{k}\right) e^{z/k}\right] = \frac{1}{\pi} e^c.$$

Therefore, $e^c = \pi$. Thus the product becomes

$$\sin \pi z = \pi z \prod_{k=-\infty, k\neq 0}^{\infty} \left(1 - \frac{z}{k}\right) e^{z/k},$$

which is a Weierstrass factorization of $\sin(\pi z)$, and that this factorization is equivalent to the factorization

$$\sin \pi z = \pi z \prod_{k=1}^{\infty} \left(1 - \frac{z}{k}\right) e^{z/k} \times \left(1 + \frac{z}{k}\right) e^{-z/k} = \pi z \prod_{k=1}^{\infty} \left(1 - \frac{z^2}{k^2}\right).$$

9.4 Counting Zeros of Analytic Functions

The rate of growth of an entire function is closely related to the density of zeros. We have quite an effective formula in this regard due to J.L.W.V. Jensen, a Danish mathematician who discovered it in the year 1899.

Theorem 9.8. [Jensen's formula]: *Let f be analytic on $\mathcal{S}[0,R]=\{z\in\mathbb{C}:|z|\leq R\}$. f has no zeros on the boundary $\partial\mathcal{S}$, $f(0)\neq 0$, and $z_1,z_2,\cdots,z_n$ are the zeros, counting multiplicity, of f in $\mathcal{S}[0,R]$. Then*

$$\log\left|f(0)\right|=\frac{1}{2\pi}\int_0^{2\pi}\log\left|f\left(Re^{i\theta}\right)\right|\,d\theta-\sum_{k=1}^{n}\log\left(\frac{R}{|z_k|}\right). \tag{9.23}$$

Proof: Let us consider a function g as

$$g(z)=f(z)\cdot\prod_{k=1}^{n}\frac{R^2-\bar{z}_kz}{R(z-z_k)}\Rightarrow\left|g(0)\right|=\left|f(0)\right|\cdot\prod_{k=1}^{n}\frac{R}{|z_k|}.$$

The zeros of the denominator of $g(z)$ are also the zeros of $f(z)$ of the same order. Hence the zeros of $f(z)$ cancels the poles z_n in the product and therefore, $g(z)$ is analytic on $\mathcal{S}[0,R]$. Also, $g(z)\neq 0$ on $\mathcal{S}[0,R]$. For, if

$$R^2-\bar{z}_kz=0,\ \text{then}\ z=R^2\bar{z}_k^{-1}$$

is the inverse point of z_k with respect to the circle $|z|=R$ and so lies outside the circle. Again,

$$\left|g(z)\right|=\left|f(z)\right|\cdot\prod_{k=1}^{n}\left|\frac{R^2-\bar{z}_kz}{R(z-z_k)}\right|.$$

Now, when $|z|=R$, we have

$$\left|\frac{R^2-\bar{z}_kz}{R(z-z_k)}\right|=\left|\frac{z\,\bar{z}-\bar{z}_kz}{R(z-z_k)}\right|=\frac{|z|}{R}\left|\frac{\bar{z}-\bar{z}_k}{z-z_k}\right|=1$$

$$\Rightarrow\quad\left|g(z)\right|=\left|f(z)\right|\ \text{on}\ \partial\mathcal{S}=\{z\in\mathbb{C}:|z|=R\}.$$

Since $g(z)$ is analytic and non-zero on $\mathcal{S}[0,R]$, $\log g(z)$ is also analytic on $\mathcal{S}[0,R]$ and consequently, $\operatorname{Re}\log g(z)=\log|g(z)|$ is harmonic on $\mathcal{S}[0,R]$. Hence by Gauss' mean value theorem (5.41),

$$\log\left|g(0)\right|=\frac{1}{2\pi}\int_0^{2\pi}\log\left|g\left(Re^{i\theta}\right)\right|\,d\theta;\ \text{as}\ \left|g(z)\right|=\left|f(z)\right|\ \text{on}\ \partial\mathcal{S}$$

$$\Rightarrow\quad\log\left[\left|f(0)\right|\cdot\prod_{k=1}^{n}\frac{R}{|z_k|}\right]=\frac{1}{2\pi}\int_0^{2\pi}\log\left|g\left(Re^{i\theta}\right)\right|\,d\theta$$

$$\Rightarrow\quad\log\left|f(0)\right|+\sum_{k=1}^{n}\frac{R}{|z_k|}=\frac{1}{2\pi}\int_0^{2\pi}\log\left|f\left(Re^{i\theta}\right)\right|\,d\theta.$$

Hence the theorem. □

Theorem 9.9. [Jensen's inequality]: *Let f be analytic on $\mathcal{S}[0,R]$, $f(0)\neq 0$ and $f(z)\neq 0$ on $\partial\mathcal{S}$. If $z_1,z_2,\cdots,z_n$ be the zeros of $f(z)$, counting multiplicity, within $\mathcal{S}[0,R]$, then*

$$\frac{R^n\left|f(0)\right|}{|z_1|\cdot|z_2|\cdots|z_n|}\leq M(R)\ \text{where}\ M(R)=\max_{|z|=R}\left|f(z)\right|. \tag{9.24}$$

Proof: As in the Theorem 9.8, we have $|g(z)| = |f(z)|$ on $\partial\mathcal{S}$ and so by the maximum modulus theorem, $|g(z)| \leq M(R)$ for $|z| \leq R$. In particular,

$$|g(0)| \leq M(R) \Rightarrow \frac{R^n|f(0)|}{|z_1|\cdot|z_2|\cdots|z_n|} < M(R).$$

Hence the result as given by Eq. (9.24) follows. □

Theorem 9.10. [Poisson–Jensen's formula] *Let f be analytic on $\mathcal{S}[0,R]$, $f(0) \neq 0$ and $f(z) \neq 0$ on $\partial\mathcal{S}$. If $z_1, z_2, \cdots, z_n$ be the zeros of $f(z)$, counting multiplicity, within $\mathcal{S}[0,R]$, then for any $z = re^{i\theta}; r < R$*

$$\log\left|f\left(re^{i\theta}\right)\right| = \frac{1}{2\pi}\int_0^{2\pi}\frac{(R^2-r^2)\log\left|f\left(Re^{it}\right)\right|}{R^2+r^2-2Rr\cos(t-\theta)}\,dt - \sum_{k=1}^{n}\log\left|\frac{R^2-\bar{z}_k re^{i\theta}}{R(re^{i\theta}-z_k)}\right|. \tag{9.25}$$

Proof: By the Jensen's formula (9.23), $|g(z)| = |f(z)$ on $\partial\mathcal{S}$. Since $g(z)$ is analytic and non-zero on $\mathcal{S}[0,R]$, $\log g(z)$ is also analytic on $\mathcal{S}[0,R]$ and consequently $\log|g(z)|$ is harmonic on $\mathcal{S}[0,R]$. So by the Poisson's integral formula (8.7)

$$\log\left|g\left(re^{i\theta}\right)\right| = \frac{1}{2\pi}\int_0^{2\pi}\frac{(R^2-r^2)}{R^2+r^2-2Rr\cos(t-\theta)}\log\left|g\left(Re^{it}\right)\right|\,dt. \tag{9.26}$$

Now, using the result

$$\log\left|g\left(re^{i\theta}\right)\right| = \log\left|f\left(re^{i\theta}\right)\right| + \sum_{k=1}^{n}\log\left|\frac{R^2-\bar{z}_k re^{i\theta}}{R(re^{i\theta}-z_k)}\right|$$

and as $\log|g(z)| = \log|f(z)|$ on $\partial\mathcal{S}$, using Eq. (9.26), we get Eq. (9.25). □

9.5 Order of an Entire Function

Recall that a polynomial $P_n(z)$ of degree n has exactly n zeros (multiple zeros repeated). Further the rate of growth as $z \to \infty$ of $|P_n(z)|$ increases as its degree n increases. Hence there exists a relationship, via the degree, between the number of zeros and the growth of the polynomial. Since the Maclaurin expansion of an entire function $f(z)$ gives an approximation to $f(z)$ by a sequence of polynomials, we might expect a similar relationship to exist for entire functions.

9.5.1 The Maximum Modulus of an Entire Function

Let $f(z)$ be a non-constant entire function. Define

$$M(r) = \max\{|f(z)| : |z| \leq r\}.$$

Since $f(z)$ is entire, by the maximum modulus principle, $|f(z)|$ reaches its maximum value $M(r)$ on the circle $|z| = r$ so that

$$M(r) = \max\{|f(z)| : |z| = r\}. \tag{9.27}$$

We sometimes use the symbol $M(r; f)$ when more than one function is involved. It is the growth rate of $M(r)$ which is most easily related to the distribution of zeros of $f(z)$. In fact, considerable information about entire functions is gained by studying how fast $M(r)$ approaches infinity. Before we define order of an entire function, we prove the following theorem.

Theorem 9.11. [Hadamard's three circle theorem]: *Let* $0 < r_1 < r < r_3$ *and suppose* f *is analytic on the closed annulus* $r_1 < |z| \leq r_3$. *If* $M(r) = \max\limits_{|z|=r} |f(z)|$, *then*

$$M(r)^{\log(r_3/r_1)} \leq M(r_1)^{\log(r_3/r)} \cdot M(r_3)^{\log(r/r_1)} \tag{9.28}$$

Proof: Let us consider the function $g(z) = z^{\alpha} f(z)$, where α is a real constant to be chosen later. If $\alpha \neq$ an integer, $g(z)$ is multi-valued in $r_1 \leq |z| \leq r_3$ (Fig. 9.1) and so we cut the annulus along the negative part of the real axis. Thus we obtain a simply-connected region $\mathcal{D}$ in which the principal branch of $g(z)$ is analytic. Hence the maximum modulus of the branch of $g(z)$ in $\mathcal{D}$ is attained on the boundary $\partial\mathcal{D}$. Since α is real, all the branches of $g(z)$ have the same modulus. If we consider another branch of $g(z)$ which is analytic in another cut annulus, it is clear that the principal branch of $g(z)$ cannot attain its maximum value on the cut.

Hence the maximum of $|g(z)|$ is attained on at least one of the bounding circles $|z| = r_1$ or $|z| = r_3$. Thus $|z^{\alpha} f(z)| \leq \max\{r_1^{\alpha} M(r_1), r_3^{\alpha} M(r_3)\}$. Hence on $|z| = r$,

$$r^{\alpha}\, M(r) \leq \max\left\{r_1^{\alpha}\, M(r_1),\ r_3^{\alpha}\, M(r_3)\right\}. \tag{9.29}$$

We now choose α such that $r_1^{\alpha}\, M(r_1) = r_3^{\alpha}\, M(r_3)$. Then

$$\alpha = -\frac{\log M(r_3) - \log M(r_1)}{\log r_3 - \log r_1}.$$

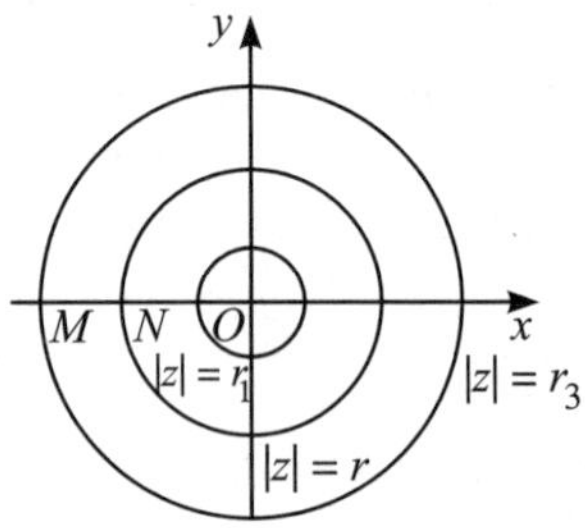

Figure 9.1 *Hadamard's three-circle theorem*

Substituting this value of α in Eq. (9.29), we get

$$M(r) \leq \left(\frac{r}{r_1}\right)^{-\alpha} M(r_1) = \left(\frac{r}{r_1}\right)^{\frac{\log M(r_3) - \log M(r_1)}{\log r_3 - \log r_1}} M(r_1)$$

$$\text{or,} \quad M(r)^{\log r_3 - \log r_1} \leq \left(\frac{r}{r_1}\right)^{\log M(r_3) - \log M(r_1)} M(r_1)^{\log r_3 - \log r_1}$$

$$\text{or,} \quad M(r)^{\log(r_3/r_1)} \leq \left[\frac{M(r_3)}{M(r_1)}\right]^{\log(r/r_1)} \cdot M(r_1)^{\log(r_3/r_1)}$$

$$\text{or,} \quad M(r)^{\log(r_3/r_1)} \leq M(r_1)^{\log(r_3/r)} \cdot M(r_3)^{\log(r/r_1)}$$

The equality of Eq.(9.28) occurs when $g(z)$ is a constant, i.e., when $f(z)$ is of the form cz^{α} for some real α, and c a constant. □

Corollary 9.5.1. $\log M(r)$ is a convex function of $\log r$.

Proof: Let $f(z)$ be analytic in the closed annulus $0 < r_1 \leq |z| \leq r_2$. If $r_1 < r < r_2$, we have, by Hadamard's three circle theorem (Eq. 9.28),

$$M(r)^{\log(r_2/r_1)} \leq M(r_1)^{\log(r_2/r)} \cdot M(r_2)^{\log(r/r_1)}$$

$$\text{or,} \quad \log(r_2/r_1)\, \log M(r) \leq \log(r_2/r)\, \log M(r_1) + \log(r/r_1)\, \log M(r_2)$$

$$\text{or,} \quad \log M(r) \leq \frac{\log r_2 - \log r}{\log r_2 - \log r_1} \log M(r_1) + \frac{\log r - \log r_1}{\log r_2 - \log r_1} \log M(r_2).$$

This inequality shows that $\log M(r)$ is a convex function of $\log r$. □

9.5.2 Entire Function of Finite Order

Definition 9.5.1. An entire function f is said to be of *finite order* if there is a positive number λ such that the inequality

$$\left|f(z)\right| \leq M(r) = e^{|z|^{\lambda}} = e^{r^{\lambda}}$$

holds for all z with $|z| = r$ sufficiently large. The infimum of all numbers λ is called the *order of the function* f.

f is said to be of *infinite order* if it is not of finite order. From the definition it is clear that order of an entire function is *non-negative*.

RESULT **71.** Let f be an entire function of order ρ and $M(r) = \max\{|f(z) : |z| = r\}$. Then

$$\rho = \limsup_{r\to\infty} \frac{\log\log M(r)}{\log r}. \tag{9.30}$$

Proof: By hypothesis, given $\varepsilon > 0$, there exists $r_0(\varepsilon) > 0$ such that

$$M(r) < e^{r^{\rho+\varepsilon}} \text{ for } r > r_0,$$

while $M(r) > e^{r^{\rho+\varepsilon}}$ for an increasing sequence $\langle r_n\rangle$ of values of r, tending to infinity. In other words,

$$\frac{\log\log M(r)}{\log r} < \rho+\varepsilon; \forall r > r_0 \text{ and } \frac{\log\log M(r)}{\log r} > \rho-\varepsilon; \text{ for } r\to\infty$$

$$\Rightarrow \rho = \limsup_{r\to\infty} \frac{\log\log M(r)}{\log r}.$$

This proves the result. □

Example 9.5.1. *Determine the order of the functions: (a)* $p(z) = a_0 + a_1z + \cdots + a_nz^n$; $a_n \neq 0$, *(b)* e^{kz}; $k \neq 0$, *and (c)* $\sin z$.

Solution. (a) Since $p(z) = a_0 + a_1z + \cdots + a_nz^n$; $a_n \neq 0$, we have

$$|p(z)| \leq |a_0| + |a_1||z| + \cdots + |a_n||z|^n.$$

Hence using relation (9.27), we get

$$\begin{aligned} M(r) &= \max_{|z|=r} |p(z)| \leq |a_0| + |a_1|r + \cdots + |a_n|r^n \\ &\leq \Big[|a_0| + |a_1| + \cdots + |a_n|\Big] r^n \end{aligned}$$

choosing $r \geq 1$. Since, ultimately $r \to \infty$, the choice is justified. Thus

$$\begin{aligned} M(r) &\leq Br^n; \text{ where } B = |a_0| + |a_1| + \cdots + |a_n| \\ \log M(r) &\leq \log B + n\log r \leq \log r + n\log r \\ &\leq (n+1)\log r \text{ (taking } r \text{ sufficiently large)}. \end{aligned}$$

Using, the relation (9.30), we get

$$\rho = \limsup_{r\to\infty} \frac{\log\log M(r)}{\log r} \leq \limsup_{r\to\infty} \frac{\log(n+1) + \log\log r}{\log r} = 0.$$

But by definition $\rho \geq 0$ and hence $\rho = 0$.

(b) Here $M(r) = e^{|k|r}$ and hence using the relation (9.30), we get

$$\rho = \limsup_{r\to\infty} \frac{\log\log M(r)}{\log r} = \limsup_{r\to\infty} \frac{\log(|k|r)}{\log r} = 1.$$

(c) Using the power series (4.9) of $\sin z$, we get

$$\begin{aligned} \left|\sin z\right| &\leq |z| + \frac{|z|^3}{3!} + \frac{|z|^5}{5!} + \cdots \\ &\leq r + \frac{r^3}{3!} + \frac{r^5}{5!} + \cdots = \frac{e^r - e^{-r}}{2} \quad \text{on } |z| \leq r. \end{aligned}$$

Also, at $z = ir$, $\sin z = \frac{e^{-r}-e^{r}}{2i}$ and so $|\sin z| = \frac{e^r - e^{-r}}{2}$. Hence, using relation (9.27), we get

$$M(r) = \max_{|z|=r} |\sin z| = \frac{e^r - e^{-r}}{2} = \frac{e^r\left(1 - e^{-2r}\right)}{2}.$$

$$\therefore \quad \log M(r) = r + \log\left(\frac{1-e^{-2r}}{2}\right) = r\left\{1 + \frac{1}{r}\log\left(\frac{1-e^{-2r}}{2}\right)\right\}.$$

Using the relation (9.30), we get

$$\rho = \limsup_{r\to\infty} \frac{\log\log M(r)}{\log r} = \lim_{r\to\infty} \frac{1}{\log r}\left[1 + \log\left\{1 + \frac{1}{r}\log\left(\frac{1-e^{-2r}}{2}\right)\right\}\right] = 1.$$

9.5.3 Estimation of the Number of Zeros

We shall denote by $n(r)$ the number of zeros of an entire function $f(z)$ in the closed disc $|z| \leq r$.

Definition 9.5.2. [The function $n(r)$]: Let $f(z)$ be an entire function with zeros at the points $z_1, z_2, \cdots$ arranged in the order of non-decreasing modulus, i.e., $|z_1| \leq |z_2| \leq \cdots$, multiple zeros being repeated according to their multiplicities. We define the function $n(r)$ to be the *number of zeros of* $f(z)$ in $|z| \leq r$. Evidently, $n(r)$ is a non-decreasing, non-negative function of r which is constant in any interval which does not contain the modulus of a zero of $f(z)$.

We observe that, if $f(0) \neq 0$, $n(r) = 0$ for $r < |z_1|$. Also, $n(r) = N$ for $|z_N| \leq r < |z_{N+1}|$.

9.5.4 Convergent Exponent

Definition 9.5.3. Let $f(z)$ be an entire function with zeros at the points $z_1, z_2, \cdots$ arranged in order of non-decreasing modulus, i.e., $|z_1| \leq |z_2| \leq \cdots$, multiple zeros being repeated according to their multiplicities and $|z_k| = r_k$; $k = 1, 2, \cdots$. We define *convergence exponent* ρ_1 of the zeros of $f(z)$ by the equation

$$\rho_1 = \limsup_{n\to\infty} \frac{\log n}{\log r_n} = \limsup_{n\to\infty} \frac{\log n(r)}{\log r}. \tag{9.31}$$

The significance of the choice of p made for the canonical product is that, with this choice, the resulting product is an entire function with order λ equal to the exponent of convergence ρ_1 of the sequence $\langle z_k \rangle$.

The convergence exponent has the following property:

Property 9.5.1. Let $f(z)$ be an entire function with zeros at the points $z_1, z_2, \cdots$ arranged in order of non-decreasing modulus, i.e., $|z_1| \leq |z_2| \leq \cdots$, multiple zeros being repeated according to their multiplicities and $|z_k| = r_k$; $k = 1, 2, \cdots$. Let ρ_1 be the convergence exponent of the zeros of $f(z)$. Then the series $\sum_{n=1}^{\infty} \frac{1}{r_n^{\alpha}}$ converges when $\alpha > \rho_1$ and diverges when $\alpha < \rho_1$.

If ρ_1 is infinite, the above series diverges for all positive values of α.

Property 9.5.2. Let $f(z)$ be an entire function with finite order ρ and $r_1, r_2, \cdots$ the moduli of the zeros of $f(z)$. Then the series $\sum_{n=1}^{\infty} \frac{1}{r_n^{\alpha}}$ converges when $\alpha > \rho$.

Property 9.5.3. Since convergence exponent ρ_1 is the lower bound of positive numbers α for which the series $\sum_{n=1}^{\infty} \frac{1}{r_n^{\alpha}}$ is convergent, it follows that $\rho_1 \leq \rho$.

Example 9.5.2. *Find the convergence exponent of the zeros of* $\cos z$.

Solution. The zeros of $\cos z$ are $(2k+1)\frac{\pi}{2}$, where $k \in \mathbb{Z}$, i.e., $\frac{\pi}{2}, -\frac{\pi}{2}, \frac{3\pi}{2}, -\frac{3\pi}{2}, \cdots$. Let $z_1 = \frac{\pi}{2}$, $z_1' = -\frac{\pi}{2}$, $z_2 = \frac{3\pi}{2}$, $z_2' = -\frac{3\pi}{2}, \cdots$, $z_k = (2k-1)\frac{\pi}{2}$, $z_k' = -(2k-1)\frac{\pi}{2}$.

Therefore, $r_1 = |z_1| = |z_1'| = \frac{\pi}{2}$, $r_2 = |z_2| = |z_2'| = \frac{3\pi}{2}, \cdots$, $r_k = |z_k| = |z_k'| = (2k-1)\frac{\pi}{2}$. Using Eq. (9.31), we get

$$\begin{aligned}
\rho_1 &= \limsup_{n\to\infty} \frac{\log n}{\log r_n} = \limsup_{n\to\infty} \frac{\log n}{\log(2n-1) + \log\frac{\pi}{2}} \\
&= \limsup_{n\to\infty} \frac{\log n}{\log\left\{n\left(2-\frac{1}{n}\right)\right\} + \log\frac{\pi}{2}} \\
&= \limsup_{n\to\infty} \frac{1}{1 + \frac{\log\left(2-\frac{1}{n}\right)}{\log n} + \frac{\log\frac{\pi}{2}}{\log n}} = 1.
\end{aligned}$$

RESULT **72.** [Borel's theorem]: The order of canonical product is equal to the convergent exponent of its zeros.

9.6 Hadamard's Factorization Theorem

Our goal in this section is to complete the characterization of entire functions of finite order. We will prove a theorem of Hadamard which asserts that every such function factors as a power of z times a canonical product of order at most λ times the exponential of a polynomial of degree at most λ. The key ingredient in the proof is Jensen's formula relating the density of the zeros of an entire function to the rate of growth at infinity of the function.

Theorem 9.12. [Hadamard's factorization theorem]: *If $f(z)$ is an entire function of finite order ρ with infinite number of zeros and $f(0) \neq 0$, then $f(z) = e^{Q(z)}\, G(z)$, where $G(z)$ is the canonical product formed with the zeros of $f(z)$ and $Q(z)$ is a polynomial of degree not greater than ρ.*

Proof: By Weierstrass factorization theorem (9.20), f has a factorization of the form

$$f(z) = e^{Q(z)}\, G(z), \tag{9.32}$$

where $G(z)$ is the canonical product with genus p formed with the zeros $z_1, z_2, \cdots$ of $f(z)$ and $Q(z)$ is an entire function. Since ρ is finite, we need to show that $Q(z)$ is a polynomial of degree $\leq \rho$. Let $m = [\rho]$, then $p \leq m$. Now

$$\begin{aligned}\log f(z) &= Q(z) + \log G(z)\\ &= Q(z) + \sum_{n=1}^{\infty} \log\left(1 - \frac{z}{z_n}\right) + \sum_{n=1}^{\infty}\left[\frac{z}{z_n} + \frac{1}{2}\left(\frac{z}{z_n}\right)^2 + \cdots + \cdots + \frac{1}{p}\left(\frac{z}{z_n}\right)^p\right]. \end{aligned} \tag{9.33}$$

Since $p \leq m$, we have,

$$\frac{d^{m+1}}{dz^{m+1}} \sum_{n=1}^{\infty}\left[\frac{z}{z_n} + \frac{1}{2}\left(\frac{z}{z_n}\right)^2 + \cdots + \frac{1}{p}\left(\frac{z}{z_n}\right)^p\right] = 0$$

and $$\frac{d^{m+1}}{dz^{m+1}} \log\left(1 - \frac{z}{z_n}\right) = \frac{d^{m+1}}{dz^{m+1}}(z_n - z) = -\frac{m!}{(z_n - z)^{m+1}}.$$

Differentiating both sides of Eq. (9.33) $(m+1)$ times, we get

$$\frac{d^{m+1}\log f(z)}{dz^{m+1}} = Q^{(m+1)}(z) + \frac{d^{m+1}}{dz^{m+1}} \sum_{n=1}^{\infty} \log\left(1 - \frac{z}{z_n}\right) + \frac{d^{m+1}}{dz^{m+1}} \sum_{n=1}^{\infty}\left[\frac{z}{z_n} + \cdots + \frac{1}{p}\left(\frac{z}{z_n}\right)^p\right]$$

or, $$\frac{d^m}{dz^m}\left(\frac{f'(z)}{f(z)}\right) = Q^{(m+1)}(z) - \sum_{n=1}^{\infty} \frac{m!}{(z_n - z)^{m+1}}.$$

Now, $Q(z)$ will be a polynomial of degree m at most if we can show that $Q^{(m+1)}(z) = 0$. Let

$$g_R(z) = \frac{f(z)}{f(0)} \prod_{|z_n| \leq R} \left(1 - \frac{z}{z_n}\right)^{-1}.$$

Since $f(z)$ is entire, $f(0) \neq 0$ and $\prod_{|z_n|\leq R}(1 - \frac{z}{z_n})^{-1}$ cancels with factors in $f(z)$. Hence, $g_R(z)$ is an entire function and $g_R(z) \neq 0$ in $|z| \leq R$. For $|z| = 2R$ and $|z_n| \leq R$, we have $|1 - \frac{z}{z_n}| \geq 1$. Hence

$$\left|g_R(z)\right| \leq \frac{|f(z)|}{|f(0)|} < Ae^{(2R)^{\rho+\varepsilon}} \text{ for } |z| = 2R. \tag{9.34}$$

By the maximum modulus theorem,

$$\left|g_R(z)\right| < Ae^{(2R)^{\rho+\varepsilon}} \text{ for } |z| < 2R. \tag{9.35}$$

Let $h_R(z) = \log g_R(z)$ such that $h_R(0) = 0$. Then $h_R(z)$ is analytic in $|z| \leq R$. Hence for Eq. (9.35),

$$\mathcal{R}]\ h_R(z) = \log\left|g_R(z)\right| < K \cdot R^{\rho+\varepsilon}; K = \text{ constant.}$$

Hence from the relation (8.22) of Borel Caratheodory theorem, we get

$$\left|h_R^{(m+1)}(z)\right| \leq \frac{2^{m+3}(m+1)!R}{(R-r)^{m+2}} K \cdot R^{\rho+\varepsilon} \text{ for } |z| = r < R.$$

Hence for $|z| = r = R/2$, $|h_R^{(m+1)}(z)| = O(R^{\rho+\varepsilon-m-1})$. But

$$h_R(z) = \log g_R(z) = \log f(z) - \log f(0) - \sum_{|z_n|\leq R} \left(1 - \frac{z}{z_n}\right)$$

$$\begin{aligned}\therefore \quad h_R^{(m+1)}(z) &= \frac{d^m}{dz^m}\left(\frac{f'(z)}{f(z)}\right) + \sum_{|z_n|\leq R} \frac{m!}{(z_n - z)^{m+1}}\\ &= O\Big(R^{\rho+\varepsilon-m-1}\Big) + O\Big(\sum_{|z_n|>R} \frac{1}{|z_n|^{m+1}}\Big)\end{aligned} \tag{9.36}$$

for $|z| = R/2$, and so also for $|z| < R/2$, by maximum modulus theorem. The first term on the right-hand side of Eq.(9.36) tends to 0 as $R \to \infty$ if $\varepsilon > 0$ is small enough since $m + 1 > \rho$. Also, the second term tends to 0 since $\sum_{n=1}^{\infty} \frac{1}{|z_n|^{m+1}}$ is convergent.

In fact, $\sum_{|z_n|>R} \frac{1}{|z_n|^{m+1}}$ becomes the remainder term for large R. Hence $Q^{(m+1)}(z) = 0$, as $Q^{(m+1)}(z)$ is independent of R. Thus $Q(z)$ is a polynomial of degree not greater than ρ. □

9.7 Meromorphic Functions

On a open connected set U, the set of analytic functions forms an integral domain — that is, it is a commutative ring with the property that the product of two elements is zero if and only if one of them is zero. The set of meromorphic functions of U forms a field — that is, a commutative ring in which every non-zero element has an inverse. The next theorem shows that the field of meromorphic function is actually the quotient field of the ring of analytic functions. That is, every meromorphic function is the quotient f/g of two analytic functions.

It is clear that the class of meromorphic functions include analytic and rational functions.

Theorem 9.13. *Every function which is meromorphic in the whole complex plane can be represented as a quotient of two entire functions.*

Proof: Let $F(z)$ be a meromorphic function on $\mathbb{C}$ for poles, say at the points $\{z_k : k \in \mathbb{N}\}$. Then by above theorem there exists an entire function $g(z)$ with prescribed zeros which are the poles of $F(z)$.

Clearly, the product

$$F(z)g(z) = f(z)(\text{say})$$

is also an entire function. Thus, $F(z) = f(z)/g(z)$. □

9.7.1 Partial Fraction Decomposition of Meromorphic Functions

Theorem 9.14. *Let $p(z)$ and $q(z)$ be two polynomials with no common zeros and $0 \leq deg\ (p) < deg\ (q)$. Let $\alpha_1, \alpha_2, \cdots, \alpha_k$ be the zeros of $q(z)$ with multipliers $m_1, m_2, \cdots, m_k$. Then $f(z) = p(z)/q(z)$ can be uniquely expressed as*

$$f(z) = \frac{p(z)}{q(z)} = \sum_{r=1}^{k}\sum_{s=1}^{m_r} \frac{A_{rs}}{(z-\alpha_r)^s}. \tag{9.37}$$

Proof [Existence]: If $f(z)$ has a pole α_1 of order m_1, then the principal part of $f(z)$ associated with α_1 can be expressed as

$$\frac{A_{11}}{z-\alpha_1} + \frac{A_{12}}{(z-\alpha_1)^2} + \cdots + \frac{A_{1m_1}}{(z-\alpha_1)^{m_1}} = \sum_{s=1}^{m_1} \frac{A_{1s}}{(z-\alpha_1)^s},$$

where A_{1s} are complex constants. Taking all the partial parts, we get

$$\begin{aligned} &\sum_{s=1}^{m_1} \frac{A_{1s}}{(z-\alpha_1)^s} + \sum_{s=1}^{m_2} \frac{A_{2s}}{(z-\alpha_2)^s} + \cdots + \sum_{s=1}^{m_k} \frac{A_{ks}}{(z-\alpha_k)^s} \\ &= \sum_{r=1}^{k}\sum_{s=1}^{m_r} \frac{A_{rs}}{(z-\alpha_r)^s} = \sum_{r=1}^{k} P_r(z), \text{ say.} \end{aligned} \tag{9.38}$$

Now, the analytic part of $\psi(z)$ given by

$$\psi(z) = f(z) - \sum_{r=1}^{k} P_r(z) = \frac{p(z)}{q(z)} - \sum_{r=1}^{k}\sum_{s=1}^{m_r} \frac{A_{rs}}{(z-\alpha_r)^s}$$

is analytic in the extended plane $\mathbb{C}_\infty$. Now each term $P_r(z)$ and $p(z)/q(z)$ converges to zero as $z \to \infty$ (as deg $(q) >$ deg (p). This shows that $\psi(z) \to 0$ for $z \to \infty$. But then ψ is necessarily bounded and hence constant by Liouville's theorem. A constant function tending to zero as $z \to \infty$ must be identically zero.

[Uniqueness]: We assume that the relation (9.37) exists. Let $\varepsilon > 0$ be small enough. Then for $z \in \mathcal{S}(\alpha_k; \varepsilon)$, Eq. (9.37) can be written as

$$\frac{p(z)}{q(z)} = g(z) + \sum_{s=1}^{m_r} \frac{A_{rs}}{(z-\alpha_r)^s}. \tag{9.39}$$

Since $\mathcal{S}(\alpha_k; \varepsilon)$ does not contain any zero of $q(z)$ other than α_k, $g(z)$ is analytic at $z = \alpha_k$. Multiplying both sides of Eq. (9.39) by $(z-\alpha_k)^{\alpha_k}$, we obtain

$$\frac{p(z)}{q(z)}\left(z-\alpha_k\right)^{\alpha_k} = g(z)\left(z-\alpha_k\right)^{\alpha_k} + \sum_{s=1}^{m_r} A_{rs}\left(z-\alpha_k\right)^{\alpha_k - r}. \tag{9.40}$$

Now the function $\frac{p(z)}{q(z)}(z-\alpha_k)^{\alpha_k}$ is analytic for all z belonging to $\mathcal{S}(\alpha_k; \varepsilon)$ and hence can be expanded in a Taylor's series in a neighbourhood of α_r in $\mathcal{S}(\alpha_k; \varepsilon)$,

$$\frac{p(z)}{q(z)}\left(z-\alpha_k\right)^{\alpha_k} = \sum_{n=0}^{\infty} C_n (z-\alpha_k)^n. \tag{9.41}$$

Combing (9.41) and (9.40), we write

$$\sum_{n=0}^{\infty} C_n(z-\alpha_k)^n = g(z)\left(z-\alpha_k\right)^{\alpha_k} + A_{k\alpha_k} + A_{k\alpha_{k-1}}\left(z-\alpha_k\right) + \cdots + A_{k1}\left(z-\alpha_k\right)^{\alpha_k-1}.$$

Comparing the coefficients, we find $c_k\alpha_k = C_0$, $A_{k\alpha_{k-1}} = C_1, \cdots, A_{k1} = C_{\alpha_k - 1}$ uniquely. Therefore, the theorem follows. □

Lemma 9.7.1. A meromorphic function f on the extended complex plane $\mathbb{C}_\infty$ is a rational.

Proof: Using the identity theorem 5.27, we note that a meromorphic function f on $\mathbb{C}_\infty$ can have only a finite number of poles.

(i) If f is analytic at ∞, we define $P_\infty(z)$ to be the constant function $f(\infty)$.

(ii) If f has a pole at ∞, we define $P_\infty(z)$ to be the principal part of f at ∞.

In any case, $P_\infty(z)$ is a polynomial and $f(z) - P_\infty(z) \longrightarrow 0$ as $z \longrightarrow \infty$. Let $\alpha_1, \alpha_2, \cdots, \alpha_m$ be the poles of $f(z)$ in the finite complex plane $\mathbb{C}$ and let $P_r(z)$ be the principal part of f at z_r given by Eq. (9.38). In particular, $P_r(z)$ is analytic at ∞ and vanishes there. Consider

$$g(z) = f(z) - P_\infty(z) - \sum_{r=1}^{m} P_r(z).$$

Since $f(z) - P_k(z)$ is analytic at z_r, and each $P_r(z)$ is analytic at z_r for $r \neq k$, $g(z)$ is analytic at each z_r. Hence $g(z)$ is an entire function and further, $g(z) \to 0$ as $z \to \infty$. By Liouville's theorem, $g(z)$ is identically zero. Thus

$$f(z) = P_\infty(z) + \sum_{r=1}^{m} P_r(z), \tag{9.42}$$

which shows that f is a rational function. The decomposition (9.42) is called *partial fraction decomposition* of rational function f. □

Example 9.7.1. *Obtain the partial fraction decomposition of* $f(z) = z^3/(z^2+1)$.

Solution. The function f is not analytic at ∞ and it has simple poles at $\pm i$. Now, f can be written in the form

$$\begin{aligned} f(z) &= \frac{z(z^2+1) - z}{z^2+1} = z - \frac{z}{z^2+1} = z - \frac{\alpha}{z-i} - \frac{\beta}{z+i} \\ &= P_\infty(z) - P_1(z) - P_2(z) \end{aligned}$$

for some constants α and β. Now,

$$\begin{aligned} \alpha &= \operatorname{Res}\{P_1; i\} = \lim_{z\to i}(z-i)\frac{z}{z^2+1} = \frac{1}{2}, \\ \beta &= \operatorname{Res}\{P_2; -i\} = \lim_{z\to -i}(z+i)\frac{z}{z^2+1} = \frac{1}{2}. \end{aligned}$$

This leads to the partial fractions decomposition as

$$f(z) = \frac{z^3}{z^2+1} = z - \frac{1}{2}\frac{1}{z-i} - \frac{1}{2}\frac{1}{z+i}.$$

9.7.2 Mittag–Leffler Theorem

In complex analysis, Mittag–Leffler's theorem (as mentioned earlier) concerns the existence of meromorphic functions with prescribed poles. Let Ω be an open subset of $\mathbb{C}$ and $\langle \alpha_r \rangle$ be a sequence of distinct points in Ω such that $\langle \alpha_r \rangle$ has no limit point in Ω. The existence of a meromorphic function f on Ω whose poles are at the points $\langle \alpha_r \rangle$ such that the singular part of f at $z = \alpha_r$ is $P_r(z)$ was proved by German Mathematician Gösta Mittag Leffler. This theorem is used to find the partial fraction of the meromorphic function with simple poles. It is a sister to the Weierstrass factorization theorem which asserts the existence of holomorphic functions with prescribed zeros.

Theorem 9.15. *Let Ω be an open set in $\mathbb{C}$. Let*

(i) $\langle \alpha_r \rangle$ be a sequence distinct non-zero complex numbers such that $|\alpha_n| < |\alpha_{n+1}|$ for $n \in \mathbb{N}$ and $\lim_{n\to\infty} |\alpha_n| = \infty$, having no points of accumulation in the finite $\mathbb{C}$-plane;

(ii) the non-constant sequence of rational polynomial functions $\langle P_r(z) \rangle$, where

$$P_r(z) = \sum_{s=1}^{m_r} \frac{A_{rs}}{(z-\alpha_r)^s}; \ r = 1, 2, \cdots$$

be assigned to each α_r; $r \in \mathbb{N}$.

Then there exists a meromorphic function f in $\mathbb{C}$ having poles at the points α_r (only there) with $P_r(z)$ as the principal part of the Laurent series expansion of f about the point α_r. Moreover, the function f having the stated property may be of the form:

$$f(z) = e(z) + \sum_{r=1}^{\infty} \Big[P_r(z) - p_r(z)\Big], \forall z \in \mathbb{C}/\{\alpha_r : r \in \mathbb{N}\}, \tag{9.43}$$

where $e(z)$ is an arbitrary entire function and $p_r(z)$ is suitable partial sum of Taylor's expansion of the singular part which is analytic in the open disc $\mathcal{S}(0; |\alpha_r|) = \{z : |z| < |\alpha_r|\} \subset \Omega$. The series converges absolutely and uniformly on any compact subset Ω of $\mathbb{C}$ not containing the poles.

Proof: Without loss of generality, we assume that $z = 0$ is not a pole of $f(z)$. Now, $P_r(z)$ is a polynomial, and so we see that $g_r(z)$ defined by

$$g_r(z) = P_r\Big(\frac{1}{z-\alpha_r}\Big) = P_r\Bigg(-\frac{1}{\alpha_r\Big(1-\frac{z}{\alpha_r}\Big)}\Bigg)$$

is analytic on the compact subset $\mathcal{S}(0; |\alpha_r|)$ of Ω. Thus the Taylor's series expansion of $g_r(z)$ is

$$\begin{aligned} g_r(z) &= -\sum_{s=1}^{m_r} \frac{A_{rs}}{\alpha_r}\Big(1-\frac{z}{\alpha_r}\Big)^{-1} = -\sum_{s=1}^{m_r} \frac{A_{rs}}{\alpha_r}\Big[1+\frac{z}{\alpha_r}+\frac{z^2}{\alpha_r^2}+\cdots\Big] \\ &= -\sum_{s=1}^{m_r} \frac{A_{rs}}{\alpha_r}\sum_{j=0}^{\infty}\frac{z^j}{\alpha_r^j} = -\sum_{j=0}^{\infty}\frac{A_{rs}}{\alpha_r^{j+1}}z^j = \sum_{j=0}^{\infty} C_j^{(r)} z^j. \end{aligned} \tag{9.44}$$

Clearly, this series is uniformly convergent in the open disc $\mathcal{S}(0; |\alpha_r|/2)$. Also, relation (9.44) can be written as

$$g_r(z) = \sum_{j=0}^{m_r} C_j^{(r)} z^j + \varphi_{m_r+1}(z) z^{m_r+1} \;\forall z \in \mathcal{S}(0; |\alpha_r|),$$

where $m_r \in \mathbb{N}$ is such that

$$2^{m_r - r} > M_r = \sup_{z \in \mathcal{S}(0;|\alpha_r|/2)} \Big|P_r(z)\Big| < \infty$$

and $$\varphi_{m_r+1}(z) = \frac{1}{2\pi i}\int_{|\zeta|=\frac{1}{2}|\alpha_r|} \frac{P_r(z)}{\zeta^{m_r+1}(\zeta - z)}\, d\zeta \;\forall z \in \mathcal{S}(0; |\alpha_r|/2).$$

Now, the function $p_r(z)$ is a polynomial obtained by expanding $P_r(z)$ into power series of z and taking a certain number of terms. So, we take

$$p_r(z) = \sum_{j=0}^{m_r} C_j^{(r)} z^j \;\forall z \in \mathbb{C}.$$

Now, for $z \in \mathcal{S}(0; |\alpha_r|/4)$, we have

$$\begin{aligned} \Big|P_r(z) - p_r(z)\Big| &= \Big|\varphi_{m_r+1}(z) z^{m_r+1}\Big| \le \Big|z\Big|^{m_r+1}\Big|\varphi_{m_r+1}(z)\Big| \\ &\le \Big|z\Big|^{m_r+1}\frac{1}{2\pi}\int_{|\zeta|=\frac{1}{2}|\alpha_r|}\frac{|P_r(z)|}{\Big|\zeta\Big|^{m_r+1}\Big|\zeta - z\Big|}\Big|d\zeta\Big| \end{aligned}$$

$$\leq \left|z\right|^{m_r+1}\frac{M_r}{2\pi}\int_{|\zeta|=\frac{1}{2}|\alpha_r|}\frac{\left|d\zeta\right|}{\left(\frac{1}{2}|\alpha_r|\right)^{m_r+1}\left(\frac{1}{2}|\alpha_r|-\frac{1}{4}|\alpha_r|\right)}$$
$$\leq \left(\frac{|\alpha_r|}{4}\right)^{m_r+1}\cdot\frac{M_r}{2\pi}\left(\frac{2}{|\alpha_r|}\right)^{m_r+1}\cdot\frac{4}{|\alpha_r|}\cdot 2\pi\frac{1}{2}|\alpha_r|$$
$$\leq 2^{-m_r}M_r < 2^{-r}\text{for } z\in\mathcal{S}(0;|\alpha_r|/2).$$

Let $R > 0$ be an arbitrarily large positive number. Since $|\alpha_n| \to \infty$ as $n \to \infty$, we can find $N_0 = N_0(R) \in \mathbb{N}$, so large that

$$|\alpha_n| > 2R \text{ when } n \geq N_0.$$

Since $\mathcal{S}(0; R) \subset \mathcal{S}(0; |\alpha_r|/2)$ for all $n > N_0$; $\mathcal{S}(0, R)$ contains none of the points $\alpha_{N_0+1}, \alpha_{N_0+2}, \cdots$. Thus by comparison test $\sum_{r=N_0}^{\infty}\left[P_r(z)-p_r(z)\right]$ converges absolutely and uniformly on every compact subset $\mathcal{S}(0; R)$. As every compact subset $\mathbb{C}/\{\alpha_r : r \in \mathbb{N}\}$ is contained in $\mathcal{S}(0; R)$, for some $R > 0$, we conclude that $\sum_{r=N_0}^{\infty}\left[P_r(z) - p_r(z)\right]$ converges uniformly on every compact subset of $\mathbb{C}/\{\alpha_r : r \in \mathbb{N}\}$.

Now the finite sum

$$f(z) - \sum_{r=N_0}^{\infty}\left[P_r(z) - p_r(z)\right] = \sum_{r=1}^{N_0}\left[P_r(z) - p_r(z)\right]$$

is a rational function whose poles in $\mathcal{S}(0; R)$ are those points of the sequence $\langle\alpha_r\rangle$ that lies in $\mathcal{S}(0; R)$. Thus, there exists a meromorphic function

$$f(z) = e(z) + \sum_{r=1}^{\infty}\left[P_r(z) - p_r(z)\right]\forall z \in \mathbb{C}/\{\alpha_r : r \in \mathbb{N}\}$$

whose singular part at α_r is $P_r(z), r \in \mathbb{N}$, where $e(z)$ is an arbitrary entire entire function. Since R can be chosen arbitrarily large, the result follows. □

For practical and simplification purpose, we prove the following simple version of the Mittag–Leffler theorem that has wider applications.

Theorem 9.16. *Let*

(i) the meromorphic function f have simple poles in the finite z-plane at $a_1, a_2, a_3, \cdots$ where $0 < |a_1| \leq |a_2| \leq \cdots \leq |a_k| \leq \cdots$.

(ii) the respective residues of f at $a_1, a_2, a_3, \cdots$ be $b_1, b_2, b_3, \cdots$ i.e., $b_k =$ Res $\{f; a_k\}$.

(iii) C_N be a positive circle of radius R_N that does not pass through any of these poles and $|f(z)| < M$ for each N for $z \in C_N$, where M is independent of N and $R_N \to \infty$ as $N \to \infty$.

Then for all z except at these poles, the Mittag–Leffer expansion of f is given by

$$f(z) = f(0) + \sum_{n=1}^{\infty} b_n\left[\frac{1}{z-a_n} + \frac{1}{a_n}\right]. \tag{9.45}$$

Proof: Let $\zeta(\notin C_N)$ be not a pole of f lying inside C_N (Fig. 9.2) of radius R_N. Then the function $\psi(z) = \frac{f(z)}{z-\zeta}$ has a simple poles at each a_k and ζ. The residue at a_k of $\psi(z)$ is

$$\text{Res}\left\{\psi; a_k\right\} = \lim_{z\to a_k}(z-a_k)\psi(z) = \frac{b_k}{a_k-\zeta};\ k = 1,2,3,\cdots,$$

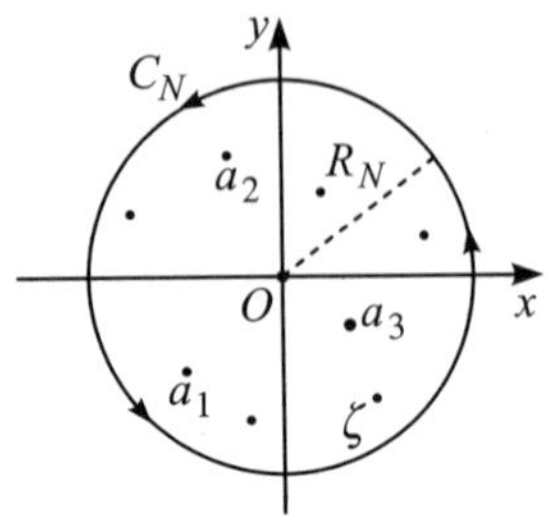

Figure 9.2 *The circle C_N of radius R_N*

and residue of $\psi(z)$ at $z = \zeta$ is

$$\text{Res}\left\{\psi; \zeta\right\} = \lim_{z\to\zeta}(z-\zeta)\psi(z) = \lim_{z\to\zeta} f(z) = f(\zeta).$$

Therefore, by Cauchy's residue theorem, we get

$$\sum \text{Res}\left\{\psi; C_N\right\} = \frac{1}{2\pi i}\oint_{C_N}\frac{f(z)}{z-\zeta}dz = f(\zeta) + \sum_k \frac{b_k}{a_k-\zeta}, \tag{9.46}$$

where the summation is taken over all poles of f inside the circle C_N and N is chosen large enough so that ζ lies inside C_N. Suppose that $f(z)$ is analytic at $z = 0$ then from Eq. (9.46) we get

$$\frac{1}{2\pi i}\oint_{C_N}\frac{f(z)}{z}dz = f(0) + \sum_k \frac{b_k}{a_k}. \tag{9.47}$$

Subtracting (9.47) from (9.46), we get

$$\begin{aligned} f(\zeta) - f(0) + \sum_k b_k\left[\frac{1}{a_k-\zeta} - \frac{1}{a_k}\right] &= \frac{1}{2\pi i}\oint_{C_N}\left[\frac{1}{z-\zeta} - \frac{1}{z}\right]f(z)dz \\ &= \frac{\zeta}{2\pi i}\oint_C \frac{f(z)}{z(z-\zeta)}dz. \end{aligned} \tag{9.48}$$

Now , since $|z-\zeta| \geq |z| - |\zeta| = R_N - |\zeta|$ and for z on $C_N : |z| = R_N$ we have $|f(z)| \leq M$.

$$\therefore \left|\oint_{C_N}\frac{f(z)}{z(z-\zeta)}dz\right| \leq \frac{M\cdot 2\pi R_N}{R_N(R_N-|\zeta|)}.$$

Now, taking $N \to \infty$ and therefore $R_N \to \infty$, it follows that

$$\lim_{N\to\infty}\oint_{C_N}\frac{f(z)}{z(z-\zeta)}dz = 0.$$

Hence from Eq. (9.48), taking $N \to \infty$, we get

$$f(\zeta) = f(0) + \sum_n b_n\left[\frac{1}{\zeta-a_n} + \frac{1}{a_n}\right].$$

This result is true for all ζ inside C_N, which are different from a_n; $n = 1,2,3,\cdots$, as obtained above. So we can write the result (9.45). □

Example 9.7.2. *Prove that* $\cot z = \frac{1}{z} + \sum_n\left(\frac{1}{z-n\pi} + \frac{1}{n\pi}\right)$, *where the summation extends over* $n = \pm1, \pm2, \cdots$.

Solution. Let $f(z) = \cot z - \frac{1}{z} = \frac{z\cos z - \sin z}{z \sin z}$. Then $f(z)$ has simple poles at $z = n\pi$; $n = \pm 1, \pm 2, \cdots$, and the residue at these poles is

$$\begin{aligned}\text{Res}[f(z), n] &= \lim_{z\to n\pi}(z-n\pi)\frac{z\cos z - \sin z}{z\sin z}\\ &= \lim_{z\to n\pi}\left(\frac{z-n\pi}{\sin z}\right)\lim_{z\to n\pi}\left(\frac{z\cos z-\sin z}{z}\right)\\ &= \lim_{z\to n\pi}\left(\frac{z-n\pi}{\sin z}\right)\lim_{z\to n\pi}\left(\cos z - \frac{\sin z}{z}\right) = 1;\ n\in\mathbb{Z}-\{0\}.\end{aligned}$$

At $z = 0$, $f(z)$ has removable singularity as

$$f(z) = \lim_{z\to 0}\left(\cot z - \frac{1}{z}\right) = \lim_{z\to 0}\frac{z\cos z-\sin z}{z\sin z} = 0.$$

Now, construct contour C_N as the square as in the Fig. 7.15. Now, $|\cot z| < 2$ and also, $(1/z)\to 0$ on C_n as $n\to\infty$. Thus, $f(z)$ is bounded on the square C_N for all values of N. Thus, all the conditions of Mittage–Leffler expansion theorem (9.45) are satisfied. Thus

$$\begin{aligned}f(z) &= f(0) + \sum_{n=1}^{\infty} b_n\left(\frac{1}{z_n}+\frac{1}{z-z_n}\right) + \sum_{n=-1}^{-\infty} b_n\left(\frac{1}{z_n}+\frac{1}{z-z_n}\right)\\ &= 0 + \sum_{n=1}^{\infty}\left(\frac{1}{n\pi}+\frac{1}{z-n\pi}\right) + \sum_{n=-1}^{-\infty}\left(\frac{1}{n\pi}+\frac{1}{z-n\pi}\right)\\ &= \sum_{n=1}^{\infty}\left[\left(\frac{1}{n\pi}+\frac{1}{z-n\pi}\right)+\left(\frac{1}{-n\pi}+\frac{1}{z+n\pi}\right)\right]\\ &= \sum_{n=1}^{\infty}\frac{2z}{z^2-n^2\pi^2};\ z\in\mathbb{C}-\mathbb{Z}.\end{aligned}$$

Therefore, $\cot z = \dfrac{1}{z} + 2z\displaystyle\sum_{n=1}^{\infty}\frac{1}{z^2-n^2\pi^2}$.

Exercises

Multiple-choice Questions

1. Suppose f and g are entire functions and $g(z)\neq 0$ for all $z\in\mathbb{C}$. If $|f(z)|\le|g(z)|$, then we conclude that

   (a) $f(z)\neq 0$ for all $z\in\mathbb{C}$ (b) f is a constant function
   (c) $f(0) = 0$ (d) for some $c\in\mathbb{C}$, $f(z) = cg(z)$

2. Let f and g be meromorphic functions on $\mathbb{C}$. If f has a zero of order k at $z = a$, then g has a pole of order m at $z = 0$. Then $g(f(z))$ has

   (a) a zero of order km at $z = a$ (b) a pole of order km at $z = a$
   (c) a zero of order $|k-m|$ at $z = a$ (d) a pole of order $|k-m|$ at $z = a$

3. Let f be an entire function such that, for some constant α, $|f(z)| \leq \alpha|z|^3$ for $|z| \geq 1$ and $f(z) = f(iz)$ for all $z \in \mathbb{C}$. Then
   (a) $f(z) = \alpha z^3$ for all $z \in \mathbb{C}$ (b) $f(z)$ is constant
   (c) $f(z)$ is a quadrilateral polynomial (d) No such $f(z)$ exists

4. Let f be an entire function. Which of the following statements are correct?
   (a) f is constant if the range of f is contained in a straight line
   (b) f is constant if f has uncountably many zeros
   (c) f is constant if f is bounded on $\{z \in \mathbb{C} : \text{Re}(z) \leq 0\}$
   (d) f is constant if the real part of f is bounded

5. Let $f(z) = \sum_{n=0}^{\infty} a_n z^n$ be an entire function and r be a positive number. Then
   (a) $\sum_{n=0}^{\infty} |a_n|^2 r^{2n} \leq \sup\limits_{|z|=r} |f(z)|^2$ (b) $\sup\limits_{|z|=r} |f(z)|^2 \leq \sum_{n=0}^{\infty} |a_n|^2 r^{2n}$
   (c) $\sum_{n=0}^{\infty} |a_n|^2 r^{2n} \leq \frac{1}{2\pi}\int_0^{2\pi} \left|f(re^{i\theta})\right|^2 d\theta$ (d) $\sup\limits_{|z|=r} |f(z)|^2 \leq \frac{1}{2\pi}\int_0^{2\pi} \left|f(re^{i\theta})\right|^2 d\theta$

6. Let f be an entire function on $\mathbb{C}$. Let $g(z) = \overline{f(\bar{z})}$. Which of the following statements is/are correct:
   (a) If $f(z) \in \mathbb{R}$ for all $z \in \mathbb{R}$, then $f = g$.
   (b) If $f(z) \in \mathbb{R}$ for all $z \in \{z : \text{Im}z = 0\} \cup \{z : \text{Im}z = a\}$, for some $a > 0$, then $f(z + ia) = f(z - ia)$ for all $z \in \mathbb{C}$.
   (c) If $f(z) \in \mathbb{R}$ for all $z \in \{z : \text{Im}z = 0\} \cup \{z : \text{Im}z = a\}$, for some $a > 0$, then $f(z + 2ia) = f(z)$ for all $z \in \mathbb{C}$.
   (d) If $f(z) \in \mathbb{R}$ for all $z \in \{z : \text{Im}z = 0\} \cup \{z : \text{Im}z = a\}$, for some $a > 0$, then $f(z+ia) = f(z)$ for all $z \in \mathbb{C}$.

7. Let $f : \mathbb{C} \to \mathbb{C}$ be an entire function. Suppose that $f = u + iv$. Then f is constant if:
   (a) $\{u(x, y) : z = x + iy \in \mathbb{C}\}$ is bounded (b) $\{v(x, y) : z = x + iy \in \mathbb{C}\}$ is bounded
   (c) $\{u(x, y) + iv(x, y) : z = x + iy \in \mathbb{C}\}$ is bounded
   (d) $\{u^2(x, y) + iv^2(x, y) : z = x + iy \in \mathbb{C}\}$ is bounded

8. Let f be an entire function on $\mathbb{C}$ and let Ω be a bounded open subset of $\mathbb{C}$. Let $S = \{\text{Re}f(z) + \text{Im}f(z) : z \in \Omega\}$. Which of the following statements is/are necessarily true?
   (a) S is an open set in $\mathbb{R}$ (b) S is an closed set in $\mathbb{R}$
   (c) S is an open set in $\mathbb{C}$ (d) S is a discrete set in $\mathbb{R}$

9. Let $K \subseteq \mathbb{C}$ be a bounded set. Let $H(\mathbb{C})$ denote the set of all entire functions and let $C(K)$ denote the set of all continuous functions on K. Consider the restriction map $r : H(\mathbb{C}) \to C(k)$ given by $r(f) = f_K$. Then r is injective if K is
   (a) compact (b) connected (c) uncountable (d) finite

10. Let f be an entire function. Suppose, for each $a \in \mathbb{R}$, there exists at least one coefficient c_n in $f(z) = \sum_{n=0}^{\infty} c_n(z - a)^n$ which is zero. Then
    (a) $f^{(n)}(0) = 0$ for infinitely many $n \geq 0$ (b) $f^{(2n)}(0) = 0$ for every $n \geq 0$
    (c) $f^{(2n+1)}(0) = 0$ for every $n \geq 0$ (d) there exists $k \geq 0$ such that $f^{(n)}(0) = 0$ for every $n \geq k$

11. Let f be a non-constant entire function. Which of the following properties is possible for f for $z \in \mathbb{C}$?
    (a) $\text{Re}f(z) = \text{Im}f(z)$ (b) $|f(z)| < 1$ (c) $\text{Im}f(z) < 0$ (d) $f(z) \neq 0$

12. Consider the functions $f, g : \mathbb{C} \to \mathbb{C}$ defined by $f(z) = e^z$, $g(z) = e^{iz}$. Let $S = \{z \in \mathbb{C} : \text{Re}z \in [-\pi, \pi]\}$. Then

    (a) f is an onto entire function (b) g is a bounded function on $\mathbb{C}$
    (c) f is bounded on S (d) g is bounded on S

13. The sum of the residues at all the poles of $f(z) = \frac{\cot \pi z}{(z+a)^2}$, where a is a constant ($a \neq 0, \pm 1, \pm 2, \cdots$) is

    (a) $\frac{1}{\pi}\sum_{n=-\infty}^{\infty} \frac{1}{(n+a)^2} - \pi \text{cosec}^2 \pi a$ (b) $-\frac{1}{\pi}\sum_{n=-\infty}^{\infty} \frac{1}{(n+a)^2} + \pi \text{cosec}^2 \pi a$
    (c) $-\frac{1}{\pi}\sum_{n=-\infty}^{\infty} \frac{1}{(n+a)^2} - \pi \text{cosec}^2 \pi a$ (d) $\frac{1}{\pi}\sum_{n=-\infty}^{\infty} \frac{1}{(n+a)^2} + \pi \text{cosec}^2 \pi a$

14. Which of the following functions f are entire functions and have simple zeros at $z = ik$ for all $k \in \mathbb{R}$?

    (a) $f(z) = a_0 + a_1 z + \cdots + a_n z^n$ for some $n \geq 1$ and some $a_0, a_1, \cdots, a_n \in \mathbb{C}$
    (b) $f(z) = a \sin 2\pi i z$, for some $a \in \mathbb{C}$
    (c) $f(z) = b \sin 2\pi(iz - \frac{1}{4})$, for some $b \in \mathbb{C}$ (d) $f(z) = e^{cz}$, for some $c \in \mathbb{C}$

15. Let f be an entire function. If $\text{Im}f \geq 0$, then

    (a) $\text{Re}f$ is constant (b) f is constant (c) $f = 0$ (d) f' is non-zero constant

16. Let f be an entire function such that $\lim_{|z| \to \infty} |f(z)| = \infty$. Then

    (a) $f(1/z)$ has an essential singularity at 0 (b) f cannot be a polynomial
    (c) f has finitely many zeros (d) $f(1/z)$ has a pole at 0

17. Let $f : \mathbb{C} \to \mathbb{C}$ be an entire function and let $g : \mathbb{C} \to \mathbb{C}$ be defined by $g(z) = y(z) - y(z+1)$ for $z \in \mathbb{C}$. Which of the following statements are true?

    (a) if $f(1/n) = 0$ for all positive integer n, then f is a constant function
    (b) if $f(n) = 0$ for all positive integers n, then f is a constant function
    (c) if $f(\frac{1}{n}) = f(1 + \frac{1}{n})$ for all positive integers n, then g is a constant function
    (d) if $f(n) = f(1+n)$ for all positive integers n, then g is a constant function

18. Let f be an entire function. If $\text{Re}f$ is bounded then

    (a) $\text{Im}f$ is constant (b) f is constant (c) $f = 0$ (d) f' is a non zero constant

Answers

1(d); 2(c); 3(b); 4(a,b,d); 5(a,c); 6(a,b,c); 7(c); 8(b); 9(c); 10(d); 11(a); 12(b); 13(a); 14(d); 15(b); 16(d); 17(a); 18(a,b)

Review Questions [Short-answer Type]

1. Show that the following infinite products converge, and find their values:

   (a) $\prod_{k=1}^{\infty}(1 + \frac{1}{k(k+2)})$ (b) $\frac{2}{1} \cdot \frac{5}{4} \cdots \frac{k^2+1}{k^2} \cdots$ (c) $\prod_{k=2}^{\infty}(1 - \frac{1}{k^2})$
   (d) $\prod_{k=3}^{\infty}(\frac{k^2-1}{k^2-4})$ (e) $\prod_{k=1}^{\infty}(1 - \frac{2}{(k+1)(k+2)})$

2. Find the regions of absolute convergence for the following products:

   (a) $\prod_{k=1}^{\infty}(1 - z^k)$ (b) $\prod_{k=0}^{\infty}(1 + z^{2k})$ (c) $\prod_{k=0}^{\infty}(1 + c_k z)$; $\sum |c_k|$ convergent

3. Show that the infinite product $\prod_{k=1}^{\infty}(1 + \frac{z^k}{k})$ is uniformly and absolutely convergent for $|z| \leq r$, where $0 < r < 1$.

4. Show that the infinite product $\prod_{k=1}^{\infty}(1 + e^{kz})$ diverges for $\text{Re}z \geq 0$ and is uniformly and converges absolutely for $\text{Re}z < 0$.

5. Prove that if w is a complex number with $|w| \leq 1/2$, then $\frac{2}{3}|w| \leq |\log(1+w)| \leq 2w$
6. Compute the (a) derivative and (b) the logarithmic derivative of $E(z;p)$
7. Find an entire function (given by a Weierstrass product) that has a zero of order
   (a) 1 at $\sqrt{n}$ for $n = 1, 2, 3, \cdots$ and no other zeros
   (b) 2 at $\sqrt{n}$ for $n = 1, 2, 3, \cdots$ and no other zeros
   (c) n at n^2 for $n = 1, 2, 3, \cdots$ and no other zeros
8. Show that a polynomial has finite order 0
9. Show that $\sin z$, $z^{-1} \sin z$ and $\cos z$ all have finite order 1.
10. Prove that an analytic function cannot be bounded in the neighbourhood of an isolated singularity
11. Show that $\sum \frac{z^n}{(n!)^\alpha}; \alpha > 0$ is an entire function of order $1/\alpha$
12. Find the canonical product of $\cos z$
13. Find the factorization for $\sinh z$ and $\cosh z$
14. Determine the order of the functions:
    (a) $e^{p(z)}$, where $p(z) = a_0 + a_1 z + \cdots + a_n z^n;\ a_n \neq 0$ (b) $e^z \sin z$ (c) $\sinh z$
    (d) e^{z^n} (e) e^{e^z} (f) $\cos\sqrt{z}$ (g) $\sum_{n=0}^{\infty} (\frac{e\alpha}{n})^{n/\alpha} z^n; \alpha > 0$ (h) $e^{\sin z}$
15. Find an entire function with finite order $1/2$
16. Find the convergence exponent of the zeros of $\sin z$
17. Show that if $a > 1$, the entire function $\prod_{n=1}^{\infty}(1 - \frac{z}{n^a})$ is of order a^{-1}
18. Find the exponent of convergence of the following sequences:
    (a) $\langle 2^k \rangle$ (b) $\langle k^r \rangle; r > 0$ (c) $\langle \log k \rangle$
19. Given an arbitrary non-negative real number μ, show that there is a sequence $\langle z_k \rangle$ with exponent of convergence μ.
20. Does the order of an entire function necessarily have to be the same as the exponent of convergence of its sequence of zeros? Justify your answer.
21. Find the infinite product representation of the following entire functions:
    (a) $e^z - 1$ (b) $e^z - e^{z_0}$ (c) $\cos \pi z$
22. Show that if $h : \mathbb{R}^2 \to \mathbb{R}$ and $f = 2h^3 + ih$ is an entire function, then h is constant.
23. What does Hadamard's factorization theorem say about an entire function of order $\lambda < 1$?
24. If a non-constant entire function of finite order λ has zeros at the points $i\sqrt{n}$, what are the possible values for λ?
25. Prove that if f is an entire function of order λ and λ is not an integer, then f has infinitely many zeros.
26. If $f(z)$ is an entire function, show that $f'(z)$ and $f(z)$ have the same order.
27. Show that the orders of the following canonical products are as given below:
    (a) $\prod_{k=1}^{\infty}(1 - \frac{z}{2^k}) = 0$ (b) (a) $\prod_{k=1}^{\infty}[1 - \frac{z}{n^5(\log n)^2}] = \frac{1}{5}$
28. Is there an entire function of order $3/2$ which has the integers as its list of zeros, counting multiplicity?
29. Suppose f is an entire function of finite order λ and μ is the exponent of convergence of the list of zeros of f. Prove that if $\mu < \lambda$, then λ is an integer.
30. Find an entire function which has a zero at each point of the sequence $\langle \log n \rangle$ for $n = 1, 2, \cdots$. Does it have finite order?

31. Given an entire function $f(z)$, show that $M(r) = \max_{|z|=r} |\{f(z)|\}$ is a continuous function of r.

32. Construct a meromorphic function on the complex plane whose poles at the Gaussian integers $m + in$ with residue 1.

33. Find all functions that are analytic everywhere in the extended complex plane $\mathbb{C}_\infty$.

Review Questions [Long-answer Type]

1. Define $\langle z_k \rangle$ by $z_k = -\frac{1}{\sqrt{k}}$ if k is odd, and $z_k = \frac{1}{\sqrt{k}} + \frac{1}{k} + \frac{1}{k\sqrt{k}}$ if k is even. Show that $\prod(1+z_k)$ converges, while $\sum z_k$ and z_k^2 diverges.

2. Does the infinite product $\prod_{k=1}^{\infty}(1+\frac{1}{k})$ converge? How about the product $\prod_{k=1}^{\infty}(1+\frac{1}{k^{3/2}})$?

3. Show that
   (a) $\prod_{k=1}^{\infty}(1+\frac{1}{k^2} = \frac{1}{2\pi}(e^{\pi}-e^{-\pi})$ (b) $\prod_{k=0}^{\infty}(1+z^{2^k}) = \frac{1}{1-z}$ for $|z|<1$

4. Show that the infinite product
   (a) $\prod_{k=1}^{\infty}(1-\frac{z}{k})e^{z/k}$ converges uniformly on a compact subsets of the plane.
   (b) $\prod_{k=1}^{\infty}(1+\frac{z}{k})e^{-z/k}$ converges absolutely and uniformly on every compact subset of $\mathbb{C}$.

5. **(Blaschke product :)** Let the sequence $\langle z_n \rangle$ be such that $0 < |z_n| < 1$ for all n and $\sum_{k=1}^{\infty}(1-|z_k|)$ converges. Show that the infinite product $f(z) = \prod_{k=1}^{\infty} \frac{z_k - z|z_k|}{1-\bar{z}_k z z_k}$ converges uniformly on the disc $\mathcal{S}[0,r]$, $0<r<1$. Hence, $f(z)$ is analytic on $\mathcal{S}(0,r)$, $z_n; n \in \mathbb{N}$ are its zeros, $f(z)$ has no zeros, and $|f(z)| \le 1$.

6. Establish one of the following formulae, and deduce from it the other using logarithmic differentiation:
$$e^z - 1 = ze^{z/2}\prod_{k=1}^{\infty}\left(1+\frac{z^2}{4\pi^2k^2}\right); \qquad \frac{1}{e^z-1} = \frac{1}{z} - \frac{1}{2} + 2z\sum_{k=1}^{\infty}\frac{1}{z^2+4\pi^2k^2}$$

7. Prove that a necessary and sufficient condition for the absolute convergence of the infinite product $\prod_{n=1}^{\infty}(1+z_n)$ is that the series $\sum_{n=1}^{\infty} z_n$ is absolute convergent.

8. Show that the infinite product
   (a) $\{(1-\frac{z}{1})e^z\}\{(1+\frac{z}{1})e^{-z}\}\{(1-\frac{z}{2})e^{z/2}\}\{(1+\frac{z}{2})e^{-z/2}\}\cdots$, and (b) $\prod_{k=1}^{\infty}(1-\frac{z^2}{k^2})$
   are uniformly and absolutely convergent in any closed and bounded region, where z is not a positive or negative integer.

9. Find the order of the zero of f at 0 and the factorization of f at 0, where $f(z) =$ (a) $\sin z - z$ and (b) $\cos z - 1 + \frac{z^2}{2}$.

10. Let $\langle z_n \rangle$ be a sequence of complex numbers with $z_k \to \infty$ as $k \to \infty$. Prove that, there exists an entire function f whose zeros are $\{z_k : k \in \mathbb{N}\}$.

11. Prove that if
    (a) $f(z) = \pi z\prod_{k=1}^{\infty}(1-\frac{z^2}{k^2})$, the logarithmic derivative of f, is an odd meromorphic function which is periodic of period 1.
    (b) Set $g(z) = \frac{\sin(\pi z)}{f(z)}$, where, $f(z)$ as in (11a). Then g is an entire function with no zeros and hence has a logarithm h which is entire. Then $\sin(\pi z) = f(z)e^{h(z)}$.
    (c) f, g and h are the functions of (11b), then the logarithmic derivative of g is $h'(z) = \pi\cot\pi z - \frac{f'(z)}{f(z)}$.

12. Given $f_1(z)$ and $f_2(z)$ are two entire functions of orders ρ_1 and ρ_2 respectively, show that
    (a) the order of $f_1(z)\cdot f_2(z)$ is $\le \max\{\rho_1,\rho_2\}$ (b) the order of $f_1(z)+f_2(z)$ is $\le \max\{\rho_1,\rho_2\}$
    and equality occurs if $\rho_1 \neq \rho_2$.
13. Suppose that $f(z)$ and $g(z)$ are entire functions such that $(f\circ g)(z)$ is a polynomial. Show that both $f(z)$ and $g(z)$ are polynomials.
14. If f is an entire function, show that $f = g^n$ for some entire function g if and only if the order of each zero of f is divisible by n.
15. Suppose f is an entire function such that $\langle z_k\rangle$ is a list of its non-zero zeros, counting multiplicity, and suppose that $\sum_{k=1}^{\infty}\frac{1}{|z_k|} < \infty$. Describe the simplest Weierstrass factorization of f.
16. Suppose f is an entire function which satisfies the inequality $|f(z)| \le |z|^{|z|}$ for $|z|$ sufficiently large. Prove that f has finite order at most 1.
17. If f is an entire function of order $\lambda(f)$, k is a non-negative integer, and $g(z) = f(z^k)$, then prove that $\lambda(g) = k\lambda(f)$, where $\lambda(g)$ is the order of g.
18. Prove that if $g(z)$ is an even entire function of finite order λ and $f(z) = g(\sqrt{\lambda})$, then f is an entire function of finite order $\lambda/2$. In particular, show that $\cos(\sqrt{z})$ has order $1/2$.
19. Prove that, if f_1 and f_2 are two entire functions with no common zeros, then there exist entire functions g_1 and g_2 such that $g_1f_1 + g_2f_2 = 1$.
20. Prove that if f and g are entire functions of finite order λ and if $f(z_k) = g(z_k)$ on a sequence which satisfies $\sum_{k=1}^{\infty}\frac{1}{|z_k|^t} = \infty$ for some $t > \lambda$, then $f(z) = g(z)$ identically.
21. Show that an even entire function of order 1 has the form $Cz^m\prod_k(1-\frac{z^2}{z_k^2})$, where m is even, C is a non-zero constant, and the sequence $\langle z_1, -z_1, z_2, -z_2, \cdots\rangle$ is a list of the zeros of f, counting multiplicity. What is the exponent of convergence for the sequence of zeros?
22. Derive the product formula of $\sin\pi z$ from the Weierstrass formula.
23. Prove that the Weierstrass product converges whenever $n_k = [\log k]$.
24. Show that if $P_k(z)$ is a polynomial of degree k such that $P_k(0) = 1$ and $P_k(z)$ has no zeros in the disc $S(0,k^3)$, then $\prod P_k(z)$ converges normally.
25. If α_n are positive roots of the equation $\tan z = z$, show that

$$\frac{z\sin z}{\sin z - z\cos z} = \frac{3}{z} + \sum_{n=1}^{\infty}\frac{2z}{z^2-\alpha^2};\ \text{where } \Big(n-\frac{1}{2}\Big)\pi < \alpha_n < \Big(n+\frac{1}{2}\Big)\pi.$$

26. Find a meromorphic function with poles of order 2 at $1, \sqrt{2}, \sqrt{3}, \cdots$ such that the residue at each pole is 0 and $\lim_{z\to\sqrt{n}}(z-\sqrt{n})^2 f(z) = 1$.
27. Let $f(z)$ be an entire function having zeros $\{z_n : n \in \mathbb{N}\}$; each zero is counted as often as its multiplicity ($z_n \neq 0\ \forall\, n$). Then prove that

$$f(z) = e^{g(z)}\prod_{n=1}^{\infty}\Big(1-\frac{z}{z_n}\Big)\exp\Big[\frac{z}{z_n}+\frac{1}{2}\Big(\frac{z}{z_n}\Big)^2+\cdots+\frac{1}{m_n}\Big(\frac{z}{z_n}\Big)^{m_n}\Big]$$

for some sequence $\langle m_n\rangle$ of positive integers and $g(z)$ is an entire function. Further, if f has a zero of order k at $z = 0$, then show that

$$f(z) = z^k e^{g(z)}\prod_{n=1}^{\infty}\Big(1-\frac{z}{z_n}\Big)\exp\Big[\frac{z}{z_n}+\frac{1}{2}\Big(\frac{z}{z_n}\Big)^2+\cdots+\frac{1}{m_n}\Big(\frac{z}{z_n}\Big)^{m_n}\Big].$$

28. **Jensen's inequality** : Let $f(z)$ be an entire function with $f(0) \neq 0$ and $z_1, z_2, \cdots$ be the zeros of $f(z)$ such that $|z_1| \leq |z_2| \leq \cdots$, multiple zeros being repeated according to their multiplicities. If $|z_N| \leq r < |z_{N+1}|$, prove that

$$\log \frac{r^N}{|z_1| \cdot |z_2| \cdots |z_N|} = \int_0^r \frac{n(t)}{t}\, dt \leq \log M(r) - \log |f(0)|$$

29. Use the infinite product expansion for $\sin(\pi z)$ to show that the Wallis's formula:

$$\prod_{k=1}^{\infty} \frac{(2k)^2}{(2k-1)(2k+1)} = \lim_{n\to\infty} \frac{2. \cdot 2 \cdot 4 \cdot 4 \cdot 6 \cdot 6 \cdots (2n) \cdot (2n)}{1 \cdot 3 \cdot 3 \cdot 5 \cdot 5 \cdot 7 \cdots (2n-1)(2n-1)}$$

converges to $\pi/2$. Use it to show that $\lim_{n\to\infty} \frac{[n!]^2}{(2n)!} \frac{2^{2n}}{\sqrt{n}} = \sqrt{\pi}$

30. Construct a meromorphic function with given zeros and poles.

31. Apply Mittage–Leffler theorem to show that

(i) $\text{cosec}\, z = \frac{1}{z} - 2z \sum_{n=1}^{\infty} \frac{(-1)^{n-1}}{z^2-(n\pi)^2}$ (ii) $\text{sech}\, z = \sum_{k=1}^{\infty} \frac{(2n-1)\pi}{z^2+(n-\frac{1}{2})^2\pi^2}$

(iii) $\tanh z = \sum\limits_{n=1}^{\infty} \frac{2z}{z^2+(n-\frac{1}{2})^2\pi^2}$ (iv) $\coth z = \frac{1}{z} + 2z \sum\limits_{n=1}^{\infty} \frac{1}{z^2+(n\pi)^2}$

(v) $\text{cosech}\, z = \frac{1}{z} - 2z \sum_{n=1}^{\infty} \frac{1}{z^2+(n\pi)^2}$ (vi) $\sec z = \sum_{k=1}^{\infty} (-1)^n \frac{(2n-1)\pi}{z^2-(n-\frac{1}{2})^2\pi^2}$

32. Prove that

(i) $\sum_{n=1}^{\infty} \frac{(-1)^n}{n^2} = -\frac{\pi^2}{12}$ (ii) $\sum_{n=1}^{\infty} \frac{1}{(n^2+1)^2} = \frac{\pi}{4} \coth\pi + \frac{\pi^2}{4} \text{cosech}^2\pi - \frac{1}{2}$

(iii) $\sum_{n=1}^{\infty} \frac{1}{n^4} = \frac{\pi^4}{90}$ (iv) $\frac{1}{1^4} + \frac{1}{3^4} + \frac{1}{5^4} + \cdots = \frac{\pi^4}{96}$ (v) $\sum_{n=1}^{\infty} \frac{1}{n^6} = \frac{\pi^6}{945}$

33. Show that if a is real and non-integer and $0 < r < 1$,

(a) $\sum_{k=-\infty}^{\infty} \frac{1}{(k+a)^2} = \pi^2 \cos^2 \pi a$ (b) $\sum_{k=-\infty}^{\infty} \frac{1}{k^2+a^2} = \frac{\pi}{a} \coth \pi a$

(c) $\sum_{k=-\infty}^{\infty} \frac{k^2-a^2}{(k^2+a^2)^2} = -\pi^2 \cosh^2 \pi a$ (d) $\sum_{k=-\infty}^{\infty} \frac{1}{(k-r)^2+a^2} = \frac{\pi}{2a} \frac{\sinh 2a\pi}{\sin^2 \pi r + \sinh^2 \pi a}$

(e) $\sum_{k=-\infty}^{\infty} \frac{(k-r)^2-a^2}{[(k-r)^2+a^2]^2} = \frac{\pi^2}{2} \frac{1-\cos 2\pi r \cosh 2a\pi}{(\sin^2 \pi r+\sinh^2 \pi a)^2}$

34. Prove that the following:

(i) $\prod_{n=1}^{\infty}(1+z^{2^n}) = (1-z)^{-1}; |z|<1$ (ii) $\prod_{n=2}^{\infty}(1+\frac{2n+1}{n^2-1}) = \frac{1}{3}$

(iii) $\prod_{n=2}^{\infty}(1-\frac{2n+1}{n^2+2n}) = 3$ (iv) $\prod_{n=2}^{\infty}[1+\frac{(-1)^n}{n}] = 1$

35. Prove that the following:

(i) $\cos z = \prod_{k=1}^{\infty}(1-\frac{4z^2}{(2k-1)^2\pi^2})$ (ii) $\cos(\frac{\pi z}{4}) - \sin(\frac{\pi z}{4}) = \prod_{n=1}^{\infty}[\frac{1+(-1)^n}{2n-1}]$

(iii) $\sin z = z\prod_{k=1}^{\infty}(1-\frac{z^2}{k^2\pi^2})$ (iv) $\cosh z = \prod_{k=1}^{\infty}(1+\frac{4z^2}{(2k-1)^2\pi^2})$

(v) $\sinh z = z\prod_{k=1}^{\infty}(1+\frac{z^2}{k^2\pi^2})$

36. Apply Mittage–Leffler theorem to show that $\tan z = -\sum_{n=1}^{\infty} \frac{2z}{z^2-(n-\frac{1}{2})^2\pi^2}$ and hence show that $\frac{1}{1^2} + \frac{1}{3^3} + \frac{1}{5^2} + \cdots = \frac{\pi^2}{8}$.

37. Prove that, given a sequence $\langle z_k \rangle$ of complex numbers converging to infinity and a sequence $\langle n_k \rangle$ of integers, there is an entire function f with given values for f and its derivatives up to order n_k at z_k for each k. [**Hint:** Use the Mittag–Leffler and Weierstrass theorems, together.]

10 Analytic Continuation

In the theory of analytic continuation, we seek to construct a function which is analytic in a larger domain and which coincides with the given function in the original domain.

In this chapter we shall introduce the idea of direct and indirect analytic continuation of an analytic function. The concepts of analytic continuation by means of power series, complete analytic function, natural boundary, and analytic continuation along a curve will be explained with the help of examples. Analytic continuation of multi-valued function and Monodromy theorem will also be discussed.

10.1 Analytic Continuation

The idea of analytic continuation rests on the notion of analytic function. Following the *uniqueness theorem,* stated as follows:

If two functions $f(z)$ and $g(z)$, analytic on a region $\mathcal{D}$, are such that $f(z) = g(z)$ on a subdomain $\mathcal{D}_1 \subset \mathcal{D}$ having a limit point in $\mathcal{D}$, then $f(z) = g(z)\ \forall z \in \mathcal{D}$.

we know that if two analytic functions agree in some small neighbourhood of a point situated in their common region of analyticity $\mathcal{D}$, they coincide everywhere in $\mathcal{D}$. Thus, an analytic continuation is a process to extend the 'domain of analyticity' i.e., to determine if there is a function $g(z)$ analytic in a larger domain whose values agree with those of $f(z)$ for $z \in \mathcal{D}$.

10.1.1 Direct Analytic Continuation

Given an analytic function f in a domain $\mathcal{D} \subset \mathbb{C}$, we are to find the largest open connected set on which it becomes, an analytic function. Suppose we have two domains $\mathcal{D}_0$ and $\mathcal{D}_1$ (Fig 10.1) in $\mathbb{C}$, whose intersection $\mathcal{D}_{01} = \mathcal{D}_0 \cap \mathcal{D}_1$ is not empty, connected and is a domain.

Definition 10.1.1. [Analytic continuation of a function]: Let $f_0 : \mathcal{D}_0 \to \mathbb{C}$ and $f_1 : \mathcal{D}_1 \to \mathbb{C}$ be two single-valued analytic functions in domains $\mathcal{D}_0$ and $\mathcal{D}_1$, respectively, satisfying that

(i) $\mathcal{D}_0$ and $\mathcal{D}_1$ has a non-empty common part $\mathcal{D}_{01}$, which is connected;

(ii) $f_0(z) \equiv f_1(z)$ for all $z \in \mathcal{D}_{01}$, i.e., $f_0|_{\mathcal{D}_{01}} = f_1|_{\mathcal{D}_{01}}$.

Then f_1 is said to be an *direct analytic continuation* or an *analytic extension* of f_0 from $\mathcal{D}_0$ to $\mathcal{D}_1$. In other words, f_0 is said to be *analytically continuable* into $\mathcal{D}_1$. In such a case, we define $g : \mathcal{D}_0 \cup \mathcal{D}_1 \to \mathbb{C}$ by $g|_{\mathcal{D}_0} = f_0$ and $g|_{\mathcal{D}_1} = f_1$, i.e.,

$$g(z) = \begin{cases} f_0(z), & z \in \mathcal{D}_0 \\ f_1(z), & z \in \mathcal{D}_1. \end{cases} \tag{10.1}$$

We see that g is analytic in $\mathcal{D}_0 \cup \mathcal{D}_1$ and we say that g is obtained by gluing f_0 and f_1 along $\mathcal{D}_{01}$. f_0 and f_1 are called the *function elements* of g and is denoted by $\langle \mathcal{D}_0, f_0 \rangle$ and $\langle \mathcal{D}_1, f_1 \rangle$ respectively.

Thus $\langle \mathcal{D}_0 \cup \mathcal{D}_1, g \rangle$ is the *direct analytic continuation* or *holomorphic extension* of both $\langle \mathcal{D}_0, f_0 \rangle$ and $\langle \mathcal{D}_1, f_1 \rangle$.

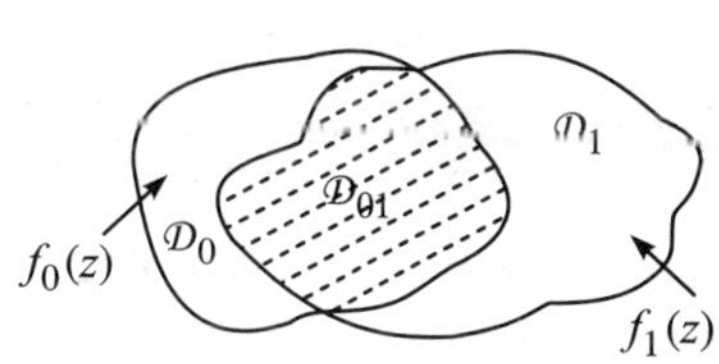

Figure 10.1 *Analytic continuation*

Theorem 10.1. [Uniqueness of direct analytic continuation]: *The direct analytic continuation, if it exists, is unique.*

Proof: Let $\mathcal{D}_0 \subset \mathbb{C}$ be a domain and let $f : \mathcal{D}_0 \to \mathbb{C}$ be an analytic function. Let us suppose that $\langle \mathcal{D}_1, g_1 \rangle$ and $\langle \mathcal{D}_1, g_2 \rangle$ be two direct analytic continuations of $\langle \mathcal{D}_0, f \rangle$ of the same function f via $\mathcal{D}_{01} = \mathcal{D}_0 \cap \mathcal{D}_1$. Then

(i) $g_1(z)$ is analytic in $\mathcal{D}_0$ and $f(z) = g_1(z)\ \forall z \in \mathcal{D}_{01}$.

(ii) $g_2(z)$ is analytic in $\mathcal{D}_1$ and $f(z) = g_2(z)\ \forall z \in \mathcal{D}_{01}$.

Thus, it follows from (i) and (ii) that $f(z) = g_1(z) = g_2(z)\ \forall z \in \mathcal{D}_{01}$, which implies

$$g_1(z) = g_2(z) \Rightarrow (g_1 - g_2)(z) = 0 \quad \forall z \in \mathcal{D}_{01}.$$

Now, g_1 and g_2 being analytic in $\mathcal{D}_1$ implies that $g_1 - g_2$ is analytic in $\mathcal{D}_1$.

$$\therefore (g_1 - g_2)(z) = 0 \text{ in } \mathcal{D}_{01} \subset \mathcal{D}_1.$$

Also, the function is analytic in $\mathcal{D}_1$. Thus,

$$(g_1 - g_2)(z) = 0\ \forall z \in \mathcal{D}_1 \Rightarrow g_1(z) = g_2(z)\ \forall z \in \mathcal{D}_1.$$

Accordingly, this maximal direct continuation is unique. □

We shall introduce the idea of direct analytic continuation by the following examples:

Example 10.1.1. (i) Let $\mathcal{D}_0 = \{z \in \mathbb{C} : |z| < 1\}$ be a domain. The infinite geometric series $f_0(z) = \sum_{n=0}^{\infty} z^n$ converges for $|z| < 1$ and its sum function $f_1(z) = (1 - z)^{-1}$ is analytic function for $z \in \mathcal{D}_1 = \mathbb{C}_\infty/\{1\}$. The geometric series diverges for $|z| \geq 1$. But we observe that

$$f_1(z) = f_0(z)\ \forall z \in \mathcal{D}_{01},$$

where, $\mathcal{D}_{01} = \{|z| < 1 \cap \mathbb{C}_\infty/\{1\}\} = \mathcal{D}_0 \cap \mathcal{D}_1$ is a domain. Thus, we may regard f_1 as determining an analytic continuation of f_0 from $\mathcal{D}_0$ into $\mathcal{D}_1$. Thus, the maximal unique direct analytic continuation is

$$\langle \mathcal{D}_1, f_1 \rangle = \langle \mathbb{C}_\infty/\{1\}, \frac{1}{1-z} \rangle.$$

It follows that f_1, which is analytic globally, is represented by a power series only locally.

(ii) Let $\mathcal{D}_0 = \{z \in \mathbb{C} : |z| < 1\}$ be a domain. The series $f_0(z) = \sum_{n=1}^{\infty} (-1)^{n-1} \frac{z^n}{n}$ converges in the disc $\mathcal{D}_0$ and its sum function $f_1(z) = \log(1 + z)$ is analytic for $z \in \mathcal{D}_1 = \mathbb{C}/\{-1, \infty\}$. Thus, the maximal unique direct analytic continuation is

$$\langle \mathcal{D}_1, f_1 \rangle = \langle \mathbb{C}/\{-1, \infty\}, \log(1 + z) \rangle.$$

(iii) Let $\mathcal{D}_0 = \{z \in \mathbb{C} : \text{Re} z > 1\}$, being a right half plane, be a domain. Let $f_0(z) = \zeta(z)$, where

$$\zeta(z) = \sum_{n=1}^{\infty} \frac{1}{n^z} = \sum_{n=1}^{\infty} \frac{1}{e^{z \ln n}},$$

where each $e^{z\ln n} = (e^z)^{\ln n}$ is an entire function so that ζ is analytic on $\mathcal{D}_0$. Let $z = x + iy$. Then,

$$\left|\frac{1}{n^z}\right| = \left|\frac{1}{e^{z\ln n}}\right| = \frac{1}{e^{x\ln n}} = \frac{1}{n^x}.$$

Since the series $\sum_{n=1}^{\infty} \frac{1}{n^x}$ converges for $x > 1$, the Riemann ζ-function converges absolutely on $\mathcal{D}_0$. Then the maximal extension of ζ is defined on $\mathbb{C}/\{1\}$ with a simple pole at $z = 1$.

(iv) Consider the function $f_0(z) = \int_0^\infty t^3 e^{-zt} dt$ which is analytic at all points of $z \in \mathcal{D}_0 = \{z \in \mathbb{C} : \text{Re}z > 0\}$. Now,

$$\begin{aligned} f_0(z) &= \lim_{M\to\infty} \int_0^M t^3 e^{-zt} dt = \lim_{M\to\infty} \left[\left(-\frac{t^3}{z} - \frac{3t^2}{z^2} - \frac{6t}{z^3} - \frac{6}{z^4}\right) e^{-zt}\right]_0^M \\ &= \lim_{M\to\infty} \left[\frac{6}{z^4} - \left(\frac{M^3}{z} + \frac{3M^2}{z^2} + \frac{6M}{z^3} + \frac{6}{z^4}\right) e^{-Mz}\right] \\ &= 6/z^4 \ \text{ if } z \in \mathcal{D}_0. \end{aligned}$$

For $\text{Re}(z) > 0$, the the integral has the value $f_1(z) = 6/z^4$. But $f_1(z)$ is analytic everywhere except $z = 0$. Since $f_1(z) = f_0(z)$ for $z \in \mathcal{D}_0$, we see that $f_1(z)$ must be the required analytic continuation.

Definition 10.1.2. [Complete analytic function]: Let $f_0(z)$ be analytic in a domain $\mathcal{D}_0$ and suppose $f_1(z), f_2(z), \cdots, f_n(z)$ are analytic continuations in the domains $\mathcal{D}_1, \mathcal{D}_2, \cdots, \mathcal{D}_n$ respectively. If there exists a function

$$g(z) = \begin{cases} f_0(z) & z \in \mathcal{D}_0 \\ f_1(z) & z \in \mathcal{D}_1 \\ \cdots & \cdots \\ f_n(z) & z \in \mathcal{D}_n, \end{cases}$$

then $g(z)$ is said to be a *complete analytic function* of the domain wider than $\mathcal{D}_0$ or $\mathcal{D}_1$, or $\cdots$ or $\mathcal{D}_n$.

This function $g(z)$, which is analytic in every one of the domains $\mathcal{D}_0, \mathcal{D}_1, \cdots, \mathcal{D}_n$, may not be analytic at some points which are outside these domains. Such points are said to be *singularities* of the complete analytic function.

Example 10.1.2. *Discuss the direct analytic continuations of the function* $f_0(z) = (1 + z^2)^{-1}$.

Solution. The function $f_0(z)$ possesses two simple poles at $z = \pm i$; otherwise it is regular throughout the whole complex plane. We first choose a point, say $z = 0$, at which f_0 is analytic and obtain its Taylor series expansion represented by

$$g(z) = \sum_{n=0}^{\infty} (-1)^n z^{2n} = 1 - z^2 + z^4 - \cdots; \ |z| < 1.$$

The series fails to converge on and beyond the unit circle $\mathcal{S}(0, 1)$, as is clear from the series for $z = 1$ even though the function $f_0(z)$ is analytic at that point. We can, in fact, continue the expansion from one region to another.

Let us consider a second expansion of $f_0(z)$ about $z = 3/4$ lying inside $\mathcal{S}(0, 1)$. We form the expansion as follows

$$\begin{aligned} f_0(z) &= \frac{1}{1+z^2} = \frac{1}{(z+i)(z-i)} = \frac{1}{2i}\left[\frac{1}{z-i} - \frac{1}{z+i}\right] \\ &= \frac{1}{2i}\left[\frac{1}{\frac{3}{4}-i}\cdot\left(1 + \frac{z-\frac{3}{4}}{\frac{3}{4}-i}\right)^{-1} - \frac{1}{\frac{3}{4}+i}\cdot\left(1 + \frac{z}{\frac{3}{4}+i}\frac{3}{4}\right)^{1}\right]; \left|z - \frac{3}{4}\right| < \frac{5}{4} \\ &= \frac{16}{25} - \frac{3}{2}\left(\frac{16}{25}\right)^2\left(z - \frac{3}{4}\right) + \frac{11}{16}\left(\frac{16}{25}\right)^3\left(z - \frac{3}{4}\right)^2 + \cdots = f_1(z), \text{ say.} \end{aligned}$$

Now $f_1(z)$, which converges in $|z-\frac{3}{4}|<\frac{5}{4}$ and coincides with $g(z)$ in the shaded region (Fig. 10.2), is clearly a direct analytic continuation of $g(z)$.

Let us construct another analytic continuation of $g(z)$. Now we consider a neighbourhood of the point $z=1$ (though it is a boundary point of the unit circle the function $f_0(z)$, is analytic there) and obtain an expansion represented by

$$f_2(z)=\frac{1}{2}-\frac{1}{2}(z-1)+\frac{1}{4}(z-1)^2-\cdots;\text{ for }|z-1|<\sqrt{2}.$$

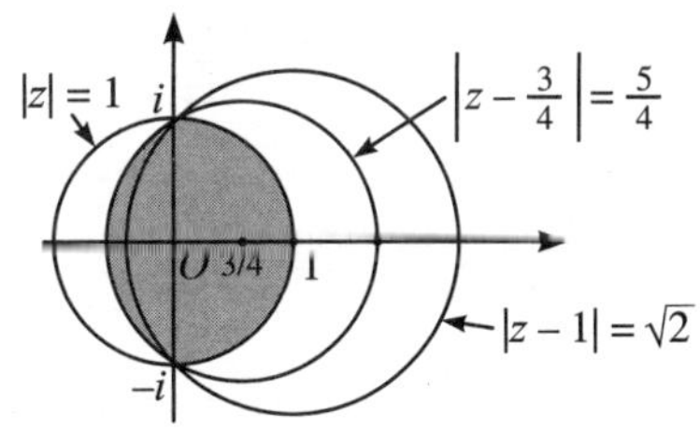

Figure 10.2 *Analytic continuation*

In this way, we can determine all possible direct analytic continuations of $g(z)$ and then continuations of these continuations and so on. Thus the complete analytic function is $(1+z^2)^{-1}$, defined in the whole complex plane barring the points $z=\pm i$.

10.1.2 Indirect Analytic Continuation

Let A be finite and have ordering, say $A=\{\alpha_1<\alpha_2<\cdots<\alpha_m\}$, and let $\langle\mathcal{D}_\alpha\rangle$ be a sequence of domains. Suppose we have pairs $\langle\mathcal{D}_\alpha,f_\alpha\rangle$ where $\mathcal{D}_\alpha$ are domains and $f_\alpha:\mathcal{D}_\alpha\to\mathbb{C}$ are analytic on a total ordering set $A=\{\alpha\}$. Thus the sequence of functions $\langle f_{\alpha_j}(z)\rangle$ will be such that

$$f_j(z)\equiv f_{j+1}(z)\text{ for } z\in\mathcal{D}_j\cap\mathcal{D}_{j+1}\neq\phi$$

and the subsequent function $f_{\alpha_{j+1}}(z)$ is the *direct analytic continuation* of the previous function $f_{\alpha_j}(z)$ from the domain $\mathcal{D}_{\alpha_j}$ to the domain $\mathcal{D}_{\alpha_{j+1}}$. Let for every j

$$\mathcal{D}_{\alpha_j}\cap\mathcal{D}_{\alpha_{j+i}}\neq\phi\text{ and } f_{\alpha_j}\Big|_{\mathcal{D}_{\alpha_j}\bigcap\mathcal{D}_{\alpha_{j+i}}}=f_{\alpha_{j+1}}\Big|_{\mathcal{D}_{\alpha_j}\bigcap\mathcal{D}_{\alpha_{j+i}}}$$

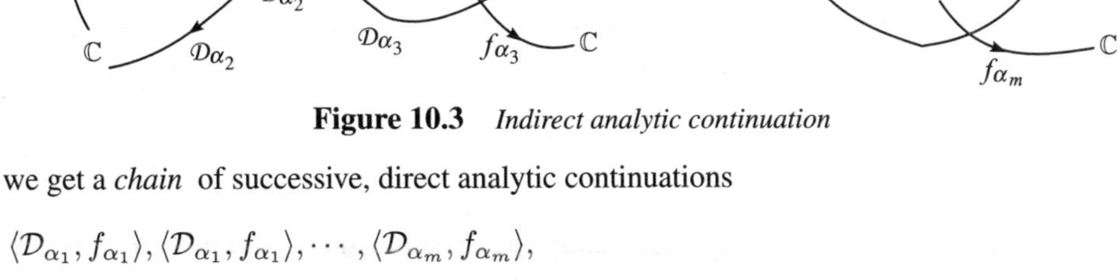

Figure 10.3 *Indirect analytic continuation*

Thus we get a *chain* of successive, direct analytic continuations

$$\langle\mathcal{D}_{\alpha_1},f_{\alpha_1}\rangle,\langle\mathcal{D}_{\alpha_1},f_{\alpha_1}\rangle,\cdots,\langle\mathcal{D}_{\alpha_m},f_{\alpha_m}\rangle,$$

where any two consecutive domains have a non-empty intersection. We say that $\langle\mathcal{D}_{\alpha_m},f_{\alpha_m}\rangle$ is an *indirect analytic continuation* of $\langle\mathcal{D}_{\alpha_1},f_{\alpha_1}\rangle$ along the chain.

Definition 10.1.3. The analytic function $g=f_{\alpha_m}:\mathcal{D}_{\alpha_m}\to\mathbb{C}$ is an *indirect analytic continuation* of $f=f_{\alpha_1}:\mathcal{D}_{\alpha_1}\to\mathbb{C}$ if g is obtained from f by a finite chain of successive direct analytic continuations. This function may be multiple-valued.

If the domains $\mathcal{D}_{\alpha_m}$ and $\mathcal{D}_{\alpha_1}$ have a non-empty intersection, the function element $\langle \mathcal{D}_{\alpha_1}, f_{\alpha_1} \rangle$ can be continued back to the initial domain $\mathcal{D}_{\alpha_1}$ along the chain, $\mathcal{D}_{\alpha_1}, \mathcal{D}_{\alpha_2}, \cdots, \mathcal{D}_{\alpha_m}, \mathcal{D}_{\alpha_1}$. In general, the function element $\langle \mathcal{D}_{\alpha_m}, f_{\alpha_m} \rangle$ obtained in the way differs from $\langle \mathcal{D}_{\alpha_1}, f_{\alpha_1} \rangle$.

10.1.3 Indirect Analytic Continuation using Power Series

One tool for studying analytic continuations is the Taylor's series. Let f_1 be analytic at a point z_1 on $\mathbb{C}$. Take the power series expansion $P(f; z_1)$ at $z_1 \in \mathcal{D}_1$ of f_1 as

$$P(f_1; z_1)(z) = \sum_{n=0}^{\infty} \frac{(z - z_1)^n}{n!} f_1^{(n)}(z_1) \tag{10.2}$$

with radius of convergence $R(z_1)$ and the domain of convergence $\mathcal{S}_1(z_1; R(z_1)) = \{z \in \mathcal{D}_1 : |z - z_1| < R(z_1)\}$. Then $f_1(z)$ is analytic in $\mathcal{S}_1(z_1; R(z_1))$. Assume f_1 does not extend to an entire function so that $R(z_1)$ must be finite. Let $z_2 \in \mathcal{S}_2(z_2; R(z_2))$.

(i) Consider $\mathcal{S}_1(z_1; R(z_1)) \supset \mathcal{S}_2(z_2; R(z_2))$, i.e., $R(z_2) < R(z_1) - |z_1 - z_2|$. In this case, the power series expansion does not go beyond the boundary $\partial\mathcal{S}_1$ of $\mathcal{S}_1$ (Fig. 10.4(a)).

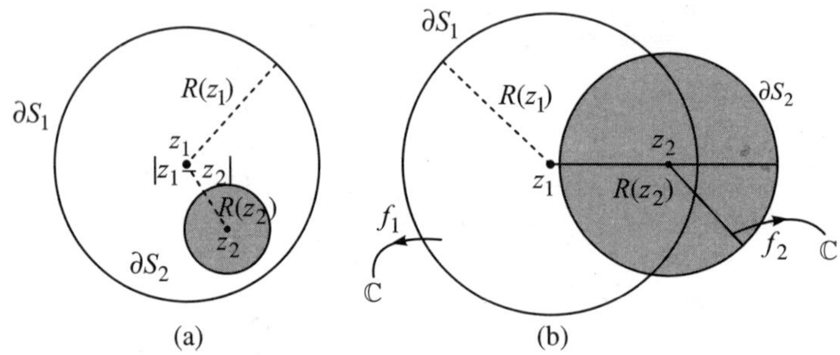

Figure 10.4 *Analytic continuation by power series*

(ii) Let $R(z_2) > R(z_1) - |z_1 - z_2|$ (Fig. 10.4 (b)). We compute the values $f_1(z_2)$, $f_1'(z_2)$, $f_1''(z_2), \cdots$, etc., by successive term-by-term differentiation of the series (10.2) and write

$$P(f_2; z_2)(z) = \sum_{n=0}^{\infty} \frac{(z - z_2)^n}{n!} f_2^{(n)}(z_2) \tag{10.3}$$

with the radius of convergence as $R(z_2)$ and domain of convergence $\mathcal{S}_2(z_2; R(z_2))$. Then $f_2(z)$ is analytic in $\mathcal{S}_2(z_2; R(z_2))$. Then $f_2(z)$ is analytic in $\mathcal{S}_2(z_2; R(z_2))$. By uniqueness theorem, $f_1(z) = f_2(z)$ for all $z \in \mathcal{S}_1 \cap \mathcal{S}_2$.

Similarly, considering a point z_3 with $R(z_3) > R(z_2) - |z_2 - z_3|$, we get an analytic function $f_3(z)$ in a circular domain $\mathcal{S}_3(z_3; R(z_3))$ such that $f_2(z) = f_3(z)$ for all $z \in \mathcal{S}_2 \cap \mathcal{S}_3$. Repeating the process we get a number of different power series representing analytic function $f_k(z)$ in their respective circular domain $\mathcal{S}_k$, which form a chain of analytic continuations of the original function $f_1(z)$ such that $\langle \mathcal{D}_k, f_k \rangle$ is a direct analytic continuation of $\langle \mathcal{D}_{k-1}, f_{k-1} \rangle$.

This process can be continued to get an element $f_n(z)$, which is analytic continuation of $f_1(z)$ along a sequence of circular domains $\mathcal{S}_1, \mathcal{S}_2, \ldots, \mathcal{S}_n$, where $\mathcal{S}_j$ is a circle given by

$$\Gamma_j : |z - z_j| < R_j;\ j = 0, 1, 2, \ldots, n$$

and where we assume that

$$R_j > R_{j-1} - |z_{j-1} - z_j|;\ j = 1, 2, \ldots, n.$$

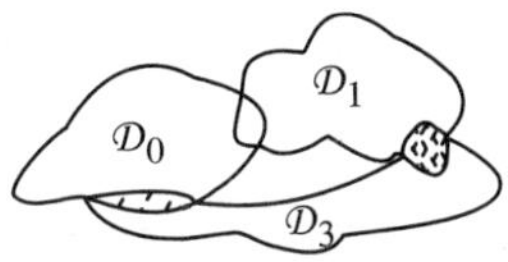

Figure 10.5 *Analytic continuation*

RESULT 73. Let f_1 be the analytic continuation of f_0 from the domain $\mathcal{D}_0$ into the domain $\mathcal{D}_1$ (Fig. 10.5). If there exists an analytic continuation f_2 of f_1 from the domain $\mathcal{D}_1$ into the domain $\mathcal{D}_2$, then for each $z \in \mathcal{D}_1 \cap \mathcal{D}_2$, it is not necessarily true that $f_2(z) = f_0(z)$. This implies that a chain of analytic continuation of a given function $f_0(z)$ on a domain $\mathcal{D}_0$ may prove a different function on a domain $\mathcal{D}_1$.

Lemma 10.1.1. Let f be analytic in a domain $\mathcal{D}$ and $z_1 \in \mathcal{D}$. $P(f, z_1)$ is the power series expansion of f around z_1 as in Eq. (10.4) with domain of convergence $|z - z_1| < R(z_1)$. $R(z_1)$ is continuous in the variable $z_1 \in \mathcal{D}$. In fact,

$$\Big|R(z_1) - R(z_2)\Big| \le |z_1 - z_2| \quad \text{for } z_1, z_2 \in \mathcal{D}.$$

Proof: We need to show that

$$R(z_2) - |z_1 - z_2| \le R(z_1) \le R(z_2) + |z_1 - z_2|.$$

Let $R(z_1) > R(z_2) + |z_1 - z_2|$. This would mean that the disc of convergence $|z - z_2| < R(z_2)$ of $P(f; z_2)$ is contained in the disc of convergence $|z - z_1| < R(z_1)$, implying that the analytic function $P(f; z_2)$ extends to the circle of convergence $|z - z_2| = R(z_2)$. This is a contradiction. □

10.1.4 Indirect Analytic Continuation along a Curve

Indirect analytic continuation along a curve is an extension of the above idea to the situation where a curve is covered by an overlapping sequence of discs and an analytic function defined on the first disc in the sequence (Fig. 10.6).

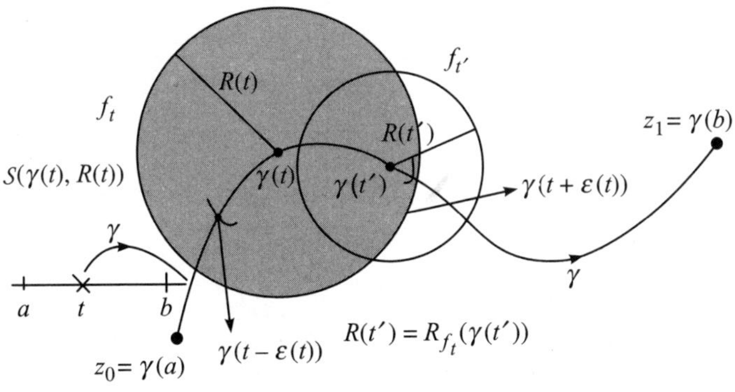

Figure 10.6 *Indirect analytic continuation along a curve*

Definition 10.1.4. Let $\gamma : [a, b] \to \mathbb{C}$ be a path in a complex plane $\mathbb{C}$ (Fig. 10.6) starting at $z_0 = \gamma(a)$. For each t, there is a convergent power series of f along γ as

$$f_t(z) = \sum_{n=0}^{\infty} a_n(t)\Big(z - \gamma(t)\Big)^n \tag{10.4}$$

with radius of convergence $R(t) > 0$ and domain of convergence $|z - \gamma(t)| < R(t)$. $f_a(z)$ is a power series representing $f(z)$ at $z_0 = \gamma(a)$. For each $t \in [a, b]$, there exists sufficiently small $\varepsilon(t) > 0$ such that we can find t' closer to t, i.e.,

$$t' \in \Big(t - \varepsilon(t), t + \varepsilon(t)\Big) \bigcap [a, b],$$

for which $f_{t'} = f_t$, by the uniqueness theorem. Thus for t' close to t, the pairs $\langle |z - \gamma(t)| < R(t), f_t \rangle$ and $\langle |z - \gamma(t')| < R(t'), f_{t'} \rangle$ are *direct analytic continuations* of one another along γ.

The final pair $\langle |z - \gamma(b)| < R(b), f_b \rangle$ is an *indirect analytic continuation* of the initial pair $\langle |z - \gamma(a)| < R(a), f_a \rangle$ along γ_1, where we regard $f_b(z)$ as either a power series or an analytic function defined near $\gamma(b)$.

Lemma 10.1.1 shows that the radius of convergence $R(t)$ of the power series (10.4) depends continuously on the parameter t.

Definition 10.1.5. We are given a power series (10.4) with radius of convergence $R(t) > 0$ and domain of convergence $|z - \gamma(t)| < R(t)$ such that for t' close to t, $f_t = f_{t'}$(as functions). We then say that f_b is the result of analytic continuation of f_a on γ.

Let $\gamma : [a, b] \to \mathbb{C}$ be a curve joining the points $z_0(= a)$ to the point b, and let $f_0(z)$ be a function analytic in domain $\mathcal{D}_0$ which contains the point z_0 in it (Fig. 10.7). Let $f_n(z)$ be the analytic continuation of $f_0(z)$ along a sequence of domains $\mathcal{D}_0, \mathcal{D}_1, \ldots, \mathcal{D}_n$ covering the the curve γ, where the domains are such that intersection of any two consecutive domains is a domain and the domain $\mathcal{D}_n$ contains a point b in it.

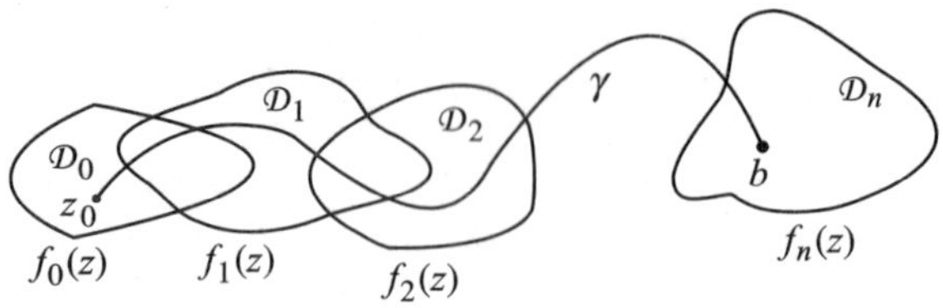

Figure 10.7 *Analytic continuation along a curve*

If $f_j(z)$ is the analytic continuation of $f_{j-1}(z)$ from $\mathcal{D}_j$ to $\mathcal{D}_j$; $j = 1, 2, \ldots, n$, then $f_0(z), f_1(z), \ldots, f_n(z)$ are elements of the analytic function $g(z)$ defined as follows:

$$g(z) = f_j(z),\ z \in \mathcal{D}_j;\ j = 1, 2, \ldots, n.$$

The element $f_n(z)$ which is analytic in domain $\mathcal{D}_n$, containing a point b in it, is called the analytic continuation along the curve Γ of the element $f_0(z)$.

Lemma 10.1.2. Given an analytic continuation $f_t(t)$ as in Eq. (10.4), the coefficients $a_n(t); n \geq 0$ and the radius of convergence $R(t)$ are continuous function of t.

Proof: Here, let $\gamma : [a, b] \to \mathbb{C}$ be a path. f_t is defined in $\mathcal{S}(\gamma(t), R(t))$, $a_n : [a, b] \to \mathbb{C}$ and $R : [a, b] \to \mathbb{R}^+ \subset \mathbb{C}$. To prove a_n and R are continuous, it is enough to show they are continuous locally.

Let $t \in [a, b]$, then $\exists\ \varepsilon(t) > 0$ such that $\forall t' \in (t - \varepsilon(t), t + \varepsilon(t)) \cap [a, b]$, we have $f_{t'} = f_t$. By Lemma 10.1.1, the radius of convergence $R(t)$ of the power series (10.4) is continuous in t. The coefficients $a_n(t')$ in Eq. (10.4) are given by

$$a_n(t') = \frac{1}{n!} f_{t'}^{(n)}(\gamma(t')) = \frac{1}{n!} f_t^{(n)}(\gamma(t')); t' \in (t - \varepsilon(t), t + \varepsilon(t)) \cap [a, b].$$

Since $f_t^{(n)}(z)$ is a continuous function of z and since γ is a continuous function, we see that $a_n(t')$ is a continuous function of t' for t' close to t.

Therefore, the coefficients $a_n(t)$ depends continuously on the parameter t. □

Theorem 10.2. *Given an analytic continuation $f_t(t)$ as in Eq. (10.4), along a path $\gamma : [a, b] \to \mathbb{C}$, f_t for $t > a$ is uniquely determined by f_a.*

Proof: Suppose the statement of the theorem is false. Then there exists

$$g_t(t) = \sum_{n=0}^{\infty} b_n(t)\Big(z - \gamma(t)\Big)^n; \quad \Big|z - \gamma(t)\Big| < \widetilde{R}(t)$$

which is also an analytic continuation along γ starting with $g_a = f_a$ and $g_t \neq f_t$ for some $t > a$. Let $t_0 = \inf\limits_{t \in [a,b]} \{f_t \neq g_t\}$. Clearly, $t_0 > a$.

Now, for $t' < t_0 - \varepsilon$, choose to $t_0 - \varepsilon$. We have

$$f_{t'} = f_{t_0 - \varepsilon} = g_{t_0 - \varepsilon} = g_{t'}.$$

But $f_{t_0 \pm \varepsilon} = f_{t_0}$ and $g_{t_0 \pm \varepsilon} = g_{t_0}$ for $\varepsilon > 0$ small enough. This implies $f_{t_0 + \varepsilon} = g_{t_0 + \varepsilon}$ for $\varepsilon > 0$ small enough, a contradiction to the definition of t_0. Hence the theorem. □

RESULT **74.** Let there be two disjoined simply-connected domains whose boundaries share a common contour. Let the domains be $\mathcal{D}_0$ and $\mathcal{D}_1$, and their common boundary ABC (Fig. 10.8) be γ. Let $f_0(z)$ be analytic in $\mathcal{D}_0$ and continuous on γ and $f_1(z)$ analytic in $\mathcal{D}_1$ and continuous in γ. Then the function $g(z)$ defined by:

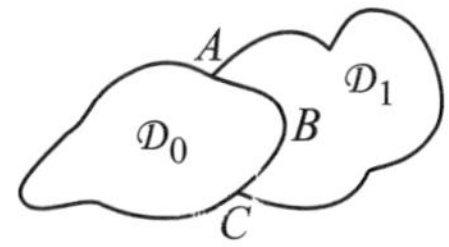

Figure 10.8 *Analytic Continuation*

$$g(z) = \begin{cases} f_0(z), & z \in \mathcal{D}_0 \\ f_0(z) \equiv f_1(z), & z \in \gamma \\ f_2(z), & z \in \mathcal{D}_1 \end{cases} \tag{10.5}$$

is called *analytic continuation* of $f_0(z)$ from $\mathcal{D}_0$ and of $f_1(z)$ from $\mathcal{D}_1$ to $g(z)$ in the extended domain $\mathcal{D}_0 \cup \mathcal{D}_1$.

Example 10.1.3. *Show that the series $\sum_{n=0}^{\infty} \frac{z^n}{2^{n+1}}$ and $\sum_{n=0}^{\infty} \frac{(z-i)^n}{(2-i)^{n+1}}$ are analytic continuation of each other.*

Solution. The first series can be written in the form $f_0(z) = \sum a_n z^n$, where $a_n = 1/2^{n+1}$. Then by ratio test the series converges if $|z| < 2$ (Fig. 10.9). Within that circle of convergence, the geometric series with ratio $z/2$ can be summed and represents the function

$$f_0(z) = \frac{1/2}{1 - z/2} = \frac{1}{2 - z}.$$

Also, $f_0(z)$ is analytic inside the circle $|z| = 2$. By the ratio test the second series $f_1(z)$ converges for

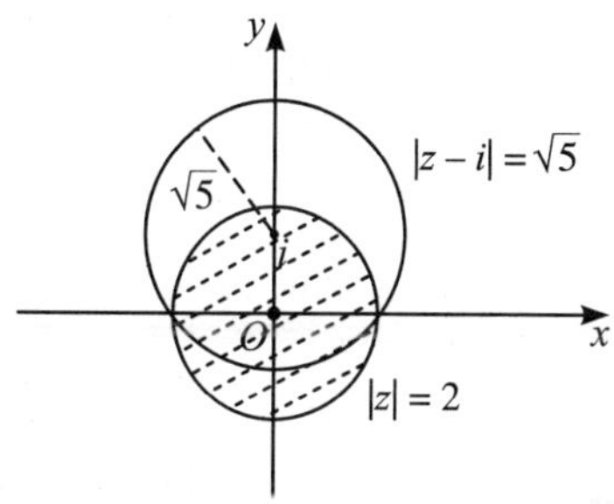

Figure 10.9 $|z| = 2, |z - i| = \sqrt{5}$

$$|(z - i)/(2 - i)| < 1 \Rightarrow |z - i| < \sqrt{5}.$$

In this circle of convergence (Fig. 10.9), the given infinite geometric series with common ratio $(z - i)/(2 - i)$ can be summed and represents the function

$$f_1(z) = \frac{1/(2-i)}{1-(z-i)/(2-i)} = \frac{1}{2-z}.$$

Also, $f_1(z)$ is analytic inside the circle $|z - i| = \sqrt{5}$. Thus, we have

(i) $f_0(z)$ is analytic within the circle $|z| = 2$.
(ii) $f_1(z)$ is analytic within the circle $|z - i| = \sqrt{5}$.
(iii) $f_0(z) = f_1(z)$ is the common area to both $|z| = 2$ and $|z - i| = \sqrt{5}$.

Thus the two series are analytic continuation of each other.

Example 10.1.4. *Show that the function $f_0(z) = \int_0^\infty e^{-zt}\,dt$ can be continued analytically. Also, construct a power series which is an analytic continuation of $f_0(z)$.*

Solution. We have

$$f_0(z) = \int_0^\infty e^{-zt}\,dt = \left[\frac{e^{-zt}}{-z}\right]_0^\infty = \begin{cases} 1/z & \text{if } \mathrm{Re}(z) > 0 \\ \infty & \text{if } \mathrm{Re}(z) \le 0. \end{cases}$$

Thus, $f_0(z)$ is analytic in the domain $\mathcal{D} : \mathrm{Re}(z) > 0$. The only singularity of $f_0(z)$ in $\mathbb{C}_\infty$ is at $z = 0$. Choose $f_1(z) = 1/z$ which is analytic everywhere except $z = 0$.

$$\therefore f_0(z) = f_1(z); \forall z \text{ for which } \mathrm{Re}(z) > 0.$$

Avoiding this singularity, $f_1(z)$ is an analytic continuation of $f_0(z)$. We use the power series again. Now, we represent $f_1(z)$ by a power series as

$$f_1(z) = \sum_{n=0}^{\infty} \frac{(-1)^n}{a^{n+1}} (z-a)^n.$$

Let R be the radius of convergence. Then

$$\frac{1}{R} = \lim_{n\to\infty} \left|\frac{(-1)^{n+1}}{a^{n+2}} \cdot \frac{a^{n+1}}{(-1)^n}\right| = \lim_{n\to\infty} \left|\frac{-1}{a}\right| = \frac{1}{|a|}.$$

$\therefore\ R = |a|$. Thus $f_1(z)$ is analytic for all z inside the circle $\Gamma_1 : |z - a| = |a|$ with radius $|a|$ and centre at a (Fig. 10.10). Now, $|a|$ is the distance of the centre from the origin so that $|a|$ is greater than the distance of the centre from the imaginary axis. This implies that the circle cuts the imaginary axis and as such has some region common with the region $\mathcal{D}$.

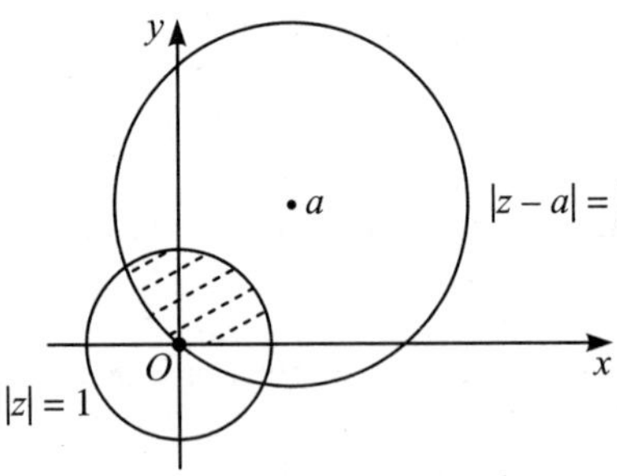

Figure 10.10 $|z| = 1, |z - a| = |a|$

$$\therefore f_0(z) = f_1(z); \forall z \text{ for which } \mathrm{Re}(z) > 0.$$

Also, the sum of this series is given by

$$\begin{aligned} f_1(z) &= \frac{1}{a} - \frac{z-a}{a^2} + \frac{(z-a)^2}{a^3} - \frac{(z-a)^3}{a^4} + \cdots ; |z-a| < |a| \\ &= \frac{1}{a}\left[1 - \frac{z-a}{a} + \frac{(z-a)^2}{a^2} - \frac{(z-a)^3}{a^3} + \cdots\right] \\ &= \frac{1}{a} \cdot \frac{1}{1 + (z-a)/a} = \frac{1}{z}, \end{aligned}$$

within the circle Γ_1. Thus $f_0(z)+f_1(z)$ in the part common to the domains $\mathcal{D}$ and $|z-a|<|a|$ shown in the Fig. 10.10. Thus, f_1 is an analytic continuation of f_0.

Example 10.1.5. *Prove that the power series* $z-\frac{z^2}{2}+\frac{z^3}{3}-\cdots$ *may be analytically continued to a wider range by means of the series*

$$\log 2-\frac{1-z}{2}-\frac{(1-z)^2}{2\cdot 2^2}-\frac{(1-z)^3}{3\cdot 2^3}-\cdots$$

Solution. Let $f_0(z)=z-\frac{z^2}{2}+\frac{z^3}{3}-\cdots$. According to the ratio test, $f_0(z)$ is convergent in $|z|<1$ and its sum function is $\log(1+z)$. Also, it is analytic inside the circle $|z|=1$. Let

$$\begin{aligned} f_1(z) &= \log 2-\frac{1-z}{2}-\frac{(1-z)^2}{2\cdot 2^2}-\frac{(1-z)^3}{3\cdot 2^3}-\cdots \\ &= \log 2-\left[\frac{1-z}{2}+\frac{1}{2}\cdot\left(\frac{1-z}{2}\right)^2+\frac{1}{3}\left(\frac{1-z}{2}\right)^3+\cdots\right] \\ &= \log 2+\log\left[1-\frac{1-z}{2}\right]=\log\left[2\frac{1+z}{2}\right]=\log(1+z). \end{aligned}$$

Evidently, $f_2(z)$ is convergent for $|(1-z)/2|<1 \Rightarrow |z-1|<2$ and the sum function is $\log(1+z)$. Also, $f_1(z)$ is analytic inside the circle $|z-1|=2$ (Fig. 10.11). Thus the two series represent the same function in the region common to the circle $|z|=1$ and $|z-1|=2$. Now

(i) $f_0(z)$ is analytic within the circle $|z|=1$.

(ii) $f_1(z)$ is analytic within the circle $|z-1|=2$.

(iii) $f_0(z)=f_1(z)$ in the area common to both $|z|=1$ and $|z-1|=2$.

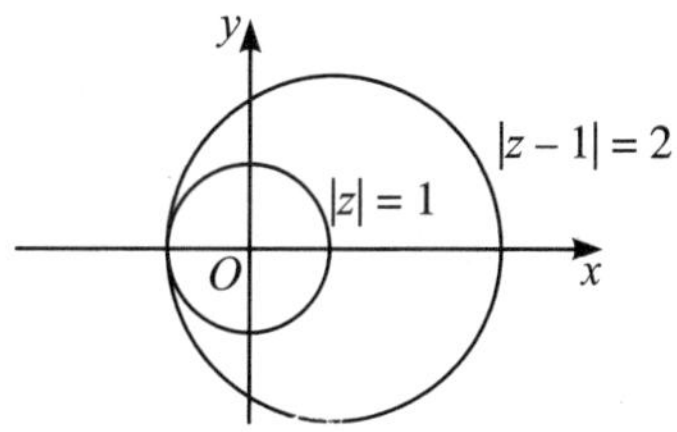

Figure 10.11 $|z|=1, |z-1|=2$

Therefore, $f_1(z)$ is analytic continuation of $f_0(z)$ from the domain $|z|<1$ to the domain $|z-1|<2$. Moreover, $|z-1|=2$ is the in compression to $|z|=1$.

Example 10.1.6. *Show that, when b is real, the series*

$$\frac{1}{2}\log(1+b^2)+i\tan^{-1}b+\frac{z-ib}{1+ib}-\frac{1}{2}\left(\frac{z-ib}{1+ib}\right)^2+\cdots$$

is an analytic continuation of the function defined by the series $z-\frac{z^2}{2}+\frac{z^3}{3}-\cdots$.

Solution. Let $f_0(z)=z-\frac{z^2}{2}+\frac{z^3}{3}-\cdots$. According to the ratio test, $f_0(z)$ is convergent for $|z|<1$ and its sum function is $\log(1+z)$. Also, it is analytic inside the circle $|z|=1$. This function has the singularity at $z=-1$. Let

$$\begin{aligned} f_1(z) &= \frac{1}{2}\log(1+b^2)+i\tan^{-1}b+\frac{z-ib}{1+ib}-\frac{1}{2}\left(\frac{z-ib}{1+ib}\right)^2+\cdots \\ &= \frac{1}{2}\log(1+b^2)+i\left[b-\frac{b^3}{3}+\frac{b^5}{5}-\frac{b^7}{7}+\cdots\right]+\frac{z-ib}{1+ib}-\frac{1}{2}\left(\frac{z-ib}{1+ib}\right)^2+\cdots \\ &= \log(1+ib)+\log\left(1+\frac{z-ib}{1+ib}\right)=\log(1+z). \end{aligned}$$

Evidently, $f_1(z)$ is convergent for $|z - ib| < \sqrt{1 + b^2}$. Also, $f_1(z)$ is analytic inside the circle $|z - ib| = |1 + ib| = \sqrt{1 + b^2}$ (Fig. 10.12). Thus, we have

(i) $f_0(z)$ is analytic within the circle $|z| = 1$.

(ii) $f_1(z)$ is analytic within the circle $|z - ib| = \sqrt{1 + b^2}$.

(iii) $f_0(z) = f_1(z)$ in the area common to both $|z| = 1$ and $|z - ib| < \sqrt{1 + b^2}$.

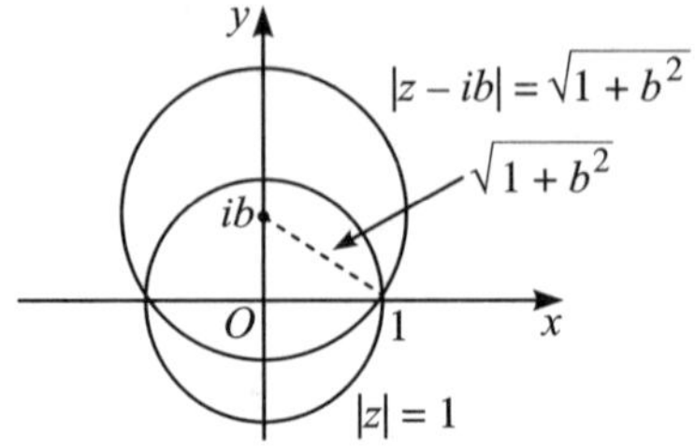

Figure 10.12 $|z| = 2, |z - 1| = 2$

Therefore, they represent analytic continuation of each other..

10.1.5 Regular and Singular Points

Let $f : \mathcal{D} \to \mathbb{C}$ be an analytic function defined in a domain $\mathcal{D} \subset \mathbb{C}$, with boundary $\partial\mathcal{D}$. A point $\zeta \in \partial\mathcal{D}$ is called a *regular point of f* if there exist a circle $\mathcal{S}(\zeta; \varepsilon)$ of the point ζ and the analytic function f_ζ such that

$$f(z) = f_\zeta(z) \ \forall z \in \mathcal{S}(\zeta; \varepsilon) \bigcap \mathcal{D}.$$

The boundary point ζ which is not a regular point is called a *singular point of f*, i.e., in any neighbourhood of the point ζ, there cannot be any analytic function coinciding with f in the part common to the neighbourhood of ζ and the domain $\mathcal{D}$.

10.1.6 Natural Boundary

In the previous examples, we have encountered with finite number of singular points situated on the boundary of the region of analyticity of the given function. It might happen that the boundary is dense with singular points. In this case analytic continuation across the boundary of the region is not possible. Such a boundary is called *natural boundary*.

Definition 10.1.6. Let $\mathcal{D} \subset \mathbb{C}$ be a domain with boundary $\partial\mathcal{D}$. If an analytic function $f : \mathcal{D} \to \mathbb{C}$ cannot be continued analytically across the boundary $\partial\mathcal{D}$, then this boundary $\partial\mathcal{D}$ is called *natural boundary*. In this case, $f(z)$ has no analytic extension to any domain which contains $\mathcal{D}$. In the power series method of analytic continuation, we have assured that the circle of convergence is not a natural boundary.

For example, the values of z given by $z = 1, z^2 = 1, z^3 = 1, \ldots$ are all singularities of $f_0(z) = \sum_{n=0}^{\infty} z^n$. These singularities all lie on the circle $|z| = 1$. Given any small arc of this circle, there will be infinitely many such singularities. These represent an impossible barrier and analytic continuation beyond $|z| = 1$ is therefore impossible. This circle $|z| = 1$ constitutes a natural boundary.

Theorem 10.3. *Every power series has at least one singular point on its circle of convergence.*

Proof: Let $f(z) = a_0 + a_1(z - z_0) + a_2(z - z_0)^2 + \cdots$ be any power series with domain of convergence $\triangle(0; R) = \{z \in \mathbb{C} : |z| < R\}$. We shall have to show that there lies at least one singular point on the circle of convergence $\partial\triangle = \{z \in \mathbb{C} : |z - z_0| = R\}$ of the function.

Suppose, on the contrary, that every point on $\partial\triangle$ are regular points of f. Then f must be analytic at all points of $\partial\triangle$. Thus f is analytic on a compact subset $\bar{\triangle}(0 : R) = \triangle(0; R) \cup \partial\triangle$.

Let $\zeta_1, \zeta_2, \cdots, \zeta_k, \cdots$ be certain number of regular points belonging to $\partial\triangle$. The definition of analyticity at a point reveals that, for each $\zeta_k \in \partial\triangle$, $\exists$ some $R_{\zeta_k} > 0$ a function f_{ζ_k}, analytic in open circles $\mathcal{S}_k(\zeta_k; R_{\zeta_k})$ and

$$f = f_{\zeta_k} \text{ on } \mathcal{S}_k\Big(\zeta_k; R_{\zeta_k}\Big) \cap \triangle(0; R).$$

In this way, if ζ_k and $\zeta_l \in \partial\triangle$ $(k \neq l)$ with $\Omega_1 = \mathcal{S}_k(\zeta_k; R_{\zeta_k}) \cap \mathcal{S}_l(\zeta_l; R_{\zeta_l}) \neq \phi$, the two functions f_{ζ_k} and f_{ζ_l}, are analytic in $\mathcal{S}_k(\zeta_k; R_{\zeta_k})$ and $\mathcal{S}_l(\zeta_l; R_{\zeta_l})$ such that

$$f = f_{\zeta_k} = f_{\zeta_l} \text{ on } \Omega_1 \cap \triangle(0; R). \tag{10.6}$$

Since Ω_1 is connected and $\Omega_1 \cap \triangle(0; R)$ is an open subset of Ω_1, by the uniqueness theorem

$$f_{\zeta_k} = f_{\zeta_l} \text{ on } \Omega_1. \tag{10.7}$$

Since $\partial\triangle$ is compact, by the Heine–Borel theorem, there exists finite number $\{\mathcal{S}_k(\zeta_k; R_{\zeta_k}); 1 \leq k \leq n\}$ from the collection $\mathcal{F} = \{\mathcal{S}_k(\zeta_k; R_{\zeta_k}); \zeta_k \in \partial\triangle\}$ such that it completely covers the boundary $\partial\triangle$ of the circle. Let

$$K = \bigcup_{k=1}^{n} \mathcal{S}_k\Big(\zeta_k; R_{\zeta_k}\Big) \text{ and } \delta = d\Big(\partial\triangle; K\Big)$$

Then, as $R_{\zeta_k} > 0$, for each k, we have $\delta > 0$. Moreover,

$$\Big\{z : z \in \mathcal{S}\Big(R; \delta\Big)\Big\} \subset K \text{ and } \triangle(0; R+\delta) \subset \mathcal{D} = K \cup \triangle(0 : R).$$

Here $\mathcal{D}$ is open connected. Now we introduce a function $g : \mathcal{D} \to \mathbb{C}$ which satisfies

$$g(z) = \begin{cases} f_{\zeta_k}(z) & \text{for } z \in \mathcal{S}_k(\zeta_k; R_{\zeta_k}); k = 1, 2, \cdots, n \\ f(z) & \text{for } z \in \triangle(0; R), \end{cases}$$

where $f_{\zeta_k}(z)$ is a direct analytic continuation of $f(z)$ in $\mathcal{S}_k(\zeta_k; R_{\zeta_k})$ of the regular point ζ_k. Due to Eq. (10.7), g is well-defined, single-valued and analytic on $\mathcal{D}$ and has same power series representation as f for $z \in \triangle(0; R)$. Thus, we observe that there exists an analytic function, say ψ, in $\triangle(0; R+\delta)$ whose power series expansion, say

$$\psi(z) = \sum_{n=0}^{\infty} b_n\, z^n \text{ for } z \in \triangle(0; R+\delta)$$

coincides with f on $\triangle(0; R)$. Since $f = g$ on $\triangle(0; R)$, by uniqueness theorem, $a_n = b_n; \forall n$. This shows that the radius of convergence of f is $R + \delta$, not R, which is a contradiction.

Hence, every point on $\partial\triangle$ cannot be regular points, i.e., there must be at least one singular point on $\partial\triangle$. □

Example 10.1.7. *Let $\mathcal{D}_1 = \{z \in \mathbb{C} : \text{Re } z > 0, \text{Im } z > 0\}$ denote the first quadrant. Set $f(z) = \log z$ for all $z \in \mathcal{D}_1$. Show that if g_1 is the analytic continuation to $\mathbb{C}/(-\infty, 0]$ of f and g_2 is the analytic continuation to $\mathbb{C}/(-\infty, 0]$ of f, then $g_1 \neq g_2$ throughout the third quadrant $\mathcal{D}_3 = \{z \in \mathbb{C} : \text{Re } z < 0, \text{Im } z < 0\}$.*

Solution. Clearly, g_1 is the principal branch of $\log z$ throughout $\mathbb{C}/(-\infty, 0]$ by the uniqueness theorem. That is,

$$g_1(z) = \int_{[1,z]} \frac{1}{\zeta}\, d\zeta$$

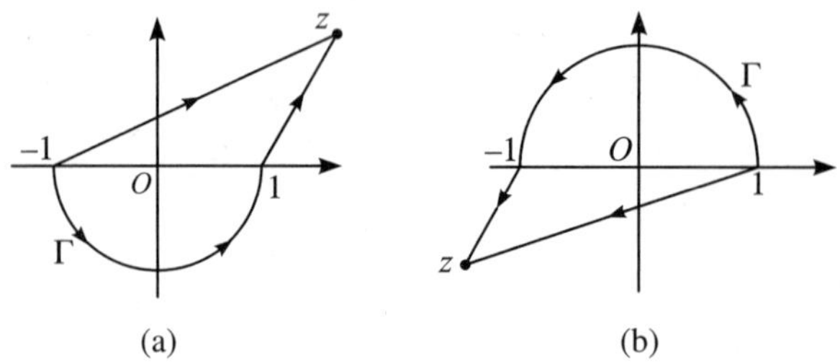

Figure 10.13

for all z barring the negative real axis including the origin.

Let γ be the closed curve (Fig. 10.13 (a)) consisting of the line segments $[1, z]$, $[z, -1]$ and the semi-circular path Γ with centre at the origin and radius of 1, where z is any point in $\mathcal{D}_1$. By Cauchy's residue theorem,

$$\int_\gamma \frac{1}{\zeta}\, d\zeta = 2\pi i \cdot 1 = 2\pi i.$$

Using the contour integral formula

$$\int_\gamma \frac{1}{\zeta}\, d\zeta = \int_{[1,z]} \frac{1}{\zeta}\, d\zeta + \int_{[z,-1]} \frac{1}{\zeta}\, d\zeta + \int_\Gamma \frac{1}{\zeta}\, d\zeta$$

$$\text{or,} \qquad 2\pi i = g_1(z) - \int_{[-1,z]} \frac{1}{\zeta}\, d\zeta + i\pi.$$

$$\text{or,} \qquad g_1(z) - \int_{[-1,z]} \frac{1}{\zeta}\, d\zeta - i\pi = g_2(z),$$

where, $g_2(z)$ is given by

$$g_2(z) = \int_{[-1,z]} \frac{1}{\zeta}\, d\zeta + i\pi \quad \text{for all } z \in \mathbb{C}/[0,\infty). \tag{10.8}$$

Hence $g_2(z) = g_1(z) = \log z$ for all $z \in \mathcal{D}_1$, that is, the mapping g_2 defined in Eq. (10.8) is the unique analytic continuation of f to $\mathbb{C}/[0,\infty)$.

Let $z \in \mathcal{D}_3$ and let γ be the curve joining the line segments $[-1, z]$, $[z, 1]$ and the unit semi-circular path Γ in the upper half plane (Fig. 10.13 (b)). Thus

$$2\pi i = \int_\gamma \frac{1}{\zeta}\, d\zeta = \int_{[-1,z]} \frac{1}{\zeta}\, d\zeta + \int_{[z,-1]} \frac{1}{\zeta}\, d\zeta + \int_\Gamma \frac{1}{\zeta}\, d\zeta$$

$$\text{or,} \qquad 2\pi i = \int_{[-1,z]} \frac{1}{\zeta}\, d\zeta - g_1(z) + i\pi$$

$$\text{or,} \qquad g_2(z) = g_1(z) + 2\pi i \text{ for all } z \in \mathcal{D}_3.$$

RESULT **75.** From the previous example we have the following observations: If γ_1 and γ_2 be two curves joining z_0 and ζ, and $\langle \mathcal{D}_0, f_0 \rangle$ be a function element at z_0, then the resulting function elements of $\langle \mathcal{D}_0, f_0 \rangle$ along the curves γ_1 and γ_2 at ζ may not be a direct analytic continuations of each other.

Definition 10.1.7. We say an analytic function f at z_0 is *analytically continuable* along a path γ starting from z_0 if $\exists$ an analytic continuation f_t, $t \in [a, b]$, where $\gamma : [a, b] \to \mathbb{C}$, such that $f_a = f$ (and $\gamma(a) = z_0$).

We consider the following examples:

(i) If f is analytic on a domain $\mathcal{D}$, then f is analytically continuable along any path γ in $\mathcal{D}$.
(ii) $f(z) = \frac{1}{z}$ cannot be analytically continued along any path passing through origin.
(iii) The analytic function $\log z$ (principal branch of the logarithm)

$$\log z = \text{Log}|z| + i \text{ Arg } z; \quad -\pi < \text{ Arg } z \leq \pi$$

has maximal domain of analyticity the slit plane $\mathbb{C}/(-\infty, 0]$.
(iv) Suppose $f(z)$ is a principal branch $\sqrt{z}$ with the series expansion about $z = 1$ as

$$f(z) = 1 + \frac{1}{2}(z-1) - \frac{1}{8}(z - 10^2 + \cdots.$$

Let $\gamma(t) = e^{it}; 0 \leq t \leq 2\pi$, be the closed path around the unit circle starting at $\gamma(0) = 1$. Then $f(z)$ has an analytic continuation along γ. The power series $f_t(z)$ in Eq. (10.4) is given explicitly by

$$f_t(z) = e^{i\pi/2} + \frac{1}{2}e^{-it/2}\left(z - e^{it}\right) - \frac{1}{8}e^{-3it/2}\left(z - e^{it}\right)^2 + \cdots.$$

Then the analytic continuation of $f(z)$ around the circle is

$$f_{2\pi}(z) = -1 - \frac{1}{2}(z-1) + \frac{1}{8}(z-1)^2 + \cdots,$$

which is just the other branch of $\sqrt{z}$.

10.2 Monodromy Theorem

If there are no singular points in between the two paths of an analytic continuation, then the result of the analytic continuation is the same for each path. Also, we have proved that if a function element can be continued analytically along every path in a simply-connected domain, then the resulting function is single-valued.

The Monodromy theorem can be stated as: if there are no singular points in between the paths of an analytic continuation, then the result of analytic continuation is the same for each path.

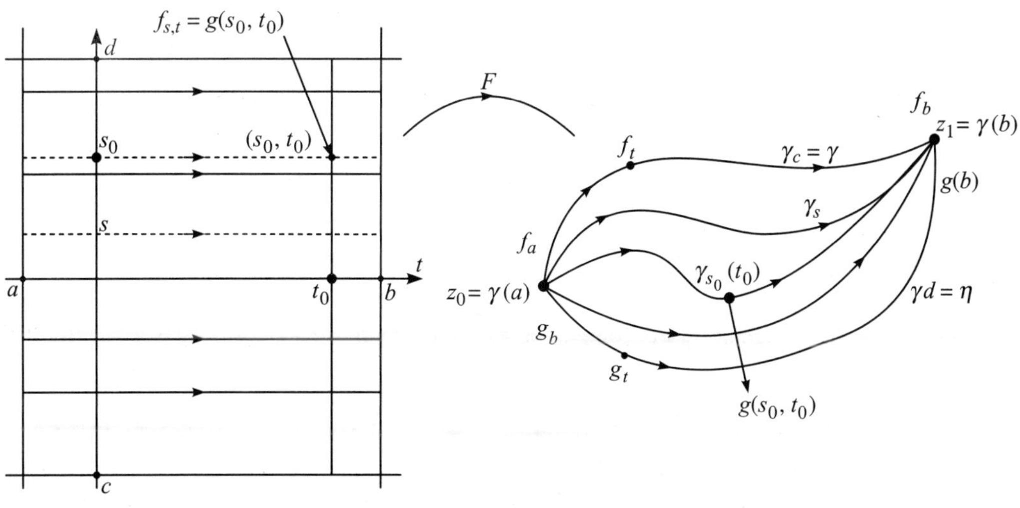

Figure 10.14 *Monodromy theorem*

Lemma 10.2.1. If f is analytic at z_0 and can be analytically continued along a path $\gamma : [a, b] \to \mathbb{C}$ from z_0 to z_1, then $\exists\, \delta > 0$ such that f can be analytically continued along any path $\eta : [a, b] \to \mathbb{C}$ from z_0 to z_1 such that

$$\left|\gamma(t) - \eta(t)\right| < \delta \ \forall t \in [a, b].$$

Moreover, the analytic continuation of f along any such path η will lead to the same function at z_1 as the one obtained via γ.

Proof: Given $f_t(z) = \sum_{n=0}^{\infty} a_n(t)(z - \gamma(t))^n$, with $|z - \gamma(t)| < R(t)$ and $f_a = f$. We have seen that $a_n : [a, b] \to \mathbb{C}$ and $R : [a, b] \to \mathbb{R}^+$ are continuous. Then $R([a, b]) = [\delta, \triangle]$, where

$$\delta = \min_{t \in [a,b]} R(t) = \inf_{t \in [a,b]} R(t).$$

Given a path $\eta : [a, b] \to \mathbb{C}$ with $\left|\eta(t) - \gamma(t)\right| < \delta\ \forall t$, and with $\eta(a) = \gamma(a) = z_0$, $\eta(b) = \gamma(b) = z_1$ (Fig. 10.14), we can define $g_t(z)$ to be the power series expansion of $f_t(z)$ about $\eta(t)$, as

$$g_t(z) = P(f_t; \eta(t)) = \sum_{n=0}^{\infty} b_n(t)\Big(z - \eta(t)\Big)^n$$

whose radius of convergence is at least $R(t) - |\eta(t) - \gamma(t)| > 0$, say $R_{g_t}(t)$. This is possible as $\eta(t)$ lies within the disc of convergence $\mathcal{S}(f_t(z); R_{gt}(t))$ of $f_t(z)$. Now we are to show that g_t is an analytic continuation along η.

Let t' be close to t. In fact, $\eta(t')$ lies within $\mathcal{S}(f_t(z); R_{gt}(t))$, and since f_t is an analytic continuation,

$$f_{t'}(z) = f_t(z) \Rightarrow g_{t'}(z) = g_t(z)$$

for t' near t. This implies that, g_t is an analytic continuation of g_a to g_b. But $g_a = f_a$ and $g_b = f_b$ at $z_1 = \eta(b) = \gamma(b)$. So any analytic continuation of f_a along η leads to f_b. This proves the lemma. □

Lemma 10.2.2. For any $t \in [a, b]$, there exists an analytic continuation along γ_s

$$f_{s,t}(z) = \sum_{n=0}^{\infty} a_n(s, t)\Big(z - \gamma_s(t)\Big)^n$$

with domain of convergence $|z - \gamma(t)| < R_{(s,t)}$ and with $f_{s,0} = f$. Then the radius of convergence $R_{(s,t)}$ of $f_{s,t}$ is a continuous function of (s, t) in $[a, b] \times [c, d]$.

Proof: Similar to the Lemma 10.1.1, it is enough to show $R_{(s,t)}$ is continuous in a neighbourhood of $(s_0, t_0)\ \forall (s_0, t_0) \in [a, b] \times [c, d]$. So, by the Lemma 10.2.1, $\exists\, \delta(s_0)$ for all s such that

$$\left|\gamma_s(t) - \gamma_{s_0}(t)\right| < \delta(s_0).$$

The analytic continuation along $f_{s,t}$ is going to be the same as along $f_{s_0,t}$, i.e., $f_{s,t} = f_{s_0,t}$. But if t' is close to t, then

$$f_{s,t'} = f_{s,t} \text{ and } f_{s_0,t'} = f_{s_0,t}$$

due to analytic continuation. In other words, there exists a neighbourhood around (s_0, t_0) where all $f_{s,t}$ represent the same analytic function, say g_{s_0,t_0}. Then $R_{g_{(s_0,t_0)}}(s, t) = R_s(t)$ is a continuous function of (s, t) in that neighbourhood by the Lemma 10.1.2. In fact, there exists a neighbourhood of $\gamma_{s_0}(t_0)$

where $g_{(s,t_0)}$ is analytic and the radius of convergence of the power series expansion $P(g_{(s_0,t_0)}, \gamma_s(t))$ is a continuous function of $\gamma_s(t)$ in that neighbourhood of $\gamma_{s_0}(t_0)$ and

$$R_s(t) = R_{g_{(s_0,t_0)}}\Big(\gamma_s(t)\Big) = R_{g_{(s_0,t_0)}}\Big(F(s,t)\Big)$$

Thus $R_{(s,t)}$ of $f_{s,t}$ varies continuously with (s,t) in a rectangle $[a,b] \times [c,d]$. □

Theorem 10.4. [Monodromy theorem (version I)]: *Let $f(z)$ be analytic at z_0. Let $\gamma : [a,b] \to \mathbb{C}$ and $\eta : [a,b] \to \mathbb{C}$ be two paths from z_0 to z_1 along which $f(z)$ can be continued analytically. If γ is homotopic (i.e., continuously deformable) to η and if there is no obstruction to analytic continuation of f_a along any intermediate path $\gamma_s(0 \le s \le 1)$, then the analytic continuation of $f(z)$ along γ and along η coincide at b.*

Proof: Let $f_{s,t}(z)$ denote the analytic continuation of $f(z)$ along γ_s. Then for any $t \in [a,b]$, the power series expansion is

$$f_{s,t}(z) = \sum_{n=0}^{\infty} a_n(s,t)\Big(z - \gamma_s(t)\Big)^n$$

with domain of convergence $|z - \gamma(t)| < R_{(s,t)}$ and with $f_{s,0} = f$.

Take the image under finite $R > 0$ of the rectangle $[a,b] \times [c,d]$, which will be $[\delta, \triangle]$ with $\delta > 0$. Thus there exists $s_0 = c < s_1 < \cdots < s_N = d$ such that $\forall t \in [a,b]$

$$\Big|\gamma_s(t) - \gamma_{s'}(t)\Big| < \delta \text{ for } s, s' \in [s_k, s_{k+1}]; k = 0, \cdots, N-1.$$

By the Lemma 10.2.1, analytic continuation of f along γ_{s_k} leads to the same power series at z_1 as that of $\gamma_{s_{k+1}}$. If we apply this now successively to $k = 0, 1, \cdots, N-1$, we conclude after N steps that along continuation along γ leads to the same power series as analytic continuation along η. Thus analytic continuation of f along any γ_s leads to one and only one analytic function at z. That proves the theorem. □

RESULT **76.** (i) Given $\langle \mathcal{D}, f\rangle$, f is an analytic function on the domain $\mathcal{D} \subset \mathbb{C}$. There is domain $\langle \mathcal{D}_1, g_1\rangle$ of maximal direct analytic continuation (extension) of $\langle \mathcal{D}, f\rangle$, i.e., $\mathcal{D}_1 \subset \mathbb{C}$ is the largest domain, $\mathcal{D}_1 \supset \mathbb{C}$ such that $\langle \mathcal{D}_1, g_1\rangle$ is the direct analytic extension or continuation of $\langle \mathcal{D}, f\rangle$, i.e., $g_1|_{\mathcal{D}} = f$.

(ii) We also have the domain $\langle \mathcal{D}_2, g_2\rangle$ of indirect analytic continuation of $\langle \mathcal{D}, f\rangle$, also called the *domain of regularity* of $\langle \mathcal{D}, f\rangle$.

We consider the following examples:

(i) Let $f(z) = 1 + z + z^2 + \cdots$ and $\mathcal{D} = \{z \in \mathbb{C} : |z| < 1\}$. Then

$$\langle \mathcal{D}_1, f_1\rangle = \langle \mathbb{C}/\{1\}, \frac{1}{1-z}\rangle.$$

Here the domain of maximal direct analytic continuation is the domain of regularity, i.e., of maximal indirect analytic continuation.

(ii) Let us consider the Riemann zeta function $f(z) = \zeta(z)$, where

$$\zeta(z) = \sum_{n=1}^{\infty} \frac{1}{n^z} = \sum_{n=1}^{\infty} \frac{1}{e^{z \ln n}}$$

and $\mathcal{D} = \{z \in \mathbb{C} : \operatorname{Re} z > 1\}$. Here $\mathcal{D}_1 = \mathbb{C}/\{1\}$ and f_1 is the extended ζ-function. $z = 1$ is a simple pole for f_1. Here the domain of maximal direct analytic continuation is the domain of regularity, i.e., of maximal indirect analytic continuation.

(iii) Let $f(z) = \text{Log}\, z$ (principal value). The domain of direct maximal analytic continuation is $\mathcal{D} = \mathbb{C}/(-\infty, 0] \subset \mathbb{C}/\{0\}$. Then $\mathcal{D}_1 = \mathcal{D}$ and $f_1 = f$. Here $\mathcal{D}$ is the domain of indirectly analytic continuation.

RESULT **77.** [Monodromy theorem (version II)]: Given a function element $\langle \mathcal{D}, f \rangle$, i.e., a holomorphic (or analytic) function f on an domain $\mathcal{D}$ if the domain of regularity of f viz., the domain of maximal indirect analytic continuation of f, $\mathcal{D}_2$, is simply connected, then $\mathcal{D}_1 = \mathcal{D}_2$, i.e., the domain of maximal direct analytic extension $\mathcal{D}_1$, is the same as the domain of regularity $\mathcal{D}_2$ (i.e., every indirect analytic extension of f is direct).

10.3 Reflection Principle

One way to extend an analytic function to a larger domain is to take advantage of some kind of symmetry operation. The *Schwarz reflection principle* does exactly this by making use of symmetry operation $z \to \bar{z}$, which practically sends a point to its reflection through the x-axis.

Definition 10.3.1. Let $f(z)$ be analytic function in a domain $\mathcal{D}$ in the complex z-plane. Then f satisfies *reflection principle* in $\mathcal{D}$ if and only if

$$\overline{f(z)} = f(\bar{z}) \ \forall z \in \mathcal{D}. \tag{10.9}$$

For example, let $f(z) = \cos z$. Then $\overline{f(z)} = \overline{\cos z}$ and

$$\begin{aligned} f(\bar{z}) &= \cos \bar{z} = \cos(x - iy) = \cos x \cos(iy) + \sin x \sin(iy) \\ &= \cos x \cosh y - i \sin x \sinh y \\ &= \overline{\cos x \cosh y + i \sin x \sinh y} \\ &= \overline{\cos x \cos(iy) - \sin x \sin(iy)} = \overline{\cos(x + iy)} = \overline{\cos z} = \overline{f(z)}. \end{aligned}$$

Hence the analytic function $f(z) = \cos z$ satisfy the reflection principle in $\mathbb{C}_\infty$. Similarly, the functions $7z, z + 11, 3z^2 + 5, z^3$ and e^z satisfy the reflection principle in the entire complex plane. The function $f(z) = z + 2i$ is analytic in $\mathcal{D}$, although it does not satisfy the reflection principle. Similarly, $2z + 5i$, $(3 + 2i) \sin z$ and e^{iz} do not satisfy the reflection principle in $\mathbb{C}$.

The following result, which is known as Schwarz reflection principle, provides sufficient conditions when the reflection of $f(z)$ in the real axis corresponds to the reflection of z.

Theorem 10.5. [Schwarz reflection principle]: *Let $f(z)$ be analytic in some domain $\mathcal{D}$ having a segment of the x-axis and is symmetric to that axis. Then f satisfies the reflection principle in $\mathcal{D}$ if and only if $f(x)$ is real for all $x \in (a, b)$ as segment on the x-axis.*

Proof: If f satisfies the reflection principle in a domain $\mathcal{D}$ which contains an interval (a, b) as segment on the x-axis, then,

$$f(x) = f(\bar{x}) = \overline{f(x)}, \quad \forall x \in (a, b)$$

so that $f(x)$ is real for all $x \in (a, b)$.

Let $f(z) = u(x, y) + iv(x, y)$ be analytic in a domain $\mathcal{D}$, and let

$$\begin{aligned} F(z) &= \overline{f(\bar{z})} = \overline{u(x, -y) + iv(x, -y)} \\ &= u(x, -y) - iv(x, -y) = u_1(x, y) + iv_1(x, y), \end{aligned}$$

say , $z \in \mathcal{D}$,

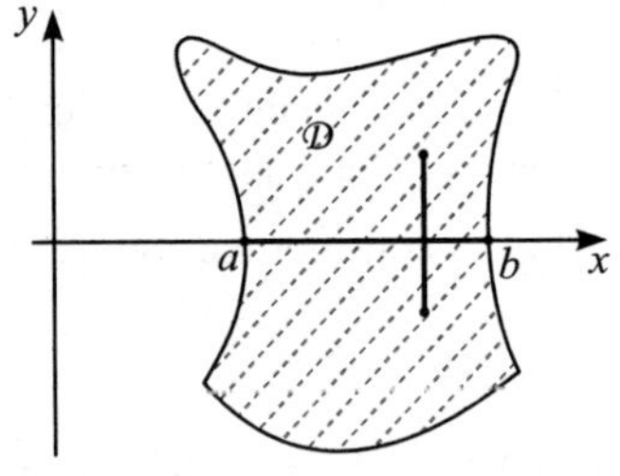

Figure 10.15 *Reflection principle*

where $u_1(x,y) = u(x,-y)$ and $v_1(x,y) = -v(x,-y)$. Since $f(z)$ is analytic in a domain $\mathcal{D}$, it satisfies C-R equations (3.22). Hence

$$\frac{\partial u_1}{\partial x} = \frac{\partial u}{\partial x} = \frac{\partial v}{\partial y} = \frac{\partial v_1}{\partial y} \text{ and } \frac{\partial u_1}{\partial y} = -\frac{\partial u}{\partial y} = \frac{\partial v}{\partial x} = -\frac{\partial v_1}{\partial x}.$$

Thus the first order partial derivatives of u_1 and v_1 are continuous (as u and v are so) and satisfies the C-R equations (3.22) in a domain $\mathcal{D}$. Hence $F(z) = \overline{f(\overline{z})}$ is analytic in $\mathcal{D}$.

Now, since $f(x)$ is real on the segment of the real axis lying in $\mathcal{D}$, we have $v(x,0) = 0$ on the segment and consequently, $f(x) = u(x,0)$. Also,

$$F(z) = u_1(x,0) + iv_1(x,0) = u(x,0) - iv(x,0) = u(x,0) \text{ as } v(x,0) = 0.$$

Therefore, $F(z) = f(z)$ at each point x on the real axis contained in $\mathcal{D}$.

$$\therefore \quad \overline{f(\overline{z})} = f(z) \Rightarrow f(\overline{z}) = \overline{f(z)} \ \forall z \in \mathcal{D}.$$

Then $u(x,-y) + iv(x,-y) = u(x,y) - iv(x,y)$, which is also true for $z = (x,0)$, a point on the real axis lying in $\mathcal{D}$.

$$\therefore u(x,0) + iv(x,0) = u(x,0) - iv(x,0) \Rightarrow v(x,0) = 0.$$

Thus $f(x)$ is real on the segment of real axis lying in $\mathcal{D}$. □

RESULT **78.** By the preceding theorem, $f(z) = x^3$ is real for all real values of x on the segment of x-axis in $\mathcal{D}$. So the reflection property (10.9) holds. But for $f(z) = z + 3i = x + i(3+y)$, $f(x) = x + 3i$ is not real for any real values of x on the segment of x-axis in $\mathcal{D}$. So the reflection property (10.9) does not hold even though $f(z) = z + 3i$ is analytic in $\mathcal{D}$.

Example 10.3.1. *Prove that $f(z)$ and $\overline{f(\overline{z})}$ are simultaneously analytic.*

Solution. Let $f(z) = u(x,y) + iv(x,y)$ be analytic in a domain $\mathcal{D}$. So it satisfies C-R equations (3.22), i.e., $u_x = v_y$ and $u_y = -v_x$. Now, since $\overline{z} = x - iy = (x,-y)$, we have

$$\overline{f(\overline{z})} = u(x,-y) - iv(x,-y) = u_1(x,y) + iv_1(x,y),$$

where $u_1(x,y) = u(x,-y)$ and $v_(x,y) = v(x,-y)$. Hence

$$\frac{\partial u_1}{\partial x} = \frac{\partial u}{\partial x} = \frac{\partial v}{\partial y} = \frac{\partial v_1}{\partial y} \text{ and } \frac{\partial u_1}{\partial y} = -\frac{\partial u}{\partial y} = \frac{\partial v}{\partial x} = -\frac{\partial v_1}{\partial x}.$$

Thus the first order partial derivatives of u_1 and v_1 are continuous (as u and v are so) and satisfies the C-R equations (3.22) in a domain $\mathcal{D}$. Hence $\overline{f(\overline{z})}$ is analytic in $\mathcal{D}$. Similarly, if $\overline{f(\overline{z})}$ is analytic in a domain $\mathcal{D}$, then $f(z)$ is also analytic in $\mathcal{D}$.

Exercises

Review Questions [Short-answer Type]

1. Prove that there cannot be two different direct analytic continuation of a function.
2. If two functions which are analytic in a domain $\mathcal{D}$ coincide in a part $\mathcal{D}_0$ of the domain, prove that they coincide in the whole domain.

3. The power series $\sum_{n=0}^{\infty}(n+1)z^n$ defines an analytic function f_0 on the open unit disc. What is the largest domain in the plane to which f_0 can be analytically continued?
4. If the identity $\sin^2 z + \cos^2 z = 1$ holds for all real values of z, show that it holds for all complex values of z.
5. If L is the line with equation $ax + by = 0$, find a formula for the map which is reflection through L, and verify that this map is conjugate analytic.
6. Let $\mathcal{H}$ be the upper half plane and let $f(z)$ be an analytic function on $\mathcal{H}$ that is continuous and real on the real axis. Prove that $f(z)$ can be extended analytically to the entire complex plane.
7. Let $f(z)$ be an entire function taking real values on the real axis and imaginary values on the imaginary axis. Prove that $f(z) + f(-z) = 0$ for all z.
8. Prove that $f(z) = z^3$ is analytic and satisfies the reflection property.
9. Verify the theorem on reflection principle with the following functions: (i) $f(z) = z+i$, (ii) $f(z) = z^2$, (iii) $f(z) = e^z$.
10. Show that the reflection through the circle $\{|z - z_0| = R\}$ of radius R centered at z_0 is given by $z^* = z_0 + R^2(z - z_0)/|z - z_0|^2$.
11. Prove that the curve in a plane with equation $y = x^2$ is a simple analytic curve.
12. Show that a reflection in a circle maps circles in the plane of circles.
13. What happens to angles between curves when they are reflected in an analytic arc?

Review Questions [Long-answer Type]

1. Define analytic continuation. Prove that direct analytic continuation is unique.
2. If $f(z)$ is analytic in a domain $\mathcal{D}$ and vanishes at all points on a domain $\mathcal{D}_0$ or line segment contained in $\mathcal{D}$, then prove that it is identically equal to 0 throughout domain $\mathcal{D}$.
3. Determine the analytic continuation of the function $f_0(z) = \int_0^{\infty} te^{-zt}\,dt$, Re$z > 0$, into the domain consisting of all the points in the complex plane except $z = 0$.
4. Show that the function defined by $f(z) = \sum z^{2^n} = z + z^2 + z^4 + \cdots$ is analytic on the open disc $|z| < 1$, and that it cannot be extended analytically to any larger open set.
5. Suppose $f(z) = \sum_{k=0}^{\infty} a_k z^k$ has radius of convergence R. Show that if $f(re^{i\theta})$ tends to ∞ as $r \to R$, then the point $Re^{i\theta}$ is a singular point of $f(z)$.
6. Suppose $f(z) = \sum_{k=0}^{\infty} a_k (z - z_0)^k$ has radius of convergence $R > 0$. Show that $f(z)$ has at least one singular point on the boundary $\partial S(z_0, R)$.
7. Prove the function $a/(z^2 + a^2)$, where $\in \mathbb{R}$, is the analytic continuation of the function $f_1(z) = \int_0^{\infty} \sin ate^{-zt}\,dt$ into the domain consisting of all the points in the complex plane except $z = \pm ai$.
8. Suppose $f(z) = \sum a_n z^n$, where the series has radius of convergence $R < \infty$. Show that there is an angle α such that $f(z)$ does not have an analytic continuation along the path $\gamma(t) = te^{it}$; $0 \leq t \leq R$. Determine the radius of convergence of the power series expansion of $f(z)$ about $te^{i\alpha}$.
9. Show that the function $f_1(z) = \frac{1}{a} + \frac{z}{a^2} + \frac{z^2}{a^3} + \frac{z^3}{a^4} + \cdots$ can be continued analytically outside the circle of convergence.
10. Show that the functions defined by the series $\sum_{n=0}^{\infty}(az)^n$ and $\sum_{n=0}^{\infty}(-1)^n \frac{(1-a)^n z^n}{(1-z)^{n+1}}$ are analytic continuation of each other.
11. Prove that the power series $\sum_{n=0}^{\infty} z^{3n}$ cannot be continued analytically beyond the circle $|z| = 1$.

12. Prove that $f_0(z) = \sum_{n=0}^{\infty} \frac{(z+i)^n}{(1+i)^{n+1}}$ is an analytic continuation of $f_1(z) = \sum_{n=0}^{\infty} z^n$, showing graphically the regions of convergence of the series.

    Show that the function $(1-z)^{-1}$ is represented by all analytic continuations of $f_1(z)$.

13. Show that the functions

$$f_1(z) = -\frac{i\pi}{2} + \sum_{j=1}^{\infty} \frac{(-1)^{j-1}}{j}(z-1)^j; |z-1| < 1$$

$$f_2(z) = \frac{i\pi}{2} + \sum_{j=1}^{\infty} \frac{1}{j}(z+1)^j; |z+1| < 1$$

    are analytic continuations of each other.

14. Consider the series $f_0(z) = \sum_{n=0}^{\infty} \frac{1}{2^n} z^{n+1}$.

    (a) Find an analytic continuation of $f_0(z)$ which converges for $z = 2+i$.

    (b) Show that the value of the analytic continuation at that point is $-2+4i$.

15. Consider the series $f_0(z) = \sum_{n=0}^{\infty} \frac{1}{3^n} z^{n+1}$.

    (a) Find an analytic continuation of $f_0(z)$ which converges for $z = 3-4i$.

    (b) Show that the value of the analytic continuation at that point is $-3 - \frac{9}{4}i$.

16. The power series $z + \frac{z^2}{2} + \frac{z^3}{3} + \cdots$ and $i\pi - (z-2) + \frac{1}{2}(z-2)^2 - \cdots$ have no common region of convergence. Prove that they are nevertheless analytic continuations of the same function.

17. If the power series $f(z) = \sum_{n=0}^{\infty} a_n(z-z_0)^n$ has a non-zero finite radius of convergence. Then the complete analytic function corresponding to $f(z)$ has at least one singularity on the circle of convergence,

18. Prove that the following series has the natural boundary $|z| = 1$.

    (a) $f_0(z) = z^{1!} + z^{2!} + z^{3!} + \ldots = \sum z^{n!}$

    (b) $f_1(z) = 1 + z + z^2 + z^4 + z^8 + \ldots = 1 + \sum z^{2^n}$.

    [**Hint:** Clearly, each root of any of the equations $z^2 = 1, z^4 = 1, z^8 = 1, \cdots$ is a singularity of $f(z)$. Hence, on any arc, however small, of the circle $|z| = 1$, the function $f(z)$ has an infinite number of singularities. Therefore, we cannot extend this function outside the circle $|z| = 1$.]

19. Prove that the unit circle $|z| = 1$ is the natural boundary of the series $\sum_{n=0}^{\infty} z^{n!}$.

20. Suppose $\langle \mathcal{D}_0, f_0 \rangle$ is a function element at z_0 and γ is a curve joining z_0 to w. If there is an open set U containing $\mathcal{D}_0$ and $\gamma(I)$, and an analytic function f on U such that $f = f_0$ on $\mathcal{D}_0$, then prove that $\langle \mathcal{D}_0, f_0 \rangle$ can be analytically continued along γ.

21. Let U be a domain and $\langle \mathcal{D}_0, g_0 \rangle$ is a function element in U which can be analytically continued along a curve γ in U to a function element $\langle \mathcal{D}, g \rangle$ in U.

    (a) If f is an entire function, prove that $\langle \mathcal{D}_0, f \circ g_0 \rangle$ can be analytically continued along γ to $\langle \mathcal{D}, f \circ g \rangle$.

    (b) If f is an entire function, h is a function analytic on U and $f \circ g_0 = h$ on $\mathcal{D}_0$, then $f \circ g = h$ on $\mathcal{D}$.

(c) If a linear differential equation on U with analytic coefficients has g_0 for a solution on $\mathcal{D}_0$, then it has g for a solution on $\mathcal{D}$.
(d) If u is harmonic function on U and $u = \text{Re}(g_0)$ on $\mathcal{D}_0$, then $u = \text{Re}(g)$ on $\mathcal{D}$.

22. Prove that
   (a) if log is the principal branch of the log function and $\mathcal{D}_0$ is any disc centered at 1 and not containing 0, then $\langle \mathcal{D}_0, \log \rangle$ can be analytically continued along any curve γ in $\mathbb{C}/\{0\}$.
   (b) any analytic continuation of the element $\langle \mathcal{D}_0, \log \rangle$ of the previous exercise, along a curve γ in $\mathbb{C}/\{0\}$, yields a function element $\langle \mathcal{D}, f \rangle$ at $\gamma(1)$ which is some branch of the log function restricted to $\mathcal{D}$. [**Hint:** That is, prove that it satisfies $e^{f(z)} = z$ on $\mathcal{D}$.]
23. Show that on the circle of convergence of a power series there is always at least one point with the following property: the function defined by the series cannot be continued analytically to the exterior of the disc along any curve passing through the point.
24. Suppose that the coefficients of the power series $f(z) = \sum_{n=0}^{\infty} a_n z^n$ are non-negative. Show that the function $f(z)$ cannot be continued analytically along the positive real axis across the circle of convergence.
25. The function $f(z)$ defined by the power series $f(z) = \sum_{n=0}^{\infty} a_n z^{n!}; a_n > 0$ cannot be continued analytically beyond its circle of convergence.
26. **(Painleve's Theorem).** Assume that $\mathcal{R}_1$ and $\mathcal{R}_2$ are two regions with an adjacent piecewise smooth boundary arc L [Figure 10.16] and $f_0(z)$ and $f_1(z)$ are analytic functions on $\mathcal{R}_1$ and $\mathcal{R}_2$ respectively. Furthermore, assume that $f_0(z)$ and $f_1(z)$ are continuous on $\mathcal{R}_1 \cup L$ and $\mathcal{R}_2 \cup L$ respectively, and $f_0(z) = f_1(z), z \in L$. Show that

$$g(z) = \begin{cases} f_0(z) & \text{if } z \in \mathcal{R}_1 \cup L \\ f_1(z) & \text{if } z \in \mathcal{R}_2 \end{cases}$$

   is analytic in $\mathcal{R}_1 \cup L \cup \mathcal{R}_2$.

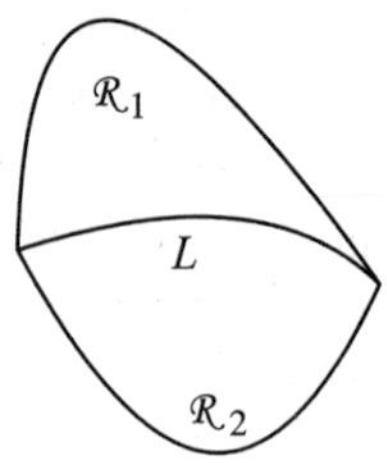

Figure 10.16 *Painleve's theorem*

27. **(Riemann's Theorem).** Let $\mathcal{D} = \mathcal{S}(z_0, r)/\{z_0\}$, and let $f(z)$ be an analytic and bounded function on $\mathcal{D}$. Show that $f(z)$ can be analytically continued to the disc $\mathcal{S}(z_0, r)$.
28. Formulate and prove a version of the Schwarz reflection principle for (a) harmonic functions, (b) meromorphic functions.
29. Prove that the composition $f \circ g$
   (a) of an analytic function f with a conjugate analytic function g is conjugate analytic.
   (b) of two conjugate analytic functions is analytic.
30. Use Schwarz reflection principle to show that
   (a) every function which is bounded and analytic in the upper half plane Im ≥ 0, and real on the real axis is constant.
   (b) $\overline{\sin z} = \sin \bar{z}$, for $z \in \mathbb{C}$.
   (c) $\overline{z^4 - 2z + 2\cos z} = \bar{z}^4 - 2\bar{z} + 2\cos \bar{z}$, for $z \in \mathbb{C}$.
31. Show that, if U is any non-empty open subset of the plane, then there is an analytic function on U which cannot be extended to be analytic on any larger open set. [**Hint:** Use the Weierstrass theorem (9.20) to construct an analytic function on U with a lot of zeros.]

Answers

3. $f_1(z) = z + 1/z, /z \neq 0$;   14. (a) $2iz \sum_{[n=0]}^{\infty} [z - (2+i)]^n i^n, \ |z - (2+i)| < 1$.

Bibliography

[1] Ablowitz, M. J. and Fokas, A. S. (2003), *Complex Variables*, Cambridge University Press.

[2] Agarwal, R. P., Perera and Pinelas, S. (2010), *An Introduction to Complex Analysis*, Springer.

[3] Ahlfors, L.V. (1979), *Complex Analysis*, McGraw-Hill Book Company, New York.

[4] Brown, J. W. and Churchill, R.V. (2009), *Complex Variables and Applications*, McGraw Hill.

[5] Conway, J. B. (1978), *Functions of One Complex Variable*, Springer International Student Edition, New York.

[6] Copson, E. T. (1935), *An introduction to the Theory of Functions of a Complex Variable*, Clarendon Press, Oxford.

[7] Gamelin, T. W. (2000), *Complex Analysis* , Springer-Verlag, New York.

[8] Henrici, P. (1974), *Applied and Computational Complex Analysis*, John Wiley & Sons, Inc., NY.

[9] Lang, S. (1977) *Complex Analysis*, Addison-Wesley.

[10] Moore, T. O. and Hadlock, E.H., (1991), *Complex Analysis*, World Scientific Publishing Co. Pte. Ltd., Singapore-9128.

[11] Nevanlinna, R. and Paatero, V. (1969), *Introduction to Complex Analysis*, Addison Wesley Publishing Company, Inc., Reading, MA.

[12] Ponnusamy, S. (2012), *Foundations of Complex Analysis*, Narosa Publishing House.

[13] Spiegel, M. R. (1964), *Theory and Problems of Complex Variables*, Schaum's Outline Series, McGraw-Hill Company, NY.

Index